Springer Proceedings in Mathematics & Statistics

Volume 183

Springer Proceedings in Mathematics & Statistics

This book series features volumes composed of selected contributions from workshops and conferences in all areas of current research in mathematics and statistics, including operation research and optimization. In addition to an overall evaluation of the interest, scientific quality, and timeliness of each proposal at the hands of the publisher, individual contributions are all refereed to the high quality standards of leading journals in the field. Thus, this series provides the research community with well-edited, authoritative reports on developments in the most exciting areas of mathematical and statistical research today.

More information about this series at http://www.springer.com/series/10533

Yoshihiro Shibata · Yukihito Suzuki
Editors

Mathematical Fluid Dynamics, Present and Future

Tokyo, Japan, November 2014

Springer

Editors
Yoshihiro Shibata
Department of Mathematics
Waseda University
Tokyo
Japan

Yukihito Suzuki
Department of Mathematics
Waseda University
Tokyo
Japan

ISSN 2194-1009 ISSN 2194-1017 (electronic)
Springer Proceedings in Mathematics & Statistics
ISBN 978-4-431-56455-3 ISBN 978-4-431-56457-7 (eBook)
DOI 10.1007/978-4-431-56457-7

Library of Congress Control Number: 2016951691

Mathematics Subject Classification (2010): 00B25

Printed on acid-free paper

This Springer imprint is published by Springer Nature
The registered company is Springer Japan KK
The registered company address is: Chiyoda First Bldg. East, 3-8-1 Nishi-Kanda, Chiyoda-ku, Tokyo 101-0065, Japan

Preface

The rapid development of computers has enabled us to design several industrial products relying on detailed numerical simulation which is based on the Navier–Stokes equation. This proves the reliability of the equation and one has become to be able to manufacture those products without knowing the mathematical details of the equation. The mathematical studies are, however, still crucial to invent completely novel way to investigate unsolved problems, because the numerical simulation does not suggest any new concepts. Moreover, even the governing equations have not been known for complicated phenomena which include several different scales. Multiphase flow is a typical and important example of such complicated multi-scale phenomena. Mathematics can play an important role to construct formulation and theory for the multiphase flows and other complex phenomena.

The research project titled *A challenge to unsolved problems in fluid engineering with modern mathematical analysis* was pursued from 2009 to 2014 (five and half years) as part of the CREST (Core Research for Evolutional Science and Technology)–SBM (Search for Breakthrough by Mathematics). The aim of the research was to tackle the above-mentioned problems enhancing the cooperation between mathematics and engineering. On the occasion of the final year of the project, the international conference was held during 11–14 November 2014 at Waseda University in Tokyo in order to announce the results obtained through the project as well as stimulate other related researches. This book is an outcome of the conference which consists of original papers offered by invited speakers. The contents range from the experimental study on cavitation jets to up-to-date mathematical analysis of the Navier–Stokes equations reflecting the feature of the conference.

This book is divided into two parts: Multiphase Flows and Other Related Topics. Both the parts contain articles on a wide range of studies from mathematics to engineering. We hope that this contribution is attractive and useful for a wide range of researchers and engineers as well.

Tokyo, Japan
December 2015

Yoshihiro Shibata
Yukihito Suzuki

Acknowledgements

We gratefully acknowledge the financial support from the Alliance for Breakthrough between Mathematics and Sciences (ABMS) supported by the Core Research for Evolutional Science and Technology (CREST) of the Japan Science and Technology Agency (JST), the Research Institute of Nonlinear Partial Differential Equations at Waseda University, the Grants-in-Aid for Scientific Research (KAKENHI, Grant No. 24224004) of the Japan Society for the Promotion of Science (JSPS) and the Institute of Mathematical Fluid Dynamics at Waseda University, for organizing the 8th CREST-SBM (Search for Breakthrough by Mathematics) International Conference titled *An International Conference on Mathematical Fluid Dynamics, Present and Future*, of which this proceedings is the outcome.

We would also like to thank Masashi Ohnawa, Linyu Peng, Hirokazu Saito, Miho Murata and Natsuko Ikezaki for their devoted work which made the smooth running of the conference possible.

Contents

Contributors

Helmut Abels Fakultät Für Mathematik, Universität Regensburg, Regensburg, Germany

Dieter Bothe Center of Smart Interfaces & Fachbereich Mathematik, Technische Universität Darmstadt, Darmstadt, Germany

Irina Vlad. Denisova Institute for Mechanical Engineering Problems, Russian Academy of Sciences, St. Petersburg, Russia

Eduard Feireisl Institute of Mathematics of the Academy of Sciences of the Czech Republic, Praha 1, Czech Republic

Giovanni Paolo Galdi Department of Mechanical Engineering, University of Pittsburgh, Pittsburgh, USA

Takeshi Gotoda Department of Mathematics, Kyoto University, Sakyo-ku, Kyoto, Japan

Matthias Hieber Technische Universität Darmstadt, Fachbereich Mathematik, Darmstadt, Germany

Toshiaki Hishida Graduate School of Mathematics, Nagoya University, Nagoya, Japan

Makoto Iima Graduate School of Science, Hiroshima University, Higashi-Hiroshima, Japan

Akane Kawaharada Department of Mathematics, Kyoto University of Education, Kyoto, Japan

Takeshi Miyazaki University of Electro-Communications, Chofu, Tokyo, Japan

Jiří Neustupa Czech Academy of Sciences, Institute of Mathematics, Prague, Czech Republic

Takaaki Nishida Kyoto University, Sakyo-ku, Kyoto, Japan

Hirofumi Notsu Faculty of Mathematics and Physics, Kanazawa University, Kakuma, Kanazawa 920-1192, Japan; Japan Science and Technology Agency (JST), PRESTO, Kawaguchi, Saitama, Japan

Toshiyuki Ogawa Graduate School of Advanced Mathematical Science, Meiji University, Nakano, Japan

Masashi Ohnawa Department of Ocean Sciences, Faculty of Marine Science, Tokyo University of Marine Science and Technology, Minato-ku, Tokyo, Japan

Mariarosaria Padula Depatment of Mathematics, Faculty of Engineering, University of Ferrara, Ferrara, Italy

Jan Prüss Institut für Mathematik, Martin-Luther-Universität Halle-Wittenberg, Halle, Germany

Keisei Saga University of Electro-Communications, Chofu, Tokyo, Japan

Takashi Sakajo Department of Mathematics, Kyoto University, Sakyo-ku, Kyoto, Japan

Takashi Okuda Sakamoto School of Science and Technology, Meiji University, Kawasaki, Japan

Stefan Schaubeck Fakultät Für Mathematik, Universität Regensburg, Regensburg, Germany

Paolo Secchi DICATAM, Mathematical Division, University of Brescia, Brescia, Italy

Adélia Sequeira Department of Mathematics and CEMAT, IST, ULisboa, Lisbon, Portugal

Yoshihiro Shibata Department of Mathematics and Research Institute of Science and Engineering, Waseda University, Tokyo, Shinjuku-ku, Japan

Yuichi Shimoda University of Electro-Communications, Chofu, Tokyo, Japan

Erika Shoji Graduate School of Science, Hiroshima University, Higashi-hiroshima, Japan

Telma Silva Department of Mathematics and CEMAT, IST, ULisboa, Lisbon, Portugal

Vsevolod Alekseevich Solonnikov St. Petersburg Department of Steklov Mathematical Institute, Russian Academy of Sciences, St.Petersburg, Russia

Yukihito Suzuki Department of Mathematics, Waseda University, Shinjuku-ku, Tokyo, Japan

Masahisa Tabata Department of Mathematics, Waseda University, Tokyo, Japan

Atusi Tani Department of Mathematics, Keio University, Yokohama, Japan

Hisasi Tani Department of Mathematics, Meiji University, Kawasaki, Japan

Yuji Tasaka Laboratory for Flow Control, Hokkaido University, Sapporo, Japan

Yoshiaki Teramoto Setsunan University, Neyagawa, Osaka-prefecture, Japan

Jorge Tiago Department of Mathematics and CEMAT, IST, ULisboa, Lisbon, Portugal

Takeshi Watanabe Rocket and Spacecraft Modeling Laboratory, School of Engineering, The University of Tokyo, Tokyo, Japan

Takayuki Yamaguchi Department of Radiation Oncology, Institute of Biomedical and Health Sciences, Hiroshima University, Higashi-hiroshima, Japan

Katsuhiro Yamamoto Waseda University, Tokyo, Japan

Satoshi Yokoyama Graduate School of Mathematical Sciences, The University of Tokyo, Tokyo, Japan

Part I
Multiphase Flows

Chapter 1
Nonconvergence of the Capillary Stress Functional for Solutions of the Convective Cahn-Hilliard Equation

Helmut Abels and Stefan Schaubeck

Abstract We show that the surface tension term $-\varepsilon \operatorname{div}(\nabla c^{\varepsilon} \otimes \nabla c^{\varepsilon})$ of the "model H" does generally not converge to the mean curvature functional of the interface as $\varepsilon \searrow 0$, where c^{ε} is the solution to a convective Cahn-Hilliard equation with mobility constant converging to 0 too fast as $\varepsilon \searrow 0$. In that case the motion of the interface is dominated by the convection term $v \cdot \nabla c^{\varepsilon}$ of the convective Cahn-Hilliard equation.

Keywords Two-phase flow · Diffuse interface model · Model H · Sharp interface limit

Mathematics Subject Classification (2000): Primary 76T99 · Secondary: 35Q30 · 35Q35 · 35R35 · 76D05 · 76D45

1.1 Introduction

In this paper we consider the limit $\varepsilon \to 0$, called sharp interface limit, for solutions of the convective Cahn-Hilliard equation

$$\partial_t c^{\varepsilon} + v \cdot \nabla c^{\varepsilon} = m(\varepsilon) \Delta \mu^{\varepsilon} \quad \text{in } \Omega \times (0, \infty), \tag{1.1}$$

$$\mu^{\varepsilon} = \varepsilon^{-1} f(c^{\varepsilon}) - \varepsilon \Delta c^{\varepsilon} \quad \text{in } \Omega \times (0, \infty), \tag{1.2}$$

$$\tfrac{\partial}{\partial n} c^{\varepsilon} = \tfrac{\partial}{\partial n} \mu^{\varepsilon} = 0 \quad \text{on } \partial\Omega \times (0, \infty), \tag{1.3}$$

$$c^{\varepsilon}|_{t=0} = c_0^{\varepsilon} \quad \text{in } \Omega, \tag{1.4}$$

where $c : \Omega \times (0, \infty) \to \mathbb{R}$ is the concentration difference of two fluids, $\mu : \Omega \times (0, \infty) \to \mathbb{R}$ is the associated chemical potential difference, $v : \Omega \times [0, \infty) \to \mathbb{R}^d$ is a given smooth velocity field, $\Omega \subseteq \mathbb{R}^d$ is a bounded domain with smooth boundary, d is the unit outer normal of $\partial\Omega$, $m(\varepsilon)$ is a mobility coefficient and $\varepsilon > 0$ is a

H. Abels (✉) · S. Schaubeck
Fakultät Für Mathematik, Universität Regensburg, 93040 Regensburg, Germany
e-mail: helmut.abels@mathematik.uni-regensburg.de

Y. Shibata and Y. Suzuki (eds.), *Mathematical Fluid Dynamics, Present and Future*, Springer Proceedings in Mathematics & Statistics 183,
DOI 10.1007/978-4-431-56457-7_1

parameter related to the "thickness" of the diffuse interface $\{x \in \Omega \,|\, |c(x,t)| < 1-\delta\}$ for $\delta \in (0,1)$, and $f = F'$, where $F\colon \mathbb{R} \to \mathbb{R}$ is a suitable double well potential e.g. $F(c) = (1-c^2)^2$.

The system (1.1) and (1.2) arises as part of the so-called "model H", which leads to system

$$\partial_t v^\varepsilon + v^\varepsilon \cdot \nabla v^\varepsilon - \operatorname{div}(\nu(c^\varepsilon)Dv^\varepsilon) + \nabla p^\varepsilon = -\varepsilon \operatorname{div}(\nabla c^\varepsilon \otimes \nabla c^\varepsilon) \quad \text{in } \Omega \times (0,\infty), \tag{1.5}$$

$$\operatorname{div} v^\varepsilon = 0 \quad \text{in } \Omega \times (0,\infty), \tag{1.6}$$

$$\partial_t c^\varepsilon + v^\varepsilon \cdot \nabla c^\varepsilon = m(\varepsilon)\Delta \mu^\varepsilon \quad \text{in } \Omega \times (0,\infty), \tag{1.7}$$

$$\mu^\varepsilon = -\varepsilon \Delta c^\varepsilon + \varepsilon^{-1} f(c^\varepsilon) \quad \text{in } \Omega \times (0,\infty), \tag{1.8}$$

where $v^\varepsilon\colon \Omega \times [0,\infty) \to \mathbb{R}^d$ is the velocity field, $Dv^\varepsilon = \frac{1}{2}(\nabla v^\varepsilon + (\nabla v^\varepsilon)^T)$, $p^\varepsilon\colon \Omega \times [0,\infty) \to \mathbb{R}$ is the pressure, and $\nu(c^\varepsilon) > 0$ is the viscosity of the mixture.

Using the method of formally matched asymptotic expansions, Abels et al. [4] showed that the solutions to the "model H" converge of the solutions to the following sharp interface models

$$\begin{aligned}
\partial_t v + v \cdot \nabla v - \operatorname{div}(\nu^\pm Dv) + \nabla p &= 0 && \text{in } \Omega^\pm(t), t > 0,\\
\operatorname{div} v &= 0 && \text{in } \Omega^\pm(t), t > 0,\\
\Delta \mu &= 0 && \text{in } \Omega^\pm(t), t > 0,\\
[v]_{\Gamma(t)} &= 0 && \text{on } \Gamma(t), t > 0,\\
-\big[\nu_{\Gamma(t)} \cdot (\nu^\pm Dv - p\mathrm{Id})\big]_{\Gamma(t)} &= \sigma\kappa\nu_{\Gamma(t)} && \text{on } \Gamma(t), t > 0,\\
V - \nu_{\Gamma(t)} \cdot v &= -\tfrac{m_0}{2}\big[\nu_{\Gamma(t)} \cdot \nabla\mu\big]_{\Gamma(t)} && \text{on } \Gamma(t), t > 0,\\
\mu &= \sigma\kappa_{\Gamma(t)} && \text{on } \Gamma(t), t > 0,
\end{aligned}$$

when $m(\varepsilon) = m_0 > 0$ and

$$\begin{aligned}
\partial_t v + v \cdot \nabla v - \operatorname{div}(\nu^\pm Dv) + \nabla p &= 0 && \text{in } \Omega^\pm(t), t > 0,\\
\operatorname{div} v &= 0 && \text{in } \Omega^\pm(t), t > 0,\\
[v]_{\Gamma(t)} &= 0 && \text{on } \Gamma(t), t > 0,\\
-\big[\nu_{\Gamma(t)} \cdot (\nu^\pm Dv - p\mathrm{Id})\big]_{\Gamma(t)} &= \sigma\kappa\nu_{\Gamma(t)} && \text{on } \Gamma(t), t > 0,\\
V - \nu_{\Gamma(t)} \cdot v &= 0 && \text{on } \Gamma(t), t > 0,
\end{aligned} \tag{1.9}$$

when $m(\varepsilon) = m_0\varepsilon, m_0 > 0$. Here $\nu^\pm > 0$ are viscosity constants. Moreover, $\Omega^\pm(t) \subset \Omega$ are open and disjoint such that $\partial\Omega^-(t) = \Gamma(t) = \partial\Omega^+(t) \cap \Omega$. The outer normal of $\partial\Omega^-(t)$ is denoted by $\nu_{\Gamma(t)}$ and the normal velocity and the mean curvature of $\Gamma(t)$ are denoted by V and κ, respectively, taken with respect to $\nu_{\Gamma(t)}$. Furthermore,

$[\,.\,]_{\Gamma(t)}$ denotes the jump of a quantity across the interface in the direction of $\nu_{\Gamma(t)}$, i.e., $[f]_{\Gamma(t)}(x) = \lim_{h\to 0}(f(x + h\nu_{\Gamma(t)}) - f(x - h\nu_{\Gamma(t)}))$ for $x \in \Gamma(t)$. To our knowledge there are only two rigorous results known so far for the sharp interface limit. These are due to Abels and Röger [6] and Abels and Lengeler [5], where convergence in the sense of varifold solutions (cf. Chen [9]) is shown provided $m(\varepsilon)^{-1}\varepsilon \to 0$ as $\varepsilon \to 0$. A proof in a stronger sense is still open.

One of the main problem is to pass to the limit in the weak formulation of the right-hand side of (1.5), which is

$$\langle H^\varepsilon, \varphi\rangle := \varepsilon \int_\Omega \nabla c^\varepsilon \otimes \nabla c^\varepsilon : \nabla\varphi\, dx \quad \varphi \in C^1(\overline{\Omega})\,.$$

Here c^ε is the solution to (1.1)–(1.4). This functional describes capillary stresses at the diffuse interface and should converge to the right-hand side of the Young-Laplace law (1.9). For $\theta = 0, 1$ we expect for $\varphi \in C^\infty_{0,\sigma}(\Omega) = \left\{f \in C_0^\infty(\Omega)^d : \operatorname{div} f = 0\right\}$ that

$$\langle H^\varepsilon, \varphi\rangle \longrightarrow 2\sigma \int_{\Gamma(t)} \nu_{\Gamma(t)} \otimes \nu_{\Gamma(t)} : \nabla\varphi\, d\mathcal{H}^{d-1} = -2\sigma \int_{\Gamma(t)} \kappa\nu_{\Gamma(t)} \cdot \varphi\, d\mathcal{H}^{d-1} \tag{1.10}$$

as $\varepsilon \to 0$. The last equality holds since $\nu_{\Gamma(t)} \otimes \nu_{\Gamma(t)} : \nabla\varphi = -\operatorname{div}_{\Gamma(t)}\varphi$ due to $\operatorname{div}\varphi = 0$. Here and in the following $\sigma \in \mathbb{R}$ is defined as

$$\sigma = \frac{1}{2}\int_{\mathbb{R}} (\theta_0'(z))^2 dz$$

and $\theta_0(x)$ is the "optimal diffuse interface profile", that is the unique solution to

$$-\theta_0'' + f(\theta_0) = 0 \text{ in } \mathbb{R}, \qquad \theta_0(0) = 0, \qquad \lim_{z\to\pm\infty} \theta_0(z) = \pm 1\,. \tag{1.11}$$

The formally matched asymptotic calculations in [4] show that the leading part of c^ε is given by $\theta_0(d(x,t)/\varepsilon)$ where d is the signed distance function to $\Gamma(t)$. Then the convergence of H^ε can be shown in the same way as in Sect. 1.3.

In the following we will simplify the problem by considering the convective Cahn-Hilliard equation (1.1) and (1.2). But we still ask, whether (1.10) holds true for the solution of (1.1) and (1.2). It was shown in [13, Chap. 6] that in the case $m(\theta) = \varepsilon$ the solutions c^ε of (1.1)–(1.4) with suitable well-prepared initial data c_0^ε converge to solutions of the geometric evolution equation

$$\begin{aligned} V &= v \cdot \nu \quad \text{on } \Gamma(t), t > 0,\\ \Gamma(0) &= \Gamma_0, \end{aligned}$$

where $\Gamma(t) \subseteq \Omega$, $t \geq 0$, is an evolving smooth hypersurface with normal velocity V and normal field ν. Moreover, one can prove (1.10) as follows: One replaces c_A^ε

defined in Sect. 1.3 by the approximate solution c_A^ε defined in [13]. Since the leading part of c_A^ε in [13] has the form $\theta_0(d(x,t)/\varepsilon)$, the arguments in Sect. 1.3 show (1.10), where one uses the convergence results for $c^\varepsilon - c_A^\varepsilon$ proven in [13, Chap. 6].

The goal of the paper is to show that (1.10) is no longer valid if $m(\varepsilon)$ tends to zero too fast as $\varepsilon \to 0$. More precisely, we will show nonconvergence in the case $m(\varepsilon) = \varepsilon^\theta$ with $\theta > 3$. Here we assume that the initial values c_0^ε have a special form defined in Sect. 1.2.

Finally, we comment on related results in the literature. Kwek [12] showed the existence of classical solutions to the convective Cahn-Hilliard equation. For the existence of weak solutions and strong solutions locally in time for the "model H" we refer to the results of Abels [1–3] and references given there. The sharp interface limit for the classical Cahn-Hilliard equation, i.e., (1.1) and (1.2) with $v = 0$ and $m(\varepsilon) = m_0 > 0$ was proved by Alikakos et al. [7] in a strong sense as long as the limit Hele-Shaw problem possesses a smooth solution and by Chen [9] in the sense of varifold solutions.

The structure of the article is as follows: In Sect. 1.2, we determine the notation and summarize the basic assumptions. In Sect. 1.3, we consider the convective Cahn-Hilliard equation with the mobility constant $m(\varepsilon) = \varepsilon^\theta$ for $\theta > 3$. For $\theta > 3$ we show that the term $-\varepsilon \operatorname{div}(\nabla c^\varepsilon \otimes \nabla c^\varepsilon)$ in general does not converge to the mean curvature of the interface, where c^ε is the solution for the convective Cahn-Hilliard equation. The reason is that the convection term $v \cdot \nabla c^\varepsilon$ dominates the motion of the interface $\Gamma(t)$. Therefore we can show that the approximate solutions do not have the form $\theta_0(d(x,t)/\varepsilon)$ as in [4] where d is the signed distance function to $\Gamma(t)$ and $\theta_0(x)$ is the "optimal diffuse interface profile".

1.2 Notation and Basic Assumptions

We denote $a \otimes b = (a_i b_j)_{i,j=1}^d$ for $a, b \in \mathbb{R}^d$ and $A : B = \sum_{i,j=1}^d A_{ij} B_{ij}$ for $A, B \in \mathbb{R}^{d\times d}$. The cofactor matrix is denoted by $\operatorname{cof}(A)$ for $A \in \mathbb{R}^{d\times d}$. We assume that $\Omega \subset \mathbb{R}^d$ is a bounded domain with smooth boundary $\partial\Omega$. For a time interval $(0,T)$, $T > 0$, we define $\Omega_T = \Omega \times (0,T)$ and $\partial_T\Omega = \partial\Omega \times (0,T)$. Moreover, n denotes the exterior unit normal on $\partial\Omega$. For a hypersurface $\Gamma(t) \subset \Omega$, $t \in [0,T]$, without boundary such that $\Gamma(t) = \partial\Omega^-(t)$ for a domain $\Omega^-(t) \subset\subset \Omega$, the interior domain is denoted by $\Omega^-(t)$ and the exterior domain by $\Omega^+(t) := \Omega \backslash (\Omega^-(t) \cup \Gamma(t))$, i.e., $\Gamma(t)$ separates Ω into an interior and an exterior domain. The exterior unit normal on $\partial\Omega^-(t) = \Gamma(t)$ is denoted by $\nu_{\Gamma(t)}$. The mean curvature of $\Gamma(t)$ is denoted by $\kappa = \kappa(t)$ with the sign convention that κ is positive, if $\Gamma(t)$ is curved in the direction of $\nu_{\Gamma(t)}$. For a signed distance function d with respect to $\Gamma(t)$, we assume $d < 0$ in $\Omega^-(t)$ and $d > 0$ in $\Omega^+(t)$. By this convention we obtain $\nabla d = \nu_{\Gamma(t)}$ on $\Gamma(t)$. Moreover, we use the definition $Q^\pm := \{(x,t) \in \Omega_T : d(x,t) \gtrless 0\}$. The "double-well" potential $F : \mathbb{R} \to \mathbb{R}$ is a smooth function taking its global minimum 0 at ± 1. For its derivative $f(c) = F'(c)$ we assume

$$f(\pm 1) = 0, \quad f'(\pm 1) > 0, \quad \int_{-1}^{u} f(s)\,ds = \int_{1}^{u} f(s)\,ds > 0 \quad \forall u \in (-1,1). \tag{1.12}$$

Moreover, we assume that there exists some constant $C_0 \geq 1$ such that $F(c)$ is monotonically increasing for $c \geq C_0$ (i.e., $f(c) \geq 0$) and monotonically decreasing for $c \leq -C_0$ (i.e., $f(c) \leq 0$). For example, this holds for $F(c) = \frac{1}{8}(1-c^2)^2$. In Eq. (1.1) the given velocity field satisfies $v \in C_b^0(\mathbb{R}; C_b^4(\overline{\Omega}))^d$ with $\operatorname{div} v = 0$ and $v \cdot n_{\partial\Omega} = 0$ on $\partial\Omega$ and the mobility constant $m(\varepsilon)$ has the form $m(\varepsilon) = \varepsilon^\theta$ for some $\theta \geq 0$. In Eq. (1.4) we choose the special initial value

$$c^\varepsilon|_{t=0} = \zeta\Big(\tfrac{d^0}{\delta}\Big)\theta_0\Big(\tfrac{d^0}{\varepsilon}\Big) + \Big(1 - \zeta\Big(\tfrac{d^0}{\delta}\Big)\Big)\Big(2\chi_{\{d^0 \geq 0\}} - 1\Big) \qquad \text{in } \Omega\,, \tag{1.13}$$

where we determine the constant $\delta > 0$ later and where d^0 is the signed distance function to an initial smooth hypersurface Γ_0 such that $\Gamma_0 = \partial\Omega_0^-$ with $\Omega_0^- \subset\subset \Omega$. Here $\zeta \in C_0^\infty(\mathbb{R})$ is a cut-off function such that

$$\zeta(z) = 1 \text{ if } |z| < \frac{1}{2}, \qquad \zeta(z) = 0 \text{ if } |z| > 1, \qquad z\zeta'(z) \leq 0 \text{ in } \mathbb{R}\,, \tag{1.14}$$

and θ_0 is the unique solution to (1.11). This choice of the initial value is natural since we can expect that $c^\varepsilon \approx \theta_0(d(x,t)/\varepsilon)$ for the model H with $m(\varepsilon) = \varepsilon^\theta, \theta = 0, 1$, see [4]. Here d is the signed distance function to the interface $\Gamma(t)$. In the following lemma we show the existence of a unique solution to the problem (1.11).

Lemma 1.1 *Let $f \in C^\infty(\mathbb{R})$ be given such that the properties (1.12) hold. Then the problem*

$$-w'' + f(w) = 0 \text{ in } \mathbb{R}, \qquad w(0) = 0, \qquad \lim_{z \to \pm\infty} w(z) = \pm 1 \tag{1.15}$$

has a unique solution. In addition, the following properties hold

$$w'(z) > 0 \qquad \forall z \in \mathbb{R}\,, \tag{1.16}$$

$$\big|w^2(z) - 1\big| + \big|w^{(n)}(z)\big| \leq C_n e^{-\alpha|z|} \qquad \forall z \in \mathbb{R},\ n \in \mathbb{N}\backslash\{0\} \tag{1.17}$$

for some constants $C_n > 0$, $n \in \mathbb{N}\backslash\{0\}$, and where α is a fixed constant such that

$$0 < \alpha < \min\Big\{\sqrt{f'(-1)}, \sqrt{f'(1)}\Big\}.$$

Proof See Remark 3.1 in [7] or [13]. □

1.3 Nonconvergence Result

First we investigate the flow of the velocity field v and prove some properties which we will need later.

Lemma 1.2 *Let $v \in C_b^0(\mathbb{R}; C_b^4(\overline{\Omega}))^d$ be a given smooth velocity field such that $v \cdot n|_{\partial\Omega} = 0$. Then there exists a unique global solution y to the problem*

$$\frac{d}{dt} y(t; y_0) = v(y(t; y_0), t), \quad y(0; y_0) = y_0$$

for all $y_0 \in \overline{\Omega}$. In particular, the flow $X(., t) =: X_t : \overline{\Omega} \to \overline{\Omega}$ defined by $X(y_0, t) = y(t; y_0)$ is a C^4-diffeomorphism for all $t \in \mathbb{R}$.

In addition, if $\operatorname{div} v = 0$ *in Ω, then it holds*

$$\det(DX_t(x)) = 1 \quad in\ \Omega_T \tag{1.18}$$

and

$$\left|DX_t^{-T} \circ X_t \nabla d^0\right|^2 = \det(\partial_{\tau_i} X_t \cdot \partial_{\tau_j} X_t)_{i,j=1}^{d-1} \quad on\ \Gamma_0\,, \tag{1.19}$$

where $\{\tau_1(x), \ldots, \tau_{d-1}(x)\}$ is an orthonormal basis of $T_x\Gamma_0$ and d^0 is the signed distance function of Γ_0.

Proof Although the results are classical, we give a proof for the convenience of the reader. By the Picard-Lindelöf theorem, there exists a unique solution $y(.; y_0) : I_{max} \to \mathbb{R}^d$ for every $y_0 \in \Omega$ where I_{max} is the maximal interval of existence. Since $v \cdot n|_{\partial\Omega} = 0$, there exists a unique global solution $y(.; y_0)$ such that $y(t; y_0) \in \partial\Omega$ for all $t \in \mathbb{R}$ when $y_0 \in \partial\Omega$, cf. [8, Sect. 35. 4. Bemerkung]. By the uniqueness of the solutions, it follows $y(t; y_0) \in \Omega$ for all $t \in I_{max}$ when $y_0 \in \Omega$. In particular, every solution $y(.; y_0)$ is bounded for $y_0 \in \overline{\Omega}$ and therefore it holds $I_{max} = \mathbb{R}$. Since $v(., t) \in C^4(\overline{\Omega})^d$ for all $t \in \mathbb{R}$, it follows from [14, III. Sect. 13 XI. Corollary] that $X_t \in C^4(\overline{\Omega})^d$. Let us show that X_t is invertible for all $t \in \mathbb{R}$. Let $t_0 \in \mathbb{R}$ be any time. Define $X_{t_0}^{-1} : \overline{\Omega} \to \overline{\Omega}$ by $X_{t_0}^{-1}(x) = \tilde{y}(-t_0; x)$ where $\tilde{y}(.; x)$ is the solution to

$$\tilde{y}'(t) = v(\tilde{y}(t), t + t_0) \text{ in } \mathbb{R}, \quad \tilde{y}(0) = x\,.$$

Claim: $X_{t_0}(X_{t_0}^{-1}(x)) = x$ for all $x \in \overline{\Omega}$.

By definition of $X_{t_0}^{-1}$, it holds

$$X_{t_0}(X_{t_0}^{-1}(x)) = y(t_0; \tilde{y}(-t_0; x))\,.$$

Since $y(. + t_0; \tilde{y}(-t_0; x)) : \mathbb{R} \to \mathbb{R}$ and $\tilde{y}$ are both solutions to

$$y'(t) = v(y(t), t + t_0) \text{ in } \mathbb{R}, \quad y(-t_0) = \tilde{y}(-t_0; x),$$

it follows $\tilde{y}(t; x) = y(t + t_0; \tilde{y}(-t_0; x))$ for all $t \in \mathbb{R}$ by uniqueness of the solution. In particular, it follows $y(t_0; \tilde{y}(-t_0; x)) = \tilde{y}(0; x) = x$. This shows the claim.

Analogously, one can show $X_{t_0}^{-1}(X_{t_0}(x)) = x$ for all $x \in \overline{\Omega}$. Hence $X_{t_0}^{-1}$ is the inverse of X_{t_0} and $X_{t_0}^{-1} \in C^4(\overline{\Omega})^d$ by the same arguments as for X_{t_0}. Due to [10, Satz 5.2], it holds

$$\frac{d}{dt} \det(DX_t(x)) = \operatorname{div} v(X, t)|_{X = X_t(x)} \det(DX_t(x)).$$

Since $\operatorname{div} v = 0$ and $X_0 = \mathrm{Id}$, we obtain

$$\det(DX_t(x)) = 1 \quad \forall (x, t) \in \Omega_T.$$

Using this property, we can verify the last assertion of the lemma. Since $X_t^{-1}(X_t(x)) = x$ for all $x \in \Omega$, it follows by differentiating with respect to x

$$\mathrm{Id} = DX_t^{-1} \circ X_t \, DX_t \quad \text{in } \Omega.$$

Due to Cramer's rule and the last equation, it follows

$$DX_t^{-T} \circ X_t = \frac{1}{\det(DX_t^T)} \operatorname{cof}(DX_t^T)^T = \operatorname{cof}(DX_t).$$

Therefore we get in a neighborhood of $\Gamma(0)$

$$\begin{aligned} |DX_t^{-T} \circ X_t \nabla d^0|^2 &= \nabla d^0 \cdot \big(\operatorname{cof}(DX_t^T)\operatorname{cof}(DX_t)\big)\nabla d^0 \\ &= \nabla d^0 \cdot \big(\operatorname{cof}(DX_t^T DX_t)\big)\nabla d^0. \end{aligned}$$

Let Q be the change-of-basis matrix taking the orthonormal basis $\{\tau_1, \ldots, \tau_{d-1}, \nu_{\Gamma_0}\}$ to the standard basis $\{e_1, \ldots, e_d\}$ in $\mathbb{R}^d$. Then Cramer's rule yields

$$Q^T = Q^{-1} = \frac{1}{\det Q} \operatorname{cof} Q^T = \operatorname{cof} Q^T,$$

and therefore it holds on Γ_0

$$\begin{aligned} \nabla d^0 \cdot \big(\operatorname{cof}(DX_t^T DX_t)\big)\nabla d^0 &= (Qe_d) \cdot \big(\operatorname{cof}(DX_t^T DX_t)\big)(Qe_d) \\ &= e_d \cdot \big(\operatorname{cof} Q^T \operatorname{cof}(DX_t^T DX_t)\operatorname{cof} Q\big)e_d \\ &= \big(\operatorname{cof}(Q^T DX_t^T DX_t Q)\big)_{dd} \\ &= \det(\partial_{\tau_i} X_t \cdot \partial_{\tau_j} X_t)_{i,j=1}^{d-1}, \end{aligned}$$

where the last equality follows due to the definition of the cofactor matrix. This completes the proof of the lemma. □

Our main result is:

Theorem 1.1 *Let $\Omega \subset \mathbb{R}^d$ be a bounded domain with smooth boundary $\partial\Omega$, Γ_0 a smooth hypersurface such that $\Gamma_0 = \partial\Omega_0^-$ for a domain $\Omega_0^- \subset\subset \Omega$ and let $(c^\varepsilon, \mu^\varepsilon)$ be the solution to the convective Cahn-Hilliard equation (1.1)–(1.3) with initial condition (1.13). Then, for every $T > 0$ and for all $\varphi \in C^\infty([0,T]; \mathscr{D}(\Omega)^d)$, it holds*

$$\int_0^T \langle H^\varepsilon, \varphi\rangle dt \longrightarrow 2\sigma \int_0^T \int_{\Gamma(t)} \left|\nabla(d^0(X_t^{-1}))\right| \nu_{\Gamma(t)} \otimes \nu_{\Gamma(t)} : \nabla\varphi \, d\mathscr{H}^{d-1}\, dt\,,$$

as $\varepsilon \searrow 0$ and where the evolving hypersurface $\Gamma(t)$ is the solution to the evolution equation

$$V(x,t) = \nu_{\Gamma(t)}(x,t)\cdot v(x,t) \text{ for } x \in \Gamma(t), t \in (0,T), \quad \Gamma(0) = \Gamma_0,$$

where V is the normal velocity of $\Gamma(t)$. Moreover, it holds

$$\left\|c^\varepsilon - (2\chi_{Q^+} - 1)\right\|^2_{L^2(\Omega_T)} = \mathscr{O}(\varepsilon)\,,$$

as $\varepsilon \searrow 0$.

Remark 1.1 In general $\left|\nabla(d^0(X_t^{-1}))\right| = \left|DX_t^{-T}\nabla d^0 \circ X_t^{-1}\right| \neq 1$. This can be shown as follows. By choosing a suitable interface Γ_0, it is sufficient to show that in general DX_t^{-T} is not length preserving. We show this by a counterexample. Let $\Omega \subset \mathbb{R}^2$ be the interior of the ellipse defined by the equation $\frac{x_1^2}{2} + \frac{x_2^2}{4} = 1$. For the velocity field $v : \overline{\Omega} \to \mathbb{R}$, we set $v(x_1, x_2) := (x_2, -2x_1)$. Note that $\operatorname{div} v = 0$ in Ω and $v \cdot n_{\partial\Omega} = 0$ on $\partial\Omega$ since $(2x_1, x_2)$ for $(x_1, x_2) \in \partial\Omega$ is a normal on $\partial\Omega$. Then the function $y : \mathbb{R} \to \Omega$ defined by $y(t) = \left(\sin(\sqrt{2}t), \sqrt{2}\cos(\sqrt{2}t)\right)$ is a solution to

$$y'(t) = v(y(t)) \text{ in } \mathbb{R}, \quad y(0) = (0, \sqrt{2})\,.$$

Since the velocity field v is independent of the time t, it follows $X_t^{-1} = X_{-t}$ where X_t is the flow of the ordinary differential equation $y' = v(y)$. Differentiating the identity $X_{-t} \circ X_t = \mathrm{Id}$ with respect to t, yields

$$0 = DX_{-t}(X_t(x))\, v(X_t(x)) - v(x) \quad \forall x \in \Omega$$

since $\partial_t X_t = v(X_t)$. Using our special solution above, we obtain $X_{\frac{\pi}{2\sqrt{2}}}(0, \sqrt{2}) = (1, 0)$. Hence we conclude

$$0 = DX_{-\frac{\pi}{2\sqrt{2}}}(1,0)\, v(1,0) - v(0,\sqrt{2}) = DX^{-1}_{\frac{\pi}{2\sqrt{2}}}(1,0)\begin{pmatrix} 0 \\ -2 \end{pmatrix} - \begin{pmatrix} \sqrt{2} \\ 0 \end{pmatrix}.$$

Thus there exists a vector $w \in \mathbb{R}^2$ and $(x,t) \in \Omega \times \mathbb{R}$ such that $\left|DX_t^{-1}(x)w\right| \neq |w|$ and therefore $DX_t^{-T}(x)$ is also not length preserving.

The strategy of the proof of the theorem is the following: First we construct a family of approximate solutions $\{c_A^\varepsilon\}_{0<\varepsilon\leq 1}$ and estimate the difference $\nabla(c^\varepsilon - c_A^\varepsilon)$. Then for an approximate functional H_A^ε, we show that $H^\varepsilon - H_A^\varepsilon \to 0$ as $\varepsilon \searrow 0$ when $\theta > 3$. Finally, we prove the assertion of the theorem for H_A^ε.

We start with the observation that $\Gamma(t) := X_t(\Gamma_0)$ is the solution to the evolution equation.

Lemma 1.3 *Let $\Gamma_0 \subset \Omega$ be a given smooth hypersurface such that $\Gamma_0 = \partial\Omega_0^-$ for a domain $\Omega_0^- \subset\subset \Omega$. Then the evolving hypersurface $\Gamma_t := \Gamma(t) := X_t(\Gamma_0) \subset \Omega$ is the solution to the problem*

$$V(x,t) = \nu_{\Gamma(t)}(x,t)\cdot v(x,t) \quad on\ \Gamma_t, t>0, \quad \Gamma(0) = \Gamma_0,$$

where V is the normal velocity and $\nu_{\Gamma(t)}$ the unit outward normal to Γ_t.

Proof The initial condition $\Gamma(0) = \Gamma_0$ is satisfied since $X_t(x)|_{t=0} = x$ for all $x \in \Omega$. Let $x_0 \in \Gamma_{t_0}$, $t_0 \in (0,T)$, be arbitrary. Then there exists $\tilde{x}_0 \in \Gamma_0$ such that $x_0 = X_{t_0}(\tilde{x}_0)$. By definition of the normal velocity, we obtain

$$\begin{aligned} V(x_0,t_0) &= \frac{d}{dt}X_t(x)\Big|_{(x,t)=(\tilde{x}_0,t_0)} \cdot \nu(x_0,t_0) = v(X_{t_0}(\tilde{x}_0),t_0)\cdot\nu(x_0,t_0) \\ &= v(x_0,t_0)\cdot\nu_{\Gamma(t)}(x_0,t_0)\,. \end{aligned}$$

This completes the proof. □

Let $d : \Omega \times [0,\infty) \to \mathbb{R}$ be the signed distance to $(\Gamma_t)_{t\in[0,\infty)}$ satisfying $d(.,t) < 0$ inside Γ_t and $d(.,t) > 0$ outside Γ_t. Note that $d^0(x) = d(x,0)$ for all $x \in \Omega$. Let $S_0(x)$ be the orthogonal projection of x on Γ_0. Then there exists a constant $\delta > 0$ such that $\Gamma_0(\delta) := \left\{x \in \Omega : \left|d^0(x))\right| < \delta\right\} \subset \Omega$ and $\tau_0 : \Gamma_0(\delta) \to (-\delta,\delta)\times\Gamma_0$ defined by $\tau_0(x) = (d^0(x), S_0(x))$ is a smooth diffeomorphism, cf. [11, Kapitel 4.6]. Furthermore, we define $\Gamma := \{(x,t) \in \Omega_T : d(x,t) = 0\}$ and $\Gamma(\delta) := \{(x,t) \in \Omega_T : |d(x,t)| < \delta\}$.

Lemma 1.4 *For* $e : \bigcup_{t\in[0,T]} X_t(\Gamma_0(\delta)) \times \{t\} \to \mathbb{R}$ *defined by* $e(x,t) := d^0\left(X_t^{-1}(x)\right)$ *the following properties hold:*

1. *$\frac{d}{dt}e(x,t) = -v(x,t)\cdot\nabla e(x,t)$ for all $(x,t) \in \bigcup_{t\in[0,T]} X_t(\Gamma_0(\delta)) \times \{t\}$.*
2. *$e(x,t)$ is a level set function for Γ_t, i.e., $e(x,t) = 0$ if and only if $x \in \Gamma_t$.*

Proof By definition of δ, the function e is differentiable with respect to x in $X_t(\Gamma_0(\delta))$ for all $t \in [0,T]$.

To 1: It holds for all $x \in \Omega$

$$X_t(X_t^{-1}(x)) = x\,.$$

Differentiating with respect to t and x, we get the identities

$$0 = DX_t(X_t^{-1}(x))\partial_t X_t^{-1}(x) + \partial_t X_t(X_t^{-1}(x)) \tag{1.20}$$

$$\text{and } \mathrm{Id} = DX_t(X_t^{-1}(x))DX_t^{-1}(x). \tag{1.21}$$

Hence we get by the definition of e

$$\begin{aligned}\frac{d}{dt}e(x,t) &= \frac{d}{dt}\left(d^0(X_t^{-1}(x))\right) = \nabla d^0(X_t^{-1}(x)) \cdot \partial_t X_t^{-1}(x)\\ &= -\nabla d^0(X_t^{-1}(x)) \cdot DX_t^{-1}\partial_t X_t(X_t^{-1}(x))\\ &= -\nabla d^0(X_t^{-1}(x)) \cdot DX_t^{-1}v(X_t(X_t^{-1}(x)),t)\\ &= -\nabla\left(d^0(X_t^{-1}(x))\right) \cdot v(x,t)\\ &= -\nabla e(x,t) \cdot v(x,t)\,,\end{aligned}$$

where we have used (1.20) and (1.21) in the third equation.
To 2: The following equivalences hold since $X_t : \Omega \to \Omega$ is a diffeomorphism

$$\begin{aligned}d^0(X_t^{-1}(x)) = 0 &\Leftrightarrow X_t^{-1}(x) \in \Gamma_0 \Leftrightarrow \exists y \in \Gamma_0 \text{ s.t. } X_t^{-1}(x) = y\\ &\Leftrightarrow \exists y \in \Gamma_0 \text{ s.t. } x = X_t(y) \Leftrightarrow x \in X_t(\Gamma_0)\,.\end{aligned}$$

This shows that e is a level set function for Γ_t. □

As mentioned in Sect. 1.2, let θ_0 be the solution to (1.11) and let ζ be a cut-off function as in (1.14). Then we define

$$c_A^\varepsilon(x,t) := \begin{cases} \pm 1 & \text{in } \overline{Q^\pm} \cap \bigcup\limits_{t\in[0,T]} \overline{X_t(\Omega\backslash\Gamma_0(\delta))} \times \{t\}\,,\\ \zeta(\frac{e}{\delta})\theta_0(\frac{e}{\varepsilon}) \pm (1-\zeta(\frac{e}{\delta})) & \text{in } Q^\pm \cap \bigcup\limits_{t\in[0,T]} X_t(\Gamma_0(\delta)\backslash\Gamma_0(\frac{\delta}{2})) \times \{t\},\\ \theta_0(\frac{e}{\varepsilon}) & \text{in } \bigcup\limits_{t\in[0,T]} X_t(\Gamma_0(\frac{\delta}{2})) \times \{t\}\,.\end{cases}$$

Note that $c_A^\varepsilon(.,0) = c^\varepsilon(.,0)$ since $e(.,0) = d^0$ and $\partial_t c_A^\varepsilon + v \cdot \nabla c_A^\varepsilon = 0$ in Ω_T since $\partial_t e + v \cdot \nabla e = 0$. Moreover, c_A^ε and Δc_A^ε satisfy Neumann boundary conditions on $\partial\Omega$ since $c_A^\varepsilon = 1$ in a neighborhood of the boundary $\partial\Omega$.

Furthermore, we define for all $\varphi \in \mathscr{D}(\Omega)^d$ the functional $H_A^\varepsilon : \mathscr{D}(\Omega)^d \to \mathbb{R}$ by

$$\langle H_A^\varepsilon, \varphi\rangle = \varepsilon \int_\Omega \nabla c_A^\varepsilon \otimes \nabla c_A^\varepsilon : \nabla\varphi\, dx\,.$$

Lemma 1.5 *Let c_A^ε be defined as above. Then there exists some constant $C > 0$ independent of ε and $\varepsilon_0 \in (0,1]$ such that the estimates*

$$\left\| \Delta c_A^\varepsilon(.,t) \right\|_{L^2(\Omega)} \le C\varepsilon^{-\frac{3}{2}}, \tag{1.22}$$

$$\left\| \nabla c_A^\varepsilon(.,t) \right\|_{L^2(\Omega)} \le C\varepsilon^{-\frac{1}{2}}, \tag{1.23}$$

$$\left\| f(c_A^\varepsilon(.,t)) \right\|_{L^2(\Omega)} \le C\varepsilon^{\frac{1}{2}}, \tag{1.24}$$

$$\left\| c_A^\varepsilon(.,t) - (2\chi_{Q^+}(.,t) - 1) \right\|_{L^2(\Omega)} \le C\varepsilon^{\frac{1}{2}} \tag{1.25}$$

hold for all $t \in [0,T]$ and $\varepsilon \in (0, \varepsilon_0)$.

Proof We obtain for all $(x,t) \in \bigcup_{t\in[0,T]} X_t(\Gamma_0(\frac{\delta}{2})) \times \{t\}$

$$\Delta c_A^\varepsilon(x,t) = \varepsilon^{-2}|\nabla e|^2 \theta_0''\left(\tfrac{e}{\varepsilon}\right) + \varepsilon^{-1}\Delta e\, \theta_0'\left(\tfrac{e}{\varepsilon}\right).$$

Hence there exists some constant $C > 0$ independent of ε and $t \in [0,T]$ such that

$$\left\| \Delta c_A^\varepsilon(.,t) \right\|_{L^2(X_t(\Gamma_0(\delta/2)))} \le C\left(\varepsilon^{-2} \left\| \theta_0''\left(\tfrac{e}{\varepsilon}\right) \right\|_{L^2(X_t(\Gamma_0(\delta/2)))} + \varepsilon^{-1}\right).$$

Using $\theta_0''(z) \le Ce^{-\alpha|z|}$ (see Lemma 1.1) for all $z \in \mathbb{R}$ and for some $C > 0$, we conclude

$$\begin{aligned}
\left\| \theta_0''(e/\varepsilon) \right\|^2_{L^2(X_t(\Gamma_0(\delta/2)))} &\le C \int_{X_t(\Gamma_0(\delta/2))} e^{-2\alpha|d^0(X_t^{-1}(x))/\varepsilon|}\, dx \\
&= C \int_{\Gamma_0(\delta/2)} e^{-2\alpha|d^0(x)/\varepsilon|} \det|DX_t(x)| dx \\
&= C \int_{\Gamma_0(\delta/2)} e^{-2\alpha|d^0(x)/\varepsilon|}\, dx,
\end{aligned}$$

where we have used (1.18). Using the identity $\int_{\Gamma_0(\delta/2)} f(x)\,dx = \int_{-\delta/2}^{\delta/2} \int_{\Gamma_0^r} f(x)\, d\mathcal{H}^{n-1}\, dr$ for all integrable functions f where $\Gamma_0^r = \{x \in \Omega : x = s + r\,\nu_{\Gamma_0}(s),\, s \in \Gamma_0\}$ for $r \in \mathbb{R}$ (we will show this identity at the end of the proof), one gets

$$\begin{aligned}
\left\| \theta_0''(e/\varepsilon) \right\|^2_{L^2(X_t(\Gamma_0(\delta/2)))} &\le C \int_{-\delta/2}^{\delta/2} \int_{\Gamma_0^r} e^{-2\alpha|d^0/\varepsilon|}\, d\mathcal{H}^{n-1}\, dr \\
&= C \int_{-\delta/2}^{\delta/2} e^{-2\alpha|r/\varepsilon|} \int_{\Gamma_0^r} 1\, d\mathcal{H}^{n-1}\, dr \\
&\le C\varepsilon
\end{aligned}$$

for some $C = C(\Gamma_0) > 0$ independent of ε and $t \in [0,T]$. Again using Lemma 1.1, we obtain in $\bigcup_{t\in[0,T]} X_t(\Gamma_0(\delta)\backslash\Gamma_0(\frac{\delta}{2})) \times \{t\}$

$$\begin{aligned}\Delta c_A^\varepsilon(x,t) &= \varepsilon^{-2}|\nabla e|^2\theta_0''(\tfrac{e}{\varepsilon})\zeta(\tfrac{e}{\delta}) + \varepsilon^{-1}\Delta e\,\theta_0'(\tfrac{e}{\varepsilon})\zeta(\tfrac{e}{\delta}) + 2\varepsilon^{-1}\theta_0'(\tfrac{e}{\varepsilon})\nabla e\cdot\nabla\big(\zeta(\tfrac{e}{\delta})\big)\\ &\quad+\big(\theta_0(\tfrac{e}{\varepsilon}) - (2\chi_{\overline{Q^+}}-1)\big)\Delta\big(\zeta(\tfrac{e}{\delta})\big)\\ &= \mathcal{O}(\varepsilon^{-2}e^{-\frac{\alpha\delta}{4\varepsilon}}).\end{aligned}$$

Altogether, this shows (1.22) for all $\varepsilon > 0$ small enough.

The second estimate can be shown by the same arguments.

Using $f(\pm1) = 0$ and the Taylor expansion, it follows

$$f(c_A^\varepsilon) = f'((2\chi_{\overline{Q^+}}-1) + \Theta(c_A^\varepsilon - 2\chi_{\overline{Q^+}}+1))\big(c_A^\varepsilon - (2\chi_{\overline{Q^+}}-1)\big)$$

for some $\Theta = \Theta(x,t) \in (0,1)$. For all $(x,t) \in \bigcup_{t\in[0,T]} X_t(\Gamma_0(\frac{\delta}{2}))\times\{t\}$, there exists some constant $C > 0$ independent of x and t such that

$$\big|c_A^\varepsilon - (2\chi_{\overline{Q^+}}-1)\big| = \big|\theta_0(\tfrac{e}{\varepsilon}) - (2\chi_{\overline{Q^+}}-1)\big| \le Ce^{-\frac{\alpha|e|}{2\varepsilon}}$$

due to Lemma 1.1. In $\bigcup_{t\in[0,T]} X_t(\Gamma_0(\delta)\backslash\Gamma_0(\frac{\delta}{2}))\times\{t\}$ we get due to the definition of c_A^ε

$$\big|c_A^\varepsilon - (2\chi_{\overline{Q^+}}-1)\big| \le Ce^{-\frac{\alpha\delta}{4}}.$$

In $\bigcup_{t\in[0,T]} \overline{X_t(\Omega\backslash\Gamma_0(\delta))}\times\{t\}$ we have $c_A^\varepsilon = 2\chi_{\overline{Q^+}}-1$. Then we can apply the same estimates as above to prove the third and also the forth assertion.

It remains to show the integral identity $\int_{\Gamma_0(\delta/2)} f(x)\,dx = \int_{-\delta/2}^{\delta/2}\int_{\Gamma_0^r} f(x)\,d\mathcal{H}^{n-1}\,dr$. To this end we choose an relatively open set $U \subset \Gamma_0$ such that U can be described as a graph, i.e., (possibly after rotation) there exists an open set $D \subset \mathbb{R}^{d-1}$ and a function $g : D \to \mathbb{R}$ such that $U = \{(y, g(y)) : y \in D\}$. Define the sets $U(\delta)$ and U_r, $r \in (-\delta,\delta)$, by

$$U(\delta) = \big\{x + r\nu_{\Gamma_0}(x) : x \in U, r \in (-\delta,\delta)\big\},\quad U_r = \big\{x + r\nu_{\Gamma_0}(x) : x \in U\big\}.$$

Then the function $\Phi : (-\delta,\delta)\times D \to U(\delta)$ defined by $\Phi(r,y) = (y, g(y)) + r\nu_{\Gamma_0}(y, g(y))$ is a smooth diffeomorphism. Let $f : \mathbb{R}^d \to \mathbb{R}$ be an arbitrary integrable function. By coordinate transformation, we obtain

$$\begin{aligned}\int_{U(\delta)} f(x)\,dx &= \int_{-\delta}^{\delta}\int_D (f\circ\Phi)(r,y)|\det(D\Phi(r,y))|\,dy\,dr\\ &= \int_{-\delta}^{\delta}\int_D (f\circ\Phi)(r,y)\big|\det(D\Phi(r,y)^T D\Phi(r,y))\big|^{\frac12}dy\,dr.\end{aligned}$$

We continue with calculating $\det(D\Phi(r,y)^T D\Phi(r,y))$. For all $i \in \{1,\dots d-1\}$, it follows

$$\partial_r\Phi(r,y)\cdot\partial_{y_i}\Phi(r,y) = \nu_{\Gamma_0}(y, g(y))\cdot\big[(e_i, \partial_i g(y)) + r\partial_i(\nu_{\Gamma_0}(y, g(y)))\big] = 0,$$

where $e_i \in \mathbb{R}^{d-1}$ is the i-th standard unit vector. Here we have used that

$$\nu_{\Gamma_0}(y, g(y)) \cdot (e_i, \partial_i g(y)) = \frac{1}{\sqrt{|\nabla g|^2+1}} \begin{pmatrix} -\nabla g \\ 1 \end{pmatrix} \cdot \begin{pmatrix} e_i \\ \partial_i g \end{pmatrix} = 0$$

and $\nu_{\Gamma_0}(y, g(y)) \cdot \partial_i(\nu_{\Gamma_0}(y, g(y))) = 0$. Hence we get

$$\det(D\Phi(r, y)^T D\Phi(r, y)) = \det\left(\begin{array}{c|c} 1 & 0 \\ \hline 0 & D_y\Phi^T D_y\Phi \end{array}\right) = \det(D_y\Phi^T(r, y) D_y\Phi(r, y)).$$

Therefore we obtain the identity

$$\begin{aligned} \int_{U(\delta)} f(x)\,dx &= \int_{-\delta}^{\delta} \int_D (f \circ \Phi)(r, y) \left|\det(D_y\Phi(r, y)^T D_y\Phi(r, y))\right|^{\frac{1}{2}} dy\,dr \\ &= \int_{-\delta}^{\delta} \int_{U_r} f(x)\,d\mathcal{H}^{d-1}\,dr \end{aligned}$$

where the last equality follows from $\Phi(r, D) = U_r$ for all $r \in (-\delta, \delta)$, that is, $\Phi(r, .) : D \to U_r$ is a chart for U_r. Using partition of the unity, the assertion follows. This completes the proof of the lemma. □

Lemma 1.6 *Let c_ε^A be defined as above and let c^ε be the unique solution to (1.1)–(1.3) with initial condition (1.13). Then, for $\theta > 3$, there exists some constant $C > 0$ independent of ε and $\varepsilon_0 > 0$ such that*

$$\varepsilon \left\|\nabla(c^\varepsilon - c_A^\varepsilon)\right\|_{L^2(\Omega_T)}^2 \le C\varepsilon^{\frac{\theta-3}{2}}, \tag{1.26}$$

$$\text{and } \left\|c^\varepsilon - c_A^\varepsilon\right\|_{L^2(\Omega_T)}^2 \le C\varepsilon^{\theta-2} \tag{1.27}$$

for all $\varepsilon \in (0, \varepsilon_0]$.

Proof Let $R = c^\varepsilon - c_A^\varepsilon$ be the remainder. Since $\partial_t c_A^\varepsilon + v \cdot \nabla c_A^\varepsilon = 0$ in Ω_T, it holds

$$\begin{aligned} \int_\Omega R(., t)\,dx &= \int_0^t \int_\Omega \partial_t R\,dx\,dt = -\int_0^t \int_\Omega v \cdot \nabla R\,dx\,dt + \varepsilon^\theta \int_0^t \int_\Omega \Delta\mu^\varepsilon\,dx\,dt \\ &= \int_0^t \int_\Omega \operatorname{div} v\, R\,dx\,dt - \int_0^t \int_{\partial\Omega} v \cdot n R\,d\mathcal{H}^{d-1}\,dt \\ &\quad + \int_0^t \int_{\partial\Omega} \frac{\partial}{\partial n}\mu^\varepsilon\,d\mathcal{H}^{d-1}\,dt = 0 \end{aligned}$$

for all $t \in [0, T]$. Hence we can find a unique solution $\Psi : \Omega_T \to \mathbb{R}$ to the problem

$$-\Delta\Psi(., t) = R(., t) \text{ in } \Omega, \qquad \frac{\partial}{\partial n}\Psi(., t) = 0 \text{ on } \partial\Omega, \qquad \int_\Omega \Psi(., t) = 0$$

for every $t \in [0, T]$. We multiply the difference of the differential equations for c^ε and c_A^ε by Ψ and integrate the resulting equation over Ω. Then we get for all $t \in (0, T)$

$$
\begin{aligned}
0 &= \int_\Omega \Psi\big[\partial_t R + v \cdot \nabla R + \varepsilon^{\theta+1} \Delta^2 R + \varepsilon^{\theta+1} \Delta^2 c_A^\varepsilon - \varepsilon^{\theta-1} \Delta f(c^\varepsilon)\big] dx \\
&= \int_\Omega \Psi(-\Delta \partial_t \Psi) - \nabla \Psi \cdot v R + \varepsilon^{\theta+1} \Delta \Psi \Delta R \, dx \\
&\quad + \int_\Omega \varepsilon^{\theta+1} \Delta \Psi \Delta c_A^\varepsilon - \varepsilon^{\theta-1} \Delta \Psi f(c^\varepsilon) \, dx \\
&= \frac{1}{2} \frac{d}{dt} \int_\Omega |\nabla \Psi|^2 dx + \int_\Omega -\nabla \Psi \cdot v R + \varepsilon^{\theta+1} |\nabla R|^2 dx \\
&\quad + \int_\Omega -\varepsilon^{\theta+1} R \Delta c_A^\varepsilon + \varepsilon^{\theta-1} R f(c^\varepsilon) \, dx ,
\end{aligned}
$$

where we have used the Neumann boundary conditions $\frac{\partial}{\partial n} \Delta c_A^\varepsilon = \frac{\partial}{\partial n} \Psi = \frac{\partial}{\partial n} f(c^\varepsilon) = 0$, $v \cdot n = 0$ and $\frac{\partial}{\partial n} \Delta c^\varepsilon = -\varepsilon^{-1} \frac{\partial}{\partial n} \mu^\varepsilon + \varepsilon^{-2} \frac{\partial}{\partial n} f(c^\varepsilon) = 0$ on $\partial \Omega$ as well as $\operatorname{div} v = 0$ in Ω.

By the assumptions $f(c^\varepsilon) \geq 0$ for $c^\varepsilon \geq C_0 \geq 1 \geq c_A^\varepsilon$ and $f(c^\varepsilon) \leq 0$ for $c^\varepsilon \leq -C_0 \leq -1 \leq c_A^\varepsilon$, we obtain

$$
\int_{\{x \in \Omega \,:\, |c^\varepsilon(x,t)| \geq C_0\}} f(c^\varepsilon) R \, dx \geq 0 .
$$

Hence Hölder's and Young's inequalities yield

$$
\begin{aligned}
&\frac{1}{2} \frac{d}{dt} \int_\Omega |\nabla \Psi|^2 dx + \varepsilon^{\theta+1} \int_\Omega |\nabla R|^2 dx + \varepsilon^{\theta-1} \int_{\{x \in \Omega \,:\, |c^\varepsilon(x,t)| \geq C_0\}} f(c^\varepsilon) R \, dx \\
&\leq \left| \int_\Omega \nabla \Psi \cdot v R \, dx \right| + \varepsilon^{\theta+1} \|R\|_{L^2(\Omega)} \left\| \Delta c_A^\varepsilon \right\|_{L^2(\Omega)} \\
&\quad + \varepsilon^{\theta-1} \left| \int_{\{x \in \Omega \,:\, |c^\varepsilon(x,t)| < C_0\}} f(c^\varepsilon) R \, dx \right| . \qquad (1.28)
\end{aligned}
$$

We estimate the right-hand side. By integration by parts and due to $v \cdot n = 0$ on $\partial \Omega$, the identity

$$
\int_\Omega \partial_{ij} \Psi v_j \partial_i \Psi \, dx = -\int_\Omega \partial_i \Psi \partial_j v_j \partial_i \Psi \, dx - \int_\Omega \partial_i \Psi v_j \partial_{ij} \Psi \, dx = -\int_\Omega \partial_i \Psi v_j \partial_{ij} \Psi \, dx
$$

yields

$$
\int_\Omega \nabla \Psi \cdot \big(D^2 \Psi \, v\big) dx = 0 .
$$

Therefore we obtain the following estimate for the first term in (1.28) on the right-hand side

$$\left|\int_\Omega \nabla\Psi\cdot v R\,dx\right| = \left|\int_\Omega \nabla\Psi\cdot v\Delta\Psi\,dx\right| = \left|\int_\Omega \nabla\Psi\cdot\left(D^2\Psi\, v\right) + \nabla v : (\nabla\Psi\otimes\nabla\Psi)\,dx\right|$$
$$= \left|\int_\Omega \nabla v : (\nabla\Psi\otimes\nabla\Psi)\,dx\right| \le \|\nabla v\|_{L^\infty(\Omega_T)}\|\nabla\Psi\|^2_{L^2(\Omega)}, \qquad (1.29)$$

for all $t\in[0,T]$ and where we have used the boundary condition $\frac{\partial}{\partial n}\Psi = 0$ on $\partial\Omega$. Using Taylor series expansion for the last term in (1.28) on the right-hand side, we get for all $\varepsilon\in(0,\varepsilon_0)$

$$\int_{\{x\in\Omega\,:\,|c^\varepsilon(x,t)|<C_0\}} f(c^\varepsilon)R\,dx = \int_{\{x\in\Omega\,:\,|c^\varepsilon(x,t)|<C_0\}} f(c_A^\varepsilon)R + f'(c_A^\varepsilon+\Theta R)R^2\,dx$$
$$\le \left\|f(c_A^\varepsilon)\right\|_{L^2(\Omega)}\|R\|_{L^2(\Omega)} + C\|R\|^2_{L^2(\Omega)}$$
$$\le C\varepsilon^{\frac{1}{2}}\|R\|_{L^2(\Omega)} + C\|R\|^2_{L^2(\Omega)} \qquad (1.30)$$

for some $\Theta=\Theta(x,t)\in(0,1)$ and some constant $C>0$ independent of ε and $t\in[0,T]$. Here we have used inequality (1.24) with the same constant $\varepsilon_0>0$. Therefore estimate (1.28) turns with (1.29), (1.30), and (1.22) into

$$\frac{1}{2}\frac{d}{dt}\int_\Omega |\nabla\Psi|^2dx + \frac{\varepsilon^{\theta+1}}{2}\int_\Omega |\nabla R|^2dx$$
$$\le C_1\Big(\|\nabla\Psi\|^2_{L^2(\Omega)} + \varepsilon^{\theta-\frac{1}{2}}\|R\|_{L^2(\Omega)} + \varepsilon^{\theta-\frac{1}{2}}\|R\|_{L^2(\Omega)} + \varepsilon^{\theta-1}\|R\|^2_{L^2(\Omega)}\Big)$$

for some $C_1>0$ independent of ε and $t\in[0,T]$. Since

$$\|R\|^2_{L^2(\Omega)} = -\int_\Omega R\Delta\Psi\,dx = \int_\Omega \nabla R\cdot\nabla\Psi\,dx \le \|\nabla R\|_{L^2(\Omega)}\|\nabla\Psi\|_{L^2(\Omega)},$$

we obtain by Young's inequality

$$\varepsilon^{\theta-\frac{1}{2}}\|R\|_{L^2(\Omega)} \le \varepsilon^{\frac{3\theta-3}{2}} + C\|\nabla\Psi\|^2_{L^2(\Omega)} + \frac{\varepsilon^{\theta+1}}{16C_1}\|\nabla R\|^2_{L^2(\Omega)}$$

and

$$\varepsilon^{\theta-1}\|R\|^2_{L^2(\Omega)} \le \frac{\varepsilon^{\theta+1}}{8C_1}\|\nabla R\|^2_{L^2(\Omega)} + C\varepsilon^{\theta-3}\|\nabla\Psi\|^2_{L^2(\Omega)}.$$

Using the last two inequalities and $\theta>3$, we get

$$\frac{1}{2}\frac{d}{dt}\int_\Omega |\nabla\Psi|^2dx + \frac{\varepsilon^{\theta+1}}{4}\int_\Omega |\nabla R|^2dx \le C\Big(\|\nabla\Psi\|^2_{L^2(\Omega)} + \varepsilon^{\frac{3\theta-3}{2}}\Big) \qquad (1.31)$$

for some $C > 0$ independent of ε. Since $R(.,0) = 0$, it follows $\Psi(.,0) = 0$. Then the Gronwall inequality yields

$$\sup_{0\le t\le T} \|\nabla\Psi(.,t)\|^2_{L^2(\Omega)} \le C\varepsilon^{\frac{3\theta-3}{2}}$$

for some $C = C(T) > 0$ independent of ε. Integrating (1.31) over $(0, T)$, yields

$$\varepsilon^{\theta+1}\|\nabla R\|^2_{L^2(\Omega_T)} \le C\Big(\|\nabla\Psi\|^2_{L^2(\Omega_T)} + \varepsilon^{\frac{3\theta-3}{2}}\Big) \le C\varepsilon^{\frac{3\theta-3}{2}}$$

for some $C > 0$ independent of ε. Furthermore, it follows

$$\|R\|^2_{L^2(\Omega_T)} \le \|\nabla\Psi\|_{L^2(\Omega_T)}\|\nabla R\|_{L^2(\Omega_T)} \le C\varepsilon^{\frac{3\theta-3}{4}}\varepsilon^{\frac{\theta-5}{4}} = C\varepsilon^{\theta-2}.$$

Hence the assertions of the lemma follow. □

Now we can show that $H^\varepsilon - H^\varepsilon_A$ converges to 0 as ε goes to zero.

Lemma 1.7 *Let H^ε and H^ε_A be defined as above and let $\theta > 3$. Then it holds for all $\varphi \in C^\infty([0,T]; \mathscr{D}(\Omega)^d)$*

$$\left|\int_0^T \langle H^\varepsilon - H^\varepsilon_A, \varphi\rangle dt\right| \longrightarrow 0,$$

as $\varepsilon \searrow 0$.

Proof We choose any $\varphi \in C^\infty([0,T]; \mathscr{D}(\Omega)^d)$ and set $R = c^\varepsilon - c^\varepsilon_A$. Then it holds by the triangle inequality

$$\begin{aligned}
&\varepsilon\left|\int_{\Omega_T} \left(\nabla c^\varepsilon \otimes \nabla c^\varepsilon - \nabla c^\varepsilon_A \otimes \nabla c^\varepsilon_A\right) : \nabla\varphi\, dx\right| \\
&\le \varepsilon\left|\int_{\Omega_T} (\nabla c^\varepsilon \otimes \nabla R) : \nabla\varphi\, dx\right| + \varepsilon\left|\int_{\Omega_T} \left(\nabla R \otimes \nabla c^\varepsilon_A\right) : \nabla\varphi\, dx\right| \\
&\le \varepsilon\|\nabla\varphi\|_{L^\infty(\Omega_T)}\|\nabla R\|_{L^2(\Omega_T)}\big(\|\nabla c^\varepsilon\|_{L^2(\Omega_T)} + \left\|\nabla c^\varepsilon_A\right\|_{L^2(\Omega_T)}\big).
\end{aligned}$$

Due to Lemma 1.5, we have

$$\left\|\nabla c^\varepsilon_A\right\|^2_{L^2(\Omega_T)} \le C\varepsilon^{-1}$$

for some $C > 0$ independent of ε. Since

$$\|\nabla c^\varepsilon\|_{L^2(\Omega_T)} \le \left\|\nabla c^\varepsilon_A\right\|_{L^2(\Omega_T)} + \|\nabla R\|_{L^2(\Omega_T)}$$

and due to Lemma 1.6, it follows

$$\left|\int_0^T \langle H^\varepsilon - H_A^\varepsilon, \varphi\rangle dt\right| \leq C\varepsilon^{\frac{\theta-3}{4}}\left(1+\varepsilon^{\frac{\theta-3}{4}}\right)$$

for some constant $C = C(\varphi) > 0$ and for all ε small enough. Since $\theta > 3$ the assertion follows. □

Lemma 1.8 *Let H_A^ε and c_A^ε be defined as above. Then it holds for all $\varphi \in \mathscr{D}(\Omega)^d$ and $t \in [0, T]$*

$$\langle H_A^\varepsilon, \varphi\rangle \longrightarrow 2\sigma \int_{\Gamma(t)} \left|\nabla(d^0(X_t^{-1}))\right| \nu_{\Gamma(t)} \otimes \nu_{\Gamma(t)} : \nabla\varphi \, d\mathscr{H}^{d-1},$$

as $\varepsilon \searrow 0$.

Proof Let $\varphi \in \mathscr{D}(\Omega)^d$ and $t \in [0, T]$ be arbitrary. Observe that

$$\left|\nabla c_A^\varepsilon\right| = \left|\varepsilon^{-1}\zeta(\tfrac{e}{\delta})\theta_0'(\tfrac{e}{\varepsilon})\nabla e + \left(\theta_0(\tfrac{e}{\varepsilon}) - \left(2\chi_{\overline{Q^+}} - 1\right)\right)\nabla(\zeta(\tfrac{e}{\delta}))\right| \leq C\varepsilon^{-1}e^{-\frac{\alpha\delta}{4\varepsilon}}$$

in $\bigcup_{t\in[0,T]} X_t(\Gamma_0(\delta)\backslash\Gamma_0(\frac{\delta}{2})) \times \{t\}$ and $\nabla c_A^\varepsilon = 0$ in $\bigcup_{t\in[0,T]} \overline{X_t(\Omega\backslash\Gamma_0(\delta))} \times \{t\}$.

Hence we can replace c_A^ε by $\theta_0(\frac{e}{\varepsilon})$ in the whole domain $\overline{\Omega_T}$ since the remainder decays exponentially as $\varepsilon \to 0$.

Since $X_t : \Omega \to \Omega$ is a diffeomorphism, we obtain by coordinate transformation

$$\begin{aligned}\langle H_A^\varepsilon, \varphi\rangle &= \varepsilon \int_\Omega \nabla c_A^\varepsilon \otimes \nabla c_A^\varepsilon : \nabla\varphi \, dx + \mathscr{O}(\varepsilon) \\ &= \varepsilon \int_\Omega \nabla c_A^\varepsilon \circ X_t \otimes \nabla c_A^\varepsilon \circ X_t : \nabla\varphi \circ X_t |\det(DX_t)| dx + \mathscr{O}(\varepsilon)\,.\end{aligned}$$

Due to Lemma 1.2, it holds

$$\det(DX_t(x)) = 1 \quad \forall (x,t) \in \Omega_T\,, \tag{1.32}$$

and the identity (1.21) yields

$$\nabla c_A^\varepsilon \circ X_t = DX_t^{-T} \circ X_t \, \nabla(c_A^\varepsilon \circ X_t)\,. \tag{1.33}$$

By Eqs. (1.32) and (1.33), we conclude

$$\langle H_A^\varepsilon, \varphi\rangle = \varepsilon \int_\Omega M\,\nabla(c_A^\varepsilon \circ X_t) \otimes M\,\nabla(c_A^\varepsilon \circ X_t) : \nabla\varphi \circ X_t \, dx + \mathscr{O}(\varepsilon)\,,$$

where $M = M(x,t) := (DX_t^{-T} \circ X_t)(x)$. Using $c_A^\varepsilon = \theta_0(d^0 \circ X_t^{-1}/\varepsilon)$ in $X_t(\Gamma(\frac{\delta}{2}))$ yields

$$\langle H_A^\varepsilon, \varphi\rangle = \varepsilon^{-1}\int_\Omega \Big(\theta_0'\Big(\tfrac{d^0}{\varepsilon}\Big)\Big)^2 M\,\nabla d^0 \otimes M\,\nabla d^0 : \nabla\varphi \circ X_t\,dx + \mathcal{O}(\varepsilon)\,. \quad (1.34)$$

Now we consider the limit $\varepsilon \searrow 0$.

Claim: Let $f \in C^1(\overline{\Omega})$ be an arbitrary function. Then it holds

$$\varepsilon^{-1}\int_\Omega \Big(\theta_0'\Big(\tfrac{d^0}{\varepsilon}\Big)\Big)^2 f\,dx \longrightarrow 2\sigma \int_{\Gamma_0} f\,d\mathcal{H}^{d-1}\,, \quad (1.35)$$

as $\varepsilon \searrow 0$.

Proof of the claim: Since there exists some constant $C > 0$ such that

$$\Big|\theta_0'\Big(\tfrac{d^0(x)}{\varepsilon}\Big)\Big| \le C e^{-\frac{\alpha|d^0(x)|}{\varepsilon}} \quad \forall x \in \Omega\,,$$

it is sufficient to consider the domain $\Gamma_0(\delta)$ instead of Ω. Hence we have the identity

$$\varepsilon^{-1}\int_{\Gamma_0(\delta)} \Big(\theta_0'\Big(\tfrac{d^0}{\varepsilon}\Big)\Big)^2 f\,dx = \varepsilon^{-1}\int_{-\delta}^{\delta} \big(\theta_0'\big(\tfrac{r}{\varepsilon}\big)\big)^2 \int_{\Gamma_0^r} f\,d\mathcal{H}^{d-1}\,dr\,,$$

where $\Gamma_0^r = \big\{x \in \Omega : x = s + r\nu_{\Gamma_0}(s),\, s \in \Gamma_0\big\}$, cf. the end of the proof of Lemma 1.5. Define $\Psi_f : \mathbb{R} \to \mathbb{R}$ by

$$\Psi_f(r) = \begin{cases} \int_{\Gamma_0^r} f\,d\mathcal{H}^{d-1} & \text{if } r \in (-\delta, \delta) \\ 0 & \text{if } r \in \mathbb{R}\backslash(-\delta,\delta)\,, \end{cases}$$

$\varphi : \mathbb{R} \to \mathbb{R}$ by

$$\varphi(r) = \frac{1}{2\sigma}\big(\theta_0'(r)\big)^2 \qquad \text{for all } r \in \mathbb{R}$$

and $\varphi_\varepsilon : \mathbb{R} \to \mathbb{R}$ by

$$\varphi_\varepsilon(r) = \varepsilon^{-1}\varphi\Big(\frac{r}{\varepsilon}\Big) \qquad \text{for all } r \in \mathbb{R}$$

for $\varepsilon \in (0, 1]$. Then φ is a one-dimensional mollifier. Therefore we obtain

$$\begin{aligned} \varepsilon^{-1}\int_{\Gamma_0(\delta)} \Big(\theta_0'\Big(\tfrac{d^0}{\varepsilon}\Big)\Big)^2 f\,dx &= 2\sigma \int_{\mathbb{R}} \varphi_\varepsilon(r)\Psi_f(r)\,dr \\ &= 2\sigma \int_{\mathbb{R}} \varphi_\varepsilon(r)\Psi_f(-(0-r))\,dr \\ &= 2\sigma\,\varphi_\varepsilon * \tilde{\Psi}_f(0)\,, \end{aligned}$$

where $\tilde{\Psi}_f(r) = \Psi_f(-r)$ and $\varphi_\varepsilon * \tilde{\Psi}_f$ denotes convolution of φ_ε and $\tilde{\Psi}_f$. To show the convergence, it is necessary to estimate $\big|\Psi_f(r) - \Psi_f(0)\big|$ for $r \le \delta$. By definition of Ψ_f, we obtain

$$
\begin{aligned}
\left|\Psi_f(r)-\Psi_f(0)\right| &= \left|\int_{\Gamma_0^r} f\,d\mathscr{H}^{d-1} - \int_{\Gamma_0} f\,d\mathscr{H}^{d-1}\right| \\
&= \left|\int_{\Gamma_0} f\circ\tau_r \left|\det(\partial_{\tau_i}\tau_r\cdot\partial_{\tau_j}\tau_r)_{i,j=1}^{d-1}\right|^{\frac{1}{2}} - f\,d\mathscr{H}^{d-1}\right| \\
&\leq \left|\int_{\Gamma_0} (f\circ\tau_r - f)\left|\det(\partial_{\tau_i}\tau_r\cdot\partial_{\tau_j}\tau_r)_{i,j=1}^{d-1}\right|^{\frac{1}{2}} d\mathscr{H}^{d-1}\right| \\
&\quad + \left|\int_{\Gamma_0} f\left(1-\left|\det(\partial_{\tau_i}\tau_r\cdot\partial_{\tau_j}\tau_r)_{i,j=1}^{d-1}\right|^{\frac{1}{2}}\right) d\mathscr{H}^{d-1}\right|,
\end{aligned}
$$

where we have used the transformation $\tau_r : \Gamma_0 \to \Gamma_0^r$ defined by $\tau_r(x) = x + r\nu_{\Gamma_0}(x)$ and where $\{\tau_1(x), \dots, \tau_{d-1}(x)\}$ is an orthonormal basis of $T_x\Gamma_0$. We estimate the two terms on the right-hand side separately. The fundamental theorem of calculus yields

$$
\begin{aligned}
&\left|\int_{\Gamma_0} (f\circ\tau_r - f)\left|\det(\partial_{\tau_i}\tau_r\cdot\partial_{\tau_j}\tau_r)_{i,j=1}^{d-1}\right|^{\frac{1}{2}} d\mathscr{H}^{d-1}\right| \\
&= \left|\int_{\Gamma_0} \left(\int_0^r \nabla f(x+s\nu_{\Gamma_0})\cdot\nu_{\Gamma_0}\,ds\right)\left|\det(\partial_{\tau_i}\tau_r\cdot\partial_{\tau_j}\tau_r)_{i,j=1}^{d-1}\right|^{\frac{1}{2}} d\mathscr{H}^{d-1}\right| \\
&\leq C|r|\|f\|_{C^1(\overline{\Omega})}
\end{aligned}
$$

for some $C = C(\Gamma_0) > 0$ independent of r. We continue with the second term. Note that

$$\partial_{\tau_i}\tau_r\cdot\partial_{\tau_j}\tau_r = \delta_{ij} + r\big(\tau_i\cdot\partial_{\tau_j}\nu_{\Gamma_0} + \tau_j\cdot\partial_{\tau_i}\nu_{\Gamma_0} + r\partial_{\tau_i}\nu_{\Gamma_0}\cdot\partial_{\tau_j}\nu_{\Gamma_0}\big).$$

Hence we can conclude

$$\left|\det(\partial_{\tau_i}\tau_r\cdot\partial_{\tau_j}\tau_r)_{i,j=1}^{d-1}\right| = 1 + \mathscr{O}(r),$$

and therefore it follows

$$1-\left|\det(\partial_{\tau_i}\tau_r\cdot\partial_{\tau_j}\tau_r)_{i,j=1}^{d-1}\right|^{\frac{1}{2}} = \frac{1-\left|\det(\partial_{\tau_i}\tau_r\cdot\partial_{\tau_j}\tau_r)_{i,j=1}^{d-1}\right|}{1+\left|\det(\partial_{\tau_i}\tau_r\cdot\partial_{\tau_j}\tau_r)_{i,j=1}^{d-1}\right|^{\frac{1}{2}}} = \mathscr{O}(r).$$

Thus we get the following estimate

$$\left|\int_{\Gamma_0} f\left(1-\left|\det(\partial_{\tau_i}\tau_r\cdot\partial_{\tau_j}\tau_r)_{i,j=1}^{d-1}\right|^{\frac{1}{2}}\right) d\mathscr{H}^{d-1}\right| \leq C|r|\|f\|_{C^0(\overline{\Omega})}$$

for some $C = C(\Gamma_0) > 0$ independent of r. Hence we obtain

$$\left|\Psi_f(r)-\Psi_f(0)\right| \leq C|r|\|f\|_{C^1(\overline{\Omega})}.$$

Applying this estimate, we can prove the assertion

$$\begin{aligned}
&\left|\varepsilon^{-1}\int_{\Gamma_0(\delta)}\left(\theta_0'\left(\tfrac{d^0}{\varepsilon}\right)\right)^2 f\,dx - 2\sigma\int_{\Gamma_0} f\,d\mathcal{H}^{d-1}\right| \\
&= 2\sigma\left|\varphi_\varepsilon * \tilde{\Psi}_f(0) - \tilde{\Psi}_f(0)\right| \\
&= 2\sigma\left|\int_{\mathbb{R}}\varphi_\varepsilon(y)\tilde{\Psi}_f(-y)\,dy - \tilde{\Psi}_f(0)\right| \\
&= 2\sigma\left|\int_{\mathbb{R}}\varphi_\varepsilon(y)\Big(\tilde{\Psi}_f(-y) - \tilde{\Psi}_f(0)\Big)dy\right| \\
&\leq 2\sigma\int_{-\sqrt{\varepsilon}}^{\sqrt{\varepsilon}}\varphi_\varepsilon(y)\left|\tilde{\Psi}_f(-y) - \tilde{\Psi}_f(0)\right|dy + 2\sigma\int_{\mathbb{R}\setminus(-\sqrt{\varepsilon},\sqrt{\varepsilon})}\varphi_\varepsilon(y)\left|\tilde{\Psi}_f(-y) - \tilde{\Psi}_f(0)\right|dy \\
&\leq C\varepsilon^{\frac{1}{2}}\|f\|_{C^1(\overline{\Omega})}\int_{\mathbb{R}}\varphi_\varepsilon(y)\,dy + C\|f\|_{C^0(\overline{\Omega})}\int_{\mathbb{R}\setminus(-\sqrt{\varepsilon},\sqrt{\varepsilon})}\varphi_\varepsilon(y)\,dy \\
&\leq C\varepsilon^{\frac{1}{2}}\|f\|_{C^1(\overline{\Omega})} + C\|f\|_{C^0(\overline{\Omega})}\int_{\mathbb{R}\setminus(-\sqrt{\varepsilon},\sqrt{\varepsilon})}\varepsilon^{-1}e^{-\frac{2\alpha|y|}{\varepsilon}}\,dy \\
&\leq C\varepsilon^{\frac{1}{2}}\|f\|_{C^1(\overline{\Omega})} + C\|f\|_{C^0(\overline{\Omega})}\varepsilon^{-1}e^{-\frac{2\alpha}{\sqrt{\varepsilon}}}
\end{aligned}$$

for some $C = C(\Gamma_0) > 0$ independent of ε and where we have used $\int_{\mathbb{R}}\varphi_\varepsilon(z)\,dz = 1$ for all $\varepsilon \in (0, 1]$. Hence the claim follows.

The relation (1.34) and the property (1.35) yield

$$\left\langle H_A^\varepsilon, \varphi\right\rangle \longrightarrow 2\sigma\int_{\Gamma_0} M\nabla d^0 \otimes M\nabla d^0 : (\nabla\varphi)\circ X_t\,d\mathcal{H}^{d-1}$$

as $\varepsilon \searrow 0$.

We apply coordinate transformation to the right-hand side. Note that due to Lemma 1.2, it holds

$$|M|^2 = \left|DX_t^{-T}\circ X_t\nabla d^0\right|^2 = \det(\partial_{\tau_i}X_t\cdot\partial_{\tau_j}X_t)_{i,j=1}^{d-1} \quad \text{on } \Gamma_0\,.$$

Then we obtain

$$\begin{aligned}
&2\sigma\int_{\Gamma_0} DX_t^{-T}\circ X_t\nabla d^0 \otimes DX_t^{-T}\circ X_t\nabla d^0 : (\nabla\varphi)\circ X_t\,d\mathcal{H}^{d-1} \\
&= 2\sigma\int_{\Gamma(t)}\left|\det(\partial_{\tau_i}X_t\cdot\partial_{\tau_j}X_t)_{i,j=1}^{d-1}\circ X_t^{-1}\right|^{\frac{1}{2}}\frac{\nabla e}{|\nabla e|}\otimes\frac{\nabla e}{|\nabla e|} : \nabla\varphi\,d\mathcal{H}^{d-1}
\end{aligned}$$

since $X_t(\Gamma_0) = \Gamma(t)$ and $DX_t^{-T}\nabla d^0 \circ X_t^{-1} = \nabla e$. Because of $\nabla e/|\nabla e| = \nu_{\Gamma(t)}$ and

$$\left|\det(\partial_{\tau_i} X_t \cdot \partial_{\tau_j} X_t)_{i,j=1}^{d-1} \circ X_t^{-1}\right|^{\frac{1}{2}} = \left|DX_t^{-T}\nabla d^0 \circ X_t^{-1}\right| = \left|\nabla\big(d^0(X_t^{-1})\big)\right|,$$

the assertion of the lemma follows.

Proof of Theorem 1.1: The first assertion of the theorem immediately follows by Lemmas 1.7 and 1.8.

The second assertion is a consequence of Lemmas 1.5 and 1.6 since $\theta > 3$. □

Acknowledgements The authors acknowledge support by the SPP 1506 "Transport Processes at Fluidic Interfaces" of the German Science Foundation (DFG) through grant AB285/3-1, AB285/4-1 and AB285/4-2. The results are part of the second author's PhD-thesis [13].

References

1. Abels, H.: Existence of weak solutions for a diffuse interface model for viscous, incompressible fluids with general densities. Comm. Math. Phys. **289**(1), 45–73 (2009)
2. Abels, H.: On a diffuse interface model for two-phase flows of viscous, incompressible fluids with matched densities. Arch. Ration. Mech. Anal. **194**(2), 463–506 (2009)
3. Abels, H.: Strong well-posedness of a diffuse interface model for a viscous, quasi-incompressible two-phase flow. SIAM J. Math. Anal. **44**(1), 316–340 (2012)
4. Abels, H., Garcke, H., Grün, G.: Thermodynamically consistent, frame indifferent diffuse interface models for incompressible two-phase flows with different densities. Math. Models Methods Appl. Sci. **22**(3), 1150013, 40 (2012)
5. Abels, H., Lengeler, D.: On sharp interface limits for diffuse interface models for two-phase flows. Interfaces Free Bound. **16**(3), 395–418 (2014)
6. Abels, H., Röger, M.: Existence of weak solutions for a non-classical sharp interface model for a two-phase flow of viscous, incompressible fluids. Ann. Inst. H. Poincaré Anal. Non Linéaire **26**(6), 2403–2424 (2009)
7. Alikakos, N.D., Bates, P.W., Chen, X.: Convergence of the Cahn-Hilliard equation to the Hele-Shaw model. Arch. Rational Mech. Anal. **128**(2), 165–205 (1994)
8. Arnold, V.I.: Gewöhnliche Differentialgleichungen -. Springer DE, Berlin, 2. aufl. edition (2001)
9. Chen, X.: Global asymptotic limit of solutions of the Cahn-Hilliard equation. J. Differential Geom. **44**(2), 262–311 (1996)
10. Eck, C., Garcke, H., Knabner, P.: Mathematische Modellierung. Springer DE, Berlin, 2. überarb. aufl. edition (2011)
11. Hildebrandt, S.: Analysis 2. Springer DE, Berlin (2003)
12. Kwek, K.H.: On the Cahn-Hilliard type equation. Ph.D. thesis, Georgia Institute of Technology (1991)
13. Schaubeck, S.: Sharp interface limits for diffuse interface models, Feb 2014. http://epub.uni-regensburg.de/29462/
14. Walter, W.: Gewöhnliche Differentialgleichungen. Springer-Lehrbuch. [Springer Textbook], 7th edn. Springer, Berlin (2000). Eine Einführung. [An introduction]

Chapter 2
On the Interface Formation Model for Dynamic Triple Lines

Dieter Bothe and Jan Prüss

Abstract This paper revisits the theory of Y. Shikhmurzaev on forming interfaces as a continuum thermodynamical model for dynamic triple lines. We start with the derivation of the balances for mass, momentum, energy and entropy in a three-phase fluid system with full interfacial physics, including a brief review of the relevant transport theorems on interfaces and triple lines. Employing the entropy principle in the form given in (Bothe and Dreyer (2015) Continuum thermodynamics of chemically reacting fluid mixtures. Acta Mech. 226, 1757–1805, [1]), but extended to this more general case, we arrive at the entropy production and perform a linear closure, except for a nonlinear closure for the sorption processes. Specialized to the isothermal case, we obtain a thermodynamically consistent mathematical model for dynamic triple lines and show that the total available energy is a strict Lyapunov function for this system.

Keywords Continuum thermodynamics · Dynamic contact line · Interfacial mass · Dynamic surface tension · Free energy Lyapunov functional

2.1 Introduction

The line at which three phases meet is called a triple line; cf. Fig. 2.1. If the phases which touch each other are all fluid phases, i.e. two immiscible liquids are in contact with another liquid or a gas, this triple line is freely deformable in space, while it is bound to move on a given surface, if one of the phases is a solid. In the latter case, one usually speaks about a dynamic contact line, while the notion of a triple

D. Bothe (✉)
Center of Smart Interfaces & Fachbereich Mathematik,
Technische Universität Darmstadt, Alarich-Weiss-Str. 10, 64287 Darmstadt, Germany
e-mail: bothe@csi.tu-darmstadt.de

J. Prüss
Institut Für Mathematik, Martin-Luther-Universität Halle-Wittenberg,
Theodor-Lieser-Str. 5, 06120 Halle (saale), Germany
e-mail: jan.pruess@mathematik.uni-halle.de

Y. Shibata and Y. Suzuki (eds.), *Mathematical Fluid Dynamics, Present and Future*, Springer Proceedings in Mathematics & Statistics 183,
DOI 10.1007/978-4-431-56457-7_2

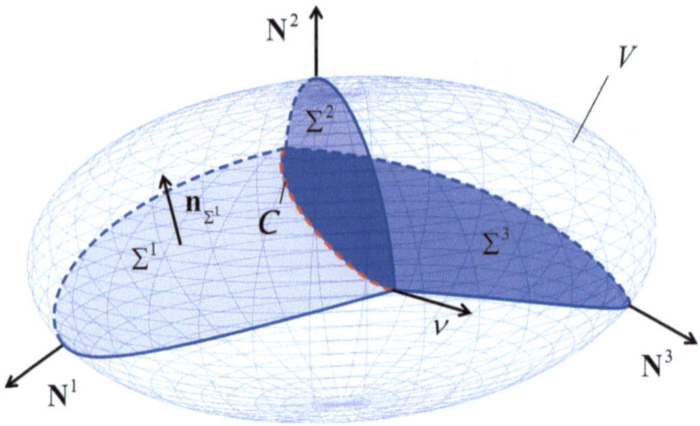

Fig. 2.1 Configuration of the phases and interfaces at the contact line

line is typically used in the former setting. Both cases share many similarities and their modeling and analysis is closely related. In applications, wetting more often appears on a solid wall, i.e. the case of a contact line is more often considered. Hence, the main body of the literature is devoted to this case. The present paper deals with dynamics triple lines, but in such a generality, that analogous results are valid for the contact line situation. Nevertheless, due to the more frequent encounter of wetting of solid supports, the brief literature survey to follow necessarily focuses on contact line dynamics.

The modeling and computation of dynamic contact lines is an active field due to the enormous relevance of wetting and dewetting phenomena in various technical and industrial applications; see [2–4] for recent surveys on the field, containing also references to experimental work. Different modeling approaches are employed, containing in particular so-called molecular-kinetic theory (MKT; see, e.g., [5, 6]) and continuum physical theories. The latter is often subsumed under the heading "hydrodynamic theory" and is mostly based on sharp-interface models, while phase field models have also been extended to cover contact lines as in [7]. The sharp-interface hydrodynamic theory started essentially with the seminal paper by Huh and Scriven [8] in which the fundamental problem of the inconsistency between a moving contact line and a no-slip condition at the fixed wall has been analyzed and shown to lead to a non-integrable stress singularity; cf. also [9] and, for a more rigorous mathematical treatment, [10, 11]. Consequently, subsequent models always rely on some "relaxation" at the contact line, and the most common way to remove the stress singularity, as already proposed in [8], is to introduce Navier-type slip close to the contact line. Besides this complication, the main extension of the standard two-phase Navier-Stokes system consists of the prescription of the dynamic contact angle θ_d, i.e. the angle which is formed between the fluid interface and the solid support, as a function of the contact line speed. At this point it is to be noted that the contact angle changes its value under dynamic conditions, while the equilibrium contact angle θ_e is usually assumed to be governed by Young's law, i.e.

$$\sigma^{gl}\cos(\theta_e) = \sigma^{gs} - \sigma^{ls} \tag{2.1}$$

in case of a liquid wetting a solid surrounded by a gas phase, where the superscripts stand for gas (g), liquid (l) and solid (s) and σ denotes the interfacial tension of the respective interface. Based on the classical experimental studies in [12], the general form of the relation between θ_d and the contact line speed is given by the heuristic relation

$$\theta_d = f_{\text{Hoff}}\big(\text{Ca} + f_{\text{Hoff}}^{-1}(\theta_e)\big), \tag{2.2}$$

where Ca denotes the Capillary number given as $\text{Ca} = \eta U/\sigma^{gl}$ with η the dynamic liquid viscosity and U the contact line speed. Several concrete correlations have been established for different materials and certain wetting scenarios like "Tanner's law" [13] or the correlation of Jiang et al. [14]. Theoretical investigations using the hydrodynamic theory identified three length-scales near the contact line: an inner region in which the fluid interface is essentially planar and touches the solid support at the equilibrium angle; a mesoscopic region in which a significant bending of the interface can occur; an outer (macroscopic) region in which the contact angle attains a different value, the so-called apparent contact angle. The hydrodynamic theory provides relations for the dependence of the contact angle on the distance from the contact line especially in the mesoscopic region; see [15–17]. Knowledge of this dependence is very useful for numerical purpose, both as a subgrid-scale model to reduce the necessary resolution at the contact-line and in order to neutralize the inherent mesh dependence of numerical solutions due to the typical under-resolution of the smallest length scales in the contact line region; for the latter, see [18, 19].

While the hydrodynamic model can describe many wetting processes at least qualitatively, in particular concerning the observed dynamical shapes of attached droplets moving on a wall, say, it does not capture the full physics of a dynamic contact line. One important deviation is the internal flow field in the wetting liquid close to an advancing contact line, which experimentally is known to be a rolling motion ([9, 20]), but is a sliding motion in numerical simulations using the above model. Moreover, there is experimental evidence that the relation between the dynamic contact angle and the contact line speed is more complicated and not of such a simple local nature; cf. [21]. For further discrepancies between experimental observations and the hydrodynamic model see [22]. A very interesting approach to overcome these short-comings has been introduced by Shikhmurzaev in [23]; see also [22]. The approach there also employs continuum physics, but accounts for the aspect of interface formation and disappearance at the contact line. A crucial point for the model development then is to include enough interfacial thermodynamics to allow for a non-constant interfacial tension, governed by a surface equations of state on all involved interfaces. For this purpose, the mass contained in the interfacial layer has to be balanced separately, since it encounters different forces compared to within the bulk phases and it is this mass density which determines the surface pressure, i.e. the surface tension. In the considered sharp interface/sharp contact line model, the interfacial mass is lumped into an area-specific mass density and the model is extended to cover the evolution of this interfacial mass density by appropriate balance equations on the moving surfaces. This model has proven a great potential to explain several physical phenomena like wetting, coalescence, cusp formation and the break-up of liquid threads; cf. [2, 22] and the introduction in [24]. The interface formation model of Shikhmurzaev has been based on the continuum thermodynamics of fluid interfaces developed in [25, 26], but with the sensible aim to formulate the most simple model which is able to describe the wetting process with dynamic contact angle and rolling motion close to the contact line with the material properties modeled via bulk and surface free energies but without a heuristic relation between contact angle and contact line speed. Several years after the fundamental paper [23] appeared, Billingham in [24] pointed out that one further condition at the contact line has to be added, and he employed a condition provided by Bedeaux in [27] which relates the rate of mass transfer from one surface into the other to the difference of the surface chemical potentials. We will come back to this point in the final remarks at the end of this paper.

In the brief survey above, the topic called "contact angle hysteresis", referring to the appearance of a full interval of possible contact angles in the static case which spans the range from the angles observed for (infinitely slowly) advancing and receding contact lines, has not been touched. This phenomenon seems to be similar to dry friction between solids and, in fact, the notion of contact line friction is also present in the literature on the molecular kinetic theory of contact lines. For this topic, we refer to [3, 4] and the references given there.

2.2 Integral Balances

We consider a region $G \subset \mathbb{R}^3$ filled with three bulk phases $\Omega^k(t)$ $(k = 1, 2, 3)$, separated by interfaces $\Sigma^k(t)$ $(k = 1, 2, 3)$ which meet at a common triple line $\mathscr{C}(t)$. As an example, imagine a liquid phase Ω^1 in the form of a water droplet sitting on another liquid, say oil, which forms bulk phase Ω^2, and being surrounded by phase Ω^3 composed of air. Then, for instance, Σ^1 denotes the oil-water interface, Σ^2 the interface between the air and the oil and Σ^3 the air-water interface. The deformable and free bounding curve at which all three interfaces meet is the so-called triple line $\mathscr{C}$. As a related but somewhat different case, consider again a liquid phase Ω^1 in the form of a droplet, but now sitting on a solid support, which forms bulk phase Ω^2, and being surrounded by gas phase Ω^3. Then two out of the three interfaces are fixed and the triple line is the set of all points where the gas-liquid interface meets the solid support. In this case, one usually calls $\mathscr{C}$ the contact line which now has reduced degrees of freedom due to the solid support. We focus on three-phase fluid systems with a common triple line and assume that the interfaces meet at angles different from 0 and π; we shall refer to this as the non-degenerate case.

We start with the integral balance of a generic extensive quantity which is present in the bulk phases with specific density ϕ, on the interfaces with specific density ϕ^Σ and on the triple line with specific density $\phi^{\mathscr{C}}$. Hence $\rho\phi$, $\rho^\Sigma\phi^\Sigma$ and $\rho^{\mathscr{C}}\phi^{\mathscr{C}}$, respectively, are the volume-, area- and line-specific densities, where ρ, ρ^Σ and $\rho^{\mathscr{C}}$

are the mass densities. If a specific bulk phase or interface is considered, we write $\rho_k \phi_k$ or $\rho_k^\Sigma \phi_k^\Sigma$, respectively, for the respective density. With this notation, the generic integral balance for a fixed control volume $V \subset G$ reads as

$$\frac{d}{dt}\left[\int_{\Omega_V} \rho\phi\, dx + \int_{\Sigma_V} \rho^\Sigma \phi^\Sigma\, do + \int_{\mathscr{C}_V} \rho^\mathscr{C} \phi^\mathscr{C}\, dl\right] = -\int_{\partial\Omega_V} (\rho\phi v + j)\cdot n\, do$$
$$-\int_{\partial\Sigma_V} (\rho^\Sigma \phi^\Sigma v^\Sigma + j^\Sigma)\cdot N\, dl - \int_{\partial\mathscr{C}_V} (\rho^\mathscr{C} \phi^\mathscr{C} v^\mathscr{C} + j^\mathscr{C})\cdot \nu\, dP \qquad (2.3)$$
$$+\int_{\Omega_V} f\, dx + \int_{\Sigma_V} f^\Sigma\, do + \int_{\mathscr{C}_V} f^\mathscr{C}\, dl.$$

In (2.3) we let dx, do and dl denote the volume, area and line measure, respectively. Moreover, dP denotes the point (i.e., counting) measure.

Here, as well as throughout the paper, we use the following condensed notation. First,

$$\Omega_V := \bigcup_{k=1}^{3} \Omega_V^k, \quad \Sigma_V := \bigcup_{k=1}^{3} \Sigma_V^k \quad \text{with } \Omega_V^k := \Omega^k \cap V, \quad \Sigma_V^k := \Sigma^k \cap V,$$

which are all time-dependent sets. We assume a single triple line, hence $\mathscr{C}_V := \mathscr{C} \cap V$. The detailed version of (2.3) then reads as

$$\frac{d}{dt}\left[\sum_{k=1}^{3}\int_{\Omega_V^k} \rho_k\phi_k\, dx + \sum_{k=1}^{3}\int_{\Sigma_V^k} \rho_k^\Sigma \phi_k^\Sigma\, do + \int_{\mathscr{C}_V} \rho^\mathscr{C} \phi^\mathscr{C}\, dl\right] =$$
$$-\sum_{k=1}^{3}\int_{\partial\Omega_V^k} (\rho_k\phi_k v_k + j_k)\cdot n^k\, do - \sum_{k=1}^{3}\int_{\partial\Sigma_V^k} (\rho_k^\Sigma \phi_k^\Sigma v_k^\Sigma + j_k^\Sigma)\cdot N^k\, dl$$
$$-\int_{\partial\mathscr{C}_V} (\rho^\mathscr{C} \phi^\mathscr{C} v^\mathscr{C} + j^\mathscr{C})\cdot \nu\, dP + \sum_{k=1}^{3}\int_{\Omega_V^k} f_k\, dx + \sum_{k=1}^{3}\int_{\Sigma_V^k} f_k^\Sigma\, do + \int_{\mathscr{C}_V} f^\mathscr{C}\, dl.$$

For better readability, we use the condensed notation whenever this is reasonable.

We apply this balancing to the extensive quantities mass, momentum, energy and entropy. The corresponding integral balances read as follows.

Mass Balance.

$$\frac{d}{dt}\left[\int_{\Omega_V} \rho\, dx + \int_{\Sigma_V} \rho^\Sigma\, do + \int_{\mathscr{C}_V} \rho^\mathscr{C}\, dl\right] = \qquad (2.4)$$
$$-\int_{\partial\Omega_V} \rho v\cdot n\, do - \int_{\partial\Sigma_V} \rho^\Sigma v^\Sigma \cdot N\, dl - \int_{\partial\mathscr{C}_V} \rho^\mathscr{C} v^\mathscr{C}\cdot \nu\, dP.$$

Momentum Balance.

$$
\begin{aligned}
&\frac{d}{dt}\Big[\int_{\Omega_V} \rho v\, dx + \int_{\Sigma_V} \rho^\Sigma v^\Sigma\, do + \int_{\mathscr{C}_V} \rho^{\mathscr{C}} v^{\mathscr{C}}\, dl\Big] = \\
&- \int_{\partial\Omega_V} \rho v (v\cdot n)\, do - \int_{\partial\Sigma_V} \rho^\Sigma v^\Sigma (v^\Sigma \cdot N)\, dl - \int_{\partial\mathscr{C}_V} \rho^{\mathscr{C}} v^{\mathscr{C}} (v^{\mathscr{C}} \cdot \nu)\, dP \qquad (2.5)\\
&+ \int_{\partial\Omega_V} S\cdot n\, do + \int_{\partial\Sigma_V} S^\Sigma \cdot N\, dl + \int_{\partial\mathscr{C}_V} S^{\mathscr{C}} \cdot \nu\, dP \\
&+ \int_{\Omega_V} \rho b\, dx + \int_{\Sigma_V} \rho^\Sigma b^\Sigma\, do + \int_{\mathscr{C}_V} \rho^{\mathscr{C}} b^{\mathscr{C}}\, dl.
\end{aligned}
$$

Here S, S^Σ and $S^{\mathscr{C}}$ denote the stress tensor in the bulk phases, on the interfaces and on the triple line, respectively, and b, b^Σ and $b^{\mathscr{C}}$ are the specific body forces.

Energy Balance.

$$
\begin{aligned}
&\frac{d}{dt}\Big[\int_{\Omega_V} \rho(e + \frac{v^2}{2})\, dx + \int_{\Sigma_V} \rho^\Sigma (e^\Sigma + \frac{(v^\Sigma)^2}{2})\, do + \int_{\mathscr{C}_V} \rho^{\mathscr{C}} (e^{\mathscr{C}} + \frac{(v^{\mathscr{C}})^2}{2})\, dl\Big] = \\
&- \int_{\partial\Omega_V} \rho(e + \frac{v^2}{2}) v\cdot n\, do - \int_{\partial\Sigma_V} \rho^\Sigma (e^\Sigma + \frac{(v^\Sigma)^2}{2}) v^\Sigma \cdot N\, dl \\
&- \int_{\partial\mathscr{C}_V} \rho^{\mathscr{C}} (e^{\mathscr{C}} + \frac{(v^{\mathscr{C}})^2}{2}) v^{\mathscr{C}} \cdot \nu\, dP \qquad (2.6)\\
&+ \int_{\partial\Omega_V} (v\cdot S - q)\cdot n\, do + \int_{\partial\Sigma_V} (v^\Sigma \cdot S^\Sigma - q^\Sigma)\cdot N\, dl + \int_{\partial\mathscr{C}_V} (v^{\mathscr{C}} \cdot S^{\mathscr{C}} - q^{\mathscr{C}})\cdot \nu\, dP \\
&+ \int_{\Omega_V} \rho v\cdot b\, dx + \int_{\Sigma_V} \rho^\Sigma v^\Sigma \cdot b^\Sigma\, do + \int_{\mathscr{C}_V} \rho^{\mathscr{C}} v^{\mathscr{C}} \cdot b^{\mathscr{C}}\, dl.
\end{aligned}
$$

Here q, q^Σ and $q^{\mathscr{C}}$ denote the heat flux in the bulk phases, on the interfaces and on the triple line, respectively. Note that energy sources due to radiation have been omitted in (2.6).

Entropy Balance.

$$
\begin{aligned}
&\frac{d}{dt}\Big[\int_{\Omega_V} \rho s\, dx + \int_{\Sigma_V} \rho^\Sigma s^\Sigma\, do + \int_{\mathscr{C}_V} \rho^{\mathscr{C}} s^{\mathscr{C}}\, dl\Big] = - \int_{\partial\Omega_V} (\rho s v + \Phi)\cdot n\, do \\
&- \int_{\partial\Sigma_V} (\rho^\Sigma s^\Sigma v^\Sigma + \Phi^\Sigma)\cdot N\, dl - \int_{\partial\mathscr{C}_V} (\rho^{\mathscr{C}} s^{\mathscr{C}} v^{\mathscr{C}} + \Phi^{\mathscr{C}})\cdot \nu\, dP \qquad (2.7)\\
&+ \int_{\Omega_V} \zeta\, dx + \int_{\Sigma_V} \zeta^\Sigma\, do + \int_{\mathscr{C}_V} \zeta^{\mathscr{C}}\, dl.
\end{aligned}
$$

Here Φ, Φ^Σ and $\Phi^\mathscr{C}$ denote the entropy flux in the bulk phases, on the interfaces and on the triple line, respectively, while ζ, ζ^Σ and $\zeta^\mathscr{C}$ are the corresponding entropy productions.

Remark Note that the internal energy density as well as the entropy density can be positive even if the area- or line-specific mass densities are considered to be zero. In other words, in the limit as $\rho^\Sigma \to 0+$ or $\rho^\mathscr{C} \to 0+$, products such as $\rho^\Sigma e^\Sigma$ or $\rho^\mathscr{C} s^\mathscr{C}$ may converge to strictly positive limit densities, i.e.

$$\rho^\Sigma e^\Sigma \to u^\Sigma, \quad \rho^\Sigma s^\Sigma \to \eta^\Sigma, \quad \rho^\mathscr{C} e^\mathscr{C} \to u^\mathscr{C}, \quad \rho^\mathscr{C} s^\mathscr{C} \to \eta^\mathscr{C}$$

with non-vanishing densities $u^\Sigma, \eta^\Sigma, u^\mathscr{C}, \eta^\mathscr{C}$ has to be allowed for. Otherwise, for instance, the surface tension for a fluid interface with zero surface mass density would automatically vanish.

2.3 Transport Theorems

The derivation of local versions of the balance equations follow by application of appropriate transport theorems and subsequent localization. The following transport theorems will be employed; see [28] as a general reference.

Volume Transport. In the general setting described above, let $V \subset \mathbb{R}^3$ be a fixed control volume in G, let Σ be short for $\bigcup_{k=1}^3 \Sigma^k$ with the time-dependent interfaces $\Sigma^k(t)$ and $n_\Sigma = n_{\Sigma^k}$ the unit normal field on $\Sigma^k(t)$ with an arbitrary fixed orientation. Let V_Σ denote the speed of normal displacement of $\Sigma^k(\cdot)$. The latter is a purely kinematic quantity, but it is related to the barycentric velocity of the interfacial mass via $V_\Sigma = v^\Sigma \cdot n_\Sigma$. Moreover, given any bulk field ϕ, the jump of ϕ at Σ is defined by the jump bracket $[\![\cdot]\!]$ according to

$$[\![\phi]\!](t,x) := \lim_{h \to 0+} \big(\phi(t, x + h n_\Sigma) - \phi(t, x - h n_\Sigma)\big). \tag{2.8}$$

With these notations and for the specific control volumes mentioned, as well as for sufficiently smooth fields, it holds that

$$\frac{d}{dt}\int_V \phi \, dx = \int_{V \setminus \Sigma} \partial_t \phi \, dx - \int_{\Sigma_V} [\![\phi]\!] \, V_\Sigma \, do, \tag{2.9}$$

where $\Sigma_V(t) := \Sigma(t) \cap V$.

Surface Transport. In the general setting described above, let $V \subset \mathbb{R}^3$ be a fixed control volume in G. Then, for sufficiently smooth fields, it holds that

$$\frac{d}{dt}\int_{\Sigma_V} \phi^\Sigma \, do = \int_{\Sigma_V} \big(\partial_t^\Sigma \phi^\Sigma - \phi^\Sigma \kappa_\Sigma V_\Sigma\big) \, do + \int_{\partial \Sigma_V} \phi^\Sigma \, V_{\partial \Sigma_V} \, dl. \tag{2.10}$$

Here ∂_t^Σ denotes the time derivative along a path that follows the normal motion of $\Sigma(\cdot)$, sometimes called Thomas-derivative, and $\kappa_\Sigma := \mathrm{div}_\Sigma(-n_\Sigma)$ is twice the mean curvature. Furthermore, $V_{\partial\Sigma_V}$ is the normal (relative to the boundary of Σ_V) speed of displacement of $\partial\Sigma_V(\cdot)$ (in the plane tangential to Σ).

Let us note in passing that the derivation of the local balance equations can be done with special control volumes such that the outer normal n_V satisfies $n_V \perp n_\Sigma$ on $\partial V \cap \Sigma$. For such control volumes the boundary contribution, i.e. the last term in (2.10), vanishes.

Line Transport. For sufficiently smooth fields, it holds that

$$\frac{d}{dt}\int_{\mathscr{C}_V} \phi^{\mathscr{C}}\, dl = \int_{\mathscr{C}_V} \Big(\frac{D^{\mathscr{C}}\phi^{\mathscr{C}}}{Dt} + \phi^{\mathscr{C}}\mathrm{div}_{\mathscr{C}} v^{\mathscr{C}}\Big)\, dl + \int_{\partial\mathscr{C}_V} \phi^{\mathscr{C}}\,\big(V_{\partial\mathscr{C}_V} - v^{\mathscr{C}}\cdot\nu\big)\, dP. \tag{2.11}$$

Here $\frac{D^{\mathscr{C}}}{Dt}$ denotes the Lagrangian derivative, following the triple line along a path with velocity $v^{\mathscr{C}}$ and $V_{\partial\mathscr{C}_V}$ is the normal (relative to the end points of $\mathscr{C}_V$) speed of displacement of $\partial\mathscr{C}_V$. Recall that ν is the outer normal to the curve $\mathscr{C}_V$ in its end points (cf. Fig. 2.1) and that dP denotes the point (i.e., counting) measure.

Remarks 1. The transport theorems above appear rather different. Actually, they can all be brought into the same form as the line transport theorem. In case of surface transport, this follows directly from the relation

$$\frac{D^\Sigma\phi^\Sigma}{Dt} = \partial_t^\Sigma\phi^\Sigma + v^\Sigma\cdot\nabla_\Sigma\phi^\Sigma$$

for the surface Lagrangian derivative. Note that $\mathrm{div}_\Sigma v^\Sigma = \mathrm{div}_\Sigma v_{||}^\Sigma - \kappa_\Sigma V_\Sigma$, hence

$$\partial_t^\Sigma\phi^\Sigma - \phi^\Sigma\kappa_\Sigma V_\Sigma = \frac{D^\Sigma\phi^\Sigma}{Dt} - \mathrm{div}_\Sigma(\phi^\Sigma v_{||}^\Sigma) + \phi^\Sigma\mathrm{div}_\Sigma v^\Sigma$$

and then, by the surface divergence theorem, Eq. (2.10) implies

$$\frac{d}{dt}\int_{\Sigma_V} \phi^\Sigma\, do = \int_{\Sigma_V} \Big(\frac{D^\Sigma\phi^\Sigma}{Dt} + \phi^\Sigma\mathrm{div}_\Sigma v^\Sigma\Big)\, do + \int_{\partial\Sigma_V} \phi^\Sigma\,\big(V_{\partial\Sigma_V} - v^\Sigma\cdot N\big)\, dl. \tag{2.12}$$

To bring the volume transport formula (2.9) into the same form, one first observes that (2.9) combines the transport formulas for both bulk phases which meet at the considered interface. If two bulk phases $\Omega^\pm(t)$ are separated by an interface $\Sigma(t)$, then a simple variant of the Reynolds transport theorem yields

$$\frac{d}{dt}\int_{\Omega_V^\pm} \phi\, dx = \int_{\Omega_V^\pm} \Big(\frac{D\phi}{Dt} + \phi\,\mathrm{div}\, v\Big)\, dx + \int_{\partial\Omega_V^\pm} \phi\,\big(V_{\partial\Omega_V^\pm} - v\cdot n\big)\, do, \tag{2.13}$$

where n is the outer unit normal to $\Omega_V^{\pm}$; note that the latter coincides with $\pm n_\Sigma$ on Σ_V.

2. An equivalent form of (2.12) reads as

$$\frac{d}{dt}\int_{\Sigma_V} \phi^\Sigma \, do = \int_{\Sigma_V} \big(\frac{D^\Sigma \phi^\Sigma}{Dt} + \phi^\Sigma \operatorname{div}_\Sigma v^\Sigma\big)\, do - \int_{\partial\Sigma_V} \phi^\Sigma \, \frac{v^\Sigma \cdot n_V}{\sqrt{1-(n_\Sigma \cdot n_V)^2}}\, dl, \tag{2.14}$$

where n_V is the outer unit normal to V. For this purpose, one first uses elementary geometry to compute $V_{\partial\Sigma_V} = -V_\Sigma \frac{n_\Sigma \cdot n_V}{\sqrt{1-(n_\Sigma \cdot n_V)^2}}$. Since $\sqrt{1-(n_\Sigma \cdot n_V)^2} = N \cdot n_V$ and $\{N, n_\Sigma, \tau\}$ with τ a unit vector tangential to $\partial\Sigma_V$ (hence also to ∂V) is a local orthonormal basis, the Eq. (2.14) follows from

$$v^\Sigma \cdot N - V_{\partial\Sigma_V} = \frac{1}{\sqrt{1-(n_\Sigma \cdot n_V)^2}} \Big((v^\Sigma \cdot N)\,(N \cdot n_V) + (v^\Sigma \cdot n_\Sigma)\,(n_\Sigma \cdot n_V) \Big).$$

The relation from (2.12) has been given in [29], while the variant (2.14) can be found in Chap. 3 in [30]; see also the appendix in [31].

3. Below, we will also use variants of the above transport theorems with built-in mass balance. These read as

$$\frac{d}{dt}\int_V \rho\phi \, dx = \int_{V\setminus\Sigma} \rho \frac{D\phi}{Dt}\, dx + \int_{\Sigma_V} [\![\dot m \phi]\!]\, do - \int_{\partial V} \rho\phi v \cdot n \, do \tag{2.15}$$

with $\dot m^{\pm} := \rho^{\pm}(v^{\pm} - v^\Sigma) \cdot n_\Sigma$ on Σ, and

$$\frac{d}{dt}\int_{\Sigma_V} \rho^\Sigma \phi^\Sigma \, do = \tag{2.16}$$
$$\int_{\Sigma_V} \big(\rho^\Sigma \frac{D^\Sigma \phi^\Sigma}{Dt} - [\![\dot m \phi^\Sigma]\!]\big)\, do + \int_{\partial\Sigma_V} \rho^\Sigma \phi^\Sigma \big(V_{\partial\Sigma_V} - v^\Sigma \cdot N\big)\, dl.$$

2.4 Local Balances

Application of the transport theorems and localization yields the following local balance equations.

Bulk Phase.

$$\partial_t \rho + \operatorname{div}(\rho v) = 0, \tag{2.17}$$
$$\partial_t(\rho v) + \operatorname{div}(\rho v \otimes v - S) = \rho b, \tag{2.18}$$
$$\partial_t(\rho e) + \operatorname{div}(\rho e v + q) = S : \nabla v, \tag{2.19}$$
$$\partial_t(\rho s) + \operatorname{div}(\rho s v + \Phi) = \zeta. \tag{2.20}$$

These are the well-known balance equations in a bulk phase.

Interface.

$$\partial_t^\Sigma \rho^\Sigma + \mathrm{div}_\Sigma(\rho^\Sigma v^\Sigma) + [\![\rho(v - v^\Sigma)\cdot n_\Sigma]\!] = 0, \tag{2.21}$$

$$\partial_t^\Sigma(\rho^\Sigma v^\Sigma) + \mathrm{div}_\Sigma(\rho^\Sigma v^\Sigma \otimes v^\Sigma - S^\Sigma) + [\![\big(\rho v \otimes (v - v^\Sigma) - S\big)\cdot n_\Sigma]\!] = \rho^\Sigma b^\Sigma, \tag{2.22}$$

$$\begin{aligned}&\partial_t^\Sigma(\rho^\Sigma e^\Sigma) + \mathrm{div}_\Sigma(\rho^\Sigma e^\Sigma v^\Sigma + q^\Sigma)\\ &\quad + [\![\Big(\rho(e + \frac{(v - v^\Sigma)^2}{2})(v - v^\Sigma) - (v - v^\Sigma)\cdot S + q\Big)\cdot n_\Sigma]\!] = S^\Sigma : \nabla_\Sigma v^\Sigma,\end{aligned} \tag{2.23}$$

$$\partial_t^\Sigma(\rho^\Sigma s^\Sigma) + \mathrm{div}_\Sigma(\rho^\Sigma s^\Sigma v^\Sigma + \Phi^\Sigma) + [\![\big(\rho s(v - v^\Sigma) + \Phi\big)\cdot n_\Sigma]\!] = \zeta^\Sigma. \tag{2.24}$$

Observe that the jump terms always appear with n_Σ as a factor. Therefore, these terms are invariant under re-orientation of the interfaces. Actually, the notion of a "jump condition" for these terms can be rather misleading. Note that $[\![f \cdot n_\Sigma]\!] = -f^+ \cdot n^+ - f^- \cdot n^-$ if the interface separates two bulk phases $\Omega^\pm$ with outer unit normals $n^\pm$. Hence, if f denotes a bulk flux, the term $-[\![f \cdot n_\Sigma]\!]$ describes the total rate of transfer from the bulk phases to the interfaces due to these fluxes. For the derivation of closure rates below, the explicit form of this term is to be used since two binary products are involved.

Triple Line.

$$\partial_t^{\mathscr{C}} \rho^{\mathscr{C}} + \mathrm{div}_{\mathscr{C}}(\rho^{\mathscr{C}} v^{\mathscr{C}}) + [\![\![\rho^\Sigma(v^\Sigma - v^{\mathscr{C}})\cdot N]\!]\!] = 0, \tag{2.25}$$

$$\begin{aligned}&\partial_t^{\mathscr{C}}(\rho^{\mathscr{C}} v^{\mathscr{C}}) + \mathrm{div}_{\mathscr{C}}(\rho^{\mathscr{C}} v^{\mathscr{C}} \otimes v^{\mathscr{C}} - S^{\mathscr{C}})\\ &\quad + [\![\![\big(\rho^\Sigma v^\Sigma \otimes (v^\Sigma - v^{\mathscr{C}}) - S^\Sigma\big)\cdot N]\!]\!] = \rho^{\mathscr{C}} b^{\mathscr{C}},\end{aligned} \tag{2.26}$$

$$\begin{aligned}&\partial_t^{\mathscr{C}}(\rho^{\mathscr{C}} e^{\mathscr{C}}) + \mathrm{div}_{\mathscr{C}}(\rho^{\mathscr{C}} e^{\mathscr{C}} v^{\mathscr{C}} + q^{\mathscr{C}})\\ &\quad + [\![\![\Big(\rho^\Sigma(e^\Sigma + \frac{(v^\Sigma - v^{\mathscr{C}})^2}{2})(v^\Sigma - v^{\mathscr{C}}) - (v^\Sigma - v^{\mathscr{C}})\cdot S^\Sigma + q^\Sigma\Big)\cdot N]\!]\!] = S^{\mathscr{C}} : \nabla_{\mathscr{C}} v^{\mathscr{C}},\end{aligned} \tag{2.27}$$

$$\partial_t^{\mathscr{C}}(\rho^{\mathscr{C}} s^{\mathscr{C}}) + \mathrm{div}_{\mathscr{C}}(\rho^{\mathscr{C}} s^{\mathscr{C}} v^{\mathscr{C}} + \Phi^{\mathscr{C}}) + [\![\![\big(\rho^\Sigma s^\Sigma(v^\Sigma - v^{\mathscr{C}}) + \Phi^\Sigma\big)\cdot N]\!]\!] = \zeta^{\mathscr{C}}. \tag{2.28}$$

Here the triple bracket $[\![\![\cdot]\!]\!]$ is defined exclusively for quantities of the form $f^\Sigma \cdot N$ by means of

$$[\![\![f^\Sigma \cdot N]\!]\!] = -\sum_{k=1}^{3} f_k^\Sigma \cdot N^k \quad \text{on } \mathscr{C}, \tag{2.29}$$

where the sum runs over all interfaces which meet at the triple line and $f_k^\Sigma := f_{|\Sigma^k}$. Let us briefly explain the appearance of such terms, e.g., for the mass balance (2.25). The transport relation (2.16) for $\phi^\Sigma \equiv 1$ yields the boundary contribution of the interfacial mass balance as

$$\int_{\partial \Sigma_V} \rho^\Sigma \left(V_{\partial \Sigma_V} - v^\Sigma \cdot N\right) dl = \sum_{k=1}^{3} \int_{\partial \Sigma_V^k} \rho_k^\Sigma \left(V_{\partial \Sigma_V^k} - v_k^\Sigma \cdot N^k\right) dl.$$

The boundary of Σ_V^k is $(\Sigma^k \cap \partial V) \cup \mathscr{C}_V$, hence

$$\int_{\partial \Sigma_V} \rho^\Sigma \left(V_{\partial \Sigma_V} - v^\Sigma \cdot N\right) dl =$$

$$\sum_{k=1}^{3} \int_{\Sigma^k \cap \partial V} \rho_k^\Sigma \left(V_{\partial \Sigma_V^k} - v_k^\Sigma \cdot N^k\right) dl + \int_{\mathscr{C}_V} \sum_{k=1}^{3} \rho_k^\Sigma \left(v^{\mathscr{C}} - v_k^\Sigma\right) \cdot N^k \, dl.$$

Employing the condensed notation, this becomes

$$\int_{\partial \Sigma_V} \rho^\Sigma (V_{\partial \Sigma_V} - v^\Sigma \cdot N) dl = \int_{\Sigma \cap \partial V} \rho^\Sigma (V_{\partial \Sigma_V} - v^\Sigma \cdot N) dl - \int_{\mathscr{C}_V} [\![\rho^\Sigma \left(v^{\mathscr{C}} - v^\Sigma\right) \cdot N]\!] \, dl.$$

2.5 Entropy Production and Closure Relations

The entropy principle states that every admissible closure for the entropy flux is such that the remaining entropy production is a sum, running over all dissipative mechanisms, of binary products. The entropy production is non-negative for any thermodynamic process, i.e. the entropy inequality holds. The system is in equilibrium, if and only if the entropy production vanishes. For more information about the employed entropy principle see [1]. We are going to apply this for bulk, interface and triple line in a fully analogous manner; the details will only be explained for the bulk case. We consider the simplest class of bulk, interface and contact line materials for which the entropy density is assumed to be a concave function of temperature and mass density, only. We hence employ constitutive relations of the form

$$\rho s = h(\rho e, \rho), \qquad \rho^\Sigma s^\Sigma = h^\Sigma(\rho^\Sigma e^\Sigma, \rho^\Sigma), \qquad \rho^{\mathscr{C}} s^{\mathscr{C}} = h^{\mathscr{C}}(\rho^{\mathscr{C}} e^{\mathscr{C}}, \rho^{\mathscr{C}}) \quad (2.30)$$

with concave functions h, h^Σ and $h^{\mathscr{C}}$. We furthermore define the (absolute) temperature and the chemical potential in the respective phase as

$$\frac{1}{T} = \frac{\partial h}{\partial(\rho e)}, \qquad \frac{1}{T^\Sigma} = \frac{\partial h^\Sigma}{\partial(\rho^\Sigma e^\Sigma)}, \qquad \frac{1}{T^{\mathscr{C}}} = \frac{\partial h^{\mathscr{C}}}{\partial(\rho^{\mathscr{C}} e^{\mathscr{C}})} \qquad (2.31)$$

and

$$-\frac{\mu}{T} = \frac{\partial h}{\partial \rho}, \qquad -\frac{\mu^\Sigma}{T^\Sigma} = \frac{\partial h^\Sigma}{\partial \rho^\Sigma}, \qquad -\frac{\mu^{\mathscr{C}}}{T^{\mathscr{C}}} = \frac{\partial h^{\mathscr{C}}}{\partial \rho^{\mathscr{C}}}. \qquad (2.32)$$

We insert the constitutive relation (2.30) for the entropy density into the respective entropy balance, use the chain rule employing the definitions (2.31) and (2.32) and eliminate all partial time derivatives by means of the other balance equations. The resulting terms are grouped in such a way that only a single full divergence appears, which contains in particular the entropy flux, all terms with the velocity divergence as a factor are collected and all remaining terms are grouped to form a sum of binary products.

Bulk Phase. The procedure above yields

$$\zeta = \operatorname{div}\,(\Phi - \frac{q}{T}) - \frac{1}{T}(\rho e + P - \rho s T - \rho\mu)\operatorname{div} v + q\cdot\nabla\frac{1}{T} + \frac{1}{T}S^\circ : \nabla v, \quad (2.33)$$

where $P := -\frac{1}{3}\operatorname{tr} S$ is the mechanical pressure and $S^\circ := S + PI$, with I denoting the identity tensor, is the traceless part of S. We will assume throughout this paper that the material in all phases does not support local densities for angular momentum (so-called couples). Hence the balance for angular momentum implies that all stress tensors which appear are symmetric; note that all stress tensors are formulated in the embedding three-dimensional Euclidean space, i.e. are symmetric 3×3-tensors.

Evidently, the simplest closure for the entropy flux in order to fulfill the entropy principle is $\Phi := \frac{q}{T}$, which is the standard choice for single component materials. This leads to the reduced entropy production, being the desired sum of binary products. Exploiting the symmetry of S, we obtain

$$\zeta = -\frac{1}{T}(\rho e + P - \rho s T - \rho\mu)\operatorname{div} v + q\cdot\nabla\frac{1}{T} + \frac{1}{T}S^\circ : D^\circ, \quad (2.34)$$

where $D := \frac{1}{2}(\nabla v + (\nabla v)^{\mathsf{T}})$ is the symmetric part of the velocity gradient and D° its traceless part. The dissipative mechanisms associated with these binary products are "volume variations", "heat conduction" and "viscous shear", in the order of their appearance in (2.34). The simplest linear (in the co-factor) closure without cross-effects leads to the relations

$$\rho e + P - \rho s T - \rho\mu = -\lambda \operatorname{div} v \qquad \text{with } \lambda \geq 0, \quad (2.35)$$

$$q = \alpha\nabla\frac{1}{T} \qquad \text{with } \alpha \geq 0, \quad (2.36)$$

$$S^\circ = 2\eta D^\circ \qquad \text{with } \eta \geq 0. \quad (2.37)$$

Note that the closure parameters λ, α, η are allowed to depend on the basic variables, say (ρ, T). Hence, in particular, the heat flux closure is equivalent to Fourier's law. For consistency with standard notation, we use 2η instead of η, above. At this point, an explanation concerning (2.35) is at order: The only quantity which requires a closure is $P = -\frac{1}{3}\operatorname{tr} S$. For a stagnant fluid, Eq. (2.35) reduces to $\rho e + p - \rho s T - \rho\mu = 0$, where p denotes the pressure at equilibrium. We therefore let the thermodynamic pressure p be defined by the Gibbs-Duhem relation, i.e. by

$$\rho e + p - \rho s T = \rho \mu. \tag{2.38}$$

Then $P = p + \pi$ with the non-equilibrium pressure contribution π and (2.35) becomes

$$\pi = -\lambda \operatorname{div} v. \tag{2.39}$$

The irreversible pressure contribution π is due to volume variations and the linear closure to model it reads as $\pi = -\lambda \operatorname{div} v$. Let us note that the thermodynamic pressure p from (2.38) satisfies the Maxwell relation $p = \rho^2 \frac{\partial \psi}{\partial \rho}$ with the free energy $\psi = \psi(T, \rho) := e - sT$. Alternatively, one can define p by the latter relation and obtain the Gibbs-Duhem relation (2.38) as a consequence. Note also that the entropy production (2.34) can now be written more concisely as

$$\zeta = q \cdot \nabla \frac{1}{T} + \frac{1}{T} S^{\mathrm{irr}} : D, \tag{2.40}$$

where the irreversible stress part is defined as $S^{\mathrm{irr}} = -\pi I + S^{\circ}$, but it is important to notice that the last term represents two independent binary products.

Interface. The same line of arguments leads to

$$\Phi^{\Sigma} = \frac{q^{\Sigma}}{T^{\Sigma}} \quad \text{and} \quad \rho^{\Sigma} e^{\Sigma} + p^{\Sigma} - \rho^{\Sigma} s^{\Sigma} T^{\Sigma} = \rho^{\Sigma} \mu^{\Sigma} \tag{2.41}$$

as well as

$$\begin{aligned} \zeta^{\Sigma} &= q^{\Sigma} \cdot \nabla_{\Sigma} \frac{1}{T^{\Sigma}} - \frac{1}{T^{\Sigma}} \pi^{\Sigma} \operatorname{div}_{\Sigma} v^{\Sigma} + \frac{1}{T^{\Sigma}} S^{\Sigma,\circ} : D^{\Sigma,\circ} \\ &+ \frac{1}{T^{\Sigma}} [\![(v - v^{\Sigma})_{||} \cdot (S \cdot n_{\Sigma})_{||}]\!] + [\![\Big(\frac{1}{T} - \frac{1}{T^{\Sigma}} \Big) \Big(\dot{m} (e + \frac{p}{\rho}) + q \cdot n_{\Sigma} \Big)]\!] \\ &- [\![\Big(\frac{\mu}{T} - \frac{\mu^{\Sigma}}{T^{\Sigma}} + \frac{1}{T^{\Sigma}} \big(\frac{(v - v^{\Sigma})^2}{2} - n_{\Sigma} \cdot \frac{S^{\mathrm{irr}}}{\rho} \cdot n_{\Sigma} \big) \Big) \dot{m}]\!]; \end{aligned} \tag{2.42}$$

recall that $\dot{m} = \rho(v - v^{\Sigma}) \cdot n_{\Sigma}$. Here π^{Σ} is the irreversible part of the interface pressure defined via $\pi^{\Sigma} + p^{\Sigma} = -\frac{1}{2} \operatorname{tr} S^{\Sigma}$ with the thermodynamic interface pressure p^{Σ} from $(2.41)_2$. Moreover, $D^{\Sigma} = \frac{1}{2} I_{\Sigma} \big(\nabla_{\Sigma} v^{\Sigma} + (\nabla_{\Sigma} v^{\Sigma})^{\mathsf{T}} \big) I_{\Sigma}$ is the symmetric interface velocity gradient, $D^{\Sigma,\circ}$ its traceless part and $I_{\Sigma} = I - n_{\Sigma} \otimes n_{\Sigma}$ denotes the surface projector, also called surface identity.

The dissipative processes associated with the binary products in (2.42) are, in the order of their appearance, interfacial heat conduction, area variation, interfacial shear, one-sided slip between the interface and a bulk phase, heat transfer to and from the interface and, finally, mass transfer to and from the interface. The following closure relations result by assuming linear relations between the corresponding co-factors with one exception: the mass transfer to or from the interface, i.e. the ad- and desorption processes $\dot{m} = \dot{m}^{\mathrm{ad}} - \dot{m}^{\mathrm{de}}$, will be modeled using a non-linear relationship in analogy to the modeling of chemical reactions; cf. [1].

$$q^{\Sigma} = \alpha^{\Sigma} \nabla_{\Sigma} \frac{1}{T^{\Sigma}} \qquad \text{with } \alpha^{\Sigma} \geq 0, \tag{2.43}$$

$$\pi^{\Sigma} = -\lambda^{\Sigma} \operatorname{div}_{\Sigma} v^{\Sigma} \qquad \text{with } \lambda^{\Sigma} \geq 0, \tag{2.44}$$

$$S^{\Sigma,\circ} = 2\eta^{\Sigma} D^{\Sigma,\circ} \qquad \text{with } \eta^{\Sigma} \geq 0, \tag{2.45}$$

$$\beta^{\Sigma}(v - v^{\Sigma})_{||} + (S n_{\Sigma})_{||} = 0 \qquad \text{with } \beta^{\Sigma} \geq 0, \tag{2.46}$$

$$\frac{1}{T} - \frac{1}{T^{\Sigma}} + \delta^{\Sigma}\big(\rho(e + \frac{p}{\rho})(v - v^{\Sigma}) + q\big) \cdot n_{\Sigma} = 0 \qquad \text{with } \delta^{\Sigma} \geq 0, \tag{2.47}$$

$$a^{\Sigma} \ln \frac{\dot{m}^{\mathrm{ad}}}{\dot{m}^{\mathrm{de}}} = \frac{\mu}{T} - \frac{\mu^{\Sigma}}{T^{\Sigma}} + \frac{1}{T^{\Sigma}}\Big(\frac{(v - v^{\Sigma})^2}{2} - n_{\Sigma} \cdot \frac{S^{\mathrm{irr}}}{\rho} \cdot n_{\Sigma}\Big) \qquad \text{with } a^{\Sigma} \geq 0. \tag{2.48}$$

The closure relation (2.48) employs the decomposition $\dot{m} = \dot{m}^{\mathrm{ad}} - \dot{m}^{\mathrm{de}}$. Note that (2.48) only fixes the ratio of ad- and desorption, while one of the rates needs to be modeled based on experimental knowledge or a micro-theory. The simplest choice is to assume a desorption rate according to $\dot{m}^{\mathrm{de}} = k^{\mathrm{de}} \rho^{\Sigma}$ with $k^{\mathrm{de}} > 0$. Observe also that (2.48) is an implicit equation regarding $\dot{m}$, since $v^{\pm} - v^{\Sigma} = (v^{\pm} - v^{\Sigma})_{||} + \dot{m}^{\pm}/\rho^{\pm}$.

At this point it should be noted that the closure relations above are given in a condensed notation: relations (2.43)–(2.45) are employed for every interface Σ^k ($k = 1, 2, 3$) separately with respective transport coefficients, while the transmission relations (2.46)–(2.48) apply to each interface in combination with any of the two adjacent bulk phases. In total, the closure relations hence yield nine conditions at each of the three interfaces.

Triple Line. Since the triple line is one-dimensional, the contact line stress tensor satisfies $S^{\mathscr{C}} = -P^{\mathscr{C}} I_{\mathscr{C}}$ with the mechanical line pressure $P^{\mathscr{C}} := -\operatorname{tr} S^{\mathscr{C}}$ and the line projector defined by $I_{\mathscr{C}} w = \langle w, \tau \rangle \tau$ with τ a unit tangent field on $\mathscr{C}$. By the same procedure as above, we obtain the following identities, where $P^{\mathscr{C}} = p^{\mathscr{C}} + \pi^{\mathscr{C}}$, and $\dot{m}^{\Sigma} = \rho^{\Sigma}(v^{\Sigma} - v^{\mathscr{C}}) \cdot N$, i.e. $\dot{m}_k^{\Sigma} = \rho_k^{\Sigma}(v_k^{\Sigma} - v^{\mathscr{C}}) \cdot N^k$ for $k = 1, 2, 3$.

$$\Phi^{\mathscr{C}} = \frac{q^{\mathscr{C}}}{T^{\mathscr{C}}} \quad \text{and} \quad \rho^{\mathscr{C}} e^{\mathscr{C}} + p^{\mathscr{C}} - \rho^{\mathscr{C}} s^{\mathscr{C}} T^{\mathscr{C}} = \rho^{\mathscr{C}} \mu^{\mathscr{C}} \tag{2.49}$$

as well as

$$\begin{aligned}\zeta^{\mathscr{C}} &= q^{\mathscr{C}} \cdot \nabla_{\mathscr{C}} \frac{1}{T^{\mathscr{C}}} - \frac{1}{T^{\mathscr{C}}} \pi^{\mathscr{C}} \operatorname{div}_{\mathscr{C}} v^{\mathscr{C}} + \frac{1}{T^{\mathscr{C}}} [\![(v^{\Sigma} - v^{\mathscr{C}})_{|||} \cdot (S^{\Sigma} \cdot N)_{|||}]\!] \\ &+ [\![\Big(\frac{1}{T^{\Sigma}} - \frac{1}{T^{\mathscr{C}}} \Big) \Big(\rho^{\Sigma} (e^{\Sigma} + \frac{p^{\Sigma}}{\rho^{\Sigma}})(v^{\Sigma} - v^{\mathscr{C}}) + q^{\Sigma} \Big) \cdot N]\!] \\ &- [\![\Big(\frac{\mu^{\Sigma}}{T^{\Sigma}} - \frac{\mu^{\mathscr{C}}}{T^{\mathscr{C}}} + \frac{1}{T^{\mathscr{C}}} \big(\frac{(v^{\Sigma} - v^{\mathscr{C}})^2}{2} - N \cdot \frac{S^{\Sigma,\mathrm{irr}}}{\rho^{\Sigma}} \cdot N \big) \Big) \dot{m}^{\Sigma}]\!] .\end{aligned} \tag{2.50}$$

Above, the notation $(\cdot)_{|||}$ denotes the component tangential to the triple line and $S^{\Sigma,\mathrm{irr}} := -\pi^{\Sigma} I_{\Sigma} + S^{\Sigma,\circ}$. In analogy to the interface we obtain the following closure relations for the dissipative processes on the triple line.

$$q^{\mathscr{C}} = \alpha^{\mathscr{C}} \nabla_{\mathscr{C}} \frac{1}{T^{\mathscr{C}}} \qquad \text{with } \alpha^{\mathscr{C}} \geq 0, \tag{2.51}$$

$$\pi^{\mathscr{C}} = -\lambda^{\mathscr{C}} \operatorname{div}_{\mathscr{C}} v^{\mathscr{C}} \qquad \text{with } \lambda^{\mathscr{C}} \geq 0, \tag{2.52}$$

$$\beta^{\mathscr{C}} (v^{\Sigma} - v^{\mathscr{C}})_{|||} + (S^{\Sigma} N)_{|||} = 0 \qquad \text{with } \beta^{\mathscr{C}} \geq 0, \tag{2.53}$$

$$\frac{1}{T^{\Sigma}} - \frac{1}{T^{\mathscr{C}}} + \delta^{\mathscr{C}} \big(\rho^{\Sigma} (e^{\Sigma} + \frac{p^{\Sigma}}{\rho^{\Sigma}})(v^{\Sigma} - v^{\mathscr{C}}) + q^{\Sigma} \big) \cdot N = 0 \qquad \text{with } \delta^{\mathscr{C}} \geq 0, \tag{2.54}$$

$$a^{\mathscr{C}} \ln \frac{\dot{m}^{\Sigma,\mathrm{ad}}}{\dot{m}^{\Sigma,\mathrm{de}}} = \frac{\mu^{\Sigma}}{T^{\Sigma}} - \frac{\mu^{\mathscr{C}}}{T^{\mathscr{C}}} + \frac{1}{T^{\mathscr{C}}} \big(\frac{(v^{\Sigma} - v^{\mathscr{C}})^2}{2} - N \cdot \frac{S^{\Sigma,\mathrm{irr}}}{\rho} \cdot N \big) \qquad \text{with } a^{\mathscr{C}} \geq 0. \tag{2.55}$$

As in the interface case, in (2.55) the decomposition of $\dot{m}^{\Sigma} = \rho^{\Sigma}(v^{\Sigma} - v^{\mathscr{C}}) \cdot N$ as $\dot{m}^{\Sigma} = \dot{m}^{\Sigma,\mathrm{ad}} - \dot{m}^{\Sigma,\mathrm{de}}$ is employed. The relation (2.55) governs the ratio of ad- and desorption at the triple line, while one of the rates needs to be modeled based on experimental knowledge or a micro-theory. Below, we assume the desorption rate to be given by $\dot{m}^{\Sigma,\mathrm{de}} = k^{\Sigma,\mathrm{de}} \rho^{\mathscr{C}}$ with $k^{\Sigma,\mathrm{de}} > 0$. The transfer relations (2.53)–(2.55) exist for every combination of the triple line with one of the interfaces, of course with individual transfer coefficients.

To complete the model it remains to fix free energy functions for the bulk phases, the interfaces and the triple line. This will only be done for a reduced model below.

2.6 Isothermal Case with Vanishing Triple Line Mass

We consider the limiting case of isothermal conditions, i.e. the internal energy balances are replaced by a known constant temperature field; in particular, we have $T_{|\Sigma} = T^{\Sigma}$ and $T^{\Sigma}_{|\mathscr{C}} = T^{\mathscr{C}}$. We also reduce the model complexity by neglecting the mass and inertia on the triple line. Moreover, we neglect any irreversible stress contributions both on the interfaces and on the triple line. For consistency with the notation in interfacial science, we do not employ the surface and line pressure, but rather let $S^\Sigma = \gamma^\Sigma I_\Sigma$ and $S^\mathscr{C} = \gamma^\mathscr{C} I_\mathscr{C}$ with the interface tensions γ^Σ and the line tension $\gamma^\mathscr{C}$. Observe that this means $S^{\Sigma,irr} = 0$. Because of zero triple line mass and isothermal conditions, we assume the line tension to be constant.

Bulk Phase.

$$\partial_t \rho + \mathrm{div}\,(\rho v) = 0, \tag{2.56}$$

$$\partial_t(\rho v) + \mathrm{div}\,(\rho v \otimes v) = \mathrm{div}\, S + \rho b, \tag{2.57}$$

where the stress is given by $S = (-p + \lambda \mathrm{div}\, v)I + 2\eta D^\circ$ according to (2.37) and (2.39). In the compressible case, an equation of state in the form $p = p(\rho)$ (with a strictly increasing function $p(\cdot)$) is to be added according to the specific fluid under consideration.

Interface. We again use the abbreviation $\dot m = \rho(v - v^\Sigma) \cdot n_\Sigma$. Then

$$\partial_t^\Sigma \rho^\Sigma + \mathrm{div}_\Sigma(\rho^\Sigma v^\Sigma) + [\![\dot m]\!] = 0, \tag{2.58}$$

$$\partial_t^\Sigma(\rho^\Sigma v^\Sigma) + \mathrm{div}_\Sigma(\rho^\Sigma v^\Sigma \otimes v^\Sigma) + [\![v\,\dot m]\!] = [\![S \cdot n_\Sigma]\!] + \mathrm{div}_\Sigma S^\Sigma + \rho^\Sigma b^\Sigma. \tag{2.59}$$

Note that

$$\mathrm{div}_\Sigma S^\Sigma = \gamma^\Sigma \kappa_\Sigma n_\Sigma + \nabla_\Sigma \gamma^\Sigma \tag{2.60}$$

in the considered case without surface viscosities.

This is complemented by the constitutive transmission conditions

$$\beta^\Sigma (v - v^\Sigma)_{||} + (S n_\Sigma)_{||} = 0, \tag{2.61}$$

$$a^\Sigma \ln \frac{\dot m^{\mathrm{ad}}}{\dot m^{\mathrm{de}}} = \mu - \mu^\Sigma + \frac{(v - v^\Sigma)^2}{2} - n_\Sigma \cdot \frac{S^{irr}}{\rho} \cdot n_\Sigma, \tag{2.62}$$

where $a^\Sigma, \beta^\Sigma \geq 0$. In addition, the material dependent interface free energy function is required. The latter determines especially the interfacial equation of state $\gamma^\Sigma = \gamma^\Sigma(\rho^\Sigma)$ and we assume that γ^Σ is a strictly decreasing function (i.e., the interface pressure depends strictly increasing on the interface mass density).

Triple Line. In analogy with the interface-related notation, we use as before the abbreviation

$$\dot m^\Sigma := \rho^\Sigma (v^\Sigma - v^\mathscr{C}) \cdot N,$$

i.e. $\dot m_k^\Sigma = \rho_k^\Sigma (v_k^\Sigma - v^\mathscr{C}) \cdot N^k$ for $k = 1, 2, 3$. Due to $\rho^\mathscr{C} \equiv 0$, the triple line mass and momentum balances become

$$[\![\dot{m}^\Sigma]\!] = 0, \tag{2.63}$$

$$[\![v^\Sigma \dot{m}^\Sigma]\!] = [\![\gamma^\Sigma N]\!] + \gamma^{\mathscr{C}} \operatorname{div}_{\mathscr{C}} I_{\mathscr{C}}. \tag{2.64}$$

This is complemented by the constitutive transmission conditions

$$v^\Sigma_{1,|||} = v^\Sigma_{2,|||} = v^\Sigma_{3,|||} =: v^{\mathscr{C}}_{|||}, \tag{2.65}$$

$$\mu^\Sigma_k - \mu^{\mathscr{C}} + ((v^\Sigma_k - v^{\mathscr{C}}) \cdot N^k)^2/2 = 0 \qquad (k = 1, 2, 3). \tag{2.66}$$

A few comments are at order: for simplicity, we consider the no-slip condition $(v^\Sigma - v^{\mathscr{C}})_{|||} = 0$, but note that the barycentric triple line velocity $v^{\mathscr{C}}$ is undefined for a triple line with zero mass. We consider $v^{\mathscr{C}}$ as the kinematic velocity of the contact line $\mathscr{C}$. The chemical potential $\mu^{\mathscr{C}}$ is determined by one of the equations in (2.66), so actually only two equations remain there. Also, observe that $v^\Sigma - v^{\mathscr{C}} \perp \tau, n_\Sigma$, hence

$$v^\Sigma - v^{\mathscr{C}} = (v^\Sigma - v^{\mathscr{C}} | N)\, N, \tag{2.67}$$

as $\mathscr{C}(t) \subset \Sigma_k(t)$ for all times.

2.7 Thermodynamical Consistency and Equilibria

For this reduced isothermal model we show that the *total available energy*, i.e. the sum of the total kinetic energy and the total free energy is a strict Lyapunov function in case of vanishing body forces. We hence let

$$\mathsf{E}_a(t) = \int_G \rho(\frac{v^2}{2} + \psi)\, dx + \int_\Sigma \rho^\Sigma(\frac{(v^\Sigma)^2}{2} + \psi^\Sigma)\, do + \int_{\mathscr{C}} \gamma^{\mathscr{C}}\, dl, \tag{2.68}$$

where G is the total domain. We are going to show

Theorem 2.1 *Let $(\rho, v, \rho^\Sigma, v^\Sigma, \Sigma, \mathscr{C})$ be a classical solution of the model from Sect. 2.6, i.e. a classical solution to* (2.56), (2.57) *with $S = (-p + \lambda \operatorname{div} v)I + 2\eta D^\circ$, where $p(\rho)$ is strictly increasing in ρ, $\lambda, \eta > 0$ and $b = 0$,* (2.58), (2.59) *with $S^\Sigma = \gamma^\Sigma I_\Sigma$, where $\gamma^\Sigma(\rho^\Sigma) > 0$ is strictly decreasing in ρ^Σ, and $b^\Sigma = 0$,* (2.61) *with $\beta^\Sigma > 0$,* (2.62) *with $a^\Sigma > 0$,* (2.63), (2.64) *with $\gamma^{\mathscr{C}}$ a positive constant,* (2.65) *and* (2.66). *We also assume that this solution is non-degenerate at the contact line, i.e. the interfaces meet at angles different from* 0 *or* π. *At the outer boundary, we assume $v \cdot n = 0$, $v \cdot Sn = 0$ on ∂G and $v^\Sigma \cdot N = 0$ on $\Sigma \cap \partial G$.*

Then the total available energy E_a from (2.68) *is a strict Lyapunov function.*

Proof (i) Let $(\rho, v, \rho^\Sigma, v^\Sigma, \Sigma, \mathscr{C})$ be a classical solution of the model from Sect. 2.6. For the bulk contribution, we first apply the transport relation (2.15) and use the momentum balance (2.18) to eliminate $\rho \frac{Dv}{Dt}$. We then exploit $\psi = \psi(\rho)$ with $\psi'(\rho) = p/\rho^2$ and use the mass balance (2.17) to eliminate $\frac{D\rho}{Dt}$. Application of the two-phase divergence theorem for partial integration in the form

$$\int_G v \cdot divS\, dx = \int_{\partial G} v \cdot Sn\, do - \int_G S : \nabla v\, dx - \int_\Sigma [\![v \cdot Sn_\Sigma]\!]\, do$$

yields

$$\begin{aligned}\frac{d}{dt}\int_G \rho(\frac{v^2}{2} + \psi)\, dx = &- \int_{\partial G} \rho(\frac{v^2}{2} + \psi) v \cdot n\, do + \int_{\partial G} v \cdot Sn\, do \\ &- \int_G S^{\text{irr}} : \nabla v\, dx - \int_\Sigma [\![v \cdot Sn_\Sigma]\!]\, do + \int_\Sigma [\![\dot m(\psi + \frac{v^2}{2})]\!]\, do.\end{aligned} \tag{2.69}$$

For the interface contribution, we first apply the transport relation (2.16) and use the momentum balance in the non-conservative form

$$\rho^\Sigma \frac{D^\Sigma v^\Sigma}{Dt} + [\![(v - v^\Sigma)\dot m]\!] = [\![S \cdot n_\Sigma]\!] + \operatorname{div}_\Sigma S^\Sigma,$$

which follows from (2.22) and (2.21), to eliminate $\rho^\Sigma \frac{D^\Sigma v^\Sigma}{Dt}$. Next, we apply the surface divergence theorem for partial integration of $v^\Sigma \cdot \operatorname{div}_\Sigma S^\Sigma$, employ $\psi^\Sigma = \psi^\Sigma(\rho^\Sigma)$ with $(\psi^\Sigma)'(\rho^\Sigma) = p^\Sigma/(\rho^\Sigma)^2 = -\gamma^\Sigma/(\rho^\Sigma)^2$, (2.21) and the interface Gibbs-Duhem relation $(2.41)_2$ to obtain

$$\begin{aligned}\frac{d}{dt}\int_\Sigma \rho^\Sigma(\frac{(v^\Sigma)^2}{2} + \psi^\Sigma)\, do = &\int_{\partial\Sigma} \rho^\Sigma(\frac{(v^\Sigma)^2}{2} + \psi^\Sigma)(V_{\partial\Sigma} - v^\Sigma \cdot N)\, dl + \int_{\partial\Sigma} v^\Sigma S^\Sigma \cdot N\, dl \\ - \int_\Sigma S^{\Sigma,\text{irr}} : D^\Sigma\, do + \int_\Sigma v^\Sigma \cdot [\![Sn_\Sigma]\!]\, do &+ \int_\Sigma [\![(\frac{(v^\Sigma)^2}{2} - v^\Sigma \cdot v - \mu^\Sigma)\dot m]\!]\, do.\end{aligned} \tag{2.70}$$

For the triple line contribution, we apply the transport relation (2.11). For constant $\gamma^{\mathscr{C}}$, this yields

$$\frac{d}{dt}\int_{\mathscr{C}} \gamma^{\mathscr{C}}\, dl = \int_{\mathscr{C}} \gamma^{\mathscr{C}} \operatorname{div}_{\mathscr{C}} v^{\mathscr{C}}\, dl = \int_{\mathscr{C}} \gamma^{\mathscr{C}} I_{\mathscr{C}} : \nabla_{\mathscr{C}} v^{\mathscr{C}}\, dl = -\gamma^{\mathscr{C}} \int_{\mathscr{C}} v^{\mathscr{C}} \cdot \operatorname{div}_{\mathscr{C}} I_{\mathscr{C}}\, dl. \tag{2.71}$$

Employing the identities (2.69), (2.70) and (2.71), we obtain

$$\begin{aligned}\dot{\mathsf{E}}_a = &\int_{\partial\Sigma} \rho^\Sigma(\frac{(v^\Sigma)^2}{2} + \psi^\Sigma)(V_{\partial\Sigma} - v^\Sigma \cdot N)\, dl + \int_{\partial\Sigma} v^\Sigma S^\Sigma \cdot N\, dl \\ &- \int_G S^{\text{irr}} : D\, dx - \int_\Sigma S^{\Sigma,\text{irr}} : D^\Sigma\, do - \int_\Sigma [\![(v - v^\Sigma)_{||} \cdot (S^{\text{irr}} n_\Sigma)_{||}]\!]\, do \\ &+ \int_\Sigma [\![(\mu - \mu^\Sigma + \frac{(v - v^\Sigma)^2}{2} - n_\Sigma \cdot \frac{S^{\text{irr}}}{\rho} \cdot n_\Sigma)\, \dot m]\!]\, do - \gamma^{\mathscr{C}} \int_{\mathscr{C}} v^{\mathscr{C}} \cdot \operatorname{div}_{\mathscr{C}} I_{\mathscr{C}}\, dl.\end{aligned} \tag{2.72}$$

Inserting the constitutive relation $S^\Sigma = \gamma^\Sigma I_\Sigma$ and exploiting the assumptions $v \cdot n = 0$, $v \cdot Sn = 0$ on ∂G and $v^\Sigma \cdot N = 0$ on $\Sigma \cap \partial G$, we get

$$\begin{aligned}\dot{\mathsf{E}}_a = &-\int_G S^{\mathrm{irr}} : D\,dx - \int_\Sigma [\![(v - v^\Sigma)_{||} \cdot (S^{\mathrm{irr}} n_\Sigma)_{||}]\!]\,do \\ &+ \int_\Sigma [\![(\mu - \mu^\Sigma + \frac{(v - v^\Sigma)^2}{2} - n_\Sigma \cdot \frac{S^{\mathrm{irr}}}{\rho} \cdot n_\Sigma)\,\dot{m}]\!]\,do \\ &+ \int_{\mathscr{C}} [\![\![(\frac{(v^\Sigma)^2}{2} + \psi^\Sigma)\dot{m}^\Sigma]\!]\!]\,dl - \int_{\mathscr{C}} [\![\![\gamma^\Sigma v^\Sigma \cdot N]\!]\!]\,dl - \gamma^{\mathscr{C}} \int_{\mathscr{C}} v^{\mathscr{C}} \cdot \mathrm{div}_{\mathscr{C}} I_{\mathscr{C}}\,dl.\end{aligned} \tag{2.73}$$

Expanding $[\![\![\gamma^\Sigma v^\Sigma \cdot N]\!]\!]$ as $[\![\![\gamma^\Sigma (v^\Sigma - v^{\mathscr{C}} + v^{\mathscr{C}}) \cdot N]\!]\!]$, exploitation of (2.64) allows to rewrite the triple line contribution as

$$\int_{\mathscr{C}} [\![\![(\frac{(v^\Sigma)^2}{2} + \mu^\Sigma)\dot{m}^\Sigma]\!]\!]\,dl - \int_{\mathscr{C}} v^{\mathscr{C}} \cdot [\![\![v^\Sigma \dot{m}^\Sigma]\!]\!]\,dl.$$

Using $(v^{\mathscr{C}})^2 [\![\![\dot{m}^\Sigma]\!]\!] = 0$ due to (2.63), where $v^{\mathscr{C}}_{|||}$ is given as the well-defined tangential part $v^\Sigma_{|||}$ by (2.65), we see that (2.73) implies

$$\begin{aligned}\dot{\mathsf{E}}_a = &-\int_G S^{\mathrm{irr}} : D\,dx - \int_\Sigma [\![(v - v^\Sigma)_{||} \cdot (S^{\mathrm{irr}} n_\Sigma)_{||}]\!]\,do \\ &+ \int_\Sigma [\![(\mu - \mu^\Sigma + \frac{(v - v^\Sigma)^2}{2} - n_\Sigma \cdot \frac{S^{\mathrm{irr}}}{\rho} \cdot n_\Sigma)\,\dot{m}]\!]\,do \\ &+ \int_{\mathscr{C}} [\![\![(\mu^\Sigma + \frac{(v^\Sigma - v^{\mathscr{C}})^2}{2})\dot{m}^\Sigma]\!]\!]\,dl.\end{aligned} \tag{2.74}$$

To come to the final representation of $\dot{\mathsf{E}}_a$, we have to write out the jump brackets $[\![\cdot]\!]$ and $[\![\![\cdot]\!]\!]$. We start with the triple line contribution and have, by (2.63),

$$[\![\![(\mu^\Sigma + \frac{(v^\Sigma - v^{\mathscr{C}})^2}{2})\dot{m}^\Sigma]\!]\!] = -\sum_{k=1}^{3} (\mu_k^\Sigma - \mu^{\mathscr{C}} + \frac{(v_k^\Sigma - v^{\mathscr{C}})^2}{2})\dot{m}_k^\Sigma.$$

Since $(v_k^\Sigma - v^{\mathscr{C}})^2 = ((v_k^\Sigma - v^{\mathscr{C}}) \cdot N^k)^2$ by (2.67), the triple line contribution vanishes due to the constitutive assumption (2.66). Insertion of the other constitutive relations, i.e. (2.37), (2.39), (2.61) and (2.62), finally leads to

$$\begin{aligned}\dot{\mathsf{E}}_a = &- \int_G \lambda(\operatorname{div} v)^2\, dx - \int_G 2\eta D^\circ : D^\circ\, dx \\ &- \int_\Sigma \beta^{\Sigma,+}(v^+ - v^\Sigma)_{||}^2\, do - \int_\Sigma \beta^{\Sigma,-}(v^- - v^\Sigma)_{||}^2\, do \\ &- \int_\Sigma a^{\Sigma,+}\big(\log \dot{m}^{+,\mathrm{ad}} - \log \dot{m}^{+,\mathrm{de}}\big)\big(\dot{m}^{+,\mathrm{ad}} - \dot{m}^{+,\mathrm{de}}\big)\, do \\ &- \int_\Sigma a^{\Sigma,-}\big(\log \dot{m}^{-,\mathrm{ad}} - \log \dot{m}^{-,\mathrm{de}}\big)\big(\dot{m}^{-,\mathrm{ad}} - \dot{m}^{-,\mathrm{de}}\big)\, do.\end{aligned} \tag{2.75}$$

Notice that, according to our condensed notation, the integrals over Σ are to be taken over the three interfaces and the notation $(\cdot)^\pm$ then denotes the respective one-sided bulk limits. Evidently, (2.75) shows that E_a is decreasing along classical solutions, i.e. E_a is a Lyapunov function.

Next we want to characterize the equilibria of the problem, proving at the same time that the total available energy E_a is a strict Lyapunov functional for the system. To this end assume that we have a solution where E_a is not strictly decreasing at all times. Then there is an interval $J = (t_1, t_2)$ where E_a is constant, hence $d\mathsf{E}_a/dt = 0$ in J. This implies, by (2.75),

$$\operatorname{div} v = 0, \quad D^\circ = 0, \quad v_{||}^+ = v_{||}^\Sigma = v_{||}^-, \quad \dot{m}^+ = \dot{m}^- = 0,$$

as $\lambda, \eta, \beta^{\Sigma,\pm}, a^{\Sigma,\pm} > 0$ by assumption. This yields $D = 0$, as well as $[\![v]\!] = 0$ on Σ, which by Lemma 1.2.1 of the monograph [32] implies $v = v_{||}^\Sigma = 0$. Next, investigating the equations for the bulk, we see that $\partial_t \rho = 0$ and $\nabla p = 0$, which implies that ρ is constant in the phases, as p_k is by assumption a strictly increasing function of ρ_k.

In the next step, we look at the equations on the interfaces. By the definition of $\dot{m}^\pm$, we obtain

$$0 = \dot{m}^\pm = \rho^\pm(v^\pm - v^\Sigma)\cdot n^\Sigma = -\rho^\pm v^\Sigma \cdot n^\Sigma,$$

hence $v^\Sigma \cdot n^\Sigma = 0$ which yields $v^\Sigma = 0$. Then the mass balance on Σ implies $\partial_t^\Sigma \rho_\Sigma = 0$ on J. Furthermore, $v = 0$ and ρ constant yield $\mu^\pm$ constant, hence $\mu^\pm = \mu^\Sigma$ is constant by (2.62). This shows that ρ^Σ is constant, as μ^Σ is strictly increasing with ρ^Σ. To see the latter, recall that $\mu^\Sigma = \psi^\Sigma + p^\Sigma/\rho^\Sigma$ and, hence, $(\mu^\Sigma)'(\rho^\Sigma) = (p^\Sigma)'(\rho^\Sigma)/\rho^\Sigma > 0$. This shows further that γ^Σ is constant. Looking at the stress transmission condition this further yields κ^Σ constant on each of the surfaces Σ_k; more precisely we obtain $\kappa^\Sigma = [\![p]\!]$, i.e. the Young-Laplace law holds on each of the surfaces Σ_k.

In the final step, we consider the equations on the contact line. Here we have $v_{|||}^{\mathscr{C}} = 0$ by (2.65), as well as

$$v^{\mathscr{C}} = (v^{\mathscr{C}}|N^k)N^k, \quad k = 1, 2, 3,$$

hence $v^{\mathscr{C}} = 0$ if $\dim \operatorname{span}\{N^k\}_{k=1}^3 = 2$, i.e. in the non-degenerate case which is assumed to hold. This further yields $\mu^{\mathscr{C}}$ constant, and there remains the force balance resembling Young's equation:

$$\sum_{k=1}^{3} \gamma^k N^k = \gamma^{\mathscr{C}} \kappa^{\mathscr{C}},$$

where $\kappa^{\mathscr{C}} = -\mathrm{div}_{\mathscr{C}} I_{\mathscr{C}} = \nabla_{\mathscr{C}} \tau$ denotes the curvature vector of the contact line $\mathscr{C}$.

So, $\dot{\mathsf{E}}_a = 0$ on (t_1, t_2) implies the following:

1. The densities are constant, and all velocities vanish.
2. The curvatures κ^{Σ_k} of the hypersurfaces Σ_k are constant.
3. $\sum_{k=1}^3 \gamma^k N^k = \gamma^{\mathscr{C}} \kappa^{\mathscr{C}}$, where the coefficients γ^j are positive constants.

But this implies that the classical solution coincides with an equilibrium of the system at any $t \in (t_1, t_2)$, hence remains at a fixed equilibrium for all $t > t_1$. Consequently, it holds that for any classical solution, E_a is strictly decreasing outside of equilibria, i.e. E_a is actually a strict Lyapunov function. □

To identify all possible equilibrium configurations is a purely geometrical problem. It appears to be a challenging problem and will not be analyzed any further, here.

Final Remarks. 1. The proof that E_a is non-increasing along classical solutions requires all interfacial and triple line conditions, in particular the condition (2.66). This confirms that the original interface formation model of Shikhmurzaev misses one contact line condition. The origin of this transmission condition is the fact that transfer of mass, here from one interface across the contact line to another interface, is a dissipative process which requires a closure relation. This is similar to the case of mass transfer across a fluid interface: even without interfacial mass, a fluid interface carries interfacial energy and, in general, entropy can be produced at the interface. In order to avoid entropy production, the condition which guarantees zero interfacial entropy production has to be added, leading in the simplest case to continuity of the chemical bulk potentials. The triple line analog is Eq. (2.66) above. In the more general case of non-trivial entropy production, thermodynamically consistent closure leads to a condition like (2.55).

2. The molecular kinetic theory of dynamic contact lines supports a friction-like dissipation term at the contact line, modeled as being proportional to the square of the contact line speed. If, instead of the non-linear closure (2.55), a linear relation is imposed, the rate of entropy production due to transfer of interfacial mass across the contact line becomes proportional to the contact line speed squared. In an isothermal setting, this entropy production is proportional to the dissipation of available energy. Hence, the so-called contact line friction can be identified with the interfacial mass transfer dissipation mechanism.

Acknowledgements The first author (D.B.) is grateful for support by the DFG within the cluster of excellence "Center of Smart Interfaces", TU Darmstadt. The second author (J.P.) thanks the DFG for continuous support in the framework of individual research projects.

References

1. Bothe, D., Dreyer, W.: Continuum thermodynamics of chemically reacting fluid mixtures. Acta Mech. **226**, 1757–1805 (2015)
2. Blake, T.D.: The physics of moving wetting lines. J. Colloid Interface Sci. **299**, 1–13 (2006)
3. Bonn, D., Eggers, J., Indekeu, J., Meunier, J., Rolley, E.: Wetting and spreading. Rev. Modern Phys. **81**, 739–805 (2009)
4. Snoeijer, J.H., Andreotti, B.: Moving contact lines: scales regimes, and dynamical transitions. Annu. Rev. Fluid Mech. **45**, 269–292 (2013)
5. Blake, T.D., Haynes, J.M.: Kinetics of liquid/liquid displacement. J. Colloid Interface Sci. **30**, 421–423 (1969)
6. De Gennes, P.-G.: Wetting: statics and dynamics. Rev. Modern Phys. **57**, 827–863 (1985)
7. Jacqmin, D.: Contact-line dynamics of a diffuse fluid interface. J. Colloid Interface Sci. **402**, 57–88 (2000)
8. Huh, C., Scriven, L.E.: Hydrodynamic model of steady movement of a solid/liquid/fluid contact line. J. Colloid Interface Sci. **35**, 85–101 (1971)
9. Dussan, E.B., Davis, V.S.H.: On the motion of a fluid-fluid interface along a solid surface. J. Fluid Mech. **65**, 71–95 (1974)
10. Pukhnachev, V.V., Solonnikov, V.A.: On the problem of dynamic contact angle. J. Appl. Math. Mech. **46**, 771–779 (1982)
11. Solonnikov, V.A.: On some free boundary problems for the Navier-Stokes equations with moving contact points and lines. Math. Ann. **302**, 743–772 (1995)
12. Hoffman, R.L.: A study of the advancing interface. I. Interface shape in liquid/gas systems. J. Colloid Interface Sci. **50**, 228–241 (1975)
13. Tanner, L.H.: The spreading of silicone oil drops on horizontal surfaces. J. Phys. D: Appl. Phys. **12**, 1473–1484 (1979)
14. Jiang, T.-S., Oh, S.-G., Slattery, J.C.: Correlation for dynamic contact angle. J. Colloid Interface Sci. **69**, 74–77 (1979)
15. Cox, R.G.: The dynamics of the spreading of liquids on a solid surface. Part 1. Viscous flow. J. Fluid Mech. **168**, 169–194 (1986)
16. Dussan, E.B.V.: The moving contact line: the slip boundary condition. J. Fluid Mech. **77**, 665–684 (1976)
17. Voinov, O.V.: Hydrodynamics of wetting. Fluid Dyn. **11**, 714–721 (1976)
18. Afkhami, S., Zaleski, S., Bussmann, M.: A mesh-dependent model for applying dynamic contact angles to VOF simulations. J. Comput. Phys. **228**, 5370–5389 (2009)
19. Fath, A., Bothe, D.: Direct Numerical Simulations of thermocapillary migration of a droplet attached to a solid wall. Int. J. Multiphase Flow **77**, 209–221 (2015)
20. Chen, Q., Ramé, E., Garoff, S.: The velocity field near moving contact lines. J. Fluid Mech. **337**, 49–66 (1997)
21. Blake, T.D., Bracke, M., Shikhmurzaev, Y.D.: Experimental evidence of nonlocal hydrodynamic influence on the dynamic contact angle. Phys. Fluids **11**, 1995–2007 (1999)
22. Shikhmurzaev, Y.D.: Capillary Flows with Forming Interfaces. Chapman & Hall/CRC, Boca Raton (2006)
23. Shikhmurzaev, Y.D.: The moving contact line on a smooth solid surface. Int. J. Multiphase Flow **19**, 589–610 (1993)
24. Billingham, J.: On a model for the motion of a contact line on a smooth solid surface. Eur. J. Appl. Math. **17**, 347–382 (2006)

25. Bedeaux, D.: Nonequilibrium thermodynamics and statistical physics of surfaces. In: Prigogine, I., Rice, S.A. (eds.) Advance in Chemical Physics, vol. 64, pp. 47–109. Jon Wiley & Sons (1986)
26. Bedeaux, D., Albano, A.M., Mazur, P.: Boundary conditions and non-equilibrium thermodynamics. Phys. A **82**, 438–462 (1976)
27. Bedeaux, D.: Nonequilibrium thermodynamic description of the three-phase contact line. J. Chem. Phys. **120**, 3744–3748 (2004)
28. Slattery, J.C., Sagis, L., Oh, E.S.: Interfacial Transport Phenomena, 2nd edn. Springer, New York (2007)
29. Cermelli, P., Fried, E., Gurtin, M.E.: Transport relations for surface integrals arising in the formulation of balance laws for evolving fluid interfaces. J. Fluid Mech. **544**, 339–351 (2005)
30. Romano, A., Marasco, A.: Continuum Mechanics—Advanced Topics and Research Trends. Modeling and Simulation in Science, Engineering and Technology (N. Bellomo, Series ed.). Birkhäuser, Boston (2010)
31. Alke, A., Bothe, D.: 3D numerical modelling of soluble surfactant at fluidic interfaces based on the Volume-of-Fluid method. Fluid Dyn. Mater. Process. **5**, 345–372 (2009)
32. Prüss, J., Simonett, G.: Moving Interfaces and Quasilinear Parabolic Problems. Monographs in Mathematics, Birkhäuser (2016)

Chapter 3
Global Solvability of the Problem on Two-Phase Capillary Fluid Motion in the Oberbeck–Boussinesq Approximation

Irina Vlad. Denisova

Abstract Unsteady motion of a drop in another incompressible fluid bounded by a rigid surface is considered in the Oberbeck–Boussinesq approximation. The liquids are separated by a closed unknown interface Γ_t where surface tension is taken into account. Global existence theorem for the problem is stated in Hölder classes of functions provided that the data have small norms and the initial configuration of the drop is close to a ball with the center in drop's barycenter. It is shown that velocity vector field and temperature deviation decay exponentially as $t \to \infty$, the interface between the liquids tending to a sphere $\{|x - h_\infty| = R_0\}$ with a center h_∞, the limiting position of drop's barycenter. It is established that if the initial data are small enough, the inner liquid will remain strictly inside the other one during all the time. The proof is based on the exponential estimate of a generalized energy and on a local existence theorem of the problem in anisotropic Hölder spaces.

Keywords Two-phase problem with unknown interface · Incompressible capillary fluid · Navier–Stokes system · Lagrangian coordinates · Hölder spaces

3.1 Statement of the Problem and the Main Result

In this paper we study unsteady motion of a drop of one viscous incompressible fluid inside another one in the Oberbeck–Boussinesq approximation. The liquids are located in a container with solid boundary Σ where the nonslip condition holds. On the unknown interface Γ_t, we take surface tension into account. Mass forces are assumed to decrease at infinity with respect to time. We prove that the Oberbeck–Boussinesq approximation gives only a small perturbation of the rest state which is damped in time. Steady fall (or uprising) of a drop in a liquid medium under

---

I.V. Denisova (✉)
Institute for Mechanical Engineering Problems, Russian Academy of Sciences, V. O., Bol'shoy prosp., 61, 199178 St. Petersburg, Russia
e-mail: denisovairinavlad@gmail.com

Y. Shibata and Y. Suzuki (eds.), *Mathematical Fluid Dynamics, Present and Future*, Springer Proceedings in Mathematics & Statistics 183,
DOI 10.1007/978-4-431-56457-7_3

gravity force was analyzed by V.A. Solonnikov in the isothermal case [11]. There fluid densities were considered to be close to each other.

We give a mathematical formulation of this two-phase problem.

Let, at the initial moment $t = 0$, a fluid with viscosity $\nu^+ > 0$ and density $\rho^+ > 0$ fill a bounded domain $\Omega_0^+ \subset \mathbb{R}^3$. Let a fluid with viscosity $\nu^- > 0$ and density $\rho^- > 0$ fill a domain Ω_0^- surrounding Ω_0^+. We denote $\partial\Omega_0^+$ by Γ_0. The boundary $\Sigma \equiv \partial(\Omega_0^+ \cup \Gamma_0 \cup \Omega_0^-)$ is a given closed surface, $\Sigma \cap \Gamma_0 = \emptyset$.

For every $t > 0$, it is necessary to find the interface Γ_t between the domains Ω_t^+ and Ω_t^-, as well as the velocity vector field $\mathbf{v}(x, t) = (v_1, v_2, v_3)$, the function p, that is the deviation from the hydrostatic pressure, and the function θ, the deviation from the average temperature value, for both fluids which satisfy the initial-boundary value problem:

$$
\begin{aligned}
&\mathscr{D}_t\mathbf{v} + (\mathbf{v}\cdot\nabla)\mathbf{v} - \nu^{\pm}\nabla^2\mathbf{v} + \frac{1}{\rho^{\pm}}\nabla p = \mathbf{f}(x,t) - \beta^{\pm}\mathbf{g}\theta, \quad \nabla\cdot\mathbf{v} = 0,\\
&\mathscr{D}_t\theta + (\mathbf{v}\cdot\nabla)\theta - k^{\pm}\nabla^2\theta = 0 \quad \text{in} \quad \Omega_t^- \cup \Omega_t^+, \quad t > 0,\\
&\mathbf{v}|_{t=0} = \mathbf{v}_0, \quad \theta|_{t=0} = \theta_0 \quad \text{in} \quad \Omega_0^- \cup \Omega_0^+, \\
&\mathbf{v}|_{\Sigma} = 0, \quad \theta|_{\Sigma} = a,
\end{aligned} \tag{3.1}
$$

$$
\begin{aligned}
&[\mathbf{v}]\big|_{\Gamma_t} \equiv \lim_{\substack{x\to x_0\in\Gamma_t,\\ x\in\Omega_t^+}} \mathbf{v}(x) - \lim_{\substack{x\to x_0\in\Gamma_t,\\ x\in\Omega_t^-}} \mathbf{v}(x) = 0, \quad [\theta]|_{\Gamma_t} = 0, \quad \Big[k^{\pm}\frac{\partial\theta}{\partial\mathbf{n}}\Big]\Big|_{\Gamma_t} = 0,\\
&[\mathsf{T}\mathbf{n}]|_{\Gamma_t} = \sigma H\mathbf{n} \quad \text{on } \Gamma_t.
\end{aligned} \tag{3.2}
$$

Here $\mathscr{D}_t = \partial/\partial t$, $\nabla = (\partial/\partial x_1, \partial/\partial x_2, \partial/\partial x_3)$, $\nu^{\pm}$, $\rho^{\pm}$ are step functions of viscosity and density, respectively, $\mathbf{f}$ is a given vector of mass forces, $\beta^{\pm} > 0$ is a step function of temperature expansion coefficient, $\mathbf{g} = g(0, 0, 1)$, g being the acceleration of free fall, $\mathbf{v}_0$ is the initial velocity, and θ_0 is the initial distribution of temperature deviation, $k^{\pm}$ is a step function of thermal conductivity, a is a given deviation of the temperature on the solid boundary, T is the stress tensor:

$$\mathsf{T}(\mathbf{v}, p) \equiv -p\mathsf{I} + \mu^{\pm}\mathsf{S}(\mathbf{v}),$$

where $\mathsf{S}(\mathbf{v})$ is the tensor with the elements $S_{ik} = \partial v_i/\partial x_k + \partial v_k/\partial x_i$, $i, k = 1, 2, 3$; $\mu^{\pm} = \nu^{\pm}\rho^{\pm}$, I is the unit matrix, $\sigma > 0$ is surface tension coefficient, $\mathbf{n}$ is the outward normal to Ω_t^+, $H(x, t)$ is twice the mean curvature of Γ_t ($H < 0$ at the points where Γ_t is convex toward Ω_t^-). We suppose that a Cartesian coordinate system $\{x\}$ is introduced in $\mathbb{R}^3$. The centered dot denotes the Cartesian scalar product. Summation from 1 to 3 over the repeated indices is implied. We mark the vectors and the vector spaces by boldface letters.

Moreover, to exclude the mass transportation through Γ_t, we assume that the liquid particles do not leave Γ_t. This means that Γ_t consists of points $x(\xi, t)$ such that the corresponding vector $\mathbf{x}(\xi, t)$ solves the Cauchy problem

$$\mathscr{D}_t\mathbf{x} = \mathbf{v}(x(\xi, t)), \quad \mathbf{x}\big|_{t=0} = \boldsymbol{\xi}, \quad \xi \in \Gamma_0, \quad t > 0. \tag{3.3}$$

Hence $\Gamma_t = \{x(\xi, t) | \xi \in \Gamma_0\}$, $\Omega_t^\pm = \{x(\xi, t) | \xi \in \Omega_0^\pm\}$.

Condition (3.3) is equivalent to the equality

$$V_{\mathbf{n}} = \mathbf{v} \cdot \mathbf{n}\big|_{\Gamma_t},$$

where $V_{\mathbf{n}}$ is the speed of the boundary Γ_t in the direction of its outward normal.

We suppose that the drop Ω_0^+ at an initial moment is close to the ball B_{R_0} whose volume equals the volume of the drop. The incompressibility of the fluids implies that the domains $\Omega_t^\pm$ conserve their volumes for all $t > 0$. In particularly, for the drop we have

$$|\Omega_t^+| = |\Omega_0^+| = \frac{4}{3}\pi R_0^3.$$

In order to simplify estimates, we introduce a new pressure function: $p_1 = p$ in Ω_t^+ and $p_1 = p + \sigma\frac{2}{R_0}$ in Ω_t^-. Then in the interface problem, only boundary condition (3.2) changes:

$$[\mathbb{T}(\mathbf{v}, p_1)\mathbf{n}]\big|_{\Gamma_t} = \sigma\Big(H + \frac{2}{R_0}\Big)\mathbf{n}. \tag{3.4}$$

To exclude the intersection between the interface Γ_t and the outer boundary Σ, we need to control the position of the barycenter of the inner fluid Ω_t^+:

$$\mathbf{h}(t) = \frac{1}{|\Omega_t^+|}\int\limits_{\Omega_t^+} \mathbf{x}\,\mathrm{d}x\ .$$

We denote by $r(\Omega, t)$ deviation function of Γ_t from the sphere $S_{R_0}(t) \equiv S_{R_0}(h(t)) = \{|x - h(t)| = R_0\}$. Next, we assume (without restriction of generality) that $\mathbf{h}(0) = 0$ and that $\Gamma \equiv \Gamma_0$ is defined by the equation

$$|x| = R_0 + r_0\Big(\frac{x}{|x|}\Big) \tag{3.5}$$

on the unit sphere S_1.

We prove for problem (3.1), (3.4), (3.3) unique solvability in anisotropic Hölder classes of functions for all $t > 0$, provided that the initial data are smooth and small enough. The same result for a problem governing the motion of single incompressible capillary fluid bounded by a free surface in the isothermal case was obtained by V.A. Solonnikov in [12]. In our proof, we follow a similar scheme. We developed this technique together with V.A. Solonnikov for proving global solvability of the problem for two incompressible capillary fluids without including mass forces [8]. The same problem with mass forces was studied in [4]. Local existence theorem for the problem in the Oberbeck–Boussinesq approximation was established in [5].

The main idea of the proof is to demonstrate an uniform exponential L_2-estimate for fluid velocity and temperature deviation from the mean value. Next, we prove step by step that they decay exponentially in the Hölder norms, pressure function tending to a step-function and the interface between the liquids doing to a sphere with the radius R_0.

We denote the anisotropic Hölder spaces of functions by $C^{k+\alpha,(k+\alpha)/2}(\mathrm{Q}_T)$, $k \in \mathbb{N}\cup\{0\}$, $\mathrm{Q}_T = \Omega\times(0,T)$, where Ω is a domain in $\mathbb{R}^n$, $n\in\mathbb{N}$; $T>0$, $\alpha\in(0,1)$. The fist exponent means the smoothness with respect to the spatial variables, the second one does the time regularity.

So, let $\mathscr{D}_x^{|\mathbf{r}|} = \partial^{(\mathbf{r})}/\partial x_1^{r_1}\dots\partial x_n^{r_n}$, $\mathbf{r}=(r_1,\dots r_n)$, $r_i\in\mathbb{N}\cup\{0\}$, $|\mathbf{r}| = r_1+\dots+r_n$, $\mathscr{D}_t^s = \partial^s/\partial t^s$, $s\in\mathbb{N}\cup\{0\}$, and $k\in\mathbb{N}$. The space $C^{k+\alpha,\frac{k+\alpha}{2}}(Q_T)$ consists of the functions with finite norm

$$|f|_{Q_T}^{(k+\alpha,\frac{k+\alpha}{2})} = \sum_{|\mathbf{r}|+2s\leqslant k} |\mathscr{D}_x^{\mathbf{r}}\mathscr{D}_t^s f|_{Q_T} + \langle f\rangle_{Q_T}^{(k+\alpha,\frac{k+\alpha}{2})},$$

where $|g|_{Q_T} = \sup_{(x,t)\in Q_T}|g(x,t)|$,

$$\langle f\rangle_{Q_T}^{(k+\alpha,\frac{k+\alpha}{2})} = \sum_{|\mathbf{r}|+2s=k} \langle\mathscr{D}_x^{\mathbf{r}}\mathscr{D}_t^s f\rangle_{Q_T}^{(\alpha,\alpha/2)} + \sum_{|\mathbf{r}|+2s=k-1} \langle\mathscr{D}_x^{\mathbf{r}}\mathscr{D}_t^s f\rangle_{t,Q_T}^{(\frac{1+\alpha}{2})},$$

$$\langle g\rangle_{Q_T}^{(\alpha,\alpha/2)} = \langle g\rangle_{x,\mathrm{Q}_T}^{(\alpha)} + \langle g\rangle_{t,\mathrm{Q}_T}^{(\alpha/2)}$$

Hölder's semi-norms with respect one variable are defined as follows:

$$\langle f\rangle_{x,\mathrm{Q}_T}^{(\alpha)} = \sup_{t\in(0,T)}\sup_{x,y\in\Omega}|f(x,t)-f(y,t)||x-y|^{-\alpha},$$

$$\langle f\rangle_{t,\mathrm{Q}_T}^{(\mu)} = \sup_{x\in\Omega}\sup_{t,\tau\in(0,T)}|f(x,t)-f(x,\tau)||t-\tau|^{-\mu}, \quad \mu\in(0,1).$$

By $C^{k+\alpha}(\varOmega)$, $k\in\mathbb{N}\cup\{0\}$, we denote the space of functions $f(x)$, $x\in\varOmega$, with the norm

$$|f|_{\varOmega}^{(k+\alpha)} = \sum_{|\mathbf{r}|\leqslant k}|\mathscr{D}_x^{\mathbf{r}} f|_{\varOmega} + \langle f\rangle_{\varOmega}^{(k+\alpha)},$$

where $|g|_{\varOmega} = \sup_{x\in\varOmega}|g(x)|$,

$$\langle f\rangle_{\varOmega}^{(k+\alpha)} = \sum_{|\mathbf{r}|=k}\langle\mathscr{D}_x^{\mathbf{r}} f\rangle_{\varOmega}^{(\alpha)} \equiv \sup_{x,y\in\varOmega}\sum_{|\mathbf{r}|=k}|\mathscr{D}_x^{\mathbf{r}} f(x) - \mathscr{D}_y^{\mathbf{r}} f(y)||x-y|^{-\alpha}.$$

We also need the following semi-norm with $\alpha,\gamma\in(0,1)$:

$$|f|_{\mathrm{Q}_T}^{(\gamma,1+\alpha)} = \langle f\rangle_{\mathrm{Q}_T}^{(\gamma,1+\alpha)} + \langle f\rangle_{t,\mathrm{Q}_T}^{(\frac{1+\alpha-\gamma}{2})},$$

where

$$\langle f\rangle_{\mathrm{Q}_T}^{(\gamma,1+\alpha)} = \sup_{t,\tau\in(0,T)}\sup_{x,y\in\Omega}\frac{|f(x,t)-f(y,t)-f(x,\tau)+f(y,\tau)|}{|x-y|^{\gamma}|t-\tau|^{(1+\alpha-\gamma)/2}}.$$

The estimate

$$\langle f\rangle_{\mathrm{Q}_T}^{(\gamma,1+\alpha)} \leqslant c_1\langle f\rangle_{\mathrm{Q}_T}^{(1+\alpha,\frac{1+\alpha}{2})}$$

is known [10].

By definition, $f\in C^{(\gamma,1+\alpha)}(\mathrm{Q}_T)$ if

$$|f|_{\mathrm{Q}_T}^{(\gamma,1+\alpha)}<\infty.$$

Finally, if a function f has finite norm

$$|f|_{\mathrm{Q}_T}^{(\gamma,\mu)} \equiv \langle f\rangle_{x,\mathrm{Q}_T}^{(\gamma)} + |f|_{t,\mathrm{Q}_T}^{(\mu)},\quad \gamma\in(0,1),\quad \mu\in[0,1),$$

where

$$|f|_{t,\mathrm{Q}_T}^{(\mu)} = \begin{cases} |f|_{\mathrm{Q}_T}+\langle f\rangle_{t,\mathrm{Q}_T}^{(\mu)} & \text{if}\quad \mu>0,\\ |f|_{\mathrm{Q}_T} & \text{if}\quad \mu=0,\end{cases}$$

then this function belongs to the Hölder space $C^{\gamma,\mu}(\mathrm{Q}_T)$.

A vector-valued function is an element of a Hölder space if all of its components belong to this space, and its norm is defined as the maximal norm of the components. So is a tensor-valued function.

In order to arrive at a fixed interface, we apply passing to the Lagrangian coordinates by the formula

$$\mathbf{x} = \boldsymbol{\xi} + \int_0^t \mathbf{u}(\xi,\tau)\,\mathrm{d}\tau \equiv \mathbf{X}_{\mathbf{u}}(\xi,t) \tag{3.6}$$

(here $\mathbf{u}(\xi,t)$ is the velocity vector field in the Lagrangian coordinates). Next, we apply the well-known relation

$$H\mathbf{n} = \Delta(t)\mathbf{x} \equiv \Delta(t)\mathbf{X}_{\mathbf{u}}(\xi,t), \tag{3.7}$$

where $\Delta(t)$ denotes the Beltrami-Laplace operator on Γ_t.

We denote:

$$\mathrm{Q}_T^{\pm}\equiv\Omega_0^{\pm}\times(0,T),\ \mathrm{D}_T\equiv \mathrm{Q}_T^{+}\cup\mathrm{Q}_T^{-},\ G_T\equiv\Gamma\times(0,T),\ \Upsilon_T\equiv\Sigma\times(0,T).$$

As a result of transformation (3.6) and of projecting boundary condition (3.4), (3.7) onto the tangent planes first to Γ_t, then to Γ, we arrive at the problem for $\mathbf{u}$, $q = p_1(X_{\mathbf{u}}, t)$, $\hat{\theta} = \theta(X_{\mathbf{u}}, t)$ with the given interface $\Gamma \equiv \Gamma_0$. If the angle between $\mathbf{n}$ and the exterior normal $\mathbf{n}_0$ to Γ is acute, this system is equivalent to the following one:

$$
\begin{aligned}
&\mathscr{D}_t\mathbf{u} - \nu^{\pm}\nabla_{\mathbf{u}}^2\mathbf{u} + \frac{1}{\rho^{\pm}}\nabla_{\mathbf{u}} q = \mathbf{f}(X_{\mathbf{u}}, t) - \beta^{\pm}\hat{\theta}\mathbf{g}(X_{\mathbf{u}}), \quad \nabla_{\mathbf{u}} \cdot \mathbf{u} = 0,\\
&\mathscr{D}_t\hat{\theta} - k^{\pm}\nabla_{\mathbf{u}}^2\hat{\theta} = 0 \ \text{ in } \mathrm{Q}_T^{\pm},\\
&\mathbf{u}|_{t=0} = \mathbf{v}_0, \qquad \hat{\theta}|_{t=0} = \theta_0 \ \text{ in } \cup\,\Omega_0^{\pm}, \quad \mathbf{u}|_{\Sigma} = 0, \quad \hat{\theta}|_{\Sigma} = a,
\end{aligned} \tag{3.8}
$$

$$
\begin{aligned}
&[\mathbf{u}]\big|_{G_T} = 0, \quad [\hat{\theta}]\big|_{G_T} = 0, \quad [k^{\pm}\mathbf{n}\cdot\nabla_{\mathbf{u}}\hat{\theta}]\big|_{G_T} = 0,\\
&[\mu^{\pm}\Pi_0\Pi\mathsf{S}_{\mathbf{u}}(\mathbf{u})\mathbf{n}]\big|_{G_T} = 0,\\
&[\mathbf{n}_0\cdot\mathsf{T}_{\mathbf{u}}(\mathbf{u}, q)\mathbf{n}]\big|_{G_T} - \sigma\mathbf{n}_0\cdot\Delta(t)\int_0^t \mathbf{u}|_{\Gamma}\,\mathrm{d}\tau = \sigma H_0 + \frac{2\sigma}{R_0} + \sigma\mathbf{n}_0\cdot\int_0^t \dot{\Delta}(\tau)\xi\big|_{\Gamma}\,\mathrm{d}\tau \ \text{on } G_T.
\end{aligned}
$$

Here we have used the notation: $\nabla_{\mathbf{u}} = \mathsf{A}\nabla$, A is the matrix of co-factors A_{ij} to the elements

$$
a_{ij}(\xi, t) = \delta_i^j + \int_0^t \frac{\partial u_i}{\partial \xi_j}\mathrm{d}t'
$$

of the Jacobian matrix of transformation (3.6), δ_i^k is the Kronecker symbol, the vector $\mathbf{n}$ is connected with $\mathbf{n}_0$ by the relation:

$$
\mathbf{n} = \frac{\mathsf{A}\mathbf{n}_0}{|\mathsf{A}\mathbf{n}_0|};
$$

$\Pi\omega = \omega - \mathbf{n}(\mathbf{n}\cdot\omega)$, $\Pi_0\omega = \omega - \mathbf{n}_0(\mathbf{n}_0\cdot\omega)$ are projections of a vector ω onto the tangent planes to Γ_t and Γ, respectively; $\mathsf{S}_{\mathbf{u}}(\mathbf{w})$ is the tensor with the elements

$$
(\mathsf{S}_{\mathbf{u}}(\mathbf{w}))_{ij} = A_{ik}\partial w_j/\partial\xi_k + A_{jk}\partial w_i/\partial\xi_k;
$$

$\mathsf{T}_{\mathbf{u}}(\mathbf{w}, q) = -q\mathsf{I} + \mu^{\pm}\mathsf{S}_{\mathbf{u}}(\mathbf{w})$, $H_0(\xi) = \mathbf{n}_0\cdot\Delta(0)\xi$ is twice the mean curvature of Γ; $\dot{\Delta}(t)$ is the operator obtained from the Beltrami–Laplace operator upon differentiation of the coefficients of the latter with respect to t.

Let be $T \in (0, \infty]$, $t, \tau > 0$. We set:

$$
\Omega = \Omega_0^- \cup \overline{\Omega_0^+} \equiv \Omega_t^- \cup \overline{\Omega_t^+}, \quad Q_T = \Omega \times (0, T);
$$

$$
\mathrm{Q}^{\pm}_{(t,\,t+\tau)} = \Omega_t^{\pm} \times (t, t+\tau), \ \mathrm{D}_{(t,\,t+\tau)} = \cup\mathrm{Q}^{\pm}_{(t,t+\tau)};
$$

and let us denote the norm $\|\cdot\|_{L_2(\Omega)}$ by $\|\cdot\|_\Omega$; $\|a\|_{W_2^1(S_1)} \equiv \|a\|_{S_1} + \|\nabla a\|_{S_1}$.

In addition, we put

$$|f|_{D_T}^{(k+\alpha,\frac{k+\alpha}{2})} \equiv |f|_{Q_T^-}^{(k+\alpha,\frac{k+\alpha}{2})} + |f|_{Q_T^+}^{(k+\alpha,\frac{k+\alpha}{2})},$$

$$|f|_{\cup\Omega^\pm}^{(k+\alpha)} \equiv |f|_{\Omega^-}^{(k+\alpha)} + |f|_{\Omega^+}^{(k+\alpha)}.$$

In the case of absence of the term $\sigma\frac{2}{R_0}\mathbf{n}_0\cdot\mathbf{n}$ in the last boundary condition, existence and uniqueness theorem for system (3.8) was proved in a bounded time interval whose value is defined by the norms of $\mathbf{v}_0$, the right-hand side functions and the curvature of the interface Γ [5, Theorem 1.1].

The term $\sigma\frac{2}{R_0}\mathbf{n}_0\cdot\mathbf{n}$ in the last boundary condition is weak with respect to the left-hand side of the equality, that is why it is easily seen that the above-mentioned Theorem 1.1 from [5] remains valid in the presence of this term.

Now we state existence theorem obtained.

Theorem 3.1 (Local existence theorem) *Suppose that* $\Gamma \in C^{3+\alpha}$, $\Sigma \in C^{2+\alpha}$, $\mathbf{f}, D_x\mathbf{f} \in \mathbf{C}^{\alpha,\frac{1+\alpha-\gamma}{2}}(\mathbb{R}^3 \times (0,T))$, $\mathbf{v}_0 \in \mathbf{C}^{2+\alpha}(\cup\Omega_0^\pm)$, $a \in C^{2+\alpha,1+\alpha/2}(\Upsilon_T)$, $\theta_0 \in C^{2+\alpha}(\cup\Omega_0^\pm)$ *with some* $\alpha \in (0,1)$, $\gamma \in (0,\alpha)$, $0 < T < \infty$. *Moreover, let the compatibility conditions are satisfied:*

$$\begin{aligned}
&\nabla\cdot\mathbf{v}_0 = 0,\ \ [\mathbf{v}_0]|_\Gamma = 0,\ \ [\theta_0]|_\Gamma = 0,\ \ \mathbf{v}_0|_\Sigma = 0,\ \ \theta_0|_\Sigma = a|_{t=0},\\
&[\mu^\pm\Pi_0\mathsf{S}(\mathbf{v}_0)\mathbf{n}_0]|_\Gamma = 0,\\
&[\Pi_0(\nu^\pm\nabla^2\mathbf{v}_0 - \frac{1}{\rho^\pm}\nabla q_0)]|_\Gamma = -[\beta^\pm\theta_0\Pi_0\mathbf{g}]|_\Gamma,\ \ k^-\nabla^2\theta_0|_\Sigma = \mathscr{D}_t a\big|_{t=0},\\
&\Pi_\Sigma(\nu^-\frac{\partial^2\mathbf{v}_0}{\partial\mathbf{n}_\Sigma^2} - \frac{1}{\rho^-}\nabla q_0)\big|_\Sigma = \Pi_\Sigma(\mathbf{f} - \beta^- a\mathbf{g})\big|_{\Sigma,\,t=0},\\
&[k^\pm\nabla^2\theta_0]|_\Gamma = [f]|_{\Gamma,\,t=0},\ \ \left[k^\pm\frac{\partial\theta_0}{\partial\mathbf{n}_0}\right]\Big|_\Gamma = 0\ \ (\frac{\partial}{\partial\mathbf{n}_0} = \mathbf{n}_0\cdot\nabla),
\end{aligned} \tag{3.9}$$

where $q_0(\xi) \equiv q(\xi,0)$ *is a solution of the diffraction problem*

$$\begin{aligned}
&\frac{1}{\rho^\pm}\nabla^2 q_0(\xi) = \nabla\cdot(\mathbf{f}(\xi,0) - \beta^\pm\theta_0\mathbf{g} - \mathscr{D}_t\mathsf{B}^*\big|_{t=0}\mathbf{v}_0(\xi)),\ \ \xi \in \Omega_0^\pm,\\
&[q_0]|_\Gamma = \left[2\mu^\pm\frac{\partial\mathbf{v}_0}{\partial\mathbf{n}_0}\cdot\mathbf{n}_0\right]\Big|_\Gamma - \sigma\Big(H_0 + \frac{2}{R_0}\Big),\\
&\left[\frac{1}{\rho^\pm}\frac{\partial q_0}{\partial\mathbf{n}_0}\right]\Big|_\Gamma = \big[\nu^\pm\mathbf{n}_0\cdot\nabla^2\mathbf{v}_0\big]\big|_\Gamma - \big[\mathbf{n}_0\cdot\beta^\pm\theta_0\mathbf{g}\big]\big|_\Gamma,\\
&\frac{1}{\rho^-}\frac{\partial q_0}{\partial\mathbf{n}_\Sigma}\Big|_\Sigma = \nu^-\mathbf{n}_\Sigma\cdot\nabla^2\mathbf{v}_0\big|_\Sigma + \mathbf{n}_\Sigma\cdot(\mathbf{f} - \beta^- a\mathbf{g})\big|_{\Sigma,\,t=0}\quad (\frac{\partial}{\partial\mathbf{n}_\Sigma} = \mathbf{n}_\Sigma\cdot\nabla).
\end{aligned}$$

Here $\mathsf{B} = \mathsf{A} - \mathsf{I}$, I *is the identity matrix,* B^* *is the transpose to* B, $\mathbf{n}_\Sigma$ *is the outward normal to* Σ, $\Pi_\Sigma\boldsymbol{\omega} \equiv \boldsymbol{\omega} - \mathbf{n}_\Sigma(\mathbf{n}_\Sigma\cdot\boldsymbol{\omega})$.

Then there exists a positive constant $T_* \leqslant T$ such that problem (3.8) has a unique solution $(\mathbf{u}, q, \hat{\theta})$ with the following properties: $\mathbf{u} \in \mathbf{C}^{2+\alpha,1+\alpha/2}(\mathrm{D}_{T_*})$, $q \in C^{(\gamma,1+\alpha)}(\mathrm{D}_{T_*})$, $\nabla q \in \mathbf{C}^{\alpha,\alpha/2}(\mathrm{D}_{T_*})$, $\hat{\theta} \in C^{2+\alpha,1+\alpha/2}(\mathrm{D}_{T_*})$, the functions $\mathbf{u}, \hat{\theta}$ being defined in a unique way, while q being done up to a bounded time dependent function. The value of T_* depends on the data norms and on the curvature of Γ.

The solution $(\mathbf{u}, q, \hat{\theta})$ is subjected to the inequality

$$\begin{aligned} &|\mathbf{u}|^{(2+\alpha,1+\alpha/2)}_{\mathrm{D}_{T_*}} + |\nabla q|^{(\alpha,\alpha/2)}_{\mathrm{D}_{T_*}} + |q|^{(\frac{1+\alpha-\gamma}{2})}_{t,\mathrm{D}_{T_*}} + \langle q\rangle^{(1+\alpha,\gamma)}_{\mathrm{D}_{T_*}} + |\hat{\theta}|^{(2+\alpha,1+\frac{\alpha}{2})}_{\mathrm{D}_{T_*}} \\ &\quad\leqslant c_4(T_*)\Big\{ |\mathbf{f}|^{(\alpha,\frac{\alpha}{2})}_{\mathrm{D}_{T_*}} + |\mathbf{f}|^{(\frac{1+\alpha-\gamma}{2})}_{t,\mathrm{D}_{T_*}} + |H_0 + \frac{2}{R_0}|^{(1+\alpha)}_{\Gamma} + |a|^{(2+\alpha,1+\frac{\alpha}{2})}_{\Upsilon_{T_*}} \\ &\qquad + c\big(T_*^{\frac{1-\gamma}{2}}, |\mathbf{v}_0|^{(1+\alpha)}_{\cup\Omega_0^\pm}\big)\Big(|\mathbf{v}_0|^{(2+\alpha)}_{\cup\Omega_0^\pm} + |\theta_0|^{(2+\alpha)}_{\cup\Omega_0^\pm}\Big)\Big\}, \qquad (3.10) \\ &|\hat{\theta}|^{(2+\alpha,1+\frac{\alpha}{2})}_{\mathrm{D}_{T_*}} \leqslant c_5(T_*)\Big\{ |a|^{(2+\alpha,1+\frac{\alpha}{2})}_{\Upsilon_{T_*}} + \big(c + T_*^{\frac{1-\gamma}{2}}|\mathbf{v}_0|^{(1)}_{\cup\Omega_0^\pm}\big)|\theta_0|^{(2+\alpha)}_{\cup\Omega_0^\pm}\Big\}, \end{aligned}$$

where $c_4(T)$, $c_5(T)$ are increasing functions of T.

Remark 3.1 Although theorem statement in [5] does not have estimate (3.10), one can easily see in the proof that it follows by passage to the limit from the solution estimates for linearized problems like (3.21), (3.23). The complete estimate was obtained in [5] as inequality (4.22).

We state now the main result of the paper.

Theorem 3.2 (Global existence theorem) *Let the hypotheses of Theorem 3.1 hold. Assume, in addition, that for $t = 0$ interface Γ is given by (3.5) on the unit sphere, and the initial data are small enough, i.e.,*

$$\begin{aligned} &|\mathrm{e}^{b_1 t}\mathbf{f}|^{(\alpha,\frac{1+\alpha-\gamma}{2})}_{\mathrm{Q}_\infty} + |\mathrm{e}^{b_1 t}\nabla\mathbf{f}|^{(\alpha,\frac{1+\alpha-\gamma}{2})}_{\mathrm{Q}_\infty} + \|\mathrm{e}^{b_1 t}\mathbf{f}\|_{\mathrm{Q}_\infty} + \big\|\mathrm{e}^{b_1 t}a\big\|_{\mathrm{W}_2^{3/2,3/4}(\Upsilon_\infty)} \\ &\quad + |\mathrm{e}^{b_1 t}a|^{(2+\alpha,1+\alpha/2)}_{\Upsilon_\infty} + |\mathbf{v}_0|^{(2+\alpha)}_{\cup\Omega_0^\pm} + |\theta_0|^{(2+\alpha)}_{\cup\Omega_0^\pm} + |r_0|^{(3+\alpha)}_{\mathrm{S}_1} \leqslant \varepsilon \ll 1. \qquad (3.11) \end{aligned}$$

Then problem (3.1), (3.4), (3.3) is uniquely solvable for all $t > 0$, and the solution $(\mathbf{v}, p_1)$ has the properties: $\mathbf{v} \in \mathbf{C}^{2+\alpha,1+\alpha/2}$, $p_1 \in C^{(\gamma,1+\alpha)}$, $\nabla p_1 \in \mathbf{C}^{\alpha,\alpha/2}$, $\theta \in C^{2+\alpha,1+\alpha/2}$, the function p_1 being defined up to a bounded time dependent function. The interface Γ_t is given for $\forall t$ by a function of $C^{3+\alpha}$:

$$|x - h(t)| = R_0 + r\left(\frac{x-h}{|x-h|}, t\right),$$

(where $h(t)$ is a position of the barycenter of Ω_t^+ at the moment t), and tending to a sphere of the radius R_0 with center in a certain point h_∞; $r_0(\omega) \equiv r(\omega, 0)$. It means that for any $t_0 \in (0, \infty)$, the solution $(\mathbf{u}, q, \hat{\theta})$ and its derivatives in local Lagrangian coordinates belong to respective Hölder spaces for a sufficiently small time interval $(t_0, t_0 + \tau)$; it is subjected to the estimate

$$|\mathbf{u}|^{(2+\alpha,1+\alpha/2)}_{D_{(t_0,t_0+\tau)}} + |\nabla q|^{(\alpha,\alpha/2)}_{D_{(t_0,t_0+\tau)}} + |q|^{(\gamma,1+\alpha)}_{D_{(t_0,t_0+\tau)}} + |\hat{\theta}|^{(2+\alpha,1+\alpha/2)}_{D_{(t_0,t_0+\tau)}} + \sup_{t\in(t_0,t_0+\tau)} |r(\cdot,t)|^{(3+\alpha)}_{S_1}$$
$$\leqslant ce^{-b_1 t_0}\Big\{|e^{b_1 t}\mathbf{f}|^{(\alpha,\frac{1+\alpha-\gamma}{2})}_{Q_\infty} + |e^{b_1 t}\nabla\mathbf{f}|^{(\alpha,\frac{1+\alpha-\gamma}{2})}_{Q_\infty} + \|e^{b_1 t}\mathbf{f}\|_{Q_\infty} + \|e^{b_1 t}a\|_{W_2^{\frac{3}{2},\frac{3}{4}}(\Upsilon_\infty)}$$
$$+|e^{b_1 t}a|^{(2+\alpha,1+\alpha/2)}_{\Upsilon_\infty} + |\mathbf{v}_0|^{(2+\alpha)}_{\cup\Omega_0^\pm} + |\theta_0|^{(2+\alpha)}_{\cup\Omega_0^\pm} + |r_0|^{(3+\alpha)}_{S_1}\Big\}, \tag{3.12}$$

where the values τ, b_1, c are independent of t_0.

One can conclude from this theorem that the stability of this solution takes place in the sense that the solution is close to zero under a small deviation of the data from zero. However, the center of the limit sphere $S_{R_0}(h_\infty)$ may be displaced with respect to the initial barycenter of Ω_0^+ for no matter how small an initial velocity $\mathbf{v}_0$ is. This displacement is evaluated in inequality (3.48) at the end of the paper. We also give there an estimate from below of the distance between the outer boundary and the initial fluid interface which guarantees the absence of the intersection between these surfaces.

Remark 3.2 We note that global solvability of a similar problem governing the motion of two fluids in the Oberbeck–Boussinesq approximation without including surface tension may obtained on the base of the results in [2, 3, 5]. In this case, one can reduce the necessary smoothness of the initial interface and take $\Gamma \in C^{2+\alpha}$.

3.2 An Energy Estimate of the Solution

In this section we prove an exponential estimate for the solution of the nonlinear problem (3.1), (3.4), (3.3) in L_2.

We will use the proposition proved in [8] for system (3.1), (3.4) in the isothermal case. We developed there the idea of constructing a function of generalized energy for the fluids proposed by Padula [9].

We suppose that the L_2-norm of mass forces decreases with respect to time in exponential way with the certain constant b and Γ_t is close to a shere.

Proposition 3.1 *Assume that a solution of problem (3.1) with $\theta = 0$, (3.4), (3.3) is defined on $[0, T]$ and $\mathbf{v}_0$ satisfies compatibility conditions (3.9). Let r be such that*

$$|r(\omega,t)|_{S_1\times(0,T)} + |\nabla_{S_1} r(\omega,t)|_{S_1\times(0,T)} \leqslant \delta_1 R_0 \ll 1 \tag{3.13}$$

and

$$\mathbf{f}(\cdot,\tau) \in L_2(\Omega), \quad \|e^{b\tau}\mathbf{f}\|_{Q_T} < \infty \tag{3.14}$$

with a big enough constant b.

Then for $\forall t \in (0, T]$

$$\|\mathbf{v}(\cdot,t)\|_{\Omega} + \|r(\cdot,t)\|_{W_2^1(S_1)} \leqslant c_1 \mathrm{e}^{-bt}\Big\{\|\mathrm{e}^{b\tau}\mathbf{f}\|_{Q_t} + \|\mathbf{v}_0\|_{\Omega} + \|r_0\|_{W_2^1(S_1)}\Big\}. \quad (3.15)$$

Proposition 3.2 *Assume that a solution of problem (3.1), (3.4), (3.3) is defined in* $[0, T]$. *Let, in addition, r satisfies smallness condition (3.13) and* $\mathbf{f}$ *does (3.14).*

Then for $\forall t \in (0, T]$

$$\|\mathbf{v}(\cdot,t)\|_{\Omega} + \|r(\cdot,t)\|_{W_2^1(S_1)} \leqslant c_2 \mathrm{e}^{-b_1 t}\Big\{\|\mathbf{v}_0\|_{\Omega} + \|\theta_0\|_{\Omega} + \|r_0\|_{W_2^1(S_1)}$$
$$+ \big\|\mathrm{e}^{b\tau}\mathbf{f}\big\|_{Q_t} + \big\|\mathrm{e}^{b_2\tau} a\big\|_{W_2^{3/2,3/4}(\Upsilon_t)}\Big\}, \quad (3.16)$$
$$\|\theta(\cdot,t)\|_{\Omega} \leqslant c_3 \mathrm{e}^{-b_2 t}\Big\{\|\theta_0\|_{\Omega} + \big\|\mathrm{e}^{b_2\tau} a\big\|_{W_2^{3/2,3/4}(\Upsilon_t)}\Big\} \quad (3.17)$$

with constants b_1, b_2, c_2 *and* c_3 *independent of* t; $b_1 = \min\{b, b_2\}$; b_2 *is the constant from (3.19).*

Proof In order to apply Proposition 3.1 to system (3.1), (3.4), we should be sure that the L_2-norm of θ decays exponentially in the certain way.

We extend the function a with preservation of class into the domain Ω so that the extension $a^* = 0$ in a neighborhood of Ω_t^+ and inside it. Let's consider the difference $\widetilde{\theta} = \theta - a^*$. It solves the problem

$$\begin{aligned}
&\mathscr{D}_t\widetilde{\theta} + (\mathbf{v}\cdot\nabla)\widetilde{\theta} - k^{\pm}\nabla^2\widetilde{\theta} = \frac{\mathrm{d}a^*}{\mathrm{d}t} - k^-\nabla^2 a^* \quad \text{in} \quad \Omega_t^- \cup \Omega_t^+, \quad t > 0,\\
&\widetilde{\theta}|_{t=0} = \widetilde{\theta}_0 - a^*|_{t=0} \quad \text{in} \quad \Omega_0^- \cup \Omega_0^+, \\
&[\widetilde{\theta}]|_{\Gamma_t} = 0, \quad \Big[k^{\pm}\frac{\partial\widetilde{\theta}}{\partial\mathbf{n}}\Big]\Big|_{\Gamma_t} = 0, \quad \widetilde{\theta}|_{\Sigma} = 0.
\end{aligned} \quad (3.18)$$

To obtain an exponential estimate for $\widetilde{\theta}$, we multiply the heat equation in (3.18) by $\widetilde{\theta}$ and integrate by parts over $\Omega_t^- \cup \Omega_t^+$:

$$\frac{1}{2}\frac{\mathrm{d}}{\mathrm{d}t}\|\widetilde{\theta}\|^2_{\cup\Omega_t^{\pm}} + \|\sqrt{k^{\pm}}\nabla\widetilde{\theta}\|^2_{\cup\Omega_t^{\pm}} = \int_{\Omega_t^-}\Big(\frac{\mathrm{d}a^*}{\mathrm{d}t} - k^-\nabla^2 a^*\Big)\widetilde{\theta}\,\mathrm{d}x.$$

And since $[\widetilde{\theta}]|_{\Gamma_t} = 0$, we have $\|\nabla\widetilde{\theta}\|^2_{\cup\Omega_t^{\pm}} = \|\nabla\widetilde{\theta}\|^2_{\Omega}$, $\Omega = \Omega_t^- \cup \overline{\Omega_t^+}$, and

$$\frac{1}{2}\frac{\mathrm{d}}{\mathrm{d}t}\|\widetilde{\theta}\|^2_{\Omega} + \min\{k^{\pm}\}\|\nabla\widetilde{\theta}\|^2_{\Omega} \leqslant \Big\|\frac{\mathrm{d}a^*}{\mathrm{d}t} - k^-\nabla^2 a^*\Big\|_{\Omega_t^-}\|\widetilde{\theta}\|_{\Omega_t^-}.$$

We apply the Poincaré inequality to the norm of $\nabla\widetilde{\theta}$ in view of $\widetilde{\theta}|_{\Sigma} = 0$, then

$$\frac{\mathrm{d}}{\mathrm{d}t}\|\widetilde{\theta}\|^2_{\Omega} + 2b_2\|\widetilde{\theta}\|^2_{\Omega} \leqslant c\Big\|\frac{\mathrm{d}a^*}{\mathrm{d}t} - k^-\nabla^2 a^*\Big\|^2_{\Omega_t^-} \quad (3.19)$$

which, due to the Gronwall lemma, gives

$$\|\widetilde{\theta}(\cdot,t)\|^2_\Omega \leqslant \mathrm{e}^{-2b_2t}\|\theta_0\|^2_\Omega + c\int_0^t \mathrm{e}^{-2b_2(t-\tau)}\Big\|\frac{\mathrm{d}a^*}{\mathrm{d}t} - k^-\nabla^2 a^*\Big\|^2_\Omega \,\mathrm{d}\tau$$
$$\leqslant c\mathrm{e}^{-2b_2t}\Big(\|\theta_0\|^2_\Omega + \big\|\mathrm{e}^{b_2\tau}a^*\big\|^2_{W_2^{2,1}(Q_t)}\Big)$$
$$\leqslant c\mathrm{e}^{-2b_2t}\Big(\|\theta_0\|^2_\Omega + \big\|\mathrm{e}^{b_2\tau}a\big\|^2_{W_2^{3/2,3/4}(\Upsilon_t)}\Big).$$

Finally,

$$\|\theta(\cdot,t)\|_\Omega \leqslant \|\widetilde{\theta}(\cdot,t)\|_\Omega + \|a^*(\cdot,t)\|_\Omega \leqslant \mathrm{e}^{-b_2t}\Big(\|\theta_0\|_\Omega + \|a(\cdot,0)\|_\Sigma + \big\|\mathrm{e}^{b_2\tau}a\big\|_{W_2^{3/2,3/4}(\Upsilon_t)}\Big).$$

Then inequality (3.16) follows from (3.17) and (3.15) for the first equation in (3.1).

□

Corollary 3.1 *The coordinates of the barycenter of Ω_t^+ satisfy the inequality*

$$|\mathbf{h}(t)| \leqslant c\Big\{\|\mathrm{e}^{b_1\tau}\mathbf{f}\|_{Q_T} + \big\|\mathrm{e}^{b_1\tau}a\big\|_{W_2^{3/2,3/4}(\Upsilon_t)} + \|\mathbf{v}_0\|_\Omega + \|\theta_0\|_\Omega + \|r_0\|_{W_2^1(S_1)}\Big\} \quad (3.20)$$

for $\forall t \in [0,T]$.

Proof Since we have the solution in the interval $[0,T]$, we know barycenter trajectory of the drop Ω_t^+: $\mathbf{h}(t) = \frac{1}{|\Omega_t^+|}\int\limits_{\Omega_t^+}\mathbf{x}\,\mathrm{d}x$. Moreover,

$$\mathbf{h}'_t(t) = |\Omega_t^+|^{-1}\int_{\Omega_t^+}\mathbf{v}(x,t)\,\mathrm{d}x.$$

We have assumed that $\mathbf{h}(0) = 0$. Inequality (3.20) follows from (3.16) and the estimate

$$|\mathbf{h}(t)| \leqslant \frac{1}{|\Omega_0^+|^{1/2}}\int_0^t \|\mathbf{v}(\cdot,\tau)\|_{\Omega_t^+}\,\mathrm{d}\tau.$$

□

3.3 Linearized Problems

Let us consider a linear problem with a given vector field $\mathbf{u}$:

$$\mathscr{D}_t\mathbf{w} - \nu^\pm\nabla^2_{\mathbf{u}}\mathbf{w} + \frac{1}{\rho^\pm}\nabla_{\mathbf{u}}s = \mathbf{f}, \quad \nabla_{\mathbf{u}}\cdot\mathbf{w} = r \ \text{ in } D_T,$$

$$\mathbf{w}\big|_{t=0} = \mathbf{w}_0 \text{ in } \Omega_0^- \cup \Omega_0^+, \tag{3.21}$$

$$[\mathbf{w}]\big|_{G_T} = 0, \quad \mathbf{w}|_\Sigma = 0, \quad [\mu^\pm \Pi_0 \Pi \mathsf{S}_\mathbf{u}(\mathbf{w})\mathbf{n}]\big|_{G_T} = \Pi_0 \mathbf{b},$$

$$[\mathbf{n}_0 \cdot \mathsf{T}_\mathbf{u}(\mathbf{w}, s)\mathbf{n}]\big|_{G_T} - \sigma \mathbf{n}_0 \cdot \Delta(t) \int_0^t \mathbf{w}\big|_\Gamma \, d\tau = b + \int_0^t B \, d\tau \text{ on } G_T.$$

The functions in the right-hand sides of all the equations, initial and boundary conditions are given.

The problem (3.21) was studied in [1, 6, 7], where the unique solvability of the problem was proved in an arbitrary finite time interval, when the surface Σ was absent and the domain $\Omega_0^- \cup \overline{\Omega_0^+}$ was the whole space $\mathbb{R}^3$. This result was obtained in the Hölder spaces with a power weight at infinity, however it is valid also in our case.

We give the formulation of the existence theorem for the problem (3.21).

Theorem 3.3 *Assume that for certain $\alpha, \ \gamma \in (0, 1), \ \gamma < \alpha, \ 0 < T < \infty$, the surfaces $\Gamma, \ \Sigma \in C^{2+\alpha}$, $\sigma > 0$, and the vector field $\mathbf{u} \in \mathbf{C}^{2+\alpha, 1+\alpha/2}(D_T)$ satisfies $[\mathbf{u}]\big|_{G_T} = 0$ and the inequality*

$$(T + T^{\gamma/2})|\mathbf{u}|_{D_T}^{(2+\alpha, 1+\alpha/2)} \leqslant \delta \tag{3.22}$$

with sufficiently small $\delta > 0$.

Moreover, we assume that the following four sets of conditions are satisfied

(1) $\mathbf{f} \in \mathbf{C}^{\alpha, \frac{\alpha}{2}}(D_T)$, $r \in C^{1+\alpha, \frac{1+\alpha}{2}}(D_T)$, $\mathbf{w}_0 \in \mathbf{C}^{2+\alpha}(\Omega_0^- \cup \Omega_0^+)$, $\mathbf{b} \in \mathbf{C}^{1+\alpha, \frac{1+\alpha}{2}}(G_T)$, $b \in C^{(\gamma, 1+\alpha)}(G_T)$, $B \in C^{\alpha, \frac{\alpha}{2}}(G_T)$*;*

(2) *The compatibility conditions*

$$\nabla \cdot \mathbf{w}_0(\xi) = r(\xi, 0) = 0, \quad [\mathbf{w}_0]\big|_\Gamma = 0, \quad \mathbf{w}_0|_\Sigma = 0,$$

$$[\mu^\pm \Pi_0 \mathsf{S}(\mathbf{w}_0(\xi))\mathbf{n}_0]\big|_{\xi \in \Gamma} = \Pi_0 \mathbf{b}(\xi, 0), \quad \xi \in \Gamma,$$

$$\left[\Pi_0 \left(\mathbf{f}(\xi, 0) - \frac{1}{\rho^\pm} \nabla s(\xi, 0) + \nu^\pm \nabla^2 \mathbf{w}_0(\xi)\right)\right]\Big|_{\xi \in \Gamma} = 0,$$

$$\Pi_\Sigma \left(\mathbf{f}(\xi, 0) - \frac{1}{\rho^-} \nabla s(\xi, 0) + \nu^- \nabla^2 \mathbf{w}_0(\xi)\right)\Big|_{\xi \in \Sigma} = 0;$$

are satisfied;

(3) *There exist a vector field $\mathbf{g} \in \mathbf{C}^{\alpha, \alpha/2}(D_T)$ and a tensor $\mathsf{G} = \{G_{ik}\}_{i,k=1}^3$, $G_{ik} \in C^{(\gamma, 1+\alpha)}(D_T) \cap C^{\gamma, 0}(D_T)$ such that the representation formulas*

$$\mathscr{D}_t r - \nabla_\mathbf{u} \cdot \mathbf{f} = \nabla \cdot \mathbf{g}, \quad \mathbf{g} = \nabla \cdot \mathsf{G} \quad (g_i = \partial G_{ik}/\partial \xi_k, \quad i = 1, 2, 3),$$

hold (in a generalized sense) and, in addition,

$$[(\mathbf{g} + \mathsf{A}^T\mathbf{f}) \cdot \mathbf{n}_0]\big|_{G_T} = 0;$$

(4) $s_0(\xi) = s(\xi, 0)$ *is a solution to the problem*

$$\begin{aligned}
&\frac{1}{\rho^\pm}\nabla^2 s_0(\xi) = \nabla \cdot \big(\mathscr{D}_t \mathsf{B}^T\big|_{t=0}\mathbf{w}_0(\xi) - \mathbf{g}(\xi, 0)\big) \equiv \nabla \cdot \mathbf{d} \text{ in } \Omega_0^- \cup \Omega_0^+,\\
&[s_0]\big|_\Gamma = \left[2\mu^\pm \frac{\partial \mathbf{w}_0}{\partial \mathbf{n}_0} \cdot \mathbf{n}_0\right]\Big|_\Gamma - b\big|_{t=0},\\
&\left[\frac{1}{\rho^\pm}\frac{\partial s_0}{\partial \mathbf{n}_0}\right]\Big|_\Gamma = \big[\mathbf{n}_0 \cdot \big(\mathbf{f}\big|_{t=0} + \nu^\pm \nabla^2 \mathbf{w}_0\big)\big]\big|_\Gamma,\\
&\frac{1}{\rho^-}\frac{\partial s_0}{\partial \mathbf{n}_\Sigma}\Big|_\Sigma = \nu^- \mathbf{n}_\Sigma \cdot \nabla^2 \mathbf{w}_0|_\Sigma + \mathbf{n}_\Sigma \cdot \mathbf{f}\big|_{\Sigma,\ t=0}.
\end{aligned}$$

Under these assumptions, the problem (3.21) has a unique solution $(\mathbf{w}, s)$, $\mathbf{w} \in \mathbf{C}^{2+\alpha,1+\alpha/2}(\mathrm{D}_T)$, $s \in C^{(\gamma,1+\alpha)}(\mathrm{D}_T)$, $\nabla s \in \mathbf{C}^{\alpha,\alpha/2}(\mathrm{D}_T)$ *(the pressure is defined up to a bounded function of time), and for arbitrary* $t' \in (0, T]$ *the inequality*

$$\begin{aligned}
N_{t'}[\mathbf{w}, s] &\equiv |\mathbf{w}|_{\mathrm{D}_{t'}}^{(2+\alpha,1+\alpha/2)} + |\nabla s|_{\mathrm{D}_{t'}}^{(\alpha,\alpha/2)} + |s|_{\mathrm{D}_{t'}}^{(\gamma,1+\alpha)}\\
&\leqslant c_1(t')\Big\{|\mathbf{f}|_{\mathrm{D}_{t'}}^{(\alpha,\alpha/2)} + |r|_{\mathrm{D}_{t'}}^{(1+\alpha,\frac{1+\alpha}{2})} + |\mathbf{w}_0|_{\cup\Omega_0^\pm}^{(2+\alpha)} + |\mathbf{g}|_{\mathrm{D}_{t'}}^{(\alpha,\alpha/2)}\\
&\quad + |\mathsf{G}|_{\mathrm{D}_{t'}}^{(\gamma,1+\alpha)} + |\mathsf{G}|_{\mathrm{D}_{t'}}^{(\gamma,0)} + |\mathbf{b}|_{\mathrm{G}_{t'}}^{(1+\alpha,\frac{1+\alpha}{2})} + |b|_{\mathrm{G}_{t'}} + |b|_{\mathrm{G}_{t'}}^{(\gamma,1+\alpha)}\\
&\quad + |\nabla_\Gamma b|_{\mathrm{G}_{t'}}^{(\alpha,\alpha/2)} + |B|_{\mathrm{G}_{t'}}^{(\alpha,\alpha/2)} + P_{t'}[\mathbf{u}]|\mathbf{w}_0|_{\cup\Omega_0^\pm}^{(1)}\Big\}
\end{aligned}$$

holds, where $c_1(t')$ *is a non-decreasing function of* $t' \leqslant T$, $\nabla_\Gamma = \Pi_0\nabla$, *and*

$$P_t[\mathbf{u}] = t^{\frac{1-\alpha}{2}}|\nabla \mathbf{u}|_{\mathrm{D}_t} + |\nabla \mathbf{u}|_{\mathrm{D}_t}^{(\alpha,\alpha/2)}.$$

Now let us consider the problem with the unknown temperature function ψ:

$$\begin{aligned}
&\mathscr{D}_t\psi - k^\pm \nabla_{\mathbf{u}}^2 \psi = f \quad \text{in } \mathrm{D}_T,\\
&\psi|_{t=0} = \psi_0 \quad \text{in } \Omega_0^- \cup \Omega_0^+,\\
&[\psi]|_{\mathrm{G}_T} = 0, \quad \psi|_{\Upsilon_T} = \varphi,\\
&[k^\pm \mathbf{n} \cdot \nabla_{\mathbf{u}}\psi]|_{\mathrm{G}_T} = d \quad \text{on } \mathrm{G}_T.
\end{aligned} \tag{3.23}$$

This problem was analyzed in [5] where the following theorem was obtained for it.

Theorem 3.4 *Assume that the surfaces* $\Gamma, \Sigma \in C^{2+\alpha}$, *and that a vector-valued function* $\mathbf{u} \in \mathbf{C}^{2+\alpha,1+\alpha/2}(\mathrm{D}_T)$, $[\mathbf{u}]|_\Gamma = 0$, *and satisfies inequality (3.22). Then for arbitrary* $f \in C^{\alpha,\alpha/2}(\mathrm{D}_T)$, $\psi_0 \in C^{2+\alpha}(\Omega_0^- \cup \Omega_0^+)$, $\varphi \in C^{2+\alpha,1+\alpha/2}(\Upsilon_T)$, $d \in C^{1+\alpha,(1+\alpha)/2}(\mathrm{G}_T)$ *which are subject to the compatibility conditions*

$$[\psi_0]|_\Gamma = 0, \quad \psi_0|_\Sigma = \varphi|_{t=0}, \quad -[k^\pm \nabla^2 \psi_0]|_\Gamma = [f|_{t=0}]|_\Gamma,$$
$$[k^\pm \frac{\partial \psi_0}{\partial \mathbf{n}_0}]|_\Gamma = d(\xi,0), \quad \xi \in \Gamma, \quad \mathscr{D}_t\varphi|_{t=0} - k^- \nabla^2 \psi_0|_\Sigma = f|_{\Sigma,\, t=0}, \tag{3.24}$$

problem (3.23) has a unique solution $\psi \in C^{2+\alpha,1+\alpha/2}(\mathrm{D}_T)$ *and the estimate*

$$|\psi|_{\mathrm{D}_T}^{(2+\alpha,1+\alpha/2)} \leqslant c_2(T)\{|f|_{\mathrm{D}_T}^{(\alpha,\alpha/2)} + |\psi_0|_{\cup\Omega_0^\pm}^{(2+\alpha)} + |\varphi|_{\Upsilon_T}^{(2+\alpha,1+\alpha/2)}$$
$$+ |d|_{\mathrm{G}_T}^{(1+\alpha,\frac{1+\alpha}{2})} + T^{\frac{1-\alpha}{2}} |\nabla \mathbf{u}|_{\mathrm{D}_T} |\psi_0|_{\cup\Omega_0^\pm}^{(1)}\} \tag{3.25}$$

holds. Here c_2 *is a nondecreasing function of* T.

3.4 Global Solvability of the Problem (3.1), (3.4), (3.3)

The aim of this section is to prove global solvability of the problem (3.1), (3.4), (3.3) in the whole time interval $\{t > 0\}$.

In the proof we use the following lemma proved in [3].

Lemma 3.1 *Let* $u \in C^{0,\frac{1+\alpha}{2}}(\mathrm{D}_{T_0})$, $T_0 > 0$, $0 < \varkappa < T_0^{1/2}$. *Then the function* u *is subject to the inequality*

$$\langle u \rangle_{t,\mathrm{D}_{T_0}}^{(\frac{1+\alpha-\gamma}{2})} \leqslant 2\varkappa^\gamma \langle u \rangle_{t,\mathrm{D}_{T_0}}^{(\frac{1+\alpha}{2})} + c\,\varkappa^{\gamma-\alpha-\frac{9}{2}} \int_0^{T_0} \|u(\cdot,\tau)\|_\Omega \, d\tau.$$

In a similar way, it is possible to prove the statement.

Lemma 3.2 *For arbitrary function*

$$u \in C^{2+\alpha,1+\frac{\alpha}{2}}(\mathrm{D}_{T_0}) \text{ and } 0 < \varkappa_1, \varkappa_2 < \min\{\operatorname{diam}\{\Omega\}, T_0^{1/2}\},$$

the inequalities

$$\langle u \rangle_{\mathrm{D}_{T_0}}^{(\alpha,\frac{\alpha}{2})} \leqslant 2\varkappa_1^2 \langle u \rangle_{\mathrm{D}_{T_0}}^{(2+\alpha,1+\frac{\alpha}{2})} + c\,\varkappa_1^{-\alpha-\frac{7}{2}} \int_0^{T_0} \|u(\cdot,\tau)\|_\Omega d\tau, \tag{3.26}$$

$$|u|_{D_{T_0}} \leqslant c\left\{\varkappa_2^{1+\alpha}\langle u\rangle_{t,D_{T_0}}^{(\frac{1+\alpha}{2})} + \varkappa_2^{-\frac{7}{2}}\int_0^{T_0}\|u(\cdot,\tau)\|_\Omega\,d\tau\right\} \tag{3.27}$$

hold.

The following proposition was proved in [4].

Proposition 3.3 *Let a solution* $(\mathbf{v}, p)$ *of problem (3.1), (3.4), (3.3) with* $\theta = 0$ *exist on the interval* $(0, T]$ *and the inequality*

$$N_{(0,T)}[\mathbf{v}, p] \equiv |\mathbf{u}|_{D_T}^{(2+\alpha,1+\alpha/2)} + |\nabla q|_{D_T}^{(\alpha,\alpha/2)} + |q|_{D_T}^{(\gamma,1+\alpha)} \leqslant \mu$$

hold. Here $(\mathbf{u}, q)$ *is the solution of the problem in the Lagrangian coordinates.*

Then

$$\begin{aligned} N_{(t_0-\tau_0,t_0)}[\mathbf{v}, p_1, r] &\equiv N_{(t_0-\tau_0,t_0)}[\mathbf{v}, p_1] + \sup_{t_0-\tau_0<\tau<t_0} |r(\cdot,\tau)|_{S_1}^{(3+\alpha)} \\ &\leqslant c_1(\delta,\tau_0)\left\{|\mathbf{f}|_{D_{(t_0-2\tau_0,t_0)}}^{(\alpha,\frac{1+\alpha-\gamma}{2})} + |\nabla\mathbf{f}|_{D_{(t_0-2\tau_0,t_0)}}^{(\alpha,\frac{1+\alpha-\gamma}{2})}\right. \\ &\quad \left. + \int_{t_0-2\tau_0}^{t_0} \left(\|\mathbf{v}(\cdot,\tau)\|_\Omega + \|r(\cdot,\tau)\|_{W_2^1(S_1)}\right) d\tau\right\}, \end{aligned} \tag{3.28}$$

where δ *is the value from (3.22),* $\tau_0 \in (0, t_0/2)$*,* τ_0 *depends on* μ*:*

$$(2\tau_0 + (2\tau_0)^{\gamma/2})\mu \leqslant \delta, \tag{3.29}$$

$c(\delta, \tau_0)$ *is a non-decreasing function.*

We use Proposition 3.3 to prove the statement as follows:

Proposition 3.4 *Assume that a solution of problem (3.1), (3.4), (3.3) is defined on the interval* $(0, T]$ *and the estimate*

$$N_{(0,T)}[\mathbf{v}, p_1, \theta] \equiv |\mathbf{u}|_{D_T}^{(2+\alpha,1+\alpha/2)} + |\nabla q|_{D_T}^{(\alpha,\alpha/2)} + |q|_{D_T}^{(\gamma,1+\alpha)} + |\widehat{\theta}|_{D_T}^{(2+\alpha,1+\alpha/2)} \leqslant \mu_1$$

holds. Here the triple $(\mathbf{u}, q, \widehat{\theta})$ *means the solution written as a function of the Lagrangian coordinates.*

Then

$$\begin{aligned} &N_{(t_0-\tau_0/2,t_0)}[\mathbf{v}, p_1, \theta, r] \equiv N_{(t_0-\tau_0/2,t_0)}[\mathbf{v}, p_1, \theta] + \sup_{t_0-\tau_0/2<\tau<t_0} |r(\cdot,\tau)|_{S_1}^{(3+\alpha)} \\ &\leqslant c(\delta,\tau_0)\left\{|\mathbf{f}|_{D_0'}^{(\alpha,\frac{1+\alpha-\gamma}{2})} + |\nabla\mathbf{f}|_{D_0'}^{(\alpha,\frac{1+\alpha-\gamma}{2})} + |a|_{\Upsilon_0'}^{(2+\alpha,1+\alpha/2)}\right. \\ &\quad \left. + \int_{t_0-2\tau_0}^{t_0} \left(\|\mathbf{v}(\cdot,\tau)\|_\Omega + \|\theta(\cdot,\tau)\|_\Omega + \|r(\cdot,\tau)\|_{W_2^1(S_1)}\right) d\tau\right\}, \end{aligned} \tag{3.30}$$

where $t_0 \in (0, T]$, $\tau_0 \in (0, t_0/2)$, τ_0 *depends on* μ *and on the constant* δ *in (3.22). Moreover, we have used the notation:* $\mathrm{D}'_\beta = \mathrm{D}_{(t_0-2\tau_0+\beta, t_0)}$, $\Upsilon'_\beta = \Sigma \times (t_0 - 2\tau_0 + \beta, t_0)$ *with* $\beta \in [0, 2\tau_0)$.

Proof We fix an arbitrary $t_0 \in (0, T]$. For $\tau_0 \in (0, t_0/2)$ we denote by $\eta_\lambda(t)$ a smooth monotone function of t such that

$$\eta_\lambda(t) = \begin{cases} 0 & \text{if } t \leqslant t_0 - 2\tau_0 + \lambda/2, \\ 1 & \text{if } t \geqslant t_0 - 2\tau_0 + \lambda, \end{cases}$$

$\lambda \in (0, \tau_0]$, and

$$\left|\dot{\eta}_\lambda(t)\right|_{\mathbb{R}} \leqslant c\lambda^{-1}, \quad \langle\dot{\eta}_\lambda(t)\rangle_{\mathbb{R}}^{(\alpha/2)} \leqslant c\lambda^{-1-\alpha/2},$$

where $\dot{\eta}_\lambda \equiv \mathrm{d}\eta_\lambda(t)/\mathrm{d}t$.

We consider the triple $\mathbf{w} = \mathbf{v}\eta_\lambda$, $s = p_1\eta_\lambda$, $\vartheta = \theta\eta_\lambda$. It satisfies the system

$$\begin{aligned}
&\mathscr{D}_t\mathbf{w} + (\mathbf{v}\cdot\nabla)\mathbf{w} - \nu^\pm\nabla^2\mathbf{w} + \frac{1}{\rho^\pm}\nabla s = \mathbf{f}\eta_\lambda + \mathbf{v}\dot{\eta}_\lambda - \beta^\pm\mathbf{g}\vartheta, \\
&\mathscr{D}_t\vartheta + (\mathbf{v}\cdot\nabla)\vartheta - k^\pm\nabla^2\vartheta = \theta\dot{\eta}_\lambda, \\
&\nabla\cdot\mathbf{w} = 0 \quad \text{in} \quad \Omega_t^- \cup \Omega_t^+, \quad t > t_0 - 2\tau_0, \\
&\mathbf{w}|_{t=t_0-2\tau_0} = 0, \quad \vartheta|_{t=t_0-2\tau_0} = 0 \quad \text{in} \quad \cup\,\Omega'_\pm \equiv \cup\Omega^\pm_{t_0-2\tau_0}, \\
&\quad \mathbf{w}|_\Sigma = 0, \quad \vartheta|_\Sigma = a\eta_\lambda, \\
&\quad [\vartheta]\big|_{\Gamma_t} = 0, \quad \left[k^\pm\frac{\partial\vartheta}{\partial\mathbf{n}}\right]\Big|_{\Gamma_t} = 0, \\
&[\mathbf{w}]\big|_{\Gamma_t} = 0, \quad [\mathrm{T}(\mathbf{w}, s)\mathbf{n}]\big|_{\Gamma_t} = \sigma\big(H + \frac{2}{R_0}\big)\mathbf{n}\eta_\lambda\big|_{\Gamma_t}.
\end{aligned} \tag{3.31}$$

We introduce the Lagrangian coordinates according to the formula

$$\mathbf{x} = \boldsymbol{\xi}' + \int\limits_{t_0-2\tau_0}^{t} \mathbf{u}(\xi', \tau)\,\mathrm{d}\tau \equiv \mathbf{X}(\xi', t), \quad \xi' \in \cup\Omega'_\pm, \quad t > t_0 - 2\tau_0, \tag{3.32}$$

where $\mathbf{u}(\xi', t) = \mathbf{v}(X(\xi', t), t)$. We transform problem (3.31) by (3.32). The functions $\mathbf{w}$, s and ϑ written in the Lagrangian coordinates will be denoted by the same symbols.

The function ϑ solves the system

$$\begin{aligned}
&\mathscr{D}_t\vartheta - k^\pm\nabla^2_{\mathbf{u}}\vartheta = \widehat{\theta}\dot{\eta}_\lambda \quad \text{in} \quad \cup\,\Omega'_\pm, \quad t > t_0 - 2\tau_0, \\
&\vartheta|_{t=t_0-2\tau_0} = 0 \quad \text{in} \quad \cup\,\Omega'_\pm, \quad [\vartheta]\big|_{\Gamma'} = 0, \\
&\quad \left[k^\pm\mathbf{n}'\cdot\nabla_{\mathbf{u}}\vartheta\right]\Big|_{\Gamma'} = 0, \quad \vartheta|_\Sigma = a\eta_\lambda.
\end{aligned} \tag{3.33}$$

Here $\Gamma' = \Gamma_{t_0-2\tau_0}$, all the notation, say $\nabla_{\mathbf{u}}$, correspond to transformation (3.32).

We verify the assumptions of Theorem 3.4 to apply it to problem (3.33). First, we choose τ_0 so small that the inequality (3.22) is satisfied. It suffices to take it such that

$$(2\tau_0 + (2\tau_0)^{\gamma/2})\mu_1 \leqslant \delta. \tag{3.34}$$

Next, compatibility conditions (3.24) are fulfilled due to (3.9). Hence, for ϑ, estimate (3.25) holds. All functions in (3.33) are equal to zero at $t = t_0 - 2\tau_0$. Thus,

$$|\widehat{\theta}|^{(2+\alpha,1+\alpha/2)}_{D'_\lambda} \leqslant |\vartheta|^{(2+\alpha,1+\alpha/2)}_{D'_{\tau_0}} \leqslant c_2(2\tau_0)\{|\widehat{\theta}\dot{\eta}_\lambda|^{(\alpha,\alpha/2)}_{D'_{\tau_0}} + |a\eta_\lambda|^{(2+\alpha,1+\alpha/2)}_{\Upsilon'_0}\}. \tag{3.35}$$

Hence for $\lambda \leqslant \tau_0 < 1$ inequality (3.35) can be extended as follows:

$$|\widehat{\theta}|^{(2+\alpha,1+\alpha/2)}_{D'_\lambda} \leqslant c_2(2\tau_0)\left\{\frac{1}{\lambda}|\widehat{\theta}|^{(\alpha,\alpha/2)}_{D'_{\lambda/2}} + \frac{1}{\lambda^{1+\frac{\alpha}{2}}}|\widehat{\theta}|_{D'_{\lambda/2}} + \frac{1}{\lambda^{1+\frac{\alpha}{2}}}|a|^{(2+\alpha,1+\alpha/2)}_{\Upsilon'_{\lambda/2}}\right\}. \tag{3.36}$$

We estimate the lower order norms of $\widehat{\theta}$ in (3.36) by Lemma 3.2, setting $\varkappa_1 = (\varepsilon\lambda)^{1/2}$ in (3.26). Next, we evaluate $|\widehat{\theta}|_{D'_{\lambda/2}}$ by the inequality (3.27) with $\varkappa_2 = (\varepsilon\lambda^{1+\frac{\alpha}{2}})^{\frac{1}{1+\alpha}}$.

As a result, we deduce the inequality

$$|\widehat{\theta}|^{(2+\alpha,1+\alpha/2)}_{D'_\lambda} \leqslant c_3(\delta)\left\{\varepsilon|\widehat{\theta}|^{(2+\alpha,1+\alpha/2)}_{D'_{\lambda/2}} + c(\varepsilon)\lambda^{-\kappa}\int\limits_{t_0-2\tau_0}^{t_0}\|\widehat{\theta}(\cdot,\tau)\|_\Omega\,d\tau + \frac{1}{\lambda^{1+\frac{\alpha}{2}}}|a|^{(2+\alpha,1+\alpha/2)}_{\Upsilon'_{\lambda/2}}\right\} \tag{3.37}$$

with $\kappa = \max\left\{\frac{11}{4} + \frac{\alpha}{2}, \left(1 + \frac{\alpha}{2}\right)\left(1 + \frac{7}{2(1+\alpha)}\right)\right\}$.

We introduce the function $\Phi(\lambda) = \lambda^\kappa|\widehat{\theta}|^{(2+\alpha,1+\alpha/2)}_{D'_\lambda}$. Since $\kappa > 1 + \frac{\alpha}{2}$, we can write (3.37) in the following way:

$$\Phi(\lambda) \leqslant c_4\varepsilon\Phi(\lambda/2) + K, \tag{3.38}$$

where $c_4 = c_3(\delta)2^\kappa$,

$$K = c_3(\delta)\left\{c(\varepsilon)\int\limits_{t_0-2\tau_0}^{t_0}\|\widehat{\theta}(\cdot,\tau)\|_\Omega\,d\tau + |a|^{(2+\alpha,1+\alpha/2)}_{\Upsilon'_0}\right\}.$$

Setting $\varepsilon = \frac{1}{2c_4}$, we deduce from (3.38) by iterations with $\lambda/2$, ..., $\lambda/2^k$ and by taking the limit as $k \to \infty$, that

$$\Phi(\lambda) \leqslant 2K.$$

This estimate with $\lambda = \tau_0$ implies the inequality

$$|\widehat{\theta}|^{(2+\alpha,1+\alpha/2)}_{D'_{\tau_0}} \leqslant c_5(\delta)\Big\{ \int\limits_{t_0-2\tau_0}^{t_0} \|\widehat{\theta}\|_\Omega \, d\tau + |a|^{(2+\alpha,1+\alpha/2)}_{\Upsilon'_0} \Big\}. \tag{3.39}$$

We apply Proposition 3.3 with $\tau_1 = \tau_0/2$ to the part of problem (3.1), (3.4) concerning $\mathbf{v}$, p_1 and consider θ as a known function in the right-hand side of the first equation in (3.1). We choose τ_0 so small that (3.29) is satisfied for τ_1. Then, by (3.28), we have

$$N_{(t_0-\tau_1,t_0)}[\mathbf{v}, p_1, r] \leqslant c_2(\delta, \tau_0)\Big\{ |\mathbf{f}|^{(\alpha, \frac{1+\alpha-\gamma}{2})}_{D'_{\tau_0}} + |\nabla \mathbf{f}|^{(\alpha, \frac{1+\alpha-\gamma}{2})}_{D'_{\tau_0}} + |\widehat{\theta}|^{(\alpha, \frac{1+\alpha-\gamma}{2})}_{D'_{\tau_0}}$$
$$+ |\nabla\widehat{\theta}|^{(\alpha, \frac{1+\alpha-\gamma}{2})}_{D'_{\tau_0}} + \int_{t_0-2\tau_1}^{t_0} \big(\|\mathbf{v}(\cdot, \tau)\|_\Omega + \|r(\cdot, \tau)\|_{W^1_2(S_1)}\big) \, d\tau \Big\}, \tag{3.40}$$

In view of (3.39), we deduce from (3.40) the inequality

$$N_{(t_0-\tau_1,t_0)}[\mathbf{v}, p_1, r] \leqslant c_2(\delta, \tau_0)\Big\{ |\mathbf{f}|^{(\alpha, \frac{1+\alpha-\gamma}{2})}_{D'_{\tau_0}} + |\nabla \mathbf{f}|^{(\alpha, \frac{1+\alpha-\gamma}{2})}_{D'_{\tau_0}} + |a|^{(2+\alpha,1+\alpha/2)}_{\Upsilon'_0}$$
$$+ \int\limits_{t_0-2\tau_0}^{t_0} \|\widehat{\theta}(\cdot, \tau)\|_\Omega \, d\tau + \int\limits_{t_0-2\tau_1}^{t_0} \big(\|\mathbf{v}(\cdot, \tau)\|_\Omega + \|r(\cdot, \tau)\|_{W^1_2(S_1)}\big) \, d\tau \Big\}$$

and estimate (3.30).

□

Lemma 3.3 *Let* $r_0 \in C^{1+\alpha}(S_1)$ *and* $\mathbf{u} \in C^{1+\alpha,0}(D_{T_0})$, $\alpha \in (0, 1)$. *Then* $r(\cdot, t) \in C^{1+\alpha}(S_1)$ *for arbitrary* $t \in (0, T_0)$ *and the inequality*

$$|r(\cdot, t)|^{(1+\alpha)}_{S_1} \leqslant c_5\big(|r_0|^{(1+\alpha)}_{S_1} + t|\mathbf{u}|^{(1+\alpha)}_{\xi, D_t}\big), \tag{3.41}$$

holds, if the norms r_0 *and* $\mathbf{u}$ *are small.*

This proposition was proved in [8] by the passage to the Lagrangian coordinates.

Now we can prove Theorem 3.2.

Proof By Theorem 3.1, there exists a local solution $(\mathbf{v}, p_1, \theta)$ on an interval $(0, T_*]$, $T_* > 1$, when ε in (3.11) is small enough (inequality (4.23) in [5]). For the norm of the solution $(\mathbf{v}, p_1, \theta)$, estimate (3.10) holds, therefore

$$N_{(0,T_*)}[\mathbf{v}, p_1, \theta] \leqslant c\Big(|\mathbf{f}|_{Q_{T_*}}^{(\alpha,\frac{1+\alpha-\gamma}{2})} + |\mathscr{D}_x\mathbf{f}|_{Q_{T_*}}^{(\alpha,\frac{1+\alpha-\gamma}{2})} + |\mathbf{v}_0|_{\cup\Omega_0^\pm}^{(2+\alpha)} + |\theta_0|_{\cup\Omega_0^\pm}^{(2+\alpha)} + |a|_{\Upsilon_{T_*}}^{(2+\alpha,1+\frac{\alpha}{2})} + |r_0|_{S_1}^{(3+\alpha)}\Big) \leqslant c_3\varepsilon \equiv \mu_1 \tag{3.42}$$

with a small $\mu_1 > 0$. In the last inequality we have taken account of the estimate

$$|H_0 + \frac{2}{R_0}|_\Gamma^{(1+\alpha)} \leqslant c|r_0|_{S_1}^{(3+\alpha)}.$$

By Proposition 3.4, there exists $\tau_0 < T_*/2$ such that (3.34) is satisfied and for $(\mathbf{v}, p_1, \theta)$, T_* estimate (3.30) holds. Lemma 3.3 guarantees the inequality

$$|r|_{S_1\times(0,T_*)}^{(1+\alpha,0)} \leqslant c_2\big(|r_0|_{S_1}^{(1+\alpha)} + c_3\varepsilon T_*\big) \leqslant \delta_1 R_0$$

($|r|_{S_1\times(0,T_*)}^{(1+\alpha,0)} \equiv \sup_{0<\tau<T_*} |r(\cdot,\tau)|_{S_1}^{(1+\alpha)}$), when ε is sufficiently small. This allows us to apply Proposition 3.2. Inequality (3.30) combined with (3.16), (3.17) leads to the estimate

$$\begin{aligned} N_{(t_0-\tau_0/2,t_0)}[\mathbf{v}, p_1, \theta, r] &\leqslant c_4\, \mathrm{e}^{-b_1(t_0-2\tau_0)}\Big\{|\mathrm{e}^{b_1 t}\mathbf{f}|_{D_0'}^{(\alpha,\frac{1+\alpha-\gamma}{2})} + |\mathrm{e}^{b_1 t}\nabla\mathbf{f}|_{D_0'}^{(\alpha,\frac{1+\alpha-\gamma}{2})} \\ &\quad + |\mathrm{e}^{b_1 t}a|_{\Upsilon_0'}^{(2+\alpha,1+\frac{\alpha}{2})} + \|\mathbf{v}_0\|_\Omega + \|\theta_0\|_\Omega + \|r_0\|_{W_2^1(S_1)}\Big\} \\ &\leqslant c_5(\tau_0)\mathrm{e}^{-b_1 t_0}\big(|\Omega|^{\frac{1}{2}} + 1\big)\varepsilon, \end{aligned} \tag{3.43}$$

where $|\Omega|$ is the measure of Ω, and $t_0 \in (2\tau_0, T_*]$.

For $t_0 = T_*$, estimate (3.42) implies that

$$|\mathbf{v}(\cdot, T_*)|_{\cup\Omega_{T_*}^\pm}^{(2+\alpha)} + |\theta(\cdot, T_*)|_{\cup\Omega_{T_*}^\pm}^{(2+\alpha)} + |r(\cdot, T_*)|_{S_1}^{(3+\alpha)} \leqslant \mu_1. \tag{3.44}$$

Next, we use Theorem 3.1 again to obtain solution in $(T_*, T_* + T_1]$ for the initial data $\mathbf{v}(\cdot, T_*)$, $\theta(\cdot, T_*)$, $r(\cdot, T_*)$. The norm of the solution is bounded:

$$N_{(T_*,T_*+T_1)}[\mathbf{v}, p_1, \theta] \leqslant \mu_2. \tag{3.45}$$

Due to Proposition 3.4, we can find $0 < \tau_1 < T_1/2$ such that satisfies (3.34) and

$$\begin{aligned} N_{(T_*+T_1-\tau_1/2,T_*+T_1)}[\mathbf{v}, p_1, \theta, r] &\leqslant c(\delta, \tau_1)\Big\{|\mathbf{f}|_{Q_{(T_*+T_1-2\tau_1,T_*+T_1)}}^{(\alpha,\frac{1+\alpha-\gamma}{2})} + |\nabla\mathbf{f}|_{Q_{(T_*+T_1-2\tau_1,T_*+T_1)}}^{(\alpha,\frac{1+\alpha-\gamma}{2})} \\ &+ |a|_{\Upsilon_{(T_*+T_1-2\tau_1,T_*+T_1)}}^{(2+\alpha,1+\alpha/2)} + \int_{T_*+T_1-2\tau_1}^{T_*+T_1} \big(\|\mathbf{v}(\cdot,\tau)\|_\Omega + \|\theta(\cdot,\tau)\|_\Omega + \|r(\cdot,\tau)\|_{W_2^1(S_1)}\big)\,d\tau\Big\}. \end{aligned} \tag{3.46}$$

Let us use again (3.41). Then in view of (3.44) and (3.45), we conclude:

$$|r|^{(1+\alpha,0)}_{S_1\times(T_*,T_*+T_1)} \leqslant c_2\big(|r(\cdot,T_*)|^{(1+\alpha)}_{S_1} + T_1|\mathbf{u}|^{(1+\alpha)}_{\xi,D(T_*,T_*+T_1)}\big) \leqslant c_2(\mu_1 + T_1\mu_2),$$

We can make μ_1 and T_1 so small that

$$c_2(\mu_1 + T_1\mu_2) \leqslant \delta_1 R_0.$$

Consequently, similarly to (3.43), relation (3.46) may be continued by virtue of Proposition 3.2 as follows

$$N_{(T_*+T_1-\tau_1,T_*+T_1)}[\mathbf{v}, p_1, \theta, r] \leqslant c_6(\delta,\tau_1)\mathrm{e}^{-b_1T_1}\big(1+|\Omega|^{1/2}\big)\varepsilon.$$

Choose ε so small that $c_6(\delta,\tau_1)\big(1+|\Omega|^{1/2}\big)\varepsilon \leqslant \mu_1$.

Hence,

$$|\mathbf{v}(\cdot,T_*+T_1)|^{(2+\alpha)}_{\cup\Omega^{\pm}_{T_*+T_1}} + |\theta(\cdot,T_*+T_1)|^{(2+\alpha)}_{\cup\Omega^{\pm}_{T_*+T_1}} + |r(\cdot,T_*+T_1)|^{(3+\alpha)}_{S_1} \leqslant \mu_1\mathrm{e}^{-b_1T_1}.$$

Thus, the norms of the initial data do not increase. Therefore we can extend the solution in the interval $(T_*+T_1, T_*+2T_1]$. This procedure may be repeated again and again as long as we like.

By repeating our argument, we should pass to the Lagrangian coordinates according to the formula

$$\mathbf{X} = \boldsymbol{\xi}^{(1)} + \int_{T_*}^{t} \widetilde{\mathbf{u}}(\xi^{(1)},\tau)\,\mathrm{d}\tau, \quad \xi^{(1)} \in \cup\Omega^{\pm}_{T_*}, \quad t \in (T_*, T_*+T_1). \tag{3.47}$$

In fact, due to the additivity of the integral, (3.47) coincides with (3.12):

$$\mathbf{X}(\xi,t) = \boldsymbol{\xi} + \int_0^{T_*} \mathbf{u}(\xi,\tau)\,\mathrm{d}\tau + \int_{T_*}^{t} \mathbf{u}(\xi,\tau)\,\mathrm{d}\tau, \quad \xi \in \cup\Omega^{\pm}_0,\ t \in (T_*, T_*+T_1),$$

because $\widetilde{\mathbf{u}}(\xi^{(1)},\tau) = \mathbf{u}(\xi,\tau)$.

The same remark is true for the coordinates of inner fluid barycenter, since the volume of the fluid is conserved:

$$\mathbf{h}(t) = \mathbf{h}(T_*) + \int_{T_*}^{t} \frac{1}{|\Omega_t^+|}\int_{\Omega_t^+} \mathbf{v}(x,\tau)\,\mathrm{d}x\,\mathrm{d}\tau = \int_0^{T_*} \frac{1}{|\Omega_t^+|}\int_{\Omega_t^+} \mathbf{v}(x,\tau)\,\mathrm{d}x\,\mathrm{d}\tau$$

$$+ \int_{T_*}^{t} \frac{1}{|\Omega_t^+|}\int_{\Omega_t^+} \mathbf{v}(x,\tau)\,\mathrm{d}x\,\mathrm{d}\tau = \frac{3}{4\pi R_0^3}\int_0^{t}\int_{\Omega_t^+} \mathbf{v}(x,\tau)\,\mathrm{d}x\,\mathrm{d}\tau, \quad t > T_*.$$

The solution of the system (3.1), (3.4), (3.3) can be extended in this way with respect to t as far as necessary. After that, inequality (3.12) follows for any finite time interval from Propositions 3.2 and 3.4.

The limiting position of the barycenter is evaluated from Corollary 3.1:

$$|h_\infty| \leqslant d \leqslant c_7\varepsilon \tag{3.48}$$

where

$$d = \frac{c_8}{|\Omega_0^+|^{1/2} b_1}\Big\{\|e^{b_1 t}\mathbf{f}\|_{Q_\infty} + \|e^{b_1 t}a\|^2_{W_2^{3/2,3/4}(\Upsilon_\infty)} + \|\mathbf{v}_0\|_\Omega + \|\theta_0\|_\Omega + \|r_0\|_{W_2^1(S_1)}\Big\}.$$

Inequality (3.48) implies an estimate concerning the distance between the surfaces Γ and Σ at initial moment. It is clear that, in order to exclude the intersection of Γ_t and Σ in the future, the data should be taken so small that the sum $d + \delta_1 R_0$ with $\delta_1,\ R_0$ from (3.13) would be strictly less than this initial distance.

Solution uniqueness follows from the uniqueness of local solutions. □

3.4.1 Conclusions

Thus, unsteady motion of a viscous incompressible two-phase fluid has been considered in a container Σ in the Oberbeck–Boussinesq approximation. The phases have been separated by a closed unknown interface Γ_t where surface tension has been taken into account. The initial interface Γ_0 has been close to a sphere with the center in drop's barycenter.

The exponential L_2–estimate for a solution has been obtained to prove global solvability of the problem. It has been noted that there holds a similar estimate for a solution to the problem without surface tension.

Global existence theorem for the problem has been stated in Hölder classes of functions provided that the data have had small norms. It has been shown that the solution in the Oberbeck–Boussinesq approximation is a small perturbation of the rest state and damped in time; the Hölder norms of velocity vector field and temperature deviation decay exponentially as $t \to \infty$, interface form tends to the sphere $\{|x - h_\infty| = R_0\}$, the limiting position of the barycenter of the inner fluid h_∞ may be displaced from initial drop's center of gravity. It has been obtained a condition under that the drop remains strictly inside the other fluid during all the time.

Acknowledgements The research leading to these results has received funding from the People Programme (Marie Curie Actions) of the European Union's Seventh Framework Programme FP7/2007-2013/under REA grant agreement n^0 319012 and from the Funds for International Co-operation under Polish Ministry of Science and Higher Education grant agreement n^0 2853/7. PR/2013/2. This work has been also partially supported by the grant NSh-1771.2014.1 for the Russian leading scientific schools.

References

1. Denisova, I.V.: The solvability in the Hölder spaces of a linear problem on the motion of two liquids separates by a closed surface. Algebra iAnaliz **5**(4), 122–148 (1993). (English trans. in St.Petersburg Math. J. **5**(4), 765–787 (1994))
2. Denisova, I.V.: Model problem connected with the motion of two incompressible fluids. Adv. Math. Sci. Appl. **17**(1), 195–223 (2007)
3. Denisova, I.V.: Global solvability of a problem on two fluid motion without surface tension. Zap. Nauchn. Sem. S.-Peterburg. Otdel. Mat. Inst. Steklov. (POMI) **348**, 19–39 (2007). (English trans. in J. Math. Sci. **152**(5), 625–637 (2008))
4. Denisova, I.V.: Global Classical Solvability of an Interface Problem on the Motion of Two Fluids. RIMS Kokyuroku Series, 1875. Kyoto University, Japan (2014)
5. Denisova, I.V., Nechasova, S.H.: Oberbeck-Boussinesq approximation for the motion of two incompressible fluids. Zap. Nauchn. Sem. S.-Peterburg. Otdel. Mat. Inst. Steklov. (POMI) **362**, 92–119 (2008). (English transl. in J. Math. Sci. **159**(4), 436–451 (2009))
6. Denisova, I.V., Solonnikov, V.A.: Solvability in Hölder spaces for a model initial boundary-value problem generated by a problem on the motion of two fluids. Zap. Nauchn. Sem. Leningrad. Otdel. Mat. Inst. Steklov. (LOMI) **188**, 5–44 (1991). (English transl. in. J. Math. Sci. **70**, 1717–1746 (1994))
7. Denisova, I.V., Solonnikov, V.A.: Classical solvability of the problem on the motion of two viscous incompressible liquids. Algebra iAnaliz **7**(5), 101–142 (1995). (English transl. in St. Petersburg Math. J. **7**(5), 755–786 (1996))
8. Denisova, I.V., Solonnikov, V.A.: Global solvability of a problem governing the motion of two incompressible capillary fluids. Zap. Nauchn. Sem. S. -Peterburg. Otdel. Mat. Inst. Steklov. (POMI) **397**, 20–52 (2011). (English transl. in J. Math. Sci. **185**(5), 668–686 (2012))
9. Padula, M.: On the exponential stability of the rest state of a viscous compressible fluid. J. Math. Fluid Mech. **1**, 62–77 (1999)
10. Solonnikov, V.A.: Estimates of the solution of an initial-boundary value problem for a linear nonstationary Navier–Stokes system. Zap. Nauchn. Sem. Leningrad. Otdel. Mat. Inst. Steklov. (LOMI) **59**, 178–254 (1976). (English transl. in J. Soviet Math. **10**(2), 336–393 (1978))
11. Solonnikov, V.A.: On a steady motion of a drop in an infinite liquid medium. Zap. Nauchn. Sem. S. -Peterburg. Otdel. Mat. Inst. Steklov. (POMI) **233**, 233–254 (1996). (English transl. in J. Math. Sci. **93**(5), 784–799 (1999))
12. Solonnikov, V.A.: Lectures on evolution free boundary problems: classical solutions. Lecture Notes in Mat. **1812**, 123–175 (2003)

Chapter 4
Stability of Steady Flow Past a Rotating Body

Giovanni Paolo Galdi and Jiří Neustupa

Abstract We study stability of a steady solution of the mathematical model describing the flow of a viscous incompressible fluid past a rotating body. We derive a sufficient condition for stability, which requires the L^1– and L^2–integrability on the time interval $(0, \infty)$ of the semigroup generated by the relevant linear operator, applied to a finite family of suitable functions, in a norm restricted to a "sufficiently large" bounded region around the body. No assumption on the smallness of the steady solution is required.

Keywords Navier–Stokes equations · Rotation · Stability

AMS math. classification (2010): 35Q30 · 35B35 · 76D05 · 76E07

4.1 Motivation and Introduction

Assume that a compact body $\mathcal{B}$ moves in a viscous Newtonian incompressible fluid so that the fluid in infinity is in the rest state and the body rotates about the x_1–axis with a constant angular velocity ω and translates in the direction of the x_1–axis with a constant velocity u_∞. (Both motions are considered with respect to the rest state in infinity.) Denoting by by $\Omega(t)$ the exterior of $\mathcal{B}$ at time t, the flow of the fluid in $Q := \{(\mathbf{x}, t);\ \mathbf{x} \in \Omega(t),\ t > 0\}$ is described by the Navier–Stokes equations

$$\partial_t \mathbf{u} - \nu \Delta \mathbf{u} + \mathbf{u} \cdot \nabla \mathbf{u} + \nabla p = \mathbf{f}, \tag{4.1}$$

$$\operatorname{div} \mathbf{u} = 0, \tag{4.2}$$

G.P. Galdi (✉)
Department of Mechanical Engineering, University of Pittsburgh, Pittsburgh, USA
e-mail: galdi@pitt.edu

J. Neustupa
Czech Academy of Sciences, Institute of Mathematics, Prague, Czech Republic
e-mail: neustupa@math.cas.cz

Y. Shibata and Y. Suzuki (eds.), *Mathematical Fluid Dynamics, Present and Future*, Springer Proceedings in Mathematics & Statistics 183,
DOI 10.1007/978-4-431-56457-7_4

where $\mathbf{u}$ and p denote the unknown velocity and the pressure, ν is the coefficient of kinematic viscosity, and $\mathbf{f}$ is an external specific body force. We assume that $\mathbf{u}$ satisfies the no–slip boundary condition on the surface of body $\mathcal{B}$, namely,

$$\mathbf{u}(\mathbf{x},t) = u_\infty \mathbf{e}_1 + \boldsymbol{\omega} \times \mathbf{x} \qquad \text{for } (\mathbf{x},t) \in \partial\Omega(t) \times \{t\}. \tag{4.3}$$

The assumption that the fluid is at rest in infinity means that

$$\mathbf{u}(\mathbf{x},t) \to \mathbf{0} \qquad \text{for } |\mathbf{x}| \to \infty. \tag{4.4}$$

The disadvantage of this model is that the domain $\Omega(t)$ occupied by the fluid depends on time. This is why it is often convenient to use the transformation

$$\mathbf{x}' = O(t)\cdot\mathbf{x} - u_\infty \mathbf{e}_1 t, \quad \mathbf{u}'(\mathbf{x}',t) = O(t)\cdot\mathbf{u}(\mathbf{x},t), \quad p'(\mathbf{x},t) = p(\mathbf{x},t), \tag{4.5}$$

where $O(t)$ is the unitary matrix of the rotation:

$$O(t) = \begin{pmatrix} 1 & 0 & 0 \\ 0 & \cos\omega t & \sin\omega t \\ 0 & -\sin\omega t & \cos\omega t \end{pmatrix}.$$

Employing these transformations into (4.1), (4.2), we obtain the equations for $\mathbf{u}'$, p' in the $\mathbf{x}'$, t coordinates:

$$\partial_t \mathbf{u}' - (\boldsymbol{\omega} \times \mathbf{x}' + u_\infty \mathbf{e}_1)\cdot\nabla'\mathbf{u}' + \boldsymbol{\omega}\times\mathbf{u}' + \mathbf{u}'\cdot\nabla'\mathbf{u}' = -\nabla' p' + \nu\Delta'\mathbf{u}' + \mathbf{f}', \tag{4.6}$$

$$\operatorname{div}'\mathbf{u}' = 0. \tag{4.7}$$

Here, $\boldsymbol{\omega} := \omega\mathbf{e}_1$ and ∇', Δ' and div$'$ are the operators acting in the $\mathbf{x}'$–space. Due to the first equation in (4.5), $(\mathbf{x},t) \in \Omega(t) \times \{t\}$ if and only if $(\mathbf{x}',t) \in \Omega(0) \times \{t\}$. Thus, the system (4.6), (4.7) can be treated for $\mathbf{x}'$ in a fixed domain $\Omega(0)$. Conditions (4.3) and (4.4) transform to

$$\mathbf{u}'(\mathbf{x}',t) = u_\infty \mathbf{e}_1 + \boldsymbol{\omega}\times\mathbf{x}' \qquad \text{for } \mathbf{x}' \in \partial\Omega(0), \tag{4.8}$$

$$\mathbf{u}'(\mathbf{x}',t) \to \mathbf{0} \qquad \text{for } |\mathbf{x}'| \to \infty. \tag{4.9}$$

Of many works studying qualitative properties of the equations (4.6), (4.7), we mention e.g. the papers [4] (by Cumsille and Tucsnak), [9] (by Farwig, Hishida and Müller), [7, 8] (by Farwig), [10–12] (by Farwig and Neustupa), [13] (by Farwig, Nečasová and Neustupa), [15, 16], (by Galdi), [20, 21] (by Galdi and Silvestre), [22] (by Geissert, Heck and Hieber), [26, 27] (by Hishida) and [28] (by Hishida and Shibata).

The existence of strong solutions on a "short" time interval $(0, T_0)$ has been studied by Hishida [26], Galdi and Silvestre [20] and Cumsille and Tucsnak [4]. In [26], the author assumes that the body force is zero and the initial velocity is in $D(A^{1/4})$, where A is the Stokes operator, and he proves the existence of a solution in the class $C([0, T_0]; D(A^{1/4})) \cap C((0, T_0]; \; D(A))$. Galdi and Silvestre [20] also deal with the case of the zero body force, they assume that the initial velocity $\mathbf{u}'_0 \in \mathbf{W}^{2,2}(\Omega)$ and they obtain a solution in $C([0, T_0]; \; \mathbf{W}^{1,2}(\Omega)) \cap C((0, T_0); \; \mathbf{W}^{2,2}(\Omega)))$. Cumsille and Tucsnak [4] formulate the result not in terms of solution $\mathbf{u}'$ of the transformed problem, but in terms of solution $\mathbf{u}$ of the original Navier–Stokes problem (4.1), (4.2) in the time–varying domain $\Omega(t)$. They consider a body force $\mathbf{f}$ locally square integrable from $(0, \infty)$ to $\mathbf{W}^{1,\infty}(\mathbb{R}^3)$, the no–slip boundary condition for the velocity on $\partial\Omega(t)$ and show that if the initial velocity $\mathbf{u}_0 \in \mathbf{W}^{1,2}(\Omega(0))$ there exists $T_0 > 0$ and a unique strong solution $\mathbf{u}$ such that

$$\begin{aligned}
&\mathbf{u} \in L^2\big(0, T_0; \; \mathbf{W}^{2,2}(\Omega(t))\big) \cap C\big([0, T_0]; \; \mathbf{W}^{1,2}(\Omega(t))\big), \\
&\partial_t \mathbf{u} \in L^2\big(0, T_0; \; \mathbf{L}^2(\Omega(t))\big).
\end{aligned}$$

Moreover, either T_0 can be extended up to infinity or the norm of $\mathbf{u}$ in $\mathbf{W}^{1,2}(\Omega(t))$ tends to infinity for $t \to T_0-$. Using the relation (4.5) between the solutions $\mathbf{u}$ and $\mathbf{u}'$, we can reformulate the result of Cumsille and Tucsnak [4] in terms of $\mathbf{u}'$ as follows: given $\mathbf{u}'_0 \in \mathbf{L}^2_\sigma(\Omega) \cap \mathbf{W}^{1,2}_{0,\sigma}(\Omega)$, there exists $T_0 > 0$ and a unique solution $\mathbf{u}'$ of the problem (4.6)–(4.9) such that

$$\begin{aligned}
&\mathbf{u}' \; \in \; L^2\big(0, T_0; \; \mathbf{W}^{2,2}(\Omega(0))\big) \cap C\big([0, T_0]; \; \mathbf{W}^{1,2}(\Omega(0))\big), \\
&\partial_t \mathbf{u}' - (\omega \times \mathbf{x}' + u_\infty \mathbf{e}_1) \cdot \nabla' \mathbf{u}' + \omega \times \mathbf{u}' \; \in \; L^2\big((0, T_0; \; \mathbf{L}^2(\Omega(0))\big).
\end{aligned} \tag{4.10}$$

We further suppose that $\Omega(0)$ is an exterior domain in $\mathbb{R}^3$ with a $C^{1,1}$ boundary $\partial\Omega(0)$ and $\mathbf{U}'$ (the velocity), Π' (the pressure) is a steady solution of the problem (4.6)–(4.9) such that

$$\mathbf{U}' \in \mathbf{L}^3(\Omega(0)), \qquad \nabla'\mathbf{U}' \in L^3(\Omega(0))^{3\times 3} \cap L^{3/2}(\Omega(0))^{3\times 3} \tag{4.11}$$

and there exists $c_1 > 0$ such that

$$|\nabla'\mathbf{U}'(\mathbf{x}')| \le \frac{c_1}{|\mathbf{x}'|} \qquad \text{for } \mathbf{x}' \in \Omega(0). \tag{4.12}$$

The existence of a steady solution with these properties for a large class of body forces $\mathbf{f}$ is known, see e.g. [14], provided that $u_\infty \neq 0$.

As we are interested in behavior of solutions $\mathbf{u}'$, p' to the problem (4.6)–(4.9) in the neighbourhood of the solution $\mathbf{U}'$, Π', it is useful to write the solutions in the form $\mathbf{u}' = \mathbf{U}' + \mathbf{v}'$, $p' = \Pi' + q'$, where $\mathbf{v}'$ are q' are "small" perturbations. The perturbations satisfy the equations

$$\partial_t \mathbf{v}' - (\omega \times \mathbf{x}' + u_\infty \mathbf{e}_1) \cdot \nabla' \mathbf{v}' + \omega \times \mathbf{v}' + \mathbf{U}' \cdot \nabla' \mathbf{v}' + \mathbf{v}' \cdot \nabla' \mathbf{U}' + \mathbf{v}' \cdot \nabla' \mathbf{v}' = -\nabla' q' + \Delta' \mathbf{v}' + \mathbf{f}, \tag{4.13}$$

$$\operatorname{div}' \mathbf{v}' = 0 \tag{4.14}$$

in $\Omega(0) \times (0, \infty)$, and the conditions

$$\mathbf{v}'(\mathbf{x}', t) = \mathbf{0} \qquad \text{for } \mathbf{x}' \in \partial\Omega(0), \tag{4.15}$$

$$\mathbf{v}'(\mathbf{x}', t) \to \mathbf{0} \qquad \text{for } |\mathbf{x}'| \to \infty. \tag{4.16}$$

Lemma 1 *Each solution $\mathbf{v}'$ to the problem (4.13)–(4.16), fulfilling (4.10), satisfies the identities*

$$\int_{\Omega(0)} \big[\partial_t \mathbf{v}' - (\omega \times \mathbf{x}' + u_\infty \mathbf{e}_1) \cdot \nabla' \mathbf{v}' + \omega \times \mathbf{v}'\big] \cdot \mathbf{v}' \, \mathrm{d}\mathbf{x}' = \frac{\mathrm{d}}{\mathrm{d}t} \frac{1}{2} \int_{\Omega(0)} |\mathbf{v}'|^2 \, \mathrm{d}\mathbf{x}', \tag{4.17}$$

$$\int_{\Omega(0)} \big[\partial_t \mathbf{v}' - (\omega \times \mathbf{x}' + u_\infty \mathbf{e}_1) \cdot \nabla' \mathbf{v}' + \omega \times \mathbf{v}'\big] \cdot \Delta' \mathbf{v}' \, \mathrm{d}\mathbf{x}' = -\frac{\mathrm{d}}{\mathrm{d}t} \frac{1}{2} \int_{\Omega(0)} |\nabla' \mathbf{v}'|^2 \, \mathrm{d}\mathbf{x}' \tag{4.18}$$

for a.a. $t \in (0, T_0)$.

Proof. We focus only on (4.18), because the proof of (4.17) is even simpler. By analogy with (4.5), we denote

$$\mathbf{v}(\mathbf{x}, t) := O(-t)\, \mathbf{v}'(\mathbf{x}', t) = O(-t) \cdot \mathbf{v}'\big(O(t)\mathbf{x} - u_\infty \, \mathbf{e}_1 t, t\big)$$

and let $o_{ij}(t)$ be the entries of the matrix $O(t)$ and v_i, v_i' be the components of $\mathbf{v}$ and $\mathbf{v}'$, respectively As $\Delta' \mathbf{v}'(\mathbf{x}', t) = O(t) \cdot \Delta \mathbf{v}(\mathbf{x}, t)$, we have

$$\begin{aligned}
&\int_{\Omega(0)} \big[\partial_t \mathbf{v}' - (\omega \times \mathbf{x}' + u_\infty \mathbf{e}_1) \cdot \nabla' \mathbf{v}' + \omega \times \mathbf{v}'\big] \cdot \Delta' \mathbf{v}' \, \mathrm{d}\mathbf{x}' \\
&\quad = \int_{\Omega(0)} \Big(O(t) \cdot \frac{\mathrm{d}}{\mathrm{d}t} \mathbf{v}\big(O(-t)\mathbf{x}' + u_\infty \, \mathbf{e}_1 t, t\big)\Big) \\
&\qquad\qquad \cdot \Big(O(t) \cdot \Delta \mathbf{v}\big(O(-t)\mathbf{x}' + u_\infty \, \mathbf{e}_1 t, t\big)\Big) \, \mathrm{d}\mathbf{x}' \\
&\quad = \int_{\Omega(0)} \Big(\frac{\mathrm{d}}{\mathrm{d}t} \mathbf{v}\big(O(-t)\mathbf{x}' + u_\infty \, \mathbf{e}_1 t, t\big)\Big) \cdot \Big(\Delta \mathbf{v}\big(O(-t)\mathbf{x}' + u_\infty \, \mathbf{e}_1 t, t\big)\Big) \, \mathrm{d}\mathbf{x}' \\
&\quad = \int_{\Omega(t)} \partial_t \mathbf{v}(\mathbf{x}, t) \cdot \Delta \mathbf{v}(\mathbf{x}, t) \, \mathrm{d}\mathbf{x} = -\int_{\Omega(t)} \partial_t \nabla \mathbf{v}(\mathbf{x}, t) : \nabla \mathbf{v}(\mathbf{x}, t) \, \mathrm{d}\mathbf{x} \\
&\quad = -\int_{\Omega(t)} \frac{\mathrm{d}}{\mathrm{d}t} \partial_j \big[o_{ik}(-t)\, v_k'\big(O(t)\mathbf{x} - u_\infty \, \mathbf{e}_1 t, t\big)\big]
\end{aligned}$$

$$
\begin{aligned}
&\qquad \partial_j\big[o_{ir}(-t)\, v'_r\big(O(t)\mathbf{x} - u_\infty\, \mathbf{e}_1 t, t\big)\big]\, \mathrm{d}\mathbf{x} \\
&= -\frac{1}{2}\int_{\Omega(t)} \frac{\mathrm{d}}{\mathrm{d}t}\Big(\partial_j\big[o_{ik}(-t)\, v'_k\big(O(t)\mathbf{x} - u_\infty\, \mathbf{e}_1 t, t\big)\big] \\
&\qquad \partial_j\big[o_{ir}(-t)\, v'_r\big(O(t)\mathbf{x} - u_\infty\, \mathbf{e}_1 t, t\big)\big]\Big)\, \mathrm{d}\mathbf{x} \\
&= -\frac{1}{2}\int_{\Omega(t)} \frac{\mathrm{d}}{\mathrm{d}t}\Big(\partial_j\big[v'_k\big(O(t)\mathbf{x} - u_\infty\, \mathbf{e}_1 t, t\big)\big] \\
&\qquad \partial_j\big[v'_k\big(O(t)\mathbf{x} - u_\infty\, \mathbf{e}_1 t, t\big)\big]\Big)\, \mathrm{d}\mathbf{x} \\
&= -\frac{1}{2}\int_{\Omega(t)} \frac{\mathrm{d}}{\mathrm{d}t}\Big(\big[\partial'_l v'_k\big(O(t)\mathbf{x} - u_\infty\, \mathbf{e}_1 t, t\big)\, o_{lj}(t)\big] \\
&\qquad \big[\partial_s v'_k\big(O(t)\mathbf{x} - u_\infty\, \mathbf{e}_1 t, t\big)\, o_{sj}(t)\big]\Big)\, \mathrm{d}\mathbf{x} \\
&= -\frac{1}{2}\int_{\Omega(t)} \frac{\mathrm{d}}{\mathrm{d}t}\Big(\partial'_l v'_k\big(O(t)\mathbf{x} - u_\infty\, \mathbf{e}_1 t, t\big)\, \partial'_l v'_k\big(O(t)\mathbf{x} - u_\infty\, \mathbf{e}_1 t, t\big)\Big)\, \mathrm{d}\mathbf{x} \\
&= -\frac{1}{2}\int_{\Omega(0)} \frac{\mathrm{d}}{\mathrm{d}t}\Big(\partial'_l v'_k(\mathbf{x}', t)\, \partial'_l v'_k(\mathbf{x}', t)\Big)\, \mathrm{d}\mathbf{x}' \\
&= -\frac{\mathrm{d}}{\mathrm{d}t}\, \frac{1}{2}\int_{\Omega(0)} |\nabla' \mathbf{v}'|^2\, \mathrm{d}\mathbf{x}'.
\end{aligned}
$$

□

In what follows, in order to simplify the notation, we shall further omit the primes and write $\mathbf{x}$, $\mathbf{v}$ and $\mathbf{U}$ instead of $\mathbf{x}'$, $\mathbf{v}'$ and $\mathbf{U}'$, respectively. We also write only Ω instead of $\Omega(0)$.

Notation. We denote by C a generic constant, i.e. a constant whose values may vary from term to term. Constants with the values fixed throughout the whole paper are denoted c_1, c_2, etc.

- $\Omega_R := \Omega \cap B_R(\mathbf{0})$ and $\Omega^R := \Omega \setminus B_R(\mathbf{0})$ (for $R > 0$)
- Vector functions and spaces of vector functions are denoted by boldface letters.
- $\mathbf{C}^\infty_{0,\sigma}(\Omega)$ is the space of infinitely differentiable divergence–free vector functions with a compact support in Ω.
- $\mathbf{L}^2_\sigma(\Omega)$ is the closure of $\mathbf{C}^\infty_{0,\sigma}(\Omega)$ in $\mathbf{L}^2(\Omega)$. The orthogonal projection of $\mathbf{L}^2(\Omega)$ onto $\mathbf{L}^2_\sigma(\Omega)$ is denoted by P_σ.
- $\|\,.\,\|_q$ and $\|\,.\,\|_{q,m}$ denote the norms of scalar– or vector– or tensor–valued functions with components in $L^q(\Omega)$ and $W^{q,m}(\Omega)$, respectively. If we consider Ω_R instead of Ω, we denote the L^q–norm by $\|\,.\,\|_{q;\,\Omega_R}$.
- $(\,.\,,\,.\,)_2$ is the scalar product in $\mathbf{L}^2_\sigma(\Omega)$.
- $\mathbf{W}^{1,2}_{0,\sigma}(\Omega)$ and $\mathbf{D}^{1,2}_{0,\sigma}(\Omega)$ are the completions of $\mathbf{C}^\infty_{0,\sigma}(\Omega)$ in the norms $\|\,.\,\|_{1,2}$ and $\|\nabla.\,\|_2$, respectively.
- We denote $A\mathbf{v} := P_\sigma \nu \Delta \mathbf{v}$ for $\mathbf{v} \in D(A) := \mathbf{W}^{1,2}_{0,\sigma}(\Omega) \cap \mathbf{W}^{2,2}(\Omega)$. Operator A (the so called Stokes operator) is a non–positive selfadjoint operator in $\mathbf{L}^2_\sigma(\Omega)$, whose

spectrum $\mathrm{Sp}(A)$ is purely continuous and coincides with the interval $(-\infty, 0]$. (See e.g. [10].) Moreover, A generates an analytic semigroup e^{At} in $\mathbf{L}^2_\sigma(\Omega)$.

- For $\mathbf{v} \in D(A)$, we set

$$\begin{aligned} B^0\mathbf{v} &= (\boldsymbol{\omega} \times \mathbf{x}) \cdot \nabla\mathbf{v} - \boldsymbol{\omega} \times \mathbf{v} \\ B^1\mathbf{v} &= u_\infty\, \partial_1\mathbf{v}, \\ B^2\mathbf{v} &= -P_\sigma\big[\mathbf{U} \cdot \nabla\mathbf{v} + \mathbf{v} \cdot \nabla\mathbf{U}\big]. \end{aligned}$$

The symmetric part of B^2 is

$$B^2_s\mathbf{v} \;:=\; -P_\sigma\big[\mathbf{v} \cdot (\nabla\mathbf{U})_s\big],$$

while the skew–symmetric (= anti–symmetric) part, labeled by subscript a, is given by

$$B^2_a\mathbf{v} \;=\; -P_\sigma\big[\mathbf{U} \cdot \nabla\mathbf{v} + \mathbf{v} \cdot (\nabla\mathbf{U})_a\big].$$

- Finally, we denote by N the nonlinear operator associated with the nonlinear term in (4.13):

$$N(\mathbf{v}) \;:=\; -P_\sigma(\mathbf{v} \cdot \nabla\mathbf{v}) \qquad \text{for } \mathbf{v} \in D(A).$$

Lemma 2 *If* $\mathbf{v}$ *is a solution to the problem (4.6)–(4.9) in the class (4.10), then* $\partial_t\mathbf{v} - B^0\mathbf{v} - B^1\mathbf{v} \in \mathbf{L}^2_\sigma(\Omega)$ *for a.a.* $t \in [0, T_0]$.

Proof One can calculate that $\partial_t\mathbf{v} - B^0\mathbf{v} - B^1\mathbf{v}$ is divergence–free in the sense of distributions. It belongs to $\mathbf{L}^2(\Omega)$ for a.a. $t \in [0, T_0]$ due to (4.10). Moreover, it is shown e.g. in [10, 41] that the normal component of $B^0\mathbf{v} + B^1\mathbf{v}$ is zero on $\partial\Omega$ (in the sense of traces). Since $\partial_t\mathbf{v}(\mathbf{x}, t) = \mathbf{0}$ for $\mathbf{x} \in \partial\Omega$, the proof is completed.

Note that the identities (4.17) and (4.18) can now be rewritten:

$$\int_\Omega \Big(\frac{\mathrm{d}\mathbf{v}}{\mathrm{d}t} - B^0\mathbf{v} - B^1\mathbf{v}\Big) \cdot \mathbf{v}\, \mathrm{d}\mathbf{x} = \frac{\mathrm{d}}{\mathrm{d}t}\,\frac{1}{2}\, \|\mathbf{v}\|_2^2, \tag{4.19}$$

$$\int_\Omega \Big(\frac{\mathrm{d}\mathbf{v}}{\mathrm{d}t} - B^0\mathbf{v} - B^1\mathbf{v}\Big) \cdot A\mathbf{v}\, \mathrm{d}\mathbf{x} = -\frac{\mathrm{d}}{\mathrm{d}t}\,\frac{1}{2}\, \|\nabla\mathbf{v}\|_2^2. \tag{4.20}$$

An operator form of system (4.13)**,** (4.14)**.** The system of equations (4.13), (4.14) can be written as an operator equation

$$\frac{\mathrm{d}\mathbf{v}}{\mathrm{d}t} - B^0\mathbf{v} - B^1\mathbf{v} \;=\; \nu A\mathbf{v} + B^2\mathbf{v} + N(\mathbf{v}) \tag{4.21}$$

in $\mathbf{L}^2_\sigma(\Omega)$. By a solution to equation (4.21) (and other equations of this type) on a time interval $[0, T)$ (where $0 < T \le \infty$), we mean a function $\mathbf{v}$ satisfying equation (4.21) a.e. in $(0, T)$ and such that

$$\mathbf{v} \in L^2\big(J;\ D(A)\big) \cap C\big(J;\ \mathbf{W}^{1,2}_{0,\sigma}(\Omega)\big),$$
$$\frac{\mathrm{d}\mathbf{v}}{\mathrm{d}t} - B^0\mathbf{v} - B^1\mathbf{v} \in L^2\big(J;\ \mathbf{L}^2_\sigma(\Omega)\big) \tag{4.22}$$

for each bounded interval $J \subset [0, T)$. This definition of strong solutions is in good agreement with (4.10).

Goals of this paper and their relation to previous results. We formulate sufficient conditions for stability of the steady solution $\mathbf{U}$ of the problem (4.6)–(4.9) without assumption on smallness of any quantity associated with $\mathbf{U}$. Obviously, stability of the solution $\mathbf{U}$ is equivalent to stability of the zero solution of equation (4.21). Note that if the operator B^2 is in an appropriate sense "sufficiently small" in comparison to νA then both the operators $\nu A + B^1 + B^2$ and $\nu A + B^0 + B^1 + B^2$ are dissipative or even essentially dissipative. (We define later what we exactly mean by this property.) Consequently, the zero solution of the equation (4.21) is stable. The condition of "sufficient smallness" of B^2 expresses the requirement that the steady flow $\mathbf{U}$ is in suitable norm small. This condition has been used e.g. in the paper [28]. Moreover, there exists a long list of papers where the authors consider just the translational motion of body $\mathcal{B}$ in a viscous incompressible fluid (which corresponds to $\boldsymbol{\omega} = \mathbf{0}$) and prove the stability of solution $\mathbf{U}$ under various assumptions on the smallness of $\mathbf{U}$. Of these papers, we can cite e.g. Heywood [23–25], Masuda [37], Maremonti [36], Galdi and Rionero [17], Galdi and Padula [18], Borchers and Miyakawa [2, 3], Kozono and Ogawa [31], Kozono and Yamazaki [32, 33], Galdi, Heywood and Shibata [19], Miyakawa [38] and Shibata [45].

Another approach to the question of stability of solutions to the differential equations of the type (4.21), based on the spectral properties of the operator $L := \nu A + B^2$ (the operators B^0 and B^1 not being considered), is presented e.g. in the papers [29], [30] (by Kielhöfer) and [43] (by Sattinger). Here, the authors show that the condition $s(L) < 0$ (where $s(L) := \sup\{\operatorname{Re}\lambda;\ \lambda \in \operatorname{Sp}(L)\}$ is the so called *spectral bound* of operator L) implies stability of the zero solution. In this case, operator L need not be dissipative. However, the condition $s(L) < 0$ can never be satisfied if the operator equation (4.21) models a flow in an exterior domain, and this holds independently of the angular velocity of rotation $\boldsymbol{\omega}$, the translational velocity $u_\infty \mathbf{e}_1$ and the steady solution $\mathbf{U}$. The reason is that the spectrum of $L := \nu A + B^0 + B^1 + B^2$ has a nonempty intersection with the imaginary axis. a) In the irrotational case $\omega = 0$, Babenko [1] provided the description of the spectrum of the so called Oseen operator $\nu A + B^1$:

$$\operatorname{Sp}(\nu A + B^1) = \operatorname{Sp}_{\mathrm{ess}}(\nu A + B^1)$$
$$= \begin{cases} \big\{\lambda \in \mathbb{C};\ \operatorname{Re}\lambda \le 0,\ \operatorname{Im}\lambda = 0\big\}, & \text{if } u_\infty = 0, \\ \big\{\lambda \in \mathbb{C};\ \operatorname{Re}\lambda \le -\nu\,(\operatorname{Im}\lambda)^2/u_\infty^2\big\} & \text{if } u_\infty \ne 0; \end{cases}$$

see also Farwig and Neustupa [10]. ($\operatorname{Sp}_{\mathrm{ess}}$ denotes the essential spectrum.) Thus, $\operatorname{Sp}_{\mathrm{ess}}(\nu A + B^1)$ coincides with the non–positive part of the real axis in the case $u_\infty = 0$, and it coincides with the parabolic region $\mathcal{P} := \{\lambda \in \mathbb{C};\ \operatorname{Re}\lambda \le -\nu\,(\operatorname{Im}\lambda)^2/u_\infty^2\}$ if $u_\infty \ne 0$. As the operator B^2 is relatively $(\nu A + B^1)$–compact (see Lemma 3 in

Sect. 2), the operator L has the same essential spectrum as $\nu A + B^1$. b) In the rotational case $\omega \neq 0$, Farwig and Neustupa [10–12], and Farwig, Nečasová and Neustupa [13] studied the spectrum of the operator $\nu A + B^0 + B^1$ and they have shown that $\mathrm{Sp}_{\mathrm{ess}}(\nu A + B^0 + B^1)$ consists of a system of infinitely many parallel half–lines $\{\lambda \in \mathbb{C};\ \mathrm{Re}\,\lambda \leq 0,\ \mathrm{Im}\,\lambda = \omega k i,\ k \in \mathbb{Z}\}$ in the case $u_\infty = 0$, and infinitely many parabolic regions of the type $\mathcal{P}$, equidistantly shifted by $k\omega i$ (where $k \in \mathbb{Z}$) if $u_\infty \neq 0$. Again, as the operator B^2 is relatively $(\nu A + B^0 + B^1)$–compact (Lemma 3 in Sect. 2), the operator $\nu A + B^0 + B^1 + B^2$ has the same essential spectrum as $\nu A + B^0 + B^1$.

The requirement $s(L) < 0$ is avoided in [41], where the stability of a steady flow around a translating body is studied. (Thus, the case considered in [41] corresponds to the problem studied in this paper if $\boldsymbol{\omega} = \mathbf{0}$.) Instead of the inequality $s(L) < 0$, the author in [41] assumes that for some "sufficiently large" $R > 0$, $\|\mathrm{e}^{Lt}\boldsymbol{\phi}\|_{2;\,\Omega_R} \leq \varphi(t)\,\|\boldsymbol{\phi}\|_2$ (or alternatively $\|\nabla \mathrm{e}^{Lt}\boldsymbol{\phi}\|_{2;\,\Omega_R} \leq \varphi(t)\,\|\boldsymbol{\phi}\|_2$) for all eigenfunctions $\boldsymbol{\phi}$ of the symmetric operator $\nu A + (1+\kappa)B_s^2$ (for some $0 < \kappa \leq 1$), associated with positive eigenvalues. (It is shown that the number of positive eigenvalues is finite.) Function φ is supposed to be in $L^1(0,\infty) \cap L^2(0,\infty)$. Applying these assumptions, the author proves the stability of the zero solution of equation (4.21) with respect to the norm $(\|\,.\,\|_2 + \|\nabla.\,\|_2)^{1/2}$ and the asymptotic decay of "small" solutions to zero in the norm $\|\nabla.\,\|_2$. Furthermore, in the case $\Omega = \mathbb{R}^3$, Deuring and Neustupa [5] have shown that the sufficient condition for stability formulated in [41] is satisfied if all the eigenvalues of operator L have negative real parts and the perturbed steady Oseen equation

$$\nu A\mathbf{u} + B^1\mathbf{u} + B^2\mathbf{u} \;=\; \mathbf{g} \tag{4.23}$$

has a unique weak solution for each $\mathbf{g} \in \mathbf{D}^{-1,2}_{\sigma,0}(\mathbb{R}^3)$ (the dual to $\mathbf{D}^{1,2}_{0,\sigma}(\mathbb{R}^3)$). Note that similar ideas as in [41] have already been applied to a general parabolic equation in a Hilbert space or to a parabolic system in an exterior domain in [39, 40], and to the linearized equation (4.21) in [42].

Let us finally mention the paper [44] by Sazonov. Here, the author also studies stability of a steady flow around a translating body (the case of $\boldsymbol{\omega} = \mathbf{0}$). He shows asymptotic stability of the zero solution of equation (4.11) in the norm of $L^3(\Omega)^3$, provided all eigenvalues of the operator $\nu A + B^1 + B^2$ (with the domain $\mathbf{W}^{2,3}(\Omega) \cap \mathbf{D}^{1,3}_{0,\sigma}(\Omega)$) have negative real parts. The used arguments are based on estimates of the semigroup, generated by Oseen's operator $\nu A + B^1$. However, we could not follow the derivation of one of the crucial estimates (Theorem 3.1 in [44]), because the author, representing the semigroup by certain line integral, uses a path crossing the essential spectrum of L. This has also been pointed out by Kobayashi and Shibata in [34].

In this paper, we assume $\boldsymbol{\omega} \neq \mathbf{0}$, do not use any assumption on smallness of solution $\mathbf{U}$, and prove a result analogous to [41]. More precisely, we denote by L the operator $\nu A + B^0 + B^1 + B^2$ and we assume that $\boldsymbol{\phi}$ is any of the eigenfunctions $\boldsymbol{\phi}$ of $\nu A + (1+\kappa)B_s^2$, for some $0 < \kappa \leq 1$, associated with positive eigenvalues. (The number of positive eigenvalues is finite.) We show that the L^1– and L^2–integrability of $\|e^{Lt}\phi\|_{2;\,\Omega_R}$ (for a "sufficiently large" $R > 0$) implies the stability of the zero solution of equation (4.21) with respect to the norm $(\|\,.\,\|_2 + \|\nabla.\,\|_2)^{1/2}$ and the asymptotic decay of a "small" solutions to zero in the norm $\|\nabla.\,\|_2$. The main result, formulated in Theorem 1, also shows that the stability of the zero solution of equation (4.21) is not affected by the nonlinear operator N.

4.2 Auxiliary Results

Recall at first that the assumptions $\nabla\mathbf{U} \in L^3(\Omega)^{3\times 3}$ and $\nabla\mathbf{U} \in L^{3/2}(\Omega)^{3\times 3}$ and Sobolev's inequality

$$\|\phi\|_r \,\leq\, c_{16}(q)\,\|\nabla\phi\|_q; \quad 1 \leq q < 3, \;\; r = \frac{3q}{3-q}, \tag{4.24}$$

see [14, p. 54], imply that $\mathbf{U} \in \mathbf{L}^a(\Omega)$ for all $3 \leq a < \infty$.

Some basic estimates. The operators B^1 and B^2 satisfy these inequalities, proved in [41]:

$$\|B^1\phi\|_2 \leq u_\infty\,\|\nabla\phi\|_2, \tag{4.25}$$

$$\|B^2_s\phi\|_2 + \|B^2_a\phi\|_2 \leq \|\nabla\mathbf{U}\|_3\,\|\phi\|_6 \leq c_3\,\|\nabla\phi\|_2, \tag{4.26}$$

$$|(B^2_s\phi, \psi)_2| \leq \|\nabla\mathbf{U}\|_3\,\|\phi\|_2\,\|\psi\|_6 \leq c_4\,\|\phi\|_2\,\|\nabla\psi\|_2 \tag{4.27}$$

for all $\phi,\ \psi \in \mathbf{L}^2_\sigma(\Omega) \cap \mathbf{D}^{1,2}_{0,\sigma}(\Omega)$. The positive constants c_3, c_4 depend only on Ω and $\mathbf{U}$. The nonlinear operator N satisfies

$$\|N(\phi)\|_2 \leq c_5\,\mathcal{Y}[\phi], \tag{4.28}$$

$$\|N(\phi)\|_2^2 \leq c_6\,\|\nabla\phi\|_2^2\,\mathcal{Y}[\phi], \tag{4.29}$$

where

$$\mathcal{Y}[\phi] \;:=\; \|A\phi\|_2^2 + \|\nabla\phi\|_2^2. \tag{4.30}$$

Estimates (4.28) and (4.29) hold for $\phi \in D(A)$.

Lemma 3 *The operator B^2 is relatively A–compact, relatively $(\nu A + B^0)$–compact and relatively $(\nu A + B^0 + B^1)$–compact in space $\mathbf{L}^2_\sigma(\Omega)$.*

The relative compactness of B^2 with respect to A and $\nu A + B^1$ is proven in [41], the statement on the relative compactness of B^2 with respect to $\nu A + B^0 + B^1$ is proven in [42].

Lemma 4 *(i) Both operators $\nu A + B^0 + B^1$ and $\nu A + B^0 + B^1 + B^2$ are closed and densely defined in $\mathbf{L}^2_\sigma(\Omega)$.*

(ii) The essential spectrum of the operator $\nu A + B^0 + B^1$ *has the form*

$$\begin{aligned}&\mathrm{Sp}_{\mathrm{ess}}(\nu A + B^0 + B^1)\\&\quad= \Big\{\lambda = \alpha + \mathrm{i}\beta + \mathrm{i}k\omega \in \mathbb{C};\ \alpha, \beta \in \mathbb{R},\ k \in \mathbb{Z},\ \alpha \le -\frac{\nu\beta^2}{u_\infty^2}\Big\}.\end{aligned}$$

(iii) $\mathrm{Sp}(\nu A + B^0 + B^1) = \mathrm{Sp}_{\mathrm{ess}}(\nu A + B^0 + B^1) \cup \Gamma$, *where* Γ *is either empty, or contains isolated eigenvalues of* $\nu A + B^0 + B^1$. *All the eigenvalues have negative real parts, finite algebraic multiplicities, and they can possibly cluster only at points on the boundary of* $\mathrm{Sp}_{\mathrm{ess}}(\nu A + B_a^1 + B_a^2)$.

(iv) If the body $\mathcal{B}$ *(and therefore also the domain* Ω*) is axially symmetric about the* x_1*–axis, then* $\Gamma = \emptyset$.

(v) The operator $\nu A + B^0 + B^1 + B^2$ *has the same essential spectrum as* $\nu A + B^0 + B^1$.

(vi) $\mathrm{Sp}(\nu A + B^0 + B^1 + B^2) = \mathrm{Sp}_{\mathrm{ess}}(\nu A + B^0 + B^1 + B^2) \cup \Gamma'$ *where set* Γ' *is either empty or it consists of isolated eigenvalues of* $\nu A + B^0 + B^1 + B^2$*, which can possibly cluster only at points on the boundary of* $\mathrm{Sp}_{\mathrm{ess}}(\nu A + B^0 + B^1 + B^2)$. *Each of the eigenvalues from* Γ' *has a finite algebraic multiplicity.*

Lemma 4 follows from [11, Theorem 1.1, Lemma 2.4] and [12, Theorem 1.2].

Lemma 5 *The operator* $\nu A + B^0 + B^1 + B^2$ *generates a* C_0*–semigroup in* $\mathbf{L}^2_\sigma(\Omega)$.

Proof The fact that $\nu A + B^0 + B^1$ generates a C_0–semigroup has already been proven by Hishida [26] (in $\mathbf{L}^2_\sigma(\Omega)$ with $u_\infty = 0$), Geisert, Heck and Hieber [22] (in $\mathbf{L}^q_\sigma(\Omega)$ for $1 < q < \infty$, with $u_\infty = 0$) and Shibata [46] (in $\mathbf{L}^q_\sigma(\Omega)$ for $1 < q < \infty$). In [22], the authors also consider the operator $\nu A + B^0 + B^1$ perturbed by B^2. However, they assume that $\mathbf{U} \in \mathbf{C}_0^\infty(\Omega)$. Nevertheless, the arguments used in [22] also hold in the case where $\mathbf{U}$ satisfies assumptions (4.11): since $\nu A + B^0 + B^1 + B^2$ is closed and densely defined in $\mathbf{L}^2_\sigma(\Omega)$ then, due to Lemma 4 and the Lumer–Phillips theorem (see e.g. [6, p. 83]), to prove the lemma it is sufficient to show that $\nu A + B^0 + B^1 + B^2 + \xi I$ is dissipative for some $\xi > 0$. This is, however, clear because

$$\big((\nu A + B^0 + B^1 + B^2)\mathbf{u} - \xi\mathbf{u}, \mathbf{u}\big)_2 = -\nu\, \|\nabla\mathbf{u}\|_2^2 + (B_s^2\mathbf{u}, \mathbf{u})_2 - \xi\, \|\mathbf{u}\|_2^2$$

and the dissipativeness of $\nu A + B^0 + B^1 + B^2 - \xi I$ now follows from estimate (4.27) and [6, Proposition 3.23].

Lemma 6 *Let* $\gamma \in \mathbb{R}$. *Then the operator* $\nu A + \gamma B_s^2$ *is selfadjoint in* $\mathbf{L}^2_\sigma(\Omega)$. *Its spectrum consists of* $\mathrm{Sp}_{\mathrm{ess}}(\nu A + \gamma B_s^2) = (-\infty, 0]$ *and at most a finite number of positive eigenvalues, each of whose has a finite multiplicity.*

This lemma is taken from [41, Lemma 3]. The algebraic and geometric multiplicities of the positive eigenvalues of $\nu A + \gamma B_s^2$ coincide due to the symmetry of $\nu A + \gamma B_s^2$. Note that there exists an interesting analogy between the operator $\nu A + \gamma B_s^2$ and the

Schrödinger operator $\Delta - g(\mathbf{x})$ in $L^2(\mathbb{R}^3)$. The so called Cwikel–Lieb–Rosenblum estimate says that the number of non–negative eigenvalues of $\Delta - g(\mathbf{x})$ is less than or equal to $C\, \|g_-\|_{3/2;\,\mathbb{R}^3}^{3/2}$ where g_- denotes the negative part of g. (See e.g. Lieb [35].) Based on this analogy, there arises a question of whether the number of the nonnegative eigenvalues of $\nu A + \gamma B_s^2$ can also be estimated by $C\, \|\nabla \mathbf{U}\|_{3/2}^{3/2}/\nu$, where C is independent of ν and $\mathbf{U}$.

Decomposition of the space $\mathbf{L}^2_\sigma(\Omega)$. Let $0 < \kappa \leq 1$ be fixed. (The assumption $\kappa \leq 1$ is used just for technical reasons.) We denote the positive eigenvalues of the operator $\nu A + (1+\kappa) B_s^2$ by $\lambda_1 \leq \lambda_2 \leq \cdots \leq \lambda_N$, each of them being repeated as many times as is its multiplicity. Let $\phi_1, \ldots, \phi_N$ be associated eigenfunctions. We can assume that the eigenfunctions have been chosen so that they constitute an orthonormal system in $\mathbf{L}^2_\sigma(\Omega)$. Denote by $\mathbf{L}^2_\sigma(\Omega)'$ the linear hull of $\phi_1, \ldots, \phi_N$ and by P' the orthogonal projection of $\mathbf{L}^2_\sigma(\Omega)$ onto $\mathbf{L}^2_\sigma(\Omega)'$. The orthogonal complement to $\mathbf{L}^2_\sigma(\Omega)'$ in $\mathbf{L}^2_\sigma(\Omega)$ is denoted by $\mathbf{L}^2_\sigma(\Omega)''$ and the orthogonal projection of $\mathbf{L}^2_\sigma(\Omega)$ onto $\mathbf{L}^2_\sigma(\Omega)''$ is denoted by P''. Then we have

$$\mathbf{L}^2_\sigma(\Omega) \;=\; \mathbf{L}^2_\sigma(\Omega)' \oplus \mathbf{L}^2_\sigma(\Omega)''$$

and the operator $\nu A + (1+\kappa) B_s^2$ is reduced on each of the subspaces $\mathbf{L}^2_\sigma(\Omega)'$ and $\mathbf{L}^2_\sigma(\Omega)''$.

The projections P' and P'' are bounded operators not only in $\mathbf{L}^2_\sigma(\Omega)$, but also in $\mathbf{W}^{1,2}_{0,\sigma}(\Omega)$, see [41]. Furthermore, using the negative definiteness of $\nu A + (1+\kappa) B_s^2$ in $\mathbf{L}^2_\sigma(\Omega)''$, one can easily derive that

$$\big((\nu A + B_s^2)\phi, \phi\big)_2 \;\leq\; \frac{\kappa}{1+\kappa}\, (\nu A\phi, \phi)_2 \;=\; -c_7\, \|\nabla\phi\|_2^2 \tag{4.31}$$

for all $\phi \in \mathbf{L}^2_\sigma(\Omega)'' \cap D(A)$, where $c_7 = \kappa\nu/(1+\kappa)$, see [41]. We call this property of $\nu A + B_s^2$ in space $\mathbf{L}^2_\sigma(\Omega)''$ the "essential dissipativity".

Lemma 7 *There exists $c_{10} > 0$ such that*

$$\int_\Omega |\mathbf{x}|^2\, |\Delta\phi_k|^2\, \mathrm{d}\mathbf{x} + \int_\Omega |\mathbf{x}|^2\, |\nabla\phi_k|^2\, \mathrm{d}\mathbf{x} \;\leq\; c_{10} \tag{4.32}$$

for all $k = 1, \ldots, N$.

Proof Assume at first that φ is an infinitely differentiable function on $[0, \infty)$ such that $\varphi(s) = 1$ for $0 \leq s \leq 1$, φ is decreasing and $|\varphi'(s)| \leq 1$, $|\varphi''(s)| \leq 2$ for $1 \leq s \leq 4$, and $\varphi(s) = 0$ for $s \geq 4$. Put $\varphi_r(s) := \varphi(s/r^2)$. Then $\varphi_r(s) = 0$ for $0 \leq s \leq r^2$, $|\varphi_r'(s)| \leq r^{-2}$, $|\varphi_r''(s)| \leq 2r^{-4}$ for $r^2 \leq s \leq 4r^2$, and $\varphi_r(s) = 0$ for $s \geq 4r^2$. Function φ_r satisfies the estimates

$$\big|\nabla\big(|\mathbf{x}|^2\varphi_r(|\mathbf{x}|^2)\big)\big| \;\leq\; c_8\, |\mathbf{x}| \qquad \text{and} \qquad \big|\nabla^2\big(|\mathbf{x}|^2\varphi_r(|\mathbf{x}|^2)\big)\big| \;\leq\; c_9. \tag{4.33}$$

The equation $[\nu A + (1+\kappa)B_s^2]\boldsymbol{\phi}_k = \lambda_k \boldsymbol{\phi}_k$ (for $k \in \{1, \ldots, N\}$) means that

$$\nu \Delta \boldsymbol{\phi}_k - (1+\kappa)\, \boldsymbol{\phi}_k \cdot (\nabla \mathbf{U})_s + \nabla p_k \;=\; \lambda_k \boldsymbol{\phi}_k, \quad \operatorname{div} \boldsymbol{\phi}_k \;=\; 0 \tag{4.34}$$

in Ω, with an appropriate function $p_k \in W^{1,2}_{loc}(\Omega)$ such that $\nabla p_k \in \mathbf{L}^2(\Omega)$. Multiplying the first equation in (4.34) by $|\mathbf{x}|^2\, \varphi_r(|\mathbf{x}|^2)\, \Delta \boldsymbol{\phi}_k$ and integrating over Ω, we get

$$\begin{aligned} \nu \int_\Omega |\mathbf{x}|^2 \varphi_r(|\mathbf{x}|^2)\, |\Delta \boldsymbol{\phi}|^2 \, \mathrm{d}\mathbf{x} - (1+\kappa) \int_\Omega \boldsymbol{\phi}_k \cdot (\nabla \mathbf{U})_s\, |\mathbf{x}|^2 \varphi_r(|\mathbf{x}|^2)\, \Delta \boldsymbol{\phi}_k \, \mathrm{d}\mathbf{x} \\ + \int_\Omega \nabla p_k \cdot |\mathbf{x}|^2 \varphi_r(|\mathbf{x}|^2) \Delta \boldsymbol{\phi}_k \, \mathrm{d}\mathbf{x} \;=\; \lambda_k \int_\Omega \boldsymbol{\phi}_k \cdot |\mathbf{x}|^2 \varphi_r(|\mathbf{x}|^2) \Delta \boldsymbol{\phi}_k \, \mathrm{d}\mathbf{x}. \end{aligned} \tag{4.35}$$

The right hand side is equal to

$$\begin{aligned} -\lambda_k \int_\Omega \nabla \boldsymbol{\phi}_k : |\mathbf{x}|^2 \varphi_r(|\mathbf{x}|^2)\, \nabla \boldsymbol{\phi}_k \, \mathrm{d}\mathbf{x} - \lambda_k \int_\Omega \nabla\big(|\mathbf{x}|^2 \varphi_r(|\mathbf{x}|^2)\big) \cdot (\nabla \boldsymbol{\phi}_k \cdot \boldsymbol{\phi}_k) \, \mathrm{d}\mathbf{x} \\ = -\lambda_k \int_\Omega |\mathbf{x}|^2\, |\nabla \boldsymbol{\phi}_k|^2 \varphi_r(|\mathbf{x}|^2) \, \mathrm{d}\mathbf{x} + \frac{\lambda_k}{2} \int_\Omega |\boldsymbol{\phi}_k|^2\, \Delta\big(|\mathbf{x}|^2 \varphi_r(|\mathbf{x}|^2)\big) \, \mathrm{d}\mathbf{x} \\ \leq -\lambda_k \int_\Omega |\mathbf{x}|^2\, |\nabla \boldsymbol{\phi}_k|^2 \varphi_r(|\mathbf{x}|^2) \, \mathrm{d}\mathbf{x} + \frac{c_9 \lambda_k}{2}\, \|\boldsymbol{\phi}_k\|_2^2. \end{aligned} \tag{4.36}$$

Due to (4.12), the modulus of the second term on the left hand side of (4.35) is less than or equal to

$$\begin{aligned} C(1+\kappa) \int_\Omega |\boldsymbol{\phi}_k|\, |\mathbf{x}|\, \varphi_r(|\mathbf{x}|^2)\, |\Delta \boldsymbol{\phi}_k| \, \mathrm{d}\mathbf{x} \\ \leq \epsilon \int_\Omega |\mathbf{x}|^2 \varphi_r(|\mathbf{x}|^2)\, |\Delta \boldsymbol{\phi}_k|^2 \, \mathrm{d}\mathbf{x} + C(\epsilon, \kappa)\, \|\boldsymbol{\phi}_k\|_2^2. \end{aligned} \tag{4.37}$$

Finally, the third term on the left hand side of (4.35) equals

$$\begin{aligned} & -\int_\Omega p_k\, \nabla\big(|\mathbf{x}|^2 \varphi_r(|\mathbf{x}|^2)\big) \cdot \Delta \boldsymbol{\phi}_k \, \mathrm{d}\mathbf{x} \\ & = \int_\Omega (\partial_j p_k)\, \nabla\big(|\mathbf{x}|^2 \varphi_r(|\mathbf{x}|^2)\big) \cdot (\partial_j \boldsymbol{\phi}_k) \, \mathrm{d}\mathbf{x} + \int_\Omega p_k\, \nabla^2\big(|\mathbf{x}|^2 \varphi_r(|\mathbf{x}|^2)\big) : \nabla \boldsymbol{\phi}_k \, \mathrm{d}\mathbf{x} \\ & = -\int_\Omega \Delta p_k\, \nabla\big(|\mathbf{x}|^2 \varphi_r(|\mathbf{x}|^2)\big) \cdot \boldsymbol{\phi}_k \, \mathrm{d}\mathbf{x} - 2 \int_\Omega \nabla p_k \cdot \nabla^2\big(|\mathbf{x}|^2 \varphi_r(|\mathbf{x}|^2)\big) \cdot \boldsymbol{\phi}_k \, \mathrm{d}\mathbf{x} \\ & \quad - \int_\Omega p_k\, \nabla \Delta\big(|\mathbf{x}|^2 \varphi_r(|\mathbf{x}|^2)\big) \cdot \boldsymbol{\phi}_k \, \mathrm{d}\mathbf{x}. \end{aligned} \tag{4.38}$$

The modulus of the first term on the right hand side is

$$(1+\kappa) \left| \int_\Omega \operatorname{div}\big(\boldsymbol{\phi}_k \cdot (\nabla \mathbf{U})_s\big)\, \nabla\big(|\mathbf{x}|^2 \varphi_r(|\mathbf{x}|^2)\big) \cdot \boldsymbol{\phi}_k \, \mathrm{d}\mathbf{x} \right|$$

$$
\begin{aligned}
&\leq (1+\kappa)\left|\int_{\Omega} \boldsymbol{\phi}_k \cdot (\nabla \mathbf{U})_s \cdot \nabla^2\big(|\mathbf{x}|^2\varphi_r(|\mathbf{x}|^2)\big) \cdot \boldsymbol{\phi}_k \, \mathrm{d}\mathbf{x}\right| \\
&\quad + (1+\kappa)\left|\int_{\Omega} [(\boldsymbol{\phi}_k \cdot (\nabla \mathbf{U})_s) \otimes \nabla\big(|\mathbf{x}|^2\varphi_r(|\mathbf{x}|^2)\big)] : \nabla \boldsymbol{\phi}_k \, \mathrm{d}\mathbf{x}\right| \\
&\leq c_9 \int_{\Omega} |\boldsymbol{\phi}_k|^2 \, |(\nabla \mathbf{U})_s| \, \mathrm{d}\mathbf{x} + c_8 \int_{\Omega} |\boldsymbol{\phi}_k| \, |(\nabla \mathbf{U})_s| \, |\mathbf{x}| \, |\nabla \boldsymbol{\phi}_k| \, \mathrm{d}\mathbf{x} \\
&\leq c_9 \, \|\boldsymbol{\phi}_k\|_6 \, \|(\nabla \mathbf{U})_s\|_{3/2} + c_1 c_8 \, \|\boldsymbol{\phi}_k\|_2 \, \|\nabla \boldsymbol{\phi}_k\|_2. \qquad (4.39)
\end{aligned}
$$

Applying (4.33), we can estimate the second and the third term on the right hand side of (4.38):

$$
\left|2\int_{\Omega} \nabla p_k \cdot \nabla^2\big(|\mathbf{x}|^2\varphi_r(|\mathbf{x}|^2)\big) \cdot \boldsymbol{\phi}_k \, \mathrm{d}\mathbf{x}\right| \;\leq\; 2c_9 \, \|\nabla p_k\|_2 \, \|\boldsymbol{\phi}_k\|_2 \;<\; \infty, \qquad (4.40)
$$

$$
\begin{aligned}
&\left|\int_{\Omega} p_k \, \nabla \Delta\big(|\mathbf{x}|^2\varphi_r(|\mathbf{x}|^2)\big) \cdot \boldsymbol{\phi}_k \, \mathrm{d}\mathbf{x}\right| \;=\; \left|\int_{\Omega} \nabla p_k \, \Delta\big(|\mathbf{x}|^2\varphi_r(|\mathbf{x}|^2)\big) \cdot \boldsymbol{\phi}_k \, \mathrm{d}\mathbf{x}\right| \\
&\leq\; c_9 \, \|\nabla p_k\|_2 \, \|\boldsymbol{\phi}_k\|_2 \;<\; \infty. \qquad (4.41)
\end{aligned}
$$

Substituting now the estimates (4.36)–(4.41) to (4.35), we obtain

$$
(\nu - \epsilon) \int_{\Omega} |\mathbf{x}|^2 \, \varphi_r(|\mathbf{x}|^2) \, |\Delta \boldsymbol{\phi}_k|^2 \, \mathrm{d}\mathbf{x} + \lambda_k \int_{\Omega} |\mathbf{x}|^2 \varphi_r(|\mathbf{x}|^2) \, |\nabla \boldsymbol{\phi}_k|^2 \, \mathrm{d}\mathbf{x} \;\leq\; C,
$$

where C depends on ϵ, κ, c_1, c_8, c_9, $\|(\nabla \mathbf{U})_s\|_{3/2}$, $\|\nabla \boldsymbol{\phi}_k\|_2$ and $\|\nabla p_k\|_2$, but it is independent of r. Choosing ϵ sufficiently small and letting $r \to \infty$, we complete the proof.

Lemma 8 *There exists* $c_{11} > 0$ *such that*

$$
\|P' B^0 \mathbf{u}\|_2 + \|P' B^1 \mathbf{u}\|_2 + \|P' B^2 \mathbf{u}\|_2 \;\leq\; c_{11} \, \|\nabla \mathbf{u}\|_2 \qquad (4.42)
$$

for all $\mathbf{u} \in D(A)$.

Proof Applying (4.12) and Lemma 7, we estimate $\|P' B^0 \mathbf{u}\|_2$ as follows:

$$
\begin{aligned}
\|P' B^0 \mathbf{u}\|_2^2 \;&=\; \sum_{k=1}^{N} \big|(B^0 \mathbf{u}, \boldsymbol{\phi}_k)_2\big|^2 \;=\; \sum_{k=1}^{N} \frac{1}{\lambda_k^2} \, \big|(B^0 \mathbf{u}, \, \nu A \boldsymbol{\phi}_k + (1+\kappa) B_s^2 \boldsymbol{\phi}_k)_2\big|^2 \\
&= \sum_{k=1}^{N} \frac{1}{\lambda_k^2} \left|\int_{\Omega} \big[(\omega \times \mathbf{x}) \cdot \nabla \mathbf{u} - \omega \times \mathbf{u}\big] \cdot \big[\nu \Delta \boldsymbol{\phi}_k - (1+\kappa) \boldsymbol{\phi}_k \cdot (\nabla U)_s\big] \, \mathrm{d}\mathbf{x}\right|^2 \\
&\leq \sum_{k=1}^{N} \frac{1}{\lambda_k^2} \Big(\omega\nu \, \|\nabla \mathbf{u}\|_2 \, \||\mathbf{x}| \, \Delta \boldsymbol{\phi}_k\|_2 + \omega(1+\kappa) \, c_1 \, \|\nabla \mathbf{u}\|_2 \, \|\boldsymbol{\phi}_k\|_2 \\
&\qquad + \omega\nu \, \|\nabla \mathbf{u}\|_2 \, \|\nabla \boldsymbol{\phi}_k\|_2 + \omega(1+\kappa) \, \|\mathbf{u}\|_6 \, \|(\nabla \mathbf{U})_s\|_{3/2} \, \|\boldsymbol{\phi}_k\|_6\Big)^2. \qquad (4.43)
\end{aligned}
$$

The right hand side is less than or equal to $C\,\|\nabla\mathbf{u}\|_2^2$ due to Sobolev's inequality (4.24) and Lemma 7. Analogous estimates of $\|P'B^1\phi\|_2$ and $\|P'B^2\phi\|_2$ directly follow from the estimates

$$\begin{aligned} \big|(B^1\mathbf{u},\,\phi_k)_2\big| \;&\le\; |u_\infty|\;\|\nabla\mathbf{u}\|_2\;\|\phi_k\|_2, \\ \big|(\mathbf{u}\cdot\nabla\mathbf{U},\,\phi_k)_2\big| \;&\le\; \|\mathbf{u}\|_6\;\|\nabla\mathbf{U}\|_{3/2}\;\|\phi_k\|_6. \end{aligned}$$

□

Obviously, $D(A)$ is dense in $\mathbf{L}^2_\sigma(\Omega)$. The next lemma states a finer result:

Lemma 9 *$D(A)\cap\mathbf{L}^2_\sigma(\Omega)''$ is dense in $\mathbf{L}^2_\sigma(\Omega)''$.*

Proof Due to the density of $\mathbf{C}^\infty_{0,\sigma}(\Omega)$ in $\mathbf{L}^2_\sigma(\Omega)$, we can find functions $\varphi_1,\,\dots,\,\varphi_N$ in $\mathbf{C}^\infty_{0,\sigma}(\Omega)$ so that they satisfy $(\varphi_i,\phi_j)_2=\delta_{ij}$ for $i,\,j=1,\dots,N$.

Let $\epsilon>0$ be given. Let $\mathbf{u}$ be an arbitrary function from $\mathbf{L}^2_\sigma(\Omega)''$. There exists $\mathbf{u}''\in\mathbf{C}^\infty_{0,\sigma}(\Omega)$ such that $\|\mathbf{u}-\mathbf{u}''\|_2<\epsilon/2$ and the numbers $\epsilon_i:=(\mathbf{u}''-\mathbf{u},\,\phi_i)_2$ satisfy $|\epsilon_i|\;\|\varphi_i\|_2<\epsilon/(2N)$ for all $i=1,\dots,N$. Now we put

$$\mathbf{v}\;:=\;\mathbf{u}''-\epsilon_1\varphi_1-\dots-\epsilon_N\varphi_N\,.$$

Then, obviously, $\mathbf{v}\in\mathbf{C}^\infty_{0,\sigma}(\Omega)$. Using the identities $(\mathbf{u},\,\phi_i)_2=0$ (for $i=1,\dots,N$) following from the inclusion $\mathbf{u}\in\mathbf{L}^2_\sigma(\Omega)''\equiv(\mathbf{L}^2_\sigma(\Omega)')^\perp$, we obtain

$$\begin{aligned} (\mathbf{v},\phi_i)_2 &= \;\epsilon_i-\epsilon_i\;=\;0\quad \text{for } i=1,\dots,N, \\ \|\mathbf{u}-\mathbf{v}\|_2 &\le\; \|\mathbf{u}-\mathbf{u}''\|_2+\sum_{i=1}^N|\epsilon_i|\;\|\varphi_i\|_2\;\le\;\frac{\epsilon}{2}+\sum_{i=1}^N\frac{\epsilon}{2N}\;=\;\epsilon. \end{aligned}$$

This completes the proof.

4.3 The Main Theorem on Stability

Let us fix number $R>0$ so large that $\mathcal{B}\subset B_R(\mathbf{0})$ and

$$c_2^2(2)\;\|\nabla\mathbf{U}\|_{3/2;\,\Omega^R}\;\le\;\frac{\nu}{8}\,. \tag{4.44}$$

(Recall the $c_2(2)$ is the constant from Sobolev's inequality (4.24).) We denote

$$L\;:=\;\nu A+B^0+B^1+B^2.$$

We impose the following condition:

(A) *there exists a function $\varphi\in L^1(0,\infty)\cap L^2(0,\infty)$ such that*

$$\left\| \mathrm{e}^{Lt} \phi \right\|_{2;\, \Omega_R} \le \varphi(t) \, \|\phi\|_2 \tag{4.45}$$

for all $\phi \in \mathbf{L}^2_\sigma(\Omega)'$ *and* $t > 0$.

Observe that (A), in fact, concerns only a finite family of functions, i.e. the functions $\phi_1, \dots, \phi_N$. The next theorem presents the main result of this paper.

Theorem 1 *Let* $\mathbf{U}$ *be a steady solution of the problem (4.6)–(4.9), satisfying conditions (4.11) and (4.12). Suppose that the operator* L *satisfies condition (A). Then there exist positive constants* δ, c_{12}, c_{13} *such that if* $\mathbf{v}_0 \in \mathbf{W}^{1,2}_{0,\sigma}(\Omega)$ *and*

$$\|\mathbf{v}_0\|_{1,2} \le \delta \tag{4.46}$$

then the problem defined by the equation (4.21) with the initial condition $\mathbf{v}(0) = \mathbf{v}_0$ *has a unique solution* $\mathbf{v}$ *on the time interval* $[0, \infty)$. *This solution satisfies the estimate*

$$\|\mathbf{v}(t)\|^2_{1,2} + c_{11} \int_0^t \left(\|\nabla \mathbf{v}(s)\|_2^2 + \|A\mathbf{v}(s)\|_2^2 \right) \mathrm{d}s \le c_{12} \, \|\mathbf{v}_0\|^2_{1,2} \tag{4.47}$$

for all $t > 0$. *Moreover,*

$$\lim_{t \to \infty} \|\nabla \mathbf{v}(t)\|_2 = 0. \tag{4.48}$$

The proof of Theorem 1 will be given later, after the next lemma and several auxiliary estimates.

Lemma 10 *Let* $T > 0$. *Function* $\mathbf{v}$ *is a solution of the equation (4.21) on the time interval* $[0, T)$ *if and only if* $\mathbf{v}$ *can be expressed as a sum* $\mathbf{w} + \mathbf{z}$, *where the functions* $\mathbf{w}$, $\mathbf{z}$ *satisfy the system*

$$\begin{aligned} \frac{\mathrm{d}\mathbf{w}}{\mathrm{d}t} - B^0\mathbf{w} - B^1\mathbf{w} = {} & \nu A\mathbf{w} + (1+\kappa) B^2_s \mathbf{w} - \kappa P'' B^2_s \mathbf{w} - P' B^0 \mathbf{w} - P' B^1 \mathbf{w} \\ & + P'' B^2_a \mathbf{w} + P'' N(\mathbf{v}), \end{aligned} \tag{4.49}$$

$$\begin{aligned} \frac{\mathrm{d}\mathbf{z}}{\mathrm{d}t} - B^0\mathbf{z} - B^1\mathbf{z} = {} & \nu A\mathbf{z} + B^2\mathbf{z} - \kappa P' B^2_s \mathbf{w} + P' B^0 \mathbf{w} + P' B^1 \mathbf{w} \\ & + P' B^2_a \mathbf{w} + P' N(\mathbf{v}) \end{aligned} \tag{4.50}$$

on the same interval $[0, T)$ *and the initial conditions*

$$\mathbf{w}(0) = P''\mathbf{v}(0), \qquad \mathbf{z}(0) = P'\mathbf{v}(0). \tag{4.51}$$

Proof If functions $\mathbf{w}$, $\mathbf{z}$ satisfy (4.49)–(4.51) then summing equations (4.49) and (4.50), we verify that $\mathbf{v}$ satisfies equation (4.11).

On the other hand, if $\mathbf{v}$ is a solution of the equation (4.11) on the interval $[0, T)$, then we at first solve the equation (4.49) with the initial condition $\mathbf{w}(0) = P''\mathbf{v}(0)$ as a linear problem for one unknown function $\mathbf{w}$. Since the "leading" terms $B^0\mathbf{w}$ and

$\nu A\mathbf{w}$ act in this equation in the same manner as in Eq. (4.6), and all other terms are of a "lower order", the existence of a strong solution can be proven in the same way as in [4]. (The information that $P'B^0\mathbf{w}$ is a "lower order" term follows from Lemma 8.) Moreover, as (4.49) is just a linear equation for the unknown function $\mathbf{w}$, we are not restricted to a "small" time interval $(0, T_0)$, and we obtain the solution on the same time interval $(0, T)$ where solution $\mathbf{v}$ does exist. Having $\mathbf{w}$, we may put $\mathbf{z} := \mathbf{v} - \mathbf{w}$ and we verify that $\mathbf{z}$ is a solution to the equation (4.50) on $(0, T)$.

Suppose, for a while, that the pair $\mathbf{w}$, $\mathbf{z}$ is a solution of the system (4.49), (4.50) on an interval $[0, T)$. We derive estimates of $\mathbf{w}$, $\mathbf{z}$, valid on the same interval $[0, T)$.

Estimate 1 *The function* $\mathbf{w}$ *satisfies the inequality*

$$\|\mathbf{w}(t)\|_2^2 + 2c_7 \int_0^t \|\nabla\mathbf{w}(s)\|_2^2 \, \mathrm{d}s \leq \|\mathbf{w}(0)\|_2^2 + \int_0^t 2\sqrt{c_5}\, \mathcal{Y}[\mathbf{v}(s)]\, \|\mathbf{w}(s)\|_2 \, \mathrm{d}s \tag{4.52}$$

where $\mathcal{Y}[\mathbf{v}(s)]$ *is defined by (4.30).*

Proof As the operator $\nu A + (1+\kappa)B_s^2$ is reduced in $\mathbf{L}^2_\sigma(\Omega)''$, (4.49) is the equation in $\mathbf{L}^2_\sigma(\Omega)''$. Consequently, since $\mathbf{w}(0)$ is also in $\mathbf{L}^2_\sigma(\Omega)''$, $\mathbf{w}(t)$ stays in $\mathbf{L}^2_\sigma(\Omega)''$ for $t > 0$. If we multiply this equation by $\mathbf{w}$, integrate over Ω, apply the identity (4.19) and the inequalities (4.31) and (4.28), we obtain

$$\begin{aligned}\frac{\mathrm{d}}{\mathrm{d}t}\frac{1}{2}\|\mathbf{w}\|_2^2 &= \big(\nu A\mathbf{w} + (1+\kappa)B_s^2\mathbf{w} - \kappa P'' B_s^2\mathbf{w}, \mathbf{w}\big)_2 + \big(N(\mathbf{v}), \mathbf{w}\big)_2 \\ &= (\nu A\mathbf{w} + B_s^2\mathbf{w}, \mathbf{w})_2 + \big(N(\mathbf{v}), \mathbf{w}\big)_2 \\ &\leq -c_7\, \|\nabla\mathbf{w}\|_2^2 + \sqrt{c_5}\, \mathcal{Y}[\mathbf{v}]\, \|\mathbf{w}\|_2.\end{aligned}$$

Integrating this estimate with respect to t, we obtain (4.52).

Estimate 2 *There exist constants* c_{14}, $c_{15} > 0$ *such that* $\mathbf{w}$ *satisfies the inequality*

$$\|\nabla\mathbf{w}(t)\|_2^2 + \nu \int_0^t \|A\mathbf{w}(s)\|_2^2 \, \mathrm{d}s \leq \|\nabla\mathbf{w}(0)\|_2^2 + c_{14} \int_0^t \|\nabla\mathbf{w}(s)\|_2^2 \, \mathrm{d}s + c_{15} \int_0^t \|\nabla\mathbf{v}(s)\|_2^2\, \mathcal{Y}[\mathbf{v}(s)] \, \mathrm{d}s. \tag{4.53}$$

Proof If we multiply equation (4.49) by $(-A\mathbf{w})$, integrate over Ω and use the identity (4.20) and the inequalities (4.25), (4.26), (4.29) and (4.42), we obtain

$$\begin{aligned}\frac{\mathrm{d}}{\mathrm{d}t}\frac{1}{2}\|\nabla\mathbf{w}\|_2^2 = {} & -\nu\, \|A\mathbf{w}\|_2^2 - (B_s^2\mathbf{w} + \kappa P' B_s^2\mathbf{w}, A\mathbf{w})_2 + (P'B^0\mathbf{w}, A\mathbf{w})_2 \\ & + (P'B^1\mathbf{w}, A\mathbf{w})_2 - (P'' B_a^2\mathbf{w}, A\mathbf{w})_2 - \big(P'' N(\mathbf{v}), A\mathbf{w}\big)_2\end{aligned}$$

$$\leq -\nu\, \|A\mathbf{w}\|_2^2 + C\, \|\nabla\mathbf{w}\|_2\, \|A\mathbf{w}\|_2 + \|N(\mathbf{v})\|_2\, \|A\mathbf{w}\|_2$$
$$\leq -\frac{\nu}{2}\, \|A\mathbf{w}\|_2^2 + \frac{c_{14}}{2}\, \|\nabla\mathbf{w}\|_2^2 + \frac{c_{15}}{2}\, \|\nabla\mathbf{v}\|_2\, \mathcal{Y}[\mathbf{v}]. \tag{4.54}$$

Integrating this with respect to time from 0 to t, we get (4.53).

Estimate 3 *Set* $c_{16} := \int_0^\infty \varphi^2(s)\,\mathrm{d}s$, $c_{17} := \left(\int_0^\infty \varphi(s)\,\mathrm{d}s\right)^2$, $c_{18} := 3c_{17}c_{17}^2(\kappa+3)^2$ *and* $c_{19} := 3c_8^2 c_{16}$. *Then*

$$\int_0^t \|\mathbf{z}(\vartheta)\|_{2;\,\Omega_R}^2\,\mathrm{d}\vartheta$$
$$\leq 3c_{16}\,\|\mathbf{z}(0)\|_2^2 + c_{18}\int_0^t \|\nabla\mathbf{w}(s)\|_2^2\,\mathrm{d}s + c_{19}\left(\int_0^t \mathcal{Y}[\mathbf{v}(s)]\,\mathrm{d}s\right)^2. \tag{4.55}$$

Proof The solution $\mathbf{z}$ of (4.50) satisfies the integral equation

$$\mathbf{z}(\vartheta) = \mathrm{e}^{L\vartheta}\mathbf{z}(0)$$
$$+ \int_0^\vartheta \mathrm{e}^{L(\vartheta-s)}\left[P'(-\kappa B_s^2 + B^0 + B^1 + B_a^2)\mathbf{w}(s) + P'N(\mathbf{v}(s))\right]\mathrm{d}s.$$

Using the inequalities (4.45) and (4.28), we obtain

$$\|\mathbf{z}(\vartheta)\|_{2;\,\Omega_R} \leq \|\mathrm{e}^{L\vartheta}\mathbf{z}(0)\|_{2;\,\Omega_R}$$
$$+ \int_0^\vartheta \left\|\mathrm{e}^{L(\vartheta-s)}\left[P'(-\kappa B_s^2 + B^0 + B^1 + B_a^2)\mathbf{w}(s) + P'N(\mathbf{v}(s))\right]\right\|_{2;\,\Omega_R}\mathrm{d}s$$
$$\leq \varphi(\vartheta)\,\|\mathbf{z}(0)\|_2 + \int_0^\vartheta \varphi(\vartheta-s)\,\left\|P'(-\kappa B_s^2 + B^0 + B^1 + B_a^2)\mathbf{w}(s)\right\|_2\,\mathrm{d}s$$
$$+ c_8\int_0^\vartheta \varphi(\vartheta-s)\,\mathcal{Y}[\mathbf{v}(s)]\,\mathrm{d}s. \tag{4.56}$$

The first integral on the right hand side is the convolution of the functions φ and $\|P'(-\kappa B_s^2 + B^0 + B^1 + B_a^2)\mathbf{w}(s)\|_2$. Due to the L^1–integrability of φ on $(0, t)$, we can estimate the L^2–norm of this integral on $(0, t)$ by the product of the L^1–norm of φ and the L^2–norm of $\|P'(-\kappa B_s^2 + B^0 + B^1 + B_a^2)\mathbf{w}(s)\|_2$. (See [14, Theorem II.11.1].) Thus, using (4.42), we get

$$\int_0^t\left(\int_0^\vartheta \varphi(\vartheta-s)\,\left\|P'(-\kappa B_s^2 + B^0 + B^1 + B_a^2)\mathbf{w}\right\|_2\,\mathrm{d}s\right)^2\mathrm{d}\vartheta$$
$$\leq c_{17}\int_0^t \left\|P'(-\kappa B_s^2 + B^0 + B^1 + B_a^2)\mathbf{w}(s)\right\|_2^2\,\mathrm{d}s$$
$$\leq c_{17}c_9^2(\kappa+3)^2\int_0^t \|\nabla\mathbf{w}\|_2^2\,\mathrm{d}s.$$

The second integral on the right hand side of (4.56) is the convolution of φ with $\mathcal{Y}[\mathbf{v}]$. Using the L^2–integrability of φ on $(0, t)$ and applying again Theorem II.11.1 in [14], we obtain

$$\int_0^t \left(\int_0^\vartheta \varphi(\vartheta - s)\, \mathcal{Y}[\mathbf{v}](s)\, \mathrm{d}s \right)^2 \mathrm{d}\vartheta \le c_{16} \left(\int_0^t \mathcal{Y}[\mathbf{v}](s)\, \mathrm{d}s \right)^2.$$

Substituting these estimates to (4.56), we obtain (4.55).

Estimate 4 *There exist positive constants c_{20}, c_{21} and c_{22} such that*

$$\begin{aligned} \|\mathbf{z}(t)\|_2^2 + \nu \int_0^t \|\nabla \mathbf{z}(s)\|_2^2\, \mathrm{d}s \;&\le\; c_{20}\, \|\mathbf{z}(0)\|_2^2 + c_{21} \int_0^t \mathcal{Y}[\mathbf{w}(s)]\, \mathrm{d}s \\ &+ c_{22} \left(\int_0^t \mathcal{Y}[\mathbf{v}(s)]\, \mathrm{d}s \right)^2 + 2c_8 \int_0^t \mathcal{Y}[\mathbf{v}(s)]\, \|\mathbf{z}(s)\|_2\, \mathrm{d}s. \end{aligned} \tag{4.57}$$

Proof Multiplying equation (4.50) by $\mathbf{z}$, integrating over Ω and using (4.19), we get

$$\begin{aligned} \frac{\mathrm{d}}{\mathrm{d}t} \frac{1}{2} \|\mathbf{z}\|_2^2 + \nu\, \|\nabla \mathbf{z}\|_2^2 = \;& \big(P'[-\kappa B_s^2 + B^0 + B^1 + B_a^2]\mathbf{w}, \mathbf{z} \big)_2 \\ &+ (B^2\mathbf{z}, \mathbf{z})_2 + \big(P' N(\mathbf{v}), \mathbf{z} \big)_2. \end{aligned} \tag{4.58}$$

The first term on the right hand side is equal to

$$\begin{aligned} &\sum_{k=1}^N \big([-\kappa B_s^2 + B^0 + B^1 + B_a^2]\mathbf{w}, \phi_k \big)_2\, (\mathbf{z}, \phi_k)_2 \\ &\quad \le \left(\sum_{k=1}^N \big([-\kappa B_s^2 + B^0 + B^1 + B_a^2]\mathbf{w}, \phi_k \big)_2^2 \right)^{1/2} \left(\sum_{k=1}^N (\mathbf{z}, \phi_k)_2^2 \right)^{1/2}. \end{aligned}$$

Applying the same approach as in the proof of Lemma 8, one can show that the first factor on the right hand side is less than or equal to $C\, \|\nabla \mathbf{w}\|_2$ and the second factor is less than or equal to $C\, \|\nabla \mathbf{z}\|_2$. Hence

$$\begin{aligned} \big(P'[-\kappa B_s^2 + B^0 + B^1 + B_a^2]\mathbf{w}, \mathbf{z} \big)_2 \le\;& C\, \|\nabla \mathbf{w}\|_2\, \|\nabla \mathbf{z}\|_2 \\ \le\;& \epsilon\, \|\nabla \mathbf{z}\|_2^2 + C(\epsilon)\, \|\mathbf{w}\|_2^2. \end{aligned} \tag{4.59}$$

Applying inequalities (4.24) and (4.44) and using the fact that $\|\nabla \mathbf{U}\|_3 < \infty$, we can estimate the second term on the right hand side of (4.58) as follows:

$$\begin{aligned} (B^2\mathbf{z}, \mathbf{z})_2 = (B_s^2\mathbf{z}, \mathbf{z})_2 \;&=\; \left(\int_{\Omega_R} + \int_{\Omega^R} \right) \mathbf{z} \cdot (\nabla \mathbf{U})_s \cdot \mathbf{z}\, \mathrm{d}\mathbf{x} \\ &\le\; \|\mathbf{z}\|_{2;\, \Omega_R}\, \|\nabla \mathbf{U}\|_{3;\, \Omega_R}\, \|\mathbf{z}\|_{6;\, \Omega_R} + \|\mathbf{z}\|_{6;\, \Omega^R}^2\, \|\nabla \mathbf{U}\|_{3/2;\, \Omega^R} \\ &\le\; \epsilon\, \|\nabla \mathbf{z}\|_2^2 + C(\epsilon)\, \|\mathbf{z}\|_{2;\, \Omega_R}^2 + c_2^2(2)\, \|\nabla \mathbf{z}\|_2^2\, \|\nabla \mathbf{U}\|_{3/2;\, \Omega^R} \end{aligned}$$

$$\leq\ \epsilon\,\|\nabla\mathbf{z}\|_2^2 + C(\epsilon)\,\|\mathbf{z}\|_{2;\,\Omega_R}^2 + \frac{\nu}{8}\,\|\nabla\mathbf{z}\|_2^2. \tag{4.60}$$

Finally, (4.28) yields

$$\big(P'N(\mathbf{v}),\mathbf{z}\big)_2\ \leq\ c_5\,\mathcal{Y}[\mathbf{v}]\,\|\mathbf{z}\|_2. \tag{4.61}$$

Estimating now the right hand side of (4.58) by means of (4.59)–(4.61), choosing $\epsilon > 0$ "sufficiently small", multiplying the whole inequality by 2, integrating with respect to time from 0 to t, and estimating the integral of $\|\mathbf{z}\|_{2;\,\Omega_R}^2$ by means of (4.55), we obtain (4.57).

Estimate 5 *There exist positive constants c_{23}, c_{24} and c_{25} such that*

$$\begin{aligned}\|\nabla\mathbf{z}(t)\|_2^2 + \nu\int_0^t \|A\mathbf{z}(s)\|_2^2\,\mathrm{d}s\ &\leq\ \|\nabla\mathbf{z}(0)\|_2^2 + c_{23}\int_0^t \|\nabla\mathbf{z}(s)\|_2^2\,\mathrm{d}s\\ &\quad + c_{24}\int_0^t \mathcal{Y}[\mathbf{w}(s)]\,\mathrm{d}s + c_{25}\int_0^t \|\nabla\mathbf{v}(s)\|_2^2\,\mathcal{Y}[\mathbf{v}(s)]\,\mathrm{d}s\end{aligned} \tag{4.62}$$

Proof Multiplying equation (4.50) by $(-A\mathbf{z})$, integrating over Ω and using (4.20), (4.26), (4.29) and (4.42), we get

$$\begin{aligned}&\frac{\mathrm{d}}{\mathrm{d}t}\,\frac{1}{2}\,\|\nabla\mathbf{z}\|_2^2\\ &\quad= -\nu\,\|A\mathbf{z}\|_2^2 - (B^2\mathbf{z},A\mathbf{z})_2 + \big(P'[\kappa B_s^2 - B^0 - B^1 - B_a^2]\mathbf{w},A\mathbf{z}\big)_2\\ &\qquad - \big(P'N(\mathbf{v}),A\mathbf{z}\big)_2\\ &\quad\leq -\nu\,\|A\mathbf{w}\|_2^2 + \|B^2\mathbf{z}\|_2\,\|A\mathbf{z}\|_2 + 4c_{11}\,\|\nabla\mathbf{w}\|_2\,\|A\mathbf{z}\|_2 + \|N(\mathbf{v})\|_2\,\|A\mathbf{z}\|_2\\ &\quad\leq -\frac{\nu}{2}\,\|A\mathbf{z}\|_2^2 + c_{23}\,\mathcal{Y}[\mathbf{z}]^2 + c_{24}\,\|\nabla\mathbf{w}\|_2^2 + c_{25}\,\|\nabla\mathbf{v}\|_2^2\,\mathcal{Y}[\mathbf{v}],\end{aligned} \tag{4.63}$$

where the constants c_{23}–c_{25} depend on ν. Now we obtain (4.62) by multiplying this inequality by 2 and integrating with respect to time from 0 to t.

Proof of Theorem 1. We can deduce from the inequality $\|\mathbf{v}_0\|_{1,2} < \delta$ and from [4] that there exists $T > 0$ such that the equation (4.21), with the initial condition $\mathbf{v}(0) = \mathbf{v}_0$, has a unique solution $\mathbf{v}$ on the interval $[0, T)$. Moreover, either $T = \infty$ or $\|\mathbf{v}(t)\|_{1,2} \to \infty$ for $t \to T-$. Assume, in the next considerations, that $t \in (0, T)$.

If we sum the estimate (4.52) with (4.53) multiplied by α, (4.57) multiplied by β and (4.62) multiplied by γ, we obtain

$$\begin{aligned}&\|\mathbf{w}(t)\|_2^2 + \alpha\,\|\nabla\mathbf{w}(t)\|_2^2 + \beta\,\|\mathbf{z}(t)\|_2^2 + \gamma\,\|\nabla\mathbf{z}(t)\|_2^2\\ &\quad + \int_0^t \Big(2c_7\,\|\nabla\mathbf{w}(s)\|_2^2 + \alpha\nu\|A\mathbf{w}(s)\|_2^2 + \beta\nu\|\nabla\mathbf{z}(s)\|_2^2 + \gamma\nu\|A\mathbf{z}(s)\|_2^2\Big)\,\mathrm{d}s\\ &\leq\ \|\mathbf{w}(0)\|_2^2 + \alpha\,\|\nabla\mathbf{w}(0)\|_2^2 + \beta c_{20}\,\|\mathbf{z}(0)\|_2^2 + \gamma\,\|\nabla\mathbf{z}(0)\|_2^2\end{aligned}$$

$$+ \int_0^t \Big[\alpha c_{14}\,\|\nabla \mathbf{w}(s)\|_2^2 + \big(\beta c_{26} + \gamma c_{24}\big)\, \mathcal{Y}[\mathbf{w}(s)] + \gamma c_{23}\,\|\nabla \mathbf{z}(s)\|_2^2\Big]\,\mathrm{d}s$$
$$+ \int_0^t \mathcal{Y}[\mathbf{v}(s)]\,\big(2\sqrt{c_5}\,\|\mathbf{w}(s)\|_2 + 2\beta c_8\,\|\mathbf{z}(s)\|_2\big)\,\mathrm{d}s$$
$$+ \beta c_{22}\left(\int_0^t \mathcal{Y}[\mathbf{v}(s)]\,\mathrm{d}s\right)^2. \tag{4.64}$$

Let us now successively choose the coefficients α, β and γ so that they satisfy the inequalities

$$\begin{aligned}
&\alpha: && \alpha \le 1, && \alpha c_{14} \le \tfrac{1}{2}c_7, && \\
&\beta: && \beta c_{20} \le 1, && \beta c_{26} \le \tfrac{3}{8}c_7 + \tfrac{1}{4}\alpha\nu, && \\
&\gamma: && \gamma \le 1, && \gamma c_{23} \le \tfrac{1}{2}\beta\nu, && \gamma c_{24} \le \tfrac{3}{8}\,c_7 + \tfrac{1}{4}\alpha\nu.
\end{aligned}$$

Then the terms on the right hand side of (4.64), quadratic in the norms of **w** and **z**, are absorbed by the left hand side, and (4.64) yields

$$\begin{aligned}
&\|\mathbf{w}(t)\|_2^2 + \alpha\,\|\nabla\mathbf{w}(t)\|_2^2 + \beta\,\|\mathbf{z}(t)\|_2^2 + \gamma\,\|\nabla\mathbf{z}(t)\|_2^2 \\
&+ \int_0^t \left(\frac{3c_7}{4}\,\|\nabla\mathbf{w}(s)\|_2^2 + \frac{\alpha\nu}{2}\,\|A\mathbf{w}(s)\|_2^2 + \frac{\beta\nu}{2}\,\|\nabla\mathbf{z}(s)\|_2^2 + \gamma\nu\,\|A\mathbf{z}(s)\|_2^2\right)\mathrm{d}s \\
&\le \|\mathbf{w}(0)\|_2^2 + \alpha\,\|\nabla\mathbf{w}(0)\|_2^2 + \beta c_{20}\,\|\mathbf{z}(0)\|_2^2 + \gamma\,\|\nabla\mathbf{z}(0)\|_2^2 \\
&+ \int_0^t \mathcal{Y}[\mathbf{u}(s)]\,\big(2\sqrt{c_5}\,\|\mathbf{w}(s)\|_2 + \beta c_8\,\|\mathbf{z}(s)\|_2\big)\,\mathrm{d}s \\
&+ \int_0^t (\alpha c_{15} + \gamma c_{25})\,\|\nabla\mathbf{v}(s)\|_2^2\,\mathcal{Y}[\mathbf{v}(s)]\,\mathrm{d}s + \beta c_{22}\left(\int_0^t \mathcal{Y}[\mathbf{v}(s)]\,\mathrm{d}s\right)^2.
\end{aligned} \tag{4.65}$$

Obviously, if we denote for simplicity by c_{27} one half of the minimum of $\frac{3}{4}c_7$, $\frac{1}{2}\alpha\nu$, $\frac{1}{2}\beta\nu$ and $\gamma\nu$, the integral on the left hand side is greater than or equal to $c_{27}\int_0^t \mathcal{Y}[\mathbf{v}(s)]\,\mathrm{d}s$. Denote

$$y(t) := \int_0^t \mathcal{Y}[\mathbf{v}(s)]\,\mathrm{d}s, \qquad z(t) := \|\mathbf{w}\|_2^2 + \alpha\,\|\nabla\mathbf{w}\|_2^2 + \beta\,\|\mathbf{z}\|_2^2 + \gamma\,\|\nabla\mathbf{z}\|_2^2.$$

Then (4.65) yields

$$\begin{aligned}
z(t) + c_{27}\,y(t) \le\ & z(0) + \int_0^t y'(s)\,\big[c_{28}\,z(s)^{1/2} + c_{29}\,z(s)\big]\,\mathrm{d}s \\
& + \beta c_{22}\,y^2(t).
\end{aligned} \tag{4.66}$$

Suppose that $z(0)$ is so small that $c_{28}\,z(0)^{1/2} + c_{29}\,z(0) < \frac{1}{4}c_{27}$. This inequality is equivalent to

$$z(0) \; < \; z_1 \; := \; \left(\frac{-c_{28} + \sqrt{c_{28}^2 + c_{27} c_{29}}}{2 c_{29}} \right)^2. \tag{4.67}$$

Denote by T' the maximum number in $[0, T]$ such that

$$c_{28}\, z(s)^{1/2} + c_{29}\, z(s) \; \le \; \frac{c_{27}}{2} \tag{4.68}$$

holds for all $s \in [0, T')$. Then (4.66) implies that

$$z(t) + \frac{c_{27}}{2}\, y(t) \; \le \; z(0) + \beta c_{22}\, y^2(t) \tag{4.69}$$

for $0 \le t < T'$. Consequently,

$$\beta c_{22}\, y^2(t) - \frac{c_{27}}{2}\, y(t) + z(0) \; \ge \; 0 \tag{4.70}$$

for $0 \le t < T'$. The discriminant of this quadratic inequality is $\frac{1}{4} c_{27}^2 - 4\beta c_{22}\, z(0)$. Assume, in addition to (4.67), that $z(0)$ is so small that

$$z(0) \; < \; \frac{{c_{27}}^2}{32 \beta c_{22}}. \tag{4.71}$$

Then the discriminant is greater than $\frac{1}{8} c_{27}^2$ and inequality (4.69) yields:

$$y(t) < \frac{c_{27} - \sqrt{c_{27}^2 - 16 \beta c_{22}\, z(0)}}{4 \beta c_{22}} \quad \text{or} \quad y(t) > \frac{c_{27} + \sqrt{c_{27}^2 - 16 \beta c_{22}\, z(0)}}{4 \beta c_{22}}. \tag{4.72}$$

If $y(0)$ satisfies the first of these two inequalities, i.e.

$$y(0) \; < \; \frac{c_{27} - \sqrt{c_{27}^2 - 16 \beta c_{22}\, z(0)}}{4 \beta c_{22}} \; = \; \frac{4 z(0)}{c_{27} + \sqrt{c_{27}^2 - 16 \beta c_{22}\, z(0)}}, \tag{4.73}$$

then, due to the continuity of $y(t)$,

$$y(t) < \frac{4 z(0)}{c_{27} + \sqrt{c_{27}^2 - 16 \beta c_{22}\, z(0)}} < \frac{4 z(0)}{c_{27} + \sqrt{\frac{1}{2} c_{27}^2}} = \frac{4 \sqrt{2}\, z(0)}{c_{27}\, (\sqrt{2} + 1)} \tag{4.74}$$

for all $0 \le t < T'$. We can now use this inequality in (4.69) in order to estimate $z(t)$:

$$z(t) \; \le \; z(0) + \left(\frac{4 \sqrt{2}}{c_{27}\, (\sqrt{2} + 1)} \right)^2 z^2(0). \tag{4.75}$$

Finally, if $z(0)$ is chosen to be less than or equal to one and so small that it also satisfies

$$z(0) + \left(\frac{4\sqrt{2}}{c_{27}\,(\sqrt{2}+1)} \right)^2 z^2(0) \;\le\; z(0) + \left(\frac{4\sqrt{2}}{c_{27}\,(\sqrt{2}+1)} \right)^2 z(0) \;<\; \frac{z_1}{2} \qquad (4.76)$$

where z_1 is defined in (4.67), then (4.75) yields $z(t) < \frac{1}{2}\, z_1$ for all $t \in [0, T')$. This implies that $T' = T$. (Otherwise, using the continuity of $z(t)$, we can derive a contradiction with the definition of T'.) Hence the estimates (4.74) and (4.75) hold for all $t \in [0, T)$. Summing (4.74) and (4.75), and taking into account that $z(0) \le 1$, we obtain

$$z(t) + y(t) \;\le\; \frac{4\sqrt{2}\, z(0)}{c_{27}\,(\sqrt{2}+1)} + z(0) + \left(\frac{4\sqrt{2}}{c_{27}\,(\sqrt{2}+1)} \right)^2 z(0) \qquad (4.77)$$

for $t \in [0, T)$.

We observe from this inequality and from the relation between $\|\mathbf{v}(t)\|_{1,2}$ and $z(t)$ that $\|\mathbf{v}(t)\|_{1,2}$ does not tend to infinity for $t \to T-$. Consequently, the solution $\mathbf{v}$ can be continued, as a solution of the equation (4.21), onto the whole time interval $[0, \infty)$ and the estimate (4.77) holds for all $t \in (0, \infty)$.

The positive number δ, used in (4.46), must be chosen so small that $\|\mathbf{v}(0)\|_{1,2} < \delta$ implies $z(0) \le 1$, $z(0) < \frac{1}{2}\, z_1$, $z(0)$ also satisfies (4.71), and $y(0)$ satisfies (4.73).

Summing the estimates (4.54) and (4.63), multiplied by 2, and applying the inequality (4.77), we deduce that

$$\frac{\mathrm{d}}{\mathrm{d}t}\, \|\nabla \mathbf{w}(t)\|_2^2 + \frac{\mathrm{d}}{\mathrm{d}t}\, \|\nabla \mathbf{z}(t)\|_2^2 + \nu\, \|A\mathbf{w}(t)\|_2^2 + \nu\, \|A\mathbf{z}(t)\|_2^2 \;\le\; C \;<\; \infty$$

for a.a. $t \in (0, \infty)$, where C is independent of t. This information, together with the information on the integrability of $\|\nabla \mathbf{w}(t)\|_2^2 + \|\nabla \mathbf{z}(t)\|_2^2$ on the interval $(0, \infty)$, implies (4.48). The proof of Theorem 1 is completed. □

Remark 1 Obviously, as $\mathrm{e}^{Lt}\phi$ is a function with values in $\mathbf{W}^{1,2}_{0,\sigma}(\Omega)$, the condition (4.45) in Assumption (A) is satisfied if there exists a function $\widetilde{\varphi} \in L^1(0, \infty) \cap L^2(0, \infty)$ such that

$$\left\| \nabla \mathrm{e}^{Lt} \phi \right\|_{2;\, \Omega_R} \;\le\; \widetilde{\varphi}(t)\, \|\phi\|_2$$

for all $\phi \in \mathbf{L}^2_\sigma(\Omega)'$ and $t > 0$.

Acknowledgements The authors thank the partial support of NSF grant DMS-1311983 (G.P. Galdi), and the Grant Agency of the Czech Republic, grant No. 13-00522S, and Academy of Sciences of the Czech Republic, RVO 67985840 (J. Neustupa). This work was also partially supported by the Department of Mechanical Engineering and Materials Science of the University of Pittsburgh, that hosted the visit of J. Neustupa in Spring 2015.

References

1. Babenko, K.I.: Spectrum of the linearized problem of flow of a viscous incompressible liquid round a body. Sov. Phys. Dokl. **27**, 25–27 (1982)
2. Borchers, W., Miyakawa, T.: L^2-decay for Navier-Stokes flows in unbounded domains, with application to exterior stationary flows. Arch. Rat. Mech. Anal. **118**, 273–295 (1992)
3. Borchers, W., Miyakawa, T.: On stability of exterior stationary Navier-Stokes flows. Acta Math. **174**, 311–382 (1995)
4. Cumsille, P., Tucsnak, M.: Wellpossedness for the Navier-Stokes flow in the exterior of a rotating obstacle. Math. Methods Appl. Sci **29**, 595–623 (2006)
5. Deuring, P., Neustupa, J.: An eigenvalue criterion for stability of a steady Navier-Stokes flow in $\mathbb{R}^3$. J. Math. Fluid Mech. **12**(2), 202–242 (2010)
6. Engel, K.J., Nagel, R.: One–Parameter semigroups for linear evolution equations. Graduate Texts in Mathematics, vol. 194. Springer (2000)
7. Farwig, R.: An L^q-analysis of viscous fluid flow past a rotating obstacle. Tohoku Math. J. **58**, 129–147 (2005)
8. Farwig, R.: Estimates of lower order derivatives of viscous fluid flow past a rotating obstacle. Banach Cent. Publ. Warsaw **70**, 73–84 (2005)
9. Farwig, R., Hishida, T., Müller, D.: L^q-Theory of a singular winding integral operator arising from fluid dynamics. Pacific J. Math. **215**, 297–312 (2004)
10. Farwig, R., Neustupa, J.: On the spectrum of a Stokes-type operator arising from flow around a rotating body. Manuscripta Math. **122**, 419–437 (2007)
11. Farwig, R., Neustupa, J.: On the spectrum of an Oseen-type operator arising from flow past a rotating body. Integr. Eqn. Oper. Theory **62**, 169–189 (2008)
12. Farwig, R., Neustupa, J.: On the spectrum of an Oseen-type operator arising from fluid flow past a rotating body in $L^q_\sigma(\Omega)$. Tohoku Math. J. **62**(2), 287–309 (2010)
13. Farwig, R., Nečasová, Š., Neustupa, J.: Spectral analysis of a Stokes-type operator arising from flow around a rotating body. J. Math. Soc. Jpn. **63**(1), 163–194 (2011)
14. Galdi, G.P.: An Introduction to the Mathematical Theory of the Navier–Stokes Equations. Steady–State Problems, 2nd edn. Springer (2011)
15. Galdi, G.P.: On the motion of a rigid body in a viscous liquid: a mathematical analysis with applications. In: Friedlander, S., Serre, D. (eds.) Handbook of Mathematical Fluid Dynamics, vol. 1. Elsevier (2002)
16. Galdi, G.P.: Steady flow of a Navier-Stokes fluid around a rotating obstacle. J. Elast. **71**, 1–31 (2003)
17. Galdi, G.P., Rionero, S.: Weighted energy methods in fluid dynamics and elasticity. Lecture Notes in Mathematics, vol. 1134. Springer, Berlin–Heidelberg–New York–Tokyo (1985)
18. Galdi, G.P., Padula, M.: A new approach to energy theory in the stability of fluid motion. Arch. Rat. Mech. Anal. **110**, 187–286 (1990)
19. Galdi, G.P., Heywood, J.G., Shibata, Y.: On the global existence and convergence to steady state of Navier-Stokes flow past an obstacle that is started from rest. Arch. Rat. Mech. Anal. **138**, 307–318 (1997)
20. Galdi, G.P., Silvestre, A.L.: Strong solutions to the Navier-Stokes equations around a rotating obstacle. Arch. Rat. Mech. Anal. **176**, 331–350 (2005)
21. Galdi, G.P., Silvestre, A.L.: The steady motion of a Navier-Stokes liquid around a rigid body. Arch. Rat. Mech. Anal. **184**, 371–400 (2007)
22. Geissert, M., Heck, H., Hieber, M.: L^p-theory of the Navier-Stokes flow in the exterior of a moving or rotating obstacle. J. Reine Angew. Math. **596**, 45–62 (2006)
23. Heywood, J.G.: On stationary solutions of the Navier-Stokes equations as limits of nonstationary solutions. Arch. Rat. Mech. Anal. **37**, 48–60 (1970)
24. Heywood, J.G.: The exterior nonstationary problem for the Navier-Stokes equations. Acta Math. **129**, 11–34 (1972)
25. Heywood, J.G.: The Navier-Stokes equations: On the existence, regularity and decay of solutions. Indiana Univ. Math. J. **29**, 639–681 (1980)

26. Hishida, T.: An existence theorem for the Navier-Stokes flow in the exterior of a rotating obstacle. Arch. Rat. Mech. Anal. **150**, 307–348 (1999)
27. Hishida, T.: L^q estimates of weak solutions to the stationary Stokes equations around a rotating body. J. Math. Soc. Jpn. **58**, 743–767 (2006)
28. Hishida, T., Shibata, Y.: $L_p - L_q$ estimate of the Stokes operator and Navier-Stokes flows in the exterior of a rotating obstacle. Arch. Rat. Mech. Anal. **193**, 339–421 (2009)
29. Kielhöfer, H.: Existenz und Regularität von Lösungen semilinearer parabolischer Anfangs-Randwertprobleme. Math. Z. **142**, 131–160 (1975)
30. Kielhöfer, H.: On the Lyapunov stability of stationary solutions of semilinear parabolic differential equations. J. Diff. Eqn. **22**, 193–208 (1976)
31. Kozono, H., Ogawa, T.: On stability of Navier-Stokes flows in exterior domains. Arch. Rat. Mech. Anal. **128**, 1–31 (1994)
32. Kozono, H., Yamazaki, M.: On stability of small stationary solutions in Morrey spaces of the Navier-Stokes equations. Indiana Univ. Math. J. **44**, 1307–1336 (1995)
33. Kozono, H., Yamazaki, M.: On a large class of stable solutions to the Navier-Stokes equations in exterior domains. Math. Z. **228**, 751–785 (1998)
34. Kobayashi, T., Shibata, Y.: On the Oseen equation in the three dimensional exterior domain. Math. Ann. **310**, 1–45 (1998)
35. Lieb, E.: The number of bound states of one–body Schrödinger operators and the Weyl problem. In: Proceedings of the AMS Symposia in Pure Mathematics, vol. 36, pp. 241–252 (1980)
36. Maremonti, P.: Asymptotic stability theorems for viscous fluid motions in exterior domains. Rend. Sem. Mat. Univ. Padova **71**, 35–72 (1984)
37. Masuda, K.: On the stability of incompressible viscous fluid motions past objects. J. Math. Soc. Jpn. **27**, 294–327 (1975)
38. Miyakawa, T.: On L^1-stability of stationary Navier-Stokes flows in $\mathbb{R}^3$. J. Math. Sci. Univ. Tokyo **4**, 67–119 (1997)
39. Neustupa, J.: Stabilizing influence of a skew-symmetric operator in semilinear parabolic equations. Rend. Mat. Sem. Univ. Padova **102**, 1–18 (1999)
40. Neustupa, J.: Stability of a steady solution of a semilinear parabolic system in an exterior domain. Far East J. Appl. Math. **15**, 307–324 (2004)
41. Neustupa, J.: Stability of a steady viscous incompressible flow past an obstacle. J. Math. Fluid Mech. **11**, 22–45 (2009)
42. Neustupa, J.: On L^2–boundedness of a C_0–semigroup generated by the perturbed Oseen–type operator arising from flow around a rotating body. Discr. Cont. Dyn. Syst.-S 758–767 (2007)
43. Sattinger, D.H.: The mathematical problem of hydrodynamic stability. J. Math. Mech. **18**, 797–817 (1970)
44. Sazonov, L.I.: Justification of the linearization method in the flow problem. Izv. Ross. Akad. Nauk Ser. Mat. **58**, 85–109 (1994). (Russian)
45. Shibata, Y.: On an exterior initial boundary value problem for Navier-Stokes equations. Q. Appl. Math. LVII **1**, 117–155 (1999)
46. Shibata, Y.: On the Oseen semigroup with rotating effect. In: Analysis, Functional, Equations, Evolution (eds.) The Günter Lumer Volume, pp. 595–611. Basel, Birkhauser Verlar (2009)

Chapter 5
Asymptotic Structure of Steady Stokes Flow Around a Rotating Obstacle in Two Dimensions

Toshiaki Hishida

Abstract This paper provides asymptotic structure at spatial infinity of plane steady Stokes flow in exterior domains when the obstacle is rotating with constant angular velocity. The result shows that there is no longer Stokes' paradox due to the rotating effect.

Keywords Plane Stokes flow · Rotating obstacle · Asymptotic representation · Stokes paradox

5.1 Introduction

Let Ω be an exterior domain in the plane $\mathbb{R}^2$ with smooth boundary $\partial\Omega$, and consider the motion of a viscous incompressible fluid around an obstacle (rigid body) $\mathbb{R}^2 \setminus \Omega$. As compared with 3D problem, we have less knowledge about exterior steady flows in 2D despite efforts of several authors mentioned below. The difficulty is to analyze the asymptotic behavior of the flow at infinity. This is related to the following hydrodynamical paradox found by Stokes [33]: The problem

$$-\Delta u + \nabla p = 0, \qquad \operatorname{div} u = 0 \qquad \text{in } \Omega, \tag{5.1}$$

$$u|_{\partial\Omega} = 0, \qquad u \to u_\infty \quad \text{as } |x| \to \infty \tag{5.2}$$

admits no solution unless $u_\infty = 0$, where $u(x) = (u_1, u_2)^T$ and $p(x)$ denote the velocity and pressure, respectively, of the fluid. Throughout this paper, all vectors are column ones and $(\cdot)^T$ denotes the transpose of vectors or matrices. Later on, Chang and Finn [6] made it clear that the Stokes paradox is interpreted in terms of the total

Dedicated to Professor Reinhard Farwig on his 60th birthday.

T. Hishida (✉)
Graduate School of Mathematics, Nagoya University, Nagoya 464-8602, Japan
e-mail: hishida@math.nagoya-u.ac.jp

Y. Shibata and Y. Suzuki (eds.), *Mathematical Fluid Dynamics, Present and Future*, Springer Proceedings in Mathematics & Statistics 183,
DOI 10.1007/978-4-431-56457-7_5

net force exerted by the fluid to the obstacle

$$N := \int_{\partial\Omega} T(u,p)\nu \, d\sigma, \tag{5.3}$$

where $T(u,p)$ is the Cauchy stress tensor given by

$$T(u,p) := \big(T_{jk}(u,p)\big) = Du - p\mathbb{I}, \quad Du := \nabla u + (\nabla u)^T, \quad \mathbb{I} = \big(\delta_{jk}\big), \tag{5.4}$$

and ν denotes the outward unit normal to $\partial\Omega$; in fact, they proved that the flow satisfying (5.1) can be bounded at infinity only if the net force (5.3) vanishes. This is an immediate consequence of asymptotic representation at infinity of solutions to (5.1) due to themselves [6], see (5.17). The original form of the Stokes paradox mentioned above follows from the result of Chang and Finn as a corollary because the net force (5.3) never vanishes provided that $\{u,p\}$ is nontrivial and satisfies (5.1) together with $u|_{\partial\Omega} = 0$. There are some other forms of the Stokes paradox, see Galdi [19, V.7], Kozono and Sohr [28, Theorem A].

For the case in which a constant velocity $u_\infty \in \mathbb{R}^2 \setminus \{0\}$ is prescribed at infinity or, equivalently, the obstacle is translating with velocity $-u_\infty$ (while the flow is at rest at infinity), Oseen [30] proposed his linearization of the Navier-Stokes system around u_∞ to get around the Stokes paradox. This works well because the fundamental solution of the Oseen operator $-\Delta u + u_\infty \cdot \nabla u + \nabla p$ possesses some decay structure with wake, while the Stokes fundamental solution grows logarithmically at infinity, see (5.18). Finn and Smith [13, 14], Smith [32] actually adopted the Oseen linearization to succeed in construction of the Navier-Stokes flow when u_∞ is not zero but small enough (and the external force is small, too, unless it is absent). Later on, Galdi [15] refined the result by means of L^q-estimates, see also [19, Sect. 12.5]. The similar existence theorem for the case $u_\infty = 0$ is still an open question even for small external force. Even before the results mentioned above, Leray [29] constructed at least one Navier-Stokes flow with finite Dirichlet integral without any smallness condition, however, the asymptotic behavior at infinity of his solution is still unclear and all the related results obtained so far are partial answers (Gilbarg and Weinberger [21, 22] and Amick [1, 2]). For details, see Galdi [16, 18, 19]. It should be noted that symmetry helps to attain the boundary condition $u \to 0$ at infinity, see [18, 31, 34] and the references therein. Among them, Yamazaki [34] employed a linearization method to construct a small Navier-Stokes flow decaying like $|x|^{-1}$ at infinity under a sort of symmetry; indeed, the symmetry he adopted enables us to avoid the Stokes paradox since the net force vanishes.

In this paper it is shown that, instead of the translation mentioned above, the rotation of the obstacle leads to the resolution of the Stokes paradox in the sense that: (i) The flow can be bounded (and even goes to a constant vector at the rate $|x|^{-1}$) at infinity even if the net force (5.3) does not vanish (Theorem 5.2.1); (ii) Given external force decaying sufficiently fast, there exists a linear flow which enjoys $u(x) = O(|x|^{-1})$ as $|x| \to \infty$ (Theorem 5.2.2). We also provide a remarkable asymptotic representation of the flow at infinity, in which the leading term involves the rotational profile $x^\perp/|x|^2$ whose coefficient is given by the torque, where $x^\perp = (-x_2, x_1)^T$, see (5.15) and (5.21). Here, the linear system arising from the flow around a rotating obstacle with constant angular velocity $a \in \mathbb{R} \setminus \{0\}$ is described as

$$-\Delta u - a\left(x^\perp \cdot \nabla u - u^\perp\right) + \nabla p = f, \qquad \operatorname{div} u = 0 \qquad \text{in } \Omega \tag{5.5}$$

in the reference frame attached to the obstacle. We recall the derivation of (5.5) in the next section.

The essential reason why there is no longer Stokes' paradox is asymptotic structure of the fundamental solution of the system (5.5) in the whole plane $\mathbb{R}^2$. Roughly speaking, the oscillation caused by rotation yields better decay structure of the fundamental solution, from which combined with some cut-off techniques we obtain the main results. It is worth while comparing with the result [9] by Farwig and the present author on the 3D Stokes flow around a rotating obstacle, in which the axis of rotation (e_3-axis without loss, where $e_3 = (0, 0, 1)^T$) plays an important role; in fact, $e_3 \cdot N$ controls the rate of decay. The result would suggest better decay studied in this paper since we do not have the axis of rotation in 2D, however, there are some difficulties compared with 3D case. Look at (5.28) below, which would be heuristically a fundamental solution, but this is by no means trivial because of lack of absolute convergence unlike 3D case. We thus employ the centering technique due to Guenther and Thomann [23], that is, we subtract the worst part, whose time-integral converges on account of oscillation, from the integrand of (5.28) such that the remaining part converges absolutely and can be treated in a usual way. This technique is also needed to justify some estimates of the fundamental solution, see Lemma 5.3.3. Asymptotic analysis of the fundamental solution to find the asymptotic representation (5.60) is similar to the argumant adopted for 3D [9], in which, however, the external force f is assumed to have a compact support. In this paper we will derive further properties of the fundamental solution and the corresponding volume potential (5.91) to deal with the external force decaying sufficiently fast for $|x| \to \infty$, see (5.11) and (5.14).

This paper is a step toward analysis of the Navier-Stokes flow around a rotating obstacle in the plane. To proceed to the nonlinear case, it is reasonable to consider the external force $f = \operatorname{div} F$ with $F(x) = O(|x|^{-2})$ in view of the nonlinear structure $u \cdot \nabla u = \operatorname{div}(u \otimes u)$, see Remark 5.3.2. This will be discussed in a forthcoming paper. As for asymptotic structure of the Navier-Stokes flow around a rotating obstacle in 3D, the leading term at infinity was found first by [10] and then the estimate of the remainder was refined by [8].

When the rotating obstacle is the two-dimensional disk and the external force is absent, the Navier-Stokes system subject to the no-slip boundary condition (5.24) admits an explicit solution (5.25) in the original frame, see Galdi [19, p. 302]. Recently, Hillairet and Wittwer [25] considered small perturbation from this solution with large angular velocity $|a|$ to find the Navier-Stokes flow decaying like $|y|^{-1}$ at infinity, whose leading profile is given by $y^\perp/|y|^2$. See also Guillod and Wittwer [24, Sect. 4], who provided among others numerical simulations of the related issue.

This paper is organized as follows. In the next section, after recalling the equations in the reference frame, we present the main theorems. Section 5.3 is essentially the central part of this paper and we carry out a detailed analysis of several asymptotic properties of the fundamental solution of (5.5) in the whole plane $\mathbb{R}^2$. Section 5.4 is devoted to decay structure of the system (5.5) to prove Theorem 5.2.1. In the final section we show the existence of a unique linear flow which goes to zero as $|x| \to \infty$ to prove Theorem 5.2.2.

5.2 Results

We begin with introducing notation. Set $B_\rho(x_0) = \{x \in \mathbb{R}^2;\ |x - x_0| < \rho\}$, where $x_0 \in \mathbb{R}^2$ and $\rho > 0$. Given exterior domain Ω with smooth boundary $\partial\Omega$, we fix $R \geqslant 1$ such that $\mathbb{R}^2 \setminus \Omega \subset B_R(0)$. For $\rho \geqslant R$ we set $\Omega_\rho = \Omega \cap B_\rho(0)$. Let D be one of Ω, $\mathbb{R}^2$ and Ω_ρ, and let $1 \leqslant q \leqslant \infty$. We denote by $L^q(D)$ the usual Lebesgue space with norm $\|\cdot\|_{L^q(D)}$. It is also convenient to introduce the weak-L^2 space $L^{2,\infty}(D)$ (one of the Lorentz spaces, see [3]) by $L^{2,\infty}(D) = \left(L^1(D), L^\infty(D)\right)_{1/2,\infty}$ with norm $\|\cdot\|_{L^{2,\infty}(D)}$, where $(\cdot,\cdot)_{1/2,\infty}$ stands for the real interpolation functor. The measurable function f belongs to $L^{2,\infty}(D)$ if and only if $\sup_{\tau>0} \tau |\{x \in D;\ |f(x)| > \tau\}|^{1/2} < \infty$, where $|\cdot|$ denotes the Lebesgue measure. Note that $L^2(D) \subset L^{2,\infty}(D)$; indeed, $|x|^{-1} \in L^{2,\infty}(D)$. By $H^k(D)$ and $H^1_0(D)$ we respectively denote the L^2-Sobolev space of k-th order ($k \geqslant 1$) with norm $\|\cdot\|_{H^k(D)}$ and the completion of $C_0^\infty(D)$ (consisting of smooth functions with compact support) in $H^1(D)$. We use the same symbol for denoting the spaces of scalar, vector and tensor valued functions.

Before stating our results, we briefly explain the derivation of the system (5.5) for the readers' convenience although that is the same as in 3D case [17, 26]. Suppose a compact obstacle (rigid body) $\mathbb{R}^2 \setminus \Omega$ is rotating about the origin in the plane with constant angular velocity $a \in \mathbb{R} \setminus \{0\}$, and let us start with the nonstationary Navier-Stokes system

$$\partial_t v + v \cdot \nabla v = \Delta v - \nabla q + g, \qquad \operatorname{div} v = 0$$

in the time-dependent exterior domain $\Omega(t) = \{y = O(at)x;\ x \in \Omega\}$ with

$$O(t) = \begin{pmatrix} \cos t & -\sin t \\ \sin t & \cos t \end{pmatrix},$$

where $v(y,t)$ and $q(y,t)$ are unknowns, while $g(y,t)$ is a given external force. The fluid velocity is assumed to attain the rotational velocity of the rigid body on the boundary $\partial\Omega(t)$ (no-slip condition), while it is at rest at infinity, that is,

$$v|_{\partial\Omega(t)} = ay^\perp, \qquad v \to 0 \quad \text{as } |y| \to \infty.$$

We take the frame attached to the obstacle by making change of variables

$$\begin{aligned} &y = O(at)x, \qquad u(x,t) = O(at)^T v(y,t), \qquad p(x,t) = q(y,t), \\ &f(x,t) = O(at)^T g(y,t), \end{aligned} \tag{5.6}$$

so that the equation of momentum is reduced to

$$\begin{aligned}\partial_t u &= O(at)^T \partial_t v + O(at)^T \left(a\,\dot{O}(at)x\right)\cdot \nabla_y v + a\,\dot{O}(at)^T v \\ &= O(at)^T(-v\cdot\nabla_y v + \Delta_y v - \nabla_y q + g) + a\left(x^\perp\cdot\nabla_x u - u^\perp\right) \\ &= -u\cdot\nabla_x u + \Delta_x u - \nabla_x p + f + a\left(x^\perp\cdot\nabla_x u - u^\perp\right)\end{aligned}$$

in Ω, where $\dot{O}(t) = \frac{d}{dt}O(t)$. If f is independent of t, then one can consider the steady problem

$$-\Delta u - a\left(x^\perp\cdot\nabla u - u^\perp\right) + \nabla p + u\cdot\nabla u = f, \qquad \operatorname{div} u = 0 \quad \text{in } \Omega \tag{5.7}$$

subject to

$$u|_{\partial\Omega} = ax^\perp, \qquad u\to 0 \ \text{ as } |x|\to\infty. \tag{5.8}$$

It is sometimes convenient to write the LHS of $(5.7)_1$ as divergence form

$$\begin{aligned}&\Delta u + a\left(x^\perp\cdot\nabla u - u^\perp\right) - \nabla p - u\cdot\nabla u \\ &= \operatorname{div}\left(S(u,p) - u\otimes u\right) = \left(\sum_{k=1}^{2}\partial_k\left\{S_{jk}(u,p) - u_j u_k\right\}\right)_{j=1,2}\end{aligned}$$

with

$$S(u,p) = \left(S_{jk}(u,p)\right) = T(u,p) + a\left(u\otimes x^\perp - x^\perp\otimes u\right) \tag{5.9}$$

where $T(u,p)$ is given by (5.4), $u\otimes v = (u_j v_k)$ stands for the matrix for given vector fields u and v, and $\partial_k = \partial_{x_k}$.

The only problem we intend to address in this paper is the associated linear system (5.5). On account of the relation

$$\int_\Omega \left[(x^\perp\cdot\nabla u - u^\perp)\cdot v + u\cdot(x^\perp\cdot\nabla v - v^\perp)\right]dx = \int_{\partial\Omega}(\nu\cdot x^\perp)(u\cdot v)\,d\sigma \tag{5.10}$$

for vector fields u and v (so long as the calculation (5.129) in Sect. 5.4 makes sense), the operator $u\mapsto x^\perp\cdot\nabla u - u^\perp$ is skew-symmetric under the homogeneous boundary condition. Also, by using the auxiliary function (5.130) below, our problem with boundary condition $(5.8)_1$ can be reduced to the problem with the homogeneous one. Hence, it is not hard to find at least one solution with $\nabla u\in L^2(\Omega)$ for (5.5) (even for the Navier-Stokes system (5.7) without restriction on the size of $|a|$) subject to the boundary condition $u|_{\partial\Omega} = ax^\perp$ (only $(5.8)_1$) along the same procedure as in Leray [29] provided $f = \operatorname{div} F$ with $F\in L^2(\Omega)$, however, we do not know whether the behavior $(5.8)_2$ at infinity is verified. The asymptotic structure of this solution for f decaying sufficiently fast at infinity and, more generally, that of $\{u,p\}$ satisfying (5.5) without assuming any boundary condition on $\partial\Omega$ are given by the following theorem. For simplicity we are concerned with smooth solutions although the result can be extended to less regular solutions (in view of Proposition 5.3.2 for the whole plane problem).

Theorem 5.2.1 *Let $a \in \mathbb{R} \setminus \{0\}$. Suppose that $\{u, p\} \in H^1_{loc}(\overline{\Omega}) \times L^2_{loc}(\overline{\Omega})$ is a smooth solution to the system (5.5) with $f \in C^\infty(\Omega)$ satisfying*

$$\int_\Omega |x||f(x)|\,dx < \infty, \qquad |f(x)| \leqslant \frac{C}{(1+|x|^3)\big(\log(e+|x|)\big)} \tag{5.11}$$

where the constant $C > 0$ is independent of $x \in \Omega$. Assume either

(i) $\nabla u \in L^r(\Omega \setminus B_R(0))$ *for some* $r \in (1, \infty)$

or

(ii) $u(x) = o(|x|)$ *as* $|x| \to \infty$.

Then there are constants $u_\infty \in \mathbb{R}^2$ and $p_\infty \in \mathbb{R}$ such that:

1. (asymptotic behavior)

$$\begin{cases} u(x) = u_\infty + (1+|a|^{-1})\,O(|x|^{-1}), \\ p(x) = -a\,u_\infty^\perp \cdot x + p_\infty + O(|x|^{-1}), \end{cases} \tag{5.12}$$

as $|x| \to \infty$.

2. (energy balance)

We have $\nabla u \in L^2(\Omega)$ (even if we do not assume (i) *with $r = 2$) and*

$$\frac{1}{2}\int_\Omega |Du|^2 dx = \int_{\partial\Omega} \left[\big(\widetilde{T}(u,p)\nu\big)\cdot(u-u_\infty) + \frac{a\,(\nu\cdot x^\perp)}{2}|u-u_\infty|^2\right] d\sigma + \int_\Omega f\cdot(u-u_\infty)\,dx \tag{5.13}$$

with $\widetilde{T}(u,p) := T(u,\ p + a\,u_\infty^\perp\cdot x - p_\infty)$, where Du and $T(\cdot,\cdot)$ are as in (5.4).

3. (asymptotic representation)

If in addition

$$f(x) = o(|x|^{-3}(\log|x|)^{-1}) \quad \text{as } |x| \to \infty, \tag{5.14}$$

then

$$u(x) - u_\infty = \frac{\alpha x^\perp - 2\beta x}{4\pi|x|^2} + (1+|a|^{-1})\,o(|x|^{-1}) \quad \text{as } |x| \to \infty, \tag{5.15}$$

where

$$\begin{aligned} \alpha &= \int_{\partial\Omega} y^\perp \cdot \big\{\big(T(u,p) + a\,u\otimes y^\perp\big)\,\nu\big\}\,d\sigma_y + \int_\Omega y^\perp\cdot f\,dy, \\ \beta &= \int_{\partial\Omega} \nu\cdot u\,d\sigma. \end{aligned} \tag{5.16}$$

If in particular $f \in C_0^\infty(\overline{\Omega})$, that is, the support of f is compact in $\mathbb{R}^2$, then the remainder decays like $O(|x|^{-2})$ in (5.15).

Note that $T(u,p)\nu$ belongs to $H^{-1/2}(\partial\Omega) := H^{1/2}(\partial\Omega)^*$ by the normal trace theorem since $T(u,p) \in L^2(\Omega_R)$ and $\operatorname{div} T(u,p) = -a\big(x^\perp \cdot \nabla u - u^\perp\big) - f \in L^2(\Omega_R)$. Therefore, the boundary integral $\int_{\partial\Omega} y^\perp \cdot \big(T(u,p)\nu\big)\, d\sigma_y$ can be understood as ${}_{H^{1/2}(\partial\Omega)}\langle y^\perp, T(u,p)\nu\rangle_{H^{-1/2}(\partial\Omega)}$ in (5.16). Since $u \in H^{1/2}(\partial\Omega)$, the same reasoning as above justifies $\int_{\partial\Omega} \big(\widetilde{T}(u,p)\nu\big) \cdot (u - u_\infty)\, d\sigma$ in (5.13). All the other integrals in (5.13) and (5.16) also make sense.

For the usual Stokes system (5.1), under the same growth condition on $u(x)$ as in Theorem 5.2.1, there is a constant $u_\infty \in \mathbb{R}^2$ such that (Chang and Finn [6, Theorem 1])

$$u(x) = u_\infty + E(x)N + O(|x|^{-1}) \tag{5.17}$$

as $|x| \to \infty$, where

$$E(x) = \frac{1}{4\pi}\left[\left(\log\frac{1}{|x|}\right)\mathbb{I} + \frac{x \otimes x}{|x|^2}\right] \tag{5.18}$$

is the Stokes fundamental solution and N denotes the net force (5.3). We observe the remarkable difference between $(5.12)_1$ and (5.17); in fact, the flow is bounded in Theorem 5.2.1 even though the net force N does not vanish. We would say that this is the resolution of the Stokes paradox.

The leading term of (5.15) is of interest since it contains the rotational profile $x^\perp/|x|^2$, which comes from the leading term of the fundamental solution of (5.5), see (5.60). The other profile $-x/(2\pi|x|^2)$ is called the flux carrier. If in particular the flux β at the boundary vanishes, then the leading term is purely rotational and that is just the case in the next theorem. Look at the coefficient (5.16) of $x^\perp/|x|^2$, where the integral $\int_{\partial\Omega} y^\perp \cdot \big(T(u,p)\nu\big)\, d\sigma_y$ stands for the torque exerted by the fluid to the obstacle. It is worth while noting that, in three dimensions, one finds the rotational profile $(e_3 \times x)/|x|^3$, whose coefficient involves the torque, in the second term after the leading one. For details, see Farwig and Hishida [9, Theorem 1.1]. It is reasonable that both $x^\perp/|x|^2 = \nabla^\perp \log|x|$ and $x/|x|^2 = \nabla \log|x|$ are solutions to (5.5) with $f = 0$ in $\mathbb{R}^2 \setminus \{0\}$ together with the constant pressure and, therefore, so is the leading term of (5.15). In fact, we observe

$$\Delta\frac{x^\perp}{|x|^2} = 0, \qquad x^\perp \cdot \nabla\frac{x^\perp}{|x|^2} = \frac{(x^\perp)^\perp}{|x|^2}, \qquad \operatorname{div}\frac{x^\perp}{|x|^2} = 0 \qquad \text{in } \mathbb{R}^2 \setminus \{0\}$$

as well as (5.118) (with $x_0 = 0$) below.

In Theorem 5.2.1 it is also possible to find the asymptotic representation of the pressure $p(x)$ without assuming any growth condition on $p(x)$ itself since it can be controlled by the growth of $u(x)$ via the Eq. $(5.5)_1$. The leading profile of $p(x) + a\,u_\infty^\perp \cdot x - p_\infty$ in $(5.12)_2$ is just the fundamental solution $Q(x) = \frac{x}{2\pi|x|^2}$ of the pressure to the Stokes system. This is because

$$\operatorname{div}\left(x^\perp\cdot\nabla u-u^\perp\right)=x^\perp\cdot\nabla\operatorname{div}u=0 \tag{5.19}$$

so that the pressure part of the fundamental solution is independent of $a\in\mathbb{R}$. Thus we are not interested in the asymptotic representation of the pressure, which the rotation of the obstacle does not affect so much. The coefficient of the leading profile $Q(x)$ is rather complicated in Theorem 5.2.1, but it becomes just the force in the next theorem, see (5.22).

The next question is whether one can actually construct a solution to (5.5) when zero velocity is prescribed at infinity as in (5.8). The following theorem gives an affirmative answer.

Theorem 5.2.2 *Let $a\in\mathbb{R}\setminus\{0\}$. Suppose that $f=\operatorname{div}F\in C^\infty(\Omega)$, with $F\in L^2(\Omega)$, satisfies* (5.11). *Then the system* (5.5) *subject to* (5.8) *admits a smooth solution $\{u,p\}$, which is of class $u\in L^{2,\infty}(\Omega)\cap H^1_{loc}(\overline{\Omega})$, $p\in L^2_{loc}(\overline{\Omega})$ as well as $\nabla u\in L^2(\Omega)$ and fulfills*

$$\begin{aligned}\|u\|_{L^{2,\infty}(\Omega)}\leqslant C\Big[1+|a|+(1+|a|^{-1})\Big(&\|F\|_{L^2(\Omega)}\\ &+\int_\Omega |x||f(x)|\,dx+\sup_{x\in\Omega}|x|^3\big(\log(e+|x|)\big)|f(x)|\Big)\Big],\end{aligned} \tag{5.20}$$

$$\|\nabla u\|_{L^2(\Omega)}\leqslant\|F\|_{L^2(\Omega)}+C|a|,$$

with some $C>0$ independent of f and $a\in\mathbb{R}\setminus\{0\}$, and $\{u,p\}$ enjoys all the assertions in Theorem 5.2.1 with $\{u_\infty,p_\infty\}=\{0,0\}$. In particular, we have

$$u(x)=\left(\int_{\partial\Omega}y^\perp\cdot\big(T(u,p)\nu\big)\,d\sigma_y+\int_\Omega y^\perp\cdot f\,dy\right)\frac{x^\perp}{4\pi|x|^2}+(1+|a|^{-1})\,o(|x|^{-1}), \tag{5.21}$$

$$p(x)=\left(\int_{\partial\Omega}T(u,p)\nu\,d\sigma+\int_\Omega f\,dy\right)\cdot\frac{x}{2\pi|x|^2}+O(|x|^{-2}), \tag{5.22}$$

as $|x|\to\infty$ under the additional condition (5.14) *(which is needed only for* (5.21)*). This is the only solution in the class $\nabla u\in L^2(\Omega)$, $\{u,p\}\in L^2_{loc}(\overline{\Omega})$ with $\{u,p\}\to\{0,0\}$ as $|x|\to\infty$.*

Note that, when $a=0$, the problem is not always solvable for given external force $f=\operatorname{div}F$ even if $F\in C_0^\infty(\overline{\Omega})$, that may be also regarded as the Stokes paradox. The L^∞-norm of $|x||u(x)|$ away from the boundary can be also estimated by the RHS of $(5.20)_1$ (see (5.135) for an approximate solution). In order to control the L^∞-norm of $u(x)$ near the boundary $\partial\Omega$, the class $H^1_{loc}(\overline{\Omega})$ is not enough. We put the term $x^\perp\cdot\nabla u-u^\perp$ in the RHS and use the regularity theory of the usual Stokes system up to the boundary to show that $u\in H^2_{loc}(\overline{\Omega})\subset L^\infty_{loc}(\overline{\Omega})$ together with a certain estimate, which enables us to obtain the similar estimate of $\sup_{x\in\Omega}(1+|x|)|u(x)|$ to $(5.20)_1$.

We conclude this section with the following exact solutions of both the Stokes and Navier-Stokes boundary value problems without external force. The Stokes flow (5.26) seems to be well known since it is found in some old literature. The Navier-Stokes flow (5.25) is found in the second edition of [19, p. 302] (I learned it from Professor Masao Yamazaki around 2008). Suppose the unit disk (rigid body) $\overline{B_1(0)}$ is rotating about the origin with constant angular velocity $a \in \mathbb{R} \setminus \{0\}$. Then the Navier-Stokes flow in the exterior $\Omega = \mathbb{R}^2 \setminus \overline{B_1(0)}$ obeys

$$-\Delta v + \nabla q + v \cdot \nabla v = 0, \qquad \operatorname{div} v = 0 \quad \text{in } \Omega \tag{5.23}$$

subject to

$$v|_{\partial\Omega} = ay^\perp, \qquad v \to 0 \ \text{ as } |y| \to \infty \tag{5.24}$$

and this problem has a solution

$$v(y) = \frac{ay^\perp}{|y|^2}, \qquad q(y) = \frac{-a^2}{2|y|^2} + \text{constant}. \tag{5.25}$$

Also, the associated Stokes problem

$$-\Delta v + \nabla q = 0, \qquad \operatorname{div} v = 0 \quad \text{in } \Omega,$$

subject to (5.24) admits a solution

$$v(y) = \frac{ay^\perp}{|y|^2}, \qquad q(y) = \text{constant}. \tag{5.26}$$

Note that the Stokes flow (5.26) does not contradict the Stokes paradox because $\int_{\partial\Omega} T(v,q)\nu\, d\sigma = 0$ due to symmetry. When the obstacle is a disk, we do not necessarily have to make the change of variables (5.6), nevertheless, we can do so and this case is not excluded in the present paper. Steady flows in the original frame correspond to time-periodic flows and are not steady in general in the reference frame via (5.6). But the Stokes flow (5.26) becomes the steady one $u(x) = ax^\perp/|x|^2$, $p(x) =$ constant in the reference frame as well and this may be regarded as a special case in Theorems 5.2.1 and 5.2.2 (when we take $p = 0$ in the latter theorem); indeed, one can verify

$$\int_{\partial\Omega} y^\perp \cdot \big(T(u,p)\nu\big)\, d\sigma_y = 4\pi a$$

in (5.21). Recently, Hillairet and Wittwer [25] proved that if the boundary value $v|_{\partial\Omega}$ is sufficiently close to $ay^\perp$ with $|a| > \sqrt{48}$ in a sense and $\int_{\partial\Omega} v \cdot \nu\, d\sigma = 0$, then the Navier-Stokes system (5.23) in the exterior $\Omega = \mathbb{R}^2 \setminus \overline{B_1(0)}$ of the disk subject to this boundary condition admits at least one smooth solution, which decays like $|y|^{-1}$ as $|y| \to \infty$. The leading profile of their solution is given by $y^\perp/|y|^2$ with some coefficient close to a.

5.3 Fundamental Solution

In this section we derive the decay structure of the fundamental solution of the linear system (5.5) in the whole plane $\mathbb{R}^2$ when $a \in \mathbb{R} \setminus \{0\}$. Because of (5.19) the pressure part of the fundamental solution is

$$Q(x-y) = \frac{x-y}{2\pi |x-y|^2}, \tag{5.27}$$

while the velocity part is given by

$$\Gamma_a(x,y) = \int_0^\infty O(at)^T \, K(O(at)x - y, t)\, dt, \tag{5.28}$$

where

$$K(x,t) = G(x,t)\mathbb{I} + H(x,t)$$

is the fundamental solution of unsteady Stokes system ($a = 0$), and it consists of the 2D heat kernel

$$G(x,t) = \frac{1}{4\pi t} e^{-|x|^2/4t}$$

and 2×2 matrix

$$H(x,t) = \int_t^\infty \nabla^2 G(x,s)\, ds = \int_t^\infty G(x,s) \left(\frac{x \otimes x}{4s^2} - \frac{\mathbb{I}}{2s} \right) ds. \tag{5.29}$$

In 2D case one can write (5.29) in terms of elementary functions

$$H(x,t) = \frac{-(x \otimes x)}{|x|^2} G(x,t) + \left(\frac{x \otimes x}{|x|^2} - \frac{\mathbb{I}}{2} \right) \frac{1 - e^{-|x|^2/4t}}{\pi |x|^2}, \tag{5.30}$$

while one cannot in 3D, see [9]. One needs more careful argument than 3D case [9] to prove that (5.28) is actually the fundamental solution, see Proposition 5.3.2.

Indeed the integral representation (5.28) does not absolutely converge, but it is convergent due to oscillation $O(at)^T$ with $a \in \mathbb{R} \setminus \{0\}$, see Lemma 5.3.2. This is a contrast to the case $a = 0$, in which (5.28) is not convergent. In this case one needs the centering technique to recover the convergence, which leads to the Stokes fundamental solution $E(x)$ given by (5.18) as follows:

$$\int_0^\infty \left(K(x,t) - \frac{e^{-e/4t}}{8\pi t} \mathbb{I} \right) dt = E(x). \tag{5.31}$$

This was clarified by Guenther and Thomann [23, Proposition 2.2]. As a part of this technique (5.31), the fundamental solution of the Laplace operator in two dimensions is recovered exactly as

$$\int_0^\infty \left(G(x,t) - \frac{e^{-1/4t}}{4\pi t} \right) dt = \frac{1}{2\pi} \log \frac{1}{|x|} \tag{5.32}$$

in terms of the heat kernel, see [23, Lemma 2.1]. Although we do not need the centering technique in the representation (5.28) itself, we will use this technique to justify some formulae in this section.

Remark 5.3.1 In [11, p. 301] Farwig, Hishida and Müller mentioned that the integral kernel $\int_0^\infty O(t)^T G(O(t)x - y, t)\, dt$ should be modified to recover the convergence in two dimensions. But this is redundant as we will see in Lemma 5.3.2 by making use of the oscillation.

For convenience we will collect a few elementary foumulae, which will be used several times. We omit the proof that is nothing but integration by parts. In the first assertion below it is possible to derive even faster decay $r^{-\{2(m-1)+2k\}}$ for every $k \in \mathbb{N}$ by k-times integration by parts, but (5.33) and (5.34) are enough for later use. Note that they are not absolutely convergent for $m \leqslant 1$ (the only case we need is $m = 1$).

Lemma 5.3.1 *Let $r > 0$.*

1. *Let $a \in \mathbb{R} \setminus \{0\}$ and $m > 0$. Then*

$$\left| \int_0^\infty e^{iat} e^{-r^2/t} \frac{dt}{t^m} \right| \leqslant \frac{C}{|a| r^{2m}}, \tag{5.33}$$

$$\left| \int_0^\infty e^{iat} \int_t^\infty e^{-r^2/s} \frac{ds}{s^{m+1}}\, dt \right| \leqslant \frac{C}{|a| r^{2m}}, \tag{5.34}$$

with some $C = C(m) > 0$ independent of $r > 0$ and $a \in \mathbb{R} \setminus \{0\}$, where $i = \sqrt{-1}$.

2. *Let $m > 1$. Then*

$$\int_0^\infty e^{-r^2/t} \frac{dt}{t^m} = \frac{\gamma(m-1)}{r^{2(m-1)}}, \tag{5.35}$$

$$\int_0^\infty \int_t^\infty e^{-r^2/s} \frac{ds}{s^{m+1}}\, dt = \frac{\gamma(m-1)}{r^{2(m-1)}}, \tag{5.36}$$

where $\gamma(\cdot)$ denotes the Euler gamma function.

We begin with the following lemma, from which the function (5.28) is well-defined.

Lemma 5.3.2 *Let $a \in \mathbb{R} \setminus \{0\}$. Then the integral $\Gamma_a(x,y)$ given by (5.28) converges for every $(x,y) \in \mathbb{R}^2 \times \mathbb{R}^2$ with $x \neq y$.*

Proof We decompose $\Gamma_a(x,y)$ as

$$\Gamma_a(x,y) = \Gamma_a^0(x,y) + \Gamma_a^1(x,y), \qquad \Gamma_a^1(x,y) = \Gamma_a^{11}(x,y) + \Gamma_a^{12}(x,y)$$

with

$$\begin{aligned}
\Gamma_a^0(x,y) &= \int_0^\infty O(at)^T G(O(at)x-y,t)\,dt,\\
\Gamma_a^{11}(x,y) &= \int_0^\infty O(at)^T \int_t^\infty G(O(at)x-y,s)\,\frac{(O(at)x-y)\otimes(O(at)x-y)}{4s^2}\,ds\,dt,\\
\Gamma_a^{12}(x,y) &= \int_0^\infty O(at)^T \int_t^\infty G(O(at)x-y,s)\,\frac{-1}{2s}\,ds\,dt.
\end{aligned} \tag{5.37}$$

We start with the convergence of $\Gamma_a^0(x,y)$ by using the centering technique as in (5.32). By (5.33) we know that

$$\left|\int_0^\infty O(at)^T e^{-1/4t}\,\frac{dt}{t}\right| \leqslant \frac{C}{|a|}. \tag{5.38}$$

Hence, it suffices to show the convergence of

$$\int_0^\infty O(at)^T \left(e^{-|O(at)x-y|^2/4t} - e^{-1/4t}\right)\frac{dt}{t}. \tag{5.39}$$

As we will see, this is absolutely convergent. For large t, we have

$$\begin{aligned}
\int_1^\infty \left|e^{-|O(at)x-y|^2/4t} - e^{-1/4t}\right|\frac{dt}{t} &\leqslant \int_1^\infty \left||O(at)x-y|^2-1\right|\frac{dt}{4t^2}\\
&\leqslant \frac{(|x|+|y|)^2+1}{4}.
\end{aligned}$$

For small t, we use the relation

$$|O(at)x-y|^2 = |x-y|^2 + 2at\big(\dot{O}(a\theta t)x\big)\cdot(O(a\theta t)x-y) \tag{5.40}$$

for some $\theta = \theta(a,t,x,y) \in (0,1)$, where $\dot{O}(t) = \frac{d}{dt}O(t)$. Then we have

$$\int_0^1 \left|e^{-|O(at)x-y|^2/4t} - e^{-1/4t}\right|\frac{dt}{t} \leqslant \int_0^1 \left(e^{-|x-y|^2/4t}\,e^{|a||x|(|x|+|y|)/2} + e^{-1/4t}\right)\frac{dt}{t} \tag{5.41}$$

with

$$\begin{aligned}\int_0^1 e^{-|x-y|^2/4t}\,\frac{dt}{t} &= \left(\int_0^1 + \int_1^{1/|x-y|^2}\right) e^{-1/4t}\,\frac{dt}{t} \\ &\leqslant 4 + \int_1^{1/|x-y|^2} \frac{dt}{t} = 4 + 2\log\frac{1}{|x-y|} \qquad (0 < |x-y| < 1),\end{aligned} \tag{5.42}$$

while

$$\int_0^1 e^{-|x-y|^2/4t}\,\frac{dt}{t} \leqslant 4 \qquad (|x-y| \geqslant 1).$$

This concludes that (5.39) is absolutely convergent.

The next integral $\Gamma_a^{11}(x, y)$ is absolutely convergent without centering as above. Given (x, y) with $x \neq y$, there is $\delta = \delta(a, x, y) > 0$ such that

$$0 < \frac{|x-y|^2}{2} \leqslant |O(at)x - y|^2 \leqslant \frac{3|x-y|^2}{2}, \qquad 0 \leqslant \forall t \leqslant \delta, \tag{5.43}$$

on account of $\lim_{t\to 0} |O(at)x - y|^2 = |x-y|^2$. This together with (5.36) implies that

$$\begin{aligned}&\int_0^\infty \int_t^\infty e^{-|O(at)x-y|^2/4s}\,\frac{|O(at)x-y|^2}{s^3}\,ds\,dt \\ &\leqslant \int_0^\delta \int_t^\infty e^{-|x-y|^2/8s}\,\frac{3|x-y|^2}{2s^3}\,ds\,dt + \int_\delta^\infty \int_t^\infty \frac{(|x|+|y|)^2}{s^3}\,ds\,dt \\ &\leqslant \frac{3|x-y|^2}{2} \int_0^\infty \int_t^\infty e^{-|x-y|^2/8s}\,\frac{ds}{s^3}\,dt + \frac{(|x|+|y|)^2}{2} \int_\delta^\infty \frac{dt}{t^2} \\ &= C + \frac{(|x|+|y|)^2}{2\delta}.\end{aligned} \tag{5.44}$$

Finally, similarly to the argument of convergence of $\Gamma_a^0(x, y)$, we can discuss $\Gamma_a^{12}(x, y)$. From (5.34) it follows that

$$\left| \int_0^\infty O(at)^T \int_t^\infty e^{-1/4s}\,\frac{ds}{s^2}\,dt \right| \leqslant \frac{C}{|a|}. \tag{5.45}$$

It thus remains to show the convergence of

$$\int_0^\infty O(at)^T \int_t^\infty \left(e^{-|O(at)x-y|^2/4s} - e^{-1/4s}\right) \frac{ds}{s^2}\,dt. \tag{5.46}$$

For large t, we have

$$\int_1^\infty\int_t^\infty \left|e^{-|O(at)x-y|^2/4s}-e^{-1/4s}\right|\frac{ds}{s^2}\,dt \leqslant \int_1^\infty\int_t^\infty \big||O(at)x-y|^2-1\big|\frac{ds}{4s^3}\,dt$$
$$\leqslant \frac{(|x|+|y|)^2+1}{8}.$$

For small t, as in (5.41), we use (5.40) to find

$$\begin{aligned}&\int_0^1\int_t^\infty \left|e^{-|O(at)x-y|^2/4s}-e^{-1/4s}\right|\frac{ds}{s^2}\,dt\\ &\leqslant \int_0^1\int_t^\infty \left(e^{-|x-y|^2/4s}\,e^{|a||x|t(|x|+|y|)/2s}+e^{-1/4s}\right)\frac{ds}{s^2}\,dt\\ &\leqslant e^{|a||x|(|x|+|y|)/2}\int_0^1\int_t^\infty e^{-|x-y|^2/4s}\frac{ds}{s^2}\,dt+4\end{aligned} \tag{5.47}$$

with

$$\begin{aligned}&\int_0^1\int_t^\infty e^{-|x-y|^2/4s}\frac{ds}{s^2}\,dt=\left(\int_0^1+\int_1^{1/|x-y|^2}\right)\int_t^\infty e^{-1/4s}\frac{ds}{s^2}\,dt\\ &\leqslant 4+4\int_1^{1/|x-y|^2}(1-e^{-1/4s})\,ds\leqslant 4+\int_1^{1/|x-y|^2}\frac{ds}{s}\\ &=4+2\log\frac{1}{|x-y|}\qquad (0<|x-y|<1),\end{aligned} \tag{5.48}$$

while

$$\int_0^1\int_t^\infty e^{-|x-y|^2/4s}\frac{ds}{s^2}\,dt\leqslant 4\qquad (|x-y|\geqslant 1).$$

This implies the absolute convergence of (5.46). We have completed the proof. □

We have concentrated ourselves only on the convergence of the integral (5.28). So the estimates appeared in the proof above are not related to the asymptotic behavior with respect to (x, y) at large distance, which will be discussed in a different way in Proposition 5.3.1, but we have tried to derive the singular behavior for $|x-y|\to 0$ as less as possible, see (5.42), (5.44) and (5.48). This behavior should be logarithmic, otherwise (5.28) cannot be the fundamental solution, but the behavior (5.44) is not clear since δ depends on x, y (probably, the part $\Gamma_a^{11}(x, y)$ would be bounded for $|x-y|\to 0$ as in the second term of the Stokes fundamental solution (5.18)). In order to ensure that the volume potential (5.91) below is well-defined, we will show the following lemma. The growth rate (5.49) with $\rho = 2|x|$ will be also used to show asymptotic representation (5.96) below.

Lemma 5.3.3 *Let $a \in \mathbb{R} \setminus \{0\}$ There is a constane $C > 0$ independent of $a \in \mathbb{R} \setminus \{0\}$ such that*

$$\int_{|y| \leqslant \rho} |\Gamma_a(x, y)|\, dy \leqslant C|a|^{-1}\rho^2 + C\rho^2 \log \rho, \tag{5.49}$$

$$\int_{|y| \leqslant \rho} |\nabla_x \Gamma_a(x, y)|\, dy \leqslant C\rho, \tag{5.50}$$

for every $x \in \mathbb{R}^2$ and $\rho \geqslant |x| + e$.

Proof To this end, it is convenient to use another representation (5.30) of $H(x, t)$ together with the centering technique (5.31) due to Guenther and Thomann [23]. But we subtract $\left(e^{-1/4t}/8\pi t\right)\mathbb{I}$ instead of $\left(e^{-e/4t}/8\pi t\right)\mathbb{I}$ since there is no need to derive $E(x)$. First of all, it follows from (5.33) that

$$\left|\int_0^\infty O(at)^T \frac{e^{-1/4t}}{8\pi t}\, dt\right| \leqslant \frac{C}{|a|}. \tag{5.51}$$

We set

$$\widetilde{\Gamma}_a(x, y) := \int_0^\infty O(at)^T \left(K(O(at)x - y, t) - \frac{e^{-1/4t}}{8\pi t}\mathbb{I}\right) dt. \tag{5.52}$$

We will see that the integral of this integrand over $(0, \infty) \times \overline{B_\rho(0)}$ with respect to (t, y) is absolutely convergent. By the transformation $y = O(at)z$ we have

$$\begin{aligned} &\int_0^\infty \int_{|y| \leqslant \rho} \left|K(O(at)x - y, t) - \frac{e^{-1/4t}}{8\pi t}\mathbb{I}\right| dy\, dt \\ &= \int_0^\infty \int_{|z| \leqslant \rho} \left|K(x - z, t) - \frac{e^{-1/4t}}{8\pi t}\mathbb{I}\right| dz\, dt. \end{aligned} \tag{5.53}$$

The useful decomposition discovered by [23] is

$$\begin{aligned} &K(x, t) - \frac{e^{-1/4t}}{8\pi t}\mathbb{I} \\ &= \frac{e^{-|x|^2/4t} - e^{-1/4t}}{8\pi t}\mathbb{I} + \left(\frac{e^{-|x|^2/4t}}{8\pi t} - \frac{1 - e^{-|x|^2/4t}}{2\pi |x|^2}\right)\left(\mathbb{I} - \frac{2x \otimes x}{|x|^2}\right) =: \frac{A + B}{8\pi}. \end{aligned} \tag{5.54}$$

Then we find

$$\int_0^\infty |A|\, dt = C\left|\log |x|\right|, \tag{5.55}$$

while we see from the transformation $\tau = |x|^2/4t$ that

$$\int_0^\infty |B|\,dt = \left|\mathbb{I} - \frac{2x\otimes x}{|x|^2}\right| \int_0^\infty \frac{-\tau e^{-\tau} + 1 - e^{-\tau}}{\tau^2}\,d\tau = \left|\mathbb{I} - \frac{2x\otimes x}{|x|^2}\right| \leqslant C \tag{5.56}$$

since

$$0 < \frac{-\tau e^{-\tau} + 1 - e^{-\tau}}{\tau^2} = \frac{d}{d\tau}\left(\frac{e^{-\tau} - 1}{\tau}\right)$$

for every $\tau > 0$. By the Fubini theorem we obtain

$$\begin{aligned}\int_{|y|\leqslant\rho} |\widetilde{\Gamma}_a(x,y)|\,dy &\leqslant C\int_{|y|\leqslant\rho} \big(1 + \big|\log|x-y|\big|\big)\,dy \\ &\leqslant C\rho^2 + C\int_{|y-x|\leqslant\rho+|x|} \big|\log|x-y|\big|\,dy \\ &\leqslant C\rho^2 + C\rho^2\log\rho\end{aligned}$$

for $\rho \geqslant |x| + e$. This together with (5.51) concludes (5.49).

For the estimate of $\nabla_x\Gamma_a(x,y)$, we first need to justify

$$\nabla_x\Gamma_a(x,y) = \int_0^\infty O(at)^T\nabla_x\big[K(O(at)x - y, t)\big]\,dt \tag{5.57}$$

with the aid of $\widetilde{\Gamma}_a(x,y)$ given by (5.52). Given $\varphi \in C_0^\infty(\mathbb{R}^2)$ arbitrarily, we have

$$\langle\widetilde{\Gamma}_a(\cdot,y), \operatorname{div}\varphi\rangle = \int_0^\infty \left\langle O(at)^T\left(K(O(at)x - y, t) - \frac{e^{-1/4t}}{8\pi t}\mathbb{I}\right), \operatorname{div}\varphi\right\rangle dt$$

because this integral over $(0,\infty)\times B_L(0)$ with respect to (t,x) is absolutely convergent by the same reasoning as in the proof of (5.49), where $L > 0$ is taken in such a way that $\operatorname{Supp}\varphi \subset B_L(0)$. We then use

$$|(\nabla K)(x,t)| \leqslant Ct^{-3/2}e^{-|x|^2/16t} + C\int_t^\infty s^{-5/2}e^{-|x|^2/16s}\,ds \tag{5.58}$$

together with (5.35)–(5.36) to get the absolute convergence

$$\begin{aligned}\int_0^\infty\int_{|x|\leqslant L} \big|\nabla_x\big[K(O(at)x - y, t)\big]\big|\,dx\,dt &\leqslant C\int_0^\infty\int_{|x|\leqslant L} |(\nabla K)(O(at)x - y, t)|\,dx\,dt \\ &= C\int_0^\infty\int_{|x|\leqslant L} |(\nabla K)(x - y, t)|\,dx\,dt \\ &\leqslant C\int_{|x|\leqslant L}\frac{dx}{|x-y|}\end{aligned}$$

as in (5.53). Hence we obtain

$$\begin{aligned}\langle \widetilde{\Gamma}_a(\cdot, y), \operatorname{div}\varphi\rangle &= -\int_0^\infty \big\langle O(at)^T\nabla_x\big[K(O(at)x-y,t)\big], \varphi\big\rangle\, dt\\ &= -\left\langle \int_0^\infty O(at)^T\nabla_x\big[K(O(at)x-y,t)\big]\, dt, \varphi\right\rangle\end{aligned}$$

for all $\varphi \in C_0^\infty(\mathbb{R}^2)$, which implies (5.57) since $\nabla_x\widetilde{\Gamma}_a(x,y) = \nabla_x\Gamma_a(x,y)$. Once we have that, by the same reasoning as above we get

$$\int_{|y|\leqslant\rho} |\nabla_x\Gamma_a(x,y)|\, dy \leqslant C\int_{|y|\leqslant\rho}\int_0^\infty |(\nabla K)(O(at)x-y,t)|\, dt\, dy \leqslant C\int_{|y|\leqslant\rho}\frac{dy}{|x-y|}$$

which leads to (5.50) for $\rho \geqslant |x| + e$. □

The following estimate provides the decay structure of $\Gamma_a(x,y)$ and plays a crucial role in this paper.

Proposition 5.3.1 *Let $a \in \mathbb{R}\setminus\{0\}$.*

1. *There is a constant $C > 0$ independent of $a \in \mathbb{R}\setminus\{0\}$ such that*

$$\left|\Gamma_a(x,y) - \frac{x^\perp\otimes y^\perp}{4\pi|x|^2}\right| \leqslant \frac{C(|a|^{-1}+|y|^2)}{|x|^2} \tag{5.59}$$

for all $(x,y)\in\mathbb{R}^2\times\mathbb{R}^2$ with $|x| > 2|y|$. In particular, we have

$$\Gamma_a(x,y) = \frac{x^\perp\otimes y^\perp}{4\pi|x|^2} + O(|x|^{-2}), \tag{5.60}$$

as $|x|\to\infty$ so long as $|y|\leqslant\rho$, where $\rho > 0$ is fixed.

2. *Similarly, there is a constant $C > 0$ independent of $a \in \mathbb{R}\setminus\{0\}$ such that*

$$\left|\Gamma_a(x,y) - \frac{x^\perp\otimes y^\perp}{4\pi|y|^2}\right| \leqslant \frac{C(|a|^{-1}+|x|^2)}{|y|^2} \tag{5.61}$$

for all $(x,y)\in\mathbb{R}^2\times\mathbb{R}^2$ with $|y| > 2|x|$.

Proof Tha latter assertion follows from the former one because $\Gamma_a(x,y) = \Gamma_{-a}(y,x)^T$ and $(y^\perp\otimes x^\perp)^T = x^\perp\otimes y^\perp$. We will show (5.59), which immediately implies (5.60). Let us start with $\Gamma_a^0(x,y)$ given by (5.37). We use the Taylor formula with respect to y around $y = 0$ to see that there is $\theta = \theta(a,t,x,y) \in (0,1)$ satisfying

$$
\begin{aligned}
&e^{-|O(at)x-y|^2/4t} \\
&= e^{-|x|^2/4t} + e^{-|x|^2/4t}\,\frac{(O(at)x)\cdot y}{2t} \\
&\quad + \frac{1}{2}\,e^{-|O(at)x-\theta y|^2/4t}\,y^T\,\frac{(O(at)x-\theta y)\otimes(O(at)x-\theta y)-2t\,\mathbb{I}}{4t^2}\,y.
\end{aligned}
\tag{5.62}
$$

According to this formula, we decompose $\Gamma_a^0(x,y)$ as

$$
\Gamma_a^0(x,y) = \Gamma_a^{01}(x) + \Gamma_a^{02}(x,y) + \Gamma_a^{03}(x,y).
$$

It follows from (5.33) that

$$
|\Gamma_a^{01}(x)| = \left|\frac{1}{4\pi}\int_0^\infty O(at)^T e^{-|x|^2/4t}\,\frac{dt}{t}\right| \leqslant \frac{C}{|a||x|^2}. \tag{5.63}
$$

Since

$$
(O(at)x)\cdot y = (x\cdot y)\cos at + (x^\perp\cdot y)\sin at \tag{5.64}
$$

and, thereby,

$$
\begin{aligned}
\{(O(at)x)\cdot y\}\,O(at)^T &= \frac{1}{2}\begin{pmatrix} x\cdot y & x^\perp\cdot y \\ -x^\perp\cdot y & x\cdot y\end{pmatrix} + \frac{\cos 2at}{2}\begin{pmatrix} x\cdot y & -x^\perp\cdot y \\ x^\perp\cdot y & x\cdot y\end{pmatrix} \\
&\quad + \frac{\sin 2at}{2}\begin{pmatrix} x^\perp\cdot y & x\cdot y \\ -x\cdot y & x^\perp\cdot y\end{pmatrix},
\end{aligned}
\tag{5.65}
$$

we have

$$
\begin{aligned}
\Gamma_a^{02}(x,y) &= \frac{1}{16\pi}\int_0^\infty e^{-|x|^2/4t}\,\frac{dt}{t^2}\begin{pmatrix} x\cdot y & x^\perp\cdot y \\ -x^\perp\cdot y & x\cdot y\end{pmatrix} + M_a^{02}(x,y) \\
&= \frac{1}{4\pi|x|^2}\begin{pmatrix} x\cdot y & x^\perp\cdot y \\ -x^\perp\cdot y & x\cdot y\end{pmatrix} + M_a^{02}(x,y)
\end{aligned}
\tag{5.66}
$$

with

$$
|M_a^{02}(x,y)| \leqslant \frac{C|y|}{|a||x|^3} \leqslant \frac{C}{|a||x|^2} \tag{5.67}
$$

for $|x| > 2|y|$, which follows from (5.33). Since $e^{-|O(at)x-\theta y|^2/4t} \leqslant e^{-|x|^2/16t}$ for $|y| < |x|/2$, it is easily seen that

$$
|\Gamma_a^{03}(x,y)| \leqslant C|y|^2\int_0^\infty \left(|x|^2 t^{-3} + t^{-2}\right) e^{-|x|^2/16t}\,dt = \frac{C|y|^2}{|x|^2} \tag{5.68}
$$

without using oscillation. Then (5.63), (5.66), (5.67) and (5.68) imply that

$$\left| \Gamma_a^0(x,y) - \frac{1}{4\pi |x|^2} \begin{pmatrix} x\cdot y & x^\perp \cdot y \\ -x^\perp \cdot y & x\cdot y \end{pmatrix} \right| \leqslant C(|a|^{-1} + |y|^2)|x|^{-2} \tag{5.69}$$

for $|x| > 2|y|$.

We proceed to the decay structure of $\Gamma_a^1(x,y)$ given by (5.37). Similarly to (5.62), we have the formula

$$\begin{aligned} &e^{-|O(at)x-y|^2/4s} \\ &= e^{-|x|^2/4s} + e^{-|x|^2/4s}\,\frac{(O(at)x)\cdot y}{2s} \\ &\quad + \frac{1}{2}\,e^{-|O(at)x-\theta y|^2/4s}\, y^T\,\frac{(O(at)x-\theta y)\otimes(O(at)x-\theta y) - 2s\,\mathbb{I}}{4s^2}\, y \end{aligned} \tag{5.70}$$

with some $\theta = \theta(a,t,s,x,y) \in (0,1)$ and, correspondingly, we decompose $\Gamma_a^{11}(x,y)$ given by (5.37) as

$$\Gamma_a^{11}(x,y) = \Gamma_a^{111}(x,y) + \Gamma_a^{112}(x,y) + \Gamma_a^{113}(x,y).$$

We write

$$\begin{aligned} &O(at)^T[(O(at)x-y)\otimes(O(at)x-y)] \\ &= (x - O(at)^T y)\otimes(O(at)x-y) \\ &= A_0 + (\cos at)\,A_c + (\sin at)\,A_s + \frac{\cos 2at}{2}\,\widetilde{A}_c + \frac{\sin 2at}{2}\,\widetilde{A}_s \end{aligned} \tag{5.71}$$

with

$$\begin{aligned} A_0 &= A_0(x,y) = \frac{-3(x\otimes y) + (x^\perp \otimes y^\perp)}{2}, \\ A_c &= A_c(x,y) = \begin{pmatrix} x_1^2 + y_1^2 & x_1x_2 + y_1y_2 \\ x_1x_2 + y_1y_2 & x_2^2 + y_2^2 \end{pmatrix}, \\ A_s &= A_s(x,y) = \begin{pmatrix} -x_1x_2 + y_1y_2 & x_1^2 + y_2^2 \\ -(x_2^2 + y_1^2) & x_1x_2 - y_1y_2 \end{pmatrix}, \\ \widetilde{A}_c &= \widetilde{A}_c(x,y) = \begin{pmatrix} -x\cdot y & x^\perp\cdot y \\ -x^\perp\cdot y & -x\cdot y \end{pmatrix}, \\ \widetilde{A}_s &= \widetilde{A}_s(x,y) = \begin{pmatrix} -x^\perp\cdot y & -x\cdot y \\ x\cdot y & -x^\perp\cdot y \end{pmatrix}. \end{aligned}$$

Using (5.34) and (5.36), we get

$$\begin{aligned} \Gamma_a^{111}(x,y) &= \frac{A_0}{16\pi}\int_0^\infty\int_t^\infty e^{-|x|^2/4s}\,\frac{ds}{s^3}\,dt + M_a^{111}(x,y) \\ &= \frac{-3(x\otimes y) + (x^\perp\otimes y^\perp)}{8\pi |x|^2} + M_a^{111}(x,y) \end{aligned} \tag{5.72}$$

with

$$|M_a^{111}(x,y)| \leqslant \frac{C}{|a||x|^2} \tag{5.73}$$

for $|x| > 2|y|$. Look at (5.64) and (5.71) to obtain

$$\{(O(at)x)\cdot y\}O(at)^T[(O(at)x-y)\otimes(O(at)x-y)] = B_0 + (\text{remainder})$$

with

$$B_0 = \frac{x\cdot y}{2}A_c + \frac{x^\perp\cdot y}{2}A_s = \frac{x\cdot y}{2}(x\otimes x) + \frac{x^\perp\cdot y}{2}(x\otimes x^\perp) + B_1 = \frac{|x|^2(x\otimes y)}{2} + B_1$$

which is independent of t, where B_1 is of degree one (resp. three) with respect to x (resp. y) and the remainder consists of all terms involving $\cos kat$ and $\sin kat$ $(1 \leqslant k \leqslant 3)$. We thus find from (5.34) and (5.36) that

$$\begin{aligned} \Gamma_a^{112}(x,y) &= \frac{|x|^2(x\otimes y)}{64\pi}\int_0^\infty\int_t^\infty e^{-|x|^2/4s}\frac{ds}{s^4}\,dt + M_a^{112}(x,y) \\ &= \frac{x\otimes y}{4\pi|x|^2} + M_a^{112}(x,y) \end{aligned} \tag{5.74}$$

with

$$|M_a^{112}(x,y)| \leqslant \frac{C\left(|a|^{-1}|y| + |y|^3\right)}{|x|^3} \leqslant \frac{C\left(|a|^{-1} + |y|^2\right)}{|x|^2} \tag{5.75}$$

for $|x| > 2|y|$. Without using oscillation, we see that

$$|\Gamma_a^{113}(x,y)| \leqslant C|y|^2|x|^2\int_0^\infty\int_t^\infty \left(|x|^2s^{-5} + s^{-4}\right)e^{-|x|^2/16s}\,ds\,dt = \frac{C|y|^2}{|x|^2} \tag{5.76}$$

for $|x| > 2|y|$. We collect (5.72), (5.73), (5.74), (5.75) and (5.76) to find

$$\left|\Gamma_a^{11}(x,y) - \frac{-(x\otimes y) + (x^\perp\otimes y^\perp)}{8\pi|x|^2}\right| \leqslant C(|a|^{-1} + |y|^2)|x|^{-2} \tag{5.77}$$

for $|x| > 2|y|$.

Finally, we decompose $\Gamma_a^{12}(x,y)$ given by (5.37) as

$$\Gamma_a^{12}(x,y) = \Gamma_a^{121}(x) + \Gamma_a^{122}(x,y) + \Gamma_a^{123}(x,y)$$

by use of (5.70) and deduce its decay structure. By (5.34) we have

$$|\Gamma_a^{121}(x)| \leqslant \frac{C}{|a||x|^2}. \tag{5.78}$$

As in the argument for $\Gamma_a^{02}(x, y)$, we employ (5.65) to obtain

$$\begin{aligned}\Gamma_a^{122}(x, y) &= \frac{-1}{32\pi}\int_0^\infty \int_t^\infty e^{-|x|^2/4s}\frac{ds}{s^3}\,dt\begin{pmatrix} x\cdot y & x^\perp\cdot y \\ -x^\perp\cdot y & x\cdot y\end{pmatrix} + M_a^{122}(x, y) \\ &= \frac{-1}{8\pi|x|^2}\begin{pmatrix} x\cdot y & x^\perp\cdot y \\ -x^\perp\cdot y & x\cdot y\end{pmatrix} + M_a^{122}(x, y)\end{aligned} \tag{5.79}$$

with

$$|M_a^{122}(x, y)| \leqslant \frac{C|y|}{|a||x|^3} \leqslant \frac{C}{|a||x|^2} \tag{5.80}$$

for $|x| > 2|y|$. Similarly to the argument for $\Gamma_a^{113}(x, y)$, it is seen that

$$|\Gamma_a^{123}(x, y)| \leqslant C|y|^2\int_0^\infty \int_t^\infty \left(|x|^2 s^{-4} + s^{-3}\right) e^{-|x|^2/16s}\,ds\,dt = \frac{C|y|^2}{|x|^2} \tag{5.81}$$

for $|x| > 2|y|$. We collect (5.78), (5.79), (5.80) and (5.81) to obtain

$$\left|\Gamma_a^{12}(x, y) - \frac{-1}{8\pi|x|^2}\begin{pmatrix} x\cdot y & x^\perp\cdot y \\ -x^\perp\cdot y & x\cdot y\end{pmatrix}\right| \leqslant C(|a|^{-1} + |y|^2)|x|^{-2} \tag{5.82}$$

for $|x| > 2|y|$. Using the simple relation

$$\begin{pmatrix} x\cdot y & x^\perp\cdot y \\ -x^\perp\cdot y & x\cdot y\end{pmatrix} = x\otimes y + x^\perp \otimes y^\perp,$$

we gather (5.69), (5.77) and (5.82) to conclude (5.60). The proof is complete. □

We next verify that (5.28) can be actually the fundamental solution. To this end, we need two lemmas.

Lemma 5.3.4 *Let $f \in L^1(\mathbb{R}^2)\cap L^\infty(\mathbb{R}^2)$ and*

$$p(x) = \int_{\mathbb{R}^2} Q(x-y)\cdot f(y)\,dy, \tag{5.83}$$

where $Q(x)$ is given by (5.27). Set

$$v^0(x, t) = O(at)^T\int_{\mathbb{R}^2} G(O(at)x - y, t) f(y)\,dy, \tag{5.84}$$

$$v^1(x, t) = O(at)^T\int_{\mathbb{R}^2} H(O(at)x - y, t) f(y)\,dy, \tag{5.85}$$

where $H(x, t)$ is given by (5.29). Then they respectively satisfy

$$\partial_t v^0 + L_a v^0 = 0, \qquad v^0(\cdot, 0) = f, \tag{5.86}$$

$$\partial_t v^1 + L_a v^1 = 0, \qquad v^1(\cdot, 0) = -\nabla p, \tag{5.87}$$

in $\mathbb{R}^2 \times (0, \infty)$, *where*

$$L_a v := -\Delta v - a\left(x^\perp \cdot \nabla v - v^\perp\right). \tag{5.88}$$

Proof The well-known estimate of singular integrals yields $\nabla p \in L^q(\mathbb{R}^2)$ for every $q \in (1, \infty)$. By the derivation (5.6) of the equation (5.7), it is obvious that $v^0(x, t)$ is a solution to the Cauchy problem (5.86), where the initial condition is understood as $\lim_{t\to 0} \|v^0(t) - f\|_{L^q(\mathbb{R}^2)} = 0$ for every $q \in (1, \infty)$. By the same reasoning, $v^1(x, t) = (v_1^1, v_2^1)^T$ with

$$\begin{aligned} v_j^1(x, t) &= -\sum_k O(at)_{kj} \int_{\mathbb{R}^2} G(O(at)x - y, t)\partial_k p(y)\, dy \\ &= -\int_{\mathbb{R}^2} \partial_{x_j} G(O(at)x - y, t) \int_{\mathbb{R}^2} Q(y - z) \cdot f(z)\, dz\, dy \qquad (j = 1, 2) \end{aligned}$$

solves (5.87). Note that the integration by parts above can be justified since $p \in L^r(\mathbb{R}^2)$ for every $r \in (2, \infty)$ by the Hardy-Littlewood-Sobolev inequality. So we have only to deduce the representation (5.85). Using the relation

$$Q(y) = \frac{y}{2\pi |y|^2} = -\int_0^\infty \nabla G(y, \tau)\, d\tau$$

and the semigroup property of the heat kernel, we find

$$\begin{aligned} v_j^1(x, t) &= \int_{\mathbb{R}^2} \sum_m \int_0^\infty \int_{\mathbb{R}^2} \partial_{x_j} G(O(at)x - y, t)(\partial_m G)(y - z, \tau)\, dy\, d\tau\, f_m(z)\, dz \\ &= -\int_{\mathbb{R}^2} \sum_m \int_0^\infty \partial_{x_j} \partial_{z_m} G(O(at)x - z, t + \tau)\, d\tau\, f_m(z)\, dz \\ &= \int_{\mathbb{R}^2} \sum_{k,m} O(at)_{kj} \int_t^\infty (\partial_k \partial_m G)(O(at)x - z, s)\, ds\, f_m(z)\, dz \end{aligned}$$

which leads us to (5.85). □

Lemma 5.3.5 *Let* $\varepsilon \geqslant 0$. *Let* $U \in \mathscr{S}'(\mathbb{R}^2)$ *fulfill*

$$\varepsilon U - \Delta U - a x^\perp \cdot \nabla U = 0 \quad \text{in } \mathbb{R}^2,$$

where $\mathscr{S}'$ *is the class of tempered distributions. Then Supp* $\widehat{U} \subset \{0\}$, *where* $\widehat{U}$ *denotes the Fourier transform of* U. *Similarly, if* $u \in \mathscr{S}'(\mathbb{R}^2)$ *and* $p \in \mathscr{S}'(\mathbb{R}^2)$ *satisfy*

$$\varepsilon u - \Delta u - a\big(x^\perp \cdot \nabla u - u^\perp\big) + \nabla p = 0, \qquad div\, u = 0 \quad in\, \mathbb{R}^2, \tag{5.89}$$

then $Supp\, \widehat{u} \subset \{0\}$ *and* $Supp\, \widehat{p} \subset \{0\}$.

Proof We will prove the second assertion along the same idea as in [11], [27, Lemma 4.2] (in which the first assertion was shown for the case $\varepsilon = 0$). By (5.19) we have $\Delta p = 0$, so that $\mathrm{Supp}\, \widehat{p} \subset \{0\}$ is obvious. We take the Fourier transform of $(5.89)_1$ to find

$$(\varepsilon + |\xi|^2)\, \widehat{u} - a\, \big(\xi^\perp \cdot \nabla_\xi \widehat{u} - \widehat{u}^\perp\big) + i\xi \widehat{p} = 0.$$

Given $\psi \in C_0^\infty(\mathbb{R}^2 \setminus \{0\})$ arbitrarily, we set

$$\phi(\xi) = \int_0^\infty O(at)\, e^{-(\varepsilon+|\xi|^2)t}\, \psi(O(at)^T \xi)\, dt \in C_0^\infty(\mathbb{R}^2 \setminus \{0\}),$$

which solves

$$(\varepsilon + |\xi|^2)\, \phi + a\, \big(\xi^\perp \cdot \nabla_\xi \phi - \phi^\perp\big) = \psi.$$

We thus obtain

$$\begin{aligned} \langle \widehat{u}, \psi \rangle &= \big\langle \widehat{u}, (\varepsilon + |\xi|^2)\, \phi + a\, \big(\xi^\perp \cdot \nabla_\xi \phi - \phi^\perp\big) \big\rangle \\ &= \big\langle (\varepsilon + |\xi|^2)\, \widehat{u} - a\, \big(\xi^\perp \cdot \nabla_\xi \widehat{u} - \widehat{u}^\perp\big), \phi \big\rangle \\ &= -\langle i\xi \widehat{p}, \phi \rangle = 0, \end{aligned}$$

which completes the proof. □

The following volume potential (5.91) is well-defined on account of (5.49) and provides a solution to (5.97) for every $f \in C_0^\infty(\mathbb{R}^2)$; that is, $\Gamma_a(x, y)$ is a fundamental solution. We also deduce several properties of (5.91) for later use, including asymptotic representation (5.96) even for less regular f, whose support is not necessarily compact but which decays sufficiently fast at infinity.

Proposition 5.3.2 *Let* $a \in \mathbb{R} \setminus \{0\}$. *Suppose*

$$f \in L^1(\mathbb{R}^2) \cap L^\infty(\mathbb{R}^2). \tag{5.90}$$

Set

$$u(x) = \int_{\mathbb{R}^2} \Gamma_a(x, y) f(y)\, dy, \tag{5.91}$$

where $\Gamma_a(x, y)$ *is given by* (5.28), *and consider* $p(x)$ *defined by* (5.83) *as well.*

1. *The function* $u(x)$ *is well-defined by* (5.91) *as an element of* $L^\infty_{loc}(\mathbb{R}^2) \cap \mathscr{S}'(\mathbb{R}^2)$.
2. *Suppose further that*

$$\int_{\mathbb{R}^2} |x||f(x)|\,dx < \infty, \qquad f(x) = O(|x|^{-3}(\log|x|)^{-1}) \ \ as\ |x| \to \infty. \tag{5.92}$$

Then the functions $u(x)$ and $p(x)$ enjoy

$$|u(x)| + |\nabla u(x)| + |p(x)| = O(|x|^{-1}) \ \ as\ |x| \to \infty \tag{5.93}$$

with estimate

$$\sup_{|x|\geqslant\rho} |x||u(x)| \leqslant C(1+|a|^{-1})\left[\int_{\mathbb{R}^2}(1+|x|)|f(x)|\,dx + \sup_{|x|\geqslant\rho/2}|x|^3(\log|x|)|f(x)|\right] \tag{5.94}$$

for every $\rho \geqslant e$, where the constant $C > 0$ is independent of $\rho \in [e, \infty)$ and $a \in \mathbb{R}\setminus\{0\}$. Furthermore, we have

$$p(x) = \int_{\mathbb{R}^2} f\,dy \cdot \frac{x}{2\pi|x|^2} + O(|x|^{-2}) \ \ as\ |x| \to \infty. \tag{5.95}$$

3. *In addition to* (5.90) *and* (5.92)*, assume* (5.14)*. Then we have*

$$u(x) = \int_{\mathbb{R}^2} y^\perp \cdot f\,dy\, \frac{x^\perp}{4\pi|x|^2} + (1+|a|^{-1})\,o(|x|^{-1}) \ \ as\ |x| \to \infty. \tag{5.96}$$

If in particular the support of f is compact, then the remainder decays like $O(|x|^{-2})$ in (5.96).

4. *Under the conditions* (5.90) *and* (5.92)*, the pair $\{u, p\}$ satisfies*

$$-\Delta u - a\left(x^\perp \cdot \nabla u - u^\perp\right) + \nabla p = f, \qquad div\, u = 0 \qquad in\ \mathbb{R}^2 \tag{5.97}$$

in the sense of distributions as well as

$$(\nabla^2 u,\ \nabla p,\ x^\perp \cdot \nabla u - u^\perp) \in L^q(\mathbb{R}^2) \quad for\ \forall\, q \in (1, \infty), \tag{5.98}$$

$$x^\perp \cdot \nabla u \in L^r(\mathbb{R}^2) \quad for\ \forall\, r \in (2, \infty). \tag{5.99}$$

If in addition $f \in C^\infty(\mathbb{R}^2)$, then we have $\{u, p\} \in C^\infty(\mathbb{R}^2)$.

Remark 5.3.2 It is also possible to deduce $\nabla u(x) = O(|x|^{-2})$ at infinity by use of similar estimates of $\nabla_x \Gamma_a(x, y)$, see (5.57), to Proposition 5.3.1 (such estimates of $\nabla_x \Gamma_a(x, y)$ are not simple consequences of Proposition 5.3.1 and one needs further several pages). Since slower decay $\nabla u(x) = O(|x|^{-1})$ in (5.93) is enough for the proof of Theorem 5.2.1, we postpone precise analysis of $\nabla_x \Gamma_a(x, y)$ until a forthcoming paper, in which the external force $f = \operatorname{div} F$ with $F(x) = O(|x|^{-2})$ will be treated by using estimates of $\nabla_y \Gamma_a(x, y)$.

Proof of Proposition 5.3.2. Let $|x| \geqslant e$, then we take $\rho = 2|x|$ in (5.49) to obtain

$$\int_{|y|\leqslant 2|x|} |\Gamma_a(x,y)||f(y)|\,dy \leqslant C(1+|a|^{-1})\|f\|_{L^\infty(\mathbb{R}^2)}|x|^2\log|x| \qquad (|x|\geqslant e).$$

By (5.61) we also have

$$\begin{aligned}\int_{|y|>2|x|} |\Gamma_a(x,y)||f(y)|\,dy &\leqslant C\int_{|y|>2|x|}\left(\frac{|x|}{|y|}+\frac{1}{|a||y|^2}\right)|f(y)|\,dy\\ &\leqslant C(1+|a|^{-1})\|f\|_{L^1(\mathbb{R}^2)} \qquad (|x|\geqslant e).\end{aligned}$$

When $|x| < e$, we similarly use (5.49) with $\rho = 2e$ and (5.61) to find

$$|u(x)| \leqslant \int_{|y|\leqslant 2e} + \int_{|y|>2e} \leqslant C(1+|a|^{-1})\big(\|f\|_{L^\infty(\mathbb{R}^2)} + \|f\|_{L^1(\mathbb{R}^2)}\big) \qquad (|x|<e). \tag{5.100}$$

We thus obtain $u \in L^\infty_{loc}(\mathbb{R}^2) \cap \mathscr{S}'(\mathbb{R}^2)$.

We next divide (5.91) into three parts:

$$\begin{aligned}u(x) &= U_1(x) + U_2(x) + U_3(x)\\ &:= \left(\int_{|y|<|x|/2} + \int_{|x|/2\leqslant|y|\leqslant 2|x|} + \int_{|y|>2|x|}\right)\Gamma_a(x,y)f(y)\,dy.\end{aligned}$$

By (5.59) and (5.92) we have

$$U_1(x) = \frac{x^\perp}{4\pi|x|^2}\int_{|y|<|x|/2} y^\perp\cdot f(y)\,dy + W(x) \tag{5.101}$$

with

$$\begin{aligned}|W(x)| &\leqslant C|a|^{-1}|x|^{-2}\int_{|y|<|x|/2}|f(y)|\,dy + C|x|^{-2}\int_{|y|<|x|/2}|y|^2|f(y)|\,dy\\ &\leqslant C|a|^{-1}|x|^{-2}\|f\|_{L^1(\mathbb{R}^2)} + C|x|^{-1}\int_{\mathbb{R}^2}|y||f(y)|\,dy.\end{aligned} \tag{5.102}$$

The second term of the first line of (5.102) can be estimated even by

$$C|x|^{-2}\int_0^{|x|/2}\big(\log(e+r)\big)^{-1}\,dr = o(|x|^{-1}) \quad \text{as } |x|\to\infty.$$

Note that this holds true under weaker assumption $f(x) = o(|x|^{-3})$ than $(5.92)_2$. This together with

$$\left|\frac{x^\perp}{4\pi|x|^2}\int_{|y|\geqslant|x|/2} y^\perp\cdot f(y)\,dy\right| \leqslant \frac{C}{|x|}\int_{|y|\geqslant|x|/2} |y||f(y)|\,dy = o(|x|^{-1})$$

implies that

$$U_1(x) = \frac{x^\perp}{4\pi|x|^2}\int_{\mathbb{R}^2} y^\perp\cdot f(y)\,dy + o(|x|^{-1}) \quad \text{as } |x|\to\infty. \tag{5.103}$$

Let $|x| \geqslant e$, then it follows from (5.49) with $\rho = 2|x|$ and (5.92) that

$$\begin{aligned}
|U_2(x)| &\leqslant \int_{|x|/2\leqslant|y|\leqslant 2|x|} |\Gamma_a(x,y)||f(y)|\,dy \\
&\leqslant C|x|^{-3}\left(\log\frac{|x|}{2}\right)^{-1}\int_{|y|\leqslant 2|x|}|\Gamma_a(x,y)|\,dy \sup_{|y|\geqslant|x|/2}|y|^3(\log|y|)|f(y)| \\
&\leqslant C(1+|a|^{-1})|x|^{-1}\sup_{|y|\geqslant|x|/2}|y|^3(\log|y|)|f(y)| \qquad (|x|\geqslant e).
\end{aligned} \tag{5.104}$$

Under stronger assumption (5.14), we see that $U_2(x) = o(|x|^{-1})$ as $|x|\to\infty$. We remark that (5.14) is needed only here. We use (5.61) to find

$$\begin{aligned}
|U_3(x)| &\leqslant C\int_{|y|>2|x|}\left(\frac{|x|}{|y|}+\frac{1}{|a||y|^2}\right)|f(y)|\,dy \\
&\leqslant C(|x|^{-1}+|a|^{-1}|x|^{-3})\int_{|y|>2|x|}|y||f(y)|\,dy = o(|x|^{-1})
\end{aligned} \tag{5.105}$$

as $|x|\to\infty$. We gather (5.101), (5.102), (5.104) and (5.105) to conclude (5.94) for every $\rho\geqslant e$. Then (5.94) with $\rho = e$ together with (5.100) for $|x| < e$ yields

$$\begin{aligned}
&\sup_{x\in\mathbb{R}^2}(1+|x|)|u(x)| \\
&\leqslant C(1+|a|^{-1})\left[\int_{\mathbb{R}^2}(1+|x|)|f(x)|\,dx + \sup_{x\in\mathbb{R}^2}(1+|x|^3)\big(\log(e+|x|)\big)|f(x)|\right].
\end{aligned} \tag{5.106}$$

Furthermore, we collect (5.103), (5.104) and (5.105) to find the asymptotic representation (5.96) as long as (5.14) is additionally imposed. If in particular $\operatorname{Supp} f\subset B_\rho(0)$ for some $\rho > 0$, then $u(x) = U_1(x)$ for $|x|\geqslant 2\rho$. In view of the first line of (5.102), we have

$$|W(x)| \leqslant C(|a|^{-1}+\rho^2)|x|^{-2}\int_{|y|<\rho}|f(y)|\,dy = O(|x|^{-2}) \quad \text{as } |x|\to\infty.$$

To show the decay of $\nabla u(x)$, consider

$$\begin{aligned} V(x) &:= \int_{\mathbb{R}^2} \nabla_x \Gamma_a(x,y) f(y)\, dy \\ &= \int_{|y|<|x|/2} + \int_{|x|/2 \leqslant |y| \leqslant 2|x|} + \int_{|y|>2|x|} =: V_1(x) + V_2(x) + V_3(x). \end{aligned}$$

Neglecting the oscillation and using (5.57)–(5.58) together with (5.35)–(5.36), we deduce

$$|\nabla_x \Gamma_a(x,y)| \leqslant \begin{cases} C|x|^{-1}, & |x| > 2|y|, \\ C|y|^{-1}, & |y| > 2|x|. \end{cases} \tag{5.107}$$

Although they are not sharp (Remark 5.3.2), they respectively yield

$$|V_1(x)| \leqslant C|x|^{-1} \|f\|_{L^1(\mathbb{R}^2)}$$

and

$$|V_3(x)| \leqslant C \int_{|y|>2|x|} |y|^{-1}|f(y)|\, dy \leqslant C|x|^{-2} \int_{|y|>2|x|} |y||f(y)|\, dy.$$

Let $|x| \geqslant e$ and use (5.50) with $\rho = 2|x|$ to find

$$|V_2(x)| \leqslant C|x|^{-3}\Big(\log \frac{|x|}{2}\Big)^{-1} \int_{|y|\leqslant 2|x|} |\nabla_x \Gamma(x,y)|\, dy \leqslant C|x|^{-2}\Big(\log \frac{|x|}{2}\Big)^{-1}.$$

We thus obtain

$$|V(x)| \leqslant \frac{C}{|x|} \qquad (|x| \geqslant e).$$

In order to conclude $\nabla u(x) = O(|x|^{-1})$ as $|x| \to \infty$, it suffices to show that

$$\nabla u = V \qquad \text{in } \mathscr{D}'(\mathbb{R}^2 \setminus \overline{B_e(0)}). \tag{5.108}$$

Given $\varphi \in C_0^\infty(\mathbb{R}^2 \setminus \overline{B_e(0)})$ arbitrarily, we have

$$\langle u, \operatorname{div} \varphi \rangle = \Big\langle \int_{\mathbb{R}^2} \Gamma_a(\cdot, y) f(y)\, dy, \operatorname{div} \varphi \Big\rangle = \int_{\mathbb{R}^2} \langle \Gamma_a(\cdot, y), \operatorname{div} \varphi \rangle f(y)\, dy,$$

in which the last equality is correct because

$$\int_{e<|x|<M} \int_{\mathbb{R}^2} |\Gamma_a(x,y)||f(y)||\operatorname{div} \varphi(x)|\, dy\, dx \leqslant C \int_{e<|x|<M} \frac{|\operatorname{div} \varphi(x)|}{|x|}\, dx < \infty$$

follows from the proof of (5.94), where $\operatorname{Supp}\varphi \subset B_M(0) \setminus \overline{B_e(0)}$. We further obtain

$$\langle u, \operatorname{div}\varphi\rangle = -\int_{\mathbb{R}^2} \langle \nabla_x \Gamma_a(\cdot, y), \varphi\rangle f(y)\, dy = -\langle V, \varphi\rangle$$

since we have

$$\int_{e<|x|<M} \int_{\mathbb{R}^2} |\nabla_x \Gamma_a(x, y)||f(y)||\varphi(x)|\, dy\, dx \leqslant C \int_{e<|x|<M} \frac{|\varphi(x)|}{|x|}\, dx < \infty$$

by computation as above. We are thus led to (5.108).

We turn to the decay property of the pressure

$$p(x) = \frac{x}{2\pi|x|^2} \cdot \int_{\mathbb{R}^2} f(y)\, dy + R(x),$$

where the remainder $R(x)$ is divided into three parts:

$$\begin{aligned} R(x) &= R_1(x) + R_2(x) + R_3(x) \\ &:= \frac{1}{2\pi}\left(\int_{|y|<|x|/2} + \int_{|x|/2\leqslant|y|\leqslant 2|x|} + \int_{|y|>2|x|}\right)\left(\frac{x-y}{|x-y|^2} - \frac{x}{|x|^2}\right) \cdot f(y)\, dy. \end{aligned}$$

We then observe

$$|R_1(x)| \leqslant \frac{1}{2\pi} \int_{|y|<|x|/2} \int_0^1 \frac{3|y|}{|x-ty|^2}\, dt\, |f(y)|\, dy \leqslant C|x|^{-2} \int_{\mathbb{R}^2} |y||f(y)|\, dy$$

and

$$\begin{aligned} |R_2(x)| &\leqslant C|x|^{-3}\Big(\log \frac{|x|}{2}\Big)^{-1} \left(\int_{|y-x|\leqslant 3|x|} \frac{1}{|x-y|}\, dy + \frac{1}{|x|} \int_{|y|\leqslant 2|x|} dy\right) \\ &= C|x|^{-2}\Big(\log \frac{|x|}{2}\Big)^{-1} \end{aligned}$$

as well as

$$\begin{aligned} |R_3(x)| &\leqslant \frac{1}{2\pi} \int_{|y|>2|x|} \left(\frac{1}{|x-y|} + \frac{1}{|x|}\right) |f(y)|\, dy \\ &\leqslant C|x|^{-2} \int_{|y|>2|x|} |y||f(y)|\, dy = o(|x|^{-2}) \end{aligned}$$

as $|x| \to \infty$. We thus obtain (5.95).

We will show that (5.91) is a solution to (5.97). We use v^0 and v^1 given by (5.84) and (5.85), which satisfy (5.86) and (5.87), respectively, by (5.90). We set

$$v(x,t) = v^0(x,t) + v^1(x,t), \qquad w(x) = \int_0^\infty v(x,t)\,dt.$$

Since neither u nor w can absolutely converge, we are unable to apply the Fubini theorem directly to them. We will show, nevertheless, that they do converge and coincide. Let us employ the centering technique as in (5.52). We consider

$$\widetilde{u}(x) = \int_{\mathbb{R}^2} \widetilde{\Gamma}_a(x,y) f(y)\,dy,$$

and

$$\widetilde{v}(x,t) = O(at)^T \int_{\mathbb{R}^2} \left(K(O(at)x - y, t) - \frac{e^{-1/4t}}{8\pi t}\mathbb{I} \right) f(y)\,dy,$$

$$\widetilde{w}(x) = \int_0^\infty \widetilde{v}(x,t)\,dt,$$

where $\widetilde{\Gamma}_a(x,y)$ is given by (5.52). Then both integrals of $\widetilde{u}$ and $\widetilde{w}$ are absolutely convergent over $(0,\infty)\times\mathbb{R}^2$ with respect to (t,y). In fact, as in (5.53), it follows from (5.54)–(5.56) together with the assumption (5.92) that

$$\begin{aligned}
&\int_0^\infty \int_{\mathbb{R}^2} \left| K(O(at)x - y, t) - \frac{e^{-1/4t}}{8\pi t}\mathbb{I} \right| |f(y)|\,dy\,dt \\
&= \int_0^\infty \int_{\mathbb{R}^2} \left| K(x-y,t) - \frac{e^{-1/4t}}{8\pi t}\mathbb{I} \right| |f(O(at)y)|\,dy\,dt \\
&\leqslant C \int_{\mathbb{R}^2} \frac{\left(\left|\log|x-y|\right| + 1 \right)}{1+|y|^3}\,dy
\end{aligned}$$

which is actually convergent. We thus obtain $\widetilde{u} = \widetilde{w}$. Since

$$u - \widetilde{u} = \int_0^\infty O(at)^T e^{-1/4t} \frac{dt}{8\pi t} \int_{\mathbb{R}^2} f(y)\,dy = w - \widetilde{w} \tag{5.109}$$

and since (5.109) does converge by (5.51), we eventually conclude that $u = w$. We now show that $\{u, p\}$ actually satisfies $(5.97)_1$ in the sense of distributions. Given $\varphi \in C_0^\infty(\mathbb{R}^2)$ arbitrarily, let us consider $\langle \widetilde{u}, L_{-a}\varphi \rangle$ since we have the adjoint relation $L_{-a} = L_a^*$, see (5.88). Then we find

$$\langle \widetilde{u}, L_{-a}\varphi \rangle = \langle \widetilde{w}, L_{-a}\varphi \rangle = \int_0^\infty \langle \widetilde{v}(t), L_{-a}\varphi \rangle\,dt,$$

in which the Fubini theorem is employed. Note that the argument does not work if $\widetilde{v}$ is replaced by v. By integration by parts we have

$$\langle \widetilde{u}, L_{-a}\varphi\rangle = \int_0^\infty \langle L_a v(t), \varphi\rangle\, dt + \int_0^\infty \big\langle L_a\big(\widetilde{v}(t) - v(t)\big), \varphi\big\rangle\, dt, \tag{5.110}$$

however, since $\widetilde{v} - v$ is independent of x and since

$$\int_{\mathbb{R}^2} (L_{-a}\varphi)(x)\, dx = -a \int_{\mathbb{R}^2} \varphi^\perp(x)\, dx,$$

we obtain

$$\begin{aligned} \int_0^\infty \big\langle L_a\big(\widetilde{v}(t) - v(t)\big), \varphi\big\rangle\, dt &= -a(u - \widetilde{u})^\perp \cdot \int_{\mathbb{R}^2} \varphi(x)\, dx \\ &= a(u - \widetilde{u}) \cdot \int_{\mathbb{R}^2} \varphi^\perp(x)\, dx \\ &= -\langle u - \widetilde{u}, L_{-a}\varphi\rangle, \end{aligned} \tag{5.111}$$

see (5.109). On the other hand, in view of (5.86) and (5.87) and by taking

$$\lim_{t\to\infty} \langle v(t), \varphi\rangle = 0, \qquad \lim_{t\to 0} \langle v(t) - (f - \nabla p), \varphi\rangle = 0,$$

into account, we have

$$\int_0^\infty \langle L_a v(t), \varphi\rangle\, dt = -\int_0^\infty \partial_t \langle v(t), \varphi\rangle\, dt = \langle f - \nabla p, \varphi\rangle. \tag{5.112}$$

We collect (5.110), (5.111) and (5.112) to obtain

$$\langle u, L_{-a}\varphi\rangle = \langle f - \nabla p, \varphi\rangle$$

for all $\varphi \in C_0^\infty(\mathbb{R}^2)$. Since $\Delta p = \operatorname{div} f$, we take the divergence of $(5.97)_1$ to see that $(\operatorname{div} u) \in \mathscr{S}'(\mathbb{R}^2)$ obeys

$$-\Delta(\operatorname{div} u) - a\, x^\perp \cdot \nabla(\operatorname{div} u) = 0$$

on account of (5.19). By Lemma 5.3.5, $\operatorname{div} u$ is a polynomial, however, from (5.93) we conclude that $\operatorname{div} u = 0$. Since $f \in L^q(\mathbb{R}^2)$ for every $q \in (1, \infty)$, the result of [11] (see also another proof given by [20]) implies (5.98). And then, (5.106) combined with (5.98) especially for $x^\perp \cdot \nabla u - u^\perp$ leads to (5.99). Finally, if $f \in C^\infty(\mathbb{R}^2)$, then we put the term $x^\perp \cdot \nabla u - u^\perp$ in the RHS together with such f to use the regularity theory of the usual Stokes system. As a consequence, we find $\{u, p\} \in C^\infty(\mathbb{R}^2)$. This completes the proof. □

For the proof of Theorem 5.2.2 we also need analysis of the system

$$\varepsilon u - \Delta u - a\big(x^\perp \cdot \nabla u - u^\perp\big) + \nabla p = f, \qquad \operatorname{div} u = 0 \quad \text{in } \mathbb{R}^2, \tag{5.113}$$

where the term εu is introduced in order to control the behavior of solutions at infinity. Indeed (5.113) is the resolvent system, but the only case we are going to consider is $\varepsilon > 0$. The velocity part of the associated fundamental solution is given by

$$\Gamma_a^{(\varepsilon)}(x, y) = \int_0^\infty e^{-\varepsilon t} O(at)^T K(O(at)x - y, t)\, dt, \tag{5.114}$$

while the pressure part is the same, see (5.27). Of course, (5.114) converges without using oscillation, however, what we need is to derive a certain estimate uniformly with respect to $\varepsilon > 0$. Therefore, we still use oscillation as well as the centering technique.

Proposition 5.3.3 *Let $a \in \mathbb{R} \setminus \{0\}$. Suppose f satisfies* (5.90) *and* (5.92). *Set*

$$u_\varepsilon(x) = \int_{\mathbb{R}^2} \Gamma_a^{(\varepsilon)}(x, y) f(y)\, dy, \qquad \varepsilon > 0, \tag{5.115}$$

where $\Gamma_a^{(\varepsilon)}(x, y)$ is given by (5.114). *Then $u_\varepsilon(x)$ enjoys* (5.94) *for every $\rho \geqslant e$, where the constant $C > 0$ is independent of $\varepsilon > 0$ (as well as $\rho \in [e, \infty)$ and $a \in \mathbb{R} \setminus \{0\}$). Furthermore, the pair $\{u_\varepsilon, p\}$ is a solution to* (5.113) *in the sense of distributions, where $p(x)$ given by* (5.83).

Proof Let $m > 0$. As in the proof of (5.33)–(5.34) by use of integration by parts, we easily find

$$\begin{aligned} &\left| \int_0^\infty e^{-\varepsilon t + iat} e^{-r^2/t} \frac{dt}{t^m} \right| + \left| \int_0^\infty e^{-\varepsilon t + iat} \int_t^\infty e^{-r^2/s} \frac{ds}{s^{m+1}}\, dt \right| \\ &\leqslant \frac{C}{\sqrt{\varepsilon^2 + a^2}\; r^{2m}} \leqslant \frac{C}{|a| r^{2m}} \end{aligned} \tag{5.116}$$

with some $C = C(m) > 0$ independent of $\varepsilon \geqslant 0$, $r > 0$ and $a \in \mathbb{R} \setminus \{0\}$. Owing to (5.116), we have the similar estimates to (5.49), (5.59) and (5.61) uniformly in $\varepsilon > 0$; namely, there is a constant $C > 0$ independent of $\varepsilon > 0$ such that

$$\begin{aligned} &\int_{|y| \leqslant 2|x|} |\Gamma_a^{(\varepsilon)}(x, y)|\, dy \leqslant C|a|^{-1}|x|^2 + C|x|^2 \log|x|, \qquad |x| \geqslant e, \\ &|\Gamma_a^{(\varepsilon)}(x, y)| \leqslant \begin{cases} C|x|^{-1}|y| + C|a|^{-1}|x|^{-2}, & |x| > 2|y|, \\ C|y|^{-1}|x| + C|a|^{-1}|y|^{-2}, & |y| > 2|x|. \end{cases} \end{aligned} \tag{5.117}$$

In fact, it follows from (5.116) that

$$\left|\int_0^\infty e^{-\varepsilon t}O(at)^T\,\frac{e^{-1/4t}}{8\pi t}\,dt\right| \leqslant \frac{C}{|a|}$$

with some $C > 0$ independent of $\varepsilon > 0$, which together with the same computing as in the proof of Lemma 5.3.3 by means of centering technique as in (5.52) yields $(5.117)_1$. Also, look at the proof of Proposition 5.3.1, in which oscillation is used in (5.63) and so on. This time, we employ (5.116) to get

$$\left|\int_0^\infty e^{-\varepsilon t}O(at)^T\,e^{-|x|^2/4t}\,\frac{dt}{t}\right| \leqslant \frac{C}{|a||x|^2}$$

and so on, where $C > 0$ is independent of $\varepsilon > 0$. The other estimates without using oscillation are obvious. For the purpose here it is enough to split the exponential function into two terms rather than (5.62) and (5.70) since we do not intend to find out the leading term. As a consequence, we obtain $(5.117)_2$. With use of (5.117) the desired estimate (5.94) for u_ε uniformly in $\varepsilon > 0$ is deduced in exactly the same way as in the proof of Proposition 5.3.2.

The proof of the latter assertion is easier than the corresponding part (the 4th assertion) of Proposition 5.3.2, in which we are forced to introduce $\widetilde{u}$. We do not need it since u_ε itself converges absolutely. Hence, we have

$$u_\varepsilon(x) = \int_0^\infty v_\varepsilon(x,t)\,dt$$

with

$$v_\varepsilon(x,t) = e^{-\varepsilon t}O(at)^T\int_{\mathbb{R}^2} K(O(at)x-y,t)f(y)\,dy,$$

which satisfies

$$\partial_t v_\varepsilon + (\varepsilon + L_a)v_\varepsilon = 0, \qquad v_\varepsilon(\cdot,0) = f - \nabla p$$

in $\mathbb{R}^2 \times (0,\infty)$, where L_a is given by (5.88). We thus obtain

$$\langle u_\varepsilon, (\varepsilon + L_{-a})\varphi\rangle = \langle f - \nabla p, \varphi\rangle$$

for all $\varphi \in C_0^\infty(\mathbb{R}^2)$. This combined with $\Delta p = \operatorname{div} f$ implies $\operatorname{div} u_\varepsilon = 0$ by Lemma 5.3.5 since $|\nabla u_\varepsilon(x)| = O(|x|^{-1})$ as $|x| \to \infty$, where this decay property is verified along the same line as the case $\varepsilon = 0$ by use of (5.50) and (5.107) for $\nabla_x \Gamma_a^{(\varepsilon)}(x,y)$ without using oscillation. The proof is complete. □

5.4 Proof of Theorem 5.2.1

To find the asymptotic representation (5.15), it would be standard to employ a potential representation formula in terms of the fundamental solution $\Gamma_a(x, y)$ as in [9] for the 3D problem, but we have to establish the decay properties (5.12) in advance in order to justify such a formula. This procedure consisting of those two steps would be also fine (and actually it works), however, there is another way, which is straightforward and leads us directly to (5.15) as well as (5.12), by means of a cut-off technique. We will adopt the latter way to prove Theorem 5.2.1. The only disadvantage compared with the former one by use of the potential representation formula is that the coefficient of the leading profile needs a bit lengthy (but elementary) calculation.

Proof of Theorem 5.2.1. We use a cut-off technique as mentioned above. In order to recover the solenoidal condition by use of the correction term with compact support, we first reduce the problem to the one with vanishing flux at the boundary $\partial\Omega$. To this end, we fix $x_0 \in \mathrm{int}\,(\mathbb{R}^2 \setminus \Omega)$ and introduce the flux carrier

$$w(x) = \beta\,\nabla\left(\frac{1}{2\pi}\log\frac{1}{|x-x_0|}\right) = \frac{-\beta\,(x-x_0)}{2\pi\,|x-x_0|^2}, \qquad \beta = \int_{\partial\Omega} \nu\cdot u\,d\sigma,$$

for given smooth solution $\{u, p\}$ of (5.5). Then we have

$$\int_{\partial\Omega} \nu\cdot w\,d\sigma = \beta,$$

$$\mathrm{div}\,w = 0, \qquad \Delta w = 0, \qquad (x-x_0)^\perp\cdot\nabla w = w^\perp \qquad \text{in } \mathbb{R}^2\setminus\{x_0\} \tag{5.118}$$

and

$$\nabla^j w(x) = \nabla^j\left(\frac{-\beta x}{2\pi\,|x|^2}\right) + O(|x|^{-(2+j)}) \qquad (j = 0, 1) \tag{5.119}$$

as $|x| \to \infty$. So the pair

$$\widetilde{u} = u - w, \qquad \widetilde{p} = p - a x_0^\perp\cdot w$$

fulfills (5.5) subject to

$$\int_{\partial\Omega} \nu\cdot\widetilde{u}\,d\sigma = 0, \tag{5.120}$$

where we note the relation

$$\partial_k(x_0^\perp\cdot w) = \sum_j (x_0^\perp)_j\partial_k\partial_j\left(\frac{\beta}{2\pi}\log\frac{1}{|x-x_0|}\right) = x_0^\perp\cdot\nabla w_k \qquad (k = 1, 2). \tag{5.121}$$

We fix $R \geqslant 1$ such that $\mathbb{R}^2 \setminus \Omega \subset B_R(0)$. Let $\psi \in C_0^\infty(B_{3R}(0); [0,1])$ be a cut-off function satisfying $\psi(x) = 1$ for $|x| \leqslant 2R$. By using the Bogovskii operator B in the annulus

$$A = \{x \in \mathbb{R}^2;\ R < |x| < 3R\},$$

see [4, 5, 19], we set

$$v = (1-\psi)\widetilde{u} + B[\widetilde{u} \cdot \nabla\psi], \qquad q = (1-\psi)\widetilde{p}.$$

It should be noted that $\int_A \widetilde{u} \cdot \nabla\psi\, dx = 0$ follows from (5.120). Then the pair $\{v, q\}$ obeys

$$-\Delta v - a\left(x^\perp \cdot \nabla v - v^\perp\right) + \nabla q = g + (1-\psi)f, \qquad \operatorname{div} v = 0 \qquad \text{in } \mathbb{R}^2 \quad (5.122)$$

for some function $g \in C_0^\infty(\mathbb{R}^2)$ whose support is a compact set in A. Here, we do not need any explicit form of g; in fact, the important quantity (5.125) below can be calculated only by taking account of the structure of the equation (5.122), that is, $\operatorname{div} S(v,q) = -g - (1-\psi)f$, see (5.9). When $u(x) = o(|x|)$, it is obvious that $v \in \mathscr{S}'(\mathbb{R}^2)$. Under the alternative assumption $\nabla u \in L^r(\Omega \setminus B_R(0))$ for some $r < \infty$, we have $\nabla v \in \mathscr{S}'(\mathbb{R}^2)$, which implies $v \in \mathscr{S}'(\mathbb{R}^2)$ by [7, Proposition 1.2.1]. Going back to (5.122), we observe $\nabla q \in \mathscr{S}'(\mathbb{R}^2)$ and thereby $q \in \mathscr{S}'(\mathbb{R}^2)$, too. Proposition 5.3.2 together with Lemma 5.3.5 concludes that

$$\begin{aligned} v(x) &= \int_{\mathbb{R}^2} \Gamma_a(x,y)\{g + (1-\psi)f\}(y)\, dy + P_v(x), \\ q(x) &= \int_{\mathbb{R}^2} Q(x-y) \cdot \{g + (1-\psi)f\}(y)\, dy + P_q(x), \end{aligned} \qquad (5.123)$$

with some polynomials P_v and P_q, however, it turns out from (5.93) and from either $\nabla v \in L^r(\mathbb{R}^2)$ with some $r \in (1,\infty)$ or $v(x) = o(|x|)$ that P_v must be a constant vector u_∞. Thus we have

$$u(x) = w(x) + \int_{\mathbb{R}^2} \Gamma_a(x,y)\{g + (1-\psi)f\}(y)\, dy + u_\infty \qquad (|x| \geqslant 3R), \quad (5.124)$$

from which combined with (5.96) and (5.119) we obtain (5.15) under the additional condition (5.14) as well as $(5.12)_1$, where the coefficient

$$\alpha = \int_{\mathbb{R}^2} y^\perp \cdot \{g + (1-\psi)f\}(y)\, dy = -\int_{\mathbb{R}^2} y^\perp \cdot \operatorname{div} S(v,q)\, dy \qquad (5.125)$$

is computed as follows.

Set

$$\alpha(\rho) := -\int_{|y|<\rho} y^\perp \cdot \operatorname{div} S(v,q)\, dy \qquad (\rho > 3R).$$

In view of (5.9) we have the relation

$$\begin{aligned} \operatorname{div}\left(y^{\perp}\cdot S(v,q)\right) &= \sum_{j,k} \partial_k\left[(y^{\perp})_j S_{jk}(v,q)\right] \\ &= y^{\perp}\cdot \operatorname{div} S(v,q) - S_{12}(v,q) + S_{21}(v,q) \\ &= y^{\perp}\cdot \operatorname{div} S(v,q) - 2a\, y\cdot v \end{aligned} \tag{5.126}$$

to find

$$\alpha(\rho) = -\int_{|y|=\rho} y^{\perp}\cdot\left(S(\widetilde{u},\widetilde{p})\,\frac{y}{\rho}\right) d\sigma - 2a\int_{|y|<\rho} y\cdot v\, dy.$$

Since $\operatorname{div} S(\widetilde{u},\widetilde{p}) = -f$ in Ω, it follows from (5.126) in which v is replaced by $\widetilde{u}$ that

$$\alpha(\rho) = \int_{\partial\Omega} y^{\perp}\cdot(S(\widetilde{u},\widetilde{p})\nu)\, d\sigma + 2a\int_{\Omega_\rho} y\cdot(\widetilde{u}-v)\, dy + \int_{\Omega_\rho} y^{\perp}\cdot f\, dy.$$

We are going to compute

$$\begin{aligned} &\int_{\partial\Omega} y^{\perp}\cdot(S(\widetilde{u},\widetilde{p})\nu)\, d\sigma \\ &= \int_{\partial\Omega} y^{\perp}\cdot\left\{\left(T(u,p) + a\,u\otimes y^{\perp}\right)\nu\right\} d\sigma \\ &\quad - \int_{\partial\Omega} y^{\perp}\cdot((Dw)\nu)\, d\sigma + a\int_{\partial\Omega} (y^{\perp}\cdot\nu)(x_0^{\perp}\cdot w)\, d\sigma \\ &\quad - a\int_{\partial\Omega} y^{\perp}\cdot\left\{(w\otimes y^{\perp})\nu\right\} d\sigma - a\int_{\partial\Omega} y^{\perp}\cdot\left\{(y^{\perp}\otimes\widetilde{u})\nu\right\} d\sigma \\ &=: \int_{\partial\Omega} y^{\perp}\cdot\left\{\left(T(u,p) + a\,u\otimes y^{\perp}\right)\nu\right\} d\sigma + J_1 + J_2 + J_3 + J_4. \end{aligned}$$

We will show that

$$J_1 = 0, \qquad J_2 + J_3 = 0, \qquad J_4 + 2a\int_{\Omega_\rho} y\cdot(\widetilde{u}-v)\, dy = 0,$$

which concludes

$$\alpha(\rho) = \int_{\partial\Omega} y^{\perp}\cdot\left\{\left(T(u,p) + a\,u\otimes y^{\perp}\right)\nu\right\} d\sigma + \int_{\Omega_\rho} y^{\perp}\cdot f\, dy.$$

Letting $\rho\to\infty$ leads us to

$$\alpha = \int_{\partial\Omega} y^{\perp}\cdot\left\{\left(T(u,p) + a\,u\otimes y^{\perp}\right)\nu\right\} d\sigma + \int_{\Omega} y^{\perp}\cdot f\, dy. \tag{5.127}$$

In fact, we observe

$$\begin{aligned}2a\int_{\Omega_\rho} y\cdot(\widetilde{u}-v)\,dy &= a\int_{\Omega_\rho}\{\psi\widetilde{u}-B[\widetilde{u}\cdot\nabla\psi]\}\cdot\nabla|y|^2\,dy\\ &= a\int_{\Omega_\rho}\mathrm{div}\left[|y|^2\{\psi\widetilde{u}-B[\widetilde{u}\cdot\nabla\psi]\}\right]dy\\ &= a\int_{\partial\Omega}|y|^2(\nu\cdot\widetilde{u})\,d\sigma = -J_4\end{aligned}$$

and

$$J_2+J_3=-a\int_{\partial\Omega}(y^\perp\cdot\nu)(y-x_0)^\perp\cdot w\,d\sigma=0$$

on account of $(y-x_0)^\perp\cdot w(y)=0$. We take account of $(\nabla w)^T=\nabla w$ and $\Delta w=0$ in $\mathbb{R}^2\setminus\{x_0\}$ to see that

$$\begin{aligned}J_1 &= -2\int_{\partial\Omega}(y-x_0)^\perp\cdot(\nu\cdot\nabla w)\,d\sigma-2x_0^\perp\cdot\int_{\partial\Omega}\nu\cdot\nabla w\,d\sigma\\ &= -2\int_{\partial\Omega}(y-x_0)^\perp\cdot(\nu\cdot\nabla w)\,d\sigma+2x_0^\perp\cdot\int_{|y-x_0|=\varepsilon}\frac{y-x_0}{\varepsilon}\cdot\nabla w\,d\sigma\end{aligned}$$

where $\varepsilon>0$ is taken in such a way that $\overline{B_\varepsilon(x_0)}\subset\mathrm{int}\,(\mathbb{R}^2\setminus\Omega)$. Using the explicit representation

$$\nabla w(y)=\frac{-\beta}{2\pi}\left(\frac{\mathbb{I}}{|y-x_0|^2}-\frac{2(y-x_0)\otimes(y-x_0)}{|y-x_0|^4}\right),$$

we find

$$\begin{aligned}\int_{\partial\Omega}(y-x_0)^\perp\cdot(\nu\cdot\nabla w)\,d\sigma &= \frac{-\beta}{2\pi}\int_{\partial\Omega}\frac{(y-x_0)^\perp\cdot\nu}{|y-x_0|^2}\,d\sigma\\ &= \frac{\beta}{2\pi}\int_{\mathbb{R}^2\setminus\left(\Omega\cup B_\varepsilon(x_0)\right)}\mathrm{div}\,\frac{(y-x_0)^\perp}{|y-x_0|^2}\,dy=0\end{aligned}$$

and

$$\int_{|y-x_0|=\varepsilon}\frac{y-x_0}{\varepsilon}\cdot\nabla w\,d\sigma=\frac{\beta}{2\pi\varepsilon^3}\int_{|y-x_0|=\varepsilon}(y-x_0)\,d\sigma=0$$

which implies that $J_1=0$. We thus obtain (5.127).

Concerning the pressure, it follows from (5.121) and (5.123) that

$$\nabla p=a x_0^\perp\cdot\nabla w+\nabla\int_{\mathbb{R}^2}Q(x-y)\cdot\{g+(1-\psi)f\}(y)\,dy+\nabla P_q\quad(|x|\geqslant 3R).$$

By (5.98) together with (5.119) we know $\nabla(p - P_q) \in L^r(\Omega)$ for every $r \in (1, \infty)$. Since $\Delta w = 0$ in $\mathbb{R}^2 \setminus \{x_0\}$, we obtain from (5.124)

$$\Delta u = \Delta \int_{\mathbb{R}^2} \Gamma_a(x, y)\{g + (1 - \psi)f\}(y)\, dy \qquad (|x| \geqslant 3R),$$

so that $\Delta u \in L^r(\Omega)$ for every $r \in (1, \infty)$ on account of (5.98). In addition, we also have

$$x^\perp \cdot \nabla u = x^\perp \cdot \nabla w + x^\perp \cdot \nabla \int_{\mathbb{R}^2} \Gamma_a(x, y)\{g + (1 - \psi)f\}(y)\, dy \qquad (|x| \geqslant 3R).$$

It thus follows from (5.99) and (5.119) that $x^\perp \cdot \nabla u \in L^r(\Omega)$ for every $r \in (2, \infty)$. Taking those as well as $(5.12)_1$ into account, we go back to (5.5) and let $|x| \to \infty$ to find that $\nabla P_q = -a u_\infty^\perp$. This implies that

$$p = a\, x_0^\perp \cdot w + \int_{\mathbb{R}^2} Q(x - y) \cdot \{g + (1 - \psi)f\}(y)\, dy - a\, u_\infty^\perp \cdot x + p_\infty \qquad (|x| \geqslant 3R)$$

for some constant p_∞. By (5.93) together with (5.119) we obtain $(5.12)_2$. We also use (5.95) and carry out a bit computation to obtain

$$\begin{aligned} p(x) + a u_\infty^\perp \cdot x - p_\infty = \Big[\int_{\partial\Omega} &\left\{ T(\widetilde{u}, \widetilde{p}) + a \left(\widetilde{u} \otimes y^\perp - y^\perp \otimes \widetilde{u} \right) \right\} \nu \, d\sigma_y \\ &+ \int_{\Omega} f\, dy - \beta a x_0^\perp \Big] \cdot \frac{x}{2\pi |x|^2} + O(|x|^{-2}) \end{aligned} \tag{5.128}$$

as $|x| \to \infty$. We stop further computation of the coefficient, however, we will recall (5.128) in Theorem 5.2.2, in which the coefficient is much simplified.

Once we have fine decay properties (5.12), we are able to justify the energy relation (5.13). We first verify (5.10) for smooth vector fields $u,\ v \in H^1_{loc}(\overline{\Omega})$ without assuming their decay properties at infinity. For each $\rho > 0$ large enough we have

$$\begin{aligned} \int_{\Omega_\rho} \left[(x^\perp \cdot \nabla u - u^\perp) \cdot v + u \cdot (x^\perp \cdot \nabla v - v^\perp) \right] dx &= \int_{\Omega_\rho} \operatorname{div} [x^\perp (u \cdot v)]\, dx \\ &= \int_{\partial\Omega} (\nu \cdot x^\perp)(u \cdot v)\, d\sigma \end{aligned} \tag{5.129}$$

since $\int_{|x|=\rho} = 0$. Letting $\rho \to \infty$, we obtain (5.10). Now, given smooth solution $\{u, p\} \in H^1_{loc}(\overline{\Omega}) \times L^2_{loc}(\overline{\Omega})$, we use the constants $\{u_\infty, p_\infty\}$ found above and set

$$u_*(x) = u(x) - u_\infty, \qquad p_*(x) = p(x) + a u_\infty^\perp \cdot x - p_\infty,$$

which satisfy

$$-\Delta u_* - a\,(x^\perp \cdot \nabla u_* - u_*^\perp) + \nabla p_* = f, \qquad \operatorname{div} u_* = 0 \qquad \text{in } \Omega.$$

We multiply u_*, perform integration by parts over Ω_ρ and use (5.129) to find the following two equalities, in which the relation $\operatorname{div} T(u_*, p_*) = \Delta u_* - \nabla p_*$ is used for the latter:

$$\int_{\Omega_\rho} |\nabla u_*|^2\, dx = \int_{\partial\Omega_\rho} \big((\nabla u_* - p_*\mathbb{I})\nu\big) \cdot u_*\, d\sigma + I,$$
$$\frac{1}{2}\int_{\Omega_\rho} |Du_*|^2\, dx = \int_{\partial\Omega_\rho} \big(T(u_*, p_*)\nu\big) \cdot u_*\, d\sigma + I,$$

where the common term I is given by

$$I = \frac{a}{2}\int_{\partial\Omega} (\nu \cdot x^\perp)|u_*|^2\, d\sigma + \int_{\Omega_\rho} f \cdot u_*\, dx.$$

Note that both $T(u_*, p_*)\nu$ and $(\nabla u_* - p_*\mathbb{I})\nu$ are understood as the normal trace being in $H^{-1/2}(\partial\Omega_\rho)$. In view of (5.124), we see from (5.93) and (5.119) that

$$\nabla u_*(x) = \nabla u(x) = O(|x|^{-1}) \quad \text{as } |x| \to \infty.$$

This together with (5.12) implies that

$$\lim_{\rho\to\infty} \int_{|x|=\rho} \left(\frac{\partial u_*}{\partial \nu} \cdot u_* - (\nu \cdot u_*)p_*\right) d\sigma = 0,$$

and that

$$\lim_{\rho\to\infty} \int_{|x|=\rho} \big(T(u_*, p_*)\nu\big) \cdot u_*\, d\sigma = 0.$$

On the other hand, we know that $f \cdot u_* \in L^1(\Omega)$ by (5.11) and (5.12) together with $u \in H^1_{loc}(\overline{\Omega}) \subset L^s_{loc}(\overline{\Omega})$ for all $s < \infty$. We thus obtain not only $\nabla u \in L^2(\Omega)$ but (5.13). This completes the proof. □

5.5 Proof of Theorem 5.2.2

Proof of Theorem 5.2.2. We begin with the proof of uniqueness. Suppose $\{u, p\}$ is a solution in the sense of distributions to (5.5) with $f = 0$ subject to $u = 0$ on $\partial\Omega$ and $\{u, p\} \to \{0, 0\}$ as $|x| \to \infty$ within the class $\nabla u \in L^2(\Omega)$, $\{u, p\} \in L^2_{loc}(\overline{\Omega})$. We put the term $x^\perp \cdot \nabla u - u^\perp$ in the RHS and use the interior regularity theory of the usual Stokes system to show that $u \in H^{k+1}_{loc}(\Omega)$, $p \in H^k_{loc}(\Omega)$ for every integer

$k \geqslant 1$ by bootstrapping argument; hence, $u,\ p \in C^\infty(\Omega)$. By Theorem 5.2.1 we have (5.13), in which the RHS vanishes. So, u is the rigid motion, but $u = 0$ on account of the boundary condition. Going back to (5.5) (with $f = 0$), we have $\nabla p = 0$, which together with $p \to 0$ at infinity yields $p = 0$. This proves the uniqueness.

We turn to the existence. It is easy to find a solution with $\nabla u \in L^2(\Omega)$ by following the method of Leray, but one cannot exclude a constant vector u_∞ at infinity even if applying Theorem 5.2.1. To get around this difficulty, we will adopt an approximation procedure specified below which brings regularizing effect at infinity. We take the auxiliary function

$$w(x) = \frac{a}{2}\,\nabla^\perp \left(\zeta(|x|)|x|^2\right) = \left\{\frac{|x|}{2}\,\zeta'(|x|) + \zeta(|x|)\right\}\left(ax^\perp\right) \tag{5.130}$$

where $\zeta \in C^\infty([0,\infty);[0,1])$ satisfies $\zeta(r) = 1\,(r \leqslant R)$ and $\zeta(r) = 0\,(r \geqslant 2R)$, where $R \geqslant 1$ is fixed such that $\mathbb{R}^2 \setminus \Omega \subset B_R(0)$. Then we have

$$w|_{\partial\Omega} = ax^\perp, \quad \operatorname{div} w = 0, \quad x^\perp \cdot \nabla w - w^\perp = \operatorname{div}\left(w \otimes x^\perp - x^\perp \otimes w\right) = 0.$$

We will find a solution of the form $u = \widetilde{u} + w$, where $\widetilde{u}$ should obey

$$\begin{cases} -\Delta\widetilde{u} - a\left(x^\perp \cdot \nabla\widetilde{u} - \widetilde{u}^\perp\right) + \nabla p = f + \Delta w, & \operatorname{div}\widetilde{u} = 0 \quad \text{in } \Omega, \\ \widetilde{u}|_{\partial\Omega} = 0, \qquad \widetilde{u} \to 0 \quad \text{as } |x| \to \infty. \end{cases} \tag{5.131}$$

As Finn and Smith performed in their paper [13] on the Oseen system (see also Galdi [19, Sect. 7.5]), for $\varepsilon \in (0,1)$, let us consider the approximate problem

$$\begin{cases} \varepsilon u_\varepsilon - \Delta u_\varepsilon - a\left(x^\perp \cdot \nabla u_\varepsilon - u_\varepsilon^\perp\right) + \nabla p_\varepsilon = f + \Delta w, & \operatorname{div} u_\varepsilon = 0 \quad \text{in } \Omega, \\ u_\varepsilon|_{\partial\Omega} = 0, \qquad u_\varepsilon \to 0 \quad \text{as } |x| \to \infty. \end{cases} \tag{5.132}$$

By $C_{0,\sigma}^\infty(\Omega)$ we denote the class of all solenoidal vector fields being in $C_0^\infty(\Omega)$. Let $H_{0,\sigma}^1(\Omega)$ be the completion of $C_{0,\sigma}^\infty(\Omega)$ in $H^1(\Omega)$. In a usual way (see, for instance, the proof of Lemma 5.3 of [27], in which the problem in each bounded domain Ω_ρ is first solved by means of the Lax-Milgram theorem and then the limit $\rho \to \infty$ is considered by using a priori estimate uniformly in ρ), one can find $u_\varepsilon \in H_{0,\sigma}^1(\Omega)$ which satisfies

$$\varepsilon\,\|u_\varepsilon\|_{L^2(\Omega)}^2 + \frac{1}{2}\|\nabla u_\varepsilon\|_{L^2(\Omega)}^2 \leqslant \frac{1}{2}\|F + \nabla w\|_{L^2(\Omega)}^2 \tag{5.133}$$

and

$$\varepsilon\langle u_\varepsilon, \varphi\rangle + \langle\nabla u_\varepsilon, \nabla\varphi\rangle - a\,\langle x^\perp \cdot \nabla u_\varepsilon - u_\varepsilon^\perp, \varphi\rangle = \langle f + \Delta w, \varphi\rangle$$

for all $\varphi \in C_{0,\sigma}^\infty(\Omega)$. We choose $p_\varepsilon \in L^2_{loc}(\overline{\Omega})$ such that $\int_{\Omega_{3R}} p_\varepsilon\, dx = 0$ and that the pair $\{u_\varepsilon, p_\varepsilon\}$ satisfies $(5.132)_1$ in the weak sense. Since $f + \Delta w \in C^\infty(\Omega)$, the interior regularity theory of the usual Stokes system implies that $u_\varepsilon,\ p_\varepsilon \in C^\infty(\Omega)$.

As in the proof of Theorem 5.2.1, we take the same cut-off function ψ together with the Bogovskii operator B in the annulus $A = \{x \in \mathbb{R}^2; R < |x| < 3R\}$ and set

$$v_\varepsilon = (1-\psi)u_\varepsilon + B[u_\varepsilon \cdot \nabla\psi], \qquad q_\varepsilon = (1-\psi)p_\varepsilon.$$

Then the pair $\{v_\varepsilon, q_\varepsilon\}$ obeys

$$\begin{aligned} \varepsilon v_\varepsilon - \Delta v_\varepsilon - a\left(x^\perp \cdot \nabla v_\varepsilon - v_\varepsilon^\perp\right) + \nabla q_\varepsilon = g_\varepsilon + (1-\psi)f \quad &\text{in } \mathbb{R}^2, \\ \operatorname{div} v_\varepsilon = 0 \quad \text{in } \mathbb{R}^2, \end{aligned}$$

where

$$\begin{aligned} g_\varepsilon = \varepsilon B[u_\varepsilon \cdot \nabla\psi] + 2\nabla\psi \cdot \nabla u_\varepsilon + (\Delta\psi + a x^\perp \cdot \nabla\psi)u_\varepsilon - \Delta B[u_\varepsilon \cdot \nabla\psi] \\ - a x^\perp \cdot \nabla B[u_\varepsilon \cdot \nabla\psi] + aB[u_\varepsilon \cdot \nabla\psi]^\perp - (\nabla\psi)p_\varepsilon. \end{aligned}$$

Here, note that $(1-\psi)\Delta w = 0$ since $\psi = 1$ ($|x| \leqslant 2R$) and since $\Delta w = 0$ ($|x| \geqslant 2R$). We use the fundamental solution (5.114) for the system (5.113). Then, by Proposition 5.3.3 with $\rho = 6R$ and Lemma 5.3.5, we find

$$v_\varepsilon(x) = \int_{\mathbb{R}^2} \Gamma_a^{(\varepsilon)}(x,y)\,\{g_\varepsilon + (1-\psi)f\}(y)\, dy$$

subject to

$$\begin{aligned} \sup_{|x|\geqslant 6R} |x||v_\varepsilon(x)| \leqslant C(1+|a|^{-1})\Big[\int_{\mathbb{R}^2} (1+|x|)\,|\{g_\varepsilon + (1-\psi)f\}(x)|\, dx \\ + \sup_{|x|\geqslant 3R} |x|^3(\log|x|)|f(x)|\Big] \end{aligned} \tag{5.134}$$

with $C > 0$ independent of ε. Here, the point is that a constant vector can be excluded since $u_\varepsilon \in L^2(\Omega)$.

By $\int_{\Omega_{3R}} p_\varepsilon\, dx = 0$ and $(5.132)_1$ we have

$$\|p_\varepsilon\|_{L^2(\Omega_{3R})} \leqslant C_R\|\nabla p_\varepsilon\|_{H^{-1}(\Omega_{3R})} \leqslant C_R\left(\|u_\varepsilon\|_{H^1(\Omega_{3R})} + \|F + \nabla w\|_{L^2(\Omega_{3R})}\right),$$

where $H^{-1}(\Omega_{3R}) := H_0^1(\Omega_{3R})^*$. This together with (5.133) and the estimate of the Bogovskii operator [4, 5, 19] lead us to

$$
\begin{aligned}
\int_A |g_\varepsilon(y)|\,dy &\leqslant 2\sqrt{2}R\|g_\varepsilon\|_{L^2(A)} \\
&\leqslant C_R\left(\|u_\varepsilon\|_{H^1(\Omega_{3R})} + \|p_\varepsilon\|_{L^2(\Omega_{3R})}\right) \\
&\leqslant C_R\left(\|\nabla u_\varepsilon\|_{L^2(\Omega_{3R})} + \|F\|_{L^2(\Omega_{3R})} + |a|\right) \\
&\leqslant C_R(\|F\|_{L^2(\Omega)} + |a|),
\end{aligned}
$$

which combined with (5.134) implies that

$$
|u_\varepsilon(x)| = |v_\varepsilon(x)| \leqslant C(1+|a|^{-1})(|a| + \|F\|_{L^2(\Omega)} + [f\,])\,|x|^{-1} \quad (|x| \geqslant 6R), \tag{5.135}
$$

where

$$
[f\,] := \int_{|x|\geqslant 2R} |x||f(x)|\,dx + \sup_{|x|\geqslant 3R} |x|^3(\log|x|)|f(x)|
$$

and $C = C(R) > 0$ is independent of $\varepsilon \in (0,1)$. By (5.133) we have

$$
\|u_\varepsilon\|_{L^{2,\infty}(\Omega_{6R})} \leqslant C\|u_\varepsilon\|_{L^2(\Omega_{6R})} \leqslant C_R\|\nabla u_\varepsilon\|_{L^2(\Omega_{6R})} \leqslant C_R(\|F\|_{L^2(\Omega)} + |a|),
$$

which together with (5.135) yields

$$
u_\varepsilon \in L^{2,\infty}(\Omega), \qquad \|u_\varepsilon\|_{L^{2,\infty}(\Omega)} \leqslant C\left\{1 + |a| + (1+|a|^{-1})\left(\|F\|_{L^2(\Omega)} + [f\,]\right)\right\}
$$

with $C = C(R) > 0$ independent of $\varepsilon \in (0,1)$. Hence, there is $\widetilde{u} \in L^{2,\infty}(\Omega) \cap H^1_{loc}(\overline{\Omega})$ with $\nabla\widetilde{u} \in L^2(\Omega)$ such that, as $\varepsilon \to 0$ along a subsequence,

$$
\begin{aligned}
&u_\varepsilon \to \widetilde{u} \;\; \text{weakly-star in}\, L^{2,\infty}(\Omega), \qquad \nabla u_\varepsilon \to \nabla\widetilde{u} \;\; \text{weakly in}\, L^2(\Omega), \\
&u_\varepsilon \to \widetilde{u} \;\; \text{weakly in}\, H^1(\Omega_\rho), \qquad u_\varepsilon \to \widetilde{u} \;\; \text{strongly in}\, L^2(\Omega_\rho),
\end{aligned}
$$

for every $\rho \geqslant R$ and, thereby,

$$
\langle \nabla\widetilde{u}, \nabla\varphi\rangle - a\,\langle x^\perp \cdot \nabla\widetilde{u} - \widetilde{u}^\perp, \varphi\rangle = \langle f + \Delta w, \varphi\rangle
$$

holds for all $\varphi \in C^\infty_{0,\sigma}(\Omega)$, as well as $\operatorname{div}\widetilde{u} = 0$. We fix ρ and use the trace inequality

$$
\|u_\varepsilon - \widetilde{u}\|_{L^2(\partial\Omega_\rho)} \leqslant C\|u_\varepsilon - \widetilde{u}\|^{1/2}_{L^2(\Omega_\rho)}\|u_\varepsilon - \widetilde{u}\|^{1/2}_{H^1(\Omega_\rho)}
$$

to see that $\widetilde{u}|_{\partial\Omega} = 0$. Since $\Delta\widetilde{u} + a\left(x^\perp\cdot\nabla\widetilde{u} - \widetilde{u}^\perp\right) + f + \Delta w \in H^{-1}(\Omega_\rho)$ for every $\rho \geqslant R$, we find an associated pressure $p \in L^2_{loc}(\overline{\Omega})$ such that the pair $\{\widetilde{u}, p\}$ solves $(5.131)_1$ in the weak sense. The interior regularity theory of the Stokes system concludes that $\{\widetilde{u}, p\}$ is smooth and, therefore, so is $u := \widetilde{u} + w$. Both estimates in (5.20) are obvious.

Let us apply Theorem 5.2.1 to $\{u, p\}$ with $\nabla u \in L^2(\Omega)$ as well as $u \in H^1_{loc}(\overline{\Omega})$, $p \in L^2_{loc}(\overline{\Omega})$. Since $u \in L^{2,\infty}(\Omega)$, we have all the properties in this theorem with

$u_\infty = 0$. We denote $p - p_\infty$ by the same symbol p so that $\{u, p\}$ is the desired solution. By $u|_{\partial\Omega} = ax^\perp$ we have

$$\beta = \int_{\partial\Omega} \nu \cdot u \, d\sigma = 0, \quad \int_{\partial\Omega} y^\perp \cdot \left\{(u \otimes y^\perp)\nu\right\} d\sigma_y = a \int_{\partial\Omega} (\nu \cdot y^\perp)|y|^2 \, d\sigma_y = 0,$$

and thereby (5.15) implies (5.21). Finally, asymptotic representation of the pressure is given by (5.128), in which $\{u_\infty, p_\infty\} = \{0, 0\}$. Since $\beta = 0$ and $u|_{\partial\Omega} = ay^\perp$, we conclude (5.22). The proof is complete. □

Acknowledgements I would like to thank Professor M. Yamazaki for stimulating discussions. Thanks are also due to Professor G.P. Galdi for informing me of [25], and to Dr. J. Guillod for sending [24] to me. This work is partially supported by Grant-in-Aid for Scientific Research, No. 24540169, from JSPS.

References

1. Amick, C.J.: On Leray's problem of steady Navier-Stokes flow past a body in the plane. Acta Math. **161**, 71–130 (1988)
2. Amick, C.J.: On the asymptotic form of Navier-Stokes flow past a body in the plane. J. Differ. Equ. **91**, 149–167 (1991)
3. Bergh, J., Löfström, J.: Interpolation Spaces. Springer, Berlin (1976)
4. Bogovskiĭ, M.E.: Solution of the first boundary value problem for the equation of continuity of an incompressible medium. Soviet Math. Dokl. **20**, 1094–1098 (1979)
5. Borchers, W., Sohr, H.: On the equations rot $v = g$ and div $u = f$ with zero boundary conditions. Hokkaido Math. J. **19**, 67–87 (1990)
6. Chang, I.-D., Finn, R.: On the solutions of a class of equations occurring in continuum mechanics, with application to the Stokes paradox. Arch. Ration. Mech. Anal. **7**, 388–401 (1961)
7. Chemin, J.-Y.: Fluides parfaits incompressibles. Astérisque 230, Société Mathématique de France (1995)
8. Farwig, R., Galdi, G.P., Kyed, M.: Asymptotic structure of a Leray solution to the Navier-Stokes flow around a rotating body. Pac. J. Math. **253**, 367–382 (2011)
9. Farwig, R., Hishida, T.: Asymptotic profile of steady Stokes flow around a rotating obstacle. Manuscripta Math. **136**, 315–338 (2011)
10. Farwig, R., Hishida, T.: Leading term at infinity of steady Navier-Stokes flow around a rotating obstacle. Math. Nachr. **284**, 2065–2077 (2011)
11. Farwig, R., Hishida, T., Müller, D.: L^q-theory of a singular "winding" integral operator arising from fluid dynamics. Pac. J. Math. **215**, 297–312 (2004)
12. Finn, R.: On the Stokes paradox and related questions, *Nonlinear Problems*, pp. 99–115. University of Wisconsin Press, Madison, Wis. (1963)
13. Finn, R., Smith, D.R.: On the linearized hydrodynamical equations in two dimensions. Arch. Ration. Mech. Anal. **25**, 1–25 (1967)
14. Finn, R., Smith, D.R.: On the stationary solution of the Navier-Stokes equations in two dimensions. Arch. Ration. Mech. Anal. **25**, 26–39 (1967)
15. Galdi, G.P.: Existence and uniqueness at low Reynolds numbers of stationary plane flow of a viscous fluid in exterior domains. In: Recent Developments in Theoretical Fluid Mechanics (Paseky 1992). Pitman Research Notes on Mathematical Series 291, pp. 1–33. Longman Science Technology, Harlow (1993)

16. Galdi, G.P.: Mathematical questions relating to the plane steady motion of a Navier-Stokes fluid past a body. In: Kozono, H., Shibata, Y. (eds.) Recent Topics on Mathematical Theory of Viscous Incompressible Fluid. Lecture Notes in Numerical and Applied Analysis, vol. 16, pp. 117–160. Kinokuniya, Tokyo (1998)
17. Galdi, G.P.: On the motion of a rigid body in a viscous liquid: a mathematical analysis with applications. In: Friedlander, S., Serre, D. (eds.) Handbook of Mathematical Fluid Dynamics, vol. I, pp. 653–791. North-Holland, Amsterdam (2002)
18. Galdi, G.P.: Stationary Navier-Stokes problem in a two-dimensional exterior domain. In: Chipot, M., Quittner, P. (eds.) Handbook of Differential Equations, Stationary Partial Differential Equations, vol. 1, pp. 71–155. Elsevier, North-Holland (2004)
19. Galdi, G.P.: An Introduction to the Mathematical Theory of the Navier-Stokes Equations. Steady-State Problems, 2nd edn. Springer (2011)
20. Galdi, G.P., Kyed, M.: A simple proof of L^q-estimates for the steady-state Oseen and Stokes equations in a rotating frame. Part I: strong solutions. Proc. Am. Math. Soc. **141**, 573–583 (2013)
21. Gilbarg, D., Weinberger, H.F.: Asymptotic properties of Leray's solution of the stationary two-dimensional Navier-Stokes equations. Russ. Math. Surv. **29**, 109–123 (1974)
22. Gilbarg, D., Weinberger, H.F.: Asymptotic properties of steady plane solutions of the Navier-Stokes equations with bounded Dirichlet integral. Ann. Scuola Norm. Sup. Pisa **5**(4), 381–404 (1978)
23. Guenther, R.B., Thomann, E.A.: Fundamental solutions of Stokes and Oseen problem in two spatial dimensions. J. Math. Fluid Mech. **9**, 489–505 (2007)
24. Guillod, J., Wittwer, P.: Asymptotic behaviour of solutions to the stationary Navier-Stokes equations in two dimensional exterior domains with zero velocity at infinity. Math. Models Methods Appl. Sci. **25**, 229–253 (2015)
25. Hillairet, M., Wittwer, P.: On the existence of solutions to the planar exterior Navier-Stokes system. J. Differ. Equ. **255**, 2996–3019 (2013)
26. Hishida, T.: An existence theorem for the Navier-Stokes flow in the exterior of a rotating obstacle. Arch. Ration. Mech. Anal. **150**, 307–348 (1999)
27. Hishida, T.: L^q estimates of weak solutions to the stationary Stokes equations around a rotating body. J. Math. Soc. Jpn. **58**, 743–767 (2006)
28. Kozono, H., Sohr, H.: On a new class of generalized solutions for the Stokes equations in exterior domains. Ann. Scuola Norm. Sup. Pisa **19**, 155–181 (1992)
29. Leray, J.: Etude de diverses equations integrales non lineaires et de quelques problemes que pose l'Hydrodynamique. J. Math. Pures Appl. **12**, 1–82 (1933)
30. Oseen, C.W.: Über die Stokessche Formel und über eine verwandte Aufgabe in der Hydrodynamik. Ark. Mat. Astron. Fys. **6**, 1–20 (1910)
31. Pileckas, K., Russo, R.: On the existence of vanishing at infinity symmetric solutions to the plane stationary exterior Navier-Stokes problem. Math. Ann. **352**, 643–658 (2012)
32. Smith, D.R.: Estimates at infinity for stationary solutions of the Navier-Stokes equations in two dimensions. Arch. Ration. Mech. Anal. **20**, 341–372 (1965)
33. Stokes, G.G.: On the effect of the internal friction of fluids on the motion of pendulums. Trans. Cambridge Phil. Soc. **9**, 8–106 (1851)
34. Yamazaki, M.: Unique existence of stationary solutions to the two-dimensional Navier-Stokes equations in exterior domains. In: Adachi, T. et al. (eds.) Mathematical Analysis on the Navier-Stokes Equations and Related Topics, Past and Future. In Memory of Professor Tetsuro Miyakawa. Gakuto International Series, Mathematical Sciences and Applications, vol. 35, pp. 220–241. Gakkotosho, Tokyo (2011)

Chapter 6
Toward Understanding Global Flow Structure

Makoto Iima, Takayuki Yamaguchi, Takeshi Watanabe, Akane Kawaharada, Yuji Tasaka and Erika Shoji

Abstract Flows in nature are generally coupled with the environment. Also, flow structures in the forms of convection rolls, vortices, boundary layers, for example, are often coupled with flow structures in other form(s) as well as the external environments such as the boundary motion and the temperature gradient. Whole flow structure in such cases is often characterized by multi-scale or hierarchy, therefore, we will term such flow structures "global flow structure". Clearly, the global flow structure is complex in both space and scales, but there are general viewpoints applicable to this category of the flow, by which we can tackle with new phenomena. In this review, we discuss the global flow structure in both views of typical problems and analysis methods itself.

Keywords Binary fluid convection · Bioconvection · Localized structure · Surface switching · Covariant Lyapunov analysis · Cellular automata

M. Iima (✉)
Graduate School of Science, Hiroshima University, 1-7-1, Kagamiyama, Higashi-hiroshima 739-8521, Japan
e-mail: makoto@mis.hiroshima-u.ac.jp

T. Yamaguchi
Department of Radiation Oncology, Institute of Biomedical and Health Sciences, Hiroshima University, Higashi-hiroshima, Japan

T. Watanabe
Rocket and Spacecraft Modeling Laboratory, School of Engineering, The University of Tokyo, Tokyo, Japan

A. Kawaharada
Department of Mathematics, Kyoto University of Education, Kyoto, Japan

Y. Tasaka
Laboratory for Flow Control, Hokkaido University, Sapporo, Japan

E. Shoji
Graduate School of Science, Hiroshima University, Higashi-hiroshima, Japan

Y. Shibata and Y. Suzuki (eds.), *Mathematical Fluid Dynamics, Present and Future*, Springer Proceedings in Mathematics & Statistics 183,
DOI 10.1007/978-4-431-56457-7_6

6.1 Introduction

Flow in many different form are ubiquitous and such forms may be categorized as flow structures according to their characteristic behaviors; waves, boundary layers, wakes, vortices, are a few examples. Such flow structure is convenient to focus on, and we have been studying single flow structure so far. In many situations, however, observed flow can not be classified to single flow structure. For the classification, the 'global flow structure' is defined [1] as the flow including the coupling of the flow structures in space and scales, which are flow patterns with the boundaries. In the following, we list four examples of the global flow structures.

1. A comprehensive example is bioconvection [2–6]. Here macroscopic flow is generated by the collective behavior of microorganisms due to their taxes, and the microscopic flow driven by an individual is influenced by the macroscopic flow; advection and diffusion change their spatial number density distribution. In terms of the governing equations, the microscopic flow governed by the Stokes equation is coupled with the macroscopic flow governed by the Navier-Stokes equations including mass flux of microorganisms. Not only single convection cell, consisting of such hierarchical structure, but also their interaction governed by a long-range interaction between convection cells adds a much more large macroscopic flow to understand.

2. Flows driven by moving boundary are also good examples. Insect's flight [7] is a typical example of heterogeneous multi-scale flow. In this case, flow structure in the form of boundary layer and separation vortices are generated by the wing motion. The generated flow is needed to generate appropriate lift and moment for insect to fly and to maneuver, which has been studied in simplified models, experiments, and direct numerical simulations [8–15]. Because the flapping flight is unstable in general, insects need to control the wing motion to generate the global flow structure in an appropriate way, which is a central question in vortex-using flapping flight.

3. Partially filled liquid in a cylinder driven by the rotating bottom causes boundary layers on the bottom and various surface motions. Their interaction generates polygonal flow [16–23], in which the surface has a polygon-like cross section. In some situations where the container is relatively smaller and the volume of the water is large, the surface shape switches among axisymmetric shape and non-axisymmetric shapes non-periodically even if the rotation speed is constant. This phenomena is called "surface switching" [19, 21, 24–27], and the flow transition is suggested a major factor to cause such switching [24, 25, 27–29]. In this case, the surface shape is an observable object, but the internal flow structure coupled with the surface shape clearly plays an essential role. Because the flow transition mechanism should be coupled with the surface shape dynamics, the idea of the global flow structure should be useful to understand this phenomena.

4. Even if the situation is much simpler, the pattern can be very complex if the elementary pattern is spatially localized. For this case, a good example is the thermal convection of binary fluid mixture, such as a mixture of water and alcohol [30–38]

(binary fluid convection). This is the same as the Rayleigh Benárd convection except for the liquid inside, which allows rich convection patterns, especially spatially localized patterns. The formation process of these localized patterns and their interaction makes a complex spatio-temporal patterns.

So far, four examples of global flow structure are introduced; their essential mechanism contain heterogeneous interaction between typical flow structure and surrounding environment, and/or homogeneous/hierchical interaction among typical flow structures. The observed spatio-temporal patterns become complicated in general. However, in some cases, their complication is just on surface and their essential mechanism can be understood simpler than it seems. In a previous review [1], we show several analysis of the global flow structure problem at that moment. After the publish, we accumulated several examples and developed analysis methods. In particular, we applied the knowledge of the global flow structure of the binary fluid convection to the bioconvection problems. Also, theoretical developments are significant in the surface switching phenomena. Analysis methods applicable for the global flow structure problem are also developing, which should be mentioned. They are the thrusts of this paper.

This review consists of two parts. One is the analysis part (Sect. 6.2), in which we mention the developments of the analysis of the global flow structure problems after the publish of Ref. [1]. We briefly summarize our recent results in the binary fluid convection [38] (Sect. 6.2.1) to focus on the bioconvection problem [5, 39] (Sect. 6.2.2). We discuss similarity between the the elementary localized bioconvection structures experimentally obtained in the bioconvection and the localized structures in binary fluid convection. For the surface switching problem (Sect. 6.2.3), we comment theoretical developments, a one-dimensional map with random noise term [40] and a theory potential flow theory [22], both of which suggests that some modification of the theory based on simple situation can describe the major part of the complicated phenomena of the global flow structure.

The other is the methodology part (Sect. 6.3), in which two new analysis methods applicable for the dynamics of the global flow structure are discussed, which may be used in the cases with and without the governing equations. We propose a new method of the analysis of dynamical systems based on the orbit property (Sect. 6.3.1). We focus on the orbit passing near the steady solution in the space defined by the phase space and the parameter space. The characteristics of the orbit can be described by the pullback vectors of the eigenvectors of the steady solution [41]. We also propose a new construction method of cellular automata (Sect. 6.3.2) which only needs measurement data and can be applied without any knowledge of target phenomena [39, 42, 43].

6.2 Phenomena

6.2.1 Localized Convection Patterns in Binary Fluid Convection

In binary fluid convection, a variety of convection patterns are observed, many of which can never seen in the ordinary Rayleigh-Bénard convection. In this review, we focus on the spatially localized convection patterns, i.e., convection rolls are only observed in a spatially localized region (active region) while the flow outside that region is almost quiet. Such structures are important because they can be unit structure(s), and their properties including interaction behavior with another structure are a convenient way to understand the global flow structure.

Two types of a variety of spatially localized convection patterns, stationary ones and moving one, have been known. Stationary localized convection patterns, or *convectons*, with different number of convection cells have been reported [32] (Fig. 6.1f, g). They can coexist even if all the parameters remain the same [32, 38]. Moving localized convection pattern, or localized traveling wave, also has convection cells in a confined region, however, the convection cells are generated at one end of the active region and transferred to the other end to disappear [31, 32, 37, 38, 44]. Further, the whole structure moves at a constant speed; they have a group velocity. The localized traveling wave has been reported in both experimentally and numerically [31, 44], but had not been characterized mathematically; it had not been obtained as a solution satisfying a mathematical condition. Watanabe et al. [37] firstly characterized the localized traveling wave as a temporally periodic solution in a moving frame with a constant velocity. In Fig. 6.1a–e, snapshots during the period is shown. Based on

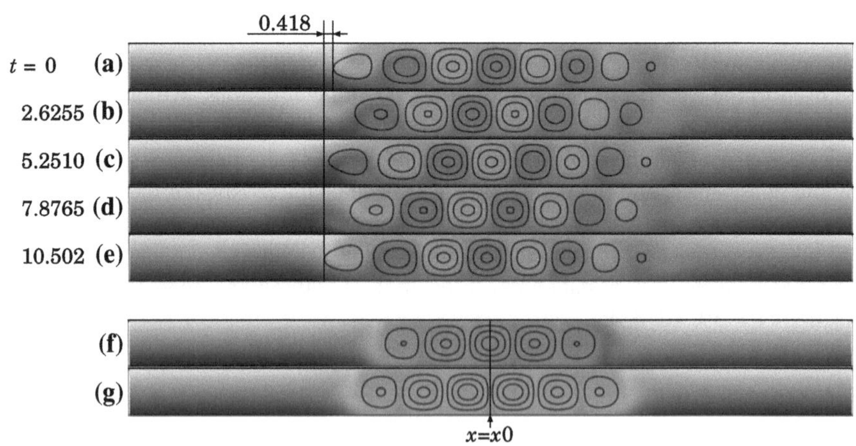

Fig. 6.1 **a–e** Localized traveling wave, which moves with changing the shapes periodically. **f, g** Convectons, which are steady. Curves and grayscale indicate streamlines and temperature field, respectively [38]

this mathematical property, they also calculated the corresponding solutions numerically and gave the bifurcation diagram [37, 38]. The shape of the bifurcation branch for the localized traveling wave is much different from those of the convectons; the bifurcation branches of the convectons are separated to two, which represent solutions with even- or odd-symmetry. Their shapes are snaky and twisted [32], while the bifurcation branch of the localized traveling wave is a Z-shape and the stable region is a function of the Rayleigh number.

The localization mechanism of these structures may be originated from the bistable structure of the conductive state where there is no flow and the convective state where convection cells covers the entire region. The coexistence of these states leads to the spatially localized solutions, which was pointed out by Pomeau [45]. Recent theory for the snaking bifurcation can be seen in Ref. [46] and references in Ref. [38].

The generation process of such localized patterns is an intriguing problem, which have been studied in both experimentally and numerically [34, 38]. In particular, Watanabe et al. studied the generation process in the parameter range where the conductive state is convectively unstable but both convectons and the localized traveling waves can exist as stable solutions [38], which is possible due to the periodic boundary condition laterally and the system size is finite. A long-time simulation starting from a pointwise initial condition added to the conductive state reveals that the perturbation initially spread to a small-amplitude wave packets and then they propagates in the opposite directions with larger velocity than that of the localized traveling wave, which is due to the convective instability (Fig. 6.2). During prop-

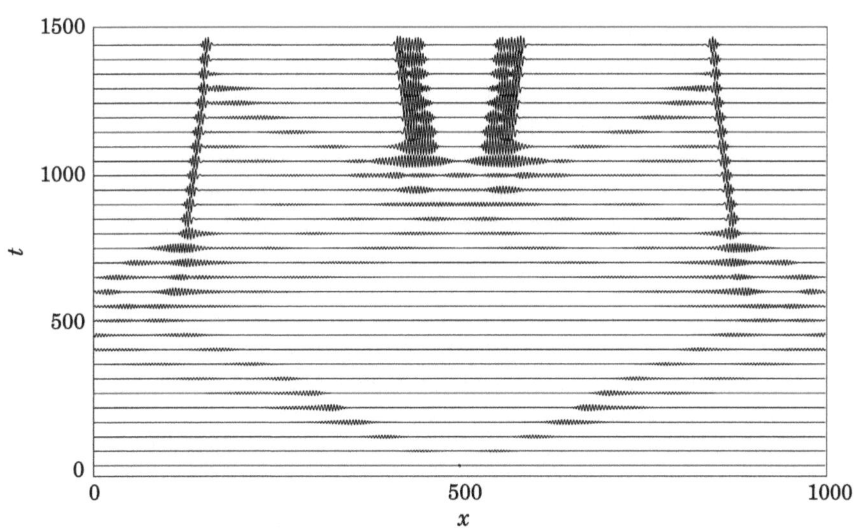

Fig. 6.2 Generation process of localized structures starting from pointwise initial condition by time evolution of the streamfunction on the middle horizontal line. The aspect ratio of the system size is 500 [38]

agation, the wave packet grows their amplitude until the nonlinear effect becomes dominant to form localized structure. Most of such localized structures are similar to the localized traveling wave, but in detail, they are different from the solution of the localized traveling wave; one of typical structure is called "cousin" of the localized traveling wave [38]. Also, they interact with each other with emitting perturbations in the 'calm' region similar to the conductive state. Perturbations can either grow to form another localized structures or absorbed in other localized structure (Fig. 6.2). After the complicated process, the whole system finally converge to co-propagating localized structures each of which is similar to the localized traveling wave.

Some part of such interactions can be reduced to a collision between counter-propagating and co-propagating localized traveling waves. We note that the collision does not occur for wave packets because of linearity. The elementary process is also analyzed using numerical solution, and their behavior depends on the parameter region (bistable region or not) greatly [37, 38, 47]. We remark that the detailed analysis of collision process requires detailed information of the localized traveling wave because setting the initial condition requires the phases of the traveling waves in addition to their distance. The detailed analysis becomes possible because of the successful of calculating solution of the localized traveling wave [37]. It is known that even unstable solutions are important to analyze the interaction process [33, 48, 49]. The relationship between such unstable solutions and collision or pattern formation process is under study.

6.2.2 Localized Convection Patterns in Bioconvection

Microorganism often has behavioral responses to stimuli, *taxis*, by which the suspension of the microorganism shows macroscopic ordered patterns, which is called bioconvection [2, 50]. Various bioconvection patterns such as polygons, lines, and dots, have been reported [6, 51, 52]. Typical formation mechanism of such patten is as follows. Microorganisms moves directionally due to some taxis. If the preferred direction is upward, they accumulate in a near-surface region to generate a layer. Because the mass density of microorganism is larger than water, the interface between the near-surface layer and the rest part becomes unstable due to the Rayleigh-Taylor instability [53]. Then, a part of the layer fall down with inducing downward flow (Fig. 6.3). The individuals in this part start to swimming upward again due to the taxis, which leads to a macroscopic convection pattern: bioconvection. More detailed reviews of bioconvection are found in Refs. [2, 50].

Here we focus on the bioconvection due to the phototaxis, the behavioral response to light [54–58]. Recently, it was reported that bioconvection pattern of *Euglena gracilis* (Fig. 6.4), a phototactic microorganism, can be spatially localized when they are strongly illuminated from the bottom [6], which contrasts other bioconvection patterns that generally cover the entire region. The localized patterns in Ref. [6] are, however, rather complex; there are spots in the central region and wavy patterns in the peripheral region. Therefore, we need to extract their essential features, or

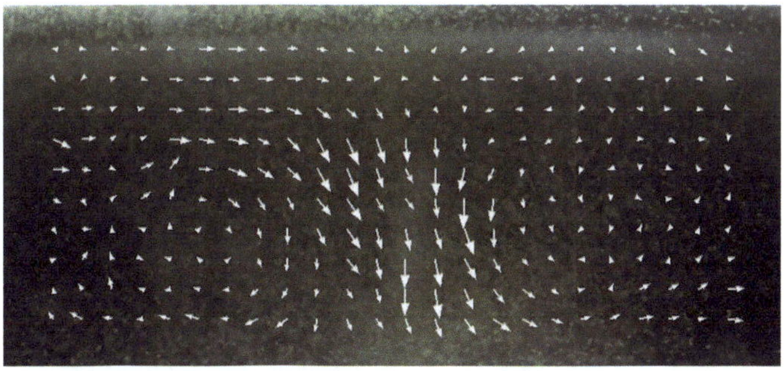

Fig. 6.3 *Side* view of the central part of the bioconvection unit. The flux of number density is also shown

Fig. 6.4 *Euglena gracilis*

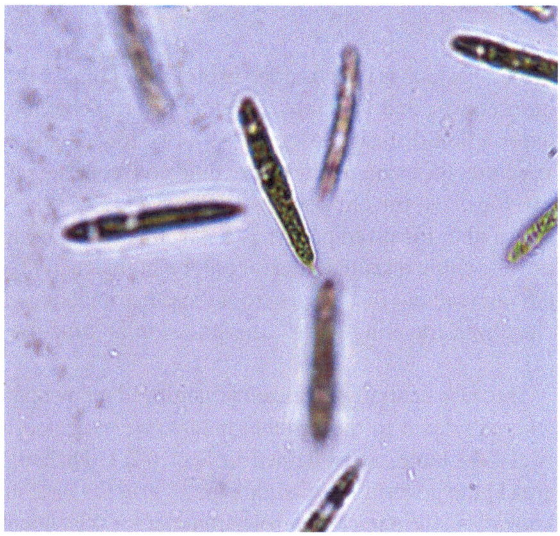

decompose the structure to prime units. For this purpose, we performed experiments using an annular container (Fig. 6.5) to suppress complex structure formation in the radial direction and exclude the wall effect in the azimuthal direction. Also the number density was tuned near the critical density of the onset of the convection. More detailed information is in Ref. [5].

We obtained the following two elementary localized convection patterns. One consists of single or several localized regions of high number density, in particular, the simplest structure with single high number density region sandwiched by two convection cells was termed "bioconvection unit" [5] (Fig. 6.3). In most cases, they do not move for hours, but occasionally the velocity changes abruptly. In other words, they has two time scales. This type of localized structure is similar to convecton in binary fluid convection [32], which are stationary localized states consisting of several convection rolls, although the present localized patterns can move with different

Fig. 6.5 Annular container and suspension of *Euglena gracilis* illuminated below

velocity, which is not observed in the case of binary fluid convection. When two structures coexist and they move, their collision behavior is similar to those observed in the reaction-diffusion systems [49]; two localized structure can be merged to single structure, for example. In binary fluid convection, similar collision behavior between two localized traveling waves is observed [37, 38].

Another localized pattern is a traveling wave of high density region, but the wave travels within a spatial interval only; a high density region is generated at one end of the interval and disappears from the other end. This localized structure was termed "localized traveling wave" after the similar structure in the binary fluid convection [5].

Because spatially localized structure in many spatially-extended dissipative systems is related to the bistable nature, we expected that this bioconvection system is also bistable, as explained in Sect. 6.2.1. Difficulties to confirm this conjecture are (1) the governing equations for *E. gracilis* bioconvection illumated below is not known; (2) the natural control parameter for this bioconvection is the number density, which is difficult to change during an experiment. Therefore we can neither find out the solutions to show the bistability bifurcation structure nor conduct experiments to show hysteresis. To overcome this difficulty, we prepared two types of the initial state: one is set so that the number density of *E. gracilis* is uniformly distributed and the other is set so that the density was distributed locally. A number of experiments to count the number of convection pattern occurrence revealed that the frequency of the occurrence clearly depended on the initial states, which suggests the bistability of the system [5].

Biological and mathematical mechanism of the localization would be the first question to answer for this system. We focus on the response behavior of *E. gracilis* to light property, and we are measuring the function of such response qualitatively, as well as calculating a model equation based on the measurements. The localized bioconvection pattern can be reproduced, but further analysis is required to discuss the localization mechanism. Figure 6.6 shows an experimental result of the typical formation process of the bioconvection units. As far as the figure and our experi-

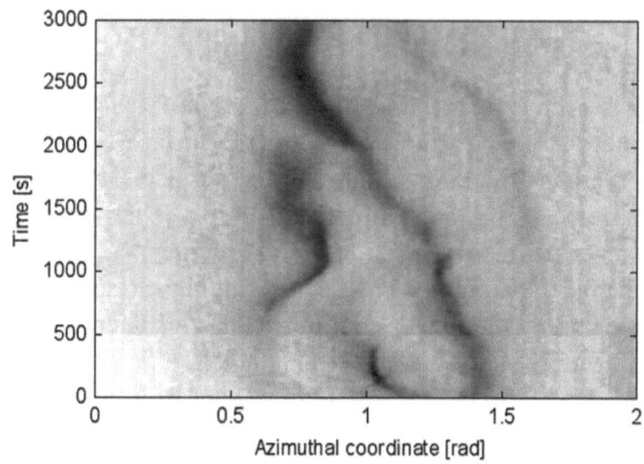

Fig. 6.6 Generation process of the localized pattern

ences are concerned, the formation process does not include a clear transition of the mechanism from the linear regime to the nonlinear regime observed in the case of binary fluid convection (Fig. 6.2). This fact suggests that the localization mechanism in the bioconvection in this system is much simpler, despite of the relatively complex hierarchy of flows from microscopic flow due to microorganism swimming to the macroscopic flow. As mentioned above, the bioconvection of *E. gracilis* and binary fluid convection shares typical localized structures in common, however, the detailed behavior are clearly different. One may expect that some mathematical structure responsible for these localized structure is shared in these systems, which should be something more than the bistability, because bistability structure alone can not account for the localized traveling wave. Comparing such similarity and difference in these two systems will help to deepen our understanding for the global flow structure problems (Fig. 6.7).

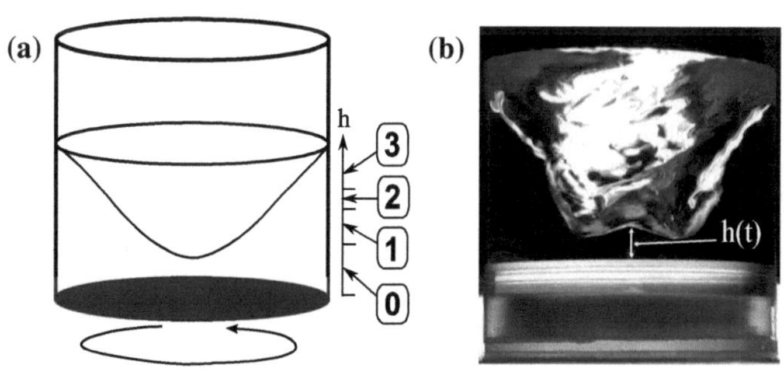

Fig. 6.7 **a** Experimental setup of the surface switching. *Bottom* of the cylinder rotates to drive the flow partially filled in the cylinder. **b** A snapshot of the surface shape. The dynamics of the surface shape is characterized by $h(t)$, the height of the *middle* of the surface shape

6.2.3 Surface Switching

When fluid is partially filled in a cylindrical container and driven by the bottom rotation, the rotating flow with free surface is also a good example of global flow structure. It is known that the free-surface shape can be polygonal despite that the system has the axisymmetry [20, 23]. If the container is relatively small and the aspect ratio is large (large liquid volume), the flow state (laminar or turbulent) can change dynamically with accompanying surface shape switching (circle and ellipse) even if the rotation speed is constant. This phenomena is termed "surface switching" [25] and is studied by experimentally [24, 26, 27, 59] and theoretically [21]. Because much of the experimental findings have been summarized in Ref. [10], our comments here are focused on the following points; a Markov process description and the recent theoretical developments.

The first comment is on the Markov process model (A. Kawaharada, M. Iima, Y. Tasaka, in preparation). A sequence of the height of the middle of the surface shape, $h(t)$, is discretized in time by taking the local average over the time interval $J\Delta t$ (J is an integer and $\Delta t = 1/30$[s] is the sampling time). The sequence of the local average of $h(t)$, $\{\overline{h}_j | j = 0, 1, ...\}$, was discretized into four states 0, 1, 2, and 3 represented by $\{m_j | j = 0, 1, ...\}$, by the following mapping:

$$m_j = \begin{cases} 0 \ (\overline{h}_j/H \leq 0.2), \\ 1 \ (0.2 < \overline{h}_j/H \leq 0.3), \\ 2 \ (0.3 < \overline{h}_j/H \leq 0.33), \\ 3 \ (0.33 < \overline{h}_j/H), \end{cases} \tag{6.1}$$

where H is the unperturbed height. The threshold values are defined by the histogram of $h(t)$ [24, 29]. After checking the robustness and reproductivity of the statistical values, we can choose appropriate value of J as 80 and the constructed Markov chain

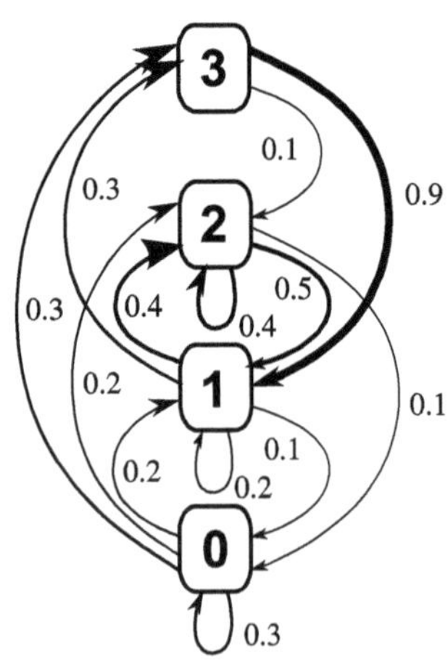

Fig. 6.8 Schematic picture of the transient process of the discretized height, 0, 1, 2, 3. As the number increases, the corresponding height increases. Arrows indicate the transition probability, their value is represented the line width

of order one reproduces the statistical properties of the original time series such as memoryless property [60]. In Fig. 6.8, the obtained Markov chain is depicted. The highest state, 3, mainly moves to the state 1. The intermediate state 2 does not move to the state 3, and the state oscillates between the state 2 and the state 1, or remains in the same state. The lower states 0 and 1 either remain the states or move to the higher states. These behavior matches our observation that the surface switching consists of a random switching among a quasi-oscillation, quasi-steady and significant surface elongation including touching the bottom [27].

Another model, the random dynamics model, was proposed by Sato et al. [40], in which the surface switching dynamics is represented by a sequence of the height of the surface shape, i.e., the local maximum and minimum of the (moving-averaged) time series of the surface height. The random dynamical system is defined by

$$h_{n+1} = F(h_n) + \xi(h_n), \tag{6.2}$$

where h_n represents the nth local maximum or minimum, $F(h_n)$ is a function defined by using the return plot, and $\xi(h_n) = \varepsilon(h_n)\xi_n$ is the stochastic term consisting of the state dependent amplitude of the noise $\varepsilon(h_n)$ and ξ_n, a random variable satisfying normal distribution $N(0, 1)$ [40]. Sato et al. succeeded in constructing one dimensional map with random fluctuations term that depends on the surface height, based on the measurements that shows a strong correlation between the surface height and the turbulent intensity [27]. Accordingly, the amplitude of the random fluctuation depends on the surface height. This random fluctuations can be regarded as the internal noise. Because the deterministic term and the stochastic term are separated in this model, we can study the effect of the stochastic term by controlling the amplitude of the stochastic term. Sato et al. revealed that the noise term essentially determines the switching behavior.

Both the Markov process model and random dynamics model share deterministic and stochastic characteristics, both of which were needed to understand the complicated behavior caused by the global flow structure. In principle, the fluid dynamics is governed by the Navier-Stokes equations, but direct description based on the Navier-Stokes equation would be too complex to understand. To extract the essential feature, the above-mentioned models will be useful if we are aware of the range of application.

The second comment is on a theoretical development. Recently, a theory was proposed to explain the polygonal flow by the instability of the axisymmetric shape [22]. Their theory assumes potential flow as the basic axisymmetric state and some resonance of centrifugal wave and gravity wave, and the effect of the boundary layers on the bottom and the side wall are much simplified to derive the resonance condition between these waves. Their theory predicts the parameter region of the polygonal flow so that the instability occurs. Mougel et al. applied this theory to the parameters corresponding to the surface switching phenomena, and the mode $m = 2$ (corresponding to the ellipse shape) has the largest growth rate [21], which may (at least partially) explain the reason why the modes $m > 2$ are not observed in the setup of the surface switching [25, 27], while other modes are observed in other setups with different container size and liquid volume.

6.3 Analysis Methods

In this section, the ideas of two analysis methods we are developing are introduced, both of which are expected applicable for the analysis and useful to understand the global flow structure.

When the governing equation is given, the bifurcation analysis is one of the powerful tools to extract the embedded mathematical structure in the global flow problems. In the bifurcation analysis, we define a class of solution which satisfies some mathematical requirements, then find the solution in the parameter space [37, 38]. In the phenomena part, the analysis of the binary fluid convection greatly owe to the bifurcation analysis, in which we define the solution of time periodic traveling solution (Sect. 6.2.1). The branch consisting of the solutions with stable/unstable information gives us characteristics of the solution or the corresponding flow structure. Such solution is useful for the analysis of dynamics. One example is the detailed study of the collision process, which became possible after obtaining the solution.

However, the formation process or interactions in the global flow structure can be understood as a dynamical process where both the initial state and the final state are often neither the same nor clearly defined. For example, the initial state for the formation process of the localized structures is often characterized by random perturbation, which is not a particular state. To directly analyze, we must treat the orbits in the phase space. In the Sect. 6.3.1, we show our recent development of the orbit analysis method that can classify the perturbations to the orbit passing near the steady solution according to the eigenvectors of the critical point.

When the governing equation is not given, the problem becomes a challenge: how can we derive the embedded rule from the observation data alone? If the time series is given, we may construct a return map to extract the embedded rule (as in the model by Sato et al. [40]), but if we desire to know the rule governing the spatio-temporal pattern from the observation, the applicable method is not known. In the Sect. 6.3.2, we propose a new method to construct the cellular automaton from the measurement data alone, without any knowledge of the data.

6.3.1 Orbit Analysis Applying Covariant Lyapunov Analysis

We assume that the governing equation of the target phenomena is known and that the orbit (in the phase space) of interest passes near a steady solution which may be unstable. Here the distance between the orbit and the solution may be measured in the space determined by the phase space and the parameter space. We focus on the dynamics of the orbit starting from a perturbed point from the point on the original orbit, which will be referred to as "perturbed orbit", hereafter.

A comprehensive example is the case that the steady solution is the saddle. In this case, the orbit and the perturbed orbit may be close to a stable manifold of the saddle, however, they may be separated to different directions after passing near the steady solution, each of which may be along (different) unstable manifold. In the collision problem of localized structure, the corresponding saddle and its unstable manifold plays a critical role [37, 38, 49]. To understand or control the relationship between the perturbation direction and the final state far before the orbit reaches the saddle, we need to know pullback vectors of the perturbation vector needed to change the orbit to the desired direction, that is, the eigenvector of the saddle. A new algorithm which pulls back vectors in tangent spaces along the orbit was proposed by applying the calculation method of covariant Lyapunov vectors [41].

Covariant Lyapunov vectors are defined on each point on an orbit as the vectors corresponding intrinsically to an exponent of growth rates of perturbations to an orbit. Further, the covariant Lyapunov vectors have an important feature that their time evolutions in both forward and backward time give also covariant Lyapunov vectors at another point on the orbit. Thus the covariant Lyapunov vectors can be used to estimate the time evolutions of orbits starting from an initial condition perturbed from the original orbit.

We remark that the covariant Lyapunov vector can not be defined for transient orbits. However, an algorithm by Ginelli et al. [61] can be applied to calculate the pullback vectors of the eigenvector. In Ref. [41], we discuss the theoretical background of this method and the algorithm was applied to one of the simplest example; a transient orbit passing near the saddle in a three-dimensional ordinary differential equations. We demonstrated that the control of the orbit far before the saddle can lead the orbit to bend to the direction of the unstable manifold of the saddle. The tested system is the following three-dimensional dynamics system:

$$\frac{d}{dt}\begin{pmatrix} x \\ y \\ z \end{pmatrix} = \begin{pmatrix} Ax\cos(\arctan(Dz)) - By\sin(\arctan(Dz)) \\ Ax\sin(\arctan(Dz)) + By\cos(\arctan(Dz)) \\ C\arctan(z) \end{pmatrix}, \tag{6.3}$$

where A, B, C and D are constants. The origin is the equilibrium of this system and the eigenvalues are A, B and C. In Fig. 6.9, the pullback of the eigenvectors of the equilibrium point (the origin) is shown in the case $A > B > 0 > C$. The pullback vector of the eigenvector tells us the appropriate direction of the perturbation to the orbit to control the perturbated orbit in the future. For example, if the orbit is perturbed along the pullback vector of the eigenvector associated with the second largest eigenvalue (B; solid arrows), then the perturbation vector near the saddle will be parallel to the eigenvector associated with the second largest eigenvalue. This is not trivial, because almost all perturbation vector become parallel to the eigenvector associated with the largest eigenvalue. Using our method, it is possible to calculate such perturbations. Detailed demonstrations are found in Ref. [41]. Further, the fact that the angle made by the two pullback vectors becomes narrow as z becomes large

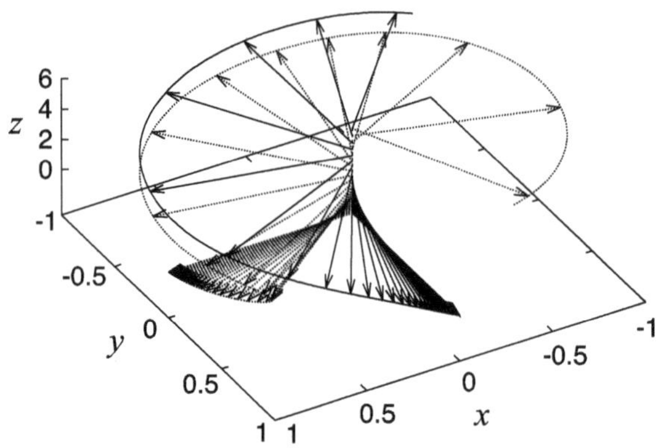

Fig. 6.9 Pullback of eigenvectors of a saddle along the orbit connecting to the saddle using covariant Lyapunov analysis. Pullback vectors of eigenvectors $(1, 0, 0)^T$ and $(0, 1, 0)^T$, which corresponds to the eigenvalues A (*dotted arrows*) and B (*solid arrows*), respectively, are shown. Endpoints of arrows are connected by curves to guide the eyes. The orbit is close to $z-$axis and the z-component of the initial state is positive

indicates that the random perturbations added at a point where z is large enough is not likely to result in an orbit passing inside the region made by these pullback vectors. Thus, the pullback vectors also give us information of the probable perturbed orbits passing near the saddle.

6.3.2 Generating Cellular Automata Rule from Measurement Data Alone

When we have a measurement data of a spatio-temporal pattern alone (e.g. video record of a phenomena), how can we extract the embedded rule of the dynamics? In some cases we may create a mathematical model based on the knowledge of physics or our understanding of the phenomena. In this section, however, we do not rely on such a priori assumption and consider the method to generate appropriate cellular automation rule, which was explained in detail in Refs. [39, 42]. The procedure to construct a rule of CA from the data is summarized as follows. First, we discretize the data in space, time, and the observed state. After discretization, the number of possible states at each site in a discretized space-time plane is finite. Second, we predetermine the interaction range and number of the states. Assuming that the rule is homogeneous, we can list all the possible interactions. Then we calculate the appearance frequency of particular state from combinations of neighbor's possible states. To define the rule, we have two choices. One is a deterministic rule by taking the state of the maximum count. The other is a stochastic rule in which each interaction can be chosen with a probability proportional to the count, for example. These definition

Fig. 6.10 *Left* Spatiotemporal pattern generated by cellular automaton (CA), Rule 90. *Right* Noise-contaminated CA, Rule 90 ($p = 0.05$)

rules are used in Refs. [39, 42, 43], but other definition rule can be applied fitted to the problem, e.g., some conservation law may be considered.

This method was firstly applied to noise contaminated CA pattern. We illustrate the definition by the elementary CA Rule 90, defined by

$$x_n(i+1) = x_{n-1}(i) + x_{n+1}(i) \pmod 2, \tag{6.4}$$

where $x_n(i) \in \{0, 1\}$ represents the state on the site (n, i); n and i are indexes of space and time, respectively. For space (indexed by n), the periodic boundary condition is applied. The noise-contaminated CA is generated by the additional procedure that the calculated state is flipped with probability p [42]. In Fig. 6.10 (Left), the spatio-temporal pattern is shown; a well-known pattern with self-similar triangles is generated. Figure 6.10 (Right) shows a noise-contaminated CA ($p = 0.05$) is shown. Although the probability is small and the CA rule is modified only slightly, the self-similar triangles are completely destroyed and we can only see a pattern with random triangles with noise. Thus, the contaminated pattern is far from typical patterns expected by the original rule (Fig. 6.10). In Ref. [42], we show that the embedded rule can be extracted by our method. Then, simple partial differential equations (PDE), the diffusion equations and the Burgers equations, are used to generate the data, and this method was applied to examine the reproductivity of the properties of the original PDE. For the diffusion equations, stochastic type model gives a robust time series of the variance if the number of the state is larger than two, but they fit the theoretical line until limited steps (20–30 steps). For the Burgers equations, stochastic type model gives a generation of shock and their merging process [39]. Also this method was tried to reproduce the experimental data of bioconvection pattern [43].

6.4 Concluding Remarks

In this review, we have outlined several phenomena related to the global flow structure in Sect. 6.3, that is, binary fluid convection, bioconvection, and surface switching, in terms of recent developments after the previous review [24]. All of these examples have multi-scale nature and heterogeneous flow structure in common. In binary fluid convection, two types of localized structures are observed and their collision dynamics based on the well controlled initial state obtained by the numerical solution reveals a heterogeneous the pattern formation process. Here the macroscopic pattern consists of localized structures composed of convection cells. In bioconvection, two types of localized structures which somehow are quite similar to those in binary fluid convection are experimentally obtained using new setup and initial states. Their interaction behaviors are similar to not only those in the binary fluid convection but also those in dissipative systems such as the reaction-diffusion systems. Similar to the case of binary fluid convection, the generated macroscopic pattern of bioconvection is heterogeneous and the multi-scale nature can be traced back to the microscopic flow around the microorganism. In the surface switching, the boundary layer due to the rotation of the bottom and the bulk flow with free surface is globally coupled, which is the main idea of the recently theory by Tophøj et al. [22].

Such multi-scale phenomena will show different aspects depending on the target scale, and they are not always fit our natural expectation. Thus we need a new analysis method applicable to such phenomena which is outlined in Sect. 6.3. When the governing equation is given, we can use analytical approach for the analysis. In addition to the bifurcation analysis which basically focus on the solution of various type, we propose a method to analyze the orbit based on the covariant Lyapunov analysis(Sect. 6.3.1). Although it is restricted to the orbit passing through near the steady solution, this method allows us to control the orbit in the direction to the unstable manifold. When the governing equation is not given, the first task to understand the phenomena mathematically is to extract the embedded rule. We proposed a framework using the idea of cellular automata to achieve this goal (Sect. 6.3.2). This method can be used without any knowledge of the target phenomena.

The merit to have the idea of the global flow structure is that we can understand different problems from a single point of view. The successful of the bioconvection analysis is clearly due to the fact that we regard the bioconvection and the binary fluid convection as a single category problem, whereby the idea used in the analysis of the binary fluid convection could be applied to the analysis of the bioconvection problems. Although some modification from the mathematical idea to the experimental setup is needed, a policy of analysis can be easily transferred. In the analysis of the surface switching, the one-dimensional random dynamics gave us insights to further flow analysis, which are now prepared by two of the authors for a series of papers [28, 29].

For the analysis method, we could only show the idea and give simple but essential examples. The analysis method based on the covariant Lyapunov analysis will be able to be directly used to control the collision problems in reaction-diffusion problems and binary fluid convection problems, because such collision can be understood as an orbit passing through the special saddle called '*scatter*'. It will be very interesting of us to know the appropriate perturbation to determine the desired result of the collision even when two localized structures are far from each other, which is one of our current interests. The method to construct the cellular automaton from the measurement data was applied to construct the bioconvection pattern, which was not spatially localized [43]. We obtained several interesting patterns, but they are not very similar to the original convection patterns. Perhaps we need much appropriate phenomena to evaluate the potential of this method.

For the model of the surface switching and the CA model has the following feature in common. That is, the major part of the dynamics is based on the deterministic simple model of single scale or single mechanics, and stochastic effect is used to represent the effect coming from other part of the dynamics. These terms are used as a convenient representation of other part of dynamics, which would be complex or have different time scales. As discussed in Ref. [24] where the Wayland test was applied to the times series data of the surface switching, this idea itself is not new; the connection between this type of representation and the mathematical structure embedded in the global flow structure problem should be clarified, then, the construction of the model will develop further.

In this review, we mentioned several phenomena of and several proposal to analyze the global flow structure. As has been discussed so far, the analysis of phenomena and the development of method can not be separated; selection of appropriate problem and method will be a great driving force for the understanding of the global flow structure.

Acknowledgements The authors were major members of the CREST project "A challenge to unsolved problems in fluid engineering with modern mathematical analysis" headed by Professor Yoshihiro Shibata in Waseda University in the JST mathematics program "Alliance for breakthrough between mathematics and sciences" governed by Professor Yasumasa Nishiura in Tohoku University from April 1st, 2010 to March 31st, 2015. We would like to appreciate both financial and academic support for this project. The part of this study was supported by CREST No. PJ74100011, and Grants-in-Aid for Scientific Research (26400396, 25289062, 23540433, 21340019.

References

1. Iima, M., Tasaka, Y., Watanabe, T.: Analysis of global flow structure. GAKUTO Int. Ser. Math. Sci. Appl. **34**, 25–38 (2011)
2. Hill, N.A., Pedley, T.J.: Bioconvection. Fluid Dyn. Res. **37**(1-2), 1–20 (2005)
3. Ishikawa, T.: Suspension biomechanics of swimming microbes. J. Royal Soc. Interface **6**(39), 815–834 (2009)
4. Platt, J.: Bioconvection patterns in cultures of free-swimming organisms. Science **133**, 1766–1767 (1961)
5. Shoji, E., Nishimori, H., Awazu, A., Izumi, S., Iima, M.: Localized bioconvection patterns and their initial state dependency in euglena gracilis suspensions in an annular container. J. Phys. Soc. Jpn. **83**, 043001 (2014)

6. Suematsu, N.J., Awazu, A., Izumi, S., Noda, S., Nakata, S., Nishimori, H.: Localized bioconvection of euglena caused by phototaxis in the lateral direction. J. Phys. Soc. Jpn. **80**(6), 064003 (2011)
7. Dudley, R.: The Biomechanics of Insect Flight: Form, Function, Evolution. Princeton University Press, Princeton (2000)
8. Iima, M., Yanagita, T.: Is a two-dimensional butterfly able to fly by symmetric flapping? J. Phys. Soc. Jpn **70**(1), 5–8 (2001)
9. Iima, M.: A paradox of hovering insect in two-dimensional space. J. Fluid. Mech. **617**, 207–229 (2008)
10. Iima, M., Yokoyama, N., Hirai, N., Senda, K.: Controlling flow structures by wing motion in a flapping-flight model. Adv. Sci. Technol. **84**, 59–65 (2013)
11. Ota, K., Suzuki, K., Inamuro, T.: Lift generation by a two-dimensional symmetric flapping wing: immersed boundary-lattice Boltzmann simulations. Fluid Dyn. Res. **44**(4), 045504 (2012)
12. Senda, K., Obara, T., Kitamura, M., Nishikata, T., Hirai, N., Iima, M., Yokoyama, N.: Modeling and emergence of flapping flight of butterfly based on experimental measurements. Robot. Auton. Syst. **60**, 670–678 (2012)
13. Senda, K., Obara, T., Kitamura, M., Yokoyama, N., Hirai, N., Iima, M.: Effects of structural flexibility of wings in flapping flight of butterfly. Bioinspiration Biomimetics **7**(2), 025002 (2012)
14. Tanaka, H., Shimoyama, I.: Forward flight of swallowtail butterfly with simple flapping motion. Bioinspiration Biomimetics **5**(2), 026003 (2010)
15. Yokoyama, N., Senda, K., Iima, M., Hirai, N.: Aerodynamic forces and vortical structures in flapping butterfly's forward flight. Phys. Fluids **25**, 021902 (2013)
16. Abderrahmane, H.A., Siddiqui, K., Vatistas, G.H.: Rotating waves within a hollow vortex core. Exp. Fluids **50**(3), 677–688 (2010)
17. Bach, B., Linnartz, E.C., Vested, M.H., Andersen, A., Bohr, T.: From Newton's bucket to rotating polygons: experiments on surface instabilities in swirling flows. J. Fluid Mech. **759**, 386–403 (2014)
18. Bergmann, R., Tophoj, L., Homan, T.A.M., Hersen, P., Andersen, A., Bohr, T.: Polygon formation and surface flow on a rotating fluid surface. J. Fluid Mech. **679**, 415–431 (2011)
19. Iga, K., Yokota, S., Watanabe, S., Ikeda, T., Niino, H., Misawa, N.: Various phenomena on a water vortex in a cylindrical tank over a rotating bottom. Fluid Dyn. Res. **46**(3), 031409 (2014)
20. Jansson, T., Haspang, M., Jensen, K.R., Hersen, P., Bohr, T.: Polygons on a rotating fluid surface. Phys. Rev. Lett. **96**(17), 1–4 (2006)
21. Mougel, J., Fabre, D., Lacaze, L.: Waves and instabilities in rotating free surface flows. Mech. Ind. **15**(2), 107–112 (2014)
22. Tophøj, L., Mougel, J., Bohr, T., Fabre, D.: Rotating polygon instability of a swirling free surface flow. Phys. Rev. Lett. **110**(19), 194502 (2013)
23. Vatistas, G.H.: A note on liquid vortex sloshing and Kelvin's equilibria. J. Fluid. Mech. **217**, 241–248 (1990)
24. Iima, M., Iijima, Y., Sato, Y., Tasaka, Y.: A time-series analysis of the free-surface motion of rotational flow. Theor. Appl. Mech. Jpn. **59**, 187–193 (2011)
25. Suzuki, T., Iima, M., Hayase, Y.: Surface switching of rotating fluid in a cylinder. Phys. Fluilds **18**, 101701 (2006)
26. Tasaka, Y., Ito, K., Iima, M.: Visualization of a rotating flow under large-deformed free surface using anisotropic flakes. J. Vis. **11**, 163–172 (2008)
27. Tasaka, Y., Iima, M.: Flow transitions in the surface switching of rotating fluid. J. Fluid. Mech. **636**, 475–484 (2009)
28. Makoto, I., Yuji, T.: Flow structures and their dynamics during surface switching of rotation fluid. Part I: Quarupole Flow Associated with Asymmetric Surface Shape (in preparation)
29. Tasaka, Y., Iima, I.: Flow structures and their dynamics during surface switching of rotating fluid. Part 2: Influence of Disk-Rim-Gap on the Switching (in preparation)

30. Barten, W., Lucke, M., Kamps, M., Schmits, R.: Convection in binary fluid mixtures. I extended traveling-wave and stationary states. Phys. Rev. E **51**, 5636–5661 (1995)
31. Barten, W., Lucke, M., Kamps, M., Schmits, R.: Convection in binary fluid mixtures. IILocalized traveilng waves. Phys. Rev. E **51**, 5662–5679 (1995)
32. Batiste, O., Knobloch, E., Alonso, A., Mercader, I.: Spatially localized binary-fluid convection. J. Fluid. Mech. **560**, 149–158 (2006)
33. Iima, M., Nishiura, Y.: Unstable periodic solution controlling collision of localized convection cells in binary fluid mixture. Phys. D **238**, 449–460 (2009)
34. Kolodner, P.: Interactions of localized pulses of traveling-wave convection with propagating disturbances. Phys. Rev. A **42**, 7504–7506 (1990)
35. Kolodner, P.: Drift, shape, and intrinsic destabilization of pulses of traveling-wave convection. Phys. Rev. A **44**, 6448–6465 (1991)
36. Niemela, J.J., Ahlers, G., Cannell, D.S.: Localized traveling-wave states in binary-fluid convection. Phys. Rev. Lett. **64**, 1365–1368 (1990)
37. Watanabe, T., Toyabe, K., Iima, M., Nishiura, Y.: Time-periodic traveling solutions of localized convection cells in binary fluid mixture. Theor. Appl. Mech. Jpn. **59**, 211–219 (2011)
38. Watanabe, T., Iima, M., Nishiura, Y.: Spontaneous formation of travelling localized structures and their asymptotic behaviour in binary fluid convection. J. Fluid Mech. **712**, 219–243 (2012)
39. Kawaharada, A., Shoji, E., Nishimori, H., Awazu, A., Izumi, S., Iima, M.: Cellular automata automatically constructed from a bioconvection pattern. In: Suzuki Y., Hagiya M. (eds.) Recent Advances in Natural Computing. Mathematics for Industry, vol. 14, pp. 15–25. Springer (2016)
40. Sato, Y., Iima, M., Tasaka, Y.: Random dynamics from a time series of rotating fluid. Hokkaido University Preprint Series in Mathematics, p. 979 (2011)
41. Yamaguchi, T., Iima, M.: Numerical analysis of transient orbits by the pullback method for covariant Lyapunov vector. Theor. Appl. Mech. Jpn. **63**, 91–96 (2015)
42. Kawaharada, A., Iima, M.: Constructing cellular automaton models from observation data. In: 2013 First International Symposium on Computing and Networking, pp. 559–562. Ieee (2013)
43. Kawaharada, A., Iima, M.: An application of data-based construction method of cellular automata to physical phenomena. J. Cell. Automata (2015) in press
44. Barten, W., Lücke, M., Kamps, M.: Localized traveling-wave convection in binary-fluid mixtures. Phys. Rev. Lett. **66**, 2621–2624 (1991)
45. Pomeau, Y.: Front motion, metastability and subcritical bifurcations in hydrodynamics. Phys. D **23**, 3–11 (1986)
46. Burke, J., Knobloch, E.: Homoclinic snaking: structure and stability. Chaos (Woodbury, N.Y.) **17**(3), 037102 (2007)
47. Kazutaka, T.: Collision dynamics of localized convection cells in binary fluid mixture: network structure and collision orbit. Master thesis, Hokkaido University (in Japanese) (2009)
48. Iima, M., Nishiura, Y.: Collision of localized traveling-wave convection cells in binary fluid. GAKUTO Int. Ser. Math. Sci. Appl. **22**, 289–303 (2005)
49. Nishiura, Y., Teramoto, T., Ueda, K.I.: Scattering of traveling spots in dissipative systems. Chaos **15**, 47509 (2005)
50. Pedley, T.J., Kessler, J.O.: Hydrodynamic phenomema in suspensions of swimming microorganisms. Ann. Rev. Fluid Mech. **24**, 313–358 (1992)
51. Jánosi, I., Kessler, J., Horváth, V.: Onset of bioconvection in suspensions of Bacillus subtilis. Phys. Rev. E **58**(4), 4793–4800 (1998). http://pre.aps.org/abstract/PRE/v58/i4/p4793_1
52. Wager, H.: On the effect of gravity upon the movements and aggregation of euglena viridis, ehrb., and other micro-organisms. Philos. Trans. Royal Soc. B: Biol. Sci. **201**(274–281), 333–390 (1911)
53. Chandrasekhar, S.: Hydrodynamic and Hydromagnetic Stability. Dover Publications (1981)
54. Ghorai, S., Panda, M.K., Hill, N.a.: Bioconvection in a suspension of isotropically scattering phototactic algae. Phys. Fluids **22**(7), 071901 (2010)
55. Ghorai, S., Panda, M.: Bioconvection in an anisotropic scattering suspension of phototactic algae. Eur. J. Mech.- B/Fluids **41**, 81–93 (2013)

56. Vincent, R., Hill, N.: Bioconvection in a suspension of phototactic algae. J. Fluid Mech. **327**, 343–371 (1996)
57. Williams, C.R., Bees, M.A.: A tale of three taxes: photo-gyro-gravitactic bioconvection. J. Exp. Biol. **214**(Pt 14), 2398–2408 (2011)
58. Williams, C.R., Bees, M.A.: Photo-gyrotactic bioconvection. J. Fluid Mech. **678**, 41–86 (2011)
59. Tasaka, Y., Iima, M., Ito, K.: Rotataing flow transition related to surface switching. J. Phys.: Conf. Ser. **137**, 12030 (2008)
60. Kawaharada, A., Iima, M., Tasaka, Y.: A Markov chain analysis of rotating flow with free surface. Abstracts of the Joint-Workshop on the Applied Mathematics 2013 (2013), pp. 132–135 (in Japanese)
61. Ginelli, F., Poggi, P.: Turchi, A., Chaté, H., Livi, R., Politi, A.: Characterizing dynamics with covariant lyapunov vectors. Phys. Rev. Lett. **99**, 1–4 (2007)

Chapter 7
Mathematical and Numerical Analysis of the Rayleigh-Plesset and the Keller Equations

Masashi Ohnawa and Yukihito Suzuki

Abstract In the present paper, we conduct mathematical analysis on the Rayleigh-Plesset and the Keller equations, ordinary differential equations of the second order widely used for describing motions of a spherically symmetric single bubble. We show that these equations admit structures of the Hamiltonian system with respect to a physically reasonable energy function perturbed by dissipation and obtain the asymptotic behavior of the solutions. Making use of this structure, we rewrite the equations into gradient systems and develop numerical codes which properly inherit conservation or dissipation of the energy from the original differential equations following the discrete gradient method.

Keywords Bubble dynamics · Rayleigh-Plesset equation · Keller equation · Hamiltonian system · Discrete gradient method

7.1 Introduction

In the present paper, we analyze mathematically the Rayleigh-Plesset and the Keller equations and based on the results, we numerically integrate the equations following the discrete gradient method which preserves important mathematical features in the discretized forms.

The Rayleigh-Plesset equation was proposed in [9, 13]. It is an ordinary differential equation of the second order which describes the radial motion of a spherical bubble immersed in a incompressible liquid and has been used extensively for exploring the dynamics of cavitation bubbles [1, 10]. Reflecting on the significance of cavitation

M. Ohnawa (✉)
Department of Ocean Sciences, Faculty of Marine Science, Tokyo University of Marine Science and Technology, 4-5-7, Konan, Minato-ku, Tokyo 108-8477, Japan
e-mail: ohnawa@m.kaiyodai.ac.jp

Y. Suzuki
Department of Mathematics, Waseda University, 3-4-1 Ohkubo, Shinjuku-ku, Tokyo 169-8555, Japan
e-mail: y_suzuki@aoni.waseda.jp

Y. Shibata and Y. Suzuki (eds.), *Mathematical Fluid Dynamics, Present and Future*, Springer Proceedings in Mathematics & Statistics 183,
DOI 10.1007/978-4-431-56457-7_7

in industry, a lot of efforts have been done to extend the Rayleigh-Plesset equation to account for phenomena in various occasions. For example, in order to model the damping oscillations of underwater explosion bubbles which the Rayleigh-Plesset equation fails to predict, Keller and Kolodner [8] introduced an effect of emission of pressure waves into the equation and proposed the Rayleigh-Plesset-Keller equation or simply the Keller equation which describes the radial motion of a spherical bubble immersed in a (nearly) incompressible, (outward) wave transmitting liquid. Depending on contexts, there are some different implications in the extensions of the Rayleigh-Plesset equation [3, 5, 11, 12, 14]. Among them, we deal with the Keller equation and the Herring equations, which we call the Keller type equations altogether.

In spite of intensive studies on modeling cavitation bubbles and derivation of the governing equations for oscillating bubbles, their mathematical aspects do not seem to have received much attention. Even to the simplest one such as the Rayleigh-Plesset equation, the existence and uniqueness of global solutions is proved only recently in [2]. Following this study, we investigate mathematical aspects of the Keller type equations. We introduce an energy function which makes physical sense and prove its preservation or dissipation. Unique existence of the global solutions and large time behaviors are obtained from this. This fact also allows us to formulate those equations as the Hamiltonian systems including dissipation or damping of energy.

As for the numerical methods, there are some peculiar methods to Hamiltonian systems, which inherit structures possessed by original differential equations. Symplectic schemes which preserve the symplectic structure of canonical Hamiltonian systems are typical examples of such kind of methods. In a different context, the so called "discrete gradient method" [4] has also been proposed, which is applicable not only to conservative systems but also to dissipative systems. This type of scheme inherits the skew-symmetry and negative semi-definiteness of the operators acting on the gradient of the Hamiltonian from the original systems and thus directly guarantees the conservation or dissipation of the discretized energy. We apply the discrete gradient method to the Keller type equations and obtain energy preserving or dissipating numerical schemes for those equations, which properly reproduce the dynamics proved by the mathematical analysis mentioned above.

The paper is organized as follows. In Sect. 7.2, we give a brief derivation of the Rayleigh-Plesset and the Keller equations from the Bernoulli equation and the conservation of mass following [9] and [8] respectively. Note that Prosreretti and Lezzi [11] extended the Rayleigh-Plesset equations in a different way starting from the compressible Navier-Stokes equations using an asymptotic expansion in terms of the Mach number of the velocity of bubble's surface. Section 7.3 is devoted to the mathematical analysis on the Keller type equations. We prove the global existence and the uniqueness of solutions and show principal properties of the dynamics such as the convergence of the solution to the equilibrium state and the exponential decay of the energy. In Sect. 7.4, the Keller type equations are reformulated as the Hamiltonian systems with dissipation or damping of energy. Based on this formulation, in Sect. 7.5 we develop a numerical scheme which properly inherits conservation or dissipation of energy following the discrete gradient method. We employ the coordinate increment method proposed by Ito and Abe [7] to get the discrete gradient operator. Finally in Sect. 7.6, we present numerical results to demonstrate the usefulness of the proposed method.

7.2 Mathematical Models for Motion of a Spherical Bubble

We consider the motions of an oscillating bubble immersed in a liquid. We assume that the bubble is spherical and the motion is spherically symmetric.

Then the radius $R = R(t)$ of the bubble as a function of time t describes the kinetics completely. The bubble is filled with vapor and a non-condensable gas. All thermal effects have been ignored and the vapor pressure p_V is assumed to be constant. The pressure of the non-condensable gas p_G follows the ideal gas law and determined by the radius of the bubble. Under these conditions, the pressure inside the bubble p_B is represented by

$$p_B(R) = p_V + p_G(R) = p_V + \bar{p}_G \left(\frac{\bar{R}}{R}\right)^{3\kappa}, \tag{7.1}$$

where a constant $\kappa \geqslant 1$ is prescribed depending on adiabatic, isothermal or intermediate nature of the process, and $\bar{p}_G$ is the reference pressure of the non-condensable gas when the bubble radius is $R = \bar{R}$ as a reference state. Particularly, if we assume the adiabatic process, κ corresponds to the heat capacity ratio.

The configuration of the bubble is illustrated in Fig. 7.1 (Table 7.1).

7.2.1 The Rayleigh-Plesset Equation

Spherically symmetric motions of the bubble induce a spherically symmetric flows in the surrounding liquid, which are irrotational. Then the radial velocity field $v = v(t, r)$ of the liquid has the velocity potential $\phi = \phi(t, r)$ such that $v = \partial\phi/\partial r$, which satisfies the Bernoulli's equation

Fig. 7.1 Spherical bubble

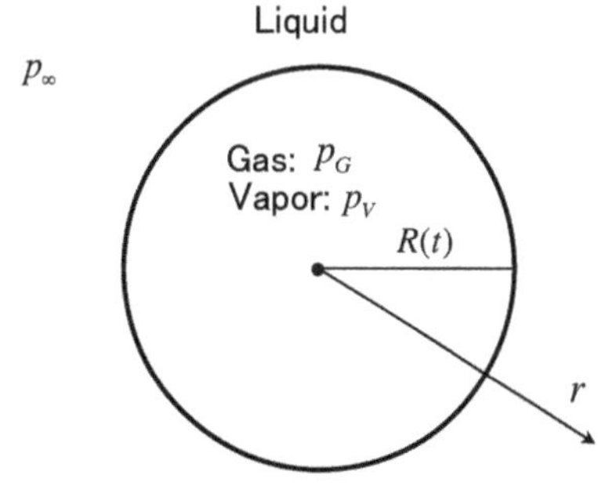

Table 7.1 List of symbols

Name	Symbol
Vapor pressure	p_V
Reference radius	$\bar{R}$
Reference pressure	$\bar{p}_G$
Exponent of the barotropic relation	κ
Coefficient of surface tension	σ
Density of the liquid	ρ_L
Coefficient of viscosity of the liquid	μ_L
Speed of sound in the liquid	c_L
External pressure	p_∞

$$\frac{\partial}{\partial r}\left[\frac{\partial \phi}{\partial t}(t,r)+\frac{1}{2}\left(\frac{\partial \phi}{\partial r}(t,r)\right)^2+\int^{p(t,r)}\frac{dp}{\rho}\right]=0, \tag{7.2}$$

where $p = p(t, r)$ and $\rho(p(t, r))$ are the pressure and density field of the liquid. We assume that the flow is barotropic, namely the density depends only on the pressure. If we further assume that the flow of the surrounding liquid is incompressible and the density is constant: $\rho = \rho_L$, then the velocity potential satisfies the Laplace equation

$$\Delta\phi = \frac{1}{r^2}\frac{\partial}{\partial r}\left(r^2\frac{\partial \phi}{\partial r}\right)=0, \tag{7.3}$$

which implies that there is a function $f = f(t)$ such that

$$\phi(t,r)=\frac{f(t)}{r}. \tag{7.4}$$

Boundary conditions at the surface of the bubble $r = R$ are the kinematic condition:

$$\left.\frac{\partial \phi}{\partial r}\right|_{r=R}=\dot{R}, \tag{7.5}$$

and the dynamic condition:

$$p_B(R(t)) - p(t,R(t)) - \frac{2\sigma}{R(t)} + 2\mu_L\frac{\partial^2\phi}{\partial r^2}(t,R(t))=0, \tag{7.6}$$

where μ_L is the coefficient of viscosity of the liquid and σ is the coefficient of surface tension. The dynamics condition represents the balance among the viscous stress, the surface tension and the pressures acting on both sides of the surface.

Evaluating Eq. (7.4) at the surface $r = R$ and appealing to the kinematic boundary condition (7.5), the function f is determined as $f(t) = -\dot{R}(t)R(t)^2$ and the velocity potential becomes

$$\phi(t, r) = -\frac{\dot{R}(t)R(t)^2}{r}. \tag{7.7}$$

We integrate Eq. (7.2) assuming both the velocity and the velocity potential vanish at infinity to get

$$p(t, r) = p_\infty - \rho_L \left[\frac{\partial \phi}{\partial t}(t, r) + \frac{1}{2} \left(\frac{\partial \phi}{\partial r}(t, r) \right)^2 \right]. \tag{7.8}$$

Substituting (7.7) into (7.8), we get

$$p(t, r) = p_\infty + \rho_L \left[\frac{\ddot{R}(t)R(t)^2 + 2\dot{R}(t)^2 R(t)}{r} - \frac{1}{2} \frac{\dot{R}(t)^2 R(t)^4}{r^4} \right], \tag{7.9}$$

which is the pressure field of the liquid induced by the motion of the bubble. Evaluating this equation at the surface $r = R$, we have

$$\frac{p(t, R(t)) - p_\infty}{\rho_L} = \ddot{R}(t)R(t) + \frac{3}{2}\dot{R}(t)^2. \tag{7.10}$$

The pressure of the liquid at the surface $p(t, R(t))$ in (7.10) is determined from Eqs. (7.6) and (7.7):

$$p(t, R(t)) = p_B(R(t)) - \frac{2\sigma}{R(t)} - 4\mu_L \frac{\dot{R}(t)}{R(t)}. \tag{7.11}$$

Substituting (7.11) into (7.10), we have

$$R\ddot{R} + \frac{3}{2}\dot{R}^2 + \frac{p_\infty - [p_B(R) - 2\sigma/R - 4\mu_L \dot{R}/R]}{\rho_L} = 0. \tag{7.12}$$

The Eq. (7.12) supplemented with (7.1) is called the Rayleigh-Plesset equation. This governs the radially symmetric motion of a spherical bubble immersed in an incompressible liquid.

7.2.2 *The Rayleigh-Plesset-Keller Equation*

As mentioned in the introduction, Keller and Kolodner [8] introduced an effect of emission of pressure waves into the Rayleigh-Plesset equation. To this end, the incompressibility condition: Eq. (7.3) is replaced with the wave equation

$$\Delta\phi - \frac{1}{c_L^2}\frac{\partial^2\phi}{\partial t^2} = 0, \tag{7.13}$$

where the wave velocity is assumed to be the speed of sound c_L in the liquid at infinity. If we consider only the outward wave, the velocity potential can be represented by

$$\phi(t, r) = \frac{f(t - r/c_L)}{r} \tag{7.14}$$

as a solution of the wave equation (7.13). Then Eq. (7.4) is replaced with Eq. (7.14). Evaluating Eq. (7.14) at the surface $r = R$ and appealing to the kinematic boundary condition (7.5), we get

$$\dot{R}(t) = -\frac{f(t - R(t)/c_L)}{R(t)^2} - \frac{1}{c_L}\frac{f'(t - R(t)/c_L)}{R(t)}. \tag{7.15}$$

On the other hand, substituting the velocity potential, Eq. (7.14) into (7.8), we get the pressure field as

$$p(t, r) = p_\infty - \rho_L \left\{ \frac{f'(t - r/c_L)}{r} + \frac{1}{2}\left[\frac{f(t - r/c_L)}{r^2} + \frac{1}{c_L}\frac{f'(t - r/c_L)}{r}\right]^2 \right\}. \tag{7.16}$$

Then the pressure at the surface is

$$p(t, R(t)) = p_\infty - \rho_L \left\{ \frac{f'(t - R(t)/c_L)}{R(t)} + \frac{1}{2}\dot{R}(t)^2 \right\}. \tag{7.17}$$

By solving (7.15) and (7.17) for $f(t - R/c_L)$ and $f'(t - R/c_L)$ in terms of the dynamical state of the bubble $(R, \dot{R})$, we have

$$f(t - R/c_L) = -R^2\dot{R} + \frac{R^2}{c_L}\left[\frac{1}{2}\dot{R}^2 - \frac{p_\infty - p(t, R)}{\rho_L}\right], \tag{7.18}$$

$$f'(t - R/c_L) = -R\left[\frac{1}{2}\dot{R}^2 - \frac{p_\infty - p(t, R)}{\rho_L}\right]. \tag{7.19}$$

Taking time derivative of Eq. (7.18) and equating it with Eq. (7.19), we can eliminate the function f to obtain the equation which governs the motion of the bubble:

$$\left(1 - \frac{\dot{R}}{c_L}\right)R\ddot{R} + \frac{3}{2}\left(1 - \frac{\dot{R}}{3c_L}\right)\dot{R}^2 = -\left(1 + \frac{\dot{R}}{c_L} + \frac{R}{c_L}\frac{d}{dt}\right)\frac{p_\infty - p(t, R)}{\rho_L}. \tag{7.20}$$

Note that the pressure of the surrounding liquid at the surface $p(t, R)$ depends only on the dynamical state of the bubble $(R, \dot{R})$ as in the case of the Rayleigh-Plesset equation. Equation (7.20) supplemented with (7.11) and (7.1) is called the Rayleigh-Plesset-Keller equation or simply the Keller equation. This governs the radially symmetric motion of a spherical bubble immersed in a (nearly) incompressible, (outward) wave transmitting liquid.

In a slightly different context, another equation called the Herring equation was derived [5, 14] and can be combined with the Keller equation as

$$\begin{aligned}\left(1-(\Lambda+1)\frac{\dot{R}}{c_L}\right)R\ddot{R}+\frac{3}{2}\left(1-\frac{3\Lambda+1}{3}\frac{\dot{R}}{c_L}\right)\dot{R}^2\\ =-\left(1+(1-\Lambda)\frac{\dot{R}}{c_L}+\frac{R}{c_L}\frac{d}{dt}\right)\frac{p_\infty-p(t,R)}{\rho_L},\end{aligned} \tag{7.21}$$

where Λ is a real parameter. The Keller and the Herring equations are recovered when $\Lambda = 0$ and $\Lambda = 1$, respectively.

7.3 Mathematical Analysis

First we quote fundamental mathematical facts from [2] for the better understanding of mathematical analysis on the Keller-Herring equation in the next theorem and the gradient system formulation in the next section.

Proposition 7.3.1 ([2])
Consider the first order system

$$\dot{R} = V, \tag{7.22}$$

$$\dot{V} = -\left(\frac{3}{2}V^2+\frac{P(R)}{\rho_L}+\frac{4\mu_L V}{\rho_L R}\right)/R, \tag{7.23}$$

where

$$P(R) := p_\infty - \left(p_B(R)-\frac{2\sigma}{R}\right) = (p_\infty - p_V) + \frac{2\sigma}{R} - \bar{p}_G\left(\frac{\bar{R}}{R}\right)^{3\kappa}. \tag{7.24}$$

This is equivalent to the Rayleigh-Plesset equation (7.10). *Assuming $p_\infty \geqslant p_V$, the followings hold.*

(i) There exists a unique equilibrium point $(R, V) = (R_*, 0)$, where R_* is the unique positive solution to the equation $P(R_*) = 0$.
(ii) For an arbitrary initial datum, there exists a unique global solution. Defining an energy function $E(R, V)$ by

$$E(R,V) := \frac{1}{2}\rho_L R^3 V^2 + G(R),\quad G(R) := \int_{R_*}^{R} P(r)r^2 dr, \tag{7.25}$$

the solution satisfies

$$\frac{d}{dt}E(R(t), V(t)) = -\mu_L R V^2.$$

(iii) *In the case* $\mu_L = 0$*, the solution is periodic for an arbitrary initial datum.*
(iv) *In the case* $\mu_L > 0$*, the solution converges to* $(R_*, 0)$ *for an arbitrary initial datum with the energy* $E(R(t), V(t))$ *decaying exponentially fast. Moreover, if* $\mu_L > \mu_C := (3\kappa\rho_L(p_\infty - p_V) + 2(3\kappa - 1)\rho_L\sigma/R_*)^{1/2} R_*/2$*, then* $V(t)$ *changes its sign only finite times while if* $\mu_L < \mu_C$*, then* $V(t)$ *changes its sign infinitely many times.*

Remark 7.3.1 Some of the statements in this proposition are not included in [2]. Since they are proved similarly to the corresponding claims in the next theorem, we omit the proof.

Remark 7.3.2 Since $p_B(R) - 2\sigma/R$ in (7.24) is the pressure of the inviscid liquid (i.e. neglecting the viscous stress) at the surface of the bubble whose radius is R (see Eq. (7.11)), $G(R)$ stands for the work done when the bubble radius increases from $r = R_*$ to $r = R$. Note that $G(R) \geqslant G(R_*) = 0$ holds for all $R > 0$ since $P(R) \gtrless 0$ for $R \gtrless R_*$. The first term of the right hand side of Eq. (7.25) is the kinetic energy of the surrounding fluid.[1]

In the same spirit, we proceed to mathematically analyze the Keller-Herring equation (7.21). Rewriting it into a first order system, we have

$$\dot{R} = V, \tag{7.26}$$

$$\left(1 - (\Lambda + 1)\frac{V}{c_L}\right)\dot{V} = -\left\{\frac{3}{2}\left(1 - \frac{3\Lambda + 1}{3}\frac{V}{c_L}\right)V^2 + \left(1 + (1 - \Lambda)\frac{V}{c_L} + \frac{R}{c_L}\frac{d}{dt}\right)\frac{P(R(t))}{\rho_L}\right\}/R. \tag{7.27}$$

Theorem 7.3.1 *Consider the Keller-Herring system (7.26) and (7.27). Assuming* $p_\infty \geqslant p_V$ *and* $V(0) < c_L/(\Lambda + 1)$*, we have the followings.*

(i) *There exists a unique equilibrium point* $(R, V) = (R_*, 0)$*, where* R_* *is the unique positive solution to the equation* $P(R_*) = 0$*.*
(ii) *For an arbitrary initial datum, there exists a unique global solution which satisfies*

$$V(t) < \frac{c_L}{\Lambda + 1} \quad \textit{for all } t \geqslant 0. \tag{7.28}$$

[1] Strictly speaking, the work and the kinetic energy have been divided by 4π.

The energy defined in (7.25) *decays monotonically as*

$$\frac{d}{dt}E(R(t),V(t))$$
$$= -\frac{R^2V^2}{c_L-(\Lambda+1)V}\left(\rho_L V^2+2(p_\infty-p_V)+(3\kappa-2)\bar{p}_G\left(\frac{\bar{R}}{R}\right)^{3\kappa}+\frac{2\sigma}{R}\right), \tag{7.29}$$

and

$$\lim_{t\to\infty}(R(t),V(t))=(R_*,0). \tag{7.30}$$

(*iii*) *Define a positive constant* c_0 *by*

$$c_0^2=\{3\kappa(p_\infty-p_V)+2(3\kappa-1)\sigma/R_*\}/4\rho_L.$$

If $c_L<c_0$, $V(t)$ *changes the sign only finite times, while if* $c_L>c_0$, $V(t)$ *changes its sign infinitely many times.*

(*iv*) *The energy decays exponentially fast, i.e. there exists positive constants* λ *and* C *such that* $E(R(t),V(t))\leqslant Ce^{-\lambda t}$ *holds for an arbitrary* $t\geqslant 0$.

Proof (i) The proof is the same as that of (i) of Proposition 7.3.1 in [2].
(ii) The uniqueness of the solution is obvious since the vector field is locally Lipschitz continuous. To obtain the energy dissipation (7.29), we multiply (7.27) by $\rho_L R^3 V$. Of the two terms containing $\dot{V}$ in the result, the first one: $\rho_L R^3 V\dot{V}$ is combined with $-3\rho_L R^2 V^3/2$ in the right hand side to make $d\left(R^3V^2/2\right)/dt$. Another one $-(\Lambda+1)\rho_L R^3V^2\dot{V}/c_L$ is transformed using (7.27) to express $\dot{V}$ in terms of R and V. Thus (7.29) is shown. Since $\lim_{R\to+0}G(R)=\lim_{R\to+\infty}G(R)=\infty$, this nonincreasing property of $E(R(t),V(t))$ assures that $(R(t),V(t))$ is contained in a certain compact set in $(0,\infty)\times\mathbb{R}$. Next we show that

$$V(t)<\frac{c_L}{\Lambda+1}\quad\text{for all } t\geqslant 0$$

provided $V(0)<c_L/(\Lambda+1)$. In fact, for an arbitrary fixed $R>0$, it is easy to see that as $V\to c_L/(\Lambda+1)-0$, the right hand side of (7.27) tends to $-\left(2P(R)+RP'(R)\right)/\left((\Lambda+1)\rho_L R\right)-c_L^2/\left((\Lambda+1)^3R\right)$. This value is negative since

$$2P(R)+RP'(R)=2(p_\infty-p_V)+(3\kappa-2)\bar{p}_G\left(\bar{R}/R\right)^{3\kappa}+2\sigma/R>0. \tag{7.31}$$

By these discussions, we see that the solution exists globally in time. The Poincaré-Bendixon theorem assures (7.30). For the detail, readers are referred to the proof of Proposition 7.3.1(iv) in [2].
(iii) Let us denote by $W(t)$ the winding number of the solution trajectory around $(R_*,0)$ by the time t. Using the argument principle, we have

$$W(t) = \mathrm{Re}\left[\frac{-1}{2\pi\sqrt{-1}}\int_0^t \frac{\dot{Z}(s)}{Z(s)-R_*}ds\right], \quad Z(s) := R(s) + \sqrt{-1}V(s),$$
$$= \frac{1}{2\pi}\int_0^t \frac{N(R(s), V(s))}{(R(s)-R_*)^2 + V(s)^2}ds, \quad N(R, V) := V^2 - \dot{V}(R - R_*). \quad (7.32)$$

Substituting (7.21) into (7.32), we have

$$N(R, V) = \frac{A}{R_*^2}(R-R_*)^2 + \frac{A}{R_* c_L}(R-R_*)V + V^2 + \mathscr{O}\left((|R-R_*| + |V|)^3\right)$$
$$\text{as } |R-R_*| + |V| \to 0, \quad \text{where } A := 3\kappa\frac{p_\infty - p_V}{\rho_L} + (3\kappa - 1)\frac{2\sigma}{\rho_L R_*} = 4c_0^2.$$

Here consider a quadratic equation $A + Ax + c_L^2 x^2 = 0$, whose determinant is $A^2 - 4Ac_L^2 = 4A(A/4 - c_L^2) = 4A(c_0^2 - c_L^2)$. In the case of $c_L > c_0$, the integrand in the second line of (7.32) is greater than a certain positive constant for sufficiently large t in view of (7.30), and the trajectory rotates around $(R_*, 0)$ infinitely many times. In the case of $c_L < c_0$, the equation $A + Ax + c_L^2 x^2 = 0$ has two negative roots. Thus $R_* V(t)/c_L(R(t) - R_*)$ converges to one of the roots of $A + Ax + c_L^2 x^2 = 0$ as $t \to \infty$ and $W(t)$ has a finite upper bound.

(iv) We may assume that $(R(0), V(0))$ is sufficiently close to but not equal to $(R_*, 0)$ without loss of generality. Choose a constant $\varepsilon \in (0, \pi/2)$ sufficiently small so that $-2c_L^{-1}R_* \tan\varepsilon$ is larger than an arbitrary real root of the equation $A + Ax + c_L^2 x^2 = 0$ if it exists, and consider the region $D := \{(R, V) \mid |\mathrm{Arctan}\,(V(t)/(R(t) - R_*))| \leqslant \varepsilon, (R, V) \in (0, \infty) \times \mathbb{R}\}$. Now we show that $(R(t), V(t)) \in D$ does not hold consecutively for longer than a certain finite time and $(R(t), V(t)) \notin D$ holds consecutively for longer than a certain positive time. In fact, by the winding number argument above, we see that $(R, V) \in D$ implies $(R - R_*)^2 + V^2 \leqslant (1 + \tan^2\varepsilon)(R - R_*)^2$ and $N(R, V) \geqslant (AR_*^{-2} - AR_*^{-1}c_L^{-1}\tan\varepsilon + \tan^2\varepsilon)(R - R_*)^2/2$. Therefore $\dot{W}(t)$ is greater than a certain positive constant ω_1 and $(R, V) \in D$ does not hold consecutively for longer than $t_1 := 2\varepsilon/2\pi\omega_1$. On the other hand, when $(R, V) \notin D$, i.e. $|\mathrm{Arctan}\,((R(t) - R_*)/V(t))| < \pi/2 - \varepsilon$ holds, we have $N(R, V) \leqslant 2\big(A/(R_* \tan\varepsilon)^2 + A/(R_* c_L \tan\varepsilon) + 1\big)V^2$ and hence $\dot{W}(t) \leqslant \omega_2$ for a certain positive constant ω_2. Therefore, the state $|\mathrm{Arctan}\,((R(t) - R_*)/V(t))| < \pi/2 - \varepsilon$ lasts for longer than $t_2 := (\pi - 2\varepsilon)/2\pi\omega_2$. Also in this case, the right hand side of (7.29) divided by $E(R(t), V(t))$ is less than $-\omega_3$ for a certain positive constant ω_3. Therefore, by defining $\lambda := \omega_3 t_2/(t_1 + t_2)$ and letting a positive constant C sufficiently large, $E(R(t), V(t)) \leqslant Ce^{-\lambda t}$ holds for an arbitrary $t \geqslant 0$. □

7.4 A Hamiltonian Formulation of the Rayleigh-Plesset-Keller Equation

7.4.1 *A Hamiltonian Formulation of the Rayleigh-Plesset Equation*

By introducing a momentum variable $Q = \rho_L R^3 \dot{R} = \rho_L R^3 V$, we rewrite the energy function (7.25) for the Rayleigh-Plesset equation as

$$E(R, Q) = \frac{Q^2}{2\rho_L R^3} + \int_{R_*}^{R} P(r) r^2 dr. \tag{7.33}$$

Theorem 7.4.1 *The inviscid Rayleigh-Plesset equation can be represented as a Hamiltonian system with Hamiltonian E defined by Eq. (7.33). Namely, the Rayleigh-Plesset equation is equivalent to the Hamilton's canonical equation:*

$$\dot{\mathbf{u}} = \mathbb{J} \nabla E \tag{7.34}$$

where $\mathbf{u} = (R, Q)^T$ *denotes a point of the phase space for the radial motion of a spherical bubble, the dot over the variable means the time derivative,*

$$\mathbb{J} = \begin{pmatrix} 0 & 1 \\ -1 & 0 \end{pmatrix} \tag{7.35}$$

is the symplectic (that is non-degenerate skew-symmetric) matrix and

$$\nabla E = \begin{pmatrix} \partial E / \partial R \\ \partial E / \partial Q \end{pmatrix},$$

is the gradient of the energy function, if we neglect the viscosity.

Moreover, adding the symmetric negative semi-definite matrix

$$\mathbf{D}_{RP}(\mathbf{u}) = \begin{pmatrix} 0 & 0 \\ 0 & -4\mu_L R \end{pmatrix} \tag{7.36}$$

to the symplectic matrix $\mathbb{J}$*, the Rayleigh-Plesset equation with viscosity can be represented in the form*

$$\dot{\mathbf{u}} = \mathbf{A}_{RP}(\mathbf{u}) \nabla E = \begin{pmatrix} 0 & 1 \\ -1 & -4\mu_L R \end{pmatrix} \begin{pmatrix} \partial E / \partial R \\ \partial E / \partial Q \end{pmatrix}, \tag{7.37}$$

where $\mathbf{A}_{RP}(\mathbf{u}) = \mathbb{J} + \mathbf{D}_{RP}(\mathbf{u})$.

Proof By definition, we have $Q = \rho_L R^3 \dot{R}$ and $\dot{Q} = \rho_L(R\ddot{R} + 3\dot{R}^2)R^2$, Then the components in the right hand side of Eq. (7.37) are

$$\frac{\partial E}{\partial Q} = \frac{Q}{\rho_L R^3} = \dot{R},$$

and

$$\begin{aligned} -\frac{\partial E}{\partial R} - 4\mu_L R\frac{\partial E}{\partial Q} &= \frac{3}{2}\frac{Q^2}{\rho_L R^4} - P(R)R^2 - 4\mu_L\frac{Q}{\rho_L R^2} \\ &= \rho_L R^2\left(\frac{3}{2}\dot{R}^2 - \frac{P(R) + 4\mu_L\dot{R}/R}{\rho_L}\right) \\ &= \rho_L R^2\left(R\ddot{R} + 3\dot{R}^2\right) = \dot{Q}, \end{aligned}$$

where the Rayleigh-Plesset equation (7.12) is used. Thus we have (7.37). If we set $\mu_L = 0$ in the computation above, (7.34) is seen to be equivalent to the inviscid Rayleigh-Plesset equation. □

According to this theorem, the time derivative of energy is

$$\begin{aligned} \frac{d}{dt}E(\mathbf{u}) &= \nabla E \cdot \dot{\mathbf{u}} = \nabla E \cdot (\mathbb{J} + \mathbf{D}_{RP})\nabla E = \nabla E \cdot \mathbf{D}_{RP}\nabla E \\ &= -4\mu_L R\dot{R}^2 \leqslant 0 \end{aligned} \tag{7.38}$$

by virtue of $\mathbb{J}$ being skew-symmetric and $\mathbf{D}_{RP}$ being negative semi-definite. Then we recover the conservation and dissipation of energy property for the inviscid and viscous Rayleigh-Plesset equation, respectively.

7.4.2 A Hamiltonian Formulation of the Keller-Herring Equation

We show that the Keller-Herring equation (7.21) can be represented as a Hamiltonian system perturbed by dissipation, as well as the Rayleigh-Plesset equation.

Theorem 7.4.2 *The Keller-Herring equation can be represented in the form*

$$\dot{\mathbf{u}} = [\mathbb{J} + \mathbf{D}_K(\mathbf{u})]\nabla E = \mathbf{A}_K(\mathbf{u})\nabla E = \begin{pmatrix} 0 & 1 \\ -1 & \alpha_K(\mathbf{u}) \end{pmatrix}\begin{pmatrix} \partial E/\partial R \\ \partial E/\partial Q \end{pmatrix}, \tag{7.39}$$

where E is the energy function defined by Eq. (7.33),

$$\mathbf{D}_K(\mathbf{u}) = \begin{pmatrix} 0 & 0 \\ 0 & \alpha_K(\mathbf{u}) \end{pmatrix}, \quad \mathbf{A}_K(\mathbf{u}) = \mathbb{J} + \mathbf{D}_K(\mathbf{u}) = \begin{pmatrix} 0 & 1 \\ -1 & \alpha_K(\mathbf{u}) \end{pmatrix} \tag{7.40}$$

and

$$\alpha_K(\mathbf{u}) = \alpha_K(R, Q) = -\frac{R^2}{c_L - (\Lambda+1)\frac{Q}{\rho_L R^3}} \left(2P(R) + P'(R)R + \frac{Q^2}{\rho_L R^6}\right). \quad (7.41)$$

Moreover, if the initial condition satisfies $(\Lambda+1)\dot{R}(0) < c_L$, *we have* $\alpha_K(\mathbf{u}(t)) < 0$ *for all* $t > 0$, *which implies* $\mathbf{D}_K$ *is negative semi-definite for all time.*

Proof The right hand side of the first equation of (7.39) is

$$\frac{\partial E}{\partial Q} = \frac{Q}{\rho_L R^3} = \dot{R}.$$

The second equation of (7.39) is

$$\begin{aligned}\dot{Q} &= -\frac{\partial E}{\partial R} + \alpha_K(\mathbf{u})\frac{\partial E}{\partial Q},\\ &= \frac{3}{2}\frac{Q^2}{\rho_L R^4} - P(R)R^2 - \frac{\frac{Q}{\rho_L R}}{c_L - (\Lambda+1)\frac{Q}{\rho_L R^3}}\left(2P(R) + P'(R)R + \frac{Q^2}{\rho_L R^6}\right).\end{aligned}$$

Since $Q = \rho_L R^3 \dot{R}$ and $\dot{Q} = \rho_L(R\ddot{R} + 3\dot{R}^2)R^2$, this is equivalent to

$$\rho_L(R\ddot{R} + \frac{3}{2}\dot{R}^2) + P(R) = -\frac{\dot{R}}{c_L - (\Lambda+1)\dot{R}}\left(2P(R) + P'(R)R + \rho_L\dot{R}^2\right). \quad (7.42)$$

This is rewritten into

$$\begin{aligned}&\left(1 - (\Lambda+1)\frac{\dot{R}}{c_L}\right)R\ddot{R} + \frac{3}{2}\left(1 - (\Lambda+1)\frac{\dot{R}}{c_L} + \frac{2}{3}\frac{\dot{R}}{c_L}\right)\dot{R}^2\\ &\qquad = -\left(1 - (\Lambda+1)\frac{\dot{R}}{c_L} + 2\frac{\dot{R}}{cL} + \frac{R}{c_L}\frac{d}{dt}\right)\frac{P(R)}{\rho_L}, \end{aligned} \quad (7.43)$$

which is the (inviscid) Keller-Herring equation itself.

Finally, recall that (7.28) and (7.31) in Theorem 7.3.1 follow from the assumptions $V(0) < c_L/(\Lambda+1)$ and $p_\infty \geqslant p_V$ with $\kappa \geqslant 1$ respectively. Therefore α_K is always negative and $\mathbf{D}_K(\mathbf{u}(t))$ is negative semi-definite for all $t \geqslant 0$. □

Similarly to (7.38), the time derivative of energy is

$$\begin{aligned}\frac{d}{dt}E(\mathbf{u}) &= \nabla E \cdot \dot{\mathbf{u}} = \nabla E \cdot (\mathbb{J} + \mathbf{D}_K)\nabla E = \nabla E \cdot \mathbf{D}_K \nabla E\\ &= \alpha_K(\mathbf{u})\dot{R}^2 \leqslant 0\end{aligned} \quad (7.44)$$

by virtue of $\mathbb{J}$ being skew-symmetric and $\mathbf{D}_K$ being negative semi-definite. Then we have recovered the dissipation of energy property for the Keller equation discussed in the previous section.

From Theorems 7.4.1 and 7.4.2, we find that the conservative part of the dynamics is governed by the same Hamiltonian both in the Rayleigh-Plesset and the Keller-Herring equations. On that account, these equations describe the same dynamics as long as the conservative part is concerned. The dissipative part of the dynamics comes from the viscosity of the liquid in the Rayleigh-Plesset equation. The same mechanism of dissipation can be incorporated in the Keller-Herring as well, though we neglected it. The dissipation in the Keller-Herring equation depends on the ratio of the velocity of the bubble surface $\dot{R}$ to the speed of sound c_L, that is, the Mach number of the velocity of the surface. If this Mach number is small, the dissipation of energy is also small. In the limit of zero Mach number, the dissipation in the Keller-Herring equation vanishes and the equation reduces to the inviscid Rayleigh-Plesset equation as expected.

In the discussion above on the conservation or dissipation of energy, it is essential to represent the equation in the form

$$\dot{\mathbf{u}} = \mathbf{A}(\mathbf{u})\nabla F(\mathbf{u}) \tag{7.45}$$

using the gradient ∇F of an energy function F and a linear operator $\mathbf{A}$ defined on a state variable $\mathbf{u}$. Indeed, the time derivative of energy is given by

$$\frac{d}{dt}F(\mathbf{u}) = \nabla F \cdot \mathbf{A}(\mathbf{u})\nabla F, \tag{7.46}$$

and the energy is conserved or dissipated if $\mathbf{A}$ is skew-symmetric or negative semi-definite, respectively. We call this formulation a gradient system.

7.5 Discrete Gradient Schemes for the Rayleigh-Plesset and Keller Equations

By applying the discrete gradient method to a gradient system (7.45), we obtain an energy-dissipative/preserving scheme. Let $[0, T]$ be a computational interval and let $\{t_n\}_{n=0}^{N}$ be a sequence of discrete time levels with $0 = t_0 < t_1 < \cdots < t_{N-1} < t_N = T$. We denote an approximate solution at $t = t_n$ by $\mathbf{U}_n$, and define $h_n := t_{n+1} - t_n$ and $h := \max_{1 \leqslant n \leqslant N} h_n$. Let $\mathbf{A}(\mathbf{x}) = \mathbf{A}(x_1, x_2)$ denote a matrix-valued function for $\mathbf{x} = (x_1, x_2)^T$. The discrete gradient scheme for a gradient system (7.45) reads

$$\frac{\mathbf{U}_{n+1} - \mathbf{U}_n}{h_n} = \mathbf{A}(\mathbf{U}_n)\widetilde{\nabla} F(\mathbf{U}_n, \mathbf{U}_{n+1}). \tag{7.47}$$

Here $\widetilde{\nabla}$ is an operator called a discrete gradient, which satisfies for $\mathbf{x} = (x_1, x_2)^T, \mathbf{y} = (y_1, y_2)^T \in \mathbb{R}^2$, and a differentiable function f that

$$f(\mathbf{y}) - f(\mathbf{x}) = \widetilde{\nabla} f(\mathbf{x}, \mathbf{y}) \cdot (\mathbf{y} - \mathbf{x}), \tag{7.48}$$

$$\widetilde{\nabla} f(\mathbf{x}, \mathbf{x}) = \nabla f(\mathbf{x}). \tag{7.49}$$

We note that the discrete gradient is not uniquely determined. In this paper, we employ the coordinate increment method [7]:

$$\widetilde{\nabla} f(\mathbf{x}, \mathbf{y}) = \begin{pmatrix} \dfrac{f(y_1, x_2) - f(x_1, x_2)}{y_1 - x_1} \\ \dfrac{f(y_1, y_2) - f(y_1, x_2)}{y_2 - x_2} \end{pmatrix}, \tag{7.50}$$

where in the case of $x_1 = y_1$ or $x_2 = y_2$, the finite difference is replaced by the partial derivative of f with respect to the first or second variable at (x_1, x_2) or (y_1, y_2), respectively. It is easy to check that this discrete gradient operator satisfies (7.48). Depending on whether the matrix $\mathbf{A}$ is skew-symmetric or negative semi-definite, the discretized equation (7.47) is energy-preserving or dissipative, because the rate of change of the energy function F is

$$\begin{aligned} \frac{F(\mathbf{U}_{n+1}) - F(\mathbf{U}_n)}{h_n} &= \widetilde{\nabla} F(\mathbf{U}_n, \mathbf{U}_{n+1})^T \frac{\mathbf{U}_{n+1} - \mathbf{U}_n}{h_n} \\ &= (\widetilde{\nabla} F)^T \mathbf{A}(\mathbf{U}_n)(\widetilde{\nabla} F) \leqslant 0. \end{aligned} \tag{7.51}$$

Let us give the concrete form of the discrete gradient for the energy function E. We apply $F = E$ defined in (7.33) and $\mathbf{A} = \mathbf{A}_{RP}$ in (7.37) or $\mathbf{A}_K$ in (7.39). Setting $\mathbf{x} = \mathbf{U}_n, \mathbf{y} = \mathbf{U}_{n+1}$ and $f = E$ in (7.50), we have

$$\begin{aligned} \widetilde{\nabla} E(\mathbf{U}_n, \mathbf{U}_{n+1}) &= \widetilde{\nabla}\left(\frac{Q^2}{2\rho_L R^3}\right)(\mathbf{U}_n, \mathbf{U}_{n+1}) + \widetilde{\nabla} G(R)(\mathbf{U}_n, \mathbf{U}_{n+1}) \\ &= \begin{pmatrix} -\dfrac{Q_n^2(R_n^2 + R_n R_{n+1} + R_{n+1}^2)}{2\rho_L R_n^3 R_{n+1}^3} + \dfrac{G(R_{n+1}) - G(R_n)}{R_{n+1} - R_n} \\ \dfrac{Q_n + Q_{n+1}}{2\rho_L R_{n+1}^3} \end{pmatrix}. \end{aligned}$$

The function G appeared in the equations above has been defined in (7.25), and the values of the function G and its derivative dG/dR can be computed if we have the value of the radius R.

The discrete gradient scheme for the Rayleigh-Plesset and the Keller equations is obtained if we set

$$\mathbf{U} = \begin{pmatrix} R \\ Q \end{pmatrix}, \quad \mathbf{A} = \begin{pmatrix} 0 & 1 \\ -1 & \alpha(R_n, Q_n) \end{pmatrix}$$

and $\tilde{\nabla}F = \tilde{\nabla}E$ in (7.47), where $\alpha(R, Q) = -4\mu_L R$ or $\alpha_K(R, Q)$ for the Rayleigh-Plesset or the Keller equation, respectively. The result is

$$\begin{aligned}\frac{R_{n+1} - R_n}{h_n} &= \frac{Q_n + Q_{n+1}}{2\rho_L R_{n+1}^3},\\ \frac{Q_{n+1} - Q_n}{h_n} &= \frac{Q_n^2(R_n^2 + R_n R_{n+1} + R_{n+1}^2)}{2\rho_L R_n^3 R_{n+1}^3} - \frac{G(R_{n+1}) - G(R_n)}{R_{n+1} - R_n}\\ &\quad + \alpha(R_n, Q_n)\frac{Q_n + Q_{n+1}}{2\rho_L R_{n+1}^3}.\end{aligned} \tag{7.52}$$

These nonlinear equations can be solved uniquely using the Newton-Raphson method such that

$$\lim_{h_n \to 0} \frac{R_{n+1} - R_n}{h_n} = \frac{Q_n}{\rho_L R_n^3},$$

is satisfied by letting h_n be sufficiently small for each n. For the Keller model, the subsonic condition

$$\frac{Q_{n+1}}{\rho_L R_{n+1}^3} < c_L$$

must also be satisfied, which holds if h_n is sufficiently small.

7.6 Numerical Results

We consider a spherical bubble collapsing under the influence of a high external pressure p_∞ in a liquid. The bubble contains saturated vapor and we assume that the vapor pressure p_V is constant and takes the value at room temperature (293K). The bubble also contains a non-condensable gas such that the bubble with the radius $\bar{R}$ is at rest under the external pressure $\bar{p}$, that is, the pressure difference between the inside and outside bubble balances with the surface tension:

$$p_B(\bar{R})(= p_V + \bar{p}_G) = \bar{p} + 2\sigma/\bar{R}.$$

This equation gives the value of $\bar{p}_G$ when we set the value of initial external pressure $\bar{p}$ which is instantaneously increased to p_∞ in the computation. The non-condensable gas is modeled by an ideal monoatomic gas and we assume that the thermodynamic processes which occur in the bubble are adiabatic. Then the exponent of the barotropic relation corresponds to the heat capacity ratio for monoatomic gases. The liquid surrounding the bubble is assumed to be inviscid. The material properties and parameters used in the computation are listed in Table 7.2. The equilibrium radius R_* in this setting is 3.73×10^{-6} [m] and the natural period T_* of the linearized system about the equilibrium point $(R_*, 0)$ in the phase plane (which is defined in Proposition

Table 7.2 Material properties and parameters

Name	Symbol	Value
Coefficient of surface tension	σ	7.275×10^{-2} [N/m]
Reference radius	$\bar{R}$	1.0×10^{-5} [m]
Vapor pressure	p_V	2337 [Pa]
Density of the liquid	ρ_L	998.2 [kg/m^3]
Exponent of the barotropic relation	κ	5/3
Coefficient of viscosity of the liquid	μ_L	0 [m^2/s]
Initial external pressure	$\bar{p}$	2500 [Pa]
External pressure	p_∞	1.0×10^5 [Pa]
Speed of sound in the liquid	c_L	1478 [m/s]

7.3.1) is

$$T_* = 2\pi R_* / \sqrt{\{3\kappa(p_\infty - p_V) + (3\kappa - 1)2\sigma/R_*\} / \rho_L} \approx 9.19 \times 10^{-7} \text{ [s]}.$$

The initial values are set to be $\mathbf{U}_0 = (R_0, Q_0)^T = (\bar{R}, 0)^T$.

7.6.1 *The Inviscid Rayleigh-Plesset Equation*

We compute the collapsing bubble described by the inviscid Rayleigh-Plesset equation ((7.12) with $\mu_L = 0$) using the discrete gradient method (DGM) for conservative systems ((7.52) with $\alpha = 0$) and fourth order Runge-Kutta method (RK4) for comparison. For each computation, we fix the time step h_n independent of n. Here and hereafter physical quantities without units are normalized using R_*, T_*, and ρ_L.

Figure 7.2 shows the time evolution of the energy defined by (7.33). It should be conserved according to Proposition 7.3.1(ii) and is indeed conserved by DGM up to the accuracy in solving (7.52). We see that the energy obtained by DGM with the time step $h_n = 5 \times 10^{-9}$ [s] (solid line) is well kept constant, while RK4 with $h_n = 5 \times 10^{-10}$ [s] (broken line) and $h_n = 2 \times 10^{-10}$ [s] (dotted line) lose energy to some extent. The energy loss by RK4 is almost invisible if we further diminishes the time step to $h_n = 1 \times 10^{-10}$ [s].

The same situation can be observed in the phase diagram. Figure 7.3 exhibits the trajectory of $(R(t), \dot{R}(t))$ over approximately six cycles computed by DGM and RK4 both with $h_n = 1 \times 10^{-9}$ [s]. Conservation of energy in the inviscid Rayleigh-Plesset equation requires that the trajectory must lie on the contour of a certain energy, which is a closed curve for the energy (7.25). We see that the trajectory by RK4 gets in and out of that by DGM, implying vasillation of the energy in the result by RK4 with an insufficient time resolution.

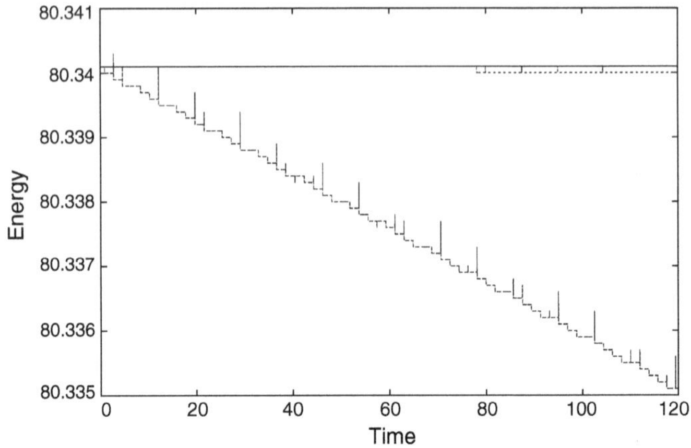

Fig. 7.2 Time evolution of the energy in the inviscid Rayleigh-Plesset equation obtained by DGM with $h_n = 5 \times 10^{-9}$ [s] (*solid*) and RK4 with $h_n = 2 \times 10^{-9}$ [s] (*broken*) and $h_n = 5 \times 10^{-10}$ [s] (*dotted*)

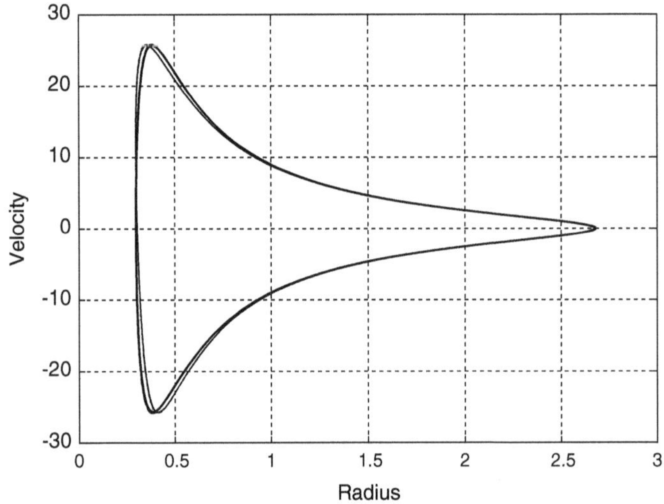

Fig. 7.3 The phase diagram of $(R(t), \dot{R}(t))$ computed by DGM (*thick line*) and RK4 (*thin line*) both with $h_n = 1 \times 10^{-9}$ [s]

Figure 7.4 presents the radius as a function of time computed by RK4 with $h_n = 1 \times 10^{-10}$ [s] (solid line), by DGM with $h_n = 5 \times 10^{-9}$ [s] (broken line), and by DGM with $h_n = 2 \times 10^{-9}$ [s] (dotted line). Although the energy is very accurately computed by DGM as seen in Fig. 7.2, the phase of the oscillation could differ from the highly accurate values obtained by RK4. The result by DGM for $h_n = 1 \times 10^{-9}$ [s] is almost identical to that by RK4.

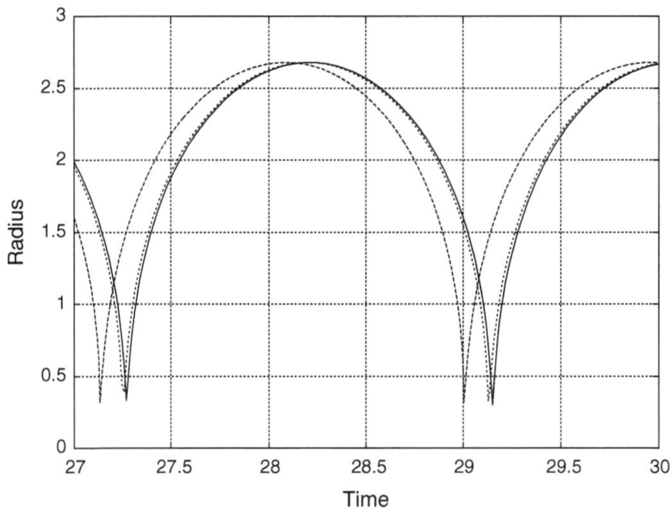

Fig. 7.4 Time evolution of the radius of the bubble in the inviscid Rayleigh-Plesset equation obtained by RK4 with $h_n = 1 \times 10^{-10}$ [s] (*solid*) and DGM with $h_n = 5 \times 10^{-9}$ [s] (*broken*) and $h_n = 2 \times 10^{-9}$ [s] (*dotted*)

7.6.2 The Keller Equation

We compute the collapsing bubble subject to the inviscid Keller equation (7.20) using DGM (7.52) with $\alpha = \alpha_K$. The time step is set to be $h_n = 2 \times 10^{-9}$ [s]. All the results presented in this subsection are compared with RK4 and confirmed to be highly accurate. Figure 7.5 shows computed radius against time. The bubble expands and shrinks almost regularly, but unlike the Rayleigh-Plesset model, the amplitude of the oscillation decreases gradually in the Keller model.

Figure 7.6 presents how the energy (7.33) of the bubble changes with time. We see that the energy decreases monotonically in accordance with the mathematical results stated in Theorem 7.3.1. The DGM well reproduces this property by virtue of (7.51), up to the accuracy in solving (7.52).

Comparing Fig. 7.5 with Fig. 7.6, the bubble drastically loses energy around the time when the bubble ceases to shrink and begins to expand. The decreasing rate of energy for the Keller equation is given in (7.29) with $\Lambda = 0$ and should correspond to the energy release rate due to the emission of pressure waves. The establishment of the relation between the loss of energy and the emission of pressure waves is left for future study.

The trajectory of $(R(t), \dot{R}(t))$ is shown in Fig. 7.7. It converges to the equilibrium point $(1, 0)$ in nondimensional coordinate in agreement with the theoretical result proved in Theorem 7.3.1.

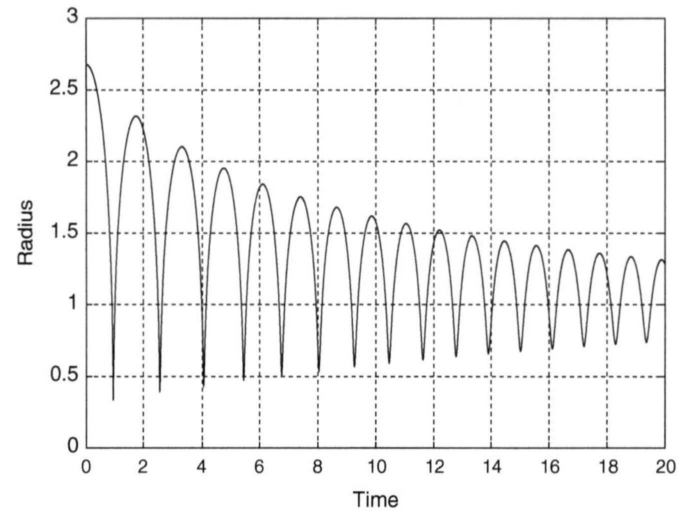

Fig. 7.5 Time evolution of the bubble radius for the Keller equation

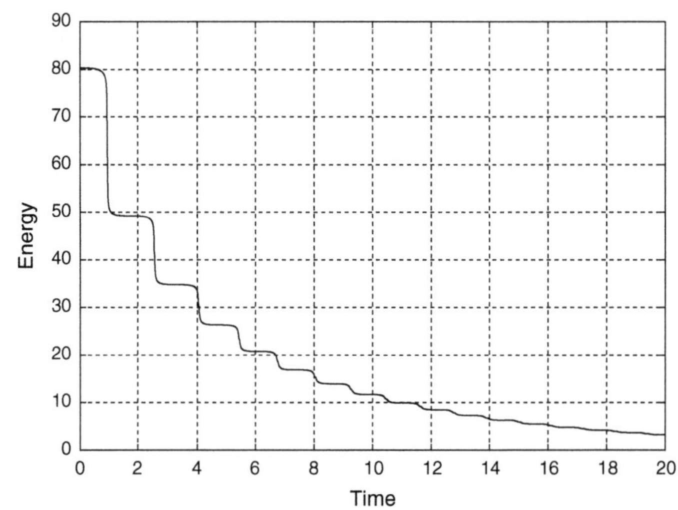

Fig. 7.6 Time evolution of the energy for the Keller equation

Finally, Fig. 7.8 shows the long time evolution of the energy in a logarithmic scale. This manifests the exponential decay of energy on the time scale proved in Theorem 7.3.1.

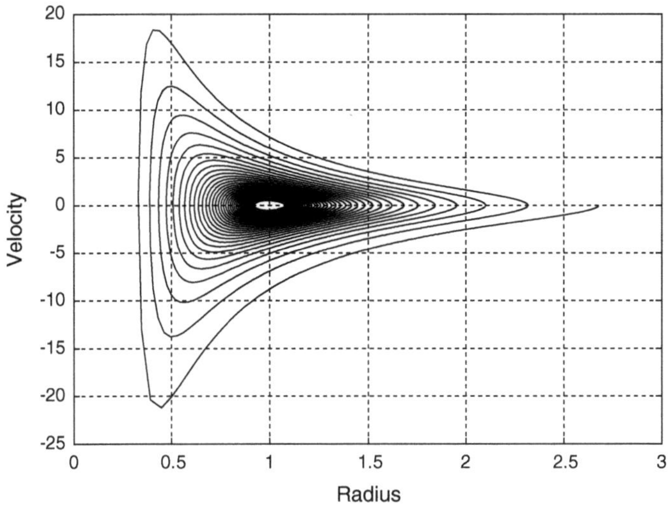

Fig. 7.7 The phase diagram of $(R(t), \dot{R}(t))$ for the Keller model

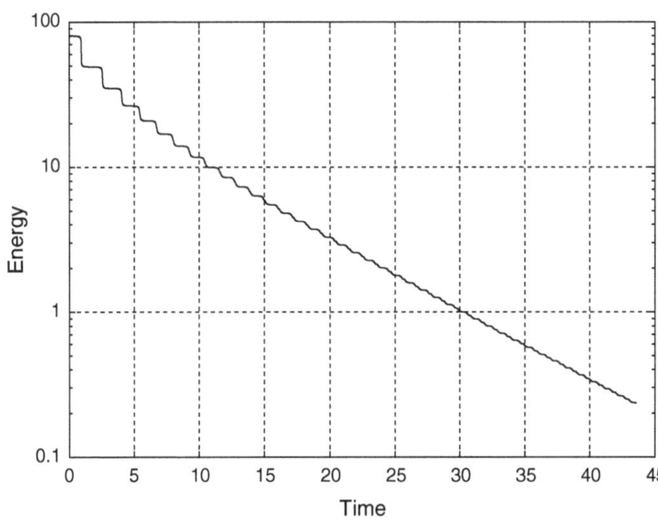

Fig. 7.8 Long time behavior of the energy for the Keller equation

7.7 Concluding Remarks

We have mathematically analyzed the Rayleigh-Plesset equation and the Keller equations and found an energy function which physically makes sense. In view of this, we reformulated them in the Hamiltonian systems. Based on this formulation, we applied the discrete gradient method to their numerical integration and reproduced precise energy conservation or dissipation. In this sense, we obtain a stable scheme without introducing numerical dissipation often associated with fully implicit schemes. Another benefit is that we can take large time step as compared to explicit schemes such as the Runge-Kutta schemes. Note however, that the time step can not be taken arbitrarily large in order to solve the nonlinear equations and that the energy conservation alone, even if it holds, does not guarantee precise computations.

Acknowledgements Financial support from the Alliance for Breakthrough between Mathematics and Sciences (ABMS) supported by the Core Research for Evolutional Science and Technology (CREST) of the Japan Science and Technology Agency (JST) are gratefully acknowledged.

References

1. Brennen, C.E.: Cavitation and Bubble Dynamics. Oxford University Press (1995)
2. Funaki, T., Ohnawa, M., Suzuki, Y., Yokoyama, S.: Existence and uniqueness of solutions to stochastic Rayleigh-Plesset equations. J. Math. Anal. Appl. **425**, 20–32 (2015)
3. Gilmore, F.R., The collapse and growth of a spherical bubble in a viscous compressible liquid. Calif. Inst. of Tech. Hydrodynamics Lab. Rep. No. 26-4 (1952)
4. Gonzalez, O.: Time integration and discrete Hamiltonian systems. J. Nonlinear Sci. **6**, 449–467 (1996)
5. Herring, C.: Theory of the pulsations of the gas bubble produced by an underwater explosion. O.S.R.D. Rep. No. 236 (1941)
6. Holzfuss, J., Rüggeberg, M., Billo, A.: Shock wave emissions of a sonoluminescing bubble. Phys. Rev. Lett. **81**, 5434–5437 (1998)
7. Ito, T., Abe, K.: Hamiltonian-conserving discrete canonical equations based on variational difference quotients. J. Comput. Phys. **76**, 85–102 (1988)
8. Keller, J.B., Kolodner, I.I.: Damping of underwater explosion bubble oscillations. J. Appl. Phys. **27**, 1152–1161 (1956)
9. Plesset, M.S.: The dynamics of cavitation bubbles. ASME J. Appl. Mech. **16**, 228–231 (1949)
10. Plesset, M.S., Prosperetti, A.: Bubble dynamics and caviation. Ann. Rev. Fluid Mech. **9**, 145–185 (1977)
11. Prosperetti, A., Lezzi, A.: Bubble dynamics in a compressible liquid. Part 1. First-order theory. J. Fluid Mech. **168**, 457–478 (1986)
12. Prosperetti, A., Lezzi, A.: Bubble dynamics in a compressible liquid. Part 2. Second-order theory. J. Fluid Mech. **185**, 289–321 (1987)
13. Rayleigh, L.: On the pressure developed in a liquid during the collapse of a spherical cavity. Philos. Mag. **34**, 94–98 (1917)
14. Trilling, L.: The collapse and rebound of a gas bubble. J. Appl. Phys. **23**, 14–17 (1952)

Chapter 8
On the Amplitude Equation of Approximate Surface Waves on the Plasma-Vacuum Interface

Paolo Secchi

Abstract In this paper we present a recent result about the propagation of weakly nonlinear surface waves on a plasma-vacuum interface. In the plasma region we consider the equations of incompressible magnetohydrodynamics, while in vacuum the magnetic and electric fields are governed by the Maxwell equations. A surface wave propagate along the plasma-vacuum interface, when it is linearly weakly stable. Following the approach of Alì and Hunter, we measure the amplitude of the surface wave by the normalized displacement of the interface in a reference frame moving with the linearized phase velocity of the wave, and obtain that it satisfies an asymptotic nonlocal, Hamiltonian evolution equation with quadratic nonlinearity. We show the local-in-time existence of smooth solutions to the Cauchy problem for the amplitude equation in noncanonical variables, and we derive the regularity of the first order corrections of the asymptotic expansion.

Keywords Incompressible magneto-hydrodynamics • Maxwell equations • Plasma-vacuum interface • Interfacial stability and instability

8.1 Introduction

Plasma-vacuum interface problems appear in the mathematical modeling of plasma confinement by magnetic fields in thermonuclear energy production (as in Tokamaks and Stellarators; see, e.g., [4]). In this model, the plasma is confined inside a perfectly conducting rigid wall and isolated from it by a region containing very low density plasma, which may qualify as vacuum, due to the effect of strong magnetic fields. In Astrophysics, the plasma-vacuum interface problem can be used for modeling the motion of a star or the solar corona when magnetic fields are taken into account.

P. Secchi (✉)
DICATAM, Mathematical Division, University of Brescia, Via Branze 43, 25123 Brescia, Italy
e-mail: paolo.secchi@unibs.it

Y. Shibata and Y. Suzuki (eds.), *Mathematical Fluid Dynamics, Present and Future*, Springer Proceedings in Mathematics & Statistics 183,
DOI 10.1007/978-4-431-56457-7_8

For the sake of simplicity, in [11] we consider the plasma-vacuum interface problem in two-dimensions, with the coupling of the incompressible MHD equations in the plasma region and the Maxwell equations in the vacuum region. The solution is close to a stationary basic state with parallel magnetic fields at the flat interface.

To study the time evolution of the plasma-vacuum interface we follow the approach of Alì and Hunter in [1] and we show that, in a unidirectional surface wave, the normalized displacement $x_2 = \varphi(t, x_1)$ of a weakly stable surface wave along the interface, in a reference frame moving with the linearized phase velocity of the wave, satisfies the quadratically nonlinear, nonlocal asymptotic equation

$$\varphi_t + \frac{1}{2}\mathbb{H}[\Phi^2]_{xx} + \Phi\varphi_{xx} = 0, \qquad \Phi = \mathbb{H}[\varphi]\,. \tag{8.1}$$

Here $\mathbb{H}$ denotes the Hilbert transform defined by

$$\mathbb{H}[\varphi](x) = \frac{1}{\pi}\,\text{p.v.}\int_{-\infty}^{+\infty} \frac{\varphi(y)}{x-y}\,dy,$$

and such that

$$\mathbb{H}[e^{ikx}] = -i\,\text{sgn(k)}\,\text{e}^{\text{ikx}}\,, \qquad \mathscr{F}[\mathbb{H}[\varphi]] = -\text{i}\,\text{sgn(k)}\,\mathscr{F}[\varphi],$$

for $\mathscr{F}$ denoting the Fourier transformation.

Equation (8.1) coincides with the amplitude equation for nonlinear Rayleigh waves [5] and current-vortex sheets in incompressible MHD [1, 2]. It is interesting that exactly the same equation appears for the incompressible plasma-vacuum interface problem, where in the vacuum part the electric and magnetic fields are ruled by the Maxwell equations. The derivation of the same equation confirms that (8.1) is a canonical model equation for nonlinear surface wave solutions of hyperbolic conservation laws, analogous to the inviscid Burgers equation for bulk waves. Equation (8.1) also admits the alternative spatial form

$$\varphi_t + [\mathbb{H}, \Phi]\Phi_{xx} + \mathbb{H}[\Phi_x^2] = 0\,, \tag{8.2}$$

where $[\mathbb{H}, \Phi]$ is the commutator of $\mathbb{H}$ with multiplication by Φ, see [8]. The alternative form (8.2) shows that (8.1) is an equation of first order, due to a cancelation of the second order spatial derivatives appearing in (8.1).

By adapting the proof of [7], in [11] we show the local-in-time existence of smooth solutions to the Cauchy problem for amplitude equation in noncanonical variables, and we derive a blow up criterion. Numerical computations [1, 5] show that solutions of (8.1) form singularities in which the derivative φ_x blows up, but φ appears to remain continuous. As far as we know, the global existence of appropriate weak solutions is an open question.

In the present note, from the previous existence result for the solution to the amplitude equation we also derive the regularity of the first order asymptotic corrections of the plasma variables and magnetic and electric fields in vacuum.

8.2 The Plasma-Vacuum Interface Problem

We consider the equations of incompressible magneto-hydrodynamics (MHD), i.e. the equations governing the motion of a perfectly conducting inviscid incompressible plasma. In the case of a homogeneous plasma (the density $\rho \equiv \text{const} > 0$), the equations in a dimensionless form read:

$$\begin{cases} \partial_t \mathbf{v} + \nabla \cdot (\mathbf{v} \otimes \mathbf{v} - \mathbf{B} \otimes \mathbf{B}) + \nabla q = 0\,, \\ \partial_t \mathbf{B} - \nabla \times (\mathbf{v} \times \mathbf{B}) = 0\,, \\ \operatorname{div} \mathbf{v} = 0\,, \ \operatorname{div} \mathbf{B} = 0\,, \end{cases} \tag{8.3}$$

where $\mathbf{v}$ denotes the plasma velocity, $\mathbf{B}$ is the magnetic field (in Alfvén velocity units), $q = p + |\mathbf{B}|^2/2$ is the total pressure, p being the pressure.

For smooth solutions, system (8.3) can be written in equivalent form as a symmetric system

$$\begin{cases} \partial_t \mathbf{v} + (\mathbf{v} \cdot \nabla)\mathbf{v} - (\mathbf{B} \cdot \nabla)\mathbf{B} + \nabla q = 0\,, \\ \partial_t \mathbf{B} + (\mathbf{v} \cdot \nabla)\mathbf{B} - (\mathbf{B} \cdot \nabla)\mathbf{v} = 0\,, \\ \operatorname{div} \mathbf{v} = 0\,. \end{cases} \tag{8.4}$$

In addition the magnetic field must satisfy the constraint

$$\operatorname{div} \mathbf{B} = 0\,,$$

which is preserved by the evolution in time if it is satisfied by the initial data.

Let $\Omega^+(t)$ and $\Omega^-(t)$ be space-time domains occupied by the plasma and the vacuum respectively, separated by an interface $\Gamma(t)$. That is, in the domain $\Omega^+(t)$ we consider system (8.4) governing the motion of the plasma and in the domain $\Omega^-(t)$ we have the Maxwell system

$$\begin{cases} \nu \partial_t \mathbf{H} + \nabla \times \mathbf{E} = 0\,, \\ \nu \partial_t \mathbf{E} - \nabla \times \mathbf{H} = 0\,, \end{cases} \tag{8.5}$$

describing the vacuum magnetic and electric fields $\mathbf{H}, \mathbf{E} \in \mathbb{R}^3$. Here, the equations are written in nondimensional form through a suitable scaling (see Mandrik–Trakhinin [9]), and $\nu = \frac{\bar{v}}{c}$, where $\bar{v}$ is the velocity of a uniform flow and c is the speed of light in vacuum. If we choose $\bar{v}$ to be the speed of sound in vacuum, we have that ν is a small, even though fixed parameter. System (8.5) is supplemented by the divergence constraints

$$\operatorname{div} \mathbf{H} = \operatorname{div} \mathbf{E} = 0$$

on the initial data.

For the sake of simplicity we consider the case of two space dimensions and write

$$\mathbf{v} = (v_1, v_2)^T, \qquad \mathbf{B} = (B_1, B_2)^T.$$

In the (three-dimensional) Maxwell equations (8.5) we assume that

$$\mathbf{H} = (H_1, H_2, 0)^T,$$

and that there is no dependence of $\mathbf{H}$ on the third space variable x_3. It follows from (8.5) that $\mathbf{E}$ takes the form

$$\mathbf{E} = (0, 0, E)^T,$$

and the Maxwell equations reduce to

$$\begin{cases} \nu\partial_t H_1 + \partial_2 E = 0\,, \\ \nu\partial_t H_2 - \partial_1 E = 0\,, \\ \nu\partial_t E - \partial_1 H_2 + \partial_2 H_1 = 0\,, \end{cases} \tag{8.6}$$

under the constraint

$$\partial_1 H_1 + \partial_2 H_2 = 0$$

on the initial data. From now on we write

$$\mathbf{H} = (H_1, H_2)^T,$$

hoping that this small abuse of notation will create no confusion for the reader.

Let us assume that the moving interface $\Gamma(t)$ takes the form

$$\Gamma(t) \doteq \{(x_1, x_2) \in \mathbb{R}^2,\ x_2 = \zeta(x_1, t)\},$$

where $t \in [0, T]$, and that $\Omega^{\pm}(t) = \{x_2 \gtrless \zeta(x_1, t)\}$.

The plasma variables are connected with the vacuum magnetic and electric fields on the interface $\Gamma(t)$ through the relations [4]

$$\partial_t\zeta = \mathbf{v} \cdot N\,, \quad q = \frac{1}{2}\big(H_1^2 + H_2^2 - E^2\big)\,, \tag{8.7}$$

$$\mathbf{B} \cdot N = 0\,, \quad \mathbf{H} \cdot N = 0\,, \quad E - \nu\partial_t\zeta H_1 = 0 \quad \text{on} \quad \Gamma(t)\,, \tag{8.8}$$

where $N = (-\partial_1\zeta, 1)$ is a normal vector and $[q] = q|_\Gamma - \frac{1}{2}|\mathbf{H}|^2_{|\Gamma} + \frac{1}{2}|\mathbf{E}|^2_{|\Gamma}$ is the jump of the total pressure across the interface.

A stationary solution of the 2-D Eqs. (8.4) and (8.6)–(8.8) with interface located at $\{x_2 = 0\}$ is given by the constant states

$$\mathbf{v}^0 = (v_1^0, 0)^T, \qquad \mathbf{B}^0 = (B_1^0, 0)^T,$$

$$\mathbf{H}^0 = (H_1^0, 0)^T, \qquad E^0 = 0, \qquad q^0 = \frac{1}{2}(H_1^0)^2.$$

We will consider the propagation of surface waves localized near the interface. The corresponding solutions must satisfy the decay conditions

$$\lim_{x_2 \to +\infty} (\mathbf{v}, \mathbf{B}, q) = U^0 \doteq (v_1^0, 0, B_1^0, 0, q^0), \tag{8.9}$$

$$\lim_{x_2 \to -\infty} (\mathbf{H}, E) = V^0 \doteq (H_1^0, 0, 0). \tag{8.10}$$

8.3 The Asymptotic Expansion

As in [1] we suppose that the perturbed interface has a slope of the order ε, where ε is a small parameter. With respect to dimensionless variables in which the wavelength of the perturbation and the velocity of the surface wave are of the order one, the time scale for quadratically nonlinear effects to significantly alter the wave profile is of the order ε^{-1}. We therefore introduce a "slow" time variable $\tau = \varepsilon t$. We also introduce a spatial variable $\theta = x_1 - \lambda t$ in a reference frame moving with the surface wave. Here, λ is the linearized phase velocity of the wave, which we will determine as part of the solution.

We write the perturbed location of the interface as

$$x_2 = \varepsilon\varphi(\theta, \tau; \varepsilon),$$

and define a new independent variable

$$\eta = x_2 - \varepsilon\varphi(\theta, \tau; \varepsilon),$$

so that the perturbed interface is located at $\eta = 0$. We look for an asymptotic expansion of the solution $U = (\mathbf{v}, \mathbf{B}, q)^T$, $V = (\mathbf{H}, E)^T$ and φ as $\varepsilon \to 0$ of the form

$$U(\theta, \eta, \tau; \varepsilon) = U^0 + \varepsilon U^{(1)}(\theta, \eta, \tau) + \varepsilon^2 U^{(2)}(\theta, \eta, \tau) + O(\varepsilon^3), \qquad \eta > 0, \tag{8.11}$$

$$V(\theta, \eta, \tau; \varepsilon) = V^0 + \varepsilon V^{(1)}(\theta, \eta, \tau) + \varepsilon^2 V^{(2)}(\theta, \eta, \tau) + O(\varepsilon^3), \qquad \eta < 0, \tag{8.12}$$

$$\varphi(\theta, \tau; \varepsilon) = \varphi^{(1)}(\theta, \tau) + \varepsilon\varphi^{(2)}(\theta, \tau) + O(\varepsilon^2). \tag{8.13}$$

We expand the partial derivatives with respect to the original time and space variables as

$$\begin{aligned} \partial_t &= -\lambda\partial_\theta + \varepsilon(\partial_\tau + \lambda\varphi_\theta\partial_\eta) - \varepsilon^2\varphi_\tau\partial_\eta\,, \\ \partial_{x_1} &= \partial_\theta - \varepsilon\varphi_\theta\partial_\eta\,, \\ \partial_{x_2} &= \partial_\eta\,. \end{aligned}$$

We substitute these expansions in (8.4) and (8.6), Taylor expand the result with respect to ε and equate coefficients of ε^1 and ε^2 to zero. In the interior the asymptotic solution satisfies at the first order

$$\begin{cases} (\lambda - v_1^0)\partial_\theta v^{(1)} + B_1^0\partial_\theta B^{(1)} - \begin{pmatrix}\partial_\theta \\ \partial_\eta\end{pmatrix} q^{(1)} = 0\,, \\ (\lambda - v_1^0)\partial_\theta B^{(1)} + B_1^0\partial_\theta v^{(1)} = 0\,, \\ \partial_\theta v_1^{(1)} + \partial_\eta v_2^{(1)} = 0\,, \qquad \text{for} \quad \eta > 0, \end{cases} \tag{8.14}$$

$$\begin{cases} \nu\lambda\partial_\theta H_1^{(1)} - \partial_\eta E^{(1)} = 0\,, \\ \nu\lambda\partial_\theta H_2^{(1)} + \partial_\theta E^{(1)} = 0\,, \\ \nu\lambda\partial_\theta E^{(1)} + \partial_\theta H_2^{(1)} - \partial_\eta H_1^{(1)} = 0\,, \qquad \text{for} \quad \eta < 0. \end{cases} \tag{8.15}$$

We expand the jump conditions in (8.7) and (8.8), with $\zeta = \varepsilon\varphi$, and equate coefficients of ε^1 and ε^2 to zero. We find that the solutions satisfy at the first order the following jump conditions

$$\begin{cases} (\lambda - v_1^0)\partial_\theta\varphi^{(1)} + v_2^{(1)} = 0\,, \\ B_1^0\partial_\theta\varphi^{(1)} - B_2^{(1)} = 0\,, \qquad H_1^0\partial_\theta\varphi^{(1)} - H_2^{(1)} = 0\,, \\ q^{(1)} = H_1^0 H_1^{(1)}\,, \qquad E^{(1)} + \nu\lambda H_1^0\partial_\theta\varphi^{(1)} = 0\,, \qquad \text{for } \eta = 0. \end{cases} \tag{8.16}$$

At the second order we obtain

$$\begin{cases} (\lambda - v_1^0)\partial_\theta v^{(2)} + B_1^0\partial_\theta B^{(2)} - \begin{pmatrix}\partial_\theta \\ \partial_\eta\end{pmatrix} q^{(2)} = p_1\,, \\ (\lambda - v_1^0)\partial_\theta B^{(2)} + B_1^0\partial_\theta v^{(2)} = p_2\,, \\ -\partial_\theta v_1^{(2)} - \partial_\eta v_2^{(2)} = p_3\,, \qquad \text{for } \eta > 0, \end{cases} \tag{8.17}$$

$$\begin{cases} \nu\lambda\partial_\theta H_1^{(2)} - \partial_\eta E^{(2)} = p_1'\,, \\ \nu\lambda\partial_\theta H_2^{(2)} + \partial_\theta E^{(2)} = p_2'\,, \\ \nu\lambda\partial_\theta E^{(2)} + \partial_\theta H_2^{(2)} - \partial_\eta H_1^{(2)} = p_3'\,, \qquad \text{for } \eta < 0, \end{cases} \tag{8.18}$$

and the jump conditions

$$\begin{cases} (\lambda - v_1^0)\partial_\theta \varphi^{(2)} + v_2^{(2)} = r_1\,, \\ B_1^0 \partial_\theta \varphi^{(2)} - B_2^{(2)} = r_2\,, \qquad H_1^0 \partial_\theta \varphi^{(2)} - H_2^{(2)} = r_3\,, \\ q^{(2)} - H_1^0 H_1^{(2)} = r_4\,, \qquad E^{(2)} + \nu\lambda H_1^0 \partial_\theta \varphi^{(2)} = r_5\,, \qquad \text{for } \eta = 0, \end{cases} \tag{8.19}$$

where we have denoted

$$\begin{aligned} p_1 &\doteq (\partial_\tau + \lambda \varphi_\theta^{(1)} \partial_\eta) v^{(1)} + (v_1^{(1)} \partial_\theta + v_2^{(1)} \partial_\eta - v_1^0 \varphi_\theta^{(1)} \partial_\eta) v^{(1)} \\ &\quad - (B_1^{(1)} \partial_\theta + B_2^{(1)} \partial_\eta - B_1^0 \varphi_\theta^{(1)} \partial_\eta) B^{(1)} - \begin{pmatrix} \varphi_\theta^{(1)} \partial_\eta q^{(1)} \\ 0 \end{pmatrix}, \\ p_2 &\doteq (\partial_\tau + \lambda \varphi_\theta^{(1)} \partial_\eta) B^{(1)} + (v_1^{(1)} \partial_\theta + v_2^{(1)} \partial_\eta - v_1^0 \varphi_\theta^{(1)} \partial_\eta) B^{(1)} \\ &\quad - (B_1^{(1)} \partial_\theta + B_2^{(1)} \partial_\eta - B_1^0 \varphi_\theta^{(1)} \partial_\eta) v^{(1)}\,, \\ p_3 &\doteq -\varphi_\theta^{(1)} \partial_\eta v_1^{(1)}\,, \end{aligned}$$

$$p_1' \doteq \nu(\partial_\tau + \lambda \varphi_\theta^{(1)} \partial_\eta) H_1^{(1)}\,, \qquad p_2' \doteq \nu(\partial_\tau + \lambda \varphi_\theta^{(1)} \partial_\eta) H_2^{(1)} + \varphi_\theta^{(1)} \partial_\eta E^{(1)}\,,$$
$$p_3' \doteq \nu(\partial_\tau + \lambda \varphi_\theta^{(1)} \partial_\eta) E^{(1)} + \varphi_\theta^{(1)} \partial_\eta H_2^{(1)}\,,$$

$$r_1 \doteq (\partial_\tau + v_1^{(1)} \partial_\theta) \varphi^{(1)}\,, \qquad r_2 \doteq -B_1^{(1)} \partial_\theta \varphi^{(1)}\,,$$
$$r_3 \doteq -H_1^{(1)} \partial_\theta \varphi^{(1)}\,, \qquad r_4 \doteq \frac{1}{2}\big(|H^{(1)}|^2 - (E^{(1)})^2\big)\,,$$
$$r_5 \doteq -\nu\lambda H_1^{(1)} \partial_\theta \varphi^{(1)} + \nu H_1^0 \partial_\tau \varphi^{(1)}\,.$$

The next step is the resolution of Eqs. (8.14)–(8.19).

8.4 The First Order Equations

We first solve system (8.14)–(8.16) by Fourier transformation. Introducing the Fourier transforms

$$\hat{U}^{(1)}(k, \eta, \tau) = \frac{1}{2\pi} \int_{-\infty}^{+\infty} U^{(1)}(\theta, \eta, \tau) e^{-ik\theta} d\theta,$$

$$\hat{V}^{(1)}(k, \eta, \tau) = \frac{1}{2\pi} \int_{-\infty}^{+\infty} V^{(1)}(\theta, \eta, \tau) e^{-ik\theta} d\theta,$$

$$\hat{\varphi}^{(1)}(k, \tau) = \frac{1}{2\pi} \int_{-\infty}^{+\infty} \varphi^{(1)}(\theta, \tau) e^{-ik\theta} d\theta,$$

and Fourier transforming (8.14)–(8.16) with respect to θ, we find a system of the form

$$ik\mathscr{A}\hat{U}^{(1)} + \mathscr{B}\partial_\eta \hat{U}^{(1)} = 0, \qquad \eta > 0, \tag{8.20}$$

with suitable real symmetric matrices $\mathscr{A}$, $\mathscr{B}$. As shown in [11], the general solution of (8.20) is

$$\hat{U}^{(1)}(k,\eta,\tau) = a(k,\tau)e^{-k\eta}\mathbf{R} + b(k,\tau)e^{k\eta}\overline{\mathbf{R}}, \qquad \eta > 0, \tag{8.21}$$

where

$$\mathbf{R} = (\lambda - v_1^0, i(\lambda - v_1^0), -B_1^0, -iB_1^0, d)^T \quad \text{and} \quad d \doteq (\lambda - v_1^0)^2 - (B_1^0)^2.$$

Here $a(k,\tau)$ and $b(k,\tau)$ are arbitrary complex-valued functions, the bar denotes a complex conjugate. The condition (8.9) at infinity implies

$$\lim_{\eta\to+\infty} \hat{U}^{(1)}(k,\eta,\tau) = 0, \tag{8.22}$$

yielding from (8.21)

$$\hat{U}^{(1)}(k,\eta,\tau) = \begin{cases} a(k,\tau)e^{-k\eta}\mathbf{R}, & \text{if } k > 0, \\ b(k,\tau)e^{k\eta}\overline{\mathbf{R}}, & \text{if } k < 0. \end{cases} \tag{8.23}$$

Then we consider problem (8.15) for $\eta < 0$. First of all we see that $\hat{E}^{(1)}$ obeys the equation

$$\partial_\eta^2 \hat{E}^{(1)} + k^2(\nu^2\lambda^2 - 1)\hat{E}^{(1)} = 0, \qquad \eta < 0. \tag{8.24}$$

Thus, in order to have

$$\lim_{\eta\to-\infty} \hat{V}^{(1)}(k,\eta,\tau) = 0 \tag{8.25}$$

(obtained from (8.10)), we need to prescribe in (8.24)

$$\nu|\lambda| < 1. \tag{8.26}$$

It is easily seen that the general solution of (8.15) is

$$\hat{H}_1^{(1)}(k,\eta,\tau) = \frac{\sigma(\lambda)}{i\nu\lambda}\left\{\alpha(k,\tau)e^{\sigma(\lambda)k\eta} - \beta(k,\tau)e^{-\sigma(\lambda)k\eta}\right\}, \tag{8.27}$$

$$\hat{H}_2^{(1)}(k,\eta,\tau) = -\frac{1}{\nu\lambda}\left\{\alpha(k,\tau)e^{\sigma(\lambda)k\eta} + \beta(k,\tau)e^{-\sigma(\lambda)k\eta}\right\}, \tag{8.28}$$

$$\hat{E}^{(1)}(k,\eta,\tau) = \alpha(k,\tau)e^{\sigma(\lambda)k\eta} + \beta(k,\tau)e^{-\sigma(\lambda)k\eta}, \tag{8.29}$$

where $\alpha(k,\tau)$ and $\beta(k,\tau)$ are arbitrary complex-valued functions and

$$\sigma(\lambda) \doteq \sqrt{1-\nu^2\lambda^2} > 0\,.$$

Finally, imposing the condition (8.25) at infinity to (8.27)–(8.29) we find that

$$\hat{V}^{(1)}(k,\eta,\tau) = \begin{cases} \alpha(k,\tau)e^{\sigma(\lambda)k\eta}\begin{pmatrix} -i\sigma(\lambda)/\nu\lambda \\ -1/\nu\lambda \\ 1 \end{pmatrix}, & \text{if } k>0\,, \\ \\ \beta(k,\tau)e^{-\sigma(\lambda)k\eta}\begin{pmatrix} i\sigma(\lambda)/\nu\lambda \\ -1/\nu\lambda \\ 1 \end{pmatrix}, & \text{if } k<0\,. \end{cases} \tag{8.30}$$

Next, we use the solution (8.23) and (8.30) in the jump conditions (8.16). First we consider the case $k>0$. Under the assumption $\lambda - v_1^0 \neq 0$ or $B_1^0 \neq 0$, the resulting equations may be written as a linear system for the unknowns $(a, \alpha, k\hat{\varphi}^{(1)})$:

$$\begin{pmatrix} 1 & 0 & 1 \\ 0 & 1 & i\nu\lambda H_1^0 \\ d & i\sigma(\lambda)H_1^0/\nu\lambda & 0 \end{pmatrix}\begin{pmatrix} a \\ \alpha \\ k\hat{\varphi}^{(1)} \end{pmatrix} = 0. \tag{8.31}$$

This system has a nontrivial solution if and only if

$$d = (\lambda - v_1^0)^2 - (B_1^0)^2 = (H_1^0)^2\sigma(\lambda). \tag{8.32}$$

Real roots λ of (8.26) and (8.32) do not exist for all stationary states (v_1^0, B_1^0, H_1^0), but only for those satisfying a suitable physical condition, as discussed in the following lemma.

Lemma 8.4.1 *1. If* $|B_1^0| > |v_1^0| + 1/\nu$, *Eq.* (8.32) *does not have any real root.*

2. *If* $|B_1^0| = |v_1^0| + 1/\nu$, *for all* $|H_1^0| > 0$ *and* $v_1^0 \neq 0$ *there exists one real root* $\lambda = -\mathrm{sgn}(v_1^0)/\nu$. *If* $v_1^0 = 0$ *then* $\lambda = \pm 1/\nu$. *Thus in any case* $|\lambda| = 1/\nu$.
3. *If* $|v_1^0| - 1/\nu \leq |B_1^0| < |v_1^0| + 1/\nu$, *for all* $|H_1^0| > 0$ *there exist one or two real roots* λ *of* (8.32) *such that* $|\lambda| < 1/\nu$.
4. *If* $|B_1^0| < |v_1^0| - 1/\nu$, *there exists* $H^* > 0$ *such that, for all* $|H_1^0| \geqslant H^*$, *there exist two real roots* λ *of* (8.32) *such that* $|\lambda| < 1/\nu$ *(coincident roots if* $|H_1^0| = H^*$*); if* $|H_1^0| < H^*$ (8.32) *does not have any real root.*

Proof See [11].

We choose λ to be one of the values found in Lemma 8.4.1 such that $|\lambda| < 1/\nu$ (cases 3 and 4). Observe that for all such $|\lambda| < 1/\nu$, from (8.32) there holds $\lambda \neq v_1^0$ and $\lambda \neq v_1^0 \pm B_1^0$, i.e. $d \neq 0$.

The solution of (8.31) is then

$$a = -k\hat{\varphi}^{(1)}, \qquad \alpha = -\nu\lambda H_1^0 ik\hat{\varphi}^{(1)} \qquad \text{if } k > 0\,.$$

For $k < 0$ the analysis is similar and gives the solution

$$b = k\hat{\varphi}^{(1)}, \qquad \beta = -\nu\lambda H_1^0 ik\hat{\varphi}^{(1)} \qquad \text{if } k < 0\,,$$

under the same condition (8.32).

Summarizing these results, we have shown that when λ satisfies (8.26) and (8.32), the solution of (8.14)–(8.16) and (8.22), (8.25) is given by

$$\hat{U}^{(1)}(k,\eta,\tau) = \begin{cases} -|k|\hat{\varphi}^{(1)}(k,\tau)e^{-k\eta}\mathbf{R}, & \text{if } k > 0\,, \\ -|k|\hat{\varphi}^{(1)}(k,\tau)e^{k\eta}\overline{\mathbf{R}}, & \text{if } k < 0\,, \end{cases} \tag{8.33}$$

$$\hat{V}^{(1)}(k,\eta,\tau) = H_1^0\hat{\varphi}^{(1)}(k,\tau)e^{\sigma(\lambda)|k|\eta}\begin{pmatrix} -\sigma(\lambda)|k| \\ ik \\ -i\nu\lambda k \end{pmatrix}. \tag{8.34}$$

This solution depends on the unknown function $\hat{\varphi}^{(1)}(k,\tau)$, which describes the profile of the surface wave, and will be determined through a solvability condition for the equations for the second order corrections.

8.5 The Second Order Equations

We introduce the Fourier transforms

$$\hat{U}^{(2)}(k,\eta,\tau) = \frac{1}{2\pi}\int_{-\infty}^{+\infty} U^{(2)}(\theta,\eta,\tau)e^{-ik\theta}d\theta,$$

$$\hat{V}^{(2)}(k,\eta,\tau) = \frac{1}{2\pi}\int_{-\infty}^{+\infty} V^{(2)}(\theta,\eta,\tau)e^{-ik\theta}d\theta,$$

$$\hat{\varphi}^{(2)}(k,\tau) = \frac{1}{2\pi}\int_{-\infty}^{+\infty} \varphi^{(2)}(\theta,\tau)e^{-ik\theta}d\theta.$$

8.5.1 The Second Order Equations in the Plasma Region

Fourier transforming (8.17) with respect to θ gives the system of equations

$$ik\mathscr{A}\hat{U}^{(2)} + \mathscr{B}\partial_\eta\hat{U}^{(2)} = \hat{p}\,, \qquad \eta > 0\,, \tag{8.35}$$

where $\mathscr{A}$, $\mathscr{B}$ are the same matrices of (8.20). From (8.9), the solution of (8.35) must satisfy the decay condition

$$\lim_{\eta \to +\infty} \hat{U}^{(2)}(k, \eta, \tau) = 0 . \tag{8.36}$$

In order to solve (8.35) and (8.36), we proceed as in [1, 11]. We look for a solution of (8.35) in the form

$$\hat{U}^{(2)}(k, \eta, \tau) = \mathbf{S}(k, \eta, \tau) + a(k, \eta, \tau)\mathbf{R} + b(k, \eta, \tau)\overline{\mathbf{R}},$$

with a suitable vector-valued function $\mathbf{S}$ and scalar functions a, b. We look for $\mathbf{S}$ such that

$$\mathbf{L} \cdot \mathscr{B}\mathbf{S} = \overline{\mathbf{L}} \cdot \mathscr{B}\mathbf{S} = 0 ,$$

where $\mathbf{L}$ is an eigenvector of

$$\mathbf{L} \cdot (i\mathscr{A} - \mathscr{B}) = 0 ,$$

normalized by

$$\mathbf{L} \cdot \mathscr{B}\mathbf{R} = \overline{\mathbf{L}} \cdot \mathscr{B}\overline{\mathbf{R}} = 1 ,$$

that is

$$\mathbf{L} = -\frac{1}{2id(\lambda - v_1^0)}\mathbf{R} .$$

In [11] we show that the solution $\mathbf{S}$ is given by

$$\mathbf{S} = \left(\frac{1}{k}\mathbf{L}_1 \cdot \hat{p}, 0, \frac{1}{k}\mathbf{L}_3 \cdot \hat{p}, \frac{1}{k}\mathbf{L}_4 \cdot \hat{p}, 0\right)^T , \tag{8.37}$$

where

$$\mathbf{L}_1 = \frac{1}{d}\left(-i(\lambda - v_1^0), 0, iB_1^0, 0, 0\right)^T , \qquad \mathbf{L}_3 = \frac{1}{d}\left(iB_1^0, 0, -i(\lambda - v_1^0), 0, 0\right)^T ,$$

$$\mathbf{L}_4 = \left(0, 0, 0, -\frac{i}{\lambda - v_1^0}, 0\right)^T .$$

As for a, b, they are given by

$$a(k, \eta, \tau) = e^{-k\eta}\left(a_0(k, \tau) + \int_0^\eta \mathbf{L} \cdot \hat{p}(k, \eta', \tau)e^{k\eta'} d\eta'\right), \tag{8.38}$$

$$b(k, \eta, \tau) = e^{k\eta}\left(b_0(k, \tau) + \int_0^\eta \overline{\mathbf{L}} \cdot \hat{p}(k, \eta', \tau)e^{-k\eta'} d\eta'\right), \tag{8.39}$$

where $a_0(k,\tau), b_0(k,\tau)$ are arbitrary functions of integration, that will be chosen later.

In order to verify the decay condition (8.36) for $\hat{U}^{(2)}(k,\eta,\tau)$, we first explicitly calculate $\mathbf{S}(k,\eta,\tau)$ from (8.37). This is a very long expression that we don't report for the sake of brevity. The fact that we need here is that $\mathbf{S}$ depends on η only through the exponentials of $-|k|\eta$ and $-(|k-\ell|+|\ell|)\eta$, so that

$$\lim_{\eta\to+\infty} \mathbf{S}(k,\eta,\tau)=0\,.$$

Thus $\hat{U}^{(2)}(k,\eta,\tau)$ satisfies (8.36) if and only if

$$\lim_{\eta\to+\infty} a(k,\eta,\tau)=0\,, \tag{8.40}$$

$$\lim_{\eta\to+\infty} b(k,\eta,\tau)=0\,. \tag{8.41}$$

From (8.38), (8.39), and the explicit calculation of $\mathbf{L}\cdot\hat{p}(k,\eta,\tau), \overline{\mathbf{L}}\cdot\hat{p}(k,\eta,\tau)$, it follows that condition (8.40) is automatically satisfied if $k>0$, and (8.41) is automatically satisfied if $k<0$. It follows that a_0 remains undetermined for $k>0$, and b_0 remains undetermined for $k<0$. Instead, (8.38)–(8.41) may be used to determine a_0 if $k<0$, and b_0 if $k>0$, as functions of $\hat{\varphi}^{(1)}$:

$$\begin{aligned} a_0(k,\tau) &= -\int_0^{+\infty} \mathbf{L}\cdot\hat{p}(k,\eta',\tau)e^{k\eta'}\,d\eta'\,, \qquad \text{if } k<0\,,\\ b_0(k,\tau) &= -\int_0^{+\infty} \overline{\mathbf{L}}\cdot\hat{p}(k,\eta',\tau)e^{-k\eta'}\,d\eta'\,, \qquad \text{if } k>0\,. \end{aligned} \tag{8.42}$$

8.5.2 The Second Order Equations in Vacuum

We take the Fourier transform of (8.18) in θ for $\eta<0$. The problem is easily solved by substitution of $ik\hat{H}_2^{(2)}=(\hat{p}_2'-ik\hat{E}^{(2)})/\nu\lambda$ in the other equations to give

$$\partial_\eta^2\hat{E}^{(2)}+k^2(\nu^2\lambda^2-1)\hat{E}^{(2)}=-P\,, \qquad \eta<0\,, \tag{8.43}$$

where $P=\nu\lambda ik\hat{p}_3'-ik\hat{p}_2'+\partial_\eta\hat{p}_1'$. We solve (8.43) with the decay condition

$$\lim_{\eta\to-\infty}\hat{E}^{(2)}(k,\eta,\tau)=0\,,$$

then we substitute in the other equations to find the other components of $\hat{V}^{(2)}(k,\eta,\tau)$ and obtain

$$\hat{V}^{(2)}(k,\eta,\tau) = \alpha'(k,\tau)e^{\sigma(\lambda)k\eta}\begin{pmatrix} \frac{\sigma(\lambda)}{i\nu\lambda} \\ -\frac{1}{\nu\lambda} \\ 1 \end{pmatrix} + \begin{pmatrix} \frac{1}{\nu\lambda ik}\hat{p}_1' \\ \frac{1}{\nu\lambda ik}\hat{p}_2' \\ 0 \end{pmatrix}$$

$$+\frac{1}{2|k|\sigma(\lambda)}\begin{pmatrix} \frac{1}{\nu\lambda ik}\left\{\int_{-\infty}^{0} e^{-\sigma(\lambda)|k||\eta-\zeta|}\partial_\zeta P(k,\zeta,\tau)\,d\zeta - e^{-\sigma(\lambda)|k\eta|}P(k,0,\tau)\right\} \\ -\frac{1}{\nu\lambda}\int_{-\infty}^{0} e^{-\sigma(\lambda)|k||\eta-\zeta|}P(k,\zeta,\tau)\,d\zeta \\ \int_{-\infty}^{0} e^{-\sigma(\lambda)|k||\eta-\zeta|}P(k,\zeta,\tau)\,d\zeta \end{pmatrix}$$

if $k > 0$,

$$\hat{V}^{(2)}(k,\eta,\tau) = \beta'(k,\tau)e^{-\sigma(\lambda)k\eta}\begin{pmatrix} -\frac{\sigma(\lambda)}{i\nu\lambda} \\ -\frac{1}{\nu\lambda} \\ 1 \end{pmatrix} + \begin{pmatrix} \frac{1}{\nu\lambda ik}\hat{p}_1' \\ \frac{1}{\nu\lambda ik}\hat{p}_2' \\ 0 \end{pmatrix}$$

$$+\frac{1}{2|k|\sigma(\lambda)}\begin{pmatrix} \frac{1}{\nu\lambda ik}\left\{\int_{-\infty}^{0} e^{-\sigma(\lambda)|k||\eta-\zeta|}\partial_\zeta P(k,\zeta,\tau)\,d\zeta - e^{-\sigma(\lambda)|k\eta|}P(k,0,\tau)\right\} \\ -\frac{1}{\nu\lambda}\int_{-\infty}^{0} e^{-\sigma(\lambda)|k||\eta-\zeta|}P(k,\zeta,\tau)\,d\zeta \\ \int_{-\infty}^{0} e^{-\sigma(\lambda)|k||\eta-\zeta|}P(k,\zeta,\tau)\,d\zeta \end{pmatrix}$$

if $k < 0$. Notice that we need to determine the arbitrary functions $\alpha'(k,\tau)$ if $k > 0$, and $\beta'(k,\tau)$ if $k < 0$.

8.5.3 The Second Order Jump Conditions

The first-order solution depends on the unknown function $\hat{\varphi}^{(1)}(k,\tau)$, which describes the profile of the surface wave, while the second order solution depends, in addition, on unknown functions $a_0(k,\tau)$, $b_0(k,\tau)$ and $\alpha'(k,\tau)$, $\beta'(k,\tau)$. Now we study the second order jump conditions and show that they reduce to a singular linear system of algebraic equations for $(a_0, b_0, \alpha', \beta', \hat{\varphi}^{(2)})$, where $\hat{\varphi}^{(2)}(k,\tau)$ is the Fourier transform of the second-order displacement of the interface. Imposing solvability conditions on this system gives the evolution equation for the function $\hat{\varphi}^{(1)}(k,\tau)$ we are looking for.

We take the Fourier transform of (8.19), where we substitute the second order corrections $\hat{U}^{(2)}$, $\hat{V}^{(2)}$ obtained in the previous sections, evaluated at $\eta = 0$.

Let us first assume $k > 0$, recalling that in this case we need to determine $a_0(k,\tau)$, $\alpha'(k,\tau)$ and $\hat{\varphi}^{(2)}(k,\tau)$ (for $k > 0$). In this case we obtain the linear system

$$\begin{pmatrix} 1 & 0 & 1 \\ 1 & 0 & 1 \\ 0 & 1/\nu\lambda & iH_1^0 \\ d & i\sigma(\lambda)H_1^0/\nu\lambda & 0 \\ 0 & 1 & i\nu\lambda H_1^0 \end{pmatrix} \begin{pmatrix} a_0 \\ \alpha' \\ k\hat{\varphi}^{(2)} \end{pmatrix} = \begin{pmatrix} \hat{r}'_1 \\ \hat{r}'_2 \\ \hat{r}'_3 \\ \hat{r}'_4 \\ \hat{r}'_5 \end{pmatrix}, \tag{8.44}$$

where $\hat{r}'_1, \dots, \hat{r}'_5$ depend on the first order corrections and b_0 given by (8.42) (in case $k > 0$ it is known). First of all we see that the first two lines of the matrix in the left-hand side of (8.44) are equal, and we can also verify that $\hat{r}'_1 = \hat{r}'_2$. Moreover, the last row of the matrix in (8.44) equals the third one multiplied by $\nu\lambda$, and actually one verifies that $\hat{r}'_5 = \nu\lambda\hat{r}'_3$. Thus (8.44) may be reduced to

$$\begin{pmatrix} 1 & 0 & 1 \\ 0 & 1/\nu\lambda & iH_1^0 \\ d & i\sigma(\lambda)H_1^0/\nu\lambda & 0 \end{pmatrix} \begin{pmatrix} a_0 \\ \alpha' \\ k\hat{\varphi}^{(2)} \end{pmatrix} = \begin{pmatrix} \hat{r}'_1 \\ \hat{r}'_3 \\ \hat{r}'_4 \end{pmatrix}. \tag{8.45}$$

The determinant of the matrix of this system is zero because of (8.32), i.e. the equation defining λ. It is easily seen that the rank of this matrix is 2. Then, the linear system (8.45) is solvable if and only if the rank of the augmented matrix is also equal to 2, and this is true if the following condition holds:

$$d\hat{r}'_3 + iH_1^0\hat{r}'_4 - iH_1^0 d\hat{r}'_1 = 0. \tag{8.46}$$

Developing the terms in (8.46) we get the solvability condition for $k > 0$

$$\mu_0\hat{\varphi}^{(1)}_\tau(k,\tau) + i\int_{-\infty}^{+\infty} \Lambda_+(k,\ell)\hat{\varphi}^{(1)}(k-\ell,\tau)\hat{\varphi}^{(1)}(\ell,\tau)\,d\ell = 0, \tag{8.47}$$

where we have denoted

$$\mu_0 = 2\frac{\lambda - v_1^0}{d} + \frac{\nu^2\lambda}{\sigma(\lambda)^2},$$

$$\begin{aligned} \Lambda_+(k,\ell) = \ell\frac{|k-\ell|(|k-\ell|-|\ell|) + (k-\ell)|\ell| - |k-\ell|\ell}{|k-\ell|+|k|+|\ell|} + \frac{(k-\ell)|\ell|(|\ell|-\ell)}{|k|+|\ell|} \\ -(k-\ell)|\ell| + \sigma(\lambda)\Big\{-k|\ell| + \frac{1}{2}\big((k+\ell)\ell - |k-\ell||\ell|\big)\Big\}. \end{aligned}$$

Thus, when $k > 0$, system (8.45) is solvable if and only if $\hat{\varphi}^{(1)}$ satisfies equation (8.47) and then the rank of the augmented matrix of the system is equal to 2. Given the solution $\hat{\varphi}^{(1)}$ of (8.47) we compute $\hat{U}^{(1)}$, $\hat{V}^{(1)}$ from (8.33) and (8.34). Thus the leading-order term of the asymptotic expansion is uniquely determined. From system (8.45) we may obtain a_0, α' in terms of an arbitrary second order wave profile $\hat{\varphi}^{(2)}$, and in turn $\hat{U}^{(2)}$, $\hat{V}^{(2)}$ from the expressions obtained in the previous sections. The wave profile $\hat{\varphi}^{(2)}$ should be determined by considering higher order terms of the asymptotic expansion, see [10].

The analysis for $k < 0$ is similar and leads to the solvability condition

$$\mu_0 \hat{\varphi}^{(1)}_\tau(k,\tau) + i \int_{-\infty}^{+\infty} \Lambda_-(k,\ell)\hat{\varphi}^{(1)}(k-\ell,\tau)\hat{\varphi}^{(1)}(\ell,\tau)\, d\ell = 0\,, \tag{8.48}$$

where we have denoted

$$\Lambda_-(k,\ell) = \ell \frac{|k-\ell|(|k-\ell|-|\ell|)-(k-\ell)|\ell|+|k-\ell|\ell}{|k-\ell|+|k|+|\ell|} + \frac{(k-\ell)|\ell|(|\ell|+\ell)}{|k|+|\ell|}$$
$$-(k-\ell)|\ell| + \sigma(\lambda)\Big\{-k|\ell| - \frac{1}{2}\big((k+\ell)\ell - |k-\ell||\ell|\big)\Big\}.$$

Given the solution $\hat{\varphi}^{(1)}$ of (8.48) we compute $\hat{U}^{(1)}$, $\hat{V}^{(1)}$ from (8.33) and (8.34). We may get b_0, β' in terms of an arbitrary second order wave profile $\hat{\varphi}^{(2)}$, and in turn $\hat{U}^{(2)}$, $\hat{V}^{(2)}$ from the expressions obtained in the previous sections. Again, also for $k < 0$ the wave profile $\hat{\varphi}^{(2)}$ should be determined by considering higher order terms of the asymptotic expansion, see [10].

8.5.4 The Kernel

The Eqs. (8.47) and (8.48) can be written in more compact form as

$$\mu_0 \hat{\varphi}^{(1)}_\tau(k,\tau) + i \int_{-\infty}^{+\infty} \Lambda_0(k,\ell)\hat{\varphi}^{(1)}(k-\ell,\tau)\hat{\varphi}^{(1)}(\ell,\tau)\, d\ell = 0\,, \quad \forall k \neq 0\,, \tag{8.49}$$

with kernel

$$\Lambda_0(k,\ell) = \begin{cases} \Lambda_+(k,\ell) & k > 0, \\ \Lambda_-(k,\ell) & k < 0, \end{cases}$$

where $\Lambda_\pm$ are defined after (8.47) and (8.48). This form is not convenient and so we look for a different formula. First we write it as

$$\Lambda_0(k,\ell) = \mathrm{sgn}(k)\, \tilde{\Lambda}_0(k-\ell,\ell),$$

so that the integral in (8.49) takes the form of a convolution product. Moreover, the new kernel $\tilde{\Lambda}_0(k,\ell)$ can be equivalently replaced in (8.49) by the symmetrized kernel

$$\tilde{\Lambda}(k,\ell) = \frac{1}{2}\Big(\tilde{\Lambda}_0(k,\ell) + \tilde{\Lambda}_0(\ell,k)\Big),$$

because the antisymmetric part of $\tilde{\Lambda}_0$ gives a vanishing integral. Thus we can write (8.49) as

$$\mu_0 \hat{\varphi}^{(1)}_\tau(k,\tau) + i\,\mathrm{sgn}(\mathrm{k}) \int_{-\infty}^{+\infty} \tilde{\Lambda}(\mathrm{k}-\ell,\ell)\,\hat{\varphi}^{(1)}(\mathrm{k}-\ell,\tau)\,\hat{\varphi}^{(1)}(\ell,\tau)\,\mathrm{d}\ell = 0\,, \tag{8.50}$$

for all $k \neq 0$. The explicit formula of $\tilde{\Lambda}$ is still rather complicated, see [11], but there is the way to simplify it. First of all we verify that $\tilde{\Lambda}$ satisfies the following properties

$$\tilde{\Lambda}(k,\ell) = \tilde{\Lambda}(\ell,k) \qquad \text{(symmetry)}, \tag{8.51}$$

$$\tilde{\Lambda}(k,\ell) = \overline{\tilde{\Lambda}(-k,-\ell)} \qquad \text{(reality)}, \tag{8.52}$$

$$\tilde{\Lambda}(\alpha k,\alpha\ell) = \alpha^2 \tilde{\Lambda}(k,\ell) \quad \forall\,\alpha > 0 \qquad \text{(homogeneity)}. \tag{8.53}$$

Considering some particular cases we can considerably simplify $\tilde{\Lambda}$ as follows

$$\tilde{\Lambda}(k,\ell) = \begin{cases} -(1+\sigma(\lambda))k\ell & \text{if } k>0,\ \ell>0\,, \\ (1+\sigma(\lambda))\ell(k+\ell) & \text{if } k+\ell>0,\ \ell<0\,. \end{cases} \tag{8.54}$$

Then the values of $\tilde{\Lambda}$ in the other regions of the (k,ℓ)-plane follow from (8.51)–(8.54). Finally $\tilde{\Lambda}$ can be written in a different way as

$$\tilde{\Lambda}(k,\ell) = -(1+\sigma(\lambda))\frac{2|k+\ell|\,|k|\,|\ell|}{|k+\ell|+|k|+|\ell|}\,, \tag{8.55}$$

and, after an appropriate rescaling in time, we write (8.50) and (8.55) as

$$\hat{\varphi}^{(1)}_\tau(k,\tau) + i\,\mathrm{sgn}(\mathrm{k}) \int_{-\infty}^{+\infty} \Lambda(\mathrm{k}-\ell,\ell)\,\hat{\varphi}^{(1)}(\mathrm{k}-\ell,\tau)\,\hat{\varphi}^{(1)}(\ell,\tau)\,\mathrm{d}\ell = 0\,, \quad \forall\,\mathrm{k} \neq 0\,, \tag{8.56}$$

with the new kernel Λ defined by

$$\Lambda(k,\ell) = \frac{2|k+\ell|\,|k|\,|\ell|}{|k+\ell|+|k|+|\ell|}\,. \tag{8.57}$$

The spacial form of (8.56) and (8.57) is, see [2, 5],

$$\varphi^{(1)}_\tau + \frac{1}{2}\mathbb{H}[\Phi^2]_{\theta\theta} + \Phi\varphi^{(1)}_{\theta\theta} = 0, \qquad \Phi = \mathbb{H}[\varphi^{(1)}]\,,$$

where $\mathbb{H}$ denotes the Hilbert transform. After renaming of variables it becomes (8.1) and (8.2).

Equations (8.56) and (8.57) is well-known as it coincides with the amplitude equation for nonlinear Rayleigh waves [5] and describes the propagation of surface waves on a tangential discontinuity (current-vortex sheet) in incompressible MHD

[1]. The derivation of the asymptotic equation (8.56) and (8.57) also for the plasma-interface problem confirms that it is a canonical model equation for nonlinear surface wave solutions of hyperbolic conservation laws, analogous to the inviscid Burgers equation for bulk waves.

Λ, defined in (8.57), is perhaps the simplest kernel arising for surface waves. It satisfies the properties

$$\Lambda(k,\ell) = \Lambda(\ell,k) \qquad \text{(symmetry)}, \tag{8.58}$$

$$\Lambda(k,\ell) = \overline{\Lambda(-k,-\ell)} \qquad \text{(reality)}, \tag{8.59}$$

$$\Lambda(\alpha k,\alpha\ell) = \alpha^2 \Lambda(k,\ell) \quad \forall\, \alpha > 0 \qquad \text{(homogeneity)}, \tag{8.60}$$

$$\Lambda(k+\ell,-\ell) = \overline{\Lambda(k,\ell)} \quad \forall k,\ell \in \mathbb{R} \qquad \text{(Hamiltonian)}. \tag{8.61}$$

The value 2 of the scaling exponent in (8.60) is consistent with the dimensional analysis in [2] for surface waves. It is shown by Alì et al. [2] that (8.61) is a sufficient condition for (8.50), in addition to (8.58) and (8.59), to admit a *Hamiltonian structure*, see also [5, 6].

The results of Sects. 8.3 to 8.5 are summarized in the following theorem.

Theorem 8.5.1 *Assume that* v_1^0, B_1^0, H_1^0 *are as in (3) or (4) of Lemma 8.4.1, and let* λ *be a real root of* (8.32). *Then the solution* $U = (\mathbf{v}, \mathbf{B}, q)^T$, $V = (\mathbf{H}, E)^T$, φ *of* (8.4), (8.6)–(8.8) *admits the asymptotic expansions* (8.11)–(8.13) *where the first order terms of the expansion are defined in* (8.33) *and* (8.34). *The location of the plasma-vacuum interface is given by*

$$x_2 = \varepsilon\varphi^{(1)}(x_1 - \lambda t, \varepsilon t) + O(\varepsilon^2),$$

as $\varepsilon \to 0$, *with* $t = O(\varepsilon^{-1})$ *and* λ *the linearized phase velocity of the surface wave. After an appropriate rescaling in time, the Fourier transform of the leading order perturbation* $\varphi^{(1)}(\theta,\tau)$ *satisfies the amplitude equation* (8.56) *and* (8.57).

We wish to stress that for the existence of surface waves propagating on the plasma-vacuum interface, it is necessary to have a real root λ of (8.32) satisfying (8.26). This is obtained if the basic state v_1^0, B_1^0, H_1^0 is as in (3) or (4) of Lemma 8.4.1.

8.6 Noncanonical Variables and Well-Posedness

As in [7] we introduce the noncanonical dependent variable $\psi(\theta,\tau)$ defined by

$$\psi(\theta,\tau) = |\partial_\theta|^{1/2}\varphi^{(1)}(\theta,\tau), \qquad \hat{\psi}(k,\tau) = |k|^{1/2}\hat{\varphi}^{(1)}(k,\tau).$$

Then rewriting Eq. (8.56) in terms of ψ gives

$$\hat{\psi}_\tau(k,\tau) + i\,k \int_{-\infty}^{+\infty} S(k-\ell,\ell)\,\hat{\psi}(k-\ell,\tau)\,\hat{\psi}(\ell,\tau)\,d\ell = 0\,, \quad \forall k \neq 0\,, \tag{8.62}$$

with kernel S given by

$$S(k,\ell) = \frac{\Lambda(k,\ell)}{|k\ell(k+\ell)|^{1/2}}.$$

The definition of S is extended by setting

$$S(k,\ell) = 0 \qquad \text{if } k\ell = 0\,.$$

The corresponding spatial form of (8.62) is

$$\partial_\tau \psi + \partial_\theta a(\psi,\psi) = 0\,, \tag{8.63}$$

where the bilinear form a is defined by

$$\widehat{a(\psi,\phi)}(k,\tau) = \int_{-\infty}^{+\infty} S(k-\ell,\ell)\,\hat{\psi}(k-\ell,\tau)\,\hat{\phi}(\ell,\tau)\,d\ell. \tag{8.64}$$

(8.63) has the form of a nonlocal Burgers equation, like (2.8) in [7], or (1.1) in [3].

8.6.1 Well-Posedness of the Amplitude Equation

We consider the initial value problem for the noncanonical equation (8.63) and (8.64), supplemented by an initial condition

$$\psi(\theta,0) = \psi_0(\theta). \tag{8.65}$$

The well-posedness of (8.63)–(8.65) easily follows by adapting the proof of Hunter [7] (given for the periodic setting) to our case.

Theorem 8.6.1 *For any $\psi_0 \in H^s(\mathbb{R})$, $s > 2$, the initial value problem* (8.63)–(8.65) *has a unique local solution*

$$\psi \in C(I; H^s(\mathbb{R})) \cap C^1(I; H^{s-1}(\mathbb{R}))$$

defined on the time interval $I = (-\tau_*, \tau_*)$, where

$$\tau_* = \frac{1}{K_s \|\psi_0\|^{1-2/s}_{L^2(\mathbb{R})} \|\psi_0\|^{2/s}_{H^s(\mathbb{R})}}, \tag{8.66}$$

for a suitable constant K_s.

For a complete proof of Theorem 8.6.1 see [11]. The well-posedness result of Theorem 8.6.1 may be easily recast as a similar result for (8.56) and (8.57).

Proof (*Sketch of proof*) The proof is based on the a priori estimate

$$\left|\frac{d}{d\tau}\|\psi\|_{H^s(\mathbb{R})}\right| \leqslant CC_s\|\psi_0\|_{L^2(\mathbb{R})}^{1-2/s}\|\psi\|_{H^s(\mathbb{R})}^{1+2/s}, \tag{8.67}$$

deduced from the properties (8.58)–(8.61) of the kernel. Using Gronwall's inequality, we deduce from (8.67) the bound

$$\|\psi(\cdot,\tau)\|_{H^s(\mathbb{R})} \leqslant \|\psi_0\|_{H^s(\mathbb{R})}\left(1-\frac{2CC_s}{s}\|\psi_0\|_{L^2(\mathbb{R})}^{1-2/s}\|\psi_0\|_{H^s(\mathbb{R})}^{2/s}|\tau|\right)^{-s/2}, \tag{8.68}$$

for $|\tau| < \tau_*$ where τ_* is given by (8.66).

The rest of the proof follows from a standard Galerkin approximation. Given any function $f \in H^s(\mathbb{R})$ with Fourier transform $\hat{f}$ we define the finite dimensional orthogonal projection

$$f^N(\theta) = P_N f(\theta) = \int_{|k|\leq N} \widehat{f}(k)e^{ik\theta}dk\,.$$

We consider the Galerkin approximation $\psi^N = P_N\psi$, defined as the solution of the approximate system of ODEs

$$\begin{cases}\partial_\tau\psi^N + P_N\partial_\theta a(\psi^N,\psi^N) = 0\,,\\ \psi^N(\theta,0) = P_N\psi_0(\theta).\end{cases} \tag{8.69}$$

The approximate solutions φ^N satisfy the a priori estimate (8.68), uniformly in N. By standard arguments we can extract a subsequence $\{\varphi^N\}$ and pass to the limit in (8.69) to obtain a solution of (8.63). The uniqueness of the solution follows by a standard argument. See [11] for details.

In [11] we also obtain the following *blow-up criterion*.

Lemma 8.6.1 *Under the assumptions of Theorem 8.6.1, if* $\psi \in C(0,T;H^s(\mathbb{R}))$ *with* $0 < T < +\infty$ *is a solution of* (8.63) *such that*

$$\int_0^T \|\psi(\cdot,\tau)\|_{s'}^{2/s'}d\tau < +\infty$$

for some $s' > 2$, *then* ψ *is continuable to a solution* $\psi \in C(0,T';H^s(\mathbb{R}))$ *with* $T' > T$.

Proof See [11].

8.6.2 Regularity of the First Order Terms $U^{(1)}$, $V^{(1)}$

The resolution of (8.56) and (8.57) via the introduction of the noncanonical variable $\psi(\theta,\tau)$, stated in Theorem 8.6.1, is the chief step. When this is achieved it is not difficult to obtain the regularity of the other first order terms (8.33) and (8.34) of the asymptotic expansions (8.11) and (8.12) as in the following lemma.

Lemma 8.6.2 *For any* $\psi_0 \in H^s(\mathbb{R})$, *with integer* $s > 2$, *let*

$$\psi \in C(I; H^s(\mathbb{R})) \cap C^1(I; H^{s-1}(\mathbb{R}))$$

be the solution of (8.63)–(8.65), *given by Theorem 8.6.1. Let* $\varphi^{(1)}$ *be defined by* $\psi(\theta,\tau) = |\partial_\theta|^{1/2}\varphi^{(1)}(\theta,\tau)$, *and* $\hat{U}^{(1)}$, $\hat{V}^{(1)}$ *be defined by* (8.33) *and* (8.34). *Then*

$$U^{(1)} \in C(I; H^s(\mathbb{R}^2_+)) \cap C^1(I; H^{s-1}(\mathbb{R}^2_+)), \quad V^{(1)} \in C(I; H^s(\mathbb{R}^2_-)) \cap C^1(I; H^{s-1}(\mathbb{R}^2_-)),$$

where $\mathbb{R}^2_\pm = \{(\theta,\eta) |\, \theta \in \mathbb{R}, \pm\eta > 0\}$.

Proof For the proof it is convenient to introduce the homogeneous space $\dot{H}^m(\mathbb{R})$ (w.r.to the independent variable θ),

$$\dot{H}^m(\mathbb{R}) = \left\{ u(\theta) : \mathbb{R} \to \mathbb{R} : \int_{-\infty}^{+\infty} |k|^{2m} |\hat{u}(k)|^2\, dk < +\infty \right\},$$

normed by

$$\|u\|_{\dot{H}^m(\mathbb{R})} = \left(\int_{-\infty}^{+\infty} |k|^{2m} |\hat{u}(k)|^2 dk \right)^{1/2}.$$

Let us consider integers $m, p \geqslant 0$ such that $m + p \leqslant s$. For every $\tau \in I$ we have

$$\|\partial_\eta^p U^{(1)}(\cdot,\tau)\|^2_{L^2(\mathbb{R}^+_\eta; \dot{H}^m(\mathbb{R}_\theta))} = \| |k|^m \partial_\eta^p \hat{U}^{(1)}(\cdot,\tau)\|^2_{L^2(\mathbb{R}^2_+)},$$

recalling that our Fourier transform is taken w.r.to θ. Substituting the definition (8.33) gives

$$\begin{aligned} \| |k|^m \partial_\eta^p \hat{U}^{(1)}(\cdot,\tau)\|^2_{L^2(\mathbb{R}^2_+)} &= \int_0^\infty \int_{\mathbb{R}} |k|^{2m+2} |\hat{\varphi}^{(1)}(k,\tau)|^2 |\partial_\eta^p e^{-|k|\eta}|^2 |R|^2 dk d\eta \\ &= \frac{1}{2}|R|^2 \int_0^\infty \int_{\mathbb{R}} |k|^{2m+2p+1} |\hat{\varphi}^{(1)}(k,\tau)|^2 2|k| e^{-2|k|\eta} dk d\eta \\ = \frac{1}{2}|R|^2 \int_{\mathbb{R}} |k|^{2m+2p} \big| |k|^{1/2} \hat{\varphi}^{(1)}(k,\tau)\big|^2 dk &= \frac{1}{2}|R|^2 \|\psi(\cdot,\tau)\|^2_{\dot{H}^{m+p}(\mathbb{R})}. \end{aligned}$$

Adding over $m, p \geqslant 0$ such that $m + p \leqslant s$ gives $U^{(1)} \in L^\infty(I; H^s(\mathbb{R}^2_+))$. The same calculation for a difference at times τ_1, τ_2 gives

$$\|\partial_\eta^p U^{(1)}(\cdot, \tau_1) - \partial_\eta^p U^{(1)}(\cdot, \tau_2)\|_{L^2(\mathbb{R}^+_\eta; \dot{H}^m(\mathbb{R}_\theta))} \leq C\|\psi(\cdot, \tau_1) - \psi(\cdot, \tau_2)\|_{\dot{H}^{m+p}(\mathbb{R})},$$

and adding again over $m + p \leqslant s$ gives the time continuity of $U^{(1)}$ in $H^s(\mathbb{R}^2_+)$, by the time continuity of ψ in $H^s(\mathbb{R})$. Thus we have obtained $U^{(1)} \in C(I; H^s(\mathbb{R}^2_+))$. The proof of $U^{(1)} \in C^1(I; H^{s-1}(\mathbb{R}^2_+))$ is similar, thanks to the time continuity of ψ_τ in $H^{s-1}(\mathbb{R})$. By the same arguments we prove the regularity of $V^{(1)}$.

Acknowledgements The author is supported by the national research project PRIN 2012 "Nonlinear Hyperbolic Partial Differential Equations, Dispersive and Transport Equations: theoretical and applicative aspects".

References

1. Alì, G., Hunter, J.K.: Nonlinear surface waves on a tangential discontinuity in magnetohydrodynamics. Q. Appl. Math. **61**(3), 451–474 (2003)
2. Alì, G., Hunter, J.K., Parker, D.F.: Hamiltonian equations for scale-invariant waves. Stud. Appl. Math. **108**(3), 305–321 (2002)
3. Benzoni-Gavage, S.: Local well-posedness of nonlocal Burgers equations. Differ. Integral Equ. **22**(3–4), 303–320 (2009)
4. Goedbloed, J.P., Poedts, S.,S.: Principles of Magnetohydrodynamics with Applications to Laboratory and Astrophysical Plasmas. Cambridge University Press, Cambridge (2004)
5. Hamilton, M.F., Il'insky, Y.A., Zabolotskaya, E.A.: Evolution equations for nonlinear Rayleigh waves. J. Acoust. Soc. Am. **97**, 891–897 (1995)
6. Hunter, J.K.: Nonlinear surface waves. In: Current progress in Hyberbolic Systems: Riemann Problems and Computations (Brunswick, ME, 1988), pp. 185–202. American Mathematical Society (1989)
7. Hunter, J.K.: Short-time existence for scale-invariant Hamiltonian waves. J. Hyperbolic Differ. Equ. **3**(2), 247–267 (2006)
8. Hunter, J.K.: Nonlinear hyperbolic surface waves. In: Nonlinear Conservation Laws and Applications. The IMA Volumes in Mathematics and its Applications, vol. **153**, pp. 303–314. Springer, New York (2011)
9. Mandrik, N., Trakhinin, Y.: Influence of vacuum electric field on the stability of a plasma-vacuum interface. Commun. Math. Sci. **12**(6), 1065–1100 (2014)
10. Marcou, A.: Rigorous weakly nonlinear geometric optics for surface waves. Asymptot. Anal. **69**(3–4), 125–174 (2010)
11. Secchi, P.: Nonlinear surface waves on the plasma-vacuum interface. Q. Appl. Math. **73**(4), 711–737 (2015)

Chapter 9
On the $\mathscr{R}$-Bounded Solution Operator and the Maximal L_p-L_q Regularity of the Stokes Equations With Free Boundary Condition

Yoshihiro Shibata

Abstract In this paper, we consider the boundary value problem of Stokes operator arising in the study of free boundary problem for the Navier-Stokes equations with surface tension in a uniform $W_r^{3-1/r}$ domain of N-dimensional Euclidean space $\mathbb{R}^N$ ($N \geqslant 2$, $N < r < \infty$). We prove the existence of $\mathscr{R}$-bounded solution operator with spectral parameter λ varying in a sector $\Sigma_{\varepsilon,\lambda_0} = \{\lambda \in \mathbb{C} \mid |\arg\lambda| \leqslant \pi - \varepsilon,\ |\lambda| \geqslant \lambda_0\}$ ($0 < \varepsilon < \pi/2$), and the maximal L_p-L_q regularity with the help of the $\mathscr{R}$-bounded solution operator and the Weis operator valued Fourier multiplier theorem. The essential assumption of this paper is the unique solvability of the weak Dirichlet-Neumann problem, namely it is assumed the unique existence of solution $\mathfrak{p} \in \mathscr{W}_q^1(\Omega)$ to the variational problem: $(\nabla\mathfrak{p}, \nabla\varphi)_\Omega = (f, \nabla\varphi)_\Omega$ for any $\varphi \in \mathscr{W}_{q'}^1(\Omega)$ with $1 < q < \infty$ and $q' = q/(q-1)$, where $\mathscr{W}_q^1(\Omega)$ is a closed subspace of $\hat{W}_{q,\Gamma}^1(\Omega) = \{\mathfrak{p} \in L_{q,\mathrm{loc}}(\Omega) \mid \nabla\mathfrak{p} \in L_q(\Omega)^N,\ \mathfrak{p}|_\Gamma = 0\}$ with respect to gradient norm $\|\nabla\cdot\|_{L_q(\Omega)}$ that contains a space $W_{q,\Gamma}^1(\Omega) = \{\mathfrak{p} \in W_q^1(\Omega) \mid \mathfrak{p}|_\Gamma = 0\}$, and Γ is one part of boundary on which free boundary condition is imposed. The unique solvability of such weak Dirichlet-Neumann problem is necessary for the unique existence of a solution to the resolvent problem with uniform estimate with respect to spectral parameter varying in (λ_0, ∞), which was proved in Shibata [13]. Our assumption is satisfied for any $q \in (1, \infty)$ by the following domains: half space, perturbed half space, bounded domains, layer, perturbed layer, straight cube, and exterior domains with $\mathscr{W}_q^1(\Omega) = \hat{W}_{q,\Gamma}^1(\Omega)$.

Keywords $\mathscr{R}$-Boundedness · Stokes equations · Free boundary condition · Surface tension · Uniform $W_r^{3-1/r}$ domain · Analytic semigroup · Maximal L_p-L_q regularity

Mathematics Subject Classification (2000) 35Q35 · 76D07

Y. Shibata (✉)
Department of Mathematics and Research Institute of Science and Engineering, Waseda University, Ohkubo 3-4-1, Tokyo, Shinjuku-ku 169-8555, Japan
e-mail: yshibata@waseda.jp

Y. Shibata and Y. Suzuki (eds.), *Mathematical Fluid Dynamics, Present and Future*, Springer Proceedings in Mathematics & Statistics 183,
DOI 10.1007/978-4-431-56457-7_9

9.1 Introduction

Let Ω be a domain in the N dimensional Euclidean space $\mathbb{R}^N$ ($N \geqslant 2$), whose boundary consists of two hypersurfaces Γ and Γ_0. We assume that the distance between Γ and Γ_0 is positive, that $\Gamma \neq \emptyset$, and that $\Gamma_0 = \emptyset$ is admissible. Let t be the time variable and $x = (x_1, \ldots, x_N) \in \mathbb{R}^N$. In this paper, we consider the maximal regularity of the non-stationary Stokes equations:

$$\begin{cases} \partial_t \mathbf{u} - \mathrm{Div}\,(\mu \mathbf{D}(\mathbf{u}) - \mathfrak{p}\mathbf{I}) = \mathbf{f}, \quad \mathrm{div}\,\mathbf{u} = g & \text{in } \Omega \times (0, T), \\ \partial_t h - \mathbf{n} \cdot \mathbf{u} = f & \text{on } \Gamma \times (0, T), \\ \{\mathbf{D}(\mathbf{u}) - \mathfrak{p}\mathbf{I} - ((\tau + \delta \Delta_\Gamma) h)\mathbf{I}\}\mathbf{n} = \mathbf{f}_b & \text{on } \Gamma \times (0, T), \\ \mathbf{u} = 0 & \text{on } \Gamma_0 \times (0, T), \\ (\mathbf{u}, h)|_{t=0} = (\mathbf{u}_0, h_0) & \text{in } \Omega \times \Gamma, \end{cases} \tag{9.1}$$

and the existence of $\mathscr{R}$-bounded solution operators for the corresponding generalized resolvent problem:

$$\begin{cases} \lambda \mathbf{u} - \mathrm{Div}\,(\mu \mathbf{D}(\mathbf{u}) - \mathfrak{p}\mathbf{I}) = \mathbf{f}, \quad \mathrm{div}\,\mathbf{u} = g & \text{in } \Omega, \\ \lambda h - \mathbf{n} \cdot \mathbf{u} = f & \text{on } \Gamma, \\ \{\mu \mathbf{D}(\mathbf{u}) - \mathfrak{p}\mathbf{I} - ((\tau + \delta \Delta_\Gamma) h)\mathbf{I}\}\mathbf{n} = \mathbf{f}_b & \text{on } \Gamma, \\ \mathbf{u} = 0 & \text{on } \Gamma_0 \end{cases} \tag{9.2}$$

with spectral parameter λ varying in $\Sigma_{\varepsilon,\lambda_0} = \{\lambda \in \Sigma_\varepsilon \mid |\lambda| \geqslant \lambda_0\}$ with $0 < \varepsilon < \pi/2$, $\lambda_0 > 0$ and $\Sigma_\varepsilon = \{\lambda \in \mathbb{C} \setminus \{0\} \mid |\arg \lambda| \leqslant \pi - \varepsilon\}$. Here, unknowns are $\mathbf{u} = {}^\top(u_1, \ldots, u_N)$, $\mathfrak{p}$ and h, where ${}^\top M$ denotes the transposed M, while $\mathbf{f} = {}^\top(f_1, \ldots, f_N)$, g, f, $\mathbf{f}_b = {}^\top(f_{b1}, \ldots, f_{bN})$ are, in the order, prescribed N-component vectors and scalar functions, and $\mathbf{u}_0$ and h_0 are also prescribed N-component vector and a scalar function. As for the remaining notations, μ and δ are positive constants (the coefficients of viscosity and surface tension; the density of the fluid is assumed to be one), $\mathbf{I}$ is the $N \times N$ identity matrix, and $\mathbf{D}(\mathbf{u}) = \nabla \mathbf{u} + {}^\top \nabla \mathbf{u}$ is the doubled deformation tensor whose (i, j) components are $D_{ij}(\mathbf{u}) = \partial_i u_j + \partial_j u_i$ with $\partial_i = \partial/\partial x_i$, $\partial_t \mathbf{u} = (\partial_t u_1, \ldots, \partial_t u_N)$ with $\partial_t = \partial/\partial t$, and Δ_Γ is the Laplace-Beltrami operator on Γ. Moreover, for any matrix field $\mathbf{K}$ with (i, j) components K_{ij}, the quantity $\mathrm{Div}\,\mathbf{K}$ is an N-vector with components $\sum_{j=1}^N \partial_j K_{ij}$. Also, for any vector of functions $\mathbf{v} = {}^\top(v_1, \ldots, v_N)$ we set $\mathrm{div}\,\mathbf{v} = \sum_{j=1}^N \partial_j v_j$. Finally, $\mathbf{n} = {}^\top(n_1, \ldots, n_N)$ stands for the unit outer normal to Γ.

Problem (9.1) arises as a linearized system of one phase free boundary problem for the Navier-Stokes equations describing the motion of incompressible viscous fluids with surface tension taken into account. To prove existence of solutions to problem (9.1), one of standard methods is to use the technique of regularizers. Therefore, we consider problem (9.1) locally in a neighbourhood of either an interior point or a boundary point. The model problem for the interior points is the Stokes equations in $\mathbb{R}^N$, and therefore we can show the maximal L_p regularity by applying the Marcinkiewicz-Mikhlin-Lizorkin multiplier theorems to the solutions written by the Fourier transform in the space-time variables. On the other hand, problem (9.1) in the boundary neighbourhood is transformed to a problem in the half-space. Applying the Fourier transform with respect to time and tangent directions, problem (9.1) becomes a system of ordinary differential equations, and we obtain the solution formula by applying the inverse Fourier transform of the solution formulas for this ordinary differential equations. To obtain the maximal L_p regularity, Solonnikov [19] applied the Marcinkiewicz-Mikhlin-Lizorkin multiplier theorems together with some Hardy type inequality to the solution formulas in the half-space. The similar idea to [19] was used for the same problem without surface tension by Mogilevskii [4, 5] and Mucha and Zajaczkowski [6]. Prüss and Simonett [7, 8] proved the maximal L_p estimate by using the $\mathscr{H}^\infty$ calculus in the two phase model problem case. Shibata and Shimizu [18] obtained even the maximal L_p-L_q (L_p in time and L_q in space) regularity by applying the Weis operator valued Fourier multiplier theorem [20] to the Laplace inverse transform of the solutions to (9.2) in the half-space written by the $\mathscr{R}$ bounded solution operators.

To prove the existence of solutions of (9.1) in the domain Ω, we construct the parametrix of the form: $u = \sum_j \phi_j u_j$, where u_j are solutions in the neighbourhood of interior points or the boundary points and the $\{\phi_j\}_j$ is the partition of unit of Ω associated with the covering $\{\Omega_j\}_j$ having the finite intersection property (cf. Proposition 9.5.1 (v) below). When the number of the covering is infinite, to prove the convergence of the infinite sum: $\sum_j \phi_j u_j$, we need the inequality

$$\sum_j \int_0^T \Big(\int_{\Omega_j} |f(x,t)|^q\,dx\Big)^{p/q} dt \leqslant C \int_0^T \Big(\int_\Omega |f(x,t)|^q\,dx\Big)^{p/q} dt. \tag{9.3}$$

But, inequality (9.3) does seem to be valid only in the $p = q$ case. Thus, in case of $p \neq q$ we need a different idea from [6–8, 19]. In fact, Shibata [14] proved the maximal L_p-L_q regularity for (9.1) with $\tau = \delta = 0$ (without surface tension case) in a general unbounded domain by constructing the $\mathscr{R}$ bounded solution operators of (9.2) with $\tau = \delta = 0$ and applying the Weis operator valued Fourier multiplier theorem to the Laplace inverse transform of solutions written by this $\mathscr{R}$ bounded solution operators.

The purpose of this paper is to prove the maximal L_p-L_q regularity of problem (9.1) by using the $\mathscr{R}$-bounded solution operators associated with problem (9.2) in a general unbounded domain, which is the continuation of the work due to Shibata [14]. To investigate solutions in the maximal L_p-L_q regularity class with $p \neq q$ is relevant for stability issues of global solutions in unbounded domains, since there, we can expect only polynomially decay properties, so that we choose p large enough freely to guarantee the global integrability in time (cf. Saito and Shibata [10] and Schonbek and Shibata [12]).

In what follows, we state our assumptions and main results. To this end, we need to recall some further notation used throughout the paper.

Notation We denote the set of all complex numbers, real numbers and natural numbers by $\mathbb{C}$, $\mathbb{R}$, and $\mathbb{N}$, respectively. Let $\mathbb{N}_0 = \mathbb{N} \cup \{0\}$. For any multi-index $\alpha = (\alpha_1, \ldots, \alpha_N) \in \mathbb{N}_0^N$ we set $\partial_x^\alpha h = \partial_1^{\alpha_1} \cdots \partial_N^{\alpha_N} h$ with $\partial_i = \partial/\partial x_i$. Especially, for scalar, θ, and N-vector, $\mathbf{u} = {}^\top(u_1, \ldots, u_N)$, functions and $n \in \mathbb{N}_0$, we set $\nabla^n \theta = (\partial_x^\alpha \theta \mid |\alpha| = n)$ and $\nabla^n \mathbf{u} = (\nabla^n u_j \mid j = 1, \ldots, N)$. In particular, $\nabla^0 \theta = \theta$, $\nabla^1 \theta = \nabla \theta$, $\nabla^0 \mathbf{u} = \mathbf{u}$ and $\nabla^1 \mathbf{u} = \nabla \mathbf{u}$. We use bold small letters to denote N-vectors and bold capital letters to denote $N \times N$ matrices, respectively. For any vectors $\mathbf{a} = {}^\top(a_1, \ldots, a_N)$ and $\mathbf{b} = {}^\top(b_1, \ldots, b_N)$, let $\mathbf{a} \cdot \mathbf{b} = <\mathbf{a}, \mathbf{b}> = \sum_{j=1}^N a_j b_j$ and let $\mathcal{T}_\mathbf{n} \mathbf{a} = \mathbf{a} - <\mathbf{a}, \mathbf{n}> \mathbf{n}$, which is the tangential part of $\mathbf{a}$ along $\mathbf{n}$. Given $1 < q < \infty$, we set $q' = q/(q-1)$. For any domain G in $\mathbb{R}^N$, let $L_q(G)$, $W_q^m(G)$, and $B_{q,p}^s(G)$ be the usual Lebesgue space, Sobolev space, and Besov space on G, while $\| \cdot \|_{L_q(G)}$, $\| \cdot \|_{W_q^m(G)}$, and $\| \cdot \|_{B_{q,p}^s(G)}$ denote their norms, respectively. We write $W_q^0(G) = L_q(G)$, and $W_q^s(G) = B_{q,p}^s(G)$. For a Banach space X with norm $\| \cdot \|_X$, let

$$X^d = \{(f_1, \ldots, f_d) \mid f_i \in X \ (i = 1, \ldots, d)\},$$

while the norm of X^d is written by $\| \cdot \|_X$ for short, that is $\|f\|_X = \sum_{j=1}^d \|f_j\|_X$ for $f = (f_1, \ldots, f_d) \in X^d$. Let $\hat{W}_q^1(G) = \{\theta \in L_{q,\mathrm{loc}}(G) \mid \nabla \theta \in L_q(G)^N\}$ with semi-norm $\|\nabla \theta\|_{L_q(G)}$. Let $X_{q,0}^1(G) = \{\theta \in X_q^1(G) \mid \theta|_{\partial G} = 0\}$ with $X \in \{\hat{W}, W\}$. Let $(\mathbf{u}, \mathbf{v})_G = \int_G \mathbf{u} \cdot \mathbf{v}\, dx$ and $(\mathbf{u}, \mathbf{v})_{\partial G} = \int_{\partial G} \mathbf{u} \cdot \mathbf{v}\, d\omega$, where $d\omega$ denotes the surface element on ∂G. For $1 \leqslant p \leqslant \infty$, $L_p((a, b), X)$ and $W_p^m((a, b), X)$ denote the usual Lebesgue space and Sobolev space of X-valued functions defined on an interval (a, b), while $\| \cdot \|_{L_p((a,b),X)}$ and $\| \cdot \|_{W_p^m((a,b),X)}$ denote their norms, respectively. Let $C_0^\infty(G)$ be the set of all C^∞ functions whose supports are compact and contained in G. For any $\gamma \in \mathbb{R}$ we set $\|e^{\gamma t} f\|_{L_p((a,b),X)} = \left(\int_a^b (e^{\gamma t} \|f(t)\|_X)^p\, dt\right)^{1/p}$. For two Banach spaces X and Y, $\mathcal{L}(X, Y)$ denotes the set of all bounded linear operators from X into Y and $\mathcal{L}(X)$ is the abbreviation of $\mathcal{L}(X, X)$. For a domain U in $\mathbb{C}$, $\mathrm{Hol}\,(U, \mathcal{L}(X, Y))$ denotes the set of all $\mathcal{L}(X, Y)$-valued holomorphic functions defined on U. Let $\Sigma_\varepsilon = \{\lambda \in \mathbb{C} \setminus \{0\} \mid |\arg \lambda| \leqslant \pi - \varepsilon\}$ and $\Sigma_{\varepsilon,\lambda_0} = \{\lambda \in \Sigma_\varepsilon \mid |\lambda| \geqslant \lambda_0\}$. Throughout the paper, the letter C denotes generic constants and $C_{a,b,c,\cdots}$ means that the constant $C_{a,b,c,\cdots}$ depends on $a, b, c, \cdots$. The value of constants C and $C_{a,b,c,\cdots}$ may change from line to line.

Now, we introduce some definitions.

Definition 9.1.1 Let $1 < r < \infty$ and $k = 2$ or 3. We say that Ω is a uniform $W_r^{k,2}$ domain, if there exist positive constants α, β and K such that the following two assertions hold:

- For any $x_0 = (x_{01}, \ldots, x_{0N}) \in \Gamma$ there exist a coordinate number j and a $W_r^{k-1/r}$ function $h(x')$ defined on $B'_\alpha(x'_0)$ such that $\|h\|_{W_r^{k-1/r}(B'_\alpha(x'_0))} \leqslant K$, and

$$\begin{aligned} \Omega \cap B_\beta(x_0) &= \{x \in \mathbb{R}^N \mid x_j > h(x'_j) \ (x'_j \in B'_\alpha(x'_{0j}))\} \cap B_\beta(x_0), \\ \Gamma \cap B_\beta(x_0) &= \{x \in \mathbb{R}^N \mid x_j = h(x'_j) \ (x'_j \in B'_\alpha(x'_{0j}))\} \cap B_\beta(x_0). \end{aligned} \tag{9.4}$$

- For any $x_0 = (x_{01}, \dots, x_{0N}) \in \Gamma_0$ there exist a coordinate number j and a $W_r^{2-1/r}$ function $h(x'_j)$ defined on $B'_\alpha(x'_{0j})$ such that $\|h\|_{W_r^{2-1/r}(B'_\alpha(x'_{0j}))} \leqslant K$, and

$$\begin{aligned}\Omega \cap B_\beta(x_0) &= \{x \in \mathbb{R}^N \mid x_j > h(x'_j)\ (x'_j \in B'_\alpha(x'_{0j}))\} \cap B_\beta(x_0),\\ \Gamma_0 \cap B_\beta(x_0) &= \{x \in \mathbb{R}^N \mid x_j = h(x'_j)\ (x'_j \in B'_\alpha(x'_{0j}))\} \cap B_\beta(x_0).\end{aligned} \tag{9.5}$$

Here, we have set

$$x'_j = (x_1, \dots, x_{j-1}, x_{j+1}, \dots, x_N),\ \ x'_{0j} = (x_{01}, \dots, x_{0j-1}, x_{0j+1}, \dots, x_{0N}),$$
$$B'_\alpha(x'_{0j}) = \{x'_j \in \mathbb{R}^{N-1} \mid |x'_j - x'_{0j}| < \alpha\},\ \ B_\beta(x_0) = \{x \in \mathbb{R}^N \mid |x - x_0| < \beta\}.$$

Remark 9.1.1 When $\Gamma_0 = \emptyset$, Ω is called a uniform $W_r^{k-1/r}$ domain. And, when $\Gamma = \emptyset$, Ω is called a uniform $W_r^{2-1/r}$ domain.

Definition 9.1.2 Let X and Y be two Banach spaces. A family of operators $\mathscr{T} \subset \mathscr{L}(X, Y)$ is called $\mathscr{R}$-bounded on $\mathscr{L}(X, Y)$, if there exist constants $C > 0$ and $q \in [1, \infty)$ such that for each $n \in \mathbb{N}$, $\{T_j\}_{j=1}^n \subset \mathscr{T}$, and $\{f_j\}_{j=1}^n \subset X$, we have

$$\int_0^1 \|\sum_{j=1}^n r_j(u) T_j f_j\|_Y^q \, du \leqslant C \int_0^1 \|\sum_{j=1}^n r_j(u) f_j\|_X^q \, du. \tag{9.6}$$

Here the Rademacher functions r_k are given by $r_k(t) = \text{sign}(\sin(2^k \pi t))$ for $t \in [0, 1]$ ($k \in \mathbb{N}$). The smallest C in (9.6) is called $\mathscr{R}$ bound of $\mathscr{T}$ on $\mathscr{L}(X, Y)$ which is written by $\mathscr{R}_{L(X,Y)}(\mathscr{T})$ in what follows.

Remark 9.1.2 The definition of $\mathscr{R}$-boundedness is independent of $q \in [1, \infty)$ (cf. [2, p. 26 3.2. Remarks (2)]).

Finally, we introduce the weak Dirichlet-Neumann problem. Let

$$\begin{aligned}\hat{W}^1_{q,\Gamma}(\Omega) &= \{\theta \in L_{q,\text{loc}}(\Omega) \mid \nabla\theta \in L_q(\Omega)^N, \theta|_\Gamma = 0\},\\ W^1_{q,\Gamma}(\Omega) &= \{\theta \in W^1_q(\Omega) \mid \theta|_\Gamma = 0\}.\end{aligned}$$

Definition 9.1.3 Let $1 < q < \infty$ and let $\mathscr{W}^1_q(\Omega)$ be a closed subspace of $\hat{W}^1_{q,\Gamma}(\Omega)$ that contains $W^1_{q,\Gamma}(\Omega)$. Then, we say that the weak Dirichlet-Neumann problem is uniquely solvable on $\mathscr{W}^1_q(\Omega)$, if the following assertion holds: For any $\mathbf{f} \in L_q(\Omega)^N$ there exists a unique $\theta \in \mathscr{W}^1_q(\Omega)$ which satisfies the variational equation:

$$(\nabla\theta, \nabla\varphi)_\Omega = (\mathbf{f}, \nabla\varphi)_\Omega \quad \text{for all } \varphi \in \mathscr{W}^1_{q'}(\Omega), \tag{9.7}$$

and the estimate: $\|\nabla\theta\|_{L_q(\Omega)} \leqslant C_q\|\mathbf{f}\|_{L_q(\Omega)}$ for some constant C_q independent of $\mathbf{f}$, θ and φ. We define a bounded linear operator $\mathscr{K}_1 \in \mathscr{L}(L_q(\Omega)^N, \mathscr{W}^1_q(\Omega))$ by $\mathscr{K}_1(\mathbf{f}) = \theta$ with $\mathbf{f} \in L_q(\Omega)^N$ and $\theta \in \mathscr{W}^1_q(\Omega)$.

Remark 9.1.3 (1) Given $\mathbf{f} \in L_q(\Omega)^N$ and $g \in W_q^{1-1/q}(\Gamma)$, there exists a unique $u \in W^1_q(\Omega) + \mathscr{W}^1_q(\Omega)$ that satisfies the variational equation:

$$(\nabla u, \nabla\theta)_\Omega = (\mathbf{f}, \nabla\varphi)_\Omega \quad \text{for any } \varphi \in \mathscr{W}^1_{q'}(\Omega) \tag{9.8}$$

subject to $u = g$ on Γ, where

$$W^1_q(\Omega) + \mathscr{W}^1_q(\Omega) = \{p_1 + p_2 \mid p_1 \in W^1_q(\Omega),\ \ p_2 \in \mathscr{W}^1_q(\Omega)\}.$$

In fact, let $\mathbf{T}^1_\Gamma : W_q^{1-1/q}(\Gamma) \to W^1_q(\Omega)$ be a map such that for any $g \in W_q^{1-1/q}(\Gamma)$, $\mathbf{T}^1_\Gamma(g) \in W^1_q(\Omega)$ satisfies the conditions:

$$\mathbf{T}^1_\Gamma(g) = g \quad \text{on } \Gamma \quad \text{and} \quad \|\mathbf{T}^1_\Gamma(g)\|_{W^1_q(\Omega)} \leqslant C\|g\|_{W_q^{1-1/q}(\Gamma)} \tag{9.9}$$

with some constant C independent of g. For $g \in W_q^{1-1/q}(\Gamma)$, let

$$u = \mathbf{T}^1_\Gamma(g) + \mathscr{K}_1(\mathbf{f} - \mathbf{T}^1_\Gamma(g)) \in W^1_q(\Omega) + \mathscr{W}^1_q(\Omega) \tag{9.10}$$

and then u satisfies (9.8). Obviously,

$$\|\nabla u\|_{L_q(\Omega)} \leqslant C_q(\|g\|_{W_q^{1-1/q}(\Gamma)} + \|\mathbf{f}\|_{L_q(\Omega)}).$$

Especially, $W^1_q(\Omega) + \mathscr{W}^1_q(\Omega)$ is the space for the pressure.

Now, we state our main result. To this end, we introduce a space $DI_q(\Omega)$ defined by

$$\begin{aligned} DI_q(\Omega) = \{g \in W^1_q(\Omega) \mid &\text{ there exists a } G \in L_q(\Omega)^N \\ &\text{such that } (g, \varphi)_\Omega = -(G, \nabla\varphi)_\Omega \text{ for any } \varphi \in W^1_{q',\Gamma}(\Omega)\}. \end{aligned} \tag{9.11}$$

Let $\mathscr{G}(g) = \{H \in L_q(\Omega)^N \mid \operatorname{div} G = \operatorname{div} H\}$ and $[\mathscr{G}(g)]$ denotes the representative elements of the set $\mathscr{G}(g)$. But $[\mathscr{G}(g)]$ is also written by $\mathscr{G}(g)$ for simplicity unless confusion may occur. The space $DI_q(\Omega)$ is the space of data for divergence equation: $\operatorname{div}\mathbf{u} = g$ in Ω with $\mathbf{n}_0 \cdot \mathbf{u} = 0$ on Γ_0,where $\mathbf{n}_0$ denotes the unit outer normal to Γ_0. We see that $\operatorname{div}\mathscr{G}(g) = g$ in Ω and $\mathbf{n}_0 \cdot \mathscr{G}(g) = 0$ on Γ_0. In fact, for any $\varphi \in C^\infty_0(\Omega)$ we have

$$(\operatorname{div}\mathscr{G}(g), \varphi)_\Omega = -(\mathscr{G}(g), \nabla\varphi)_\Omega = (g, \varphi)_\Omega,$$

which furnishes that $\operatorname{div}\mathscr{G}(g) = g$ in Ω. Moreover, for any $\psi \in C_0^1(\Gamma_0)$ we choose $\varphi \in W^1_{q',\Gamma}(\Omega)$ in such a way that $\varphi|_{\Gamma_0} = \psi$, and then,

$$(\mathbf{n}_0 \cdot \mathscr{G}(g), \psi)_{\Gamma_0} = (\operatorname{div}\mathscr{G}(g), \varphi)_\Omega + (\mathscr{G}(g), \nabla\varphi)_\Omega = (g, \varphi)_\Omega - (g, \varphi)_\Omega = 0,$$

which furnishes that $\mathbf{n}_0 \cdot \mathscr{G}(g) = 0$ on Γ_0.

Let

$$\|g\|_{DI_q(\Omega)} = \|g\|_{W^1_q(\Omega)} + \inf\{\|H\|_{L_q(\Omega)} \mid H \in \mathscr{G}(g)\}$$

for $g \in DI_q(\Omega)$, and then $DI_q(\Omega)$ is a Banach space with norm $\|\cdot\|_{DI_q(\Omega)}$.

In this paper, we say that $\mathbf{u} \in W^1_q(\Omega)^N$ satisfies

$$\operatorname{div}\mathbf{u} = g \quad \text{in } \Omega, \quad \mathbf{n}_0 \cdot \mathbf{u}|_{\Gamma_0} = 0 \tag{9.12}$$

if it holds that

$$(\mathbf{u}, \nabla\varphi)_\Omega = (\mathscr{G}(g), \nabla\varphi) \quad \text{for any } \varphi \in \mathscr{W}^1_{q'}(\Omega). \tag{9.13}$$

Note that when $W^1_{q',\Gamma}(\Omega)$ is dense in $\mathscr{W}^1_{q'}(\Omega)$, assertions (9.12) and (9.13) are equivalent. Although (9.13) implies (9.12), the opposite direction can not be proved in general.

Concerning the existence of $\mathscr{R}$ bounded solution operator for problem (9.2), we have the following theorem.

Theorem 9.1.4 *Let $1 < q < \infty$, $0 < \varepsilon < \pi/2$, $N < r < \infty$ and $max(q, q') \leqslant r$. Assume that Ω is a uniform $W^{3,2}_r$ domain, and that the weak Dirichlet-Neumann problem is uniquely solvable on $\mathscr{W}^1_m(\Omega)$ for $m = q$ and q'. Let $J_q(\Omega)$ be the solenoidal space defined by*

$$J_q(\Omega) = \{\mathbf{f} \in L_q(\Omega)^N \mid (\mathbf{f}, \nabla\varphi)_\Omega = 0 \ \textit{for any } \varphi \in \mathscr{W}^1_{q'}(\Omega)\}, \tag{9.14}$$

and let $W^1_{q,\mathbf{n}}(\Omega)$ be the space for the boundary data defined by

$$W^1_{q,\mathbf{n}}(\Omega) = \{\mathbf{f}_b \in W^1_q(\Omega)^N \mid <\mathbf{f}_b, \mathbf{n}>= 0 \quad on\ \Gamma\}.$$

Let

$$\begin{aligned} X_q(\Omega) &= \{(\mathbf{f}, f, \mathbf{f}_b, g) \mid \mathbf{f} \in J_q(\Omega),\ f \in W^2_q(\Omega), \\ &\qquad\qquad \mathbf{f}_b \in W^1_{q,\mathbf{n}}(\Omega),\ g \in DI_q(\Omega)\}, \\ \mathscr{X}_q(\Omega) &= \{(F_1, \ldots, F_7) \mid F_1, F_3, F_5 \in L_q(\Omega)^N,\ F_2 \in W^{2-1/q}_q(\Gamma), \\ &\qquad\qquad F_4 \in W^1_q(\Omega)^N,\ F_6 \in L_q(\Omega),\ F_7 \in W^1_q(\Omega)\}. \end{aligned} \tag{9.15}$$

Then, there exist a constant $\lambda_0 \geqslant 1$ and operator families: $\mathbf{A}(\lambda)$, $\mathbf{P}(\lambda)$ and $\mathbf{H}(\lambda)$ with

$$\mathbf{A}(\lambda) \in \mathrm{Hol}\,(\Sigma_{\varepsilon,\lambda_0}, \mathscr{L}(\mathscr{X}_q(\Omega), W^2_q(\Omega)^N)),$$
$$\mathbf{P}(\lambda) \in \mathrm{Hol}\,(\Sigma_{\varepsilon,\lambda_0}, \mathscr{L}(\mathscr{X}_q(\Omega), W^1_q(\Omega) + \mathscr{W}^1_q(\Omega))),$$
$$\mathbf{H}(\lambda) \in \mathrm{Hol}\,(\Sigma_{\varepsilon,\lambda_0}, \mathscr{L}(\mathscr{X}_q(\Omega), W^{3-1/q}_q(\Gamma)))$$

such that for any $\lambda \in \Sigma_{\varepsilon,\lambda_0}$ *and* $(\mathbf{f}, f, \mathbf{f}_b, g) \in X_q(\Omega)$,

$$\mathbf{u} = \mathbf{A}(\lambda)\mathscr{F}_\lambda(\mathbf{f}, f, \mathbf{f}_b, g), \quad \mathfrak{p} = \mathbf{P}(\lambda)\mathscr{F}_\lambda(\mathbf{f}, f, \mathbf{f}_b, g), \quad h = \mathbf{H}(\lambda)\mathscr{F}_\lambda(\mathbf{f}, f, \mathbf{f}_b, g),$$

are unique solutions to (9.2), *where*

$$\mathscr{F}_\lambda(\mathbf{f}, f, \mathbf{f}_b, g) = (\mathbf{f}, f, \lambda^{1/2}\mathbf{f}_b, \mathbf{f}_b, \lambda\mathscr{G}(g), \lambda^{1/2}g, g), \tag{9.16}$$

and

$$\mathscr{R}_{\mathscr{L}(\mathscr{X}_q(\Omega), W^{2-j}_q(\Omega)^N)}(\{(\tau\partial_\tau)^\ell(\lambda^{j/2}\mathbf{A}(\lambda)) \mid \lambda \in \Sigma_{\varepsilon,\lambda_0}\}) \leqslant \gamma_* \quad (j = 0, 1, 2),$$
$$\mathscr{R}_{\mathscr{L}(\mathscr{X}_q(\Omega), L_q(\Omega)^N)}(\{(\tau\partial_\tau)^\ell(\nabla\mathbf{P}(\lambda)) \mid \lambda \in \Sigma_{\varepsilon,\lambda_0}\}) \leqslant \gamma_*,$$
$$\mathscr{R}_{\mathscr{L}(\mathscr{X}_q(\Omega), W^{3-j}_q(\Omega)^{N+1})}(\{(\tau\partial_\tau)^\ell(\lambda^j\mathbf{H}(\lambda)) \mid \lambda \in \Sigma_{\varepsilon,\lambda_0}\}) \leqslant \gamma_* \quad (j = 0, 1)$$

with some constant $\gamma_* > 0$ *for* $\ell = 0, 1$. *Here and hereafter,* λ *represents a complex number with* $\lambda = \gamma + i\tau \in \mathbb{C}$.

Remark 9.1.4 (1) Here, F_1, F_2, F_3, F_4, F_5, F_6, and F_7 are corresponding variables to $\mathbf{f}, f, \lambda^{1/2}\mathbf{f}_b, \mathbf{f}_b, \lambda\mathscr{G}(g), \lambda^{1/2}g$, and g, respectively. The norms $\|\cdot\|_{X_q(\Omega)}$ and $\|\cdot\|_{\mathscr{X}_q(\Omega)}$ of the spaces $X_q(\Omega)$ and $\mathscr{X}_q(\Omega)$ are defined by

$$\begin{aligned}\|(\mathbf{f}, f, \mathbf{f}_b, g)\|_{X_q(\Omega)} &= \|\mathbf{f}\|_{L_q(\Omega)} + \|g\|_{DI_q(\Omega)} + \|\mathbf{f}_b\|_{W^1_q(\Omega)},\\ \|(F_1, \ldots, F_7)\|_{\mathscr{X}_q(\Omega)} &= \|(F_1, F_3, F_5, F_6)\|_{L_q(\Omega)} + \|(F_4, F_7)\|_{\tilde{W}^1_q(\Omega)}\\ &\quad + \|F_2\|_{W^{2-1/q}_q(\Gamma)}.\end{aligned} \tag{9.17}$$

(2) Given $\mathbf{f} \in L_q(\Omega)$ and $\mathbf{f}_b \in W^1_q(\Omega)^N$, let $\theta \in W^1_q(\Omega) + \mathscr{W}^1_q(\Omega)$ be a solution to the variational equation:

$$(\nabla\theta, \nabla\varphi)_\Omega = (\mathbf{f}, \nabla\varphi)_\Omega \quad \text{for any } \varphi \in \mathscr{W}^1_{q'}\Omega)$$

subject to $\theta = \mathbf{f}_b \cdot \mathbf{n}$ on Γ. If $\mathbf{u}$, $\mathfrak{p}$ and h satisfy the Eqs. (9.1), then $\mathbf{u}$, $\mathfrak{p} - \theta$ and h satisfy the Eqs. (9.1) replacing $\mathbf{f}$ and $\mathbf{f}_b$ by $\mathbf{f} - \nabla\theta$ and $\mathbf{f}_b - < \mathbf{f}_b, \mathbf{n} > \mathbf{n}$. Thus, without loss of generality we may assume that $\mathbf{f} \in J_q(\Omega)$ and $\mathbf{f}_b \in W^1_{q,\mathbf{n}}(\Omega)$.

To state the maximal L_p-L_q regularity theorem for problem (9.1), we introduce the space $\mathbf{W}^{-1}_q(\Omega)$ and its norm $\|\cdot\|_{\mathbf{W}^{-1}_q(\Omega)}$ as follows: Let $(1 - \Delta)^{-1/2}$ be the operator defined by $(1 - \Delta)^{-1/2}f = \mathscr{F}^{-1}[(1 + |\xi|^2)^{-1/4}\mathscr{F}[f]]$, where $\mathscr{F}$ and $\mathscr{F}^{-1}$ are Fourier transform and Fourier inverse transform defined by

$$\mathscr{F}[f](\xi) = \int_{\mathbb{R}^N} e^{-ix\cdot\xi} f(x)\,dx, \quad \mathscr{F}^{-1}[g(\xi)] = \frac{1}{(2\pi)^N}\int_{\mathbb{R}^N} e^{ix\cdot\xi} g(\xi)\,d\xi.$$

Let ι be an extension map from $L_{1,\mathrm{loc}}(\Omega)$ into $L_{1,\mathrm{loc}}(\mathbb{R}^N)$ having the following properties:
(e-1) For any $1 < q < \infty$ and $f \in W_q^1(\Omega)$, $\iota f \in W_q^1(\mathbb{R}^N)$, $\iota f = f$ in Ω and

$$\|\iota f\|_{W_q^i(\mathbb{R}^N)} \leqslant C_q \|f\|_{W_q^i(\Omega)}$$

for $i = 0, 1$ with some constant C_q depending on q, r and Ω.
(e-2) For any $1 < q < \infty$ and $f \in W_q^1(\Omega)$,

$$\|(1-\Delta)^{-1/2}\iota(\nabla f)\|_{L_q(\mathbb{R}^N)} \leqslant C_q \|f\|_{L_q(\Omega)}$$

with some constant C_q depending on q, r and Ω.

In the following, such extension map ι is fixed. Then, we define the space $\mathbf{W}_q^{-1}(\Omega)$ and its norm $\|\cdot\|_{\mathbf{W}_q^{-1}(\Omega)}$ by

$$\mathbf{W}_q^{-1}(\Omega) = \{f \in L_{1,\mathrm{loc}}(\Omega) \mid (1-\Delta)^{-1/2}\iota f \in L_q(\mathbb{R}^N)\},$$
$$\|f\|_{\mathbf{W}_q^{-1}(\Omega)} = \|(1-\Delta)^{-1/2}\iota f\|_{L_q(\mathbb{R}^N)}.$$

Our maximal L_p-L_q regularity theorem for problem (9.1) is the following.

Theorem 9.1.5 *Let $1 < q < \infty$, $0 < \varepsilon < \pi/2$, $N < r < \infty$, $T > 0$. Assume that $\max(q, q') \leqslant r$, that Ω is a uniform $W_r^{3,2}$ domain, and that the weak Dirichlet-Neumann problem is uniquely solvable on $\mathscr{W}_m^1(\Omega)$ for $m = q$ and q'. Then, the following unique existence theorem holds:*

Let $\mathbf{u}_0 \in B_{q,p}^{2(1-1/p)}(\Omega)^N$ and $h_0 \in W_{q,p}^{3-1/p-1/q}(\Gamma)$ be initial data for (9.1) *and let* $\mathbf{f}$, g, f, *and* $\mathbf{f}_b$ *be right members of* (9.1) *with*

$$\begin{aligned}
&\mathbf{f} \in L_p((0,T), J_q(\Omega)), \ g \in L_p((0,T), DI_q(\Omega)^N) \cap W_p^1((0,T), \mathbf{W}_q^{-1}(\Omega)) \\
&\mathscr{G}(g) \in W_p^1((0,T), L_q(\Omega)^N), \quad f \in L_p((0,T), W_q^{2-1/q}(\Gamma)), \\
&\mathbf{f}_b \in L_p((0,T), W_{q,\mathbf{n}}^1(\Omega)) \cap W_p^1((0,T), \mathbf{W}_q^{-1}(\Omega)^N).
\end{aligned}$$

We assume that the initial data and right-members satisfy the compatibility condition:

$$\mathbf{u}_0 = 0 \ on\ \Gamma_0, \ \mathbf{u}_0 - \mathscr{G}(g)|_{t=0} \in J_q(\Omega), \ \mu\mathscr{T}_{\mathbf{n}}\mathbf{D}(\mathbf{u}_0)\mathbf{n} = \mathscr{T}_{\mathbf{n}}\mathbf{f}_b|_{t=0} \ on\ \Omega. \quad (9.18)$$

Then, problem (9.1) *admits unique solutions* $\mathbf{u}$, $\mathfrak{p}$ *and* h *with*

$$\begin{aligned}
&\mathbf{u} \in L_p((0,T), W_q^2(\Omega)^N \cap J_q(\Omega)) \cap W_p^1((0,T), L_q(\Omega)^N), \\
&\mathfrak{p} \in L_p((0,T), W_q^1(\Omega) + \mathscr{W}_q^1(\Omega)), \\
&h \in L_p((0,T), W_q^{2-1/q}(\Gamma)) \cap W_p^1((0,T), W_q^{3-1/q}(\Gamma)),
\end{aligned} \quad (9.19)$$

possessing the estimate:

$$\mathscr{I}((0,t),\mathbf{u},\mathfrak{p},h) \leqslant Ce^{\gamma t}\mathscr{M}(t,\mathbf{u}_0,h_0,\mathbf{f},f,\mathbf{f}_b,g) \tag{9.20}$$

for any $t \in (0,T]$ *with some positive constants* C *and* γ*, where we have set*

$$\begin{aligned}
&\mathscr{I}((0,t),\mathbf{u},\mathfrak{p},h) = \|\mathbf{u}\|_{L_p((0,t),W^2_q(\Omega))} + \|\partial_t\mathbf{u}\|_{L_p((0,t),L_q(\Omega))}\\
&\quad + \|\nabla\mathfrak{p}\|_{L_p((0,t),L_q(\Omega))} + \|h\|_{L_p((0,t),W^{3-1/q}_q(\Gamma))} + \|\partial_t h\|_{L_p((0,t),W^{2-1/q}_q(\Gamma))},\\
&\mathscr{M}(t,\mathbf{u}_0,h_0,\mathbf{f},f,\mathbf{f}_b,g) = \|\mathbf{u}_0\|_{B^{2(1-1/p)}_{q,p}(\Omega)} + \|h_0\|_{B^{3-1/p-1/q}_{q,p}(\Gamma)}\\
&\quad + \|\mathbf{f}\|_{L_p((0,t),L_q(\Omega))} + \|\partial_t\mathscr{G}(g)\|_{L_p((0,t),L_q(\Omega))} + \|\partial_t(g,\mathbf{f}_b)\|_{L_p((0,t),\mathbf{W}^{-1}_q(\Omega))}\\
&\quad + \|(g,\mathbf{f}_b)\|_{L_p((0,t),W^1_q(\Omega))}.
\end{aligned} \tag{9.21}$$

Remark 9.1.5 (1) The third compatibility condition in (9.18) usually should read as

$$\mu\mathscr{T}_{\mathbf{n}}\mathbf{D}(\mathbf{u}_0)\mathbf{n} = \mathscr{T}_{\mathbf{n}}\mathbf{f}_b|_{t=0} \quad \text{on } \Gamma.$$

Both of $\mu\mathbf{D}(\mathbf{u}_0)$ and $\mathbf{f}_b|_{t=0}$ belong to $B^{1-2/p}_{q,p}(\Omega)$, but according to the trace theorem due to Schneider [11], we do not have their boundary trace when $1-2/p \leqslant 1/q$. Thus, in this case, we can not impose the compatibility condition: $\mu\mathscr{T}_{\mathbf{n}}\mathbf{D}(\mathbf{u}_0)\mathbf{n} = \mathscr{T}_{\mathbf{n}}\mathbf{f}_b|_{t=0}$ on Γ. On the other hand, in our approach, it is necessary to assume that $\mu\mathscr{T}_{\mathbf{n}}\mathbf{D}(\mathbf{u}_0)\mathbf{n} = \mathscr{T}_{\mathbf{n}}\mathbf{f}_b|_{t=0}$, so that we assume that it holds on Ω, where $\mathbf{n}$ is suitably extended to Ω. For the application to the nonlinear problem, the third compatibility condition is naturally satisfied on Ω.
(2) As was discussed in Remark 9.1.3 (2), considering the pressure term $\mathfrak{p}-\theta$ instead of $\mathfrak{p}$ and noting the fact that $\mathscr{T}_{\mathbf{n}}(\mathbf{f}_b)$ is independent of the value of $< \mathbf{f}_b, \mathbf{n} >$, without loss of generality we may assume that $\mathbf{f} \in L_p((0,T),J_q(\Omega))$ and $\mathbf{f}_b \in L_p((0,T),W^1_{q,\mathbf{n}}(\Omega) \cap W^1_p((0,T),\mathbf{W}^{-1}_q(\Omega)^N)$.

Example 9.1.1 (1) When Ω is a bounded domain or an exterior domain with $\Gamma_0 = \emptyset$, $\mathscr{W}^1_q(\Omega) = \hat{W}^1_{q,\Gamma}(\Omega)$ (cf. [9, 17]).
(2) When Ω is a half-space and a bent half-space with $\Gamma_0 = \emptyset$, we take $\mathscr{W}^1_q(\Omega) = \hat{W}^1_{q,0}(\Omega)$.
(3) Let $\varphi(x')$ $(x' = (x_1,\ldots,x_{N-1}))$ be a function in $W^{3-1/r}_r(\mathbb{R}^{N-1})$ with $N < r < \infty$. Let

$$\begin{aligned}
H_\varphi &= \{x = (x_1,\ldots,x_N) \in \mathbb{R}^N \mid x_N > \varphi(x')\ (x' \in \mathbb{R}^{N-1})\},\\
\Gamma_\varphi &= \{x = (x_1,\ldots,x_N) \in \mathbb{R}^N \mid x_N = \varphi(x')\ (x' \in \mathbb{R}^{N-1})\}.
\end{aligned}$$

Assume that there exists an $R > 0$ such that $\Omega \cap B^R = H_\varphi \cap B^R$ and $\Gamma \cap B^R = \Gamma_\varphi \cap B^R$. Moreover, we assume that Γ is a hypersurface of $W^{3-1/r}_r$ class and that $\Gamma_0 = \emptyset$. In this case, we take $\mathscr{W}^1_q(\Omega) = \hat{W}^1_{q,0}(\Omega)$. Here, we have set $B^L = \{x \in \mathbb{R}^N \mid |x| > L\}$ and $B_L = \{x \in \mathbb{R}^N \mid |x| < L\}$ for any $L > 0$.

(4) Let $\varphi_1(x') \in W_r^{3-1/r}(\mathbb{R}^{N-1})$ and let $\varphi_2(x') \in W_r^{2-1/r}(\mathbb{R}^{N-1})$, respectively. Assume that $\varphi_2(x') < a < b < \varphi_1(x')$ for any $x' \in \mathbb{R}^{N-1}$ with some real numbers a and b. Let

$$H_{\varphi_1,\varphi_2} = \{x = (x_1, \ldots, x_N) \in \mathbb{R}^N \mid \varphi_2(x') < x_N < \varphi_1(x') \ (x' \in \mathbb{R}^{N-1})\},$$
$$\Gamma_{\varphi_i} = \{x = (x_1, \ldots, x_N) \in \mathbb{R}^N \mid x_N = \varphi_i(x') \ (x' \in \mathbb{R}^{N-1})\} \ (i = 1, 2).$$

Assume that there exists an $R > 0$ such that $\Omega \cap B^R = H_{\varphi_1,\varphi_2} \cap B^R$, $\Gamma \cap B^R = \Gamma_{\varphi_2} \cap B^R$ and $\Gamma_0 \cap B^R = \Gamma_{\varphi_1} \cap B^R$. Moreover, we assume that Γ is the hypersurface of $W_r^{3-1/r}$ class and Γ_0 the hypersurface of $W_r^{2-1/r}$ class, respectively. In this case, we take $\mathscr{W}_q^1(\Omega) = \hat{W}^1_{q,\Gamma}(\Omega)$.
(5) Let D be a compact domain with $W_r^{3-1/r}$ boundary in $\mathbb{R}^{N-1}$ and let $\Omega = \{x = (x', x_N) \mid x' \in D,\ x_N \in \mathbb{R}\} = D \times \mathbb{R}$. In this case, we take $\mathscr{W}_q^1(\Omega) = \hat{W}^1_{q,0}(\Omega)$.

Remark 9.1.6 In examples (2) (3) and (4), $\hat{W}^1_{q,\Gamma}(\Omega)$ coincides with the closure of $W^1_{q,\Gamma}(\Omega)$ with semi-norm $\|\nabla \cdot\|_{L_q(\Omega)}$ (cf. Shibata [13, Appendix]).

The paper is organized as follows: In Sect. 9.2, we introduce the reduced Stokes operators by eliminating the pressure term and the divergence equations and state Theorem 9.2.1 concerning the existence of $\mathscr{R}$-bounded solution operator for the generalized resolvent problem of the reduced Stokes operators and Theorem 9.2.2 concerning the maximal L_p-L_q regularity for the initial boundary value problem of the reduced Stokes operators. In Sect. 9.3, the existence of $\mathscr{R}$ bounded solution operators is proved in the case of model problems in $\mathbb{R}^N$ and in the half space with free boundary condition and non-slip condition. In Sect. 9.4, the problems in bent half-space are discussed. In Sect. 9.5, we prove Theorem 9.2.1 in a general domain by constructing a parametrix. In Sect. 9.6, we prove Theorem 9.2.2 by applying the Weis operator valued Fourier multiplier theorem [20] to the inverse Laplace transform of solutions to (9.2) written by the $\mathscr{R}$-bounded solution operator obtained in Theorem 9.1.4. Finally, in Sect. 9.7, according to what is pointed out in Sect. 9.2, we prove Theorems 9.1.4 and 9.1.5 with the help of Theorems 9.2.1 and 9.2.2.

9.2 Reduced Stokes Problem

9.2.1 Equivalence Between Stokes and Reduced Stokes Equations

In this section, eliminating the pressure term $\mathfrak{p}$ and the divergence equation: $\operatorname{div}\mathbf{u} = g$ in (9.2), we deduce the reduced Stokes equations equivalent to the Eqs. (9.2). Let $K_1 = K_1(\mathbf{u})$ be a unique solution of the variational problem:

$$(\nabla K_1(\mathbf{u}), \nabla\varphi)_\Omega = (\operatorname{Div}(\mu\mathbf{D}(\mathbf{u})) - \nabla\operatorname{div}\mathbf{u}, \nabla\varphi)_\Omega \quad \text{for any } \varphi \in \mathscr{W}^1_{q'}(\Omega), \quad (9.22)$$

subject to $K_1(\mathbf{u}) = \mu < \mathbf{D}(\mathbf{u})\mathbf{n}, \mathbf{n} > -\mathrm{div}\,\mathbf{u}$ on Γ, while $K_2(h)$ is a unique solution of the variational problem:

$$(\nabla K_2(h), \nabla\varphi)_\Omega = 0 \quad \text{for any } \varphi \in \mathscr{W}^1_{q'}(\Omega), \tag{9.23}$$

subject to $K_2(h) = -(\tau + \delta\Delta_\Gamma)h$ on Γ. In fact, as was seen in Remark 9.1.3, $K_1(\mathbf{u})$ and $K_2(h)$ are defined by

$$\begin{aligned} K_1(\mathbf{u}) &= g_1 + \mathscr{K}_1(\mathrm{Div}\,(\mu\mathbf{D}(\mathbf{u})) - \nabla\mathrm{div}\,\mathbf{u} - \nabla g_1), \\ K_2(h) &= -g_2 + \mathscr{K}_1(\nabla g_2). \end{aligned} \tag{9.24}$$

with $g_1 = \mathbf{T}^1_\Gamma(\mu < \mathbf{D}(\mathbf{u})\mathbf{n}, \mathbf{n} > -\mathrm{div}\,\mathbf{u})$ and $g_2 = \mathbf{T}^1_\Gamma\{(\tau + \delta\Delta_\Gamma)h\}$. Obviously, $\mathbf{K}_1(\mathbf{u})$ and $\mathbf{K}_2(h)$ belong to $W^1_q(\Omega) + \mathscr{W}^1_q(\Omega)$ and satisfy the estimates:

$$\begin{aligned} \|\nabla K_1(\mathbf{u})\|_{L_q(\Omega)} &\leqslant C\|\nabla\mathbf{u}\|_{W^1_q(\Omega)}, \\ \|\nabla K_2(h)\|_{L_q(\Omega)} &\leqslant C\|h\|_{W^{3-1/q}_q(\Omega)}. \end{aligned} \tag{9.25}$$

We consider the reduced Stokes equations:

$$\begin{cases} \lambda\mathbf{u} - \mathrm{Div}\,(\mu\mathbf{D}(\mathbf{u}) - (K_1(\mathbf{u}) + K_2(h))\mathbf{I}) = \mathbf{f} & \text{in } \Omega, \\ \lambda h - \mathbf{n}\cdot\mathbf{u} = f & \text{on } \Gamma, \\ \mathscr{T}_\mathbf{n}(\mu\mathbf{D}(\mathbf{u})\mathbf{n}) = \mathscr{T}_\mathbf{n}\mathbf{f}_b, \quad \mathrm{div}\,\mathbf{u} = \mathbf{n}\cdot\mathbf{f}_b & \text{on } \Gamma, \\ \mathbf{u} = 0 & \text{on } \Gamma_0. \end{cases} \tag{9.26}$$

Note that the third equations in (9.26) are equivalent to

$$(\mu\mathbf{D}(\mathbf{u}) - (K_1(\mathbf{u}) + K_2(\eta))\mathbf{I})\mathbf{n} - ((\tau + \delta\Delta_\Gamma)h)\mathbf{n} = \mathbf{f}_b \quad \text{on } \Gamma.$$

In what follows, we discuss the equivalence between (9.2) and (9.26). First, we assume that (9.2) is uniquely solvable. Let $\mathbf{f} \in L_q(\Omega)^N$, $f \in W^2_q(\Omega)$, and $\mathbf{f}_b \in W^2_q(\Omega)^N$. Let $g \in W^1_q(\Omega)$ be a unique solution of the variational equation:

$$\lambda(g, \varphi)_\Omega + (\nabla g, \nabla\varphi)_\Omega = -(\mathbf{f}, \nabla\varphi)_\Omega \text{ for any } \varphi \in W^1_{q',\Gamma}(\Omega), \tag{9.27}$$

subject to $g = \mathbf{n}\cdot\mathbf{f}_b$ on Γ. The unique existence of g is guaranteed for large $\lambda > 0$ (cf. Theorem 9.6.2 in Sect. 9.6.2). In this case, we see that

$$\mathscr{G}(g) = \lambda^{-1}(\nabla g + \mathbf{f}). \tag{9.28}$$

Let $\mathbf{u} \in W^2_q(\Omega)^N$, $\mathfrak{p} \in W^1_q(\Omega) + \mathscr{W}^1_q(\Omega)$ and $h \in W^{3-1/q}_q(\Gamma)$ be unique solutions of (9.2), and then by (9.12), (9.13) and (9.28)

$$\operatorname{div} \mathbf{u} = g \text{ in } \Omega, \quad (\mathbf{u}, \nabla\varphi)_\Omega = \lambda^{-1}(\nabla g + \mathbf{f}, \nabla\varphi) \quad \text{for any } \varphi \in \mathscr{W}^1_{q'}(\Omega). \tag{9.29}$$

On the other hand, by (9.22) and (9.23) for any $\varphi \in \mathscr{W}^1_{q'}(\Omega)$ we have

$$\begin{aligned}
&(\mathbf{f}, \nabla\varphi)_\Omega = (\lambda\mathbf{u} - \operatorname{Div}(\mu\mathbf{D}(\mathbf{u}) - \mathfrak{p}\mathbf{I}), \nabla\varphi)_\Omega \\
&= \lambda(\mathbf{u}, \nabla\varphi)_\Omega - (\nabla\operatorname{div}\mathbf{u}, \nabla\varphi)_\Omega - (\operatorname{Div}(\mu\mathbf{D}(\mathbf{u})) - \nabla\operatorname{div}\mathbf{u}, \nabla\varphi)_\Omega + (\nabla\mathfrak{p}, \nabla\varphi)_\Omega \\
&= \lambda(\mathbf{u}, \nabla\varphi)_\Omega - (\nabla\operatorname{div}\mathbf{u}, \nabla\varphi)_\Omega + (\nabla(\mathfrak{p} - (K_1(\mathbf{u}) + K_2(h))), \nabla\varphi)_\Omega,
\end{aligned}$$

which, combined with (9.29), furnishes that

$$(\nabla(\mathfrak{p} - (K_1(\mathbf{u}) + K_2(h))), \nabla\varphi)_\Omega = 0 \quad \text{for any } \varphi \in \mathscr{W}^1_{q'}(\Omega).$$

Moreover, by (9.22) and (9.23)

$$\begin{aligned}
&\mathfrak{p} - (K_1(\mathbf{u}) + K_2(\mathbf{u})) \\
&\quad =< \mu\mathbf{D}(\mathbf{u})\mathbf{n}, \mathbf{n} > -(\tau + \delta\Delta_\Gamma)h - < \mathbf{f}_b, \mathbf{n} > -K_1(\mathbf{u}) - K_2(h) \\
&\quad = \operatorname{div}\mathbf{u} - \mathbf{n}\cdot\mathbf{f}_b = g - g = 0 \quad \text{on } \Gamma,
\end{aligned}$$

because $\operatorname{div}\mathbf{u} = g$ in Ω and $g = \mathbf{n}\cdot\mathbf{f}_b$ on Γ. Thus, the uniqueness of solutions implies that $\mathfrak{p} = K_1(\mathbf{u}) + K_2(h)$, which yields that $\mathbf{u}$ and h satisfy the Eqs. (9.26).

Conversely, we assume that problem (9.26) is uniquely solvable. In what follows, we assume that

$$\mathbf{n}\cdot\mathbf{f}_b = 0 \text{ on } \Gamma_0, \quad (\mathbf{f}, \nabla\varphi)_\Omega = 0 \text{ for any } \varphi \in \mathscr{W}^1_{q'}(\Omega). \tag{9.30}$$

Given $g \in DI_q(\Omega)$, let $K \in W^1_q(\Omega) + \mathscr{W}^1_q(\Omega)$ be a solution to the variational problem:

$$(\nabla K, \nabla\varphi)_\Omega = (\lambda\mathscr{G}(g) - \nabla g, \nabla\varphi)_\Omega \quad \text{for any } \varphi \in \mathscr{W}^1_{q'}(\Omega), \tag{9.31}$$

subject to $K = -g$ on Γ. Let $\mathbf{u}$ and h be solutions of problem:

$$\begin{cases}
\lambda\mathbf{u} - \operatorname{Div}(\mu\mathbf{D}(\mathbf{u}) - (K_1(\mathbf{u}) + K_2(h))\mathbf{I}) = \mathbf{f} + \nabla K & \text{in } \Omega, \\
\lambda h - \mathbf{n}\cdot\mathbf{u} = f & \text{on } \Gamma, \\
\mathscr{T}_\mathbf{n}(\mu\mathbf{D}(\mathbf{u})\mathbf{n}) = \mathscr{T}_\mathbf{n}(\mathbf{f}_b + g\mathbf{n}) = \mathscr{T}_\mathbf{n}(\mathbf{f}_b) & \text{on } \Gamma, \\
\operatorname{div}\mathbf{u} = \mathbf{n}\cdot(\mathbf{f}_b + g\mathbf{n}) = g & \text{on } \Gamma, \\
\mathbf{u} = 0 & \text{on } \Gamma_0.
\end{cases} \tag{9.32}$$

By (9.30), (9.31) and (9.32),

$$\begin{aligned}
&(\lambda\mathscr{G}(g) - \nabla g, \nabla\varphi)_\Omega = (\nabla K, \nabla\varphi)_\Omega \\
&\quad = (\lambda\mathbf{u} - \operatorname{Div}(\mu\mathbf{D}(\mathbf{u}) - (K_1(\mathbf{u}) + K_2(h))\mathbf{I}), \nabla\varphi) \\
&\quad = (\lambda\mathbf{u}, \nabla\varphi)_\Omega - (\nabla\operatorname{div}\mathbf{u}, \nabla\varphi)_\Omega \quad \text{for any } \varphi \in \mathscr{W}^1_{q'}(\Omega).
\end{aligned} \tag{9.33}$$

Since $W^1_{q',\Gamma}(\Omega) \subset \mathscr{W}^1_{q'}(\Omega)$, by (9.33), (9.11) and the divergence theorem of Gauß

$$\begin{aligned}&\lambda(\mathrm{div}\,\mathbf{u}, \varphi)_\Omega + (\nabla \mathrm{div}\,\mathbf{u}, \nabla\varphi)_\Omega \\ &\quad = \lambda(g, \varphi)_\Omega + (\nabla g, \nabla\varphi)_\Omega \quad \text{for any } \varphi \in W^1_{q',\Gamma}(\Omega),\end{aligned}$$

so that

$$\lambda(g - \mathrm{div}\,\mathbf{u}, \varphi)_\Omega + (\nabla(g - \mathrm{div}\,\mathbf{u}), \nabla\varphi)_\Omega \quad \text{for any } \varphi \in W^1_{q',\Gamma}(\Omega).$$

Since $\mathrm{div}\,\mathbf{u} = g$ on Γ, the uniqueness of solutions implies that $\mathrm{div}\,\mathbf{u} = g$ in Ω, which inserted into (9.33) yields that $(\mathbf{u}, \nabla\varphi)_\Omega = (\mathscr{G}(g), \nabla\varphi)_\Omega$ for any $\varphi \in \mathscr{W}^1_{q'}(\Omega)$. Thus, in view of (9.12) and (9.13) we conclude that $\mathbf{u}$, h and $\mathfrak{p} = K_1(\mathbf{u}) + K_2(h) - K$ satisfy the Eqs. (9.2) under the assumptions (9.30).

9.2.2 On the $\mathscr{R}$ Bounded Solution Operators for the Reduced Stokes Problem

From the observation in Sect. 9.2.1, in what follows we consider the reduced Stokes equations (9.26) instead of Eqs. (9.2). For any domain $D \in \mathbb{R}^N$, we set

$$\begin{aligned}Y_q(D) &= \{(\mathbf{f}, f, \mathbf{f}_b) \mid \mathbf{f} \in L_q(D)^N,\ f \in W^{2-1/q}_q(\partial D),\ \mathbf{f}_b \in W^1_q(D)^N\},\\ \mathscr{Y}_q(D) &= \{(F_1, F_2, F_3, F_4) \mid F_1, F_3 \in L_q(D)^N,\ F_2 \in W^{2-1/q}_q(\partial D),\\ &\qquad F_4 \in W^1_q(\Omega)^N\},\end{aligned} \tag{9.34}$$

where ∂D is the boundary of D. And we set

$$\begin{aligned}\|(\mathbf{f}, f, \mathbf{f}_b)\|_{Y_q(D)} &= \|\mathbf{f}\|_{L_q(D)} + \|f\|_{W^{2-1/q}_q(\partial D)} + \|\mathbf{f}_b\|_{W^1_q(D)},\\ \|(F_1, F_2, F_3, F_4)\|_{\mathscr{Y}_q(D)} &= \|(F_1, F_3)\|_{L_q(D)} + \|F_2\|_{W^{2-1/q}_q(\partial D)} + \|F_4\|_{W^1_q(D)},\end{aligned}$$

for any $(\mathbf{f}, f, \mathbf{f}_b) \in Y_q$ and $(F_1, F_2, F_3, F_4) \in \mathscr{Y}_q(D)$, respectively. We prove the following theorem instead of Theorem 9.1.4.

Theorem 9.2.1 *Let $1 < q < \infty$, $0 < \varepsilon < \pi/2$, $N < r < \infty$ and $\max(q, q') \leqslant r$. Assume that Ω is a uniform $W^{3,2}_r$ domain, and that the weak Dirichlet-Neumann problem is uniquely solvable on $\mathscr{W}^1_m(\Omega)$ for $m = q$ and q'. Then, the following two assertions hold.*
(1) **Existence** *There exist a constant $\lambda_0 \geqslant 1$ and operator families $\mathscr{A}(\lambda)$ and $\mathscr{H}(\lambda)$ with*

$$\begin{aligned}\mathscr{A}(\lambda) &\in \mathrm{Hol}\,(\Sigma_{\varepsilon,\lambda_0}, \mathscr{L}(\mathscr{Y}_q(\Omega), W^2_q(\Omega)^N)),\\ \mathscr{H}(\lambda) &\in \mathrm{Hol}\,(\Sigma_{\varepsilon,\lambda_0}, \mathscr{L}(\mathscr{Y}_q(\Omega), W^{3-1/q}_q(\Omega)))\end{aligned}$$

such that for any $\lambda \in \Sigma_{\varepsilon,\lambda_0}$ *and* $(\mathbf{f}, f, \mathbf{f}_b) \in Y_q(\Omega)$,

$$\mathbf{u} = \mathscr{A}(\lambda)F_\lambda(\mathbf{f}, f, \mathbf{f}_b), \quad h = \mathscr{H}(\lambda)F_\lambda(\mathbf{f}, f, \mathbf{f}_b),$$

where $F_\lambda(\mathbf{f}, f, \mathbf{f}_b) = (\mathbf{f}, f, \lambda^{1/2}\mathbf{f}_b, \mathbf{f}_b)$, *are solutions to* (9.26), *and*

$$\mathscr{R}_{\mathscr{L}(\mathscr{Y}_q(\Omega), W_q^{2-j}(\Omega)^N)}(\{(\tau\partial_\tau)^\ell(\lambda^{j/2}\mathscr{A}(\lambda)) \mid \lambda \in \Sigma_{\varepsilon,\lambda_0}\}) \leqslant \gamma_{*R} \quad (j = 0, 1, 2),$$
$$\mathscr{R}_{\mathscr{L}(\mathscr{Y}_q(\Omega), W_q^{3-j}(\Omega))}(\{(\tau\partial_\tau)^\ell(\lambda^{j}\mathscr{H}(\lambda)) \mid \lambda \in \Sigma_{\varepsilon,\lambda_0}\}) \leqslant \gamma_{*R} \quad (j = 0, 1)$$

with some constant $\gamma_{*R} > 0$ *for* $\ell = 0, 1$.

(2) **Uniqueness:** *There exists a* $\lambda_0 \geqslant 1$ *such that if* $\mathbf{u} \in W_q^2(\Omega)^N$ *and* $h \in W_q^{3-1/q}(\Gamma)$ *satisfy the homogeneous equations:*

$$\begin{cases} \lambda\mathbf{u} - \mathrm{Div}\,(\mu\mathbf{D}(\mathbf{u}) - (K_1(\mathbf{u}) + K_2(h))\mathbf{I}) = 0 & \text{in } \Omega, \\ \lambda h - \mathbf{n}\cdot\mathbf{u} = 0 & \text{on } \Gamma, \\ \{\mu\mathbf{D}(\mathbf{u}) - (K_1(\mathbf{u}) + K_2(h))\mathbf{I} - ((\tau + \delta\Delta_\Gamma)h)\mathbf{I}\}\mathbf{n} = 0 & \text{on } \Gamma, \\ \mathbf{u} = 0 & \text{on } \Gamma_0. \end{cases} \tag{9.35}$$

with $\lambda \in \Sigma_{\varepsilon,\lambda_0}$, *then* $\mathbf{u} = 0$ *and* $h = 0$.

9.2.3 Time Dependent Reduced Stokes Equations

In this subsection, we consider the time dependent problem:

$$\begin{cases} \partial_t\mathbf{u} - \mathrm{Div}\,(\mu\mathbf{D}(\mathbf{u}) - (K_1(\mathbf{u}) + K_2(h))\mathbf{I}) = \mathbf{f} & \text{in } \Omega \times (0, T), \\ \partial_t h - \mathbf{n}\cdot\mathbf{u} = f & \text{on } \Gamma \times (0, T), \\ \mathscr{T}_\mathbf{n}(\mu\mathbf{D}(\mathbf{u})\mathbf{n}) = \mathscr{T}_\mathbf{n}\mathbf{f}_b, \quad \mathrm{div}\,\mathbf{u} = \mathbf{n}\cdot\mathbf{f}_b & \text{on } \Gamma \times (0, T), \\ \mathbf{u} = 0 & \text{on } \Gamma_0 \times (0, T), \\ (\mathbf{u}, h)|_{t=0} = (\mathbf{u}_0, h_0) & \text{on } \Omega \times \Gamma. \end{cases} \tag{9.36}$$

Concerning the maximal L_p-L_q regularity for problem (9.36), we prove the following theorem.

Theorem 9.2.2 *Let* $1 < q < \infty$, $0 < \varepsilon < \pi/2$, $N < r < \infty$, $T > 0$. *Assume that* $max(q, q') \leqslant r$, *that* Ω *is a uniform* $W_r^{3,2}$ *domain, and that the weak Dirichlet-Neumann problem is uniquely solvable on* $\mathscr{W}_m^1(\Omega)$ *for* $m = q$ *and* q'. *Then, we have the following unieque existence theorem:*

Let $\mathbf{u}_0 \in B^{2(1-1/p)}_{q,p}(\Omega)^N$ *and* $h_0 \in W^{3-1/p-1/q}_{q,p}(\Gamma)$ *be initial data for* (9.1) *and let* $\mathbf{f}$, f, *and* $\mathbf{f}_b$ *be right members of* (9.1) *with*

$$\mathbf{f} \in L_p((0,T), J_q(\Omega)), \quad f \in L_p((0,T), W^{2-1/q}_q(\Gamma)),$$
$$\mathbf{f}_b \in L_p((0,T), W^1_{q,\mathbf{n}}(\Omega)) \cap W^1_p((0,T), \mathbf{W}^{-1}_q(\Omega)^N).$$

We assume that the following compatibility conditions are satisfied:

$$\mathbf{u}_0 = 0 \text{ on } \Gamma_0, \ \mu\mathscr{T}_\mathbf{n}\mathbf{D}(\mathbf{u}_0)\mathbf{n} = \mathscr{T}_\mathbf{n}\mathbf{f}_b|_{t=0}, \ \operatorname{div}\mathbf{u}_0 = \mathbf{n}\cdot\mathbf{f}_b|_{t=0} \ \text{ on } \Omega. \tag{9.37}$$

Then, problem (9.1) *admits unique solutions* $\mathbf{u}$ *and* h *with*

$$\mathbf{u} \in L_p((0,T), W^2_q(\Omega)^N \cap J_q(\Omega)) \cap W^1_p((0,T), L_q(\Omega)^N),$$
$$h \in L_p((0,T), W^{3-1/q}_q(\Gamma)) \cap W^1_p((0,T), W^{2-1/q}_q(\Gamma)), \tag{9.38}$$

possessing the estimate:

$$\mathscr{I}_R(t, \mathbf{u}, h) \leqslant Ce^{\gamma t}\mathscr{M}_R(t, \mathbf{u}_0, h_0, \mathbf{f}, f, \mathbf{f}_b) \tag{9.39}$$

for any $t \in (0, T]$ *with some positive constants* C *and* γ, *where we have set*

$$\mathscr{I}_R(t, \mathbf{u}, h) = \|\mathbf{u}\|_{L_p((0,t),W^2_q(\Omega))} + \|\partial_t\mathbf{u}\|_{L_p((0,t),L_q(\Omega))}$$
$$+ \|h\|_{L_p((0,t),W^{3-1/q}_q(\Gamma))} + \|\partial_t h\|_{L_p((0,t),W^{2-1/q}_q(\Gamma))}$$
$$\mathscr{M}_R(t, \mathbf{u}_0, h_0, \mathbf{f}, f, \mathbf{f}_b) = \|\mathbf{u}_0\|_{B^{2(1-1/p)}_{q,p}(\Omega)} + \|h_0\|_{B^{3-1/p-1/q}_{q,p}(\Gamma)} + \|\mathbf{f}\|_{L_p((0,t),L_q(\Omega))}$$
$$+ \|f\|_{L_p((0,t),W^{2-1/q}_q(\Gamma))} + \|\partial_t\mathbf{f}_b\|_{L_p((0,t),\mathbf{W}^{-1}_q(\Omega))} + \|\mathbf{f}_b\|_{L_p((0,t),W^1_q(\Omega))}. \tag{9.40}$$

Next, we consider the generation of analytic semigroup associated with the Eqs. (9.36) with $\mathbf{f} = 0$, $f = 0$ and $\mathbf{f}_b = 0$. Let

$$\mathscr{H}_q(\Omega) = \{(\mathbf{u}, h) \mid \mathbf{u} \in L_q(\Omega), \quad h \in W^{2-1/q}_q(\Gamma)\},$$
$$\mathscr{D}_q(\mathscr{A}) = \{(\mathbf{u}, h) \mid \mathbf{u} \in W^2_q(\Omega), \ h \in W^{3-1/q}_q(\Omega), \mathbf{u}|_{\Gamma_0} = 0,$$
$$\mathscr{T}_\mathbf{n}(\mu\mathbf{D}(\mathbf{u})\mathbf{n}) = 0, \ \operatorname{div}\mathbf{u} = 0 \quad \text{on } \Gamma\},$$
$$\mathscr{A}(\mathbf{u}, h) = (\operatorname{Div}(\mu\mathbf{D}(\mathbf{u}) - (K_1(\mathbf{u}) + K_2(h))\mathbf{I}), \mathbf{n}\cdot\mathbf{u}) \quad \text{for } (\mathbf{u}, h) \in \mathscr{D}_q(\mathscr{A}).$$

Then, we have the following theorem.

Theorem 9.2.3 *Let* $1 < q < \infty$, $0 < \varepsilon < \pi/2$, $N < r < \infty$, $T > 0$. *Assume that* $max(q, q') \leqslant r$, *that* Ω *is a uniform* $W^{3,2}_r$ *domain, and that the weak Dirichlet-Neumann problem is uniquely solvable on* $\mathscr{W}^1_m(\Omega)$ *for* $m = q$ *and* q'. *Then, the operator* $\mathscr{A}$ *generates a continuous semigroup* $\{T(t)\}_{t\geqslant 0}$ *on* $\mathscr{H}_q(\Omega)$ *which is analytic.*

Moreover, if $\mathbf{u}_0 \in J_q(\Omega)$, *then* $T(t)(\mathbf{u}_0, h_0) \in J_q(\Omega)$ *for any* $t \in (0, \infty)$.

9.3 Reduced Stokes Equations in $\mathbb{R}^N$ and $\mathbb{R}^N_+$

9.3.1 Reduced Stokes Equations in $\mathbb{R}^N$

In this subsection, let

$$DI_{0,q}(\mathbb{R}^N) = \{g \in W^1_q(\mathbb{R}^N) \mid \text{there exists a } G \in L_q(\mathbb{R}^N)^N \\ \text{such that } (g, \varphi)_{\mathbb{R}^N} = -(G, \nabla\varphi)_{\mathbb{R}^N} \text{ for any } \varphi \in W^1_{q'}(\mathbb{R}^N)\}.$$

Let $\mathscr{G}_0(g) = \{H \in L_q(\mathbb{R}^N)^N \mid \operatorname{div} G = \operatorname{div} H\}$ and $[\mathscr{G}_0(g)]$ denotes the representative elements of the set $\mathscr{G}_0(g)$. But $[\mathscr{G}_0(g)]$ is also written by $\mathscr{G}_0(g)$ for simplicity unless confusion may occur. We see that $\operatorname{div} \mathscr{G}_0(g) = g$ in $\mathbb{R}^N$. For $g \in DI_{0,q}(\mathbb{R}^N)$, let

$$\|g\|_{W^{-1}_q(\mathbb{R}^N)} = \inf\{\|H\|_{L_q(\mathbb{R}^N)} \mid H \in \mathscr{G}_0(g)\}. \tag{9.41}$$

We know that the weak Laplace equation:

$$(\nabla u, \nabla\varphi)_{\mathbb{R}^N} = (\mathbf{f}, \nabla\varphi)_{\mathbb{R}^N} \quad \text{for any } \varphi \in \hat{W}^1_{q'}(\mathbb{R}^N)$$

is uniquely solvable for any $\mathbf{f} \in L_q(\mathbb{R}^N)$. In fact, we have $u = \Delta^{-1}\operatorname{div}\mathbf{f}$, so that

$$\|\nabla u\|_{L_q(\mathbb{R}^N)} \leqslant C_q \|\mathbf{f}\|_{L_q(\mathbb{R}^N)}.$$

For any $\mathbf{u} \in W^2_q(\mathbb{R}^N)$, let $K(\mathbf{u})$ be a unique solution of the variational equation:

$$(\nabla K(\mathbf{u}), \nabla\varphi)_{\mathbb{R}^N} = (\operatorname{Div}(\mu\mathbf{D}(\mathbf{u})) - \nabla\operatorname{div}\mathbf{u}, \nabla\varphi)_{\mathbb{R}^N} \quad \text{for any } \varphi \in \hat{W}^1_{q'}(\mathbb{R}^N).$$

Then, we consider the reduced Stokes equations:

$$\lambda\mathbf{u} - \operatorname{Div}(\mu\mathbf{D}(\mathbf{u}) - K(\mathbf{u})\mathbf{I}) = \mathbf{f} \quad \text{in } \mathbb{R}^N, \tag{9.42}$$

and we prove the following theorem.

Theorem 9.3.1 *Let $1 < q < \infty$ and $0 < \varepsilon < \pi/2$. Then, there exists an operator family $\mathscr{S}_0(\lambda) \in \operatorname{Hol}(\Sigma_\varepsilon, \mathscr{L}(L_q(\mathbb{R}^N)^N, W^2_q(\mathbb{R}^N)^N))$ such that for any $\lambda \in \Sigma_\varepsilon$ and $\mathbf{f} \in L_q(\mathbb{R}^N)^N$, $\mathbf{u} = \mathscr{S}_0(\lambda)\mathbf{f}$ is a unique solution of the reduced Stokes equation* (9.42), *and*

$$\mathscr{R}_{\mathscr{L}(L_q(\mathbb{R}^N)^N, W^{2-j}_q(\mathbb{R}^N)^N)}\{(\tau\partial_\tau)^\ell(\lambda^{j/2}\mathscr{S}_0(\lambda)) \mid \lambda \in \Sigma_{\varepsilon,\lambda_0}\}) \leqslant c_{\lambda_0}$$

for $\ell = 0, 1$, $j = 0, 1, 2$ and any λ_0. Here and hereafter, c_{λ_0} is some constant such that $\sup_{\lambda_0 \geqslant 1} c_{\lambda_0}$ is bounded and $c_{\lambda_0} \to \infty$ as $\lambda_0 \to 0$.

In what follows, we prove Theorem 9.3.1. Let g be a solution of the resolvent problem for weak Laplace operator:

$$(\lambda g, \varphi)_{\mathbb{R}^N} + (\nabla g, \nabla\varphi)_{\mathbb{R}^N} = (\mathbf{f}, \nabla\varphi)_{\mathbb{R}^N} \quad \text{for any } \varphi \in W^1_{q'}(\mathbb{R}^N). \tag{9.43}$$

And then, according to what was pointed out in Sect. 9.2.1, a solution $\mathbf{u}$ of the Stokes equations:

$$\lambda\mathbf{u} - \mathrm{Div}\,(\mu\mathbf{D}(\mathbf{u}) - \mathfrak{p}\mathbf{I}) = \mathbf{f}, \quad \mathrm{div}\,\mathbf{u} = g \quad \text{in } \mathbb{R}^N, \tag{9.44}$$

with suitable pressure term $\mathfrak{p}$, is also a solution of the Eq. (9.42). Thus, we start with the following lemma.

Lemma 9.3.2 *Let* $1 < q < \infty$ *and* $0 < \varepsilon < \pi/2$. *Then, there exists an operator family* $\mathcal{G}_0(\lambda) \in \mathrm{Hol}\,(\Sigma_\varepsilon, \mathcal{L}(L_q(\mathbb{R}^N)^N, DI_{0,q}(\mathbb{R}^N)))$ *such that for any* $\lambda \in \Sigma_\varepsilon$ *and* $\mathbf{f} \in L_q(\mathbb{R}^N)^N$, $g = \mathcal{G}_0(\lambda)\mathbf{f}$ *is a unique solution of problem* (9.43), *and*

$$\mathcal{R}_{\mathcal{L}(L_q(\mathbb{R}^N)^N, W^{1-j}_q(\mathbb{R}^N))}(\{(\tau\partial_\tau)^\ell(\lambda^{j/2}\mathcal{G}_0(\lambda)) \mid \lambda \in \Sigma_{\varepsilon,\lambda_0}\}) \leqslant c_{\lambda_0} \tag{9.45}$$

for $\ell = 0, 1$, $j = 0, 1, 2$ *and any* $\lambda_0 > 0$.

In what follows, $\mathcal{F}$ and $\mathcal{F}^{-1}_\xi$ denote the Fourier transform and the inverse Fourier transform defined by

$$\mathcal{F}[f](\xi) = \int_{\mathbb{R}^N} e^{-i\xi\cdot x} f(x)\,dx, \quad \mathcal{F}^{-1}_\xi[g(\xi)](x) = \frac{1}{(2\pi)^N}\int_{\mathbb{R}^N} e^{ix\cdot\xi} g(\xi)\,d\xi,$$

respectively. To prove Lemma 9.3.2, we use the following three lemmas.

Lemma 9.3.3 (Theorem 3.3 in [3]) *Let* $1 < q < \infty$ *and let* Λ *be a set in* $\mathbb{C}$. *Let* $m = m(\lambda, \xi)$ *be a function defined on* $\Lambda \times (\mathbb{R}^N \setminus \{0\})$ *which is infinitely differentiable with respect to* $\xi \in \mathbb{R}^N \setminus \{0\}$ *for each* $\lambda \in \Lambda$. *Assume that for any multi-index* $\alpha \in \mathbb{N}^N_0$ *there exists a constant* C_α *depending on* α *and* Λ *such that*

$$|\partial^\alpha_\xi m(\lambda, \xi)| \leqslant C_\alpha|\xi|^{-|\alpha|} \tag{9.46}$$

for any $(\lambda, \xi) \in \Lambda \times (\mathbb{R}^N \setminus \{0\})$. *Let* K_λ *be an operator defined by*

$$K_\lambda f = \mathcal{F}^{-1}_\xi[m(\lambda, \xi)\mathcal{F} f(\xi)].$$

Then, the family of operators $\{K_\lambda \mid \lambda \in \Lambda\}$ *is* $\mathcal{R}$*-bounded on* $\mathcal{L}(L_q(\mathbb{R}^N))$ *and*

$$\mathcal{R}_{\mathcal{L}(L_q(\mathbb{R}^N))}(\{K_\lambda \mid \lambda \in \Lambda\}) \leqslant C_{q,N} \max_{|\alpha|\leqslant N+1} C_\alpha \tag{9.47}$$

with some constant $C_{q,N}$ *depending only on* q *and* N.

Lemma 9.3.4 *(a) Let* X *and* Y *be Banach spaces, and let* $\mathcal{T}$ *and* $\mathcal{S}$ *be* $\mathcal{R}$*-bounded families in* $\mathcal{L}(X, Y)$. *Then,* $\mathcal{T} + \mathcal{S} = \{T + S \mid T \in \mathcal{T}, \ S \in \mathcal{S}\}$ *is also an* $\mathcal{R}$*-bounded family in* $\mathcal{L}(X, Y)$ *and*

$$\mathcal{R}_{\mathcal{L}(X,Y)}(\mathcal{T} + \mathcal{S}) \leqslant \mathcal{R}_{\mathcal{L}(X,Y)}(\mathcal{T}) + \mathcal{R}_{\mathcal{L}(X,Y)}(\mathcal{S}).$$

(b) Let X, Y and Z be Banach spaces, and let $\mathscr{T}$ *and* $\mathscr{S}$ *be* $\mathscr{R}$*-bounded families in* $\mathscr{L}(X, Y)$ *and* $\mathscr{L}(Y, Z)$*, respectively. Then,* $\mathscr{S}\mathscr{T} = \{ST \mid T \in \mathscr{T},\ S \in \mathscr{S}\}$ *also an* $\mathscr{R}$*-bounded family in* $\mathscr{L}(X, Z)$ *and*

$$\mathscr{R}_{\mathscr{L}(X,Z)}(\mathscr{S}\mathscr{T}) \leqslant \mathscr{R}_{\mathscr{L}(X,Y)}(\mathscr{T})\mathscr{R}_{\mathscr{L}(Y,Z)}(\mathscr{S}).$$

(c) Let $1 < p,\ q < \infty$ *and let D be a domain in* $\mathbb{R}^N$*. Let* $m = m(\lambda)$ *be a bounded function defined on a subset* Λ *in* $\mathbb{C}$ *and let* $M_m(\lambda)$ *be a map defined by* $M_m(\lambda)f = m(\lambda)f$ *for any* $f \in L_q(D)$*. Then,* $\mathscr{R}_{\mathscr{L}(L_q(D))}(\{M_m(\lambda) \mid \lambda \in \Lambda\}) \leqslant C_{N,q,D}\|m\|_{L_\infty(\Lambda)}$.

(d) Let $n = n(\tau)$ *be a* C^1*-function defined on* $\mathbb{R} \setminus \{0\}$ *that satisfies the conditions* $|n(\tau)| \leqslant \gamma$ *and* $|\tau n'(\tau)| \leqslant \gamma$ *with some constant* $c > 0$ *for any* $\tau \in \mathbb{R} \setminus \{0\}$*. Let* T_n *be the operator-valued Fourier multiplier defined by* $T_n f = \mathscr{F}^{-1}(n\mathscr{F}[f])$ *for any* f *with* $\mathscr{F}[f] \in \mathscr{D}(\mathbb{R}, L_q(D))$*. Then,* T_n *is extended to a bounded linear operator from* $L_p(\mathbb{R}, L_q(D))$ *into itself. Moreover, denoting this extension also by* T_n*, we have*

$$\|T_n\|_{\mathscr{L}(L_p(\mathbb{R}, L_q(D)))} \leqslant C_{p,q,D}\gamma.$$

Here, $\mathscr{D}(\mathbb{R}, L_q(D))$ *denotes the set of all* $L_q(D)$*-valued* C^∞ *functions on* $\mathbb{R}$ *with compact support.*

Proof The assertions (a) and (b) follow from [2, p. 28, Proposition 3.4], and the assertions (c) and (d) follow from [2, p. 27, Remarks 3.2] (see also Bourgain [1]). □

Lemma 9.3.5 *Let* $0 < \varepsilon < \pi/2$*. Then, for any* $\lambda \in \Sigma_\varepsilon$ *and* $x \geqslant 0$*, we have*

$$|\lambda + x| \geqslant (\sin\varepsilon)(|\lambda| + |x|). \tag{9.48}$$

Proof Let $\lambda = |\lambda|e^{i\theta}$ with $|\theta| \leqslant \pi - \varepsilon$, and then we have

$$\begin{aligned}|\lambda + x|^2 &= |\lambda|^2 + 2x|\lambda|\cos\theta + x^2 \geqslant |\lambda|^2 - 2x|\lambda|\cos\varepsilon + x^2\\ &= \cos\varepsilon(|\lambda| - x)^2 + (1 - \cos\varepsilon)(|\lambda|^2 + x^2) \geqslant (1 - \cos\varepsilon)(|\lambda|^2 + x^2)\\ &\geqslant 2\sin^2\varepsilon(|\lambda|^2 + x^2) \geqslant \{\sin\varepsilon(|\lambda| + x)\}^2\end{aligned}$$

which yields (9.48). □

A proof of Lemma 9.3.2. Since $C_0^\infty(\mathbb{R}^N)$ is dense in $L_q(\mathbb{R}^N)$, we may assume that $\mathbf{f} \in C_0^\infty(\mathbb{R}^N)^N$. Let

$$\begin{aligned}\mathscr{G}_0(\lambda)\mathbf{f} &= -\mathscr{F}_\xi^{-1}\Big[\frac{\mathscr{F}[\operatorname{div}\mathbf{f}](\xi)}{\lambda + |\xi|^2}\Big](x) = -\mathscr{F}_\xi^{-1}\Big[\frac{i\xi\cdot\mathscr{F}[\mathbf{f}](\xi)}{\lambda + |\xi|^2}\Big](x)\\ &= -\sum_{k=1}^N \frac{\partial}{\partial x_k}\mathscr{F}_\xi^{-1}\Big[\frac{\mathscr{F}[f_k](\xi)}{\lambda + |\xi|^2}\Big](x).\end{aligned} \tag{9.49}$$

Obviously, $g = \mathscr{G}_0(\lambda)$ satisfies the equation $(\lambda - \Delta)g = -\mathrm{div}\,\mathbf{f}$ in $\mathbb{R}^N$, so that by the divergence theorem of Gauß g also satisfies the variational Eq. (9.43). Let

$$\mathscr{B}_0(\lambda) f = \mathscr{F}_\xi^{-1}\Big[\frac{\mathscr{F}[f](\xi)}{\lambda + |\xi|^2}\Big], \tag{9.50}$$

and then by Lemmas 9.3.3, 9.3.4 and 9.3.5, we have

$$\begin{aligned} &\mathscr{B}_0(\lambda) \in \mathrm{Hol}\,(\Sigma_\varepsilon, \mathscr{L}(L_q(\mathbb{R}^N), W_q^2(\mathbb{R}^N))), \\ &\mathscr{R}_{\mathscr{L}(L_q(\mathbb{R}^N), W_q^{2-j}(\mathbb{R}^N))}(\{(\tau\partial_\tau)^\ell(\lambda^{j/2}\mathscr{B}_0(\lambda)) \mid \lambda \in \Sigma_{\varepsilon,\lambda_0}\}) \leqslant c_{\lambda_0} \end{aligned} \tag{9.51}$$

for $\ell = 0, 1$, $j = 0, 1, 2$ and any $\lambda_0 > 0$. Thus, in view of (9.49), we define $\mathscr{G}_0(\lambda)$ acting on $\mathbf{f} = {}^\top(f_1, \ldots, f_N)$ by

$$\mathscr{G}_0(\lambda)\mathbf{f} = -\sum_{k=1}^N \frac{\partial}{\partial x_k}\mathscr{B}_0(\lambda) f_k,$$

and then, by (9.51) $\mathscr{G}_0$ has the properties stated in Lemma 9.3.2. This completes the proof of Lemma 9.3.2. □

Next, we consider the divergence equation:

$$\mathrm{div}\,\mathbf{v} = g \quad \text{in } \mathbb{R}^N, \tag{9.52}$$

where g is a solution of (9.43). Since $\mathbf{v}$ is represented by

$$\mathbf{v} = -\mathscr{F}^{-1}\Big[\frac{i\xi\mathscr{F}[g](\xi)}{|\xi|^2}\Big],$$

in view of (9.49), we define an operator $\mathscr{S}_1(\lambda)$ by

$$\mathscr{S}_1(\lambda)\mathbf{f} = \mathscr{F}_\xi^{-1}\Big[\frac{\xi\xi\cdot\mathscr{F}[\mathbf{f}](\xi)}{|\xi|^2(\lambda + |\xi|^2)}\Big],$$

and then by Lemmas 9.3.3, 9.3.4 and 9.3.5, we have the following lemma.

Lemma 9.3.6 *Let $1 < q < \infty$ and $0 < \varepsilon < \pi/2$. Then, there exists a $\mathscr{S}_1(\lambda) \in \mathrm{Hol}\,(\Sigma_\varepsilon, \mathscr{L}(L_q(\mathbb{R}^N)^N, W_q^2(\mathbb{R}^N)^N))$ such that for any $\lambda \in \Sigma_\varepsilon$ and $\mathbf{f} \in L_q(\mathbb{R}^N)^N$, $\mathbf{v} = \mathscr{S}_1(\lambda)\mathbf{f}$ is a solution of (9.52), where g is a solution of (9.43), and*

$$\mathscr{R}_{\mathscr{L}(L_q(\mathbb{R}^N)^N, W_q^{2-j}(\mathbb{R}^N)^N)}(\{(\tau\partial_\tau)^\ell(\lambda^{j/2}\mathscr{S}_1(\lambda)) \mid \lambda \in \Sigma_{\varepsilon,\lambda_0}\}) \leqslant c_{\lambda_0}$$

for $\ell = 0, 1$, $j = 0, 1, 2$ and any $\lambda_0 > 0$.

Let $\mathbf{u} = \mathbf{v} + \mathbf{w}$ in (9.44) with $\mathbf{v} = \mathscr{S}_1(\lambda)\mathbf{f}$, and then $\mathbf{w}$ satisfies the equations:

$$\lambda \mathbf{w} - \mathrm{Div}\,(\mu \mathbf{D}(\mathbf{w}) - \mathfrak{p}\mathbf{I}) = \tilde{\mathbf{f}} \quad \mathrm{div}\,\mathbf{w} = 0 \quad \text{in } \mathbb{R}^N \tag{9.53}$$

with $\tilde{\mathbf{f}} = \mathbf{f} - (\lambda \mathbf{v} - \mathrm{Div}\,(\mu \mathbf{D}(\mathbf{v})))$.

Theorem 9.3.7 *Let $1 < q < \infty$ and $0 < \varepsilon < \pi/2$. Then, there exists $\mathscr{S}_2(\lambda) \in \mathrm{Hol}\,(\Sigma_\varepsilon, \mathscr{L}(L_q(\mathbb{R}^N)^N, W_q^2(\mathbb{R}^N)^N))$ such that for any $\lambda \in \Sigma_\varepsilon$ and $\tilde{\mathbf{f}} \in L_q(\mathbb{R}^N)^N$, $\mathbf{w} = \mathscr{S}_2(\lambda)\tilde{\mathbf{f}}$ is a unique solution of* (9.53), *and*

$$\mathscr{R}_{\mathscr{L}(L_q(\mathbb{R}^N)^N, W_q^{2-j}(\mathbb{R}^N)^N)}(\{(\tau \partial_\tau)^\ell (\lambda^{j/2}\mathscr{S}_2(\lambda)) \mid \lambda \in \Sigma_{\varepsilon,\lambda_0}\}) \leqslant c_{\lambda_0}$$

for $\ell = 0, 1$, $j = 0, 1, 2$ and any $\lambda_0 > 0$.

Proof Applying divergence to (9.53), we have $\mathfrak{p} = \Delta^{-1}\mathrm{div}\,\mathbf{f}$, so that $(\lambda - \Delta)\mathbf{w} = \tilde{\mathbf{f}} - \nabla \Delta^{-1}\mathrm{div}\,\tilde{\mathbf{f}}$. Thus, we have $\mathbf{w} = \mathscr{S}_2(\lambda)\tilde{\mathbf{f}}$ with

$$\mathscr{S}_2(\lambda)\tilde{\mathbf{f}} = \mathscr{F}_\xi^{-1}\Big[\frac{\mathscr{F}[\tilde{\mathbf{f}}](\xi) - \xi\xi \cdot \mathscr{F}[\tilde{\mathbf{f}}](\xi)|\xi|^{-2}}{\lambda + |\xi|^2}\Big].$$

By Lemmas 9.3.3, 9.3.4 and 9.3.5, we have Theorem 9.3.7, which completes the proof of Theorem 9.3.7. □

Since $\mathbf{u} = \mathbf{v} + \mathbf{w}$ is a solution of the Eqs. (9.44), we define an operator family by

$$\mathscr{S}_0(\lambda)\mathbf{f} = \mathscr{S}_1(\lambda)\mathbf{f} + \mathscr{S}_2(\lambda)\mathbf{f} - \mathscr{S}_2(\lambda)(\lambda \mathscr{S}_1(\lambda)\mathbf{f} - \mathrm{Div}\,(\mu \mathbf{D}(\mathscr{S}_1(\lambda)\mathbf{f})))$$

and then, by Lemmas 9.3.4, 9.3.6 and Theorem 9.3.7, we see that $\mathscr{S}_0(\lambda)$ satisfies the properties stated in Theorem 9.3.1, which completes the proof of Theorem 9.3.1.

9.3.2 Reduced Stokes Equations in $\mathbb{R}^N_+$ with Free Boundary Condition

Let

$$\mathbb{R}^N_+ = \{x = (x_1, \ldots, x_N) \mid x_N > 0\}, \quad \mathbb{R}^N_0 = \{x = (x_1, \ldots, x_N) \mid x_N = 0\},$$

and $\mathbf{n}_0 = (0, \ldots, 0, -1)$. Recall that

$$\hat{W}^1_{q,0}(\mathbb{R}^N_+) = \{u \in L_{q,\mathrm{loc}}(\mathbb{R}^N_+) \mid \nabla u \in L_q(\mathbb{R}^N_+)^N, \; u|_{x_N=0} = 0\},$$
$$W^1_{q,0}(\mathbb{R}^N_+) = \{u \in W^1_q(\mathbb{R}^N_+) \mid u|_{x_N=0}\},$$

In this subsection, let

$$DI_{F,q}(\mathbb{R}^N_+) = \{g \in W^1_q(\Omega) \mid \text{there exists a } G \in L_q(\mathbb{R}^N)^N \text{ such that } (g, \varphi)_{\mathbb{R}^N} = -(G, \nabla\varphi)_{\mathbb{R}^N} \text{ for any } \varphi \in W^1_{q',0}(\mathbb{R}^N)\}.$$

Let $\mathscr{G}_F(g) = \{H \in L_q(\mathbb{R}^N)^N \mid \operatorname{div} G = \operatorname{div} H\}$ and $[\mathscr{G}_F(g)]$ denotes the representative elements of the set $\mathscr{G}_F(g)$. But $[\mathscr{G}_F(g)]$ is also written by $\mathscr{G}_F(g)$ for simplicity unless confusion may occur. We see that $\operatorname{div} \mathscr{G}_F(g) = g$ in $\mathbb{R}^N_+$. For $g \in DI_{F,q}(\mathbb{R}^N)$, let

$$\|g\|_{W^{-1}_q(\mathbb{R}^N_+)} = \inf\{\|H\|_{L_q(\mathbb{R}^N_+)} \mid H \in \mathscr{G}_F(g)\}. \tag{9.54}$$

The weak Dirichlet problem:

$$(\nabla u, \nabla\varphi)_{\mathbb{R}^N_+} = (\mathbf{f}, \nabla\varphi)_{\mathbb{R}^N_+} \quad \text{for any } \varphi \in \hat{W}^1_{q',0}(\mathbb{R}^N)$$

is uniquely solvable for any $\mathbf{f} \in L_q(\mathbb{R}^N)^N$ with $u \in \hat{W}^1_{q,0}(\mathbb{R}^N_+)$. For any $\mathbf{u} \in W^2_q(\mathbb{R}^N_+)$, let $K_{F1}(\mathbf{u}) \in W^1_q(\mathbb{R}^N_+) + \mathscr{W}^1_q(\mathbb{R}^N_+)$ be a unique solution to the variational problem:

$$(\nabla K_{F1}(\mathbf{u}), \nabla\varphi)_{\mathbb{R}^N_+} = (\operatorname{Div}(\mu\mathbf{D}(\mathbf{u})) - \nabla \operatorname{div} \mathbf{u}, \nabla\varphi)_{\mathbb{R}^N_+} \tag{9.55}$$

for any $\varphi \in \hat{W}^1_{q',0}(\mathbb{R}^N_+)$, subject to $K_{F1}(\mathbf{u}) =< \mu\mathbf{D}(\mathbf{u})\mathbf{n}_0, \mathbf{n}_0 > -\operatorname{div} \mathbf{u}$ on $\mathbb{R}^N_0$, while for $h \in W^{3-1/q}_q(\mathbb{R}^{N-1})$ $K_{F2}(h) \in W^1_q(\mathbb{R}^N_+) + \hat{W}^1_{q,0}(\mathbb{R}^N_+)$ be a unique solution to the variational problem:

$$(\nabla K_{F2}(h), \nabla\varphi)_{\mathbb{R}^N_+} = 0 \quad \text{for any } \varphi \in \hat{W}^1_{q',0}(\mathbb{R}^N_+) \tag{9.56}$$

subject to $K_{F2}(h) = -(\tau + \delta\Delta' h)$ on $\mathbb{R}^N_0$, where

$$\Delta' h = \sum_{j=1}^{N-1} \frac{\partial^2}{\partial x_j^2} h.$$

We see that

$$\begin{aligned} \|\nabla K_{F1}(\mathbf{u})\|_{L_q(\mathbb{R}^N_+)} &\leqslant C\|\nabla \mathbf{u}\|_{W^1_q(\mathbb{R}^N_+)}, \\ \|\nabla K_{F2}(h)\|_{L_q(\mathbb{R}^N_+)} &\leqslant C\|h\|_{W^{3-1/q}_q(\mathbb{R}^N_+)}. \end{aligned} \tag{9.57}$$

In this subsection, we consider the following reduced Stokes equations:

$$\begin{cases} \lambda\mathbf{u} - \operatorname{Div}(\mu\mathbf{D}(\mathbf{u}) - (K_{F1}(\mathbf{u}) + K_{F2}(h))\mathbf{I}) = \mathbf{f} & \text{in } \mathbb{R}^N_+, \\ \lambda h - \mathbf{n}_0 \cdot \mathbf{u} = f & \text{on } \mathbb{R}^N_0, \\ \mathscr{T}_{\mathbf{n}_0}(\mu\mathbf{D}(\mathbf{u})\mathbf{n}_0) = \mathscr{T}_{\mathbf{n}_0}\mathbf{f}_b, \quad \operatorname{div} \mathbf{u} = \mathbf{n}_0 \cdot \mathbf{f}_b & \text{on } \mathbb{R}^N_0. \end{cases} \tag{9.58}$$

The purpose of this subsection is to prove the following theorem.

Theorem 9.3.8 *Let $1 < q < \infty$ and $0 < \varepsilon < \pi/2$. Let $Y_q(\mathbb{R}^N_+)$ and $\mathscr{Y}_q(\mathbb{R}^N_+)$ be the spaces defined in* (9.34) *with $D = \mathbb{R}^N_+$. Then, there exist operator families $\mathscr{S}_{F0}(\lambda)$ and $\mathscr{T}_{F0}(\lambda)$ with*

$$\mathscr{S}_{F0}(\lambda) \in \mathrm{Hol}\,(\Sigma_\varepsilon, \mathscr{L}(\mathscr{Y}_q(\mathbb{R}^N_+), W^2_q(\mathbb{R}^N_+)^N)),$$
$$\mathscr{T}_{F0}(\lambda) \in \mathrm{Hol}\,(\Sigma_\varepsilon, \mathscr{L}(\mathscr{Y}_q(\mathbb{R}^N_+), W^{3-1/q}_q(\mathbb{R}^{N-1})))$$

such that for any $\lambda \in \Sigma_\varepsilon$ and $(\mathbf{f}, f, \mathbf{f}_b) \in Y_q(\mathbb{R}^N_+)$, problem (9.58) *admits unique solutions $\mathbf{u} = \mathscr{S}_{F0}(\lambda)F_\lambda(\mathbf{f}, f, \mathbf{f}_b)$ and $h = \mathscr{T}_{F0}(\lambda)F_\lambda(\mathbf{f}, f, \mathbf{f}_b)$, where $F_\lambda(\mathbf{f}, f, \mathbf{f}_b) = (\mathbf{f}, f, \lambda^{1/2}\mathbf{f}_b, \mathbf{f}_b)$, and*

$$\mathscr{R}_{\mathscr{L}(\mathscr{Y}_q(\mathbb{R}^N_+), W^{2-j}_q(\mathbb{R}^N_+)^N)}(\{(\tau\partial_\tau)^\ell(\lambda^{j/2}\mathscr{S}_{F0}(\lambda)) \mid \lambda \in \Sigma_{\varepsilon,\lambda_0}\}) \leqslant c_{\lambda_0} \quad (j = 0, 1, 2),$$
$$\mathscr{R}_{\mathscr{L}(\mathscr{Y}_q(\mathbb{R}^N_+), W^{3-1/q-j}_q(\mathbb{R}^N_+))}(\{(\tau\partial_\tau)^\ell(\lambda^{j}\mathscr{T}_{F0}(\lambda)) \mid \lambda \in \Sigma_{\varepsilon,\lambda_0}\}) \leqslant c_{\lambda_0} \quad (j = 0, 1)$$

for $\ell = 0, 1$ and any $\lambda_0 > 0$.

In what follows, we prove Theorem 9.3.8. For this purpose, first we consider the following resolvent problem for the weak Laplace-Dirichlet operator:

$$(\lambda g, \varphi)_{\mathbb{R}^N_+} + (\nabla g, \nabla\varphi)_{\mathbb{R}^N_+} = (\mathbf{f}, \nabla\varphi)_{\mathbb{R}^N_+} \quad \text{for any } \varphi \in \mathscr{W}^1_{q'}(\mathbb{R}^N_+) \tag{9.59}$$

subject to $g = \mathbf{n}_0 \cdot \mathbf{f}_b$ on $\mathbb{R}^N_0$. Then, according to what was pointed out in Sect. 9.2.1, solutions $\mathbf{u}$ and h of the following equations:

$$\begin{cases} \lambda\mathbf{u} - \mathrm{Div}\,(\mu\mathbf{D}(\mathbf{u}) - \mathfrak{p}\mathbf{I}) = \mathbf{f}, \quad \mathrm{div}\,\mathbf{u} = g & \text{in } \mathbb{R}^N_+, \\ \lambda h - \mathbf{n}_0 \cdot \mathbf{u} = f & \text{on } \mathbb{R}^N_0, \\ (\mu\mathbf{D}(\mathbf{u}) - \mathfrak{p}\mathbf{I})\mathbf{n}_0 - ((\tau + \delta\Delta')h)\mathbf{n}_0 = \mathbf{f}_b & \text{on } \mathbb{R}^N_0. \end{cases} \tag{9.60}$$

are also solutions of problem (9.58).

In case of $g = 0$ in (9.60), we know the following theorem.

Theorem 9.3.9 *Let $1 < q < \infty$ and $0 < \varepsilon < \pi/2$. Let $Y_q(\mathbb{R}^N_+)$ and $\mathscr{Y}_q(\mathbb{R}^N_+)$ be the same spaces as in Theorem 9.3.8. Then, there exist operator families $\mathscr{S}_{F1}(\lambda)$, $\mathscr{P}_{F1}(\lambda)$ and $\mathscr{T}_{F1}(\lambda)$ with*

$$\mathscr{S}_{F1}(\lambda) \in \mathrm{Hol}\,(\Sigma_\varepsilon, \mathscr{L}(\mathscr{Y}_q(\mathbb{R}^N_+), W^2_q(\mathbb{R}^N_+)^N)),$$
$$\mathscr{P}_{F1}(\lambda) \in \mathrm{Hol}\,(\Sigma_\varepsilon, \mathscr{L}(\mathscr{Y}_q(\mathbb{R}^N_+), W^1_q(\mathbb{R}^N_+) + \mathscr{W}^1_q(\mathbb{R}^N_+))),$$
$$\mathscr{T}_{F1}(\lambda) \in \mathrm{Hol}\,(\Sigma_\varepsilon, \mathscr{L}(\mathscr{Y}_q(\mathbb{R}^N_+), W^{3-1/q}_q(\mathbb{R}^N_+)^N))$$

such that for any $\lambda \in \Sigma_\varepsilon$ and $(\mathbf{f}, f, \mathbf{f}_b) \in Y_q(\mathbb{R}^N_+)$, problem (9.60) *with $g = 0$ admits unique solutions $\mathbf{u} = \mathscr{S}_{F1}(\lambda)F_\lambda(\mathbf{f}, f, \mathbf{f}_b)$, $\mathfrak{p} = \mathscr{P}_{F1}(\lambda)F_\lambda(\mathbf{f}, f, \mathbf{f}_b)$ and $h = \mathscr{T}_{F1}(\lambda)F_\lambda(\mathbf{f}, f, \mathbf{f}_b)$, and*

$$\mathcal{R}_{\mathcal{L}(\mathcal{Y}_q(\mathbb{R}^N_+), W^{2-j}_q(\mathbb{R}^N_+)^N)}(\{(\tau\partial_\tau)^\ell(\lambda^{j/2}\mathcal{S}_{F1}(\lambda)) \mid \lambda \in \Sigma_{\varepsilon,\lambda_0}\}) \leqslant c_{\lambda_0} \quad (j = 0, 1, 2),$$
$$\mathcal{R}_{\mathcal{L}(\mathcal{Y}_q(\mathbb{R}^N_+), L_q(\mathbb{R}^N_+)^N)}(\{(\tau\partial_\tau)^\ell(\nabla\mathcal{P}_{F1}(\lambda)) \mid \lambda \in \Sigma_{\varepsilon,\lambda_0}\}) \leqslant c_{\lambda_0},$$
$$\mathcal{R}_{\mathcal{L}(\mathcal{Y}_q(\mathbb{R}^N_+), W^{3-1/q-j}_q(\mathbb{R}^N_+))}(\{(\tau\partial_\tau)^\ell(\lambda^j\mathcal{T}_{F1}(\lambda)) \mid \lambda \in \Sigma_{\varepsilon,\lambda_0}\}) \leqslant c_{\lambda_0} \quad (j = 0, 1)$$

for $\ell = 0, 1$ *and any* $\lambda_0 > 0$.

Proof Since Theorem 9.3.9 was essentially proved in Shibata and Shimizu [18], we may omit the proof. □

Thus, we consider the divergence equation:

$$\operatorname{div}\mathbf{v} = g \quad \text{in } \mathbb{R}^N_+, \tag{9.61}$$

where g is a solution of (9.59). The $\mathbf{n}_0 \cdot \mathbf{f}_b$ being renamed ρ in what follows, we prove the following lemma.

Lemma 9.3.10 *Let* $1 < q < \infty$ *and* $0 < \varepsilon < \pi/2$. *Let*

$$Y^1_q(\mathbb{R}^N_+) = \{(\mathbf{f}, \rho) \mid \mathbf{f} \in L_q(\mathbb{R}^N_+)^N,\ \ \rho \in W^1_q(\mathbb{R}^N_+)\},$$
$$\mathcal{Y}^1_q(\mathbb{R}^N_+) = \{(F_1, F_8, F_9) \mid F_1 \in L_q(\mathbb{R}^N_+)^N,\ \ F_8 \in L_q(\mathbb{R}^N_+),\ \ F_9 \in W^1_q(\mathbb{R}^N_+)\}.$$

Then, we have the following assertions.

(a) There exists an operator family $\mathcal{G}_{F1}(\lambda) \in \mathrm{Hol}\,(\Sigma_\varepsilon, \mathcal{L}(\mathcal{Y}^1_q(\mathbb{R}^N_+), \dot{DI}_{F,q}(\mathbb{R}^N_+)))$ *such that for any* $\lambda \in \Sigma_\varepsilon$ *and* $(\mathbf{f}, \rho) \in Y^1_q(\mathbb{R}^N_+)$, $g = \mathcal{G}_{F1}(\lambda)(\mathbf{f}, \lambda^{1/2}\rho, \rho)$ *be a unique solution of problem* (9.59) *with* $\rho = \mathbf{n}_0 \cdot \mathbf{f}_b$, *and*

$$\mathcal{R}_{\mathcal{L}(\mathcal{Y}^1_q(\mathbb{R}^N_+), W^{1-j}_q(\mathbb{R}^N_+))}(\{(\tau\partial_\tau)^\ell(\lambda^{j/2}\mathcal{G}_{F1}(\lambda)) \mid \lambda \in \Sigma_{\varepsilon,\lambda_0}\}) \leqslant c_{\lambda_0}$$

for $\ell = 0, 1$ *and* $j = 0, 1, 2$ *and any* $\lambda_0 > 0$.

(b) Let g be the function given in (a). Then, there exists an operator family $\mathcal{G}_{F2}(\lambda) \in \mathrm{Hol}\,(\Sigma_\varepsilon, \mathcal{L}(\mathcal{Y}^1_q(\mathbb{R}^N_+), W^2_q(\mathbb{R}^N_+)^N))$ *such that for any* $\lambda \in \Sigma_\varepsilon$ *and* $(\mathbf{f}, \rho) \in Y^1_q(\mathbb{R}^N_+)$, *problem* (9.61) *admits a solution* $\mathbf{v} = \mathcal{G}_{F2}(\lambda)(\mathbf{f}, \lambda^{1/2}\rho, \rho)$, *and*

$$\mathcal{R}_{\mathcal{L}(\mathcal{Y}^1_q(\mathbb{R}^N_+), W^{2-j}_q(\mathbb{R}^N_+)^N)}(\{(\tau\partial_\tau)^\ell(\lambda^{j/2}\mathcal{G}_{F2}(\lambda)) \mid \lambda \in \Sigma_{\varepsilon,\lambda_0}\}) \leqslant c_{\lambda_0}$$

for $\ell = 0, 1$, $j = 0, 1, 2$ *and any* $\lambda_0 > 0$.

Remark 9.3.1 (1) The F_8 and F_9 are variables corresponding to $\lambda^{1/2}\rho$ and ρ.

Proof (a) Since $C^\infty_0(\mathbb{R}^N_+)$ is dense in $L_q(\mathbb{R}^N_+)$, we may assume that $\mathbf{f} \in C^\infty_0(\mathbb{R}^N_+)^N$. In this case, we construct a solution g satisfying the Laplace equations of the strong form

$$\lambda g - \Delta g = -\operatorname{div}\mathbf{f} \quad \text{in } \mathbb{R}^N_+, \quad g|_{x_N=0} = \rho|_{x_N=0}. \tag{9.62}$$

For any function a defined on $\mathbb{R}^N_+$, let

$$a^e(x) = \begin{cases} a(x) & (x_N > 0), \\ a(x', -x_n) & (x_N < 0), \end{cases} \quad a^o(x) = \begin{cases} a(x) & (x_N > 0), \\ -a(x', -x_n) & (x_N < 0). \end{cases}$$

The a^e and a^o are even and odd extensions of a to $x_N < 0$, respectively. Let g_1 be a solution of the equation:

$$(\lambda - \Delta)g_1 = -(\mathrm{div}\,\mathbf{f})^o \quad \text{in } \mathbb{R}^N, \quad g_1|_{x_N=0},$$

which is defined by

$$g_1 = \mathscr{G}_{F3}(\lambda)\mathbf{f} \quad \text{with} \quad \mathscr{G}_{F3}(\lambda)\mathbf{f} = -\mathscr{F}^{-1}\Big[\frac{\mathscr{F}[(\mathrm{div}\,\mathbf{f})^o](\xi)}{\lambda + |\xi|^2}\Big](x).$$

Since $(\mathrm{div}\,\mathbf{f})^o = \sum_{j=1}^{N-1} \partial_j(f_j^o) + \partial_N(f_N^e)$ as follows from $\mathbf{f}|_{x_N=0} = 0$, we have

$$\begin{aligned} g_1 = \mathscr{G}_{F3}(\lambda)\mathbf{f} &= -\mathscr{F}^{-1}\Big[\frac{\sum_{j=1}^{N-1} i\xi_j \mathscr{F}[f_j^o](\xi) + i\xi_N \mathscr{F}[f_N^e](\xi)}{\lambda + |\xi|^2}\Big] \\ &= -\sum_{j=1}^{N-1} \frac{\partial}{\partial x_j}\mathscr{F}^{-1}\Big[\frac{\mathscr{F}[f_j^o](\xi)}{\lambda + |\xi|^2}\Big] - \frac{\partial}{\partial x_N}\mathscr{F}^{-1}\Big[\frac{\mathscr{F}[f_N^e](\xi)}{\lambda + |\xi|^2}\Big]. \end{aligned} \tag{9.63}$$

Moreover, $g_1|_{x_N=0} = 0$. In fact,

$$\begin{aligned} \hat{g}_1(\xi', 0) &= -\frac{1}{2\pi}\int_{-\infty}^{\infty} \frac{d\xi_N}{\lambda + |\xi|^2}\int_0^{\infty} (e^{-iy_N\xi_N} - e^{iy_N\xi_N})\mathscr{F}'[(\mathrm{div}\,\mathbf{f})](\xi', y_N)\,dy_N \\ &= -\int_0^{\infty} \mathscr{F}'[\mathrm{div}\,\mathbf{f}](\xi', y_N)\Big(\frac{1}{2\pi}\int_{-\infty}^{\infty} \frac{e^{-iy_N\xi_N} - e^{iy_N\xi_N}}{\lambda + |\xi|^2}\,d\xi_N\Big)\,dy_N. \end{aligned}$$

Here and hereafter, the partial Fourier transform with respect to $x' = (x_1, \ldots, x_{N-1})$ and its inversion formula are defined by

$$\hat{f}(\xi', x_N) = \mathscr{F}'[f](\xi', x_N) = \int_{\mathbb{R}^{N-1}} e^{-ix'\cdot\xi'} f(x', x_N)\,dx',$$
$$\mathscr{F}^{-1}_{\xi'}[g(\xi', x_N)](x') = \frac{1}{(2\pi)^{N-1}}\int_{\mathbb{R}^{N-1}} e^{ix'\cdot\xi'} g(\xi', x_N)\,d\xi'$$

with $\xi' = (\xi_1, \ldots, \xi_{N-1})$. By the residue theorem in the theory of one complex varible, we have

$$\frac{1}{2\pi}\int_{-\infty}^{\infty}\frac{e^{ia\xi_N}}{\lambda+|\xi|^2}\,d\xi_N=\frac{e^{-|a|B}}{2B},\quad \frac{1}{2\pi}\int_{-\infty}^{\infty}\frac{\xi_N e^{ia\xi_N}}{\lambda+|\xi|^2}\,d\xi_N=-\mathrm{sign}(a)\frac{e^{-|a|B}}{2} \tag{9.64}$$

for any $a\in\mathbb{R}\setminus\{0\}$. Here and hereafte, $B=\sqrt{\lambda+A^2}$ with $A=|\xi'|$ and $\mathrm{Re}\,B>0$. Thus, we have $g_1(x',0)=0$.

By Lemmas 9.3.3 and 9.3.5, $\mathscr{G}_{F3}(\lambda)\in \mathrm{Hol}\,(\Sigma_\varepsilon, \mathscr{L}(L_q(\mathbb{R}^N_+), W^1_q(\mathbb{R}^N_+)))$, and

$$\mathscr{R}_{\mathscr{L}(L_q(\mathbb{R}^N_+)^N, W^{1-j}_q(\mathbb{R}^N_+))}(\{(\tau\partial_\tau)^\ell(\lambda^{j/2}\mathscr{G}_{F3}(\lambda))\mid \lambda\in\Sigma_{\varepsilon,\lambda_0}\})\leqslant c_{\lambda_0}$$

for $\ell=0,1$, $j=0,1,2$ and any $\lambda_0>0$.

Next, we consider the equation:

$$(\lambda-\Delta)g_2=0\quad\text{in }\mathbb{R}^N_+,\quad g_2|_{x_N=0}=\rho|_{x_N=0}. \tag{9.65}$$

Applying the partial Fourier transform to (9.65) we have

$$D_N^2\hat g_2-(\lambda+|\xi'|^2)\hat g_2=0\quad\text{for }x_N>0,\quad \hat g_2|_{x_N=0}=\hat\rho|_{x_N=0}. \tag{9.66}$$

Let

$$g_2(x)=\mathscr{F}^{-1}_{\xi'}[e^{-Bx_N}\hat\rho(\xi',0)](x'). \tag{9.67}$$

We see that g_2 satisfies (9.66). By the Volevich trick

$$a(x_N)b(0)=-\int_0^\infty\frac{\partial}{\partial y_N}(a(x_N+y_N)b(y_N))\,dy_N$$

and the identity:

$$1=(\lambda+A^2)B^{-2}=\lambda B^{-2}-\sum_{j=1}^{N-1}(i\xi_j)(i\xi_j)B^{-2},$$

we have

$$\begin{aligned}
g_2(x)&=-\int_0^\infty\mathscr{F}^{-1}_{\xi'}[e^{-B(x_N+y_N)}\mathscr{F}'[\partial_N\rho](\xi',y_N)](x')\,dy_N\\
&\quad+\int_0^\infty\mathscr{F}^{-1}_{\xi'}[Be^{-B(x_N+y_N)}\mathscr{F}'[\rho](\xi',y_N)](x')\,dy_N\\
&=-\int_0^\infty\mathscr{F}^{-1}_{\xi'}\Big[\frac{\lambda^{1/2}}{B^2}\lambda^{1/2}e^{-B(x_N+y_N)}\mathscr{F}'[\partial_N\rho](\xi',y_N)\Big](x')\,dy_N\\
&\quad-\int_0^\infty\mathscr{F}^{-1}_{\xi'}\Big[\frac{A}{B^2}Ae^{-B(x_N+y_N)}\mathscr{F}'[\partial_N\rho](\xi',y_N)\Big](x')\,dy_N\\
&\quad+\int_0^\infty\mathscr{F}^{-1}_{\xi'}\Big[\frac{1}{B}\lambda^{1/2}e^{-B(x_N+y_N)}\mathscr{F}'[\lambda^{1/2}\rho](\xi',y_N)\Big](x')\,dy_N
\end{aligned}$$

$$- \sum_{j=1}^{N-1} \int_0^\infty \mathscr{F}_{\xi'}^{-1}\Big[\frac{i\xi_j}{A}\frac{1}{B} A e^{-B(x_N+y_N)} \mathscr{F}'[\partial_j \rho](\xi', y_N)\Big](x')\, dy_N.$$

Moreover, using the identity:

$$e^{-B(x_N+y_N)} = -B^{-1}\frac{\partial}{\partial x_N} e^{-B(x_N+y_N)},$$

we have

$$\begin{aligned} g_2(x) = \frac{\partial}{\partial x_N}\Big\{ &\int_0^\infty \mathscr{F}_{\xi'}^{-1}\Big[\frac{\lambda^{1/2}}{B^3}\lambda^{1/2} e^{-B(x_N+y_N)} \mathscr{F}'[\partial_N \rho](\xi', y_N)\Big](x')\, dy_N \\ &+ \int_0^\infty \mathscr{F}_{\xi'}^{-1}\Big[\frac{A}{B^3} A e^{-B(x_N+y_N)} \mathscr{F}'[\partial_N \rho](\xi', y_N)\Big](x')\, dy_N \\ &- \int_0^\infty \mathscr{F}_{\xi'}^{-1}\Big[\frac{1}{B^2}\lambda^{1/2} e^{-B(x_N+y_N)} \mathscr{F}'[\lambda^{1/2} \rho](\xi', y_N)\Big](x')\, dy_N \\ &+ \sum_{j=1}^{N-1} \int_0^\infty \mathscr{F}_{\xi'}^{-1}\Big[\frac{i\xi_j}{A}\frac{1}{B^2} A e^{-B(x_N+y_N)} \mathscr{F}'[\partial_j \rho](\xi', y_N)\Big](x')\, dy_N\Big\}. \end{aligned}$$

Let $\mathscr{G}_{F4}(\lambda)(F_8, F_9) = \frac{\partial}{\partial x_N}\mathscr{G}_{F5}(\lambda)(F_8, F_9)$ with

$$\begin{aligned} &\mathscr{G}_{F5}(\lambda)(F_8, F_9) \\ &= \int_0^\infty \mathscr{F}_{\xi'}^{-1}\Big[\frac{\lambda^{1/2}}{B^3}\lambda^{1/2} e^{-B(x_N+y_N)} \mathscr{F}'[\partial_N F_9](\xi', y_N)\Big](x')\, dy_N \\ &+ \int_0^\infty \mathscr{F}_{\xi'}^{-1}\Big[\frac{A}{B^3} A e^{-B(x_N+y_N)} \mathscr{F}'[\partial_N F_9](\xi', y_N)\Big](x')\, dy_N \\ &- \int_0^\infty \mathscr{F}_{\xi'}^{-1}\Big[\frac{1}{B^2}\lambda^{1/2} e^{-B(x_N+y_N)} \mathscr{F}'[F_8](\xi', y_N)\Big](x')\, dy_N \\ &+ \sum_{j=1}^{N-1} \int_0^\infty \mathscr{F}_{\xi'}^{-1}\Big[\frac{i\xi_j}{A}\frac{1}{B^2} A e^{-B(x_N+y_N)} \mathscr{F}'[\partial_j F_9](\xi', y_N)\Big](x')\, dy_N. \end{aligned}$$

Obviously, $g_2(x) = \mathscr{G}_4(\lambda)(\lambda^{1/2}\rho, \rho)$. Let

$$\mathscr{Z}_q^1(\mathbb{R}_+^N) = \{(F_8, F_9) \mid F_8 \in L_q(\mathbb{R}_+^N),\ \ F_9 \in W_q^1(\mathbb{R}_+^N)\},$$

and then, we have

$$\mathscr{R}_{\mathscr{L}(\mathscr{Z}_q^1(\mathbb{R}_+^N), W_q^{2-j}(\mathbb{R}_+^N))}(\{(\tau\partial_\tau)^\ell(\lambda^{j/2}\mathscr{G}_{F5}(\lambda)) \mid \lambda \in \Sigma_{\varepsilon,\lambda_0}\}) \leqslant c_{\lambda_0} \tag{9.68}$$

for $\ell = 0, 1$ and $j = 0, 1, 2$ and any $\lambda_0 > 0$. If we prove (9.68), then, we have

$$\mathscr{R}_{\mathscr{L}(\mathscr{X}_q^1(\mathbb{R}_+^N), W_q^{1-j}(\mathbb{R}_+^N))}(\{(\tau\partial_\tau)^\ell(\lambda^{j/2}\mathscr{G}_{F4}(\lambda)) \mid \lambda \in \Sigma_{\varepsilon,\lambda_0}\}) \leqslant c_{\lambda_0} \tag{9.69}$$

for $\ell = 0, 1$ and $j = 0, 1, 2$ and any $\lambda_0 > 0$.

To prove (9.68), we use a lemma concerning the $\mathscr{R}$ boundedness of operators defined on $\mathbb{R}_+^N$. To state this lemma, we introduce some classes of multipliers.

Definition 9.3.11 Let V be a domain in $\mathbb{C}$, let $\Xi = (\mathbb{R}^{N-1} \setminus \{0\}) \times V$, and let $m\colon \Xi \to \mathbb{C}$, $(\xi', \lambda) \mapsto m(\xi', \lambda)$ be C^1 with respect to τ (where $\lambda = \gamma + i\tau$) and C^∞ with respect to ξ'.

(1) $m(\xi', \lambda)$ is called a multiplier of order s with type 1 on Ξ if there hold the estimates:

$$\begin{aligned} |\partial_{\xi'}^{\kappa'} m(\xi', \lambda)| &\leqslant C_{\kappa'}(|\lambda|^{1/2} + |\xi'|)^{s-|\kappa'|}, \\ |\partial_{\xi'}^{\kappa'}(\tau\partial_\tau m(\xi', \lambda))| &\leqslant C_{\kappa'}(|\lambda|^{1/2} + |\xi'|)^{s-|\kappa'|} \end{aligned} \tag{9.70}$$

for any multi-index $\kappa' \in \mathbb{N}_0^{N-1}$ and $(\xi', \lambda) \in \Xi$ with some constant $C_{\kappa'}$ depending solely on κ' and Ξ.

(2) $m(\xi', \lambda)$ is called a multiplier of order s with type 2 on Ξ if there hold the estimates:

$$\begin{aligned} |\partial_{\xi'}^{\kappa'} m(\xi', \lambda)| &\leqslant C_{\kappa'}(|\lambda|^{1/2} + |\xi'|)^{s}|\xi'|^{-|\kappa'|}, \\ |\partial_{\xi'}^{\kappa'}(\tau\partial_\tau m(\xi', \lambda))| &\leqslant C_{\kappa'}(|\lambda|^{1/2} + |\xi'|)^{s}|\xi'|^{-|\kappa'|} \end{aligned} \tag{9.71}$$

for any multi-index $\kappa' \in \mathbb{N}_0^{N-1}$ and $(\xi', \lambda) \in \Xi$ with some constant $C_{\kappa'}$ depending solely on κ' and Ξ.

Let $\mathbb{M}_{s,i}(V)$ be the set of all multipliers of order s with type i on Ξ $(i = 1, 2)$.

Obviously, $\mathbb{M}_{s,i}(V)$ are complex vector spaces. Moreover, the following lemma follows from the inequality $(|\lambda|^{1/2} + |\xi'|)^{-|\alpha'|} \leqslant |\xi'|^{-|\alpha'|}$ and the Leibniz rule immediately.

Lemma 9.3.12 *Let s_1, s_2 be two real numbers. Then, the following three assertions hold.*

(a) Given $m_i \in \mathbb{M}_{s_i,1}(V)$ $(i = 1, 2)$, we have $m_1 m_2 \in \mathbb{M}_{s_1+s_2,1}(V)$.
(b) Given $\ell_i \in \mathbb{M}_{s_i,i}(V)$ $(i = 1, 2)$, we have $\ell_1\ell_2 \in \mathbb{M}_{s_1+s_2,2}(V)$.
(c) Given $n_i \in \mathbb{M}_{s_i,2}(V)$ $(i = 1, 2)$, we have $n_1 n_2 \in \mathbb{M}_{s_1+s_2,2}(V)$.

The estimate (9.68) follows immediately from the following lemma.

Lemma 9.3.13 *Let $1 < q < \infty$, $0 < \varepsilon < \pi/2$ and $\lambda_0 > 0$. Let*

$$\ell_0(\xi', \lambda) \in \mathbb{M}_{-2,1}(\Sigma_{\varepsilon,\lambda_0}), \quad \ell_1(\xi', \lambda) \in \mathbb{M}_{-2,2}(\Sigma_{\varepsilon,\lambda_0}).$$

If we define the operators $L_j(\lambda)$ $(j = 1, 2)$ by

$$[L_1(\lambda)f](x) = \int_0^\infty \mathscr{F}_{\xi'}^{-1}[\ell_0(\xi', \lambda)\lambda^{1/2}e^{-B(x_N+y_N)}\mathscr{F}'[f](\xi', y_N)](x')\,dy_N,$$

$$[L_2(\lambda)f](x) = \int_0^\infty \mathscr{F}_{\xi'}^{-1}[\ell_1(\xi,\lambda)Ae^{-B(x_N+y_N)}\mathscr{F}'[f](\xi',y_N)](x')\,dy_N,$$

then

$$\mathscr{R}_{\mathscr{L}(L_q(\mathbb{R}^N_+),W_q^{2-j}(\mathbb{R}^N))}(\{(\tau\partial_\tau)^\ell(\lambda^{j/2}L_i(\lambda)) \mid \lambda \in \Sigma_\vartheta\}) \leqslant \gamma$$

for $s = 0, 1$, $i = 1, 2$ *and* $j = 0, 1, 2$ *with some constant* γ *depending essentially on* ε *and* λ_0.

Proof We can prove the lemma with the help of Lemma 5.4 in Shibata and Shimizu [18] immediateley, so that we may omit the proof. □

Let $\mathscr{G}_{F1}(\lambda)(F_1, F_8, F_9) = \mathscr{G}_{F3}(\lambda)F_1 + \mathscr{G}_{F4}(\lambda)(F_8, F_9)$, and then $\mathscr{G}_{F1}(\lambda)$ possesses the required properties. This completes the proof of the assertion (a).

(b) First, we construct a $\mathbf{v}_1$ satisfying the relation: $\operatorname{div}\mathbf{v}_1 = g_1$ in $\mathbb{R}^N_+$. Again, we may assume that $\mathbf{f} \in C_0^\infty(\mathbb{R}^N_+)^N$. Recalling the definition of g_1 given in (9.63), we define $\mathbf{v}_1 = {}^\top(v_1, \ldots, v_N)$ by

$$v_{1j} = \mathscr{F}^{-1}\Big[\frac{-i\xi_j\mathscr{F}[g_1](\xi)}{|\xi|^2}\Big] = \mathscr{V}_{1j}(\lambda)\mathbf{f}$$

with

$$\mathscr{V}_{1j}(\lambda)\mathbf{f} = \mathscr{F}^{-1}\Big[\frac{\sum_{k=1}^{N-1}\xi_j\xi_k\mathscr{F}[f_k^o](\xi) + \xi_j\xi_N\mathscr{F}[f_N^e](\xi)}{|\xi|^2(\lambda+|\xi|^2)}\Big].$$

Let $\mathscr{G}_{F6}(\lambda) = {}^\top(\mathscr{V}_{11}(\lambda), \ldots, \mathscr{V}_{1N}(\lambda))$. By Lemmas 9.3.3 and 9.3.5, we see that $\mathscr{G}_{F6}(\lambda) \in \operatorname{Hol}(\Sigma_\varepsilon, \mathscr{L}(L_q(\mathbb{R}^N_+)^N, W_q^2(\mathbb{R}^N_+)^N))$, that $\mathbf{v}_1 = \mathscr{G}_{F6}(\lambda)\mathbf{f}$ satisfies the relation: $\operatorname{div}\mathbf{v}_1 = g_1$ in $\mathbb{R}^N_+$, and that

$$\mathscr{R}_{\mathscr{L}(L_q(\mathbb{R}^N_+)^N,W_q^{2-j}(\mathbb{R}^N_+)^N)}(\{(\tau\partial_\tau)^\ell(\lambda^{j/2}\mathscr{G}_{F6}(\lambda)) \mid \lambda \in \Sigma_{\varepsilon,\lambda_0}\}) \leqslant c_{\lambda_0}$$

for $\ell = 0, 1$, $j = 0, 1, 2$ and any $\lambda_0 > 0$.

Next, we construct a $\mathbf{v}_2$ satisfying the relation: $\operatorname{div}\mathbf{v}_2 = g_2$ in $\mathbb{R}^N_+$. Recall that $g_2 = \partial_N h$ with $h = \mathscr{G}_{F5}(\lambda)(\lambda^{1/2}\rho, \rho)$. Thus, we define $\mathbf{v}_2 = {}^\top(v_{21}, \ldots, v_{2N})$ by

$$v_{2j}(\lambda)\rho = -\mathscr{F}^{-1}\Big[\frac{i\xi_j\mathscr{F}[(\partial_N h)^o](\xi)}{|\xi|^2}\Big].$$

Since $(\partial_N h)^o = \partial_N(h^e)$, we have

$$v_{2j} = \mathscr{F}^{-1}\Big[\frac{\xi_j\xi_N\mathscr{F}[h^e](\xi)}{|\xi|^2}\Big].$$

Let

$$\mathscr{G}_{F7j}(\lambda)(F_8, F_9) = \mathscr{F}^{-1}\Big[\frac{\xi_j\xi_N\mathscr{F}[\{\mathscr{G}_{F5}(\lambda)(F_8, F_9)\}^e](\xi)}{|\xi|^2}\Big],$$

and then for any $n \in \mathbf{N}$, $\{a_\ell\}_{\ell=1}^n \subset \mathbb{C}$, $\{\lambda_\ell\}_{\ell=1}^n \subset \Sigma_\varepsilon$, $\{(F_{8\ell}, F_{9\ell})\}_{\ell=1}^n \subset \mathscr{X}_q^1(\mathbb{R}^N_+)$, and $m = 0, 1, 2$, we have

$$\sum_{\ell=1}^n a_\ell \mathscr{G}_{F7j}(\lambda_\ell)(F_{8\ell}, F_{9\ell}) = \mathscr{F}^{-1}\Big[\frac{\xi_j\xi_N}{|\xi|^2}\mathscr{F}[\{\sum_{\ell=1}^n a_\ell \mathscr{G}_{F5}(\lambda_\ell)(F_{8\ell}, F_{9\ell})\}^e](\xi)\Big],$$

and therefore,

$$\|\sum_{\ell=1}^n a_\ell \mathscr{G}_{F7j}(\lambda_\ell)(F_{8\ell}, F_{9\ell})\|_{L_q(\mathbb{R}^N)} \leqslant C\|\sum_{\ell=1}^n a_\ell \mathscr{G}_{F5}(\lambda_\ell)(F_{8\ell}, F_{9\ell})\|_{L_q(\mathbb{R}^N_+)}.$$

Let $\mathscr{G}_{F7}(\lambda) = {}^\top(\mathscr{G}_{F71}, \ldots, \mathscr{G}_{F7N}(\lambda))$, and then, by (9.68)

$$\mathscr{R}_{\mathscr{L}(\mathscr{X}_q^1(\mathbb{R}^N_+), W_q^{2-j}(\mathbb{R}^N_+))}(\{(\tau\partial_\tau)^\ell(\lambda^{j/2}\mathscr{G}_{F7}(\lambda)) \mid \lambda \in \Sigma_{\varepsilon,\lambda_0}\}) \leqslant c_{\lambda_0}$$

for $\ell = 0, 1$, $j = 0, 1, 2$ and any $\lambda_0 > 0$. Let $\mathscr{G}_{F2}(\lambda)(F_1, F_8, F_9) = \mathscr{G}_{F6}(\lambda)F_1 + \mathscr{G}_{F7}(\lambda)(F_8, F_9)$ and then $\mathscr{G}_{F2}(\lambda)$ satisfies the properties stated in Lemma 9.3.10 (b), which completes the proof of Lemma 9.3.10. □

Let $\mathbf{u}$ be a solution of the Eqs. (9.60) and let $\mathbf{w} = \mathbf{u} - \mathbf{v}$, where $\mathbf{v}$ is a solution of the divergence equation (9.61). And then, $\mathbf{w}$, $\mathfrak{p}$ and h satisfy the equations:

$$\begin{cases} \lambda\mathbf{w} - \mathrm{Div}\,(\mu\mathbf{D}(\mathbf{w}) - \mathfrak{p}\mathbf{I}) = \mathbf{f} - \lambda\mathbf{v} + \mathrm{Div}\,(\mu\mathbf{D}(\mathbf{v})), \quad \mathrm{div}\,\mathbf{w} = 0 & \text{in } \mathbb{R}^N_+, \\ \lambda h - \mathbf{n}_0 \cdot \mathbf{w} = f + \mathbf{n}_0 \cdot \mathbf{v} & \text{on } \mathbb{R}^N_0, \\ (\mu\mathbf{D}(\mathbf{w}) - \mathfrak{p}\mathbf{I})\mathbf{n}_0 - ((\tau + \delta\Delta')h)\mathbf{n}_0 = \mathbf{f}_b - \mu\mathbf{D}(\mathbf{v})\mathbf{n}_0 & \text{on } \mathbb{R}^N_0. \end{cases}$$

By Theorem 9.3.9 and Lemma 9.3.10, we have

$$\mathbf{u} = \mathbf{v} + \mathscr{S}_{F1}(\lambda)F, \quad \mathfrak{p} = \mathscr{P}_{F1}(\lambda)F, \quad h = \mathscr{T}_{F1}(\lambda)F$$

with

$$F = (\mathbf{f} - \lambda\mathbf{v} + \mathrm{Div}\,(\mu\mathbf{D}(\mathbf{v})), \mathbf{n}_0 \cdot \mathbf{v} + f, \lambda^{1/2}(\mathbf{f}_b - \mu\mathbf{D}(\mathbf{v})\mathbf{n}_0), \mathbf{f}_b - \mu\mathbf{D}(\mathbf{v})\mathbf{n}_0),$$
$$\mathbf{v} = \mathscr{G}_{F2}(\lambda)(\mathbf{f}, \lambda^{1/2}\mathbf{n}_0 \cdot \mathbf{f}_b, \mathbf{n}_0 \cdot \mathbf{f}_b).$$

As was discussed in Sect. 9.2, $\mathfrak{p} = K_{F1}(\mathbf{u}) + K_{F2}(h)$, and then $\mathbf{u}$ and h are the unique solutions of problem (9.58). Thus, we define operators $\mathscr{S}_{F0}(\lambda)$ and $\mathscr{T}_{F0}(\lambda)$ by

$$\begin{aligned} \mathscr{S}_{F0}(\lambda)(F_1, F_2, F_3, F_4) &= \mathscr{G}_{F2}(\lambda)(F_1, \mathbf{n}_0 \cdot F_3, \mathbf{n}_0 \cdot F_4) + \mathscr{S}_{F1}(\lambda)(F_1, F_2, F_3, F_4) \\ &\quad + \mathscr{S}_{F1}(\lambda)(\mathscr{F}^1(\lambda)(F_1, F_3, F_4)) \\ \mathscr{T}_{F0}(\lambda)(F_1, F_2, F_3, F_4) &= \mathscr{T}_{F1}(\lambda)(F_1, F_2, F_3, F_4) + \mathscr{T}_{F1}(\lambda)(\mathscr{F}^1(\lambda)(F_1, F_3, F_4)) \end{aligned}$$

with

$$\mathcal{F}^1(\lambda)F' = (\mathcal{F}_1(\lambda)F', \mathcal{F}_2(\lambda)F', \lambda^{1/2}\mathcal{F}_3(\lambda)F', \mathcal{F}_3(\lambda)F') \ (F' = (F_1, F_3, F_4)),$$
$$\mathcal{F}_1(\lambda)F' = -\lambda\mathcal{G}_{F2}(\lambda)F'' + \mathrm{Div}\,(\mu\mathbf{D}(\mathcal{G}_{F2}(\lambda)F'')) \quad (F'' = (F_1, \mathbf{n}_0\cdot F_3, \mathbf{n}_0\cdot F_4)),$$
$$\mathcal{F}_2(\lambda)F' = \mathbf{n}_0\cdot\mathcal{G}_{F2}(\lambda)F'', \quad \mathcal{F}_3(\lambda)F' = -\mu\mathbf{D}(\mathcal{G}_{F2}(\lambda)F'').$$

By Theorem 9.3.9, Lemmas 9.3.10 and 9.3.4, operators $\mathcal{S}_{F0}(\lambda)$ and $\mathcal{T}_{F0}(\lambda)$ satisfy the required properties in Theorem 9.3.8, which completes the proof of Theorem 9.3.8.

9.3.3 Reduced Stokes Equations in $\mathbb{R}^N_+$ with Non-slip Boundary Condition

In this subsection, let

$$DI_{D,q}(\mathbb{R}^N_+) = \{g \in W^1_q(\mathbb{R}^N_+) \mid \text{there exists a } G \in L_q(\mathbb{R}^N_+)^N \text{ such that } (g,\varphi)_{\mathbb{R}^N_+} = -(G, \nabla\varphi)_{\mathbb{R}^N_+} \text{ for any } \varphi \in W^1_{q'}(\mathbb{R}^N_+)\}.$$

Let $\mathcal{G}_D(g) = \{H \in L_q(\mathbb{R}^N_+)^N \mid \mathrm{div}\, G = \mathrm{div}\, H\}$ and $[\mathcal{G}_D(g)]$ denotes the representative elements of the set $\mathcal{G}_D(g)$. But $[\mathcal{G}_D(g)]$ is also written by $\mathcal{G}_D(g)$ for simplicity unless confusion may occur. We see that $\mathrm{div}\,\mathcal{G}_D(g) = g$ in $\mathbb{R}^N_+$ and $\mathcal{G}_D(g)\cdot\mathbf{n}_0|_{x_N=0} = 0$. For $g \in DI_{D,q}(\mathbb{R}^N)$, let

$$\|g\|_{W^{-1}_q(\mathbb{R}^N_+)} = \inf\{\|H\|_{L_q(\mathbb{R}^N_+)} \mid H \in \mathcal{G}_D(g)\}. \tag{9.72}$$

We know that the weak Neumann problem:

$$(\nabla u, \nabla\varphi)_{\mathbb{R}^N_+} = (\mathbf{f}, \nabla\varphi)_{\mathbb{R}^N_+} \quad \text{for any } \varphi \in \hat{W}^1_{q'}(\mathbb{R}^N)$$

is uniquely solvable for any $\mathbf{f} \in L_q(\mathbb{R}^N)^N$ with $u \in \hat{W}^1_q(\mathbb{R}^N_+)$. For any $\mathbf{u} \in W^2_q(\mathbb{R}^N_+)$, let $K_D(\mathbf{u}) \in \hat{W}^1_q(\mathbb{R}^N_+)$ be a unique solution to the variational problem:

$$(\nabla K_D(\mathbf{u}), \nabla\varphi)_{\mathbb{R}^N_+} = (\mathrm{Div}\,(\mu\mathbf{D}(\mathbf{u})) - \nabla\mathrm{div}\,\mathbf{u}, \nabla\varphi)_{\mathbb{R}^N_+} \tag{9.73}$$

for any $\varphi \in \hat{W}^1_{q'}(\mathbb{R}^N_+)$. We have

$$\|\nabla K_D(\mathbf{u})\|_{L_q(\mathbb{R}^N_+)} \leqslant C\|\nabla^2\mathbf{u}\|_{L_q(\mathbb{R}^N_+)}. \tag{9.74}$$

In this subsection, we consider the following reduced Stokes equations:

$$\lambda\mathbf{u} - \mathrm{Div}\,(\mu\mathbf{D}(\mathbf{u}) - K_D(\mathbf{u})\mathbf{I}) = \mathbf{f} \quad \text{in } \mathbb{R}^N_+, \quad \mathbf{u} = 0 \quad \text{on } \mathbb{R}^N_0. \tag{9.75}$$

The purpose of this subsection is to prove.

Theorem 9.3.14 *Let $1 < q < \infty$ and $0 < \varepsilon < \pi/2$. Then, there exists an operator family $\mathscr{S}_{D0}(\lambda) \in \mathrm{Hol}\,(\Sigma_\varepsilon, \mathscr{L}(L_q(\mathbb{R}^N_+)^N, W^2_q(\mathbb{R}^N_+)^N))$ such that for any $\lambda \in \Sigma_\varepsilon$ and $\mathbf{f} \in L_q(\mathbb{R}^N_+)^N$, problem (9.75) admits a unique solution $\mathbf{u} = \mathscr{S}_{D0}(\lambda)\mathbf{f}$, and*

$$\mathscr{R}_{\mathscr{L}(L_q(\mathbb{R}^N_+)^N, W^{2-j}_q(\mathbb{R}^N_+)^N)}(\{(\tau\partial_\tau)^\ell(\lambda^{j/2}\mathscr{S}_{D0}(\lambda)) \mid \lambda \in \Sigma_{\varepsilon,\lambda_0}\}) \leqslant c_{\lambda_0}$$

for $\ell = 0, 1$, $j = 0, 1, 2$ and any $\lambda_0 > 0$.

In what follows, we prove Theorem 9.3.14. For this purpose, we consider the resolvent problem for the weak Laplace-Neumann operator:

$$(\lambda g, \varphi)_{\mathbb{R}^N_+} + (\nabla g, \nabla\varphi)_{\mathbb{R}^N_+} = (\mathbf{f}, \nabla\varphi)_{\mathbb{R}^N_+} \quad \text{for any } \varphi \in \hat{W}^1_{q'}(\mathbb{R}^N_+), \tag{9.76}$$

and then, according to what was pointed out in Sect. 9.2.1, a solution $\mathbf{u}$ of the equations:

$$\lambda\mathbf{u} - \mathrm{Div}\,(\mu\mathbf{D}(\mathbf{u}) - \mathfrak{p}\mathbf{I}) = \mathbf{f}, \quad \mathrm{div}\,\mathbf{u} = g \quad \text{in } \mathbb{R}^N_+, \quad \mathbf{u} = 0 \quad \text{on } \mathbb{R}^N_0, \tag{9.77}$$

is also a solution of (9.75). Thus, we start with the following lemma.

Lemma 9.3.15 *Let $1 < q < \infty$ and $0 < \varepsilon < \pi/2$. Then, we have the following assertions.*

(a) There exists an operator family $\mathscr{G}_{D1}(\lambda) \in \mathrm{Hol}\,(\Sigma_\varepsilon, \mathscr{L}(L_q(\mathbb{R}^N_+)^N, DI_{D,q}(\mathbb{R}^N_+)))$ such that for any $\lambda \in \Sigma_\varepsilon$ and $\mathbf{f} \in L_q(\mathbb{R}^N_+)^N$, problem (9.76) admits a unique solution $g = \mathscr{G}_{D1}(\lambda)\mathbf{f}$, and

$$\mathscr{R}_{\mathscr{L}(L_q(\mathbb{R}^N_+)^N, W^{1-j}_q(\mathbb{R}^N_+))}(\{(\tau\partial_\tau)^\ell(\lambda^{j/2}\mathscr{G}_{D1}(\lambda)) \mid \lambda \in \Sigma_{\varepsilon,\lambda_0}\}) \leqslant c_{\lambda_0}$$

for $\ell = 0, 1$, $j = 0, 1, 2$ and any $\lambda_0 > 0$.

(b) Let g be the function given in (a). We consider the divergence equation:

$$\mathrm{div}\,\mathbf{v} = g \quad in\ \mathbb{R}^N_+, \quad \mathbf{n}_0 \cdot \mathbf{v} = 0 \quad on\ \mathbb{R}^N_0. \tag{9.78}$$

Then, there exists an operator family $\mathscr{G}_{D2}(\lambda) \in \mathrm{Hol}\,(\Sigma_\varepsilon, \mathscr{L}(L_q(\mathbb{R}^N_+)^N, W^2_q(\mathbb{R}^N_+)^N))$ such that for any $\lambda \in \Sigma_\varepsilon$ and $\mathbf{f} \in L_q(\mathbb{R}^N_+)^N$, problem (9.78) admits a solution $\mathbf{v} = \mathscr{G}_{D2}(\lambda)\mathbf{f}$, and

$$\mathscr{R}_{\mathscr{L}(L_q(\mathbb{R}^N_+)^N, W^{2-j}_q(\mathbb{R}^N_+)^N)}(\{(\tau\partial_\tau)^\ell(\lambda^{j/2}\mathscr{G}_{D2}(\lambda)) \mid \lambda \in \Sigma_{\varepsilon,\lambda_0}\}) \leqslant c_{\lambda_0}$$

for $\ell = 0, 1$, $j = 0, 1, 2$ and any $\lambda_0 > 0$.

Proof (a) Since $C^\infty_0(\mathbb{R}^N_+)$ is dense in $L_q(\mathbb{R}^N_+)$, we may assume that $\mathbf{f} \in C^\infty_0(\mathbb{R}^N_+)^N$. Instead of (9.76), we consider a solution of the equations of the strong form:

$$\lambda g - \Delta g = -\mathrm{div}\,\mathbf{f} \quad \text{in } \mathbb{R}^N_+, \quad \partial_N g = 0 \quad \text{on } \mathbb{R}^N_0. \tag{9.79}$$

Let $g = \mathscr{G}_{D1}(\lambda)\mathbf{f}$ with

$$\mathscr{G}_{D1}(\lambda)\mathbf{f} = -\mathscr{F}^{-1}\Big[\frac{\mathscr{F}[(\mathrm{div}\,\mathbf{f})^e](\xi)}{\lambda + |\xi|^2}\Big](x).$$

Since $(\mathrm{div}\,\mathbf{f})^e = \sum_{j=1}^{N-1} \partial_j f_j^e + \partial_N f_N^o$, we have

$$\begin{aligned}\mathscr{G}_{D1}(\lambda)\mathbf{f} &= -\mathscr{F}^{-1}\Big[\frac{\sum_{j=1}^{N-1} i\xi_j \mathscr{F}[f_j^e](\xi) + i\xi_N \mathscr{F}[f_N^o](\xi)}{\lambda + |\xi|^2}\Big] \\ &= -\sum_{j=1}^{N-1} \frac{\partial}{\partial x_j}\mathscr{B}_0(\lambda) f_j^e - \frac{\partial}{\partial x_N}\mathscr{B}_0(\lambda) f_N^o,\end{aligned} \tag{9.80}$$

where $\mathscr{B}_0(\lambda)$ is the operator defined in (9.50).

Moreover, $\partial_N g|_{x_N=0} = 0$ and $\mathscr{B}_0(\lambda) f_N^o|_{x_N=0} = 0$. In fact,

$$\begin{aligned}&\mathscr{F}'[\partial_N g](\xi', 0) \\ &= -\frac{i}{2\pi}\int_{-\infty}^{\infty} \frac{\xi_N}{\lambda + |\xi|^2}\,d\xi_N \int_0^{\infty} (e^{-iy_N\xi_N} + e^{iy_N\xi_N})\mathscr{F}'[(\mathrm{div}\,\mathbf{f})](\xi', y_N)\,dy_N \\ &= -i\int_0^{\infty} \mathscr{F}'[\mathrm{div}\,\mathbf{f}](\xi', y_N)\Big(\frac{1}{2\pi}\int_{-\infty}^{\infty} \frac{\xi_N(e^{-iy_N\xi_N} + e^{iy_N\xi_N})}{\lambda + |\xi|^2}\,d\xi_N\Big)\,dy_N; \\ &\mathscr{F}'[\mathscr{B}_0(\lambda) f_N^o](\xi', 0) \\ &= -\frac{i}{2\pi}\int_{-\infty}^{\infty} \frac{1}{\lambda + |\xi|^2}\,d\xi_N \int_0^{\infty} (e^{-iy_N\xi_N} - e^{iy_N\xi_N})\mathscr{F}'[f_N](\xi', y_N)\,dy_N \\ &= -i\int_0^{\infty} \mathscr{F}'[f_N](\xi', y_N)\Big(\frac{1}{2\pi}\int_{-\infty}^{\infty} \frac{e^{-iy_N\xi_N} - e^{iy_N\xi_N}}{\lambda + |\xi|^2}\,d\xi_N\Big)\,dy_N.\end{aligned}$$

Thus, by (9.64), we have $\partial_N g|_{x_N=0} = 0$ and $\mathscr{B}_0(\lambda) f_N^o|_{x_N=0} = 0$. Since $(\mathrm{div}\,\mathbf{f})^e = \mathrm{div}\,\mathbf{f}$ in $\mathbb{R}^N_+$, g satisfies (9.79).

By (9.51), we have

$$\begin{aligned}&\mathscr{G}_{D1}(\lambda) \in \mathrm{Hol}\,(\Sigma_\varepsilon, \mathscr{L}(L_q(\mathbb{R}^N_+), W^1_q(\mathbb{R}^N_+))), \\ &\mathscr{R}_{\mathscr{L}(L_q(\mathbb{R}^N_+)^N, W^{1-j}_q(\mathbb{R}^N_+))}(\{(\tau\partial_\tau)^\ell(\lambda^{j/2}\mathscr{G}_{D1}(\lambda) \mid \lambda \in \Sigma_{\varepsilon,\lambda_0}\}) \leqslant c_{\lambda_0}\end{aligned}$$

for $\ell = 0, 1$, $j = 0, 1, 2$ and any $\lambda_0 > 0$. This completes the proof of the assertion (a).

(b) Let $g = \mathscr{G}_{D1}(\lambda)\mathbf{f}$ be a function constructed in the proof of (a) above. Noting that g is defined in $\mathbb{R}^N$, we define $\mathbf{v}(x) = {}^\top(v_1(x), \ldots, v_N(x))$ with

$$v_j(x) = -\mathscr{F}^{-1}\Big[\frac{i\xi_j\mathscr{F}[g](\xi)}{|\xi|^2}\Big](x).$$

And then, inserting the formula (9.80), we have $v_j = \mathscr{G}_{D2j}(\lambda)\mathbf{f}$ with

$$\mathscr{G}_{D2j}(\lambda)\mathbf{f} = -\mathscr{F}^{-1}\Big[\frac{\sum_{k-1}^{N-1}\xi_j\xi_k\mathscr{F}[f_j^e](\xi) + \xi_j\xi_N\mathscr{F}[f_N^o](\xi)}{|\xi|^2(\lambda+|\xi|^2)}\Big].$$

Thus, letting $\mathscr{G}_{D2}(\lambda)\mathbf{f} = {}^\top(\mathscr{G}_{D21}(\lambda)\mathbf{f}, \ldots, \mathscr{F}_{D2N}(\lambda)\mathbf{f})$, we see that $\mathscr{G}_{D2}(\lambda)$ has the required properties in the assertion (b). □

Next, we consider the Stokes equations:

$$\lambda\mathbf{w} - \mathrm{Div}\,(\mu\mathbf{D}(\mathbf{w}) - \mathfrak{p}\mathbf{I}) = \mathbf{f}, \quad \mathrm{div}\,\mathbf{w} = 0 \quad \text{in } \mathbb{R}^N_+, \quad \mathbf{w} = 0 \quad \text{on } \mathbb{R}^N_0, \tag{9.81}$$

and we prove the following theorem.

Theorem 9.3.16 *Let $1 < q < \infty$ and $0 < \varepsilon < \pi/2$. Then, there exists an operator family $\mathscr{S}_{D1}(\lambda)$ and $\mathscr{P}_{D1}(\lambda)$ with*

$$\mathscr{S}_{D1}(\lambda) \in \mathrm{Hol}\,(\Sigma_\varepsilon, \mathscr{L}(L_q(\mathbb{R}^N_+)^N, W^2_q(\mathbb{R}^N_+)^N)),$$
$$\mathscr{P}_{D1}(\lambda) \in \mathrm{Hol}\,(\Sigma_\varepsilon, \mathscr{L}(L_q(\mathbb{R}^N_+)^N, W^1_q(\mathbb{R}^N_+) + \mathscr{W}^1_q(\mathbb{R}^N_+))),$$

such that for any $\lambda \in \Sigma_\varepsilon$ and $\mathbf{f} \in L_q(\mathbb{R}^N_+)^N$, problem (9.81) *admits unique solutions $\mathbf{w} = \mathscr{S}_{D1}(\lambda)\mathbf{f}$ and $\mathfrak{p} = \mathscr{P}_{D1}(\lambda)\mathbf{f}$, and*

$$\mathscr{R}_{\mathscr{L}(L_q(\mathbb{R}^N_+)^N, W^{2-j}_q(\mathbb{R}^N_+)^N)}(\{(\tau\partial_\tau)^\ell(\lambda^{j/2}\mathscr{S}_{D1}(\lambda)) \mid \lambda \in \Sigma_{\varepsilon,\lambda_0}\}) \leqslant c_{\lambda_0},$$
$$\mathscr{R}_{\mathscr{L}(L_q(\mathbb{R}^N_+)^N)}(\{(\tau\partial_\tau)^\ell(\nabla\mathscr{P}_{D1}(\lambda)) \mid \lambda \in \Sigma_{\varepsilon,\lambda_0}\}) \leqslant c_{\lambda_0}$$

for $\ell = 0, 1$, $j = 0, 1, 2$ and any $\lambda_0 > 0$.

Proof Since $\mathrm{Div}\,(\mathbf{D}(\mathbf{w})) = \Delta\mathbf{w}$ if $\mathrm{div}\,\mathbf{w} = 0$, the Eqs. (9.81) are rewritten as

$$\lambda\mathbf{w} - \mu\Delta\mathbf{w} + \nabla\mathfrak{p} = \mathbf{f}, \quad \mathrm{div}\,\mathbf{w} = 0 \quad \text{in } \mathbb{R}^N_+, \quad \mathbf{w} = 0 \quad \text{on } \mathbb{R}^N_0. \tag{9.82}$$

Let $\mathbf{w}_1 = \mathscr{S}_{D2}(\lambda)\mathbf{f} = {}^\top(\mathscr{S}_{D21}(\lambda)\mathbf{f}, \ldots, \mathscr{S}_{D2N}(\lambda)\mathbf{f})$ and $\mathfrak{p}_1 = \mathscr{P}_{D2}\mathbf{f}$ with

$$\mathscr{S}_{D2j}(\lambda)\mathbf{f} = \mathscr{F}^{-1}\Big[\frac{\mathscr{F}[\tilde{\mathbf{f}}](\xi) - |\xi|^{-2}\xi_j\xi\cdot\mathscr{F}[\tilde{\mathbf{f}}](\xi)}{\lambda+\mu|\xi|^2}\Big](x),$$
$$\mathscr{P}_{D2}\mathbf{f} = \mathscr{F}\Big[\frac{\xi\cdot\mathscr{F}[\tilde{\mathbf{f}}](\xi)}{|\xi|^2}\Big](x),$$

where $\tilde{\mathbf{f}}$ has been defined by $\tilde{\mathbf{f}} = {}^\top(f_1^e, \ldots, f_{N-1}^e, f_N^o))$ for $\mathbf{f} = {}^\top(f_1, \ldots, f_N)$. We have $w_N = \mathscr{S}_{D2N}(\lambda)\mathbf{f} = 0$ on $\mathbb{R}^N_0$. In fact, we observe that

$$
\begin{aligned}
\mathscr{F}'[\mathscr{S}_{D2N}(\lambda)\mathbf{f}](\xi', 0) &= \int_0^\infty \hat{f}_N(\xi', y_N)\Big(\frac{1}{2\pi}\int_{-\infty}^\infty \frac{e^{-iy_N\xi_N} - e^{iy_N\xi_N}}{\lambda + \mu|\xi|^2}\,d\xi_N\Big)\,dy_N \\
&- \sum_{k=1}^{N-1}\int_0^\infty \xi_k\hat{f}_k(\xi', y_N)\Big(\frac{1}{2\pi}\int_{-\infty}^\infty \frac{\xi_N(e^{-iy_N\xi_N} + e^{i\xi_N y_N})}{(\lambda + \mu|\xi|^2)|\xi|^2})\,d\xi_N\Big)\,dy_N \\
&- \int_0^\infty \hat{f}_N(\xi', y_N)\Big(\frac{1}{2\pi}\int_{-\infty}^\infty \frac{\xi_N^2(e^{-iy_N\xi_N} - e^{i\xi_N y_N})}{(\lambda + \mu|\xi|^2)|\xi|^2}\,d\xi_N\Big)\,dy_N
\end{aligned}
$$

By (9.64),

$$
\int_0^\infty \hat{f}_N(\xi', y_N)\Big(\frac{1}{2\pi}\int_{-\infty}^\infty \frac{e^{-iy_N\xi_N} - e^{iy_N\xi_N}}{\lambda + \mu|\xi|^2}\,d\xi_N\Big)\,dy_N = 0.
$$

Moreover, by the residue theorem in the theory of one complex variable, we have

$$
\begin{aligned}
\frac{1}{2\pi i}\int_{-\infty}^\infty \frac{e^{ia\xi_N}}{|\xi|^2(\lambda + \mu|\xi|^2)}\,d\xi_N &= \mu^{-1}\frac{e^{-|a|A}}{\tilde{B}^2 - A^2} + \frac{e^{-|a|B}}{A^2 - \tilde{B}^2}, \\
\frac{1}{2\pi i}\int_{-\infty}^\infty \frac{e^{ia\xi_N}\xi_N}{|\xi|^2(\lambda + \mu|\xi|^2)}\,d\xi_N &= \mu^{-1}\text{sign}\,(a)\Big[\frac{iAe^{-|a|A}}{\tilde{B}^2 - A^2} + \frac{iBe^{-|a|B}}{A^2 - \tilde{B}^2}\Big], \\
\frac{1}{2\pi i}\int_{-\infty}^\infty \frac{e^{ia\xi_N}\xi_N^2}{|\xi|^2(\lambda + \mu|\xi|^2)}\,d\xi_N &= -\mu^{-1}\Big[\frac{A^2e^{-|a|A}}{\tilde{B}^2 - A^2} + \frac{B^2e^{-|a|B}}{A^2 - \tilde{B}^2}\Big]
\end{aligned}
$$

for any $a \in \mathbb{R} \setminus \{0\}$, where we have set $\tilde{B} = \sqrt{\lambda\mu^{-1} + |\xi'|^2}$ with $\mathrm{Re}\,\tilde{B} > 0$, so that we have $\mathbf{w}_1 \cdot \mathbf{n}_0|_{x_N=0} = \mathscr{S}_{D2N}(\lambda)\mathbf{f}|_{x_N=0} = 0$.

Moreover, by Lemmas 9.3.3 and 9.3.5, we have

$$
\begin{aligned}
&\mathscr{S}_{D2}(\lambda) \in \mathrm{Hol}\,(\Sigma_\varepsilon, \mathscr{L}(L_q(\mathbb{R}^N_+)^N, W^2_q(\mathbb{R}^N_+)^N)), \\
&\mathscr{P}_{D2} \in \mathscr{L}(L_q(\mathbb{R}^N_+)^N, \mathscr{W}^1_q(\mathbb{R}^N_+)), \\
&\mathscr{R}_{\mathscr{L}(L_q(\mathbb{R}^N_+)^N, W^{2-j}_q(\mathbb{R}^N_+)^N)}(\{(\tau\partial_\tau)^\ell(\lambda^{j/2}\mathscr{S}_{D2}(\lambda)) \mid \lambda \in \Sigma_{\varepsilon,\lambda_0}\}) \leqslant c_{\lambda_0}, \\
&\mathscr{R}_{\mathscr{L}(L_q(\mathbb{R}^N_+)^N)}(\{(\tau\partial_\tau)^\ell(\nabla\mathscr{P}_{D2}) \mid \lambda \in \Sigma_{\varepsilon,\lambda_0}\}) \leqslant c_{\lambda_0},
\end{aligned}
$$

for $\ell = 0, 1$, $j = 0, 1, 2$ and any $\lambda_0 > 0$.

Next, we consider the equations:

$$
\begin{cases}
\lambda\mathbf{w}_2 - \mathrm{Div}\,(\mu\mathbf{D}(\mathbf{w}_2) - \mathfrak{p}_2\mathbf{I}) = 0, \quad \mathrm{div}\,\mathbf{w}_2 = 0 & \text{in } \mathbb{R}^N_+, \\
w_{2j} = h_j \ \ (j = 1, \ldots, N-1), \quad w_{2N} = 0 & \text{on } \mathbb{R}^N_0
\end{cases}
\tag{9.83}
$$

with $\mathbf{w}_2 = {}^\top(w_{21}, \ldots, w_{2N})$. Recalling that $\mathrm{Div}\,(\mu\mathbf{D}(\mathbf{w}_2) - \mathfrak{p}_2) = \mu\Delta\mathbf{w}_2 - \nabla\mathfrak{p}_2$ when $\mathrm{div}\,\mathbf{w}_2 = 0$ and applying the partial Fourier transform, we have

$$\begin{cases} \partial_N^2 \hat w_{2j} - \tilde B^2 \hat w_{2j} + \mu^{-1} i\xi_j \hat{\mathfrak p}_2 = 0 \quad (j = 1, \ldots, N-1) & \text{for } x_N > 0, \\ \partial_N^2 \hat w_{2N} - \tilde B^2 \hat w_{2N} + \mu^{-1}\partial_N \hat{\mathfrak p}_2 = 0, \quad \sum_{j=1}^{N-1} i\xi_j \hat w_{2j} + \partial_N \hat w_{2N} = 0 & \text{for } x_N > 0, \\ \hat w_{2j} = \hat h_j \quad (j = 1, \ldots, N-1), \quad \hat w_{2N} = 0 & \text{for } x_N = 0. \end{cases} \tag{9.84}$$

For unknown complex numbers α_j, β_j and γ, we set

$$\hat w_{2j} = \alpha_j(e^{-\tilde B x_N} - e^{-Ax_N}) + \beta_j e^{-\tilde B x_N} (j = 1, \ldots, N), \quad \hat{\mathfrak p}_2 = \gamma e^{-Ax_N},$$

and inserting these formulas into (9.84) yields

$$\begin{aligned} &-(A^2 - \tilde B^2)\alpha_j + \mu^{-1} i\xi_j \gamma = 0 \quad (j = 1, \ldots, N-1), \\ &-(A^2 - \tilde B^2)\alpha_N - \mu^{-1} A\gamma = 0, \\ &i\sum_{j=1}^{N-1} \xi_j(\alpha_j + \beta_j) - \tilde B(\alpha_N + \beta_N) = 0, \quad i\sum_{j=1}^{N-1} \xi_j \alpha_j - A\alpha_N = 0, \\ &\beta_j = \hat h_j \quad (j = 1, \ldots, N-1), \quad \beta_N = 0. \end{aligned}$$

Therefore, we have

$$\begin{aligned} \hat w_{2j}(\xi', x_N) &= \tilde{\mathcal M}(x_N)\frac{i\xi_j}{A}\sum_{k=1}^{N-1} \xi_k \hat h_k(\xi', 0) + e^{-\tilde B x_N}\hat h_j(\xi', 0) \quad (j = 1, \ldots, N-1), \\ \hat w_{2N}(\xi', x_N) &= \tilde{\mathcal M}(x_N)\sum_{k=1}^{N-1} \xi_k \hat h_k(\xi', 0), \quad \tilde{\mathcal M}(x_N) = \frac{e^{-\tilde B x_N} - e^{-Ax_N}}{\tilde B - A}, \\ \hat{\mathfrak p}_2(\xi', x_N) &= -\mu(A + \tilde B)\sum_{k=1}^{N-1} \frac{\xi_k}{A}\hat h_k(\xi', 0). \end{aligned}$$

Thus, by the Volevich trick and the identity: $1 = \frac{\lambda\mu^{-1} + A^2}{\tilde B^2}$, we have

$$\begin{aligned} &w_{2j}(x) \\ &= -\sum_{k=1}^{N-1}\int_0^\infty \mathcal F^{-1}_{\xi'}\Big[\frac{i\xi_j\xi_k}{\mu A^2\tilde B^2}\lambda^{\frac12} A\tilde{\mathcal M}(x_N + y_N)\mathcal F'[\lambda^{\frac12}\partial_N h_k](\xi', y_N)\Big](x')\,dy_N \\ &\quad + i\sum_{k=1}^{N-1}\int_0^\infty \mathcal F^{-1}_{\xi'}\Big[\frac{i\xi_j}{A\tilde B^2}A^2\tilde{\mathcal M}(x_N + y_N)\mathcal F'[\partial_k\partial_N h_k](\xi', y_N)\Big](x')\,dy_N \\ &\quad + \sum_{k=1}^{N-1}\int_0^\infty \mathcal F^{-1}_{\xi'}\Big[\frac{i\xi_j\xi_k}{A^2\tilde B^2}Ae^{-\tilde B(x_N + y_N)}\mathcal F'[(\lambda\mu^{-1} - \Delta')h_k](\xi', y_N)\Big](x')\,dy_N \\ &\quad + \sum_{k=1}^{N-1}\int_0^\infty \mathcal F^{-1}_{\xi'}\Big[\frac{i\xi_j\xi_k}{A^2\tilde B^2}A^2\tilde{\mathcal M}(x_N + y_N)\mathcal F'[(\lambda\mu^{-1} - \Delta')h_k](\xi', y_N)\Big](x')\,dy_N \end{aligned}$$

$$
\begin{aligned}
&- \int_0^\infty \mathscr{F}_{\xi'}^{-1}\Big[\frac{1}{\mu \tilde{B}^2}\lambda^{\frac12} e^{-\tilde{B}(x_N+y_N)}\mathscr{F}'[\lambda^{\frac12}\partial_N h_j](\xi', y_N)\Big](x')\,dy_N \\
&+ \sum_{k=1}^{N-1}\int_0^\infty \mathscr{F}_{\xi'}^{-1}\Big[\frac{i\xi_k}{A\tilde{B}^2} A e^{-\tilde{B}(x_N+y_N)}\mathscr{F}'[\partial_k\partial_N h_j](\xi', y_N)\Big](x')\,dy_N \\
&+ \int_0^\infty \mathscr{F}_{\xi'}^{-1}\Big[\frac{\lambda^{\frac12}}{\mu \tilde{B}^3}\lambda^{\frac12} e^{-\tilde{B}(x_N+y_N)}\mathscr{F}'[(\lambda\mu^{-1}-\Delta')h_j](\xi', y_N)\Big](x')\,dy_N \\
&+ \int_0^\infty \mathscr{F}_{\xi'}^{-1}\Big[\frac{A}{\mu \tilde{B}^3} A e^{-\tilde{B}(x_N+y_N)}\mathscr{F}'[(\lambda\mu^{-1}-\Delta')h_j](\xi', y_N)\Big](x')\,dy_N
\end{aligned}
$$

$$
\begin{aligned}
&w_{2N}(x) \\
&= - \sum_{k=1}^{N-1}\int_0^\infty \mathscr{F}_{\xi'}^{-1}\Big[\frac{\xi_k}{A\tilde{B}^2}\lambda^{\frac12} A\tilde{\mathscr{M}}(x_N+y_N)\mathscr{F}'[\lambda^{\frac12}\partial_N h_k](\xi', y_N)\Big](x')\,dy_N \\
&+ i\sum_{k=1}^{N-1}\int_0^\infty \mathscr{F}_{\xi'}^{-1}\Big[\frac{1}{\tilde{B}^2} A^2\tilde{\mathscr{M}}(x_N+y_N)\mathscr{F}'[\partial_k\partial_N h_k](\xi', y_N)\Big](x')\,dy_N \\
&+ \sum_{k=1}^{N-1}\int_0^\infty \mathscr{F}_{\xi'}^{-1}\Big[\frac{\xi_k}{A\tilde{B}^2} A e^{-\tilde{B}x_N}\mathscr{F}'[(\lambda\mu^{-1}-\Delta')h_k](\xi', y_N)\Big](x')\,dy_N \\
&+ \sum_{k=1}^{N-1}\int_0^\infty \mathscr{F}_{\xi'}^{-1}\Big[\frac{\xi_k}{A\tilde{B}^2} A^2\tilde{\mathscr{M}}(x_N+y_N)\mathscr{F}'[(\lambda\mu^{-1}-\Delta')h_k](\xi', y_N)\Big](x')\,dy_N
\end{aligned}
$$

$$
\begin{aligned}
&\mathfrak{p}_2(x_N) \\
&= i\sum_{k=1}^{N-1}\int_0^\infty \mathscr{F}_{\xi'}^{-1}\Big[\mu e^{-A(x_N+y_N)}\mathscr{F}'[\partial_k\partial_N h_k](\xi', y_N)\Big](x')\,dy_n \\
&- \sum_{k=1}^{N-1}\int_0^\infty \mathscr{F}_{\xi'}^{-1}\Big[\frac{\xi_k\lambda^{\frac12}}{A\tilde{B}} e^{-A(x_N+y_N)}\mathscr{F}'[\lambda^{\frac12}\partial_N h_k](\xi', y_N)\Big](x')\,dy_n \\
&+ i\sum_{k=1}^{N-1}\int_0^\infty \mathscr{F}_{\xi'}^{-1}\Big[\frac{\mu A}{\tilde{B}} e^{-A(x_N+y_N)}\mathscr{F}'[\partial_k\partial_N h_k](\xi', y_N)\Big](x')\,dy_n \\
&- \sum_{k,\ell=1}^{N-1}\int_0^\infty \mathscr{F}_{\xi'}^{-1}\Big[\frac{\mu\xi_\ell}{A} e^{-A(x_N+y_N)}\mathscr{F}'[\partial_k\partial_\ell h_k](\xi', y_N)\Big](x')\,dy_n \\
&+ \sum_{k=1}^{N-1}\int_0^\infty \mathscr{F}_{\xi'}^{-1}\Big[\frac{\lambda^{\frac12}}{\tilde{B}} e^{-A(x_N+y_N)}\mathscr{F}'[\lambda^{\frac12}\partial_k h_k](\xi', y_N)\Big](x')\,dy_n \\
&- \sum_{k,\ell=1}^{N-1}\int_0^\infty \mathscr{F}_{\xi'}^{-1}\Big[\frac{\mu\xi_\ell}{\tilde{B}} e^{-A(x_N+y_N)}\mathscr{F}'[\partial_k\partial_\ell h_k](\xi', y_N)\Big](x')\,dy_n \qquad (9.85)
\end{aligned}
$$

Let $\mathscr{S}_{D3j}(\lambda)$, $\mathscr{S}_{D3N}(\lambda)$ and $\mathscr{P}_{D3}(\lambda)$ be operators acting on (F_{10}, F_{11}, F_{12}) with

$$F_{10} = (F_{10,1}, \ldots, F_{10,N-1}) \in L_q(\mathbb{R}^N_+)^{N-1}, \quad F_{11} = (F_{11,1}, \ldots F_{11,N-1}) \in W^1_q(\mathbb{R}^N_+)^{N-1},$$
$$F_{12} = (F_{12,11}, \ldots, F_{12,N-1}) \in W^2_q(\mathbb{R}^N_+)^{N-1}$$

defined by replacing $\lambda^{1/2}\partial_N h_k$ by $\partial_N F_{11.k}$, $\partial_k\partial_N h_k$ by $\partial_k\partial_N F_{12,k}$, λh_k by $F_{10,k}$, and $\Delta' h_k$ by $\Delta' F_{12,k}$ in (9.85), respectively. Let $\mathscr{S}_{D3}(\lambda) = {}^\top(\mathscr{S}_{D31}(\lambda), \ldots, \mathscr{S}_{D3N}(\lambda))$ and then, for any $\mathbf{h}' = (h_1, \ldots, h_{N-1}) \in W^2_q(\mathbb{R}^N_+)^{N-1}$ and $\lambda \in \Sigma_\varepsilon$, $\mathbf{w}_2 = \mathscr{S}_{D3}(\lambda) F_{D\lambda}\mathbf{h}'$ and $\mathfrak{p}_2 = \mathscr{P}_{D3}(\lambda) F_{D\lambda}\mathbf{h}'$ with $F_{D\lambda}\mathbf{h}' = (\lambda\mathbf{h}', \lambda^{1/2}\mathbf{h}', \mathbf{h}')$ are unique solutions of problem (9.83). Moreover, let

$$\begin{aligned}&\mathscr{X}^2_q(\mathbb{R}^N_+)\\ &= \{(F_{10}, F_{11}, F_{12}) \mid F_{10} \in L_q(\mathbb{R}^N_+)^{N-1}, \ F_{11} \in W^1_q(\mathbb{R}^N_+)^{N-1}, \ F_{12} \in W^2_q(\mathbb{R}^N_+)^{N-1}\},\end{aligned}$$

and then, we have

$$\begin{aligned}&\mathscr{R}_{\mathscr{L}(\mathscr{X}^2_q(\mathbb{R}^N_+), W^{2-j}_q(\mathbb{R}^N_+)^N)}(\{(\tau\partial_\tau)^\ell(\lambda^{j/2}\mathscr{S}_{D3}(\lambda)) \mid \lambda \in \Sigma_{\varepsilon,\lambda_0}\}) \leqslant c_{\lambda_0},\\ &\quad\mathscr{R}_{\mathscr{L}(\mathscr{X}^2_q(\mathbb{R}^N_+), L_q(\mathbb{R}^N_+)^N)}(\{(\tau\partial_\tau)^\ell(\nabla\mathscr{P}_{D3}(\lambda)) \mid \lambda \in \Sigma_{\varepsilon,\lambda_0}\}) \leqslant c_{\lambda_0},\end{aligned} \tag{9.86}$$

for $\ell = 0, 1$, $j = 0, 1, 2$ and $\lambda_0 > 0$. To prove (9.86), we use Lemma 9.3.13 and the following lemma due to Shibata and Shimizu [18].

Lemma 9.3.17 *Let* $1 < q < \infty$, $0 < \varepsilon < \pi/2$ *and* $\lambda_0 > 0$. *Given* $\ell_1(\xi', \lambda) \in \mathbb{M}_{-2,2}(\Sigma_{\varepsilon,\lambda_0})$, *we define the operators* $L_i(\lambda)$ $(i = 3, 4, 5)$ *by*

$$\begin{aligned}[L_3(\lambda)f](x) &= \int_0^\infty \mathscr{F}^{-1}_{\xi'}[\ell_1(\xi', \lambda)Ae^{-A(x_N+y_N)}\mathscr{F}'[f](\xi', y_N)](x')\,dy_N,\\ [L_4(\lambda)f](x) &= \int_0^\infty \mathscr{F}^{-1}_{\xi'}[\ell_1(\xi', \lambda)A^2\mathscr{M}(x_N + y_N)\mathscr{F}[f](\xi', y_N)](x')\,dy_N,\\ [L_5(\lambda)f](x) &= \int_0^\infty \mathscr{F}^{-1}_{\xi'}[\ell_1(\xi', \lambda)\lambda^{1/2}A\mathscr{M}(x_N + y_N)\mathscr{F}[f](\xi', y_N)](x')\,dy_N.\end{aligned}$$

Then,

$$\mathscr{R}_{\mathscr{L}(L_q(\mathbb{R}^N_+), W^{2-j}_q(\mathbb{R}^N_+))}\big(\{(\tau\partial_\tau)^\ell(\lambda^{j/2}L_i(\lambda)) \mid \lambda \in \Sigma_\vartheta\}\big) \leqslant \gamma$$

for $s = 0, 1$, $j = 0, 1, 2$, *and* $i = 3, 4, 5$ *with some positive constant* γ *depending on* λ_0, N, q, *and* ε.

If we set

$$\begin{aligned}\mathscr{S}_{D1}(\lambda)\mathbf{f} &= \mathscr{S}_{D2}(\lambda)\mathbf{f} - \mathscr{S}_{D3}(\lambda)(\lambda\mathscr{S}'_{D2}(\lambda)\mathbf{f}, \lambda^{\frac12}\mathscr{S}'_{D2}(\lambda)\mathbf{f}, \lambda\mathscr{S}'_{D2}(\lambda)\mathbf{f}),\\ \mathscr{P}_{D1}(\lambda)\mathbf{f} &= \mathscr{P}_{D2}\mathbf{f} - \mathscr{P}_{D3}(\lambda)(\lambda\mathscr{S}'_{D2}(\lambda)\mathbf{f}, \lambda^{\frac12}\mathscr{S}'_{D2}(\lambda)\mathbf{f}, \lambda\mathscr{S}'_{D2}(\lambda)\mathbf{f}),\end{aligned}$$

with $\mathscr{S}'_{D2}(\lambda)\mathbf{f} = (\mathscr{S}_{D21}(\lambda)\mathbf{f}, \ldots, \mathscr{S}_{D2N-1}(\lambda)\mathbf{f})$, then $\mathscr{S}_{D1}(\lambda)$ and $\mathscr{P}_{D1}(\lambda)$ satisfy the required properties in Theorem 9.3.16, which completes the proof of Theorem 9.3.16. □

9.4 On the $\mathscr{R}$ Bounded Solution Operators in a Bent Half-Space

9.4.1 Unit Outer Normal and Laplace-Beltrami Operator in a Bent-Half Space

In this section we consider (9.2) in a bent half-space. Let $\Phi : \mathbb{R}^N \to \mathbb{R}^N$ be a bijective map of C^1 class and let Φ^{-1} be its inverse map. We assume that $\nabla\Phi$ and $\nabla\Phi^{-1}$ have the forms: $\nabla\Phi(x) = \mathscr{A} + B(x)$ and $\nabla\Phi^{-1}(\xi) = \mathscr{A}_{-1} + B_{-1}(\xi)$, where $\mathscr{A}$ and $\mathscr{A}_{-1}$ are orthonormal matrices with constant coefficients and $B(x)$ and $B_{-1}(\xi)$ are matrices of functions in $W^2_r(\mathbb{R}^N)$ with $N < r < \infty$ such that

$$\|(B, B_{-1})\|_{L_\infty(\mathbb{R}^N)} \leqslant M_1, \quad \|\nabla(B, B_{-1})\|_{W^1_r(\mathbb{R}^N)} \leqslant M_2. \tag{9.87}$$

We will choose M_1 small enough eventually, so that we may assume that $0 < M_1 \leqslant 1 \leqslant M_2$. Let $\Omega_+ = \Phi(\mathbb{R}^N_+)$ and let $\Gamma_+ = \Phi(\mathbb{R}^N_0)$, which is the boundary of Ω_+. Let $\mathbf{n}_+$ be the unit outer normal to Γ_+. Setting $\Phi^{-1} = (\Phi_{-1,1}, \ldots, \Phi_{-1,N})$, we see that Γ_+ is represented by $\Phi_{-1,N}(\xi) = 0$, which furnishes that

$$\mathbf{n}_+(x) = \frac{(\nabla\Phi_{-1,N}) \circ \Phi(x)}{|\nabla\Phi_{-1,N}| \circ \Phi(x)} = \frac{(a_{N1} + b_{N1}(x), \ldots, a_{NN} + b_{NN}(x))}{(\sum_{j=1}^N (a_{Nj} + b_{Nj}(x))^2)^{1/2}}, \tag{9.88}$$

where we have set $\mathscr{A}_{-1} = (a_{ij})$ and $B_{-1} \circ \Phi(x) = (b_{ij}(x))$. The $\mathbf{n}_+$ is defined on $\mathbb{R}^N$ and

$$\|\mathbf{n}_+\|_{L_\infty(\mathbb{R}^N)} \leqslant C_N, \quad \|\nabla\mathbf{n}_+\|_{W^1_q(\mathbb{R}^N)} \leqslant C_{M_2}. \tag{9.89}$$

Let g_{ij} be the (i, j) component of the first fundamental matrix G on Γ_+, which is defined by

$$g_{ij} = \sum_{k=1}^N \frac{\partial\Phi_k(x', 0)}{\partial x_i}\frac{\partial\Phi_k(x', 0)}{\partial x_j} = \delta_{ij} + \tilde{g}_{ij}(x', 0) \tag{9.90}$$

with $\tilde{g}_{ij}(x) = \sum_{k=1}^N (b_{ki}(x)a_{kj} + a_{ki}b_{kj}(x) + b_{ki}(x)b_{kj}(x))$, where δ_{ij} is the Kronecker's delta symbol. By (9.87),

$$\|\tilde{g}_{ij}\|_{L_\infty(\mathbb{R}^N)} \leqslant CM_1, \quad \|\nabla\tilde{g}_{ij}\|_{W^1_r(\mathbb{R}^N)} \leqslant C_{M_2}. \tag{9.91}$$

Let $\mathfrak{g}_{\Gamma_+} = \sqrt{\det G}$ and $G^{-1} = (g^{ij}_{\Gamma_+})$. Choosing $M_1 > 0$ suitably small, by (9.91) we may write $\mathfrak{g}_{\Gamma_+} = 1 + \tilde{\mathfrak{g}}$ and $g^{ij}_{\Gamma_+} = \delta_{ij} + \tilde{g}^{ij}$ with

$$\|(\tilde{\mathfrak{g}}, \tilde{g}^{ij})\|_{L_\infty(\mathbb{R}^N)} \leqslant CM_1, \quad \|\nabla(\tilde{\mathfrak{g}}, \tilde{g}^{ij})\|_{W^1_r(\mathbb{R}^N)} \leqslant C_{M_2}. \tag{9.92}$$

The Laplace-Beltrami operator Δ_{Γ_+} is defined by

$$\Delta_{\Gamma_+} f(x') = \sum_{i,j=1}^{N-1} \frac{1}{\mathfrak{g}_{\Gamma_+}(x',0)} \frac{\partial}{\partial x_i}\Big(\mathfrak{g}_{\Gamma_+}(x',0) g^{ij}_{\Gamma_+}(x',0) \frac{\partial f(x')}{\partial x_j}\Big).$$

Let

$$\begin{aligned} \Delta' f &= \sum_{j=1}^{N-1} \frac{\partial^2 f}{\partial x_j^2}, \quad \mathscr{D}_+ f = \sum_{i,j=1}^{N-1} \tilde{g}^{ij}(x',0) \frac{\partial^2 f}{\partial x_i \partial x_j} + \sum_{j=1}^{N-1} \tilde{g}^j(x',0) \frac{\partial f}{\partial x_j}, \\ \tilde{g}^j &= \sum_{i=1}^{N-1} \Big(\frac{\partial \tilde{g}^{ij}}{\partial x_i} + \frac{g^{ij}_{\Gamma_+}}{\mathfrak{g}_{\Gamma_+}} \frac{\partial \tilde{\mathfrak{g}}}{\partial x_i}\Big), \end{aligned} \tag{9.93}$$

and then $\Delta_{\Gamma_+} f = \Delta' f + \mathscr{D}_+ f$. By (9.92) and the Sobolev imbedding theorem,

$$\|\tilde{g}^j\|_{W^1_r(\mathbb{R}^N)} \leqslant C_{M_2}. \tag{9.94}$$

9.4.2 Reduced Stokes Equations with Free Boundary Condition in a Bent-Half Space

First, we consider the generalized resolvent problem for the reduced Stokes equations with free boundary condition. For $\mathbf{v} \in W^2_q(\Omega_+)^N$, let $K^0_{+1}(\mathbf{v})$ be a unique solution to the variational problem:

$$(\nabla K^0_{+1}(\mathbf{v}), \nabla\varphi)_{\Omega_+} - (\mathrm{Div}\,(\mu\mathbf{D}(\mathbf{v})) - \nabla \mathrm{div}\,\mathbf{v}, \nabla\varphi)_{\Omega_+} \quad \text{for any } \varphi \in \hat{W}^1_{q',0}(\Omega_+)$$

subject to $K^0_{+1}(\mathbf{v}) =< \mu\mathbf{D}(\mathbf{v})\mathbf{n}_+, \mathbf{n}_+ > -\mathrm{div}\,\mathbf{v}$ on Γ_+, while for $h \in W^{3-1/q}_q(\Gamma_+)$, let $K^0_{+2}(h)$ be a unique solution to the variational problem:

$$(\nabla K^0_{+2}(h), \nabla\varphi)_{\Omega_+} = 0 \quad \text{for any } \varphi \in \hat{W}^1_{q',0}(\Omega_+)$$

subject to $K^0_{+2}(h) = -(\tau + \delta\Delta_{\Gamma_+})h$ on Γ_+. And then, we consider the generalized resolvent problem for the reduced Stokes operator with free boundary condition:

$$\begin{cases} \lambda\mathbf{v} - \mathrm{Div}\,(\mu\mathbf{D}(\mathbf{v}) - (K^0_{+1}(\mathbf{v}) + K^0_{+2}(h))\mathbf{I}) = \mathbf{g} & \text{in } \Omega_+, \\ \lambda h - \mathbf{n}_+ \cdot \mathbf{v} = g & \text{on } \Gamma_+, \\ (\mu\mathbf{D}(\mathbf{v}) - (K^0_{+1}(\mathbf{v}) + K^0_{+2}(h))\mathbf{I} - ((\tau + \delta\Delta_{\Gamma_+})h)\mathbf{I})\mathbf{n}_+ = \mathbf{g}_b & \text{on } \Gamma_+. \end{cases} \tag{9.95}$$

The purpose of this subsection is to prove.

Theorem 9.4.1 *Let $1 < q < \infty$ and $0 < \varepsilon < \pi/2$. Let $Y_q(\Omega_+)$ and $\mathscr{Y}_q(\Omega_+)$ be the spaces defined in (9.34) with $D = \Omega_+$. Then, there exist positive numbers $M_1 > 0$, $\lambda_0 \geqslant 1$, and operator families $\mathscr{S}_{F+}(\lambda)$ and $\mathscr{T}_{F+}(\lambda)$ with*

$$\mathscr{S}_{F+}(\lambda) \in \mathrm{Hol}\,(\Sigma_{\varepsilon,\lambda_0}, \mathscr{L}(\mathscr{Y}_q(\Omega_+), W^2_q(\Omega_+)^N)),$$
$$\mathscr{T}_{F+}(\lambda) \in \mathrm{Hol}\,(\Sigma_{\varepsilon,\lambda_0}, \mathscr{L}(\mathscr{Y}_q(\Omega_+), W^{3-1/q}_q(\Gamma_+)^N))$$

such that for any $\lambda \in \Sigma_{\varepsilon,\lambda_0}$ and $(\mathbf{g}, g, \mathbf{g}_b) \in Y_q(\Omega_+)$, problem (9.95) admits unique solutions $\mathbf{v} = \mathscr{S}_{F+}(\lambda)(\mathbf{g}, g, \lambda^{1/2}\mathbf{g}_b, \mathbf{g}_b)$ and $h = \mathscr{T}_{F+}(\lambda)(\mathbf{g}, g, \lambda^{1/2}\mathbf{g}_b, \mathbf{g}_b)$.

Moreover, we have

$$\mathscr{R}_{\mathscr{L}(\mathscr{Y}_q(\Omega_+), W^{2-j}_q(\Omega_+)^N)}(\{(\tau\partial_\tau)^\ell(\lambda^{j/2}\mathscr{S}_{F+}(\lambda)) \mid \lambda \in \Sigma_{\varepsilon,\lambda_0}\}) \leqslant \gamma_{+1},$$
$$\mathscr{R}_{\mathscr{L}(\mathscr{Y}_q(\Omega_+), W^{3-1/q-k}_q(\Gamma_+))}(\{(\tau\partial_\tau)^\ell(\lambda^k\mathscr{T}_{F+}(\lambda)) \mid \lambda \in \Sigma_{\varepsilon,\lambda_0}\}) \leqslant \gamma_{+1}$$

for $\ell = 0, 1$, $j = 0, 1, 2$, $k = 0, 1$ with some positive constant γ_{+1} depending on λ_0, M_2, N, q and ε.

In what follows, we prove Theorem 9.4.1. For this purpose, we use the symbols given in Sect. 9.4.1. By the change of variable: $\xi = \Phi(x)$, we have

$$\frac{\partial}{\partial\xi_j} = \sum_{k=1}^N (a_{kj} + b_{kj}(x))\frac{\partial}{\partial x_k}. \tag{9.96}$$

Thus, the variational equation:

$$(\nabla u, \nabla\varphi)_{\Omega_+} = (\mathbf{g}, \nabla\varphi)_{\Omega_+} \quad \text{for any } \varphi \in \hat{W}^1_{q',0}(\Omega_+)$$

subject to $u = g$ on Γ_+ is transformed to the variational equation:

$$(\nabla v, \nabla\psi)_{\mathbb{R}^N_+} + (\mathscr{B}^0\nabla v, \nabla\psi)_{\mathbb{R}^N_+} = (\mathbf{h}, \nabla\psi)_{\mathbb{R}^N_+} \quad \text{for any } \psi \in W^1_{q',0}(\mathbb{R}^N_+) \tag{9.97}$$

subject to $v = f$ on $\mathbb{R}^N_0$, where $\mathbf{h} = \det G(\mathscr{A}_{-1} + B_{-1} \circ \Phi)\mathbf{g} \circ \Phi$, $f = g \circ \Phi$, $v = u \circ \Phi$, $\psi = \varphi \circ \Phi$, and $\mathscr{B}^0$ is an $N \times N$ matrix whose (k, ℓ) component $p_{k\ell}$ is given by

$$p_{k\ell} = \delta_{k\ell}(\det G - 1) + \sum_{j=1}^N (b_{kj}a_{\ell j} + a_{kj}b_{\ell j} + b_{kj}b_{\ell j})\det G. \tag{9.98}$$

Since $N < r < \infty$, by the Sobolev imbedding theorem, $\|\nabla b_{ij}\|_{L_\infty(\mathbb{R}^N)} \leqslant \|\nabla b_{ij}\|_{W_r^1(\mathbb{R}^N)}$, and therefore by (9.87) we have

$$\|(\det G - 1, p_{k\ell})\|_{L_\infty(\mathbb{R}^N)} \leqslant CM_1, \quad \|(\nabla \det G, \nabla p_{k\ell})\|_{W_r^1(\mathbb{R}^N)} \leqslant C_{M_2}. \tag{9.99}$$

Choosing $M_1 > 0$ small enough and using the Banach fixed point theorem, we can easily prove the following lemma.

Lemma 9.4.2 *Let* $1 < q < \infty$. *Then, there exist* $M_1 \in (0, 1)$ *and an operator* $\mathscr{K}_2$ *with*

$$\mathscr{K}_2 \in \mathscr{L}(L_q(\mathbb{R}^N_+)^N, W^1_q(\mathbb{R}^N_+) + \hat{W}^1_{q,0}(\mathbb{R}^N_+))$$

such that for any $\mathbf{f} \in L_q(\mathbb{R}^N_+)^N$ *and* $f \in W_q^{1-1/q}(\mathbb{R}^{N-1})$, $v = \mathscr{K}_2(\mathbf{f}, f)$ *is a unique solution to the variational problem* (9.97) *possessing the estimate:*

$$\|\nabla \mathscr{K}_2(\mathbf{f}, f)\|_{L_q(\mathbb{R}^N_+)} \leqslant C\{\|\mathbf{f}\|_{L_q(\mathbb{R}^N_+)} + \|f\|_{W_q^{1-1/q}(\mathbb{R}^{N-1})}\}.$$

Next, we transform problem (9.95) to problem in the half-space by the change of variable: $\xi = \Phi(x)$. Given $\mathbf{v} \in W^2_q(\Omega_+)$, we set $\mathbf{u} = \mathscr{A}_{-1}\mathbf{v} \circ \Phi$, and then

$$\frac{\partial v_i}{\partial \xi_j} + \frac{\partial v_j}{\partial \xi_i} = \sum_{k,\ell=1}^N (a_{ki}a_{\ell j} + a_{kj}a_{\ell i})\frac{\partial u_\ell}{\partial x_k} + b^d_{ij} : \nabla \mathbf{u} \tag{9.100}$$

where b^d_{ij} is the $N \times N$ matrix whose (k, ℓ) component is $b_{ki}a_{\ell j} + b_{kj}a_{\ell i}$ and $b^d_{ij} : \nabla \mathbf{u}$ means that

$$b^d_{ij} : \nabla \mathbf{u} = \sum_{k,\ell=1}^N (b_{ki}a_{\ell j} + b_{kj}a_{\ell i})\frac{\partial u_\ell}{\partial x_k}. \tag{9.101}$$

By (9.87), we have

$$\|b^d_{ij}\|_{L_\infty(\mathbb{R}^N)} \leqslant CM_1, \quad \|\nabla b^d_{ij}\|_{W^1_r(\mathbb{R}^N)} \leqslant CM_2. \tag{9.102}$$

Since $\sum_{k=1}^N a^2_{Nk} = 1$, setting $\mathscr{B}_\mathbf{n} = \{\sum_{k=1}^N (a_{Nk} + b_{Nk}(x))^2\}^{-1} - 1$, by (9.87) we have

$$\|\mathscr{B}_\mathbf{n}\|_{L_\infty(\mathbb{R}^N)} \leqslant CM_1, \quad \|\nabla \mathscr{B}_\mathbf{n}\|_{W^1_q(\mathbb{R}^N)} \leqslant C_{M_2}, \tag{9.103}$$

and therefore, we have

$$< \mathbf{D}(\mathbf{v})\mathbf{n}_+, \mathbf{n}_+ >=< \mathbf{D}(\mathbf{u})\mathbf{n}_0, \mathbf{n}_0 > +\mathscr{B}^1 : \nabla \mathbf{u} \tag{9.104}$$

where $\mathbf{n}_0 = (0, \ldots, 0, -1)$ and $\mathscr{B}^1$ is an $N \times N$ matrix whose (k, ℓ) componet is

$$2\mathscr{B}_{\mathbf{n}}\delta_{NN} + (1 + \mathscr{B}_{\mathbf{n}})[\sum_{i,j=1}^{N} \{a_{Nj}b_{Ni} + b_{Nj}(a_{Ni} + b_{Ni})\}(a_{ki}a_{\ell j} + a_{kj}a_{\ell i})$$
$$+ \sum_{i,j=1}^{N} (a_{Nj} + b_{Nj})(a_{Ni} + b_{Ni})(b_{ki}a_{\ell j} + b_{kj}a_{\ell i})].$$

By (9.87) we have

$$\|\mathscr{B}^1 : \nabla\mathbf{u}\|_{W^1_q(\mathbb{R}^N_+)} \leqslant CM_1\|\nabla^2\mathbf{u}\|_{L_q(\mathbb{R}^N_+)} + C_{M_2}\|\nabla\mathbf{u}\|_{L_q(\mathbb{R}^N_+)}.$$

Let $b_{\rm div}$ be an $N \times N$ matrix whose (k, ℓ) component is $\sum_{j=1}^N a_{kj}b_{\ell j}$, and then we have

$$\operatorname{div}\mathbf{v} = \operatorname{div}\mathbf{u} + b_{\rm div} : \nabla\mathbf{u}.$$

Thus, there exists an operator $\mathscr{R}^1$ acting on $\mathbf{u}$ such that

$$\begin{aligned} &< \mu\mathbf{D}(\mathbf{v})\mathbf{n}_+, \mathbf{n}_+ > -\operatorname{div}\mathbf{v} =< \mu\mathbf{D}(\mathbf{u})\mathbf{n}_0, \mathbf{n}_0 > -\operatorname{div}\mathbf{u} + \mathscr{R}^1\mathbf{u}, \\ &\|\mathscr{R}^1\mathbf{u}\|_{L_q(\mathbb{R}^N_+)} \leqslant CM_1\|\nabla\mathbf{u}|_{L_q(\mathbb{R}^N_+)}, \\ &\|\mathscr{R}^1\mathbf{u}\|_{W^1_q(\mathbb{R}^N_+)} \leqslant CM_1\|\nabla^2\mathbf{u}\|_{L_q(\mathbb{R}^N_+)} + C_{M_2}\|\nabla\mathbf{u}\|_{L_q(\mathbb{R}^N_+)}. \end{aligned} \tag{9.105}$$

Analogously, we see that there exist operators $\mathscr{R}^i$ $(i = 2, 3, 4, 5)$ acting on $\mathbf{u}$ such that

$$\begin{aligned} &\mathscr{A}_{-1}\operatorname{Div}\mathbf{D}(\mathbf{v}) = \operatorname{Div}\mathbf{D}(\mathbf{u}) + \mathscr{R}^2\mathbf{u}, \\ &\det G(\mathscr{A}_{-1} + B_{-1} \circ \Phi)(\operatorname{Div}(\mu\mathbf{D}(\mathbf{v})) - \nabla\operatorname{div}\mathbf{v}) = \operatorname{Div}(\mu\mathbf{D}(\mathbf{u})) - \nabla\operatorname{div}\mathbf{u} + \mathscr{R}^3\mathbf{u}, \\ &\mathscr{A}_{-1}(\mu\mathbf{D}(\mathbf{v})\mathbf{n}_+) = \mu\mathbf{D}(\mathbf{u})\mathbf{n}_0 + \mathscr{R}^4\mathbf{u}, \\ &\mathbf{n}_+ \cdot \mathbf{v} = \mathbf{n}_0 \cdot \mathbf{u} + \mathscr{R}^5\mathbf{u}, \\ &\|\mathscr{R}^i\mathbf{u}\|_{L_q(\mathbb{R}^N_+)} \leqslant CM_1\|\nabla^2\mathbf{u}\|_{L_q(\Omega)} + C_{M_2}\|\nabla\mathbf{u}\|_{L_q(\mathbb{R}^N_+)} \quad (i = 2, 3), \\ &\|\mathscr{R}^4\mathbf{u}\|_{L_q(\mathbb{R}^N_+)} \leqslant CM_1\|\nabla\mathbf{u}\|_{L_q(\mathbb{R}^N_+)}, \\ &\|\mathscr{R}^4\mathbf{u}\|_{W^1_q(\mathbb{R}^N_+)} \leqslant CM_1\|\nabla^2\mathbf{u}\|_{L_q(\mathbb{R}^N_+)} + C_{M_2}\|\mathbf{u}\|_{W^1_q(\mathbb{R}^N_+)}, \\ &\|\mathscr{R}^5\mathbf{u}\|_{W^2_q(\mathbb{R}^N_+)} \leqslant CM_1\|\nabla^2\mathbf{u}\|_{L_q(\mathbb{R}^N_+)} + C_{M_2}\|\mathbf{u}\|_{W^1_q(\mathbb{R}^N_+)}. \end{aligned} \tag{9.106}$$

Let $\mathfrak{p}_1 = K^0_{+1}(\mathbf{v}) \circ \Phi$ and $\mathfrak{p}_2 = K^0_{+2}(h) \circ \Phi$, and then by (9.93), (9.97), (9.105), and (9.106), $\mathfrak{p}_1$ and $\mathfrak{p}_2$ satisfy the variational equations:

$$(\nabla\mathfrak{p}_1, \nabla\psi)_{\mathbb{R}^N_+} + (\mathscr{B}^0\nabla\mathfrak{p}_1, \nabla\psi)_{\mathbb{R}^N_+} = (\operatorname{Div}(\mu\mathbf{D}(\mathbf{u})) - \nabla\operatorname{div}\mathbf{u} + \mathscr{R}^3\mathbf{u}, \nabla\psi)_{\mathbb{R}^N_+} \tag{9.107}$$

for any $\psi \in W^1_{q',0}(\mathbb{R}^N_+)$ subject to $\mathfrak{p}_1 =< \mu\mathbf{D}(\mathbf{u})\mathbf{n}_0, \mathbf{n}_0 > -\mathrm{div}\,\mathbf{u} + \mathscr{R}^1\mathbf{u}$ on $\mathbb{R}^N_0$, and

$$(\nabla\mathfrak{p}_2, \nabla\psi)_{\mathbb{R}^N_+} + (\mathscr{B}^0\nabla\mathfrak{p}_2, \nabla\psi)_{\mathbb{R}^N_+} = 0 \quad \text{for any } \psi \in W^1_{q',0}(\mathbb{R}^N_+) \tag{9.108}$$

subject to $\mathfrak{p}_2 = -(\tau + \delta\Delta')\eta - \delta\mathscr{D}_+\eta$ on $\mathbb{R}^N_0$, respectively. Here, we have set $\eta = h \circ \Phi$. Moreover, by (9.88), (9.104) and (9.106) problem (9.95) is transformed to the equations:

$$\begin{aligned}
&\lambda\mathbf{u} - \mathrm{Div}\,(\mu\mathbf{D}(\mathbf{u})) - \mathscr{R}^2\mathbf{u} + (\mathbf{I} + \mathscr{B}^2)\nabla(\mathfrak{p}_1 + \mathfrak{p}_2) = \mathbf{f} && \text{in } \mathbb{R}^N_+,\\
&\lambda\eta - \mathbf{n}_0\cdot\mathbf{u} - \mathscr{R}^5\mathbf{u} = f && \text{on } \mathbb{R}^N_0,\\
&(\mu\mathbf{D}(\mathbf{u})\mathbf{n}_0 + \mathscr{R}^4\mathbf{u} - (\mathfrak{p}_1 + \mathfrak{p}_2 + (\tau + \delta\Delta' + \delta\mathscr{D}_+)\eta)(\mathbf{n}_0 + \mathscr{B}^3) = \mathbf{f}_b && \text{on } \mathbb{R}^N_0,
\end{aligned}$$

where we have set $\mathbf{f} = \mathscr{A}_{-1}\mathbf{g}\circ\Phi$, $f = g\circ\Phi$, $\mathbf{f}_b = \mathscr{A}_{-1}\mathbf{g}_b\circ\Phi$, $\mathscr{B}^2 = \mathscr{A}_{-1}{}^\top(B_{-1}\circ\Phi)$, and $\mathscr{B}^3 = \mathscr{B}_\mathbf{n}{}^\top(a_{N1},\ldots,a_{NN}) + (1 + \mathscr{B}_\mathbf{n})^\top(b_{N1},\ldots,b_{NN})$. By (9.87), we have

$$\|\mathscr{B}^i\|_{L_\infty(\mathbb{R}^N)} \leqslant CM_1, \quad \|\nabla\mathscr{B}^i\|_{W^1_q(\mathbb{R}^N)} \leqslant C_{M_2} \quad (i = 2, 3). \tag{9.109}$$

By (9.55), (9.56) and Lemma 9.4.2, we have

$$\begin{aligned}
\mathfrak{p}_1 &= K^0_1(\mathbf{u}) + \mathscr{K}_2(\mathscr{R}^3\mathbf{u} - \mathscr{B}^0\nabla K^0_1(\mathbf{u}), \mathscr{R}^1\mathbf{u}),\\
\mathfrak{p}_2 &= K^0_2(\eta) - \mathscr{K}_2(\mathscr{B}^0\nabla K^0_2(\eta), \delta\mathscr{D}_+\eta),
\end{aligned}$$

and therefore, finally we arrive at

$$\begin{cases}
\lambda\mathbf{u} - \mathrm{Div}\,(\mu\mathbf{D}(\mathbf{u}) - (K^0_1(\mathbf{u}) + K^0_2(\eta))\mathbf{I}) + \mathscr{R}^6(\mathbf{u}, \eta) = \mathbf{f} & \text{in } \mathbb{R}^N_+,\\
\lambda\eta - \mathbf{n}_0\cdot\mathbf{u} - \mathscr{R}^5\mathbf{u} = f & \text{on } \mathbb{R}^N_0,\\
\mathscr{T}_{\mathbf{n}_0}(\mu\mathbf{D}(\mathbf{u})\mathbf{n}_0) + \mathscr{T}_{\mathbf{n}_0}(\mathscr{R}^7(\mathbf{u})) = \mathscr{T}_{\mathbf{n}_0}(\mathbf{f}_b) & \text{on } \mathbb{R}^N_0,\\
\mathrm{div}\,\mathbf{u} + \mathscr{R}^7(\mathbf{u})\cdot\mathbf{n}_0 = \mathbf{n}_0\cdot\mathbf{f}_b & \text{on } \mathbb{R}^N_0
\end{cases} \tag{9.110}$$

with

$$\begin{aligned}
\mathscr{R}^6(\mathbf{u}, \eta) &= -\mathscr{R}^2\mathbf{u} + \nabla\{\mathscr{K}_2(\mathscr{R}^3\mathbf{u} - \mathscr{B}^0\nabla K^0_1(\mathbf{u}), \mathscr{R}^1\mathbf{u})\\
&\quad - \mathscr{K}_2(\mathscr{B}^0\nabla K^0_2(\eta), \delta\mathscr{D}_+\eta)\} + \mathscr{B}^2\nabla\{K^0_1(\mathbf{u}) + K^0_2(\eta)\\
&\quad + \mathscr{K}_2(\mathscr{R}^3\mathbf{u} - \mathscr{B}^0\nabla K^0_1(\mathbf{u}), \mathscr{R}^1\mathbf{u}) - \mathscr{K}_2(\mathscr{B}^0\nabla K^0_2(\eta), \delta\mathscr{D}_+\eta)\},\\
\mathscr{R}^7(\mathbf{u}) &= \mathscr{R}^4\mathbf{u} - (\mathscr{R}^1\mathbf{u})\mathbf{n}_0 - \{< \mathbf{D}(\mathbf{u}), \mathbf{n}_0, \mathbf{n}_0 > -\mathrm{div}\,\mathbf{u} + \mathscr{R}^1\mathbf{u}\}\mathscr{B}^3.
\end{aligned} \tag{9.111}$$

Let $\mathscr{S}_{F0}(\lambda)$ and $\mathscr{T}_{F0}(\lambda)$ be the solution operators of problem (9.58) given in Theorem 9.3.8, and set $\mathbf{u} = \mathscr{S}_{F0}(\lambda)F_\lambda(\mathbf{f}, f, \mathbf{f}_b)$ and $\eta = \mathscr{T}_{F0}(\lambda)F_\lambda(\mathbf{f}, f, \mathbf{f}_b)$ in (9.110), where $F_\lambda(\mathbf{f}, f, \mathbf{f}_b) = (\mathbf{f}, f, \lambda^{1/2}\mathbf{f}_b, \mathbf{f}_b)$. Then, the Eqs. (9.110) are rewritten as follows:

$$\begin{cases} \lambda\mathbf{u} - \mathrm{Div}\,(\mu\mathbf{D}(\mathbf{u}) - (K_1^0(\mathbf{u}) + K_2^0(\eta))\mathbf{I}) = \mathbf{f} - \mathscr{R}^8(\lambda)F_\lambda(\mathbf{f}, f, \mathbf{f}_b) & \text{in } \mathbb{R}^N_+, \\ \lambda\eta - \mathbf{n}_0\cdot\mathbf{u} = f - \mathscr{R}^9(\lambda)F_\lambda(\mathbf{f}, f, \mathbf{f}_b) & \text{on } \mathbb{R}^N_0, \\ \mathscr{T}_{\mathbf{n}_0}(\mu\mathbf{D}(\mathbf{u})\mathbf{n}_0) = \mathscr{T}_{\mathbf{n}_0}(\mathbf{f}_b - \mathscr{R}^{10}(\lambda)F_\lambda(\mathbf{f}, f, \mathbf{f}_b)) & \text{on } \mathbb{R}^N_0, \\ \mathrm{div}\,\mathbf{u} = \mathbf{n}_0\cdot(\mathbf{f}_b - \mathscr{R}^{10}(\lambda)F_\lambda(\mathbf{f}, f, \mathbf{f}_b)) & \text{on } \mathbb{R}^N_0, \end{cases} \tag{9.112}$$

where we have set

$$\begin{aligned} \mathscr{R}^8(\lambda)(F_1, F_2, F_3, F_4) &= \mathscr{R}^6(\mathscr{S}_1(\lambda)(F_1, F_2, F_3, F_4), \mathscr{T}(\lambda)(F_1, F_2, F_3, F_4)), \\ \mathscr{R}^9(\lambda)(F_1, F_2, F_3, F_4) &= -\mathscr{R}^5\mathscr{S}_1(\lambda)(F_1, F_2, F_3, F_4), \\ \mathscr{R}^{10}(\lambda)(F_1, F_2, F_3, F_4) &= \mathscr{R}^7(\mathscr{S}_1(\lambda)(F_1, F_2, F_3, F_4)). \end{aligned}$$

For any $\delta > 0$, we have

$$\begin{aligned} \|\mathscr{R}^6(\mathbf{u}, \eta)\|_{L_q(\mathbb{R}^N_+)} &\leqslant C(M_1+\delta)(\|\nabla^2\mathbf{u}\|_{L_q(\mathbb{R}^N_+)} + \|\eta\|_{W_q^{3-1/q}(\mathbb{R}^{N-1})}) \\ &\quad + C_{M_2,\delta}(\|\mathbf{u}\|_{W_q^1(\mathbb{R}^N_+)} + \|\eta\|_{W_q^{2-1/q}(\mathbb{R}^{N-1})}); \\ \|\mathscr{R}^7(\mathbf{u})\|_{L_q(\mathbb{R}^N_+)} &\leqslant CM_1\|\nabla\mathbf{u}\|_{L_q(\mathbb{R}^N_+)}; \\ \|\mathscr{R}^7(\mathbf{u})\|_{W_q^1(\mathbb{R}^N_+)} &\leqslant C(M_1+\delta)\|\nabla^2\mathbf{u}\|_{L_q(\mathbb{R}^N_+)} + C_{M_2,\delta}\|\mathbf{u}\|_{W_q^1(\mathbb{R}^N_+)}. \end{aligned} \tag{9.113}$$

To prove (9.113), we use the following lemma which follows immediately from the Sobolev imbedding theorem.

Lemma 9.4.3 *Let $1 < q < r < \infty$ and $N < r < \infty$. Then, there exists a constant $C_{N,q,r}$ such that*

$$\|ab\|_{L_q(\mathbb{R}^N_+)} \leqslant C_{N,q,r}\|a\|_{L_r(\mathbb{R}^N_+)}\|b\|^{1-N/r}_{L_q(\mathbb{R}^N_+)}\|\nabla b\|^{N/r}_{L_q(\mathbb{R}^N_+)}. \tag{9.114}$$

Moroever, for any $\delta > 0$ we have

$$\|ab\|_{L_q(\mathbb{R}^N_+)} \leqslant \delta\|\nabla b\|_{L_q(\mathbb{R}^N_+)} + C_{N,q,r}\delta^{-\frac{N}{r-N}}\|a\|^{\frac{r}{r-N}}_{L_r(\mathbb{R}^N_+)}\|b\|_{L_q(\mathbb{R}^N_+)}. \tag{9.115}$$

By Lemma 9.4.3, (9.91), (9.93), and (9.94), we have

$$\|\mathscr{D}_+ f\|_{W_q^{1-1/q}(\mathbb{R}^{N-1})} \leqslant C(M_1+\delta)\|f\|_{W_q^{3-1/q}(\mathbb{R}^{N-1})} + C_{M_2,\delta}\|f\|_{W_q^{2-1/q}(\mathbb{R}^{N-1})}. \tag{9.116}$$

Thus, by Lemma 9.4.2, (9.57), (9.87), (9.98), (9.99), (9.105), (9.106), and (9.116), we have the estimate for $\mathscr{R}^6(\mathbf{u}, \eta)$ in (9.113). Analogously, by (9.105), (9.106) and (9.109), we have the estimates for $\mathscr{R}^7(\mathbf{u})$ in (9.113).

Let

$$\begin{aligned} Q_0(\lambda)(\mathbf{f}, f, \mathbf{f}_b) &= (\mathscr{R}^8(\lambda)F_\lambda(\mathbf{f}, f, \mathbf{f}_b), \mathscr{R}^9(\lambda)F_\lambda(\mathbf{f}, f, \mathbf{f}_b), \mathscr{R}^{10}(\lambda)F_\lambda(\mathbf{f}, f, \mathbf{f}_b)), \\ \mathscr{Q}_0(\lambda)F &= (\mathscr{R}^8(\lambda)F, \mathscr{R}^9(\lambda)F, \lambda^{1/2}\mathscr{R}^{10}(\lambda)F, \mathscr{R}^{10}(\lambda)F) \quad (F = (F_1, F_2, F_3, F_4)), \end{aligned}$$

$$\|(\mathbf{f}, f, \mathbf{f}_b)\|_{Y_q(\mathbb{R}^N_+)} = \|\mathbf{f}\|_{L_q(\mathbb{R}^N_+)} + \|f\|_{W^{2-1/q}_q(\mathbb{R}^{N-1})} + \|\mathbf{f}_b\|_{W^1_q(\mathbb{R}^N_+)},$$
$$\|(F_1, F_2, F_3, F_4)\|_{\mathscr{Y}_q(\mathbb{R}^N_+)} = \|(F_1, F_3)\|_{L_q(\mathbb{R}^N_+)} + \|F_2\|_{W^{2-1/q}_q(\mathbb{R}^{N-1})} + \|F_4\|_{W^1_q(\mathbb{R}^N_+)},$$

where we have set $F_\lambda(\mathbf{f}, f, \mathbf{f}_b) = (\mathbf{f}, f, \lambda^{1/2}\mathbf{f}_b, \mathbf{f}_b)$. Obviously,

$$F_\lambda Q_0(\mathbf{f}, f, \mathbf{f}_b) = \mathscr{Q}_0(\lambda) F_\lambda(\mathbf{f}, f, \mathbf{f}_b). \tag{9.117}$$

By the definition of $\mathscr{R}$-boundedness (cf. Definition 9.1.2), Lemma 9.3.4, Theorem 9.3.8 with $\lambda_0 = 1$, (9.106), and (9.113), we have

$$\mathscr{R}_{\mathscr{L}(\mathscr{Y}_q(\mathbb{R}^N_+))}(\{(\tau\partial_\tau)^\ell \mathscr{Q}_0(\lambda) \mid \lambda \in \Sigma_{\varepsilon,\lambda_0}\}) \leqslant C(M_1 + \delta) + C_{M_2,\delta}\lambda_0^{-1/2} \quad (\ell = 0, 1)$$

for any $\lambda_0 \geqslant 1$ with some costant C independent of $\lambda_0 \geqslant 1$. Thus, we choose $M_1 > 0$ and $\delta > 0$ so small that $CM_1 \leqslant 1/4$, $C\delta \leqslant 1/4$, and then we choose $\lambda_0 > 1$ so large that $C_{M_2,\delta}\lambda_0^{-1/2} \leqslant 1/4$. Thus, we have

$$\mathscr{R}_{\mathscr{L}(\mathscr{Y}_q(\mathbb{R}^N_+))}(\{(\tau\partial_\tau)^\ell \mathscr{Q}_0(\lambda) \mid \lambda \in \Sigma_{\varepsilon,\lambda_0}\}) \leqslant 3/4 \ (\ell = 0, 1), \tag{9.118}$$

which, combined with (9.117), furnishes that

$$\|F_\lambda Q_0(\lambda)(\mathbf{f}, f, \mathbf{f}_b)\|_{\mathscr{Y}_q(\mathbb{R}^N_+)} \leqslant (3/4)\|F_\lambda(\mathbf{f}, f, \mathbf{f}_b)\|_{\mathscr{Y}_q(\mathbb{R}^N_+)}. \tag{9.119}$$

Since $\|F_\lambda(\mathbf{f}, f, \mathbf{f}_b)\|_{\mathscr{Y}_q(\mathbb{R}^N_+)}$ is an equivalent norm to $\|(\mathbf{f}, f, \mathbf{f}_b)\|_{Y_q(\mathbb{R}^N_+)}$ provided that $\lambda \neq 0$, by (9.119) there exists an inverse operator $(\mathbf{I} - Q_0(\lambda))^{-1}$ of $\mathbf{I} - Q_0(\lambda)$ in $\mathscr{L}(Y_q(\mathbb{R}^N_+))$ for any $\lambda \in \Sigma_{\varepsilon,\lambda_0}$. Thus, in view of (9.112),

$$\mathbf{u} = \mathscr{S}_1(\lambda) F_\lambda(\mathbf{I} - Q_0(\lambda))^{-1}(\mathbf{f}, f, \mathbf{f}_b), \quad \eta = \mathscr{T}_1(\lambda) F_\lambda(\mathbf{I} - Q_0(\lambda))^{-1}(\mathbf{f}, f, \mathbf{f}_b)$$

are unique solutions of problem (9.110).

On the other hand, by (9.118)

$$(\mathbf{I} - \mathscr{Q}_0(\lambda))^{-1} = \sum_{j=0}^{\infty} \mathscr{Q}_0(\lambda)^j$$

exists in $\mathscr{L}(\mathscr{Y}_q(\mathbb{R}^N_+))$ and

$$\mathscr{R}_{\mathscr{L}(\mathscr{Y}_q(\mathbb{R}^N_+))}(\{(\tau\partial_\tau)^\ell (\mathbf{I} - \mathscr{Q}_0(\lambda))^{-1} \mid \lambda \in \Sigma_{\varepsilon,\lambda_0}\}) \leqslant c \quad (\ell = 0, 1) \tag{9.120}$$

with some positive constant c. Since

$$F_\lambda(\mathbf{I} - Q_0(\lambda))^{-1} = \sum_{j=1}^{\infty} F_\lambda Q_0(\lambda)^j = (\sum_{j=0}^{\infty} \mathscr{Q}_0(\lambda)^j) F_\lambda = (\mathbf{I} - \mathscr{Q}_0(\lambda))^{-1} F_\lambda$$

as follows from (9.117), setting

$$\tilde{\mathscr{S}}(\lambda) = \mathscr{S}_{F0}(\lambda)(\mathbf{I} - \mathscr{Q}_0(\lambda))^{-1}, \quad \tilde{\mathscr{T}}(\lambda) = \mathscr{T}_{F0}(\lambda)(\mathbf{I} - \mathscr{Q}_0(\lambda))^{-1},$$

we have

$$\mathbf{u} = \tilde{\mathscr{S}}(\lambda) F_\lambda(\mathbf{f}, f, \mathbf{f}_b), \quad \eta = \tilde{\mathscr{T}}(\lambda) F_\lambda(\mathbf{f}, f, \mathbf{f}_b).$$

Moreover, by Theorem 9.3.8, Lemma 9.3.4, and (9.119), we have

$$\begin{aligned}
&\mathscr{R}_{\mathscr{L}(\mathscr{Y}_q(\mathbb{R}^N_+), W^{2-j}_q(\mathbb{R}^N_+)^N)}(\{(\tau\partial_\tau)^\ell(\lambda^{j/2}\tilde{\mathscr{S}}(\lambda)) \mid \lambda \in \Sigma_{\varepsilon,\lambda_0}\}) \leqslant c,\\
&\mathscr{R}_{\mathscr{L}(\mathscr{Y}_q(\mathbb{R}^N_+), W^{3-1/q-k}_q(\mathbb{R}^N_+)^N)}(\{(\tau\partial_\tau)^\ell(\lambda^k\tilde{\mathscr{T}}(\lambda)) \mid \lambda \in \Sigma_{\varepsilon,\lambda_0}\}) \leqslant c
\end{aligned} \tag{9.121}$$

for $j = 0, 1, 2$, $k = 0, 1$ and $\ell = 0, 1$ with some positive constant c.

Recalling that

$$\mathbf{u} = \mathscr{A}_{-1}\mathbf{v}_\circ\Phi, \ \eta = h \circ \Phi, \ \mathbf{f} = \mathscr{A}_{-1}\mathbf{g} \circ \Phi, \ f = g \circ \Phi, \ \mathbf{f}_b = \mathscr{A}_{-1}\mathbf{g} \circ \Phi,$$

we define operators $\mathscr{S}_{F+}(\lambda)$ and $\mathscr{T}_{F+}(\lambda)$ by

$$\begin{aligned}
&\mathscr{S}_{F+}(\lambda)(F_1, F_2, F_3, F_4)\\
&\quad = [{}^\top\mathscr{A}_{-1}\tilde{\mathscr{S}}(\lambda)(\mathscr{A}_{-1}F_1 \circ \Phi, F_2 \circ \Phi, \mathscr{A}_{-1}F_3 \circ \Phi, \mathscr{A}_{-1}F_4 \circ \Phi)] \circ \Phi^{-1},\\
&\mathscr{T}_{F+}(\lambda)(F_1, F_2, F_3, F_4)\\
&\quad = [\tilde{\mathscr{T}}(\lambda)(\mathscr{A}_{-1}F_1 \circ \Phi, F_2 \circ \Phi, \mathscr{A}_{-1}F_3 \circ \Phi, \mathscr{A}_{-1}F_4 \circ \Phi)] \circ \Phi^{-1},
\end{aligned}$$

and then by (9.121) we see easily that $\mathscr{S}_{F+}(\lambda)$ and $\mathscr{T}_{F+}(\lambda)$ satisfy the required properties in Theorem 9.4.1, which completes the proof of Theorem 9.4.1.

9.4.3 Reduced Stokes Equations with Non-slip Boundary Condition in a Bent-Half Space

In this subsection, we consider the generalized resolvent problem for the reduced Stokes equations with non-slip boundary condition. For $\mathbf{v} \in W^2_q(\Omega_+)^N$, let $K^1_+(\mathbf{v})$ be a unique solution to the variational problem:

$$(\nabla K^1_+(\mathbf{v}), \nabla\varphi)_{\Omega_+} - (\mathrm{Div}\,(\mu\mathbf{D}(\mathbf{v})) - \nabla\mathrm{div}\,\mathbf{v}, \nabla\varphi)_{\Omega_+} \quad \text{for any } \varphi \in \hat{W}^1_{q'}(\Omega_+).$$

And, we consider the generalized resolvent problem for the reduced Stokes operator with non-slip boundary condition:

$$\lambda\mathbf{v} - \mathrm{Div}\,(\mu\mathbf{D}(\mathbf{v}) - K^1_+(\mathbf{v})\mathbf{I}) = \mathbf{g} \quad \text{in } \Omega_+, \quad \mathbf{v} = 0 \quad \text{on } \Gamma_+. \tag{9.122}$$

Then, we have the following theorem.

Theorem 9.4.4 *Let $1 < q < \infty$ and $0 < \varepsilon < \pi/2$. Then, there exist positive numbers $M_1 > 0$, $\lambda_0 \geqslant 1$ and an operator family $\mathcal{S}_{D+}(\lambda)$ with*

$$\mathcal{S}_{D+}(\lambda) \in \mathrm{Hol}\,(\Sigma_{\varepsilon,\lambda_0}, \mathcal{L}(L_q(\Omega_+)^N, W_q^2(\Omega_+)^N))$$

such that for any $\lambda \in \Sigma_{\varepsilon,\lambda_0}$ and $\mathbf{g} \in L_q(\Omega_+)^N$, problem (9.122) admits a unique solution $\mathbf{v} = \mathcal{S}_{D+}(\lambda)\mathbf{g}$.

Moreover, we have

$$\mathcal{R}_{\mathcal{L}(L_q(\Omega_+)^N, W_q^{2-j}(\Omega_+)^N)}(\{(\tau\partial_\tau)^\ell(\lambda^{j/2}\mathcal{S}_{D+}(\lambda)) \mid \lambda \in \Sigma_{\varepsilon,\lambda_0}\}) \leqslant \gamma_{+2},$$

for $\ell = 0, 1$ and $j = 0, 1, 2$ and $k = 0, 1$ with some positive constant γ_{+2} depending on λ_0, M_2, N, q and ε.

Employing the same argumentation as in Sect. 9.4.2, we can prove Theorem 9.4.4 with the help of Theorem 9.3.14, so that we may omit the proof.

9.5 Proof of Theorem 9.2.1

9.5.1 Some Preparation for the Proof of Theorem 9.2.1

First, we state several properties of a uniform $W_r^{3,2}$ domain (cf. [3]).

Proposition 9.5.1 *Let $N < r < \infty$, let Ω be a uniform $W_r^{3,2}$ domain in $\mathbb{R}^N$ and let $\Gamma_1 = \Gamma$. Then, for any $M_1 \in (0, 1)$ there exist $M_2 \geqslant 1$, $d^i \in (0, 1)$ $(i = 0, 1, 2)$, and at most countably many N-vector of functions $\Phi_j^0 \in W_r^2(\mathbb{R}^N)^N$, $\Phi_j^1 \in W_r^3(\mathbb{R}^N)^N$, points $x_j^i \in \Gamma_i$ $(i = 0, 1)$, and $x_j^2 \in \Omega$ such that the following assertions hold.*

(i) *For $i = 0, 1$ and $j \in \mathbb{N}$, the maps: $\mathbb{R}^N \ni x \mapsto \Phi_j^i(x) \in \mathbb{R}^N$ are bijective.*

(ii) $\Omega = \Big(\bigcup_{i=0}^1 \bigcup_{j=1}^\infty (\Phi_j^i(\mathbb{R}_+^N) \cap B_{d^i}(x_j^i))\Big) \cup \Big(\bigcup_{j=1}^\infty B_{d^2}(x_j^2)\Big)$, $\quad B_{d^2}(x_j^2) \subset \Omega$, $\Phi_j^i(\mathbb{R}_+^N) \cap B_{d^i}(x_j^i) = \Omega \cap B_{d^i}(x_j^i)$, $\quad \Phi_j^i(\mathbb{R}_0^N) \cap B_{d^i}(x_j^i) = \Gamma_i \cap B_{d^i}(x_j^i)$ $(i = 0, 1)$.

(iii) *There exist C^∞ functions ζ_j^i and $\tilde\zeta_j^i$ $(i = 0, 1, 2, \quad j \in \mathbb{N})$ such that*

$$0 \leqslant \zeta_j^i,\ \tilde\zeta_j^i \leqslant 1, \quad \mathrm{supp}\,\zeta_j^i,\ \mathrm{supp}\,\tilde\zeta_j^i \subset B_{d^i}(x_j^i), \quad \|(\zeta_j^i, \tilde\zeta_j^i)\|_{W_\infty^3(\mathbb{R}^N)} \leqslant c,$$

$$\tilde\zeta_j^i = 1 \text{ on } \mathrm{supp}\,\zeta_j^i, \quad \sum_{i=0}^2\sum_{j=1}^\infty \zeta_j^i = 1 \text{ on } \overline{\Omega}, \quad \sum_{j=1}^\infty \zeta_j^i = 1 \text{ on } \Gamma_i \ (i = 0, 1).$$

Here, c is a constant independent of $j \in \mathbb{N}$.

(iv) *For* $i = 0, 1$ *and* $j \in \mathbb{N}$, $\nabla \Phi^i_j = \mathscr{A}^i_j + B^i_j$, $\nabla(\Phi^i_j)^{-1} = \mathscr{A}^i_{j,-} + B^i_{j,-}$, *where* $\mathscr{A}^i_j$ *and* $\mathscr{A}^i_{j,-}$ *are* $N \times N$ *constant orthonormal matrices, and* B^i_j *and* $B^i_{j,-}$ *are* $N \times N$ *matrices of* $W^{1+i}_r(\mathbb{R}^N)$ *functions which satisfy the conditions:* $\|(B^i_j, B^i_{j,-})\|_{L_\infty(\mathbb{R}^N)} \leqslant M_1$ *and* $\|\nabla(B^i_j, B^i_{j,-})\|_{W^{1+i}_r(\mathbb{R}^N)} \leqslant M_2$.

(v) *There exists a natural number* $L \geqslant 2$ *such that any* $L+1$ *distinct sets of* $\{B_{d^i}(x^i_j) \mid i = 0, 1, 2, \ j \in \mathbb{N}\}$ *have an empty intersection.*

Proof Employing the same argument as in Appendix A of Enomoto and Shibata [3], we can prove Proposition 9.5.1, so that we may omit the proof. □

Next, we prepare some propositions used to construct a parametrix. In the following, we write $B^i_j = B_{d^i}(x^i_j)$ for the sake of simplicity. By the finite intersection property stated in Proposition 9.5.1 (v), for any $r \in [1, \infty)$ there exists a constant $C_{r,L}$ such that

$$\Big[\sum_{j=1}^\infty \|f\|^r_{L_r(\Omega \cap B^i_j)}\Big]^{\frac{1}{r}} \leqslant C_{r,L}\|f\|_{L_r(\Omega)} \quad \text{for any } f \in L_r(\Omega). \tag{9.123}$$

Proposition 9.5.2 *Let* X *be a Banach space and* X^* *its dual space, while* $\|\cdot\|_X$, $\|\cdot\|_{X^*}$, *and* $<\cdot,\cdot>$ *be the norm of* X, *the norm of* X^*, *and the duality pairing between of* X *and* X^*, *respectively. Let* $n \in \mathbb{N}$, $l = 1, \ldots, n$, *and* $\{a_l\}_{l=1}^n \subset \mathbb{C}$, *and let* $\{f^l_j\}_{j=1}^\infty$ *be sequences in* X^* *and* $\{g^l_j\}_{j=1}^\infty$, $\{h_j\}_{j=1}^\infty$ *be sequences of positive numbers. Assume that there exist maps* $\mathscr{N}_j : X \to [0, \infty)$ *such that*

$$|< f^l_j, \varphi >| \leqslant M_3 g^l_j \mathscr{N}_j(\varphi) \quad (l = 1, \ldots, n), \quad \Big|\Big\langle \sum_{l=1}^n a_l f^l_j, \varphi\Big\rangle\Big| \leqslant M_3 h_j \mathscr{N}_j(\varphi)$$

for any $\varphi \in X$ *with some positive constant* M_3 *independent of* $j \in \mathbb{N}$ *and* $l = 1, \ldots, n$. *If*

$$\sum_{j=1}^\infty (g^l_j)^q < \infty, \quad \sum_{j=1}^\infty (h_j)^q < \infty, \quad \sum_{j=1}^\infty (\mathscr{N}_j(\varphi))^{q'} \leqslant (M_4\|\varphi\|_X)^{q'}$$

with $1 < q < \infty$ *and* $q' = q/(q-1)$ *for some positive constant* M_4, *then the infinite sum* $f^l = \sum_{j=1}^\infty f^l_j$ *exists in the strong topology of* X^* *and*

$$\|f^l\|_{X^*} \leqslant M_3 M_4\Big(\sum_{j=1}^\infty (g^l_j)^q\Big)^{1/q}, \quad \Big\|\sum_{l=1}^n a_l f^l\Big\|_{X^*} \leqslant M_3 M_4\Big(\sum_{j=1}^\infty (h_j)^q\Big)^{1/q}. \tag{9.124}$$

Proof Let $F^l_m = \sum_{j=1}^m f^l_j$. We can show that $\{F_m\}_{m=1}^\infty$ is a Cauchy sequence in X^*, which implies the existence of f^l. Then the estimates (9.124) follow immediately. □

In the following, we write $\mathscr{H}^i_j = \Phi^i_j(\mathbb{R}^N_+)$, $\Gamma^i_j = \Phi^i_j(\mathbb{R}^N_0)$ $(i = 0, 1)$ and $\mathscr{H}^2_j = \mathbb{R}^N$ for the sake of simplicity. Let $n \in \mathbb{N}_0$, $f \in W^n_q(\Omega)$, and let η^i_j be functions in $C^\infty_0(B^i_j)$ with $\|\eta^i_j\|_{W^n_\infty(\mathbb{R}^N)} \leqslant c_0$ for some constant c_0 independent of $j \in \mathbb{N}$. Then, since $\Omega \cap B^i_j = \mathscr{H}^i_j \cap B^i_j$, by (9.123)

$$\sum_{j=1}^\infty \|\eta^i_j f\|^q_{W^n_q(\mathscr{H}^i_j)} \leqslant C_q\|f\|_{W^n_q(\Omega)}. \tag{9.125}$$

The following propositions are used to define the infinite sum of $\mathscr{R}$-bounded operator families defined on $\mathscr{H}^i_j$.

Proposition 9.5.3 *Let* $1 < q < \infty$, $i = 0, 1, 2$, *and* $n \in \mathbb{N}_0$. *Let* η^i_j *be a function in* $C^\infty_0(B^i_j)$ *such that* $\|\eta^i_j\|_{W^n_\infty(\mathbb{R}^N)} \leqslant c_1$ *for any* $j \in \mathbb{N}$ *with some constant* c_1 *independent of* $j \in \mathbb{N}$. *Let* f_j $(j \in \mathbb{N})$ *be elements in* $W^n_q(\mathscr{H}^i_j)$ *such that* $\sum_{j=1}^\infty \|f_j\|^q_{W^n_q(\mathscr{H}^i_j)} < \infty$. *Then,* $\sum_{j=1}^\infty \eta^i_j f_j$ *converges some* $f \in W^n_q(\Omega)$ *strongly in* $W^n_q(\Omega)$, *and*

$$\|f\|_{W^n_q(\Omega)} \leqslant C_q\{\sum_{j=1}^\infty \|f_j\|^q_{W^n_q(\mathscr{H}^i_j)}\}^{1/q}.$$

Proof For any $\alpha \in \mathbb{N}^N_0$ with $|\alpha| \leqslant n$, we have

$$|(\partial^\alpha_x(\eta^i_j f_j), \varphi)_\Omega| \leqslant C_N c_1\|f_j\|_{W^n_q\mathscr{H}^i_j)}\|\varphi\|_{L_{q'}(\Omega\cap B^i_j)}.$$

By (9.123),

$$\sum_{j=1}^\infty \|\varphi\|^{q'}_{L_{q'}(\Omega\cap B^i_j)} \leqslant C_{q'}\|\varphi\|^{q'}_{L_{q'}(\Omega)},$$

so that by Proposition 9.5.2, $\sum_{j=1}^\infty \eta^i_j f_j$ converges to some $f \in W^n_q(\Omega)$ strongly in $W^n_q(\Omega)$ and $\|f\|^q_{W^n_q(\Omega)} \leqslant C_q \sum_{j=1}^\infty \|f_j\|^q_{W^n_q(\mathscr{H}^i_j)}$. □

Proposition 9.5.4 *Let* $1 < q < \infty$ *and* $n = 2, 3$. *Then we have the following assertions.*

(1) There exist extension maps $\mathbf{T}^n_j : W^{n-1/q}_q(\Gamma^1_j) \to W^n_q(\mathscr{H}^1_j)$ *such that for any* $h \in W^{n-1/q}_q(\Gamma^1_j)$, $\mathbf{T}^n_j h = h$ *on* Γ^1_j *and* $\|\mathbf{T}^n_j h\|_{W^n_q(\mathscr{H}^1_j)} \leqslant C\|h\|_{W^{n-1/q}_q(\Gamma^1_j)}$ *with some constant* $C > 0$ *independent of* $j \in \mathbb{N}$.

(2) There exists an extension map $\mathbf{T}^n_\Gamma : W^{n-1/q}_q(\Gamma) \to W^n_q(\Omega)$ *such that for* $h \in W^{n-1/q}_q(\Gamma)$, $\mathbf{T}^n_\Gamma h = h$ *on* Γ *and* $\|\mathbf{T}^n_\Gamma h\|_{W^n_q(\Omega)} \leqslant C\|h\|_{W^{n-1/q}_q(\Gamma)}$ *with some constant* $C > 0$.

Proof (1) In view of Sect. 9.4.1, by the small perturbation from $\mathbb{R}^N_+$, we can solve the Dirichlet problem: $(\lambda - \Delta)\tilde{h} = 0$ in $\mathscr{H}^1_j$ subject to $\tilde{h} = h$ on Γ^1_j for large λ independent of $j \in \mathbb{N}$. Moreover, we have the estimate: $\|\tilde{h}\|_{W^n_q(\mathscr{H}^i_j)} \leqslant C\|h\|_{W^{n-1/q}_q(\Gamma^1_j)}$ with some $C > 0$ independent of $j \in \mathbb{N}$. Let $\mathbf{T}^n_j h = \tilde{h}$, and then $\mathbf{T}^n_j$ are the required extension maps.
(2) Let Ω_1 be a domain such that $\Omega_1 \supset \Omega$ and the boundary of $\Omega_1 = \Gamma$. Then, there exists a large $\lambda > 0$ such that the Dirichlet problem $(\lambda - \Delta)u = 0$ in Ω_1 subject to $u = h$ on Γ admits a unique solution $u \in W^n_q(\Omega_1)$ for any $h \in W^{n-1/q}_q(\Gamma)$ possessing the estimate: $\|u\|_{W^n_q(\Omega_1)} \leqslant C\|h\|_{W^{n-1/q}_q(\Gamma)}$. Let $\mathbf{T}^n h = u|_\Omega$, and then $\mathbf{T}^n$ is the required extension map. □

Proposition 9.5.5 *Let $1 < q < \infty$ and $n = 2, 3$ and let $\eta_j \in C^\infty_0(B^1_j)$ $(j \in \mathbb{N})$ with $\|\eta_j\|_{W^n_\infty(\mathbb{R}^N)} \leqslant c_2$ for some constant c_2 independent of $j \in \mathbb{N}$. Then, we have the following two assertions:*
(1) Let f_j $(j \in \mathbb{N})$ be functions in $W^{n-1/q}_q(\Gamma^1_j)$ such that $\sum_{j=1}^\infty \|f_j\|^q_{W^{n-1/q}_q(\Gamma^1_j)} < \infty$, and then the infinite sum $\sum_{j=1}^\infty \eta_j f_j$ converges to some $f \in W^{n-1/q}_q(\Gamma)$ strongly in $W^{n-1/q}_q(\Gamma)$ and

$$\|f\|_{W^{n-1/q}_q(\Gamma)} \leqslant C_q\{\sum_{j=1}^\infty \|f_j\|^q_{W^{n-1/q}_q(\Gamma^1_j)}\}^{1/q}.$$

(2) For any $h \in W^{n-1/q}_q(\Gamma)$,

$$\sum_{j=1}^\infty \|\eta_j h\|^q_{W^{n-1/q}_q(\Gamma^1_j)} \leqslant C\|h\|^q_{W^{n-1/q}_q(\Gamma)}.$$

Proof (1) Let $\mathbf{T}^n_j$ be operators given in Proposition 9.5.4 (1), and then $\sum_{j=1}^N \eta_j f_j = \sum_{j=1}^N \eta_j \mathbf{T}^n_j f_j$ on Γ. Since $\sum_{j=1}^\infty \|\mathbf{T}^n_j f_j\|^q_{W^n_q(\mathscr{H}^1_j)} \leqslant C_q \sum_{j=1}^\infty \|f_j\|^q_{W^{n-1/q}_q(\Gamma_j)} < \infty$, by Proposition 9.5.3, $\sum_{j=1}^N \eta_j \mathbf{T}^n_j f_j$ converges to some $\tilde{f} \in W^n_q(\Omega)$ as $N \to \infty$ strongly in $W^n_q(\Omega)$ and

$$\|\tilde{f}\|_{W^n_q(\Omega)} \leqslant C_q\{\sum_{j=1}^\infty \|T^n_j f_j\|^q_{W^n_q(\mathscr{H}^1_j)}\}^{1/q} \leqslant C_q\{\sum_{j=1}^\infty \|f_j\|^q_{W^{n-1/q}_q(\Gamma^1_j)}\}^{1/q},$$

which furnishes that for any $1 \leqslant N < M$

$$\begin{aligned} &\lim_{N,M\to\infty} \|\sum_{j=1}^M \eta_j f_j - \sum_{j=1}^N \eta_j f_j\|_{W^{n-1/q}_q(\Gamma)} \\ &\leqslant C_q \lim_{N,M\to\infty} \|\sum_{j=1}^M \eta_j T^n_j f_j - \sum_{j=1}^N \eta_j T^n_j f_j\|_{W^n_q(\Omega)} = 0. \end{aligned}$$

Thus, $\sum_{j=1}^N \eta_j f_j$ is a Cauchy sequence in $W_q^{n-1/q}(\Gamma)$, so that $\sum_{j=1}^N \eta_j f_j$ converges to some $f \in W_q^{n-1/q}(\Gamma)$ strongly in $W_q^{n-1/q}(\Gamma)$. Since the trace operator is continuous from $W_q^{n-1/q}(\Omega)$ to $W_q^{n-1/q}(\Gamma)$, we have $\tilde f|_\Gamma = f$, and so that

$$\|f\|_{W_q^{n-1/q}(\Gamma)} \leqslant C_q\|\tilde f\|_{W_q^n(\Omega)} \leqslant C_q\{\sum_{j=1}^\infty \|f_j\|^q_{W_q^{n-1/q}(\Gamma_j^1)}\}^{1/q}.$$

(2) By Proposition 9.5.4 (2),

$$\|\eta_j h\|_{W_q^{n-1/q}(\Gamma_j^1)} = \|\eta_j \mathbf{T}_\Gamma^n h\|_{W_q^{n-1/q}(\Gamma)} \leqslant C\|\eta_j\mathbf{T}_\Gamma^n h\|_{W_q^n(\Omega)} \leqslant Cc_2\|\mathbf{T}_\Gamma^n h\|_{W_q^n(\Omega\cap B_j^1)}.$$

Since $\mathbf{T}_\Gamma^n h \in W_q^n(\Omega)$, by (9.125)

$$\sum_{j=1}^\infty \|\eta_j h\|^q_{W_q^{n-1/q}(\Gamma_j^1)} \leqslant C_q c_2^q\|\mathbf{T}_\Gamma^n h\|^q_{W_q^n(\Omega)} \leqslant C_q c_2^q \|h\|^q_{W_q^{n-1/q}(\Gamma)}.$$

This completes the proof of Proposition 9.5.5. □

9.5.2 Local Solutions

To costruct a parametrix for reduced Stokes equations (9.26), we consider the following problems:

$$\lambda\mathbf{u}_j^0 - \mathrm{Div}\,\mu\mathbf{D}(\mathbf{u}_j^0) - K_j^0(\mathbf{u}_j^0)\mathbf{I}) = \mathbf{f}_j^0 \quad \text{in } \mathcal{H}_j^0, \quad \mathbf{u}_j^0|_{\Gamma_j^0} = 0; \tag{9.126}$$

$$\begin{cases} \lambda\mathbf{u}_j^1 - \mathrm{Div}\,(\mu\mathbf{D}(\mathbf{u}_j^1) - (K_j^1(\mathbf{u}_j^1) + L_j(h_j))\mathbf{I}) = \mathbf{f}_j^1 & \text{in } \mathcal{H}_j^1, \\ \lambda h_j - \mathbf{n}_j^1\cdot\mathbf{u}_j^1 = f_j & \text{on } \Gamma_j^1, \\ (\mu\mathbf{D}(\mathbf{u}_j^1) - (K_j^1(\mathbf{u}_j^1) + L_j(h_j))\mathbf{I})\mathbf{n}_j^1 - ((\tau + \delta\Delta_{\Gamma_j^1})h_j^1)\mathbf{n}_j^1 = \mathbf{f}_{bj}, & \text{on } \Gamma_j^1 : \end{cases} \tag{9.127}$$

$$\lambda\mathbf{u}_j^2 - \mathrm{Div}\,(\mu\mathbf{D}(\mathbf{u}_j^2) - K_j^2(\mathbf{u}_j^2)\mathbf{I}) = \mathbf{f}_j^2 \quad \text{in } \mathcal{H}_j^2. \tag{9.128}$$

Here, $\mathbf{n}_j^1$ stands for the unit outer normal to Γ_j^1 and $\Delta_{\Gamma_j^1}$ is the Laplace-Beltrami operator on Γ_j^1. According to what was discussed in Sect. 9.4.1, we may assume that $\mathbf{n}_j^1$ are defined in $\mathbb{R}^N$ and satisfies the estimates:

$$\|\mathbf{n}_j^1\|_{L_\infty(\mathbb{R}^N)} \leqslant C, \quad \|\nabla\mathbf{n}_j^1\|_{W_r^1(\mathbb{R}^N)} \leqslant C_{M_2}, \tag{9.129}$$

while $\Delta_{\Gamma_j^1}$ has the representation of the form:

$$\Delta_{\Gamma_j^1} f = \Delta' f + \mathscr{D}_{\Gamma_j^1} f$$

with $\Delta' f = \sum_{j=1}^{N-1} \partial_j^2 f$ and $\mathscr{D}_{\Gamma_j^1} f = \sum_{i,j=1}^{N-1} a_{ij}\partial_i\partial_j f + \sum_{i=1}^{N-1} a_i\partial_i f$ in a local chart, where a_{ij} and a_i possess the estimates:

$$\|a_{ij}\|_{L_\infty(\mathbb{R}^N)} \leqslant CM_1, \quad \|(\nabla a_{ij}, a_i)\|_{W_r^1(\mathbb{R}^N)} \leqslant C_{M_2}. \tag{9.130}$$

Here and hereafter, C denotes a generic constant independent of M_2 and $j \in \mathbb{N}$ and C_{M_2} a generic constant independent of $j \in \mathbb{N}$ but depending on M_2.

As for the remaining notations, for $\mathbf{u}_j^0 \in W_q^2(\mathscr{H}_j^0)$, let $K_j^0(\mathbf{u}_j^0) \in \hat{W}_q^1(\mathscr{H}_j^0)$ be a solution to the variational problem:

$$(\nabla K_j^0(\mathbf{u}_j^0), \nabla\varphi)_{\mathscr{H}_j^0} = (\mathrm{Div}\,(\nu_j^0 D(\mathbf{u}_j^0)) - \nabla\mathrm{div}\,\mathbf{u}_j^0, \nabla\varphi)_{\mathscr{H}_j^0} \tag{9.131}$$

for any $\varphi \in \hat{W}_{q'}^1(\mathscr{H}_j^0)$. For $\mathbf{u}_j^1 \in W_q^2(\mathscr{H}_j^1)$, let $K_j^1(\mathbf{u}_j^1) \in W_q^1(\mathscr{H}_j^1) + \hat{W}_{q,0}^1(\mathscr{H}_j^1)$ be a unique solution of the variational equation:

$$(\nabla K_j^1(\mathbf{u}_j^1), \nabla\varphi)_{\mathscr{H}_j^1} = (\mathrm{Div}\,(\mu D(\mathbf{u}_j^1)) - \nabla\mathrm{div}\,\mathbf{u}_j^1, \nabla\varphi)_{\mathscr{H}_j^1} \tag{9.132}$$

for any $\varphi \in \hat{W}_{q',0}^1(\mathscr{H}_j^1)$, subject to $K_j^1(\mathbf{u}_j^1) =< \nu_j^1 D(\mathbf{u}_j^1)\mathbf{n}_j^1, \mathbf{n}_j^1 > -\mathrm{div}\,\mathbf{u}_j^1$ on Γ_j^1, while for $h_j \in W_q^{3-1/q}(\Gamma_j^1)$, let $L_j(h_j) \in W_q^1(\mathscr{H}_j^1) + \hat{W}_{q,0}^1(\mathscr{H}_j^1)$ be a unique solution of the variational equation:

$$(\nabla L_j(h_j), \nabla\varphi)_{\mathscr{H}_j^1} = 0 \quad \text{for any } \varphi \in \hat{W}_{q',0}^1(\mathscr{H}_j^1), \tag{9.133}$$

subject to $L_j(h_j) = -(\tau + \delta\Delta_{\Gamma_j^1})h_j$ on Γ_j^1. For $\mathbf{u}_j^2 \in W_q^2(\mathscr{H}_j^2)$, let $K_j^2(\mathbf{u}_j^2) \in \hat{W}_q^1(\mathscr{H}_j^2)$ be a unique solution of the variational equation:

$$(\nabla K_j^2(\mathbf{u}_j^2), \nabla\varphi)_{\mathscr{H}_j^2} = (\mathrm{Div}\,(\mu D(\mathbf{u}_j^2)) - \nabla\mathrm{div}\,\mathbf{u}_j^2, \nabla\varphi)_{\mathscr{H}_j^2} \tag{9.134}$$

for any $\varphi \in \hat{W}_{q'}^1(\mathscr{H}_j^2)$.

Choosing $M_1 \in (0, 1)$ suitably small, we have the unique existence of solutions to (9.131), (9.132) (9.133), and (9.134) possessing the estimates:

$$\begin{aligned} \|\nabla K_j^i(\mathbf{u}_j^i)\|_{L_q(\mathscr{H}_j^i)} &\leqslant c_3\|\nabla\mathbf{u}_j^i\|_{W_q^1(\mathscr{H}_j^i)}, \\ \|\nabla L_j(h_j)\|_{L_q(\mathscr{H}_j^i)} &\leqslant c_3\|h_j\|_{W_q^{3-1/q}(\Gamma_j^1)} \end{aligned} \tag{9.135}$$

with some constant c_3 independent of $j \in \mathbb{N}$. Let

$$Z_q(\mathscr{H}_j^i) = \begin{cases} L_q(\mathscr{H}_j^i)^N & (i = 0, 2), \\ Y_q(\mathscr{H}_j^1), \end{cases} \qquad \mathscr{Z}_q(\mathscr{H}_j^i) = \begin{cases} L_q(\mathscr{H}_j^i)^N & (i = 0, 2), \\ \mathscr{Y}_q(\mathscr{H}_j^1), \end{cases}$$

for the notational simplicity. And then, by Theorems 9.3.1, 9.4.1 and 9.4.4 there exist constant $\lambda_0 \geqslant 1$ and operators families

$$\mathcal{S}_j^i(\lambda) \in \mathrm{Hol}\,(\Sigma_{\varepsilon,\lambda_0}, \mathcal{L}(\mathcal{Z}_q(\mathcal{H}_j^i)^N, W_q^2(\mathcal{H}_j^i)^N)) \quad (i = 0, 1, 2),$$
$$\mathcal{T}_j(\lambda) \in \mathrm{Hol}\,(\Sigma_{\varepsilon,\lambda_0}, \mathcal{L}(\mathcal{Z}_q(\mathcal{H}_j^1), W_q^{3-1/q}(\Gamma_j^1)))$$

such that problems (9.126), (9.127) and (9.128) admit unique solutions

$$\mathbf{u}_j^i = \mathcal{S}_j^i(\lambda)\mathbf{f}_j^i, \quad \mathbf{u}_j^1 = \mathcal{S}_j^1(\lambda)F_\lambda(\mathbf{f}_j, f_j, \mathbf{f}_{bj}), \quad h_j = \mathcal{T}_j(\lambda)F_\lambda(\mathbf{f}_j, f_j, \mathbf{f}_{bj})$$

for $i = 0, 2$ and $j \in \mathbb{N}$. Moreover, we have

$$\begin{aligned}
&\mathcal{R}_{\mathcal{L}(\mathcal{Z}_q(\mathcal{H}_j^i)^N, W_q^{2-m}(\mathcal{H}_j^i)^N)}(\{(\tau\partial_\tau)^\ell(\lambda^{m/2}\mathcal{S}_j^i(\lambda)) \mid \lambda \in \Sigma_{\varepsilon,\lambda_0}\}) \leqslant \omega_0, \\
&\mathcal{R}_{\mathcal{L}(\mathcal{Z}_q(\mathcal{H}_j^1), W_q^{3-n-1/q}(\Gamma_j^1))}(\{(\tau\partial_\tau)^\ell(\lambda^n\mathcal{T}_j(\lambda) \mid \lambda \in \Sigma_{\varepsilon,\lambda_0}\}) \leqslant \omega_0
\end{aligned} \tag{9.136}$$

for $\ell = 0, 1$, $m = 0, 1, 2$ and $n = 0, 1$ with some positive constant ω_0 independent of $j \in \mathbb{N}$. By (9.136), we have

$$\begin{aligned}
&\|(\lambda\mathbf{u}_j^i, \lambda^{1/2}\nabla\mathbf{u}_j^i, \nabla^2\mathbf{u}_j^i)\|_{L_q(\mathcal{H}_j^i)} \leqslant C\|\mathbf{f}_j^i\|_{L_q(\mathcal{H}_j^i)} \quad (i = 0, 2), \\
&\|(\lambda\mathbf{u}_j^1, \lambda^{1/2}\nabla\mathbf{u}_j^1, \nabla^2\mathbf{u}_j^1)\|_{L_q(\mathcal{H}_j^1)} + \|(\lambda h_j, \nabla h_j)\|_{W_q^{2-1/q}(\Gamma_j^1)} \\
&\quad \leqslant C(\|(\mathbf{f}_j^1, \lambda^{1/2}\mathbf{f}_{bj}\|_{L_q(\mathcal{H}_j^1)} + \|f_j\|_{W_q^{2-1/q}(\Gamma_j^1)} + \|\mathbf{f}_{bj}\|_{W_q^1(\mathcal{H}_j^1)})
\end{aligned} \tag{9.137}$$

for any $j \in \mathbb{N}$ and $\lambda \in \Sigma_{\varepsilon,\lambda_0}$ with some constant C independent of $j \in \mathbb{N}$.

9.5.3 Construction of a Parametrix

For $\mathbf{f} \in L_q(\Omega)^N$, $f \in W_q^{2-1/q}(\Gamma)$ and $\mathbf{f}_b \in W_q^1(\Omega)^N$, we consider the reduced Stokes equations (9.26). We use the symbols given in Proposition 9.5.1. Let us define $\mathbf{u}$ and h by

$$\mathbf{u} = \sum_{i=0}^{2}\sum_{j=1}^{\infty} \zeta_j^i \mathbf{u}_j^i, \quad h = \sum_{j=1}^{\infty} \zeta_j^1 h_j \tag{9.138}$$

with

$$\begin{aligned}
&\mathbf{u}_j^i = \mathcal{S}_j^i(\lambda)\tilde{\zeta}_j^i\mathbf{f} \quad (i = 0, 2), \\
&\mathbf{u}_j^1 = \mathcal{S}_j^1(\lambda)F_\lambda(\tilde{\zeta}_j^1\mathbf{f}, \tilde{\zeta}_j^1 f, \tilde{\zeta}_j^1\mathbf{f}_b), \quad h_j = \mathcal{T}_j(\lambda)F_\lambda(\tilde{\zeta}_j^1\mathbf{f}, \tilde{\zeta}_j^1 f, \tilde{\zeta}_j^1\mathbf{f}_b),
\end{aligned} \tag{9.139}$$

where we have set $F_\lambda(\mathbf{g}, g, \mathbf{g}_b) = (\mathbf{g}, g, \lambda^{1/2}\mathbf{g}, \mathbf{g})$. The right hand sides of (9.138) converge strongly in $W^2_q(\Omega)$ and $W^{3-1/q}_q(\Gamma)$, respectively. In fact, by (9.125), (9.137), and Proposition 9.5.5 (2),

$$\begin{aligned}
&\sum_{j=1}^\infty \|h_j\|^q_{W^{2-1/q}_q(\Gamma^1_j)}\\
&\quad\leqslant C\sum_{j=1}^\infty\{\|\tilde\zeta^1_j(\mathbf{f}, \lambda^{1/2}\mathbf{f}_b)\|^q_{L_q(\mathscr{H}^1_j)} + \|\tilde\zeta^1_j f\|^q_{W^{2-1/q}_q(\Gamma^1_j)} + \|\tilde\zeta^1_j\mathbf{f}_b\|^q_{W^1_q(\mathscr{H}^1_j)}\}\\
&\quad\leqslant C\{\|(\mathbf{f}, \lambda^{1/2}\mathbf{f}_b)\|^q_{L_q(\Omega)} + \|f\|^q_{W^{2-1/q}_q(\Gamma)} + \|\mathbf{f}_b\|^q_{W^1_q(\Omega)}\} < \infty,
\end{aligned}$$

by Proposition 9.5.5, the infinite sum $\sum_{j=1}^\infty \zeta^1_j h_j$ converges to h in $W^{3-1/q}_q(\Gamma)$ strongly and

$$\|h\|_{W^{3-1/q}_q(\Gamma)} \leqslant C_q(\|(\mathbf{f}, \lambda^{1/2}\mathbf{f}_b)\|_{L_q(\Omega)} + \|f\|_{W^{2-1/q}_q(\Gamma)} + \|\mathbf{f}_b\|_{W^1_q(\Omega)}).$$

Analogously, the infinite sum $\sum_{i=0}^2\sum_{j=1}^\infty \mathbf{u}^i_j$ strongly converges to $\mathbf{u}$ in $W^2_q(\Omega)^N$ and

$$\|\mathbf{u}\|_{W^2_q(\Omega)} \leqslant C_q(\|(\mathbf{f}, \lambda^{1/2}\mathbf{f}_b)\|_{L_q(\Omega)} + \|f\|_{W^{2-1/q}_q(\Gamma)} + \|\mathbf{f}_b\|_{W^1_q(\Omega)}).$$

Inserting $\mathbf{u}$ and h into (9.26) and the fact that $\mathbf{n} = \mathbf{n}^1_j$ on $\operatorname{supp}\zeta^1_j \cap \Gamma$, we have

$$\begin{cases}
\lambda\mathbf{u} - \operatorname{Div}(\mu\mathbf{D}(\mathbf{u}) - (K_1(\mathbf{u}) + K_2(h)) = \mathbf{f} - \mathbf{V}^1(\lambda)(\mathbf{f}, f, \mathbf{f}_b) & \text{in } \Omega,\\
\lambda h - \mathbf{n}\cdot\mathbf{u} = f & \text{on } \Gamma,\\
(\mu\mathbf{D}(\mathbf{u}) - (K_1(\mathbf{u}) + K_2(h)\mathbf{I})\mathbf{n} - ((\tau + \delta\Delta_\Gamma)h)\mathbf{n} & \\
\qquad\qquad = \mathbf{f}_b - \mathbf{V}^2(\lambda)(\mathbf{f}, f, \mathbf{f}_b)) & \text{on } \Gamma,
\end{cases} \tag{9.140}$$

where we have set

$$\begin{aligned}
\mathbf{V}^1(\lambda)(\mathbf{f}, f, \mathbf{f}_b) &= \mu\sum_{i=0}^2\sum_{j=1}^\infty\{\operatorname{Div}(\mathbf{D}(\zeta^i_j\mathbf{u}^i_j)) - \zeta^i_j\operatorname{Div}(\mathbf{D}(u^i_j))\}\\
&\quad - \sum_{i=0}^2\sum_{j=1}^\infty\{\nabla K_1(\zeta^i_j\mathbf{u}^i_j) - \zeta^i_j\nabla K^i_j(\mathbf{u}^i_j)\} - \sum_{j=1}^\infty\{\nabla K_2(\zeta^1_j h_j) - \zeta^1_j\nabla L_j(h_j)\},\\
\mathbf{V}^2(\lambda)(\mathbf{f}, f, \mathbf{f}_b) &= \sum_{i=0}^2\sum_{j=1}^\infty \mu\{\mathbf{D}(\zeta^i_j\mathbf{u}^i_j)) - \zeta^i_j\mathbf{D}(\mathbf{u}^i_j)\}\mathbf{n}^i_j\\
&\quad - \sum_{j=1}^\infty \mu(< \mathbf{D}(\zeta^1_j\mathbf{u}^1_j)\mathbf{n}^1_j, \mathbf{n}^1_j > - \zeta^1_j < \mathbf{D}(\mathbf{u}^1_j)\mathbf{n}^1_j, \mathbf{n}^1_j >)\mathbf{n}^1_j
\end{aligned} \tag{9.141}$$

$$+\sum_{j=1}^{\infty}(\operatorname{div}(\zeta_j^1\mathbf{u}_j^1)-\zeta_j^1\operatorname{div}\mathbf{u}_j^1)\mathbf{n}_j^1. \tag{9.142}$$

Here, we have used the fact that

$$\nabla K_1(\mathbf{u})=\sum_{i=0}^{2}\sum_{j=1}^{\infty}\nabla K(\zeta_j^i\mathbf{u}_j^i),\quad \nabla K_2(h)=\sum_{j=1}^{\infty}\nabla K_2(\zeta_j^1 h_j) \tag{9.143}$$

which follows from the strong convergence in (9.138) and the continuity of $\nabla\mathcal{K}_1$ in Definition 9.1.3. Recalling (9.138) and (9.139), we define the operators $\mathcal{S}(\lambda)$ and $\mathcal{T}(\lambda)$ acting on $F=(F_1,F_2,F_3,F_4)\in\mathcal{Y}_q(\Omega)$ by

$$\mathcal{S}(\lambda)F=\sum_{i=0,2}\sum_{j=1}^{\infty}\zeta_j^i\mathcal{S}_j^i(\lambda)\tilde{\zeta}_j^iF_1+\sum_{j=1}^{\infty}\zeta_j^1\mathcal{S}_j^1(\lambda)\tilde{\zeta}_j^1F,$$

$$\mathcal{T}(\lambda)F=\sum_{j=1}^{\infty}\zeta_j^1\mathcal{T}_j(\lambda)\tilde{\zeta}_j^1F.$$

And then, we have

$$\begin{aligned}
&\mathcal{S}(\lambda)\in \operatorname{Hol}(\Sigma_{\varepsilon,\lambda_0},\mathcal{L}(\mathcal{Y}_q(\Omega),W_q^2(\Omega)^N)),\\
&\mathcal{T}(\lambda)\in \operatorname{Hol}(\Sigma_{\varepsilon,\lambda_0},\mathcal{L}(\mathcal{Y}_q(\Omega),W_q^{3-1/q}(\Gamma)),\\
&\mathcal{R}_{\mathcal{L}(\mathcal{Y}_q(\Omega),W_q^{2-j}(\Omega)^N}(\{(\tau\partial_\tau)^\ell(\lambda^{j/2}\mathcal{S}(\lambda))\mid\lambda\in\Sigma_{\varepsilon,\lambda_0}\})\leqslant C_q\omega_0,\\
&\mathcal{R}_{\mathcal{L}(\mathcal{Y}_q(\Omega),W_q^{3-k}(\Omega)^N}(\{(\tau\partial_\tau)^\ell(\lambda^{k}\mathcal{T}(\lambda))\mid\lambda\in\Sigma_{\varepsilon,\lambda_0}\})\leqslant C_q\omega_0
\end{aligned} \tag{9.144}$$

for $\ell=0,1$, $j=0,1,2$ and $k=0,1$.

In fact, for any $m\in\mathbb{N}$, $\{\lambda_\ell\}_{\ell=1}^m\subset\Sigma_{\varepsilon,\lambda_0}$, $\{F_\ell\}_{\ell=1}^m\subset\mathcal{Y}_q(\Omega)$ and $\{a_\ell\}_{\ell=1}^m\subset\mathbb{C}$, by (9.136)

$$\begin{aligned}
\|\sum_{\ell=1}^m a_\ell\lambda_\ell^n\mathcal{T}_j(\lambda_\ell)\tilde{\zeta}_j^1F_\ell\|_{W^{3-n-1/q}(\Gamma)}&\leqslant\sum_{\ell=1}^m|a_\ell|\|\lambda_\ell^n\mathcal{T}_j(\lambda_\ell)\tilde{\zeta}_j^1F_\ell\|_{W_q^{3-n-1/q}(\Gamma_j^1)}\\
&\leqslant\omega_0\sum_{\ell=1}^m|a_\ell|\|\tilde{\zeta}_j^1F_\ell\|_{\mathcal{Y}_q(\mathcal{H}_j^1)}\quad(n=0,1).
\end{aligned}$$

Thus, employing the same argument as in the proof of Proposition 5.3 in [14], by (9.125) and Proposition 9.5.5, we see that

$$\mathcal{T}(\lambda)\in\operatorname{Hol}(\Sigma_{\varepsilon,\lambda_0},\mathcal{L}(\mathcal{Y}_q(\Omega),W_q^{3-n-1/q}(\Gamma))),$$

$$\|\sum_{\ell=1}^m a_\ell\lambda_\ell^n\mathcal{T}(\lambda_\ell)F_\ell\|_{W^{3-n-1/q}(\Gamma)}^q$$

$$\leqslant C_q \sum_{j=1}^{\infty} \| \sum_{\ell=1}^{m} a_\ell \lambda_\ell^n \zeta_j^1 \mathscr{T}_j(\lambda_\ell) \tilde{\zeta}_j^1 F_\ell \|^q_{W^{3-n-1/q}(\Gamma_j^1)}\} \quad (n = 0, 1).$$

Thus, by (9.136), the monotone congergence theorem in the theory of Lebesgue integral, (9.125) and Proposition 9.5.5 (2) we have

$$\begin{aligned}
&\int_0^1 \| \sum_{\ell=1}^{m} r_\ell(u) \lambda_\ell^n \mathscr{T}(\lambda_\ell) F_\ell \|^q_{W_q^{3-n-1/q}(\Gamma)} \, du \\
&\leqslant C_q \sum_{j=1}^{\infty} \int_0^1 \| \sum_{\ell=1}^{m} r_\ell(u) \lambda_\ell^n \mathscr{T}_j(\lambda_\ell) \tilde{\zeta}_j^1 F_\ell \|^q_{W_q^{3-n-1/q}(\Gamma_j^1)} \, du \\
&\leqslant C_q \omega_0^q \int_0^1 \sum_{j=1}^{\infty} \| \tilde{\zeta}_j^1 \sum_{\ell=1}^{m} r_\ell(u) F_\ell \|^q_{\mathscr{Y}_q(\mathscr{H}_j^1)} \, du \\
&\leqslant C_q \omega_0 \int_0^1 \| \sum_{\ell=1}^{m} r_\ell(u) F_\ell \|^q_{\mathscr{Y}_q(\Omega)} \, du \quad (n = 0, 1).
\end{aligned}$$

Thus, we have proved the assertion for $\mathscr{T}(\lambda)$. Analogously, we can prove the assertion for $\mathscr{S}(\lambda)$.

9.5.4 Representation of the Remainder Terms $\mathbf{V}^i(\lambda)(\mathbf{f}, f, \mathbf{f}_b)$

In this subsection, we prove the following lemma.

Lemma 9.5.6 *Let λ_0 and ω_0 be the same constants as in* (9.136). *Let $\mathbf{V}^1(\lambda)$ and $\mathbf{V}^2(\lambda)$ be the operators defined in* (9.142) *and set*

$$\mathbf{V}(\lambda)(\mathbf{f}, f, \mathbf{f}_b) = (\mathbf{V}^1(\lambda)(\mathbf{f}, f, \mathbf{f}_b), 0, \mathbf{V}^2(\lambda)(\mathbf{f}, f, \mathbf{f}_b)).$$

Then, there exists an operator family $\mathscr{V}(\lambda) \in \mathrm{Hol}\,(\Sigma_{\varepsilon,\lambda_0}, \mathscr{L}(\mathscr{Y}_q(\Omega)))$ such that

$$F_\lambda \mathbf{V}(\lambda)(\mathbf{f}, f, \mathbf{f}_b) = \mathscr{V}(\lambda) F_\lambda(\mathbf{f}, f, \mathbf{f}_b) \quad ((\mathbf{f}, f, \mathbf{f}_b) \in Y_q(\Omega)), \tag{9.145}$$

$$\mathscr{R}_{\mathscr{L}(\mathscr{Y}_q(\Omega))}(\{(\tau\partial_\tau)^\ell \mathscr{V}(\lambda) : \lambda \in \Sigma_{\varepsilon,\lambda_1}\}) \leqslant (\sigma + C_\sigma \lambda_1^{-1/2})\omega_0 \tag{9.146}$$

($\ell = 0, 1$) *for any $\sigma > 0$ and $\lambda_1 \geqslant \max(\lambda_0, 1)$, where C_σ is some constant depending on σ but independent of λ_1. Here and hereafter, F_λ is the operator defined by*

$$F_\lambda(\mathbf{f}, f, \mathbf{f}_b) = (\mathbf{f}, f, \lambda^{1/2}\mathbf{f}_b, \mathbf{f}_b).$$

Proof First we consider the term:

$$\mathbf{V}^{11}(\lambda)(\mathbf{f}, f, \mathbf{f}_b) = \sum_{i=0}^{2}\sum_{j=1}^{\infty} \mu\{\mathrm{Div}\,(\mathbf{D}(\zeta^i_j\mathbf{u}^i_j)) - \zeta^i_j \mathrm{Div}\,(\mathbf{D}(\mathbf{u}^i_j))\}.$$

Let $\mathrm{Div}\,(\mathbf{D}(\varphi\mathbf{u})) - \varphi\mathrm{Div}\,(\mathbf{D}(\mathbf{u})) = \mathscr{I}_1(\varphi, \mathbf{u})$ for any scalar function φ and N-vector functions u, where we have set

$$\mathscr{I}_1(\varphi, \mathbf{u}) = (\nabla\varphi)\mathrm{div}\,\mathbf{u} + 2(\nabla\varphi)\nabla\mathbf{u} + (\nabla\varphi)^\top\nabla\mathbf{u} + (\nabla^2\varphi)\mathbf{u} + (\Delta\varphi)\mathbf{u}.$$

Using this symbol, we define an operator $\mathscr{V}^{11}(\lambda)$ acting on $F = (F_1, F_2, F_3, F_4) \in \mathscr{Y}_q(\Omega)$ by

$$\mathscr{V}^{11}(\lambda)F = \sum_{i=0,2}\sum_{j=1}^{\infty} \mu\mathscr{I}_1(\zeta^i_j, \mathscr{S}^i_j(\lambda)\tilde\zeta^i_j F_1)) + \sum_{j=1}^{\infty}\mu\mathscr{I}_1(\zeta^1_j, \mathscr{S}^1_j(\lambda)\tilde\zeta^1_j F).$$

And then, we have

$$\begin{aligned}
&\mathscr{V}^{11}(\lambda) \in \mathrm{Hol}\,(\Sigma_{\varepsilon,\lambda_0}, \mathscr{L}(\mathscr{Y}_q(\Omega))),\\
&\mathbf{V}^{11}(\lambda)(\mathbf{f}, f, \mathbf{f}_b) = \mathscr{V}^{11}(\lambda)F_\lambda(\mathbf{f}, f, \mathbf{f}_b), \quad ((\mathbf{f}, f, \mathbf{f}_b) \in Y_q(\Omega)),\\
&\mathscr{R}_{\mathscr{L}(\mathscr{Y}_q(\Omega), L_q(\Omega)^N)}(\{(\tau\partial_\tau)^\ell\mathscr{V}^{11}(\lambda) \mid \lambda \in \Sigma_{\varepsilon,\lambda_1}\}) \leqslant C_{q,L}\omega_0\lambda_1^{-1/2}
\end{aligned} \tag{9.147}$$

($\ell = 0, 1$) for any $\lambda_1 \geqslant \lambda_0 \geqslant 1$. Here and hearafter, λ_1 denotes any number with $\lambda_1 \geqslant max(\lambda_0, 1)$.

In fact, for any $m \in \mathbb{N}$, $\{\lambda_\ell\}_{\ell=1}^m \subset \Sigma_{\varepsilon,\lambda_0}$, $\{F_\ell\}_{\ell=1}^m \subset \mathscr{Y}_q(\Omega)$ and $\{a_\ell\}_{\ell=1}^m \subset \mathbb{C}$, by (9.136)

$$\begin{aligned}
\|\sum_{\ell=1}^m a_\ell \mathrm{div}\,\mathscr{S}^1_j(\lambda_\ell)\tilde\zeta^1_j F_\ell\|_{L_q(\mathscr{H}^1_j)} &\leqslant \sum_{\ell=1}^m |a_\ell|\|\mathscr{S}^1_j(\lambda_\ell)\tilde\zeta^1_j F_\ell\|_{W^1_q(\mathscr{H}^1_j)}\\
&\leqslant \lambda_0^{-1/2}\omega_0\sum_{\ell=1}^m \|\tilde\zeta^1_j F_\ell\|_{\mathscr{Y}_q(\mathscr{H}^1_j)}.
\end{aligned}$$

Let $\mathscr{V}^{111}(\lambda) = \sum_{j=1}^\infty(\nabla\zeta^1_j)\mathrm{div}\,\mathscr{S}^1_j(\lambda)\tilde\zeta^1_j F$, and then employing the same argument as in the proof of Proposition 5.3 in [14], by (9.125) and Proposition 9.5.5, we see that

$$\begin{aligned}
&\mathscr{V}^{111}(\lambda) \in \mathrm{Hol}\,(\Sigma_{\varepsilon,\lambda_0}, \mathscr{L}(\mathscr{Y}_q(\Omega), L_q(\Gamma)^N)),\\
&\|\sum_{\ell=1} a_\ell\mathscr{V}^{111}(\lambda_\ell)F_\ell\|_\Omega^q \leqslant C_q\sum_{j=1}^\infty\|\sum_{\ell=1}^m a_\ell \mathrm{div}\,\mathscr{S}^1_j(\lambda_\ell)\tilde\zeta^1_j F_\ell\|^q_{L_q(\mathscr{H}^1_j)}.
\end{aligned}$$

Thus, by (9.136), the monotone congergence theorem in the theory of Lebesgue integral, Lemma 9.3.4, (9.125) and Proposition 9.5.5 (2) we have

$$\begin{aligned}
&\int_0^1 \|\sum_{\ell=1}^m r_\ell(u)\mathscr{V}^{111}(\lambda_\ell)F_\ell\|_{L_q(\Omega)}^q\,du \\
&\quad\leqslant C_q\sum_{j=1}^\infty\int_0^1 \|\sum_{\ell=1}^m r_\ell(u)\mathscr{S}_j^1(\lambda_\ell)\tilde\zeta_j^1F_\ell\|_{W_q^1(\mathscr{H}_j^1)}^q\,du \\
&\quad\leqslant C_q\sum_{j=1}^\infty \lambda_1^{-1/2}\int_0^1 \|\sum_{\ell=1}^m r_\ell(u)\lambda_\ell^{1/2}\mathscr{S}_j^1(\lambda_\ell)\tilde\zeta_j^1F_\ell\|_{W_q^1(\mathscr{H}_j^1)}^q\,du \\
&\quad\leqslant C_q\omega_0^q\lambda_1^{-1/2}\int_0^1 \sum_{j=1}^\infty\|\tilde\zeta_j^1\sum_{\ell=1}^m r_\ell(u)F_\ell\|_{\mathscr{Y}_q(\mathscr{H}_j^1)}^q\,du \\
&\quad\leqslant C_q\omega_0\lambda_1^{-1/2}\int_0^1 \|\sum_{\ell=1}^m r_\ell(u)F_\ell\|_{\mathscr{Y}_q(\Omega)}^q\,du,
\end{aligned}$$

where $\{\lambda_\ell\}_{\ell=1}^m \subset \Sigma_{\varepsilon,\lambda_1}$ and we have used Propsotion 9.3.4. Thus, we have

$$\mathscr{R}_{\mathscr{L}(\mathscr{Y}_q(\Omega),L_q(\Omega)^N)}(\{\mathscr{V}^{111}(\lambda)\mid \lambda\in\Sigma_{\varepsilon,\lambda_1}\}) \leqslant C_{q,L}\omega_0\lambda_1^{-1/2}.$$

Analogously, we can prove the assertions for other terms, so that we have (9.147).

Second, we consider the terms:

$$\nabla K_1(\zeta_j^i\mathbf{u}_j^i) - \zeta_j^i\nabla K_j^i(\mathbf{u}_j^i),\quad \nabla K_2(\zeta_j^1 h) - \zeta_j^1\nabla L_j(h_j).$$

The former was treated in Sect. 5 in Shibata [14], and employing the same argumentation as in Sect. 5 in [14] we see that

$$\mathbf{V}^{12}(\lambda)(\mathbf{f}, f, \mathbf{f}_b) = \sum_{i=0}^2\sum_{j=1}^\infty\{\nabla K_1(\zeta_j^i\mathbf{u}_j^i) - \zeta_j^i\nabla K_j^i(\mathbf{u}_j^i)\}$$

exists strongly in $L_q(\Omega)^N$ for any $(\mathbf{f}, f, \mathbf{f}_b) \in Y_q(\Omega)$. Moreover, there exists an operator $\mathscr{V}^{12}(\lambda) \in \mathrm{Hol}\,(\Sigma_{\varepsilon,\lambda_0}, \mathscr{L}(\mathscr{Y}_q(\Omega), L_q(\Omega)^N)))$ such that

$$\begin{aligned}
&\mathbf{V}^{12}(\lambda)(\mathbf{f}, f, \mathbf{f}_b) = \mathscr{V}^{12}(\lambda)F_\lambda(\mathbf{f}, f, \mathbf{f}_b)\quad ((\mathbf{f}, f, \mathbf{f}_b) \in Y_q(\Omega)),\\
&\mathscr{R}_{\mathscr{L}(\mathscr{Y}_q(\Omega),L_q(\Omega)^N)}(\{(\tau\partial_\tau)^\ell\mathscr{V}^{12}(\lambda)\mid \lambda\in\Sigma_{\varepsilon,\lambda_1}\}) \leqslant (\sigma + C_\sigma\lambda_1^{-1/2})\omega_0
\end{aligned}\tag{9.148}$$

($\ell = 0, 1$) for any $\sigma > 0$ and $\lambda_1 \geqslant \max(\lambda_0, 1)$, where C_σ is some constant depending on σ. Thus, in the following, we treat $\nabla K_2(\zeta_j^1 h) - \zeta_j^1\nabla L_j(h_j)$. But, employing the same argument as in the following, we can also prove (9.148).

We start with the following inequalities of Poincaré type with uniform constant, which was proved in Shibata [13, Lemmas 3.4, 3.5].

Lemma 9.5.7 *Let $N < r < \infty$ and let Ω be a $W^{2,2}_r$ uniform domain in $\mathbb{R}^N$. Then, there exists a constant $c_4 > 0$ independent of $j = 1, 2, 3, \ldots$ such that*

$$\|\varphi\|_{W^1_q(\mathcal{H}^1_j \cap B^1_j)} \leqslant c_4 \|\nabla\varphi\|_{L_q(\mathcal{H}^1_j \cap B^1_j)} \qquad \text{for any } \varphi \in \hat{W}^1_{q,0}(\mathcal{H}^1_j),$$
$$\|\psi\|_{W^1_q(\Omega \cap B^1_j)} \leqslant c_4 \|\nabla\psi\|_{L_q(\Omega \cap B^1_j)} \qquad \text{for any } \psi \in \mathcal{W}^1_q(\Omega).$$

To handle $(\nabla\zeta^1_j)L_j(h_j)$, we use the following lemma.

Lemma 9.5.8 *Let $1 < q < \infty$. Then, there exists a constant c_5 independent of $j \in \mathbb{N}$ such that*

$$\|L_j(h)\|_{L_q(\mathcal{H}^1_j \cap B^1_j)} \leqslant c_5(\|h\|_{W^{2-1/q}_q(\Gamma^1_j)} + \|h\|^{1-1/q}_{W^{2-1/q}_q(\Gamma^1_j)} \|h\|^{1/q}_{W^{3-1/q}_q(\Gamma^1_j)}) \tag{9.149}$$

for any $h \in W^{3-1/q}_q(\Gamma^1_j)$.

Remark 9.5.1 Applying Young's inequality to (9.149), we have

$$\|L_j(h)\|_{L_q(\mathcal{H}^1_j \cap B^1_j)} \leqslant \sigma\|h\|_{W^{3-1/q}_q(\Gamma^1_j)} + C_\sigma\|h\|_{W^{2-1/q}_q(\Gamma^1_j)} \tag{9.150}$$

for any $\sigma > 0$ and $h \in W^{3-1/q}_q(\Gamma^1_j)$ with some constant C_σ depending on σ but is independent of h and $j \in \mathbb{N}$.

Proof In the following, C stands for generic constants independent of $j \in \mathbb{N}$. Let $\mathbf{T}^3_j$ be the operator given in Proposition 9.5.4. We write $L_j(h) = -\mathbf{T}^3_j(\tau + \delta\Delta)h + f$, where $f \in \hat{W}^1_{q,0}(\mathcal{H}^1_j)$ is a unique solution to the variational problem:

$$(\nabla f, \nabla\varphi)_{\mathcal{H}^1_j} = (\nabla\mathbf{T}^3_j(\tau + \delta\Delta_{\Gamma^1_j})h, \nabla\varphi)_{\mathcal{H}^1_j} \quad \text{for any } \varphi \in \hat{W}^1_{q',0}(\mathcal{H}^1_j). \tag{9.151}$$

Let $Y_{q',j} = \{\psi \in L_{q'}(\mathcal{H}^1_j) \mid \operatorname{supp}\psi \subset B^1_j \cap \mathcal{H}^1_j\}$ and let ψ be an arbitrary element of $Y_{q',j}$. By Lemma 9.5.7,

$$|(\psi, \varphi)_{\mathcal{H}^1_j}| \leqslant \|\psi\|_{L_{q'}(\mathcal{H}^1_j)}\|\varphi\|_{L_q(\mathcal{H}^1_j \cap B^1_j)} \leqslant c_4\|\psi\|_{L_{q'}(\mathcal{H}^1_j)}\|\nabla\varphi\|_{L_q(\mathcal{H}^1_j)}$$

for any $\varphi \in \hat{W}^1_{q,0}(\mathcal{H}^1_j)$. By the Hahn-Banach theorem, there exists a $\mathbf{g} \in L_{q'}(\mathcal{H}^1_j)^N$ such that $\mathbf{g}$ satisfies the variational equation:

$$(\psi, \varphi)_{\mathcal{H}^1_j} = (\mathbf{g}, \nabla\varphi)_{\mathcal{H}^1_j} \quad \text{for any } \varphi \in \hat{W}^1_{q,0}(\mathcal{H}^1_j) \tag{9.152}$$

and the estimate:

$$\|\mathbf{g}\|_{L_{q'}(\mathcal{H}^1_j)} \leqslant C\|\psi\|_{L_{q'}(\mathcal{H}^1_j)}. \tag{9.153}$$

Let $X_{q',j} = \{\Psi \in \hat{W}^1_{q',0} \mid \nabla\Psi \in W^1_{q'}(\mathscr{H}^1_j)^N\}$, and then in view of (9.152) and (9.153), there exists a unique $\Psi \in X_{q',j}$ such that

$$(\nabla\Psi, \nabla\varphi)_{\mathscr{H}^1_j} = (\mathbf{g}, \nabla\varphi)_{\mathscr{H}^1_j} \quad \text{for any } \varphi \in \hat{W}^1_{q,0}(\mathscr{H}^1_j), \tag{9.154}$$
$$\|\nabla\Psi\|_{W^1_{q'}(\mathscr{H}^1_j)} \leqslant C(\|\mathbf{g}\|_{L_{q'}(\mathscr{H}^1_j)} + \|\psi\|_{L_{q'}(\mathscr{H}^1_j)}) \leqslant C\|\psi\|_{L_{q'}(\mathscr{H}^1_j)}. \tag{9.155}$$

Since $f \in \hat{W}^1_{q,0}(\mathscr{H}^1_j)$, by (9.151), (9.152), (9.154), and the divergence theorem of Gauß

$$\begin{aligned}(f, \psi)_{\mathscr{H}^1_j} &= (\nabla f, \mathbf{g})_{\mathscr{H}^1_j} = (\nabla f, \nabla\Psi)_{\mathscr{H}^1_j} = (\nabla\mathbf{T}^3_j(\tau + \delta\Delta_{\Gamma^1_j})h, \nabla\Psi)_{\mathscr{H}^1_j} \\ &= (\mathbf{T}^3_j(\tau + \delta\Delta_{\Gamma^1_j})h, \mathbf{n}^1_j \cdot \nabla\Psi)_{\Gamma^1_j} - (\mathbf{T}^3_j(\tau + \delta\Delta_{\Gamma^1_j})h, \Delta\Psi)_{\mathscr{H}^1_j},\end{aligned}$$

which, combined with (9.155) and classical interpolation inequality:

$$\|v\|_{L_q(\mathscr{H}^1_j)} \leqslant C\|v\|^{1-1/q}_{L_q(\mathscr{H}^1_j)}\|\nabla v\|^{1/q}_{L_q(\mathscr{H}^1_j)}, \tag{9.156}$$

furnishes that

$$\begin{aligned}|(f, \psi)_{\mathscr{H}^1_j}| \leqslant C\{&\|\mathbf{T}^3_j(\tau + \delta\Delta_{\Gamma^1_j})h\|_{L_q(\mathscr{H}^1_j)} \\ &+ \|\nabla\mathbf{T}^3_j(\tau + \delta\Delta_{\Gamma^1_j})h\|^{1/q}_{L_q(\mathscr{H}^1_j)}\|\mathbf{T}^3_j(\tau + \delta\Delta_{\Gamma^1_j})h\|^{1-1/q}_{L_q(\mathscr{H}^1_j)}\}\|\psi\|_{L_{q'}(\mathscr{H}^1_j)}.\end{aligned}$$

Since ψ is chosen arbitrarily,

$$\begin{aligned}\|f\|_{L_q(\mathscr{H}^1_j \cap B^1_j)} \leqslant C\{&\|\mathbf{T}^3_j(\tau + \delta\Delta_{\Gamma^1_j})h\|_{L_q(\mathscr{H}^1_j)} \\ &+ \|\nabla\mathbf{T}^3_j(\tau + \delta\Delta_{\Gamma^1_j})h\|^{1/q}_{L_q(\mathscr{H}^1_j)}\|\mathbf{T}^3_j(\tau + \delta\Delta_{\Gamma^1_j})h\|^{1-1/q}_{L_q(\mathscr{H}^1_j)}\},\end{aligned}$$

which, combined with Proposition 9.5.4, completes the proof of Lemma 9.5.8. □

Write $\nabla K_2(\zeta^1_j h_j) - \zeta^1_j \nabla L_j(h_j) = \nabla(K_2(\zeta^1_j h_j) - \zeta^1_j L_j(h_j)) + (\nabla\zeta^1_j)L_j(h_j)$ and first we consider $(\nabla\zeta^1_j)L_j(h_j)$. And then, $\mathbf{V}^{13}(\lambda)(\mathbf{f}, f, \mathbf{f}_b) = \sum_{j=1}^\infty(\nabla\zeta^1_j)L_j(h_j)$ exists strongly in $L_q(\Omega)^N$ for any $(\mathbf{f}, f, \mathbf{f}_b) \in Y_q(\Omega)$. Moreover, there exists an operator family $\mathscr{V}^{13}(\lambda) \in \mathrm{Hol}\,(\Sigma_{\varepsilon,\lambda_0}, \mathscr{L}(\mathscr{Y}_q(\Omega), L_q(\Omega)^N))$ such that

$$\begin{aligned}&\mathbf{V}^{13}(\lambda)(\mathbf{f}, f, \mathbf{f}_b) = \mathscr{V}^{13}(\lambda)F_\lambda(\mathbf{f}, f, \mathbf{f}_b) \quad ((\mathbf{f}, f, \mathbf{f}_b) \in Y_q(\Omega)), \\ &\mathscr{R}_{\mathscr{L}(\mathscr{Y}_q(\Omega), L_q(\Omega)^N)}(\{(\tau\partial_\tau)^\ell\mathscr{V}^{13}(\lambda) \mid \lambda \in \Sigma_{\varepsilon,\lambda_1}\}) \leqslant (\sigma + C_\sigma\lambda_1^{-1})\omega_0\end{aligned} \tag{9.157}$$

($\ell = 0, 1$) for any $\sigma > 0$ and $\lambda_1 \geqslant \max(\lambda_0, 1)$ with some constant C_σ depending on σ but independent of λ_1.

In fact, in view of (9.139), we define operators $\mathscr{L}_j(\lambda)$ acting on $F \in \mathscr{Y}_q(\mathscr{H}_j^1)$ by $\mathscr{L}_j(\lambda)F = (\nabla\zeta_j^1)L_j(\mathscr{T}_j(\lambda)F)$, and then

$$\mathscr{L}_j(\lambda) \in \mathrm{Hol}\,(\Sigma_{\varepsilon,\lambda_0}, \mathscr{L}(\mathscr{Y}_q(\mathscr{H}_j^1), L_q(\mathscr{H}_j^1)^N)),$$
$$(\nabla\zeta_j^1)L_j(h_j) = \mathscr{L}_j(\lambda)F_\lambda(\tilde\zeta_j^1\mathbf{f}, \tilde\zeta_j^1 f, \tilde\zeta_j^1\mathbf{f}_b) \quad \text{for } (\mathbf{f}, f, \mathbf{f}_b) \in Y_q(\Omega).$$

Moreover, by Lemma 9.5.8

$$\|\mathscr{L}_j(\lambda)F\|^q_{L_q(\mathscr{H}_j^1)} \leqslant \sigma\|\mathscr{T}_j(\lambda)F\|^q_{W_q^{3-1/q}(\Gamma_j^1)} + C_\sigma\|\mathscr{T}_j(\lambda)F\|^q_{W_q^{2-1/q}(\Gamma_j^1)}$$

for any $F \in \mathscr{Y}_q(\mathscr{H}_j^1)$ and any $\sigma > 0$ with some constant C_σ depending on σ. Noting that $\nabla\zeta_j^1 = \tilde\zeta_j^1(\nabla\zeta_j^1)$, we define an operator $\mathscr{V}^{13}(\lambda)$ acting on $F \in \mathscr{Y}_q(\Omega)$ by

$$\mathscr{V}^{13}(\lambda)F = \sum_{j=1}^\infty \tilde\zeta_j^1\mathscr{L}_j(\lambda)\tilde\zeta_j^1 F.$$

And then, by (9.136), Proposition 9.5.5 and (9.125), we see that $\mathscr{V}^{13}(\lambda)$ is the operator family satisfying the properties (9.157).

Next, we consider $\nabla(K_2(\zeta_j^1 h_j) - \zeta_j^1 L_j(h_j))$. In this case, we see that

$$\mathbf{V}^{14}(\lambda)(\mathbf{f}, f, \mathbf{f}_b) = \sum_{j=1}^\infty \nabla(K_2(\zeta_j^1 h_j) - \zeta_j^1 L_j(h_j))$$

exists strongly in $L_q(\Omega)^N$ and there exists an operator family $\mathscr{V}^{14}(\lambda)$ with

$$\mathscr{V}^{14}(\lambda) \in \mathrm{Hol}\,(\Sigma_{\varepsilon,\lambda_0}, \mathscr{L}(\mathscr{Y}_q(\Omega), L_q(\Omega)^N))$$

such that

$$\mathbf{V}^{14}(\lambda)(\mathbf{f}, f, \mathbf{f}_b) = \mathscr{V}^{14}(\lambda)F_\lambda(\mathbf{f}, f, \mathbf{f}_b),$$
$$\mathscr{R}_{\mathscr{L}(\mathscr{Y}_q(\Omega), L_q(\Omega)^N)}(\{(\tau\partial_\tau)^\ell\mathscr{V}^{14}(\lambda) \mid \lambda \in \Sigma_{\varepsilon,\lambda_1}\}) \leqslant (\sigma + C_\sigma\lambda_1^{-1/2})\omega_0 \tag{9.158}$$

($\ell = 0, 1$) for any $\sigma > 0$ and $\lambda_1 \leqslant \max(\lambda_0, 1)$ with some constant depending on σ but independent of λ_1.

In fact, noting that $\Omega \cap B_j^1 = \mathscr{H}_j^1 \cap B_j^1$ and $\Gamma \cap B_j^1 = \Gamma_j^1 \cap B_j^1$, by (9.23) and (9.133) we have

$$\begin{aligned}
&(\nabla(K_2(\zeta_j^1 h_j) - \zeta_j^i L_j(h_j)), \nabla\varphi)_\Omega \\
&= -((\nabla\zeta_j^1)L_j(h_j), \nabla\varphi)_{\mathscr{H}_j^1} + (\nabla L_j(h_j), (\nabla\zeta_j^1)\varphi)_{\mathscr{H}_j^1} \\
&= -(2(\nabla\zeta_j^1)L_j(h_j), \nabla\varphi)_{\mathscr{H}_j^1} - (L_j(h_j), (\Delta\zeta_j^1)\varphi)_{\mathscr{H}_j^1}
\end{aligned} \tag{9.159}$$

for any $\varphi \in \mathscr{W}^1_{q'}(\Omega)$, subject to

$$K_2(\zeta^1_j h_j) - \zeta^1_j L_j(h_j) = -\delta\{2(\nabla_{\Gamma^1_j}\zeta^1_j)\cdot(\nabla_{\Gamma^1_j}h_j) + (\Delta\zeta^1_j)h_j\} \quad \text{on } \Gamma.$$

Let $\mathscr{W}^{-1}_q(\Omega)$ be the dual space of $\mathscr{W}^1_{q'}(\Omega)$ and let $< \cdot, \cdot >_\Omega$ be the duality pairing between $\mathscr{W}^{-1}_q(\Omega)$ and $\mathscr{W}^1_{q'}(\Omega)$. If we define $I_j(\lambda)(\mathbf{f}, f, \mathbf{f}_b) \in \mathscr{W}^{-1}_q(\Omega)$ and $\mathscr{I}_j(\lambda) \in \mathrm{Hol}\,(\Sigma_{\varepsilon,\lambda_0}, \mathscr{L}(\mathscr{Y}_q(\Omega), \mathscr{W}^{-1}_q(\Omega)))$ acting on $F \in \mathscr{Y}_q(\Omega)$ by

$$\begin{aligned}
&< I_j(\lambda)(\mathbf{f}, f, \mathbf{f}_b), \varphi >_\Omega = -(2(\nabla\zeta^1_j)L_j(h_j), \nabla\varphi)_\Omega - ((\Delta\zeta^1_j)L_j(h_j), \varphi)_\Omega,\\
&< \mathscr{I}_j(\lambda)F, \varphi >_\Omega = -(2(\nabla\zeta^1_j)L_j(\mathscr{T}_j(\lambda)\tilde{\zeta}^1_j F), \nabla\varphi)_\Omega\\
&\qquad\qquad - ((\Delta\zeta^1_j)L_j(\mathscr{T}_j(\lambda)\tilde{\zeta}^1_j F), \varphi)_\Omega
\end{aligned} \tag{9.160}$$

for any $\varphi \in \mathscr{W}^1_{q'}(\Omega)$. By Lemmas 9.5.7, 9.5.8, (9.136), and (9.137), we have

$$\begin{aligned}
&| < I_j(\lambda)(\mathbf{f}, f, \mathbf{f}_b), \varphi >_\Omega | \leqslant (\sigma + C_\sigma\lambda_1^{-1})\omega_0\|F_\lambda(\mathbf{f}, f, \mathbf{f}_b)\|_{Y_q(\Omega\cap B^1_j)}\|\nabla\varphi\|_{L_q(\Omega\cap B^1_j)},\\
&| < \mathscr{I}_j(\lambda)F, \varphi >_\Omega | \leqslant C(\sigma\|\mathscr{T}_j(\lambda)\tilde{\zeta}^1_j F\|_{W^{3-1/q}_q(\Gamma^1_j)}\\
&\qquad\qquad + C_\sigma\|\mathscr{T}_j(\lambda)\tilde{\zeta}^1_j F\|_{W^{2-1/q}_q(\Gamma^1_j)})\|\nabla\varphi\|_{L_q(\Omega\cap B^1_j)}.
\end{aligned}$$

By Proposition 9.5.2, (9.125) and Proposition 9.5.5, there exist $I(\lambda)(\mathbf{f}, f, \mathbf{f}_b) \in \mathscr{W}^{-1}_q(\Omega)$ and $\mathscr{I}(\lambda) \in \mathrm{Hol}\,(\Sigma_{\varepsilon,\lambda_0}, \mathscr{L}(\mathscr{Y}_q(\Omega), \mathscr{W}^{-1}_q(\Omega)))$ such that the infinite sums $\sum_{j=1}^\infty I_j(\lambda)(\mathbf{f}, f, \mathbf{f}_b)$ and $\sum_{j=1}^\infty \mathscr{I}_j(\lambda)F$ converge to I and $\mathscr{I}(\lambda)F$ in $\mathscr{W}^{-1}_q(\Omega)$ strongly for any $\lambda \in \Sigma_{\varepsilon,\lambda_0}$, respectively, $I(\lambda)(\mathbf{f}, f, \mathbf{f}_b) = \mathscr{I}(\lambda)F_\lambda(\mathbf{f}, f, \mathbf{f}_b)$, and

$$\begin{aligned}
&\|I(\lambda)(\mathbf{f}, f, \mathbf{f}_b)\|_{\mathscr{W}^{-1}_q(\Omega)} \leqslant (\sigma + C_\sigma\lambda_1^{-1})\omega_0\|F_\lambda(\mathbf{f}, f, \mathbf{f}_b)\|_{Y_q(\Omega)},\\
&\|\sum_{\ell=1}^m a_\ell\mathscr{I}(\lambda)F\|^q_{\mathscr{W}^{-1}_q(\Omega)} \leqslant \sum_{j=1}^\infty\{\sigma\|\sum_{\ell=1}^m a_\ell\mathscr{T}_j(\lambda_\ell)\tilde{\zeta}^1_j F_\ell\|^q_{W^{3-1/q}_q(\Gamma^1_j)}\\
&\qquad\qquad + C_\sigma\|\sum_{\ell=1}^m a_\ell\mathscr{T}_j(\lambda_\ell)\tilde{\zeta}^1_j F_\ell\|_{W^{2-1/q}_q(\Gamma^1_j)}\}.
\end{aligned} \tag{9.161}$$

By the second formula in (9.161), the monotone congergence theorem in theory of Lebesgue integral, Lemma 9.3.4, and (9.136) we have

$$\mathscr{R}_{\mathscr{L}(\mathscr{Y}_q(\Omega), \mathscr{W}^{-1}_q(\Omega))}(\{(\tau\partial_\tau)^\ell\mathscr{I}(\lambda) \mid \lambda \in \Sigma_{\varepsilon,\lambda_1}\}) \leqslant (\sigma + C_\sigma\lambda_1^{-1})\omega_0 \tag{9.162}$$

($\ell = 0, 1$) for any $\sigma > 0$ and $\lambda_1 \geqslant \max(\lambda_0, 1)$ with some constant C_σ depending on σ but independent of λ_1. Let $\mathbf{G}$ be an operator in $\mathscr{L}(\mathscr{W}^{-1}_q(\Omega), L_q(\Omega)^N)$ such that for any $\theta \in \mathscr{W}^{-1}_q(\Omega)$

$$< \theta, \varphi >_\Omega = (\mathbf{G}(\theta), \nabla\varphi)_\Omega \quad \text{for any } \varphi \in \mathscr{W}^1_{q'}(\Omega),$$

$$\|\mathbf{G}(\theta)\|_{L_q(\Omega)} \leqslant \sup\{| < \theta, \varphi >_\Omega | : \varphi \in \mathscr{W}^1_{q'}(\Omega), \ \|\nabla\varphi\|_{L_{q'}(\Omega)} = 1\}. \tag{9.163}$$

Such an operator $\mathbf{G}$ can be constructed by the Hahn-Banach theorem. By (9.159) and (9.163),

$$(\nabla(\sum_{j=1}^\infty K_2(\zeta^1_j h_j) - \zeta^1_j L_j(h_j)), \nabla\varphi)_\Omega = (\mathbf{G}(\mathscr{I}(\lambda)F_\lambda(\mathbf{f}, f, \mathbf{f}_b)), \nabla\varphi)_\Omega \tag{9.164}$$

for any $\varphi \in \mathscr{W}^1_{q'}(\Omega)$. Moreover, by (9.162) and (9.163), we have

$$\mathscr{R}_{\mathscr{L}(\mathscr{Y}_q(\Omega), L_q(\Omega)^N)}(\{(\tau\partial_\tau)^\ell \mathbf{G}\mathscr{I}(\lambda) \mid \lambda \in \Sigma_{\varepsilon,\lambda_1}\}) \leqslant (\sigma + C_\sigma\lambda_1^{-1})\omega_0 \tag{9.165}$$

($\ell = 0, 1$) for any $\sigma > 0$ and $\lambda_1 \geqslant \max(\lambda_0, 1)$ with some constant C_σ depending on σ but independent of λ_1.

On the other hand, by (9.137)

$$\|2(\nabla_{\Gamma^1_j}\zeta^1_j)\cdot\nabla_{\Gamma^1_j}h_j + (\Delta\zeta^1_j)h_j\|_{W^{1-1/q}_q(\Gamma)} \leqslant C\omega_0|\lambda|^{-1}\|F_\lambda(\tilde\zeta^1_j\mathbf{f}, \tilde\zeta^1_j f, \tilde\zeta^1_j\mathbf{f}_b)\|_{\mathscr{Y}_q(\mathscr{H}^1_j)}$$

for any $j \in \mathbb{N}$ and $\lambda \in \Sigma_{\varepsilon,\lambda_0}$ with some constant C independent of j and λ, so that by Proposition 9.5.2, (9.125) and Proposition 9.5.5, we see that there exists a $\mathfrak{h}(\lambda)(\mathbf{f}, f, \mathbf{f}_b) \in W^{1-1/q}_q(\Gamma)$ such that the infinite sum

$$-\sigma\sum_{j=1}^\infty(2(\nabla_{\Gamma^1_j}\zeta^1_j)\cdot\nabla_{\Gamma^1_j}h_j + (\Delta_{\Gamma^1_j}\zeta^1_j)h_j)$$

converges to $\mathfrak{h}(\lambda)(\mathbf{f}, f, \mathbf{f}_b)$ strongly in $W^{1-1/q}_q(\Gamma)$ and

$$\|\mathfrak{h}(\lambda)(\mathbf{f}, f, \mathbf{f}_b)\|_{W^{1-1/q}_q(\Gamma)} \leqslant C|\lambda|^{-1}\|F_\lambda(\mathbf{f}, f, \mathbf{f}_b)\|_{\mathscr{Y}_q(\Omega)}. \tag{9.166}$$

Moreover, there exists a $\mathscr{N}(\lambda) \in \mathrm{Hol}\,(\Sigma_{\varepsilon,\lambda_0}, \mathscr{L}(\mathscr{Y}_q(\Omega), W^{1-1/q}_q(\Gamma)))$ such that

$$\begin{aligned}&\mathscr{N}(\lambda)F_\lambda(\mathbf{f}, f, \mathbf{f}_b) = \mathfrak{h}(\lambda)(\mathbf{f}, f, \mathbf{f}_b) \quad \text{on } \Gamma,\\ &\mathscr{R}_{\mathscr{L}(\mathscr{Y}_q(\Omega), W^{1-1/q}_q(\Gamma))}(\{(\tau\partial_\tau)^\ell\mathscr{N}(\lambda) \mid \lambda \in \Sigma_{\varepsilon,\lambda_1}\}) \leqslant C\lambda_1^{-1}\omega_0\end{aligned} \tag{9.167}$$

($\ell = 0, 1$) for any $\lambda_1 \geqslant \max(\lambda_0, 1)$. Since

$$\sum_{j=1}^\infty(K_2(\zeta^1_j h_j) - \zeta^1_j L_j(h_j)) = \mathfrak{h}(\lambda)(\mathbf{f}, f, \mathbf{f}_b) \quad \text{on } \Gamma,$$

by Remark 9.1.3 and (9.164) we have

$$
\begin{aligned}
&\nabla\Big(\sum_{j-1}^{\infty} K_2(\zeta_j^1 h_j) - \zeta_j^1 L_j(h_j)\Big) \\
&\quad = \mathbf{T}_\Gamma^1 \mathfrak{h}(\lambda)(\mathbf{f}, f, \mathbf{f}_b) + \mathscr{K}_1(I(\lambda)(\mathbf{f}, f, \mathbf{f}_b) - \mathbf{T}_\Gamma^1 \mathfrak{h}(\lambda)(\mathbf{f}, f, \mathbf{f}_b)). \qquad (9.168)
\end{aligned}
$$

Let

$$
\mathscr{V}^{14}(\lambda)F = \mathbf{T}_\Gamma^1 \mathscr{N}(\lambda)F + \mathscr{K}_1(\mathscr{I}(\lambda)F - \mathbf{T}_\Gamma^1 \mathscr{N}(\lambda)F)
$$

for $F \in \mathscr{Y}_q(\Omega)$, and then by (9.168), (9.167), and (9.165), $\mathscr{V}^{14}(\lambda)$ satisfies the properties stated in (9.158). Summing up, we have proved that $\mathbf{V}^1(\lambda) = \sum_{i=1}^4 \mathbf{V}^{1i}(\lambda)$ and morever, if we set $\mathscr{V}^1(\lambda) = \sum_{i=1}^4 \mathscr{V}^{1i}(\lambda)$, then

$$
\begin{aligned}
&\mathscr{V}^1(\lambda) \in \mathrm{Hol}\,(\Sigma_{\varepsilon,\lambda_0}, \mathscr{L}(\mathscr{Y}_q(\Omega), L_q(\Omega)^N)), \\
&\mathscr{V}^1(\lambda)F_\lambda(\mathbf{f}, f, \mathbf{f}_b) = \mathbf{V}^1(\lambda)(\mathbf{f}, f, \mathbf{f}_b) \quad ((\mathbf{f}, f, \mathbf{f}_b) \in Y_q(\Omega)), \\
&\mathscr{R}_{\mathscr{L}(\mathscr{Y}_q(\Omega), L_q(\Omega)^N)}(\{(\tau\partial_\tau)^\ell \mathscr{V}^1(\lambda) \mid \lambda \in \Sigma_{\varepsilon,\lambda_1}\}) \leqslant (C_q\sigma + C_{\sigma,q}\lambda_1^{-1/2})\omega_0 \qquad (9.169)
\end{aligned}
$$

for $\ell = 0, 1$ and for any $\sigma > 0$ and $\lambda_1 \geqslant \max(1, \lambda_0)$.

Finally, we consider $\mathbf{V}^2(\lambda)(\mathbf{f}, f, \mathbf{f}_b)$. Write $\mathbf{D}(\varphi\mathbf{u}) = (\mathscr{A}\nabla\varphi)\mathbf{u}$ with some constant matrix $\mathscr{A}$, and we have

$$
\begin{aligned}
&\mathbf{V}^2(\lambda)(\mathbf{f}, f, \mathbf{f}_b) \\
&= \sum_{i=0}^{2}\sum_{j=1}^{\infty} \mu(\mathscr{A}\nabla\zeta_j^i)\mathbf{u}_j^i - \sum_{j=1}^{\infty} \mu < (\mathscr{A}\nabla\zeta_j^1)\mathbf{u}_j^1\mathbf{n}_j^1, \mathbf{n}_j^1 > \mathbf{n}_j^1 + \sum_{j=1}^{\infty}((\nabla\zeta_j^1)\cdot\mathbf{u}_j^1)\mathbf{n}_j^1,
\end{aligned}
$$

where the right hand side converges strongly in $W_q^1(\Omega)$. Moreover, we define an operator $\mathscr{V}^2(\lambda)$ acting on $F = (F_1, F_2, F_3, F_4) \in \mathscr{Y}_q(\Omega)$ by

$$
\begin{aligned}
\mathscr{V}^2(\lambda)F &= \sum_{i=0,1}\sum_{j=1}^{\infty} \mu(\mathscr{A}\nabla\zeta_j^i)\mathscr{S}_j^i(\lambda)\tilde{\zeta}_j^i F_1 + \sum_{j=1}^{\infty} \mu(\mathscr{A}\nabla\zeta_j^1)\mathscr{S}_j^1(\lambda)\tilde{\zeta}_j^1 F \\
&\quad - \sum_{j=1}^{\infty} \mu < (\mathscr{A}\nabla\zeta_j^1)\mathscr{S}_j^1(\lambda)\tilde{\zeta}_j^1 F, \mathbf{n}_j^1 > \mathbf{n}_j^1 + \sum_{j=1}^{\infty}((\nabla\zeta_j^1)\cdot\mathscr{S}_j^1(\lambda)\tilde{\zeta}_j^1 F)\mathbf{n}_j^1,
\end{aligned}
$$

by (9.137), (9.136), Proposition 9.5.2 and (9.125), we see that

$$
\begin{aligned}
&\mathscr{V}_2(\lambda) \in \mathrm{Hol}\,(\Sigma_{\varepsilon,\lambda_0}, \mathscr{L}(\mathscr{Y}_q(\Omega), W_q^1(\Omega)^N)), \\
&\mathscr{V}_2(\lambda)F_\lambda(\mathbf{f}, f, \mathbf{f}_b) = \mathbf{V}^2(\lambda)(\mathbf{f}, f, \mathbf{f}_b) \quad ((\mathbf{f}, f, \mathbf{f}_b) \in Y_q(\Omega)), \\
&\mathscr{R}_{\mathscr{L}(\mathscr{Y}_q(\Omega), W_q^{1-n}(\Omega)^N)}(\{(\tau\partial_\tau)^\ell(\lambda^{n/2}\mathscr{V}_2(\lambda)) \mid \lambda \in \Sigma_{\varepsilon,\lambda_1}\}) \leqslant C\omega_0\lambda_1^{-1/2} \qquad (9.170)
\end{aligned}
$$

for $\ell = 0, 1$ and $n = 0, 1$ and for any $\lambda_1 \geqslant \max(\lambda_0, 1)$.

Let

$$\mathscr{V}(\lambda) = (\mathscr{V}_1(\lambda), 0, \lambda^{1/2}\mathscr{V}_2(\lambda), \mathscr{V}_2(\lambda))$$

and then by (9.169) and (9.170), $\mathscr{V}(\lambda)$ is the operator satisfying the properties in Lemma 9.5.6, which completes the proof of Lemma 9.5.6.

9.5.5 Proof of Theorem 9.2.1

Recall that

$$\|(\mathbf{f}, f, \mathbf{f}_b)\|_{Y_q(\Omega)} = \|\mathbf{f}\|_{L_q(\Omega)} + \|f\|_{W_q^{2-1/q}(\Gamma)} + \|\mathbf{f}_b\|_{W_q^1(\Omega)}, \quad \text{for } (\mathbf{f}, f, \mathbf{f}_b) \in Y_q(\Omega),$$
$$\|(F_1, F_2, F_3, F_4)\|_{\mathscr{Y}_q(\Omega)} = \|(F_1, F_3)\|_{L_q(\Omega)} + \|F_2\|_{W_q^{2-1/q}(\Gamma)} + \|F_4\|_{W_q^1(\Omega)}$$

for $F = (F_1, F_2, F_3, F_4) \in \mathscr{Y}_q(\Omega)$, and then

$$\|F_\lambda(\mathbf{f}, f, \mathbf{f}_b)\|_{\mathscr{Y}_q(\Omega)} = \|(\mathbf{f}, \lambda^{1/2}\mathbf{f}_b)\|_{L_q(\Omega)} + \|f\|_{W_q^{2-1/q}(\Gamma)} + \|\mathbf{f}_b\|_{W_q^1(\Omega)},$$

so that $\|F_\lambda(\mathbf{f}, f, \mathbf{f}_b)\|_{L_q(\Omega)}$ $(\lambda \neq 0)$ give equivalent norms of $Y_q(\Omega)$. By Lemma 9.5.6,

$$\begin{aligned}\|F_\lambda\mathbf{V}(\lambda)(\mathbf{f}, f, \mathbf{f}_b)\|_{\mathscr{Y}_q(\Omega)} &= \|\mathscr{V}(\lambda)F_\lambda(\mathbf{f}, f, \mathbf{f}_b)\|_{\mathscr{Y}_q(\Omega)} \\ &\leqslant (\sigma + C_\sigma\lambda_1^{-1/2})\omega_0\|\mathbf{F}_\lambda(\mathbf{f}, f, \mathbf{f}_b)\|_{L_q(\Omega)}.\end{aligned}$$

Choosing σ and λ_1 in such a way that $0 < \sigma\omega_0 \leqslant 1/4$ and $C_\sigma\lambda_1^{-1/2}\omega_0 \leqslant 1/4$, we have

$$\|F_\lambda\mathbf{V}(\lambda)(\mathbf{f}, f, \mathbf{f}_b)\|_{L_q(\Omega)} \leqslant (1/2)\|\mathbf{F}_\lambda(\mathbf{f}, f, \mathbf{f}_b)\|_{L_q(\Omega)}$$

for any $\lambda \in \Sigma_{\varepsilon,\lambda_1}$, so that $(I - \mathbf{V}(\lambda))^{-1} = \sum_{j=0}^\infty \mathbf{V}(\lambda)^j$ exists in $\mathscr{L}(Y_q(\Omega))$. Moreover, by (9.146) $(I - \mathscr{V}(\lambda))^{-1} = \sum_{j=0}^\infty (F_\lambda\mathscr{V}(\lambda))^j$ exists and

$$\mathscr{R}_{\mathscr{L}(\mathscr{X}_{\mathscr{R},q}(\Omega))}(\{(\tau\partial_\tau)^\ell(I - \mathscr{V}(\lambda))^{-1} \mid \lambda \in \Sigma_{\varepsilon,\lambda_1}\}) \leqslant 2 \quad (\ell = 0, 1). \tag{9.171}$$

By (9.140), (9.142), and (9.144), and (9.140),

$$\mathbf{u} = \mathscr{S}(\lambda)F_\lambda(I - \mathbf{V}(\lambda))^{-1}(\mathbf{f}, f, \mathbf{f}_b), \quad h = \mathscr{T}(\lambda)F_\lambda(I - \mathbf{V}(\lambda))^{-1}(\mathbf{f}, f, \mathbf{f}_b)$$

are solutions of the equations (9.26). Sine $F_\lambda\mathbf{V}(\lambda) = \mathscr{V}(\lambda)F_\lambda$, we have

$$F_\lambda(I - \mathbf{V}(\lambda))^{-1} = F_\lambda\sum_{j=1}^\infty \mathbf{V}(\lambda)^j = \sum_{j=1}^\infty \mathscr{V}(\lambda)^j F_\lambda = (I - \mathscr{V}(\lambda))^{-1}F_\lambda,$$

so that

$$\mathbf{u} = \mathscr{S}(\lambda)(I - \mathscr{V}(\lambda))^{-1} F_\lambda(\mathbf{f}, f, \mathbf{f}_b), \quad h = \mathscr{T}(\lambda)(I - \mathscr{V}(\lambda))^{-1} F_\lambda(\mathbf{f}, f, \mathbf{f}_b).$$

Thus, setting $\mathscr{A}(\lambda) = \mathscr{S}(\lambda)(I - \mathscr{V}(\lambda))^{-1}$ and $\mathscr{H}(\lambda) = \mathscr{T}(\lambda)(I - \mathscr{V}(\lambda))^{-1}$, by (9.144), (9.171), and Proposition 9.3.4, we see that $\mathscr{A}(\lambda)$ and $\mathscr{H}(\lambda)$ satisfy the properties stated in Theorem 9.2.1, which completes the proof of the existence part of Theorem 9.2.1.

Next, we prove the uniqueness part of Theorem 9.2.1. Let $\mathbf{u} \in W_q^2(\Omega)^N$ and $h \in W_q^{3-1/q}(\Gamma)$ satisfy the homogeneous equations (9.35). Let $\mathbf{g}$ be any element in $J_{q'}(\Omega)$ and let $\mathbf{v} \in W_{q'}^2(\Omega)^N$ and $\rho \in W_{q'}^{3-1/q'}(\Gamma)$ be solutions to the equations:

$$\begin{cases} \lambda\mathbf{v} - \mathrm{Div}\,(\mu\mathbf{D}(\mathbf{v}) - (K_1(\mathbf{v}) + K_2(\rho))\mathbf{I}) = \mathbf{g} & \text{in } \Omega, \\ \lambda\rho - \mathbf{n}\cdot\mathbf{v} = 0 & \text{on } \Gamma, \\ (\mu\mathbf{D}(\mathbf{v}) - (K_1(\mathbf{v}) + K_2(\rho))\mathbf{I})\mathbf{n} - ((\tau + \delta\Delta_\Gamma)\rho)\mathbf{n} = 0 & \text{on } \Gamma, \\ \mathbf{v} = 0 & \text{on } \Gamma_0. \end{cases} \tag{9.172}$$

First, we observe that $\mathbf{v} \in J_{q'}(\Omega)$. In fact, for any $\varphi \in \hat{W}_q^1(\Omega)$, we have

$$\begin{aligned} 0 = (\mathbf{g}, \nabla\varphi)_\Omega &= \lambda(\mathbf{v}, \nabla\varphi)_\Omega - (\mathrm{Div}\,(\mu\mathbf{D}(\mathbf{v})) - \nabla\mathrm{div}\,\mathbf{v}, \nabla\varphi)_\Omega \\ &= \lambda(\mathbf{v}, \nabla\varphi)_\Omega - (\nabla\mathrm{div}\,\mathbf{v}, \nabla\varphi)_\Omega. \end{aligned} \tag{9.173}$$

Since $W_{q,\Gamma}^1(\Omega) \subset \hat{W}_q^1(\Omega)$, for any $\varphi \in W_{q,\Gamma}^1(\Omega)$ we have

$$0 = \lambda(\mathrm{div}\,\mathbf{v}, \varphi)_\Omega + (\nabla\mathrm{div}\,\mathbf{v}, \nabla\varphi)_\Omega,$$

so that the uniqueness guaranteed by Theorem 9.6.1 in Sect. 9.6.2 implies that $\mathrm{div}\,\mathbf{v} = 0$, which inserted into (9.173) yields that $(\mathbf{v}, \nabla\varphi)_\Omega = 0$ for any $\varphi \in \hat{W}_q^1(\Omega)$, that is $\mathbf{v} \in J_{q'}(\Omega)$. Analogously, we have $\mathbf{u} \in J_q(\Omega)$.

Since $K_1(\mathbf{v}),\ K_2(\rho) \in W_{q'}^1(\Omega) + \hat{W}_{q'}^1(\Omega)$, we write $K_1(\mathbf{v}) = A_1 + A_2$ and $K_2(\rho) = B_1 + B_2$ with $A_1,\ B_1 \in W_{q'}^1(\Omega)$ and $A_2,\ B_2 \in \hat{W}_{q'}^1(\Omega)$. Noting that $A_1 = K_1(\mathbf{v})$ and $B_1 = K_2(\rho)$ on Γ and that $(\mathbf{u}, \nabla A_2)_\Omega = (\mathbf{u}, \nabla B_2)_\Omega = 0$, by the divergence theorem of Gauß we have

$$\begin{aligned} (\mathbf{u}, \mathbf{g})_\Omega &= \lambda(\mathbf{u}, \mathbf{v}) - (\mathbf{u}, \mathrm{Div}\,(\mu\mathbf{D}(\mathbf{v}) - (A_1 + B_1)\mathbf{I}))_\Omega \\ &= \lambda(\mathbf{u}, \mathbf{v}) - (\mathbf{n}\cdot\mathbf{u}, (\tau + \delta\Delta_\Gamma)\rho)_\Gamma + \frac{\mu}{2}(\mathbf{D}(\mathbf{u}), \mathbf{D}(\mathbf{v}))_\Omega - (\mathrm{div}\,\mathbf{u}, A_1 + B_1)_\Omega. \end{aligned}$$

Finally, using the second equations in (9.172), we have

$$(\mathbf{u}, \mathbf{g})_\Omega = \lambda(\mathbf{u}, \mathbf{v})_\Omega - \lambda\tau(h, \rho)_\Gamma + \lambda\delta(\nabla_\Gamma h, \nabla_\Gamma\rho)_\Gamma + \frac{\mu}{2}(\mathbf{D}(\mathbf{u}), \mathbf{D}(\mathbf{v}))_\Omega. \tag{9.174}$$

Analogously, we have

$$\begin{aligned}0 &= (\lambda\mathbf{u} - \mathrm{Div}\,(\mu(\mathbf{D}(\mathbf{u}) - K_1(\mathbf{u}) + K_2(h))\mathbf{I}), \mathbf{v})_\Omega \\ &= \lambda(\mathbf{u}, \mathbf{v})_\Omega - \lambda\tau(h, \rho)_\Gamma + \lambda\delta(\nabla_\Gamma h, \nabla_\Gamma \rho)_\Gamma + \frac{\mu}{2}(\mathbf{D}(\mathbf{u}), \mathbf{D}(\mathbf{v}))_\Omega,\end{aligned}$$

which, combined with (9.174), furnishes that

$$(\mathbf{u}, \mathbf{g})_\Omega = 0 \quad \text{for any } \mathbf{g} \in J_{q'}(\Omega). \tag{9.175}$$

For any $\mathbf{f} \in C_0^\infty(\Omega)^N$, let $\psi \in \mathscr{W}^1_{q'}(\Omega)$ be a solution to the variational equation $(\mathbf{f}, \nabla\varphi)_\Omega = (\nabla\psi, \nabla\varphi)_\Omega$ for any $\varphi \in \mathscr{W}^1_q(\Omega)$. Let $\mathbf{g} = \mathbf{f} - \nabla\psi$, and then $\mathbf{g} \in J_{q'}(\Omega)$ and $(\mathbf{u}, \nabla\psi)_\Omega = 0$. Thus, by (9.175), $(\mathbf{u}, \mathbf{f})_\Omega = (\mathbf{u}, \mathbf{g})_\Omega = 0$, which, combined with the arbitrariness of the choice of $\mathbf{f}$, furnishes that $\mathbf{u} = 0$. And then, by the second equation of (9.35) yields that $h = 0$. This completes the proof of the uniqueness part of Theorem 9.2.1.

9.6 Proof of Theorem 9.2.2

9.6.1 Existence Part

In this section, we prove the unique existence of solutions to time dependent problem (9.36). In this subsection, we prove the existence part. For this purpose, we transform the problem to the zero initial data case. To this end, we take a domain Ω_1 such that $\partial\Omega_1 = \Gamma_0$ and $\Omega \subset \Omega_1$. The Ω_1 is a uniform $W_r^{2-1/r}$ $(N < r < \infty)$ domain. Let $\mathbf{u}_0 \in B^{2(1-1/p)}_{q,p}(\Omega)$ be an initial velocity field for problem (9.1) and let $\tilde{\mathbf{u}}_0 = (\tilde{u}_{01}, \ldots, \tilde{u}_{0N})$ be an extension of $\mathbf{u}_0$ to Ω_1 such that $\mathbf{u}_0 = \tilde{\mathbf{u}}_0$ on Ω and $\|\tilde{\mathbf{u}}_0\|_{B^{2(1-1/p)}_{q,p}(\Omega_1)} \leqslant C\|\mathbf{u}_0\|_{B^{2(1-1/p)}_{q,p}(\Omega)}$. We consider the time-shifted heat equations:

$$\partial_t v_j + \lambda_0 v_j - \mu\Delta v_j = 0 \quad \text{in } \Omega_1 \times (0, \infty), \quad v_j|_{\Gamma_0} = 0, \quad v_j|_{t=0} = \tilde{u}_{0j} \tag{9.176}$$

$(j = 1, \ldots, N)$. Since $\tilde{u}_{0j}$ satisfies the compatibility condition: $\tilde{u}_{0j}|_{\Gamma_0} = u_{0j}|_{\Gamma_0} = 0$ as follows from (9.18), employing the similar argumentation to that in Shibata [14, 15], we see that there exist v_j $(j = 1, \ldots, N)$ such that

$$\begin{aligned}&v_j \in L_p((0, \infty), W^2_q(\Omega_1)) \cap W^1_p((0, \infty), L_q(\Omega_1)), \\ &\|\partial_t v_j\|_{L_p((0,\infty),L_q(\Omega_1))} + \|v_j\|_{L_p((0,\infty),W^2_q(\Omega_1))} \\ &\quad \leqslant C\|\tilde{u}_{0j}\|_{B^{2(1-1/p)}_{q,p}(\Omega_1)} \leqslant C\|\mathbf{u}_0\|_{B^{2(1-1/p)}_{q,p}(\Omega)}.\end{aligned} \tag{9.177}$$

Thus, we define $\mathbf{v}$ by $\mathbf{v} = (v_1, \ldots, v_N)$.

Concerning the initial data $h_0 \in B^{3-1/p-1/q}_{q,p}(\Gamma)$ for the height function of problem (9.1), let $\tilde h_0 \in B^{3-1/p}_{q,p}(\mathbb{R}^N)$ be an extension of h_0 to $\mathbb{R}^N$ such that $\tilde h_0 = h_0$ on Γ and $\|\tilde h\|_{B^{3-1/p}_{q,p}(\mathbb{R}^N)} \leqslant C\|h\|_{B^{3-1/p-1/q}_{q,p}(\Gamma)}$. We define the function d by

$$d = e^{-At}\tilde h_0 = \mathscr{F}^{-1}[e^{-t\sqrt{1+|\xi|^2}}\mathscr{F}[\tilde h_0](\xi)](x).$$

We have

$$\begin{aligned}
&d \in L_p((0,\infty), W^3_q(\mathbb{R}^N)) \cap W^1_p((0,\infty), W^2_q(\mathbb{R}^N)),\\
&\|d\|_{L_p((0,\infty),W^3_q(\mathbb{R}^N))} + \|\partial_t d\|_{L_p((0,\infty),W^2_q(\mathbb{R}^N))}\\
&\quad \leqslant C\|\tilde h_0\|_{B^{3-1/p}_{q,p}(\mathbb{R}^N)} \leqslant C\|h_0\|_{B^{3-1/p-1/q}_{q,p}(\Gamma)}.
\end{aligned} \tag{9.178}$$

Setting $\mathbf{u} = \mathbf{v} + \mathbf{w}$ and $h = d + \rho$, we have

$$\begin{cases}
\partial_t\mathbf{w} - \mathrm{Div}\,(\mu\mathbf{D}(\mathbf{u}) - (K_1(\mathbf{w}) + K_2(\rho))\mathbf{I}) = \tilde{\mathbf{f}}, & \text{in } \Omega \times (0,T),\\
\partial_t\rho - \mathbf{n}\cdot\mathbf{w} = \tilde f & \text{on } \Gamma \times (0,T),\\
\mathscr{T}_\mathbf{n}(\mu\mathbf{D}(\mathbf{w})\mathbf{n}) = \mathscr{T}_\mathbf{n}(\tilde{\mathbf{f}}_b), \quad \mathrm{div}\,\mathbf{w} = \mathbf{n}\cdot\tilde{\mathbf{f}}_b & \text{on } \Gamma \times (0,T),\\
\mathbf{w} = 0 & \text{on } \Gamma_0 \times (0,T),\\
(\mathbf{w},\rho)|_{t=0} = (0,0) & \text{in } \Omega \times \Gamma,
\end{cases} \tag{9.179}$$

where

$$\begin{aligned}
&\tilde{\mathbf{f}} = \mathbf{f} - \partial_t\mathbf{v} + \mathrm{Div}\,(\mu(\mathbf{D}(\mathbf{v}) - (K_1(\mathbf{v}) + K_2(d))\mathbf{I}), \quad \tilde f = f - \partial_t d + \mathbf{n}\cdot\mathbf{v},\\
&\tilde{\mathbf{f}}_b = \mathbf{f}_b - \mathscr{T}_\mathbf{n}(\mu\mathbf{D}(\mathbf{v})\mathbf{n}) - (\mathrm{div}\,\mathbf{v})\mathbf{n}
\end{aligned}$$

By the compatibility condition (9.37), we have

$$\mathscr{T}_\mathbf{n}\tilde{\mathbf{f}}_b|_{t=0} = 0, \quad \mathbf{n}\cdot\tilde{\mathbf{f}}_b|_{t=0} = 0 \quad \text{in } \Omega.$$

Moreover, by (9.177) and (9.178), we have

$$\mathscr{M}_R(t,0,0,\tilde{\mathbf{f}},\tilde f,\tilde{\mathbf{f}}_b) \leqslant C\mathscr{M}_R(t,\mathbf{u}_0,h_0,\mathbf{f},f,\mathbf{f}_b).$$

Thus, from now on we consider problem (9.36) with $\mathbf{u}_0 = 0$ and $h_0 = 0$ under the condition:

$$\mathbf{f}_b|_{t=0} = 0 \quad \text{in } \Omega. \tag{9.180}$$

Given any function $f(\cdot,t)$ defined on $(0,T)$, let f_0 denotes the zero extension of f to $(-\infty,0)$, namely $f_0(\cdot,t) = f(\cdot,t)$ for $t \in (0,T)$ and $f_0(\cdot,t) = 0$ for $t \in (-\infty,0)$. Let E_t be an operator defined by

$$[E_t f](\cdot, s) = \begin{cases} f_0(\cdot, s) & \text{for } s < t, \\ f_0(\cdot, 2t - s) & \text{for } s > t. \end{cases} \tag{9.181}$$

Obviously, $[E_t f](\cdot, s) = 0$ for $s \notin (0, 2t)$, Moreover, if $f|_{t=0} = 0$, then we have

$$\partial_s [E_t f](\cdot, s) = \begin{cases} 0 & \text{for } s \notin (0, 2t), \\ (\partial_s f)(\cdot, s) & \text{for } s \in (0, t), \\ -(\partial_s f)(\cdot, 2t - s) & \text{for } s \in (t, 2t). \end{cases} \tag{9.182}$$

Instead of (9.36) with $\mathbf{u}_0 = 0$ and $h_0 = 0$, we consider the equations:

$$\begin{cases} \partial_t \mathbf{u} - \mathrm{Div}\,(\mu \mathbf{D}(\mathbf{u}) - (K_1(\mathbf{u}) + K_2(h))\mathbf{I}) = E_t[\mathbf{f}] & \text{in } \Omega \times \mathbb{R}, \\ \partial_t h - \mathbf{n} \cdot \mathbf{u} = E_t[f] & \text{on } \Gamma \times \mathbb{R}, \\ \mathcal{T}_\mathbf{n}(\mu \mathbf{D}(\mathbf{u})\mathbf{n}) = \mathcal{T}_\mathbf{n}(E_t[\mathbf{f}_b]), \quad \mathrm{div}\, \mathbf{u} = \mathbf{n} \cdot E_t[\mathbf{f}_b] & \text{on } \Gamma \times \mathbb{R}, \\ \mathbf{u} = 0 & \text{on } \Gamma_0 \times \mathbb{R}. \end{cases} \tag{9.183}$$

Let $\mathcal{L}$ and $\mathcal{L}^{-1}$ denote the Laplace-Fourier transform and the inverse Laplace-Fourier transform with respect to t defined by

$$\mathcal{L}[f](\tau) = \int_{-\infty}^{\infty} e^{-(\gamma + i\tau)t} f(t)\, dt, \quad \mathcal{L}^{-1}[g](t) = \frac{1}{2\pi} \int_{-\infty}^{\infty} e^{(\gamma + i\tau)t} g(\tau)\, d\tau.$$

Let $\mathcal{F}_t$ and $\mathcal{F}_\tau^{-1}$ be the Fourier transform with respect to t and the inverse Fourier transform with respect to τ, and then $\mathcal{L}[f](\lambda) = \mathcal{F}_t[e^{-\gamma t} f(t)]$ and $\mathcal{L}^{-1}[g](t) = e^{\gamma t} \mathcal{F}_\tau^{-1}[g(\tau)](t)$ with $\lambda = \gamma + i\tau$ ($\gamma, \tau \in \mathbb{R}$). Applying the Laplace-Fourier transform to (9.183), we have

$$\begin{cases} \lambda \mathbf{v} - \mathrm{Div}\,(\mu \mathbf{D}(\mathbf{v}) - (K_1(\mathbf{v}) + K_2(\eta))\mathbf{I}) = \hat{\mathbf{f}} & \text{in } \Omega, \\ \lambda \eta - \mathbf{n} \cdot \mathbf{v} = \hat{f} & \text{on } \Gamma, \\ \mathcal{T}_\mathbf{n}(\mu \mathbf{D}(\mathbf{v})\mathbf{n}) = \mathcal{T}_\mathbf{n}(\hat{\mathbf{f}}_b), \quad \mathrm{div}\, \mathbf{v} = \mathbf{n} \cdot \hat{\mathbf{f}}_b & \text{on } \Gamma, \\ \mathbf{v} = 0 & \text{on } \Gamma_0. \end{cases} \tag{9.184}$$

Here,

$$\hat{\mathbf{f}} = \mathcal{L}[E_t \mathbf{f}](x, \lambda), \quad \hat{f} = \mathcal{L}[E_t f](x, \lambda), \quad \hat{\mathbf{f}}_b = \mathcal{L}[E_t \mathbf{f}_b](x, \lambda).$$

Applying the solution operators $\mathcal{A}(\lambda)$ and $\mathcal{H}(\lambda)$ given in Theorem 9.2.1, we have

$$\mathbf{v} = \mathcal{A}(\lambda)\mathbf{D}_\lambda, \quad \eta = \mathcal{H}(\lambda)\mathbf{D}_\lambda, \quad \mathbf{D}_\lambda = (\hat{\mathbf{f}}, \hat{f}, \lambda^{1/2}\hat{\mathbf{f}}_b, \nabla \hat{\mathbf{f}}_b). \tag{9.185}$$

Let $\Lambda_\gamma f$ be the operator defined by $\Lambda_\gamma f = \mathcal{L}^{-1}[\lambda^{1/2}\mathcal{L}[f](\lambda)]$. Note that $\lambda^{1/2}\hat{\mathbf{f}} = \mathcal{L}[\Lambda_\gamma^{1/2} E_t \mathbf{f}]$ and $\lambda^{1/2}\hat{\mathbf{f}}_b = \mathcal{L}[\Lambda_\gamma^{1/2} E_t \mathbf{f}_b]$. Applying the inverse Laplace-Fourier transform, the solutions $\mathbf{u}$ and h of problem (9.183) are given by $\mathbf{u} = \mathcal{L}^{-1}[\mathcal{A}(\lambda)\mathbf{D}_\lambda]$ and

$h = \mathscr{L}^{-1}[\mathscr{H}(\lambda)\mathbf{D}_\lambda)$. Since $(\partial_t, \Lambda_\gamma^{1/2})\mathbf{u}(\cdot, t) = \mathscr{L}^{-1}[(\lambda, \lambda^{1/2})\mathscr{A}(\lambda)\mathbf{D}_\lambda]$ and $\partial_t h = \mathscr{L}^{-1}[\lambda\mathscr{H}(\lambda)\mathbf{D}_\lambda]$, by the Weis operator valued Fourier multiplier theorem [20] and Theorem 9.2.1,

$$\begin{aligned}
&\|e^{-\gamma s}\partial_s \mathbf{u}\|_{L_p(\mathbb{R}, L_q(\Omega))} + \|e^{-\gamma t}\Lambda_\gamma^{1/2}\mathbf{u}\|_{L_p(\mathbb{R}, W_q^1(\Omega))} + \|e^{-\gamma t}\mathbf{u}\|_{L_p(\mathbb{R}, W_q^2(\Omega))} \\
&\quad + \|e^{-\gamma s}\partial_t h\|_{L_p(\mathbb{R}, W_q^{2-1/q}(\Gamma))} + \|e^{-\gamma t} h\|_{L_p(\mathbb{R}, W_q^{3-1/q}(\Gamma))} \\
&\leqslant C\{\|e^{-\gamma s}(E_t\mathbf{f}, \Lambda_\gamma^{1/2}E_t\mathbf{f}_b)\|_{L_p(\mathbb{R}, L_q(\Omega))} + \|e^{-\gamma s}E_t\mathbf{f}_b\|_{L_p(\mathbb{R}, W_q^1(\Omega))} \\
&\quad + \|e^{-\gamma s}E_t f\|_{L_p(\mathbb{R}, W_q^{2-1/q}(\Gamma))}\}
\end{aligned} \tag{9.186}$$

for any $\gamma \geqslant \gamma_0$ with some constants C and γ_0. From Shibata [16, Appendix] we have

$$\begin{aligned}
&\|e^{\gamma s}\Lambda_\gamma^{1/2} f\|_{L_p(\mathbb{R}, L_q(\Omega))} \leqslant C\{\|e^{-\gamma s}\partial_s f\|_{L_p(\mathbb{R}, \mathbf{W}_q^{-1}(\Omega))} + \|e^{-\gamma s} f\|_{L_p(\mathbb{R}, W_q^1(\Omega))}\}, \\
&\gamma\|e^{-\gamma s} f\|_{L_p(\mathbb{R}, X)} \leqslant C\|e^{-\gamma s}\partial_s f\|_{L_p(\mathbb{R}, X)} \quad (X \in \{L_q(\Omega), W_q^{2-1/q}(\Gamma)\}),
\end{aligned} \tag{9.187}$$

which, combined with (9.186), furnishes that

$$\begin{aligned}
&\|e^{-\gamma s}\partial_t \mathbf{u}\|_{L_p(\mathbb{R}, L_q(\Omega))} + \|e^{-\gamma s}\mathbf{u}\|_{L_p(\mathbb{R}, W_q^2(\Omega))} + \gamma\|e^{-\gamma s}\mathbf{u}\|_{L_p(\mathbb{R}, L_q(\Omega))} \\
&+ \|e^{-\gamma s}\partial_t h\|_{L_p(\mathbb{R}, W_q^{2-1/q}(\Gamma))} + \|e^{-\gamma s} h\|_{L_p(\mathbb{R}, W_q^{3-1/q}(\Gamma))} + \gamma\|e^{-\gamma s} h\|_{L_p(\mathbb{R}, W_q^{2-1/q}(\Gamma))} \\
&\leqslant C\tilde{\mathscr{M}}_\gamma
\end{aligned} \tag{9.188}$$

with

$$\begin{aligned}
\tilde{\mathscr{M}}_\gamma &= \|e^{-\gamma s}E_t\mathbf{f}\|_{L_p(\mathbb{R}, L_q(\Omega))} + \|e^{-\gamma s}\partial_s E_t\mathbf{f}_b\|_{L_p(\mathbb{R}, \mathbf{W}_q^{-1}(\Omega))} + \|e^{-\gamma s}E_t\mathbf{f}_b\|_{L_p(\mathbb{R}, W_q^1(\Omega))} \\
&\quad + \|e^{-\gamma s}E_t f\|_{L_p(\mathbb{R}, W_q^{2-1/q}(\Gamma)}
\end{aligned}$$

By (9.181) and (9.182), we have

$$\tilde{\mathscr{M}}_\gamma \leqslant C\mathscr{M}_R(t, 0, 0, \mathbf{f}, f, \mathbf{f}_b). \tag{9.189}$$

On the other hand, by (9.188) and (9.189) we have

$$\begin{aligned}
&\|\mathbf{u}\|_{L_p((-\infty,0), L_p(\Omega))} + \|h\|_{L_p((-\infty,0), W_q^2(\Omega))} \\
&\leqslant \|e^{-\gamma s}\mathbf{u}\|_{L_p(\mathbb{R}, L_p(\Omega))} + \|e^{-\gamma s} h\|_{L_p(\mathbb{R}, W_q^2(\Omega))} \\
&\leqslant \gamma^{-1}C\mathscr{M}_R((0, t), 0, 0, \mathbf{f}, f, \mathbf{f}_b) \to 0 \quad \text{letting } \gamma \to \infty,
\end{aligned}$$

which furnishes that $\|\mathbf{u}\|_{L_p((-\infty,0), L_p(\Omega))} + \|h\|_{L_p((-\infty,0), W_q^2(\Omega))} = 0$. This implies that $\mathbf{u}(\cdot, s) = 0$ and $h(\cdot, s) = 0$ for $s \in (-\infty, 0)$. In particular, $(\mathbf{u}, h)|_{s=0} = (0, 0)$ in $\Omega \times \Gamma$. Moreover, we have

$$\mathscr{I}_R(t, \mathbf{u}, h) \leqslant Ce^{\gamma t}\mathscr{M}_R(t, 0, 0, \mathbf{f}, f, \mathbf{f}_b). \tag{9.190}$$

Since $[E_t f](\cdot, s) = f(\cdot, s)$ for $s \in (0, t)$, $\mathbf{u}$ and h satisfy the equations (9.36) with $T = t$ and $(\mathbf{u}, h)|_{s=0} = (0, 0)$. For $0 < t_1 < t_2 \leqslant T$, let $\mathbf{u}^{t_i}$ and h^{t_i} be solutions of equations (9.36) with $T = t_i$ and $(\mathbf{u}_0, h_0) = (0, 0)$. By the uniqueness of solutions proved in Sect. 9.6.3, we have $(\mathbf{u}^{t_1}, \mathfrak{p}^{t_1}, h^{t_1}) = (\mathbf{u}^{t_2}, \mathfrak{p}^{t_2}, h^{t_2})$ for $s \in (0, t_1)$, so that if we set $(\mathbf{u}, \mathfrak{p}, h) = (\mathbf{u}^T, \mathfrak{p}^T, h^T)$, then we have $(\mathbf{u}, \mathfrak{p}, h) = (\mathbf{u}^t, \mathfrak{p}^t, h^t)$ for any $t \in (0, T)$. This completes the proof of the existence part of Theorem 9.2.2.

9.6.2 The Weak Laplace Problem with Dirichlet Condition

In this subsection, let $W^{-1}_q(\Omega)$ be the dual space of $W^1_{q',\Gamma}(\Omega)$. By the Hahn-Banach theorem there exists a map $\mathscr{F}_0 : W^{-1}_q(\Omega) \to L_q(\Omega)^{N+1}$ such that for any $\mathscr{I} \in W^{-1}_q(\Omega)$

$$\mathscr{I}(\psi) = (\mathscr{F}_{00}(\mathscr{I}), \psi)_\Omega + (\mathscr{F}_0'(\mathscr{I}), \nabla\psi)_\Omega$$

for any $\psi \in W^1_{q',\Gamma}(\Omega)$, where $\mathscr{F}_0(\mathscr{I}) = (\mathscr{F}_{00}(\mathscr{I}), \mathscr{F}_0'(\mathscr{I}))$, and

$$\|\mathscr{F}_0(\mathscr{I})\|_{L_q(\Omega)} \leqslant C_q\|\mathscr{I}\|_{W^{-1}_q(\Omega)}.$$

Especially, if g is represented by $g = g_0 + \operatorname{div}\mathbf{g}$ with $(g_0, \mathbf{g}) \in L_q(\Omega)^{N+1}$, then $g \in W^{-1}_q(\Omega)$ and

$$\|g\|_{W^{-1}_q(\Omega)} \leqslant \|(g_0, \mathbf{g})\|_{L_q(\Omega)}.$$

First, we consider the following resolvent problem:

$$\lambda(g, \varphi)_\Omega + (\nabla g, \nabla\varphi)_\Omega = (\mathbf{f}, \nabla\varphi)_\Omega + (f, \varphi)_\Omega \quad \text{for any } \varphi \in W^1_{q',\Gamma}(\Omega) \tag{9.191}$$

subject to $g = h$ on Γ. We have the following theorem.

Theorem 9.6.1 *Let* $1 < q < \infty$, $0 < \varepsilon < \pi/2$, $N < r < \infty$ *and* $\max(q, q') \leqslant r$. *Assume that* Ω *is a uniform* $W^{2,2}_r(\Omega)$ *domain. Let*

$$Y^2_q(\Omega) = \{(\mathbf{f}, f, h) \mid (\mathbf{f}, f) \in L_q(\Omega)^{N+1},\ h \in W^1_q(\Omega)\},$$
$$\mathscr{Y}^2_q(\Omega) = \{(F_0, F_8, F_9) \mid F_0 \in L_q(\Omega)^{N+1},\ F_8 \in L_q(\Omega),\ F_9 \in W^1_q(\Omega)\},$$

where F_0, F_8 *and* F_9 *are variables corresponding to* $(\mathbf{f}, f)$, $\lambda^{1/2}h$ *and* h. *Then, there exist a constant* $\lambda_0 > 0$ *and an operator family* $\mathscr{G}_\Omega(\lambda)$ *with*

$$\mathscr{G}_\Omega(\lambda) \in \mathrm{Hol}\,(\Sigma_{\varepsilon,\lambda_0}, \mathscr{L}(\mathscr{Y}^2_q(\Omega), W^1_q(\Omega) \cap W^{-1}_q(\Omega)))$$

such that for any $\lambda \in \Sigma_{\varepsilon,\lambda_0}$ *and* $(\mathbf{f}, f, h) \in Y_q^2(\Omega)$, $g = \mathscr{G}_\Omega(\lambda)(\mathbf{f}, f, \lambda^{1/2}h, h)$ *is a unique solution of problem* (9.191), *and*

$$\mathscr{R}_{\mathscr{L}(\mathscr{Y}_q^2(\Omega), W_q^{1-j}(\Omega))}(\{(\tau\partial_\tau)^\ell(\lambda^{j/2}\mathscr{G}_\Omega(\lambda)) \mid \lambda \in \Sigma_{\varepsilon,\lambda_0}\}) \leqslant c$$

for $\ell = 0, 1$ *and* $j = 0, 1, 2$.

Proof In the case where $\Omega = \mathbb{R}^N$ and $\Omega = \mathbb{R}^N_+$ employing the same arguments as in the proof of Lemma 9.3.2, Lemma 9.3.10 (a) and Lemma 9.3.15 (a), we can prove Theorem 9.6.1. When Ω is the bent half space Ω_+ given in Sect. 9.4, we set $W_q^{-1}(\Omega_+)$ is the dual space of $W^1_{q',0}(\Omega_+)$. In this case, we consider the variational equation:

$$\lambda(g, \varphi)_{\Omega_+} + (\nabla g, \nabla\varphi)_{\Omega_+} = (\mathbf{f}, \nabla\varphi)_{\Omega_+} + (f, \varphi)_{\Omega_+} \tag{9.192}$$

for any $\varphi \in W^1_{q',0}(\Omega_+)$ subject to $g = h$ on Γ_+. As was seen in Sect. 9.4.2, by the change of variable: $\xi = \Phi(x)$ the variational problem (9.192) is transformed to the variational problem:

$$\lambda(u \det G, \psi)_{\mathbb{R}^N_+} + (\nabla u, \nabla\psi)_{\mathbb{R}^N_+} + (\mathscr{B}^0\nabla u, \nabla\psi)_{\mathbb{R}^N_+} = (\tilde{\mathbf{f}}, \nabla\psi)_{\mathbb{R}^N_+} + (\tilde{f}, \psi)_{\mathbb{R}^N_+}$$

subject to $u = \tilde{h}$. Setting $v = u \det G$, we have

$$\begin{aligned}\lambda(v, \psi)_{\mathbb{R}^N_+} &+ (\nabla v, \nabla\psi)_{\mathbb{R}^N_+} + (\nabla((\det G)^{-1} - 1)v, \nabla\psi)_{\mathbb{R}^N_+} \\ &+ (\mathscr{B}^0\nabla((\det G)^{-1}v), \nabla\psi)_{\mathbb{R}^N_+} = (\tilde{\mathbf{f}}, \nabla\psi)_{\mathbb{R}^N_+} + (\tilde{f}, \psi)_{\mathbb{R}^N_+}\end{aligned}$$

subject to $v = (\det G)^{-1}\tilde{h}$. Since

$$\|(1 - (\det G)^{-1})\|_{L_\infty(\mathbb{R}^N)} \leqslant CM_1, \quad \|\nabla(1 - (\det G)^{-1})\|_{L_r(\mathbb{R}^N_+)} \leqslant C_{M_2},$$

by (9.98) and (9.99), the terms:

$$(\nabla((\det G)^{-1} - 1)v, \nabla\psi)_{\mathbb{R}^N_+} + (\mathscr{B}^0\nabla((\det G)^{-1}v), \nabla\psi)_{\mathbb{R}^N_+}$$

can be regarded as a small perturbation from the main part: $\lambda(v, \psi)_{\mathbb{R}^N_+} + (\nabla v, \nabla\psi)_{\mathbb{R}^N_+}$ Thus, using the Banach fixed point argument, we can prove Theorem 9.6.1 when $\Omega = \Omega_+$ with Dirichlet boundary condition.

When $\Omega = \Omega_+$ with Neumann boundary condition, we set $W_q^{-1}(\Omega_+)$ is the dual space of $W^1_{q'}(\Omega_+)$. In this case, we consider the variational equation:

$$\lambda(g, \varphi)_{\Omega_+} + (\nabla g, \nabla\varphi)_{\Omega_+} = (\mathbf{f}, \nabla\varphi)_{\Omega_+} + (f, \varphi)_{\Omega_+}$$

for any $\varphi \in W^1_{q'}(\Omega_+)$. Employing the same argumentation as above, we can prove Theorem 9.6.1.

When Ω is a uniform $W^{2,2}_r$ domain, we use the same argumentation as in Sect. 9.5. Let g^i_j be solutions to the following variational equations of three different cases:

$$\lambda(g^m_j, \psi)_{\mathcal{H}^m_j} + (\nabla g^m_j, \nabla\psi)_{\mathcal{H}^m_j} = (\tilde\zeta^m_j\mathbf{f}, \nabla\psi)_{\mathcal{H}^m_j} + (\tilde\zeta^m_j f, \psi)_{\mathcal{H}^m_j}$$

for any $\psi \in W^1_{q'}(\mathcal{H}^m_j)$ $(m = 0, 2)$;

$$\lambda(g^1_j, \psi)_{\mathcal{H}^1_j} + (\nabla g^1_j, \nabla\psi)_{\mathcal{H}^1_j} = (\tilde\zeta^1_j\mathbf{f}, \nabla\psi)_{\mathcal{H}^1_j} + (\tilde\zeta^1_j f, \psi)_{\mathcal{H}^1_j} \tag{9.193}$$

for any $\psi \in W^1_{q',0}(\mathcal{H}^1_j)$, subject to $g^1_j = \tilde\zeta^1_j h$ on Γ^1_j. By the results for the whole space and bent half space, there exist operator families $\mathcal{G}^m_j(\lambda)$ with

$$\mathcal{G}^m_j(\lambda) \in \mathrm{Hol}\,(\Sigma_{\varepsilon,\lambda_0}, \mathcal{L}(\mathcal{Y}^2_q(\mathcal{H}^m_j), W^1_q(\mathcal{H}^m_j) \cap W^{-1}_q(\mathcal{H}^m_j))),$$
$$\mathcal{R}_{\mathcal{L}(\mathcal{Y}^2_q(\mathcal{H}^m_j), W^{1-k}_q(\mathcal{H}^m_j))}(\{(\tau\partial_\tau)^\ell(\lambda^{k/2}\mathcal{G}^m_j(\lambda) \mid \lambda \in \Sigma_{\varepsilon,\lambda_0}\}) \leqslant c$$

for $\ell = 0, 1$ and $k = 0, 1, 2$, where c is some constant independent of $j \in \mathbb{N}$. Here, $W^{-1}_q(\mathcal{H}^m_j)$ are the dual space of $W^1_{q'}(\mathcal{H}^m_j)$ for $m = 0, 2$ and $W^1_{q',0}(\mathcal{H}^1_j)$, respectively. From (9.193), we have

$$\begin{aligned}
&\lambda(\sum_{i=0}^2\sum_{j=1}^\infty \zeta^i_j g^i_j, \psi)_\Omega + (\nabla(\sum_{i=0}^2\sum_{j=1}^\infty \zeta^i_j g^i_j), \nabla\psi)_\Omega \\
&= (\mathbf{f}, \nabla\psi)_\Omega + (f, \psi)_\Omega + 2(\sum_{i=0}^2\sum_{j=1}^\infty (\nabla\zeta^i_j)g^i_j, \nabla\psi)_\Omega \\
&+ (\sum_{i=0}^2\sum_{j=1}^\infty ((\Delta\tilde\zeta^i_j)g^i_j, \psi)_\Omega - (\sum_{j=1}^\infty \{(\nabla\tilde\zeta^0_j)\cdot\mathbf{n}_0\}g^0_j, \psi)_{\Gamma_0}.
\end{aligned}$$

Let I be an element of $W^{-1}_q(\Omega)$ defined by

$$< I, \psi >= (\sum_{j=1}^\infty \{(\nabla\tilde\zeta^0_j)\cdot\mathbf{n}_0\}g^0_j, \psi)_{\Gamma_0},$$

where $< \cdot, \cdot >$ denotes the dual paring between $W^{-1}_q(\Omega)$ and $W^1_{q',\Gamma}(\Omega)$, and then by (9.125) and the classical interplation inequality about the boundary trace to Γ_0 like (9.156), we have

$$| < I, \psi > | \leqslant C_q\{\sum_{j=1}^\infty (\varepsilon\|g^0_j\|^q_{W^1_q(\mathcal{H}^0_j)} + C_\varepsilon\|g^0_j\|_{L_q(\mathcal{H}^0_j)}\}^{1/q}\|\psi\|_{W^1_{q'}(\Omega)}.$$

Thus, there exists an $N+1$ vector of functions $\mathbf{G} = (G_0, \mathbf{G}') \in L_q(\Omega)^{N+1}$ such that

$$< I, \psi >_\Omega = (G_0, \psi)_\Omega + (\mathbf{G}', \nabla\psi)_\Omega,$$
$$\|\mathbf{G}\|_{L_q(\Omega)} \leqslant C_q\{\sum_{j=1}^{\infty}(\varepsilon\|g_j^0\|^q_{W^1_q(\mathscr{H}_j^0)} + C_\varepsilon\|g_j^0\|_{L_q(\mathscr{H}_j^0)}\}^{1/q},$$

and then, we have

$$\begin{aligned}
&\lambda(\sum_{i=0}^{2}\sum_{j=1}^{\infty}\zeta_j^i g_j^i, \psi)_\Omega + (\nabla(\sum_{i=0}^{2}\sum_{j=1}^{\infty}\zeta_j^i g_j^i), \nabla\psi)_\Omega \\
&= (\mathbf{f}, \nabla\psi)_\Omega + (f, \psi)_\Omega + (2\sum_{i=0}^{2}\sum_{j=1}^{\infty}(\nabla\zeta_j^i)g_j^i + \mathbf{G}', \nabla\psi)_\Omega \\
&+ (\sum_{i=0}^{2}\sum_{j=1}^{\infty}((\Delta\tilde{\zeta}_j^i)g_j^i + G_0, \psi)_\Omega.
\end{aligned}$$

Employing the similar argumentation to that in Sect. 9.5, we can prove Theorem 9.6.1. This completes the proof of Theorem 9.6.1. □

Next, we consider the following time dependent problem corresponding to (9.191):

$$\begin{cases} \partial_t g - \Delta g = f & \text{in } \Omega \times (0, T), \\ g = h & \text{on } \Gamma \times (0, T), \\ g = g_0 & \text{in } \Omega, \end{cases} \tag{9.194}$$

where the Laplace operator Δ is defined for $g \in W^1_q(\Omega)$ by

$$< \Delta g, \psi > = (\nabla g, \nabla\psi)_\Omega \quad \text{for any } \psi \in W^1_{q',0}(\Omega).$$

Here and hereafter, $< \cdot, \cdot >$ denotes the dual pairing between $W^{-1}_m(\Omega)$ and $W^1_{m',\Gamma}(\Omega)$ for $m = q$ and q' with $q'' = q$. We have the following theorem.

Theorem 9.6.2 *Let* $1 < q < \infty$, $T > 0$, $N < r < \infty$ *and* $\max(q, q') \leqslant r$. *Assume that* Ω *is a uniform* $W^{2,2}_r(\Omega)$ *domain. Then, for any initial data* $g_0 \in B^{1-1/p}_{q,p}(\Omega)$ *and right members*

$$f \in L_p((0, T), W^{-1}_q(\Omega)), \;\; h \in W^1_p((0, T), \mathbf{W}^{-1}_q(\Omega)) \cap L_p((0, T), W^1_q(\Omega)),$$

satisfying the compatibility condition: $g_0 = h|_{t=0}$ *on* Ω, *problem* (9.194) *admits a unique solution* g *with*

$$g \in W^1_p((0, T), W^{-1}_q(\Omega)) \cap L_p((0, T), W^1_q(\Omega))$$

possessing the estimate:

$$\|g\|_{L_p((0,T),W_q^1(\Omega))} + \|\partial_t g\|_{L_p((0,t),W_q^{-1}(\Omega))} \leqslant Ce^{\gamma T}\{\|g_0\|_{B_q^{1-1/p}(\Omega)}$$
$$+ \|f\|_{L_p((0,T),W_q^{-1}(\Omega))} + \|g\|_{L_p((0,T),W_q^1(\Omega))} + \|\partial_t g\|_{L_p((0,T),\mathbf{W}_q^{-1}(\Omega))}\}.$$

Proof Theorem 9.6.1 enable us to prove the existence part of Theorem 9.6.2 in the same manner as in the proof of the existence part of Theorem 9.2.2 in Sect. 9.6.1.

The uniqueness follows from the existence of the dual problem: In fact, let g satisfy the regularity condition:

$$g \in L_p((0,T), W_q^1(\Omega)) \cap W_p^1((0,T), W_q^{-1}(\Omega))$$

and the homogeneous equations:

$$\begin{cases} \partial_t g - \Delta g = 0 & \text{in } \Omega \times (0,T), \\ g = 0 & \text{on } \Gamma \times (0,T), \\ g|_{t=0} = 0 & \text{in } \Omega. \end{cases}$$

Note that $g \in L_p((0,T), W_{q,0}^1(\Omega))$. For any $f \in L_{p'}((0,T), W_{q'}^{-1}(\Omega))$, let u with

$$u \in L_{p'}((0,T, W_{q'}^1(\Omega)) \cap W_{p'}^1((0,T), W_{q'}^{-1}(\Omega))$$

be a solution to the time reversed equations:

$$\begin{cases} \partial_t u + \Delta u = f & \text{in } \Omega \times (0,T), \\ u = 0 & \text{on } \Gamma \times (0,T), \\ u|_{t=T} = 0 & \text{in } \Omega. \end{cases}$$

Note that $u \in L_{p'}((0,T), W_{q',0}^1(\Omega))$. Then, we have

$$\int_0^T < g(\cdot,s), f(\cdot,s) > ds = \int_0^T < g(\cdot,s), (\partial_s u + \Delta u)(\cdot,s) > ds$$
$$= -\int_0^T < (\partial_s g - \Delta g)(\cdot,s), u(\cdot,s) > ds = 0,$$

which furnishes that $g = 0$. This completes the proof of Theorem 9.6.2. □

As an application of Theorem 9.6.2, we prove that if $\mathbf{u}$ satisfies the regularity condition:

$$\mathbf{u} \in W_p^1((0,T), L_q(\Omega)) \cap L_p((0,T), W_q^2(\Omega))$$

and the equations:

$$\begin{cases} \partial_t \mathbf{u} - \mathrm{Div}\,(\mu \mathbf{D}(\mathbf{u}) - (K_1(\mathbf{u}) + K_2(h))\mathbf{I}) = \mathbf{f} & \text{in } \Omega \times (0, T), \\ \mathrm{div}\,\mathbf{u} = 0 & \text{on } \Gamma \times (0, T), \\ \mathbf{u} = 0 & \text{on } \Gamma_0 \times (0, T), \\ \mathbf{u}|_{t=0} = \mathbf{u}_0 & \text{in } \Omega, \end{cases}$$

and if $\mathbf{f}(\cdot, t) \in J_q(\Omega)$ for any $t \in (0, T)$ and $\mathrm{div}\,\mathbf{u}_0 = 0$ in Ω, then $\mathrm{div}\,\mathbf{u} = 0$ in $\Omega \times (0, T)$. Moreover, if $\mathbf{u}_0 \in J_q(\Omega)$, then $\mathbf{u} \in J_q(\Omega)$ for any $t \in (0, T)$.

In fact, for any $\varphi \in \mathscr{W}^1_{q'}(\Omega)$, we have

$$0 = (\mathbf{f}, \nabla\varphi)_\Omega = (\partial_t \mathbf{u}, \nabla\varphi)_\Omega + (\nabla \mathrm{div}\,\mathbf{u}, \nabla\varphi)_\Omega. \tag{9.195}$$

Since $\varphi \in W^1_{q',\Gamma}(\Omega) \subset \mathscr{W}^1_{q'}(\Omega)$, for any $\varphi \in W^1_{q',\Gamma}(\Omega)$ we have

$$(\partial_t \mathrm{div}\,\mathbf{u}, \varphi)_\Omega + (\nabla \mathrm{div}\,\mathbf{u}, \nabla\varphi)_\Omega = 0.$$

Namely, $\mathrm{div}\,\mathbf{u} \in W^1_p((0, T), W^{-1}_q(\Omega)) \cap L_p((0, T), W^1_q(\Omega))$ and $\mathrm{div}\,\mathbf{u}$ satisfies the equations:

$$\begin{cases} \partial_t \mathrm{div}\,\mathbf{u} - \Delta \mathrm{div}\,\mathbf{u} = 0 & \text{in } \Omega \times (0, T), \\ \mathrm{div}\,\mathbf{u} = 0 & \text{on } \Gamma \times (0, T), \\ \mathrm{div}\,\mathbf{u} = 0 & \text{in } \Omega, \end{cases}$$

so that the uniqueness, guaranteed by Theorem 9.6.2, furnishes that $\mathrm{div}\,\mathbf{u} = 0$ for any $t \in (0, T)$.

Moreover, inserting this formula into (9.195) yields that $(\partial_t \mathbf{u}(\cdot, t), \nabla\varphi)_\Omega = 0$ for any $\varphi \in \mathscr{W}^1_{q'}(\Omega)$. Thus, if $(\mathbf{u}|_{t=0}, \nabla\varphi)_\Omega = (\mathbf{u}_0, \nabla\varphi)_\Omega = 0$, then $(\mathbf{u}(\cdot, t), \nabla\varphi)_\Omega = 0$ for any $\varphi \in \mathscr{W}^1_{q'}(\Omega)$ and $t \in (0, T)$.

9.6.3 Uniqueness Part

Finally, we prove the uniqueness part of Theorem 9.1.5. Let $\mathbf{u}$ and h with

$$\begin{aligned} &\mathbf{u} \in L_p((0, T), W^2_q(\Omega)^N) \cap W^1_p((0, T), L_q(\Omega)^N), \\ &h \in L_p((0, T), W^{3-1/q}_q(\Gamma)) \cap W^1_p((0, T), W^{2-1/q}_q(\Gamma)) \end{aligned}$$

satisfy the homogeneous equations:

$$\begin{cases} \partial_t \mathbf{u} - \mathrm{Div}\,(\mu \mathbf{D}(\mathbf{u}) - (K_1(\mathbf{u}) + K_2(\mathbf{u}))\mathbf{I}) = 0 & \text{in } \Omega \times (0, T), \\ \partial_t h - \mathbf{n} \cdot \mathbf{u} = 0 & \text{on } \Gamma \times (0, T), \\ \mathscr{T}_{\mathbf{n}}(\mu \mathbf{D}(\mathbf{u})\mathbf{n}) = 0, \quad \mathrm{div}\, \mathbf{u} = 0 & \text{on } \Gamma \times (0, T), \\ \mathbf{u} = 0 & \text{on } \Gamma_0 \times (0, T), \\ (\mathbf{u}, h)|_{t=0} = (0, 0) & \text{in } \Omega \times \Gamma, \end{cases} \tag{9.196}$$

For any $\mathbf{g} \in L_{p'}((0, T), J_{q'}(\Omega))$, let $\mathbf{v}$ and ρ with

$$\mathbf{v} \in L_{p'}((0, T), W^2_{q'}(\Omega)^N) \cap W^1_{p'}((0, T), L_{q'}(\Omega)^N),$$
$$\rho \in L_{p'}((0, T), W^{3-1/q'}_{q'}(\Gamma)) \cap W^1_{p'}((0, T), W^{2-1/q'}_{q'}(\Gamma))$$

be solutions to the time reversed equations:

$$\begin{cases} \partial_t \mathbf{v} + \mathrm{Div}\,(\mu \mathbf{D}(\mathbf{v}) - (K_1(\mathbf{v}) + K_2(\rho))\mathbf{I}) = \mathbf{g} & \text{in } \Omega \times (0, T), \\ \partial_t \rho + \mathbf{n} \cdot \mathbf{v} = 0 & \text{on } \Gamma \times (0, T), \\ \mathscr{T}_{\mathbf{n}}(\mu \mathbf{D}(\mathbf{v})\mathbf{n}) = 0, \quad \mathrm{div}\, \mathbf{v} = 0 & \text{on } \Gamma \times (0, T), \\ \mathbf{v} = 0 & \text{on } \Gamma_0 \times (0, T), \\ (\mathbf{v}, \rho)|_{t=T} = (0, 0) & \text{in } \Omega \times \Gamma, \end{cases} \tag{9.197}$$

From the fact mentioned in the last part of Sect. 9.6.2 it follows that

$$\mathbf{u}(\cdot, t) \in J_q(\Omega), \quad \mathbf{v}(\cdot, t) \in J_{q'}(\Omega) \quad \text{for any } t \in (0, T). \tag{9.198}$$

Since

$$K_1(\mathbf{u}) + K_2(h) \in L_p((0, T), W^1_q(\Omega) + \mathscr{W}^1_q(\Omega)),$$
$$K_1(\mathbf{v}) + K_2(\rho) \in L_{p'}((0, T), W^1_{q'}(\Omega) + \mathscr{W}^1_{q'}(\Omega)),$$

we write

$$K_1(\mathbf{u}) + K_2(h) = A_1 + A_2, \quad K_1(\mathbf{v}) + K_2(\rho) = B_1 + B_2$$

with $A_1 \in L_p((0, T), W^1_q(\Omega))$, $B_1 \in L_{p'}((0, T), W^1_{q'}(\Omega))$, $A_2 \in L_p((0, T), \mathscr{W}^1_q(\Omega))$, and $B_2 \in L_{p'}((0, T), \mathscr{W}^1_{q'}(\Omega))$. Since $K_1(\mathbf{u}) + K_2(h) = A_1$ and $K_1(\mathbf{v}) + K_2(\rho) = B_1$ on Γ, we have

$$(\mu \mathbf{D}(\mathbf{u}) - A_1 \mathbf{I})\mathbf{n} = (\tau + \delta \Delta_\Gamma) h, \quad (\mu \mathbf{D}(\mathbf{v}) - B_1 \mathbf{I})\mathbf{n} = (\tau + \delta \Delta_\Gamma) \rho$$

on Γ. Moreover, by (9.198), we have

$$\int_0^T (\mathbf{u}(\cdot, t), \nabla B_2(\cdot, t))_\Omega \, dt = 0, \quad \int_0^T (\nabla A_2(\cdot, t), \mathbf{v}(\cdot, t))_\Omega \, dt = 0.$$

Thus, using (9.196), (9.197) and the divergence theorem of Gauß we have

$$\int_0^T (\mathbf{u}(\cdot, t), \mathbf{g}(\cdot, t))_\Omega \, dt = 0. \tag{9.199}$$

In fact,

$$\begin{aligned}
\int_0^T (\mathbf{u}, \mathbf{g})_\Omega \, dt &= \int_0^T (\mathbf{u}, \partial_t \mathbf{v} + \mathrm{Div}\,(\mu \mathbf{D}(\mathbf{v}) - B_1 \mathbf{I}))_\Omega \, dt \\
= -\int_0^T (\mathbf{u}_t, \mathbf{v})_\Omega \, dt &+ \int_0^T (\mathbf{u} \cdot \mathbf{n}, (\tau + \delta \Delta_\Gamma)\rho)_\Gamma \, dt - \frac{\mu}{2} \int_0^T (\mathbf{D}(\mathbf{u}), \mathbf{D}(\mathbf{v}))_\Omega \, dt \\
= -\int_0^T (\mathrm{Div}\,(\mu \mathbf{D}(\mathbf{u}) - A_1 \mathbf{I}), \mathbf{v})_\Omega \, dt &+ \int_0^T (h_t, (\tau + \delta \Delta_\Gamma)\rho)_\Gamma \, dt \\
&- \frac{\mu}{2} \int_0^T (\mathbf{D}(\mathbf{u}), \mathbf{D}(\mathbf{v}))_\Omega \, dt \\
= \int_0^T ((\tau + \delta \Delta_\Gamma) h, \rho_t) \, dt &+ \int_0^T (h_t, (\tau + \delta \Delta_\Gamma)\rho)_\Gamma \, dt = 0.
\end{aligned}$$

For any $\mathbf{f} \in C^\infty((0, T), C_0^\infty(\Omega)^N)$, let θ be a function in $C^\infty((0, T), \mathscr{W}^1_{q'}(\Omega))$ satisfying the variational equation:

$$(\nabla \theta(\cdot, t), \nabla \psi)_\Omega = (\mathbf{f}(\cdot, t), \nabla \psi)_\Omega \quad \text{for any } \psi \in \mathscr{W}^1_q(\Omega).$$

Since $\mathbf{f} - \nabla \theta \in L_{p'}((0, T), J_q(\Omega))$, by (9.199)

$$\int_0^T (\mathbf{u}, \mathbf{f})_\Omega \, dt = \int_0^T (\mathbf{u}, \mathbf{f} - \nabla \theta)_\Omega \, dt + \int_0^T (\mathbf{u}, \nabla \theta)_\Omega \, dt = 0.$$

Since $C^\infty((0, T), C_0^\infty(\Omega)^N)$ is dense in $L_{p'}((0, T), L_{q'}(\Omega)^N)$, we have $\mathbf{u} = 0$. By the second equation in (9.196), $h_t = 0$, which, combined with $h|_{t=0} = 0$, furnishes that $h = 0$. This completes the proof of the uniquness part of Theorem 9.2.2.

9.7 Proofs of Theorems 9.1.4 and 9.1.5

First, according to what was pointed out in Sect. 9.2.1, under the assumption (9.30), we prove Theorem 9.1.4 with the help of Theorem 9.2.1. Given $g \in DI_q(\Omega)$, let $K \in W^1_q(\Omega) + \mathscr{W}^1_q(\Omega)$ be a solution to the variational problem (9.31). Let $\mathbf{u} \in W^2_q(\Omega)^N$ and $h \in W^{3-1/q}_q(\Gamma)$ be solutions to the equations (9.32), the unique existence of which is guaranteed by Theorem 9.2.1. Then, by (9.32) and (9.33), we have

$$\lambda(\mathbf{u} - \mathscr{G}(g), \nabla \varphi)_\Omega - (\nabla(\mathrm{div}\, \mathbf{u} - g), \nabla \varphi)_\Omega = 0 \tag{9.200}$$

for any $\varphi \in \mathscr{W}^1_{q'}(\Omega)$ subject to $\operatorname{div}\mathbf{u} - g = 0$ on Γ. Since $\varphi \in W^1_{q',\Gamma}(\Omega) \subset \mathscr{W}^1_{q'}(\Omega)$, applying the divergence theorem of Gauß to the first term in (9.200) and using (9.13) we have

$$\lambda((\operatorname{div}\mathbf{u} - g), \varphi)_\Omega - (\nabla(\operatorname{div}\mathbf{u} - g), \nabla\varphi)_\Omega = 0$$

for any $\varphi \in W^1_{q',\Gamma}(\Omega)$ subject to $\operatorname{div}\mathbf{u} - g = 0$ on Γ. Thus, by Theorem 9.6.1, $\operatorname{div}\mathbf{u} = g$ in Ω provided that λ_0 is chosen larger if necessary, which, inserted into (9.200), furnishes that $(\mathbf{u}, \nabla\varphi)_\Omega = (\mathscr{G}(g), \nabla\varphi)_\Omega$ for any $\varphi \in \mathscr{W}^1_{q'}(\Omega)$. Thus, setting $\mathfrak{p} = K_1(\mathbf{u}) + K_2(h) + K$, we see that $\mathbf{u}$, $\mathfrak{p}$ and h are required solutions of the equations (9.2). Therefore, the existence part of Theorem 9.1.4 immmediately follows from Theorem 9.2.1.

To prove the uniqueness part, let $\mathbf{u}$, $\mathfrak{p}$ and h be solutions to the homogeneous equations:

$$\begin{cases} \lambda\mathbf{u} - \operatorname{Div}(\mu\mathbf{D}(\mathbf{u}) - \mathfrak{p}\mathbf{I}) = 0, \quad \operatorname{div}\mathbf{u} = 0 & \text{in } \Omega, \\ \lambda h - \mathbf{n}\cdot\mathbf{u} = 0 & \text{on } \Gamma, \\ \{\mu\mathbf{D}(\mathbf{u}) - \mathfrak{p}\mathbf{I} - ((\tau + \delta\Delta_\Gamma)h)\mathbf{I}\}\mathbf{n} = 0 & \text{on } \Gamma, \\ \mathbf{u} = 0 & \text{on } \Gamma_0 \end{cases} \tag{9.201}$$

with

$$\mathbf{u} \in W^2_q(\Omega)^N, \quad \mathfrak{p} \in W^1_q(\Omega) + \mathscr{W}^1_q(\Omega), \quad h \in W^{3-1/q}_q(\Gamma).$$

For any $\varphi \in \mathscr{W}^1_{q'}(\Omega)$, we have

$$0 = (\lambda\mathbf{u} - \operatorname{Div}(\mu\mathbf{D}(\mathbf{u}) - \mathfrak{p}\mathbf{I}), \nabla\varphi)_\Omega = (\nabla(\mathfrak{p} - (K_1(\mathbf{u}) + K_2(h)), \nabla\varphi)_\Omega.$$

Moreover, $\mathfrak{p} = K_1(\mathbf{u}) + K_2(h)$ on Γ, so that the uniqueness implies that $\mathfrak{p} = K_1(\mathbf{u}) + K_2(h)$. Thus, $\mathbf{u}$ and h satisfy the homogeneous equations (9.35), so that $\mathbf{u} = 0$ and $h = 0$, which completes the proof of the uniqueness of solutions to (9.2).

Next, we prove Theorem 9.1.5 with the help of Theorem 9.2.2 under the assumption (9.30). Given $g \in L_p((0,T), DI_q(\Omega)) \cap W^1_p((0,T), \mathbf{W}^{-1}_q(\Omega))$ with $\mathscr{G}(g) \in W^1_p((0,T), L_q(\Omega)^N)$, let $K \in L_p((0,T), W^1_q(\Omega) + \mathscr{W}^1_q(\Omega))$ be a solution to the variational problem

$$(\nabla K, \nabla\varphi)_\Omega = (\partial_t\mathscr{G}(g) - \nabla g, \nabla\varphi)_\Omega \tag{9.202}$$

for any $\varphi \in \mathscr{W}^1_{q'}(\Omega)$ subject to $K = -g$ on Γ, which possesses the estimate:

$$\|\nabla K\|_{L_q(\Omega)} \leqslant C(\|g\|_{W^1_q(\Omega)} + \|\partial_t\mathscr{G}(g)\|_{L_q(\Omega)}).$$

Let $\mathbf{u}$ and h be solutions of the equations:

$$\begin{cases} \partial_t\mathbf{u} - \mathrm{Div}\,(\mu\mathbf{D}(\mathbf{u}) - (K_1(\mathbf{u}) + K_2(h))\mathbf{I}) = \mathbf{f} + \nabla K & \text{in } \Omega \times (0, T), \\ \partial_t h - \mathbf{n}\cdot\mathbf{u} = f & \text{on } \Gamma \times (0, T), \\ \mathscr{T}_{\mathbf{n}}(\mu\mathbf{D}(\mathbf{u})\mathbf{n}) = \mathscr{T}_{\mathbf{n}}(\mathbf{f}_b + g\mathbf{n}) = \mathscr{T}_{\mathbf{n}}(\mathbf{f}_b) & \text{on } \Gamma \times (0, T), \\ \mathrm{div}\,\mathbf{u} = \mathbf{n}\cdot(\mathbf{f}_b + g\mathbf{n}) = g & \text{on } \Gamma \times (0, T), \\ \mathbf{u} = 0 & \text{on } \Gamma_0 \times (0, T), \\ (\mathbf{u}, h)|_{t=0} = (\mathbf{u}_0, h_0) & \text{on } \Omega \times \Gamma, \end{cases} \tag{9.203}$$

with

$$\mathbf{u} \in W^1_p((0, T), L_q(\Omega)^N) \cap L_p((0, T), W^2_q(\Omega)^N),$$
$$h \in W^1_p((0, T), W^{2-1/q}_q(\Gamma)) \cap L_p((0, T), W^{3-1/q}_q(\Gamma)),$$

which possess the estimate:

$$\begin{aligned} \mathscr{I}_R(t, \mathbf{u}, h) &\leqslant Ce^{\gamma t}\{\mathscr{M}_R(t, \mathbf{f}, f, \mathbf{f}_b) \\ &+ \|\partial_t\mathscr{G}(g)\|_{L_p((0,T),L_q(\Omega))} + \|g\|_{L_p((0,T),W^1_q(\Omega))} + \|\partial_t g\|_{L_p((0,T),\mathbf{W}^{-1}_q(\Omega))}\} \end{aligned} \tag{9.204}$$

for any $t \in (0, T]$ with some positive constants C and γ. Here, we have used the the assumption that $\mathbf{f}_b\cdot\mathbf{n} = 0$ on Γ. Since we assume that $\mathbf{f} \in L_p((0, T), J_q(\Omega))$, for any $\varphi \in \mathscr{W}^1_{q'}(\Omega)$

$$(\nabla K, \nabla\varphi)_\Omega = (\partial_t\mathbf{u}, \nabla\varphi)_\Omega - (\nabla\mathrm{div}\,\mathbf{u}, \nabla\varphi)_\Omega,$$

which, combined with (9.202), furnishes that

$$\partial_t(\mathbf{u} - \mathscr{G}(g), \nabla\varphi) - (\nabla(\mathrm{div}\,\mathbf{u} - g), \nabla\varphi)_\Omega = 0 \quad \text{for any } \varphi \in \mathscr{W}^1_{q'}(\Omega). \tag{9.205}$$

Since

$$(\mathrm{div}\,\mathbf{u}, \varphi)_\Omega = -(\mathbf{u}, \nabla\varphi)_\Omega, \quad (g, \varphi)_\Omega = -(\mathscr{G}(g), \nabla\varphi)_\Omega$$

for any $\varphi \in W^1_{q',\Gamma}(\Omega)$ and since $\mathbf{u}$ and $\mathscr{G}(g)$ belong to $L_p((0, T), L_q(\Omega)^N)$, we have

$$\mathrm{div}\,\mathbf{u} - g \in L_p((0, T), W^1_q(\Omega)) \cap W^1_p((0, T), W^{-1}_q(\Omega)),$$

where $W^{-1}_q(\Omega)$ is the dual space of $W^1_{q',\Gamma}(\Omega)$ Moreover, since $\varphi \in W^1_{q',\Gamma}(\Omega) \subset \mathscr{W}^1_{q'}(\Omega)$, applying the divergence theorem of Gauß to the first term in (9.205) and using the boundary condition: $\mathrm{div}\,\mathbf{u} = g$ on $\Gamma \times (0, T)$ in (9.203) and the compatibility condition: $\mathrm{div}\,\mathbf{u}_0 = g|_{t=0}$ in Ω, we see that $\mathrm{div}\,\mathbf{u} - g$ satisfies the equations:

$$\begin{cases} \partial_t(\operatorname{div}\mathbf{u} - g) - \Delta(\operatorname{div}\mathbf{u} - g) = 0 & \text{in } \Omega \times (0, T), \\ \operatorname{div}\mathbf{u} - g = 0 & \text{on } \Gamma \times (0, T), \\ (\operatorname{div}\mathbf{u} - g)|_{t=0} = 0 & \text{in } \Omega. \end{cases}$$

Thus, the uniqueness guaranteed by Theorem 9.6.2 implies that $\operatorname{div}\mathbf{u} = g$ in $\Omega \times (0, T)$. Inserting this fact into (9.205), we have

$$(\mathbf{u}, \nabla\varphi)_\Omega = (\mathcal{G}(g), \nabla\varphi)_\Omega$$

for any $\varphi \in \hat{W}^1_{q'}(\Omega)$ provided that the compatibility condition: $\mathbf{u}_0 - \mathcal{G}(g)|_{t=0} \in J_q(\Omega)$ holds. Thus, setting $\mathfrak{p} = K + K_1(\mathbf{u}) + K_2(h)$ and noting that

$$0 = \operatorname{div}\mathbf{u} - g =< \mu\mathbf{D}(\mathbf{u})\mathbf{n}, \mathbf{n} > -(K_1(\mathbf{u}) + K_2(h) + K) - (\tau + \delta\Delta_\Gamma)h \quad \text{in } \Gamma,$$

we see that $\mathbf{u}$, $\mathfrak{p}$ and h are the required solutions of the equations (9.1).

To prove the uniqueness part, let $\mathbf{u}$, $\mathfrak{p}$ and h be solutions to the homogeneous equations:

$$\begin{cases} \partial_t\mathbf{u} - \operatorname{Div}(\mu\mathbf{D}(\mathbf{u}) - \mathfrak{p}\mathbf{I}) = 0 & \text{in } \Omega \times (0, T), \\ \lambda h - \mathbf{n}\cdot\mathbf{u} = 0 & \text{on } \Gamma \times (0, T), \\ \{\mu\mathbf{D}(\mathbf{u}) - \mathfrak{p}\mathbf{I} - ((\tau + \delta\Delta_\Gamma)h)\mathbf{I}\}\mathbf{n} = 0 & \text{on } \Gamma \times (0, T), \\ \mathbf{u} = 0 & \text{on } \Gamma_0 \times (0, T), \\ (\mathbf{u}, h)|_{t=0} = (0, 0) & \text{in } \Omega \times \Gamma, \end{cases} \tag{9.206}$$

with

$$\begin{aligned} &\mathbf{u} \in L_p((0, T), W^2_q(\Omega)^N \cap W^1_p((0, T), J_q(\Omega)^N), \\ &\mathfrak{p} \in L_p((0, T), W^1_q(\Omega) + \hat{W}^1_q(\Omega)), \\ &h \in L_p((0, T), W^{3-1/q}_q(\Gamma)) \cap W^1_p((0, T), W^{2-1/q}_q(\Gamma)). \end{aligned}$$

For any $\varphi \in \hat{W}^1_{q'}(\Omega)$, we have

$$0 = (\partial_t\mathbf{u} - \operatorname{Div}(\mu\mathbf{D}(\mathbf{u}) - \mathfrak{p}\mathbf{I}), \nabla\varphi)_\Omega = (\nabla(\mathfrak{p} - (K_1(\mathbf{u}) + K_2(h)), \nabla\varphi)_\Omega.$$

Moreover, $\mathfrak{p} = K_1(\mathbf{u}) + K_2(h)$ on Γ, so that the uniqueness implies that $\mathfrak{p} = K_1(\mathbf{u}) + K_2(h)$. Thus, $\mathbf{u}$ and h satisfy the homogeneous equations (9.196), so that $\mathbf{u} = 0$ and $h = 0$, which completes the proof of the uniqueness of solutions to (9.1).

Acknowledgements The author would like to express his gratitude to Dr. Hirokazu Saito for his finding a gap in constructing the solution operators in the model problems in Shibata [14, Sect. 3]. In this paper, such a gap is filled completely.

References

1. Bourgain, J.: Vector-valued singular integrals and the H^1-BMO duality. In: Borkholder, D. (ed.) Probability Theory and Harmonic Analysis, pp. 1–19. Marcel Dekker, New York (1986)
2. Denk, R., Hieber, M., Prüß, J.: $\mathscr{R}$-boundedness, Fourier multipliers and problems of elliptic and parabolic type. Mem. AMS **166**(788) (2003)
3. Enomoto, Y., Shibata, Y.: On the $\mathscr{R}$-sectoriality and its application to some mathematical study of the viscous compressible fluids. Funkcial Ekvac. **56**(3), 441–505 (2013)
4. Mogilevskii, I.Sh.: Estimates of solutions of a general intial-boundary value problem for the linear nonstationary system of Navier-Stokes equations in a half-space. Zap Nauchn. Sem. LOMI **84**, 147–173 (1979). in Russian
5. Mogilevskii, ISh: Solvability of a general boundary value problem for a linearized nonstationary system of Navier-Stokes equations. Zap Nauchn. Sem. LOMI **110**, 105–119 (1981). in Russian
6. Mucha, P.B., Zajączkowski, W.: On the existence for the Cauchy-Neumann problem for the Stokes system in the L_p-framework. Studia Math. **143**(1), 75–101 (2000)
7. Prüss, J., Simonett, G.: On the two-phase Navier-Stokes equations with surface tension. Interfaces Free Bound. **12**, 311–345 (2010)
8. Prüss, J., Simonett, G.: Analytic solutions for the two-phase Navier-Stokes equations with surface tension and gravity. In: Parabolic Problems, pp. 507–540. Birkhäuser, Basel (2011)
9. Prüss, J., Simonett, G.: Moving interface and quasilinear parabolic equations. In: Manuscript (2015)
10. Saito, H., Shibata, Y.: On the global wellposedness of free boundary problem for the Naveir Stokes systems with surface tension. J. Math. Soc. Japan. (to appear)
11. Schneider, C.: Trace operators in Besov and Triebel-Lizorkin spaces. J. Anal. Appl. **29**, 275–302 (2010). doi:10.4171/ZAA/1409
12. Schonbek, M., Shibata, Y.: On a global well-posedness of strong dynamics of incompressible nematic liquid crystals in $\mathbb{R}^N$. J. Evolution Eqns. (to appear)
13. Shibata, Y.: Generalaized resolvent estimates of the Stokes equations with first order boundary condition in a general domain. J. Math. Fluid Mech. **15**(1), 1–40 (2013)
14. Shibata, Y.: On the $\mathscr{R}$-boundedness of solution operators for the stokes equations with free boundary condition. Diff. Int. Eqns. **27**, 313–368 (2014)
15. Shibata, Y.: On the $\mathscr{R}$-boundedness of solution operators for the weak Dirichlet-Neumann problem. In: Hishida, T. (ed.) Mathematical Analysis of Incompressible Flow February 4–6, 2013, pp. 1–18. RIMS Kôkyûroku 1875, Kyoto University (2014)
16. Shibata, Y.: On some free boundary problem of the Navier-Stokes equations in the maximal L_p-L_q regularity class. J. Diff. Equ. **258**, 4127–4155 (2015)
17. Shibata, Y.: On some free boundary problem for the Navier-Stokes equations in an exterior domain, I local well-posedness. In: Manuscript (2016)
18. Shibata, Y., Shimizu, S.: On the L_p-L_q maximal regularity of the Stokes problem with first order boundary condition; Model Problem. J. Math. Soc. Jpn. **64**(2), 561–626 (2012)
19. Solonnikov, V.: L_p-estimates for a linear problem arising in the study of the motion of an isolated liqued mass. J. Math. Sci. **189** (2013), 699–732 (translated from Problemy Matematicheskogo Analyiza **69** (2013), 137–166)
20. Weis, L.: Operator-valued Fourier multiplier theorems and maximal L_p -regularity. Math. Ann. **319**, 735–758 (2001)

Chapter 10
On the Solvability of Free Boundary Problem for Viscous Compressible Fluids in an Infinite Time Interval

Vsevolod Alekseevich Solonnikov

Abstract We consider evolution free boundary problem for two viscous compressible fluids contained in a bounded vessel and separated by a free (unknown) variable surface. We prove that this problem is uniquely solvable in the anisotropic Sobolev spaces, and under certain assumptions the solution is defined for $t > 0$ and decays exponentially as $t \to \infty$. In the proof we use the estimate of "modified energy" obtained by M. Padula.

Keywords Compressible fluids · Exponential stability · Sobolev spaces

10.1 Introduction

The paper is concerned with the free boundary problem

$$\begin{cases} \rho(x,t)(\mathscr{D}_t v + (v\cdot\nabla)v)) - \nabla\cdot T(v) + \nabla p(\rho) = 0, \\ \mathscr{D}_t\rho + \nabla\cdot(\rho v) = 0, \quad x \in \Omega_t^+ \cup \Omega_t^-, \quad t > 0, \\ [v] = 0, \quad [-p(\rho)n + T(v)n] = 0, \quad V_n = v\cdot n, \quad x \in \Gamma_t, \\ v^-(x,t) = 0, \quad x \in S, \\ v(x,0) = v_0(x), \quad \rho(x,0) = \rho_0(x), \quad x \in \Omega_0^+ \cup \Omega_0^-, \end{cases} \tag{10.1}$$

where $v(x,t) = v^\pm(x,t)$, $\rho(x,t) = \rho^\pm(x,t)$ for $x \in \Omega_t^\pm$, Ω_t^+ and Ω_t^- are bounded domains separated by a free interface $\Gamma_t = \partial\Omega_t^+$ that is given for $t = 0$ and should be found for $t > 0$. The domain $\Omega = \Omega_t^+ \cup \Gamma_t \cup \Omega_t^-$ is fixed; the surface $S = \partial\Omega$ is bounded away from Γ_t. By $T(v) \equiv T^\pm(v)$ we mean the viscous part of the stress tensor:

$$T^\pm(v) = \mu^\pm S(v) + \mu_1^\pm I\nabla\cdot v, \quad x \in \Omega_t^\pm,$$

V.A. Solonnikov (✉)
St. Petersburg Department of Steklov Mathematical Institute, Russian Academy of Sciences, 27, Fontanka, 191023 St.Petersburg, Russia
e-mail: solonnik@pdmi.ras.ru

Y. Shibata and Y. Suzuki (eds.), *Mathematical Fluid Dynamics, Present and Future*, Springer Proceedings in Mathematics & Statistics 183,
DOI 10.1007/978-4-431-56457-7_10

where $S(v) = \nabla v + (\nabla v)^T$ is the doubled rate-of-strain tensor, n is the normal to Γ_t exterior with respect to Ω^+ and V_n is the velocity of evolution of Γ_t in the direction n, $\mu^\pm, \mu_1^\pm = const > 0$, $[u]|_{\Gamma_t} = u^+ - u^-$ is the jump of the function u on Γ_t.

The pressure functions $p^\pm(\rho)$ are positive strictly increasing functions of a positive argument possessing Lipshitz continuous derivatives.

By the rest state we mean the solution of (10.1) with $v^\pm(x,t) = 0$. Then the domains $\Omega_t^\pm$ are independent of t and $\rho^\pm(x,t) = \overline{\rho}^\pm = M^\pm/|\Omega^\pm|$, where $M^\pm$ are total masses of the fluids and $|\Omega^\pm| = mes\Omega^\pm$. The jump conditions on Γ_t reduce to

$$p^+(\overline{\rho}^+) = p^-(\overline{\rho}^-). \tag{10.2}$$

Our aim is to prove the stability of the rest state. Here we restrict ourselves with the case

$$p^+(\rho) = p^-(\rho) \equiv p(\rho), \quad \overline{\rho}^+ = \overline{\rho}^- = \overline{\rho} = M/|\Omega|, \quad M = M^+ + M^-,$$

hence (10.2) is fulfilled automatically; moreover, $\Theta = \rho - \overline{\rho}$ satisfies the condition

$$\int_\Omega \Theta(x,t)dx = 0. \tag{10.3}$$

We work in the Lagrangian coordinates, in which (10.1) takes the form

$$\begin{cases} r(\xi,t)\mathscr{D}_t u - \nabla_u \cdot T_u(u) + \nabla_u p(r) = 0, \\ \mathscr{D}_t r + r\nabla_u \cdot u = 0, \quad \xi \in \Omega_0^+ \cup \Omega_0^-, \quad t > 0, \\ [u] = 0, \quad [-p(r)n + T_u(u)n] = 0, \quad \xi \in \Gamma_t, \\ u^- = 0, \quad \xi \in S, \\ u(\xi,0) = v_0(\xi), \quad r(\xi,0) = \rho_0(\xi), \quad \xi \in \Omega_0^+ \cup \Omega_0^-, \end{cases} \tag{10.4}$$

where $r = \rho(X(\xi,t),t)$, $u(\xi,t) = v(X,t)$,

$$X(\xi,t) = \xi + \int_0^t u(\xi,\tau)d\tau \equiv \xi + U(\xi,t), \quad \xi \in \Omega_0^- \cup \Omega_0^+.$$

We set $\nabla_u = J_u^{-1}A\nabla$, $S_u(u) = (\nabla_u u) + (\nabla_u u)^T$, $T_u(u) = \mu S_u(u) + \mu_1 I\nabla_u \cdot u$, $J_u = det\mathscr{L}$, $\mathscr{L} = \left(\frac{\partial X}{\partial \xi}\right)$. The elements of the matrix A are co-factors of the elements $l_{ij} = \delta_{ij} + \int_0^t \frac{\partial u_i}{\partial \xi_j}d\tau$ of the matrix $\mathscr{L}$. The normal $n(X)$ to Γ_t is connected with the normal $n_0(\xi)$ to Γ_0 by

$$n(X) = \frac{An_0}{|An_0|}.$$

By virtue of (10.3), the function $\theta(\xi,t)=\Theta(X(\xi,t),t)$ satisfies the orthogonality condition

$$\int_{\Omega}\theta(\xi,t)J_u(\xi,t)d\xi=0. \tag{10.5}$$

Problem (10.4) can be written in the form

$$\begin{cases}\overline{\rho}\mathscr{D}_t u-\nabla\cdot T(u)+p'(\overline{\rho})\nabla\theta=l_1(u,\theta),\\ \mathscr{D}_t\theta+\overline{\rho}\nabla\cdot u=l_2(u,\theta),\quad \xi\in\Omega_0^+\cup\Omega_0^-,\\ u(\xi,t)=0,\quad \xi\in S,\\ [u]=0,\quad [\mu\Pi_0 S(u)n_0]=l_3(u),\\ [-p'(\overline{\rho})\theta+n_0\cdot T(u)n_0]=l_4(u,\theta),\quad \xi\in\Gamma_0,\\ u(\xi,0)=v_0(\xi),\quad \theta(\xi,0)=\theta_0(\xi)=\rho_0(\xi)-\overline{\rho},\quad \xi\in\Omega_0^+\cup\Omega_0^-,\end{cases} \tag{10.6}$$

where

$$\begin{aligned}&l_1(u,\theta)=-\vartheta\mathscr{D}_t u+(\nabla_u\cdot T_u(u)-\nabla\cdot T(u))+p'(\overline{\rho})\nabla\theta-p'(\overline{\rho}+\theta)\nabla_u\theta,\\ &l_2(u,\theta)=-r\nabla_u\cdot u+\overline{\rho}\nabla\cdot u,\\ &l_3(u)=[\mu(\Pi_0^2 S(u)n_0-\Pi_0\Pi S_u(u)n)],\\ &l_4(u,\theta)=-[n\cdot T_u(u)n-n_0\cdot T(u)n_0]+[p(\overline{\rho}+\theta)-p(\overline{\rho})-p'(\overline{\rho})\theta],\end{aligned} \tag{10.7}$$

$$\Pi f=f-n(f\cdot n),\quad \Pi_0 f=f-n_0(f\cdot n_0).$$

Our main result is as follows.

Theorem 1 *Let* $S,\Gamma_0\in W_2^{3/2+l}$, $1/2<l<1$, $p\in C^2(\overline{\rho}/2,3\overline{\rho}/2)$, $\overline{\rho},p'(\overline{\rho})>0$. *Then for arbitrary* $v_0^\pm\in W_2^{l+1}(\Omega_0^\pm)$, $\rho_0^\pm\in W_2^{l+1}(\Omega_0^\pm)$ *such that*

$$\begin{aligned}&v_0^-(\xi,0)|_S=0,\quad \int_{\Omega_0}\theta_0(\xi)d\xi=0,\\ &[v_0]=0,\quad [-p(\rho_0)n_0+T(u_0)n_0]=0,\quad \xi\in\Gamma_0,\\ &|\rho_0|\geqslant c_0>0,\end{aligned} \tag{10.8}$$

$$\sum_{\pm}\|v_0\|_{W_2^{l+1}(\Omega_0^\pm)}+\sum_{\pm}\|\theta_0\|_{W_2^{l+1}(\Omega_0^\pm)}\leqslant\varepsilon\ll 1, \tag{10.9}$$

problem (10.4) has a unique solution defined for $t \geqslant 0$, and

$$\begin{aligned}&\sum_{\pm}(\|e^{at}u\|_{W_2^{l+2,l/2+1}(Q_\infty^\pm)} + \|e^{at}\theta\|_{W_2^{l+1,0}(Q_\infty^\pm)} + \|e^{at}\mathscr{D}_t\theta\|_{W_2^{l+1,0}(Q_\infty^\pm)})\\ &\leqslant c\sum_{\pm}(\|v_0\|_{W_2^{l+1}(\Omega_0^\pm)} + \|\theta_0\|_{W_2^{l+1}(\Omega_0^\pm)}),\quad a>0,\quad Q_\infty^\pm = \Omega_0^\pm \times (0,\infty).\end{aligned} \tag{10.10}$$

For problem (10.1) this means that the velocity decays exponentially, the density tends to its mean value $\overline{\rho}$, and $\Gamma_t \to \Gamma_\infty = X(\cdot,\infty)\Gamma_0$ as $t \to \infty$.

Theorem 1 is proved in [1] for the Dirichlet problem with no-slip condition on the fixed boundary S, see also [2, 3]. The proof of the estimate (10.10) is also outlined in [1] but it is not complete. Here (10.10) is obtained by the method that enables one to avoid technicalities connected with the analysis of model problems.

We recall the definition of the Sobolev spaces used in this paper. By $W_2^l(D)$, $D \subset \mathbb{R}^n$, we mean the space with the norm given by

$$\|u\|^2_{W_2^l(D)} = \sum_{|j|\leqslant l} \|\mathscr{D}^j u\|^2_{L_2(D)},$$

if l is an integer, or

$$\|u\|^2_{W_2^l(D)} = \|u\|^2_{W_2^{[l]}(D)} + \sum_{|j|=[l]}\int_D\int_D |\mathscr{D}^j u(x) - \mathscr{D}^j u(y)|^2 \frac{dxdy}{|x-y|^{n+2\lambda}},$$

if $l = [l]+\lambda$, $0<\lambda<1$. We work in anisotropic spaces

$$W_2^{r,r/2}(D_T) = W_2^{r,0}(D_T) \cap W_2^{0,r/2}(D_T),$$

where $D_T = D\times(0,T)$, $T \leqslant \infty$,

$$W_2^{r,0}(D_T) = L_2(0,T;W_2^r(D)),\quad W_2^{0,s}(D_T) = W_2^s(0,T;L_2(D)).$$

It is convenient to define the norm in $W_2^{0,l}(D_T)$, $l = [l]+\lambda$, by

$$\|u\|^2_{W_2^{0,l}(D_T)} = \int_0^1 \frac{dh}{h^{1+2\lambda}}\int_h^T \|\mathscr{D}_t^{[l]}u(\cdot,t) - \mathscr{D}_t^{[l]}u(\cdot,t-h)\|^2_{L_2(D)}dt + \|u\|^2_{W_2^{0,[l]}(D_T)}$$

(as a rule, $T>1$ in the sequel).

We deal with functions decaying exponentially as $t\to\infty$ and belonging to the Sobolev spaces with exponential weight $e^{\beta t}$, $\beta>0$. For functions vanishing as $t\leqslant 0$, the norm

$$||e^{\beta t}u\|_{W_2^{0,l}(D_T)}$$

is equivalent to

$$\Big(\int_{-\infty}^{\infty}(1+|s|)^{2l}\|\widetilde{u}(\cdot,s)\|^2_{L_2(D)}ds_2\Big)^{1/2}$$

where $\widetilde{u}(x,s)$ is the Laplace transform of the function u and $s=s_1+is_2$, $s_1=-\beta$.

10.2 Linear Problem

Along with (10.4), we consider a linear problem

$$\begin{cases} \overline{\rho}\mathscr{D}_t v-\nabla\cdot T(v)+p'(\overline{\rho})\nabla\theta=f(x,t),\\ \mathscr{D}_t\theta+\overline{\rho}\nabla\cdot v=h(x,t), \quad x\in\Omega_0^+\cup\Omega_0^-, \quad t>0,\\ v(x,t)=0, \quad x\in S,\\ [v]=0, \quad [-p'(\overline{\rho})\theta n_0+T(u)n_0]=b(x,t), \quad x\in\Gamma_0,\\ v(x,0)=v_0(x), \quad \theta(x,0)=\theta_0(x), \quad x\in\Omega_0^+\cup\Omega_0^-, \end{cases} \tag{10.11}$$

Theorem 2 *1. For arbitrary $v_0^\pm\in W_2^{l+1}(\Omega_0^\pm)$, $\theta_0^\pm\in W_2^{l+1}(\Omega_0^\pm)$, $f^\pm\in W_2^{l,l/2}(Q_T^\pm)$, $h^\pm\in W_2^{l+1,0}(Q_T^\pm)$, $b\in W_2^{l+1/2,l/2+1/4}(G_T)$ satisfying the compatibility conditions*

$$\begin{aligned} &v_0^-(x)=0, \quad x\in S,\\ &[v_0]=0, \quad [-p'(\overline{\rho})\theta_0 n_0+T(u_0)n_0]=b(x,0), \quad x\in\Gamma_0, \end{aligned} \tag{10.12}$$

problem (10.11) has a unique solution $v\in W_2^{2+l,1+l/2}(Q_T^\pm)$, $\theta,\theta_t\in W_2^{l+1,0}(Q_T^\pm)$, $T<\infty$ defined for $t\in(0,T)$ and satisfying the inequality

$$\begin{aligned} &\sum_{\pm}(\|v\|_{W_2^{l+2,l/2+1}(Q_T^\pm)}+\|\theta\|_{W_2^{l+1,0}(Q_T^\pm)}+\|\mathscr{D}_t\theta\|_{W_2^{l+1,0}(Q_T^\pm)})\\ &\leqslant c(T)\Big(\sum_{\pm}(\|v_0\|_{W_2^{l+1}(\Omega_0^\pm)}+\|\theta_0\|_{W_2^{l+1}(\Omega_0^\pm)}\\ &+\|f\|_{W_2^{l,l/2}(Q_T^\pm)}+\|h\|_{W_2^{l+1,0}(Q_T^\pm)})+\|b\|_{W_2^{l+1/2,l/2+1/4}(G_T)}\Big), \end{aligned} \tag{10.13}$$

where $Q_T^\pm=\Omega_0^\pm\times(0,T)$, $G_T=\Gamma_0\times(0,T)$. The number T grows without limits as the sum of norms in the right hand side of (10.13) tends to zero.

2. Moreover, if

$$\int_\Omega\theta_0(x)dx=0, \quad \int_\Omega h(x,t)dx=0, \quad \forall t>0, \tag{10.14}$$

then the solution is defined for all $t>0$, the condition $\int_\Omega\theta(x,t)dx=0$ holds and

$$\sum_{\pm}(\|e^{\beta t}v\|_{W_2^{l+2,l/2+1}(Q_T^{\pm})} + \|e^{\beta t}\theta\|_{W_2^{l+1,0}(Q_T^{\pm})} + \|e^{\beta t}\mathscr{D}_t\theta\|_{W_2^{l+1,0}(Q_T^{\pm})})$$
$$\leqslant c\Big(\sum_{\pm}(\|v_0\|_{W_2^{l+1}(\Omega_0^{\pm})} + \|\theta_0\|_{W_2^{l+1}(\Omega_0^{\pm})} + \|e^{\beta t}f\|_{W_2^{l,l/2}(Q_T^{\pm})} + \|e^{\beta t}h\|_{W_2^{l+1,0}(Q_T^{\pm})})$$
$$+ \|e^{\beta t}b\|_{W_2^{1/2+l,1/4+l/2}(G_T)}\Big) \tag{10.15}$$

with $\beta > 0$ *and* c *independent of* $T \leqslant \infty$.

If $b = 0$, then the problem (10.11) can be written as

$$\mathscr{D}_t U + \mathfrak{A}U = F, \quad U|_{t=0} = U_0,$$

where $U = (v, \theta)^T$, $U_0 = (v_0, \theta_0)^T$, $F = (f, h)^T$. The operator $\mathfrak{A}$ is defined by

$$\mathfrak{A}U = \begin{pmatrix} -\frac{1}{\overline{\rho}}\nabla \cdot T(\cdot) & \frac{p'(\overline{\rho})}{\overline{\rho}}\nabla \\ \overline{\rho}\nabla\cdot & 0 \end{pmatrix} U,$$

and the domain of $\mathfrak{A}$ is the set of $U = (v, \theta)^T$ such that

$$v \in W_2^{2+l}(\Omega_0^+) \cap W_2^{2+l}(\Omega_0^-),$$
$$\theta \in W_2^{1+l}(\Omega_0^+) \cap W_2^{1+l}(\Omega_0^-), \quad \int_{\Omega} \theta(x)dx = 0,$$
$$v|_{x\in S} = 0, \quad [v]|_{x\in\Gamma_0} = 0, \quad [-p'(\overline{\rho})\theta n + T(v)n]|_{x\in\Gamma_0} = 0.$$

Since

$$\overline{\rho}sI + \mathfrak{A}: \ (W_2^{2+l}(\Omega_0^+) \cap W_2^{2+l}(\Omega_0^-)) \times (W_2^{1+l}(\Omega_0^+) \cap W_2^{1+l}(\Omega_0^-))$$
$$\to (W_2^{l}(\Omega_0^+) \cap W_2^{l}(\Omega_0^-)) \times (W_2^{1+l}(\Omega_0^+) \cap W_2^{1+l}(\Omega_0^-)),$$

the operator $(\overline{\rho}sI + \mathfrak{A})^{-1}$ is not compact, which is the main difficulty of the problem.

The first statement of Theorem 2 is a consequence of local in time existence theorem for the parabolic problem

$$\begin{cases} \overline{\rho}v_t - \nabla \cdot T(v) = f'(x,t), & x \in \Omega_0^+ \cup \Omega_0^-, \\ v(x,t) = 0, & x \in S, \\ [v] = 0, \quad [T(u)n_0] = b'(x,t), & x \in \Gamma_0, \\ v(x,0) = v_0(x), & x \in \Omega_0^- \cup \Omega_0^+, \end{cases} \tag{10.16}$$

established in [4]. The proof of the estimate (10.15) is based on the ideas of M. Padula who has estimated a Lyapunov type function, so called modified energy, for the complete equations of motion of the compressible viscous fluid ([5, 6], see also [7]). We obtain similar estimate for problem (10.11).

Proposition 1 *If (10.14) hold, then the solution of the problem (10.11) satisfies the inequality*

$$\sup_{t<T} e^{2\beta t}(\|v(\cdot,t)\|^2_{L_2(\Omega)} + \|\theta(\cdot,t)\|^2_{L_2(\Omega)}) + \int_0^T e^{2\beta t}(\|v\|^2_{W^1_2(\Omega)} + \|\theta\|^2_{L_2(\Omega)})dt$$
$$\leqslant c\Big(\|v_0\|^2_{L_2(\Omega)} + \|\theta_0\|^2_{L_2(\Omega)} + \int_0^T e^{2\beta t}(\|f\|^2_{L_2(\Omega)} + \|h\|^2_{L_2(\Omega)} + \|b\|^2_{L_2(\Gamma_0)})dt\Big) \tag{10.17}$$

with $T \leqslant \infty$, $\beta > 0$ and c independent of T.

Proof We recall that the conditions (10.14) imply $\int_\Omega \theta(x,t)dx = 0, \quad t > 0$. By standard calculation we obtain

$$\frac{1}{2}\frac{d}{dt}\overline{\rho}\|v\|^2_{L_2(\Omega)} + \sum_{\pm}\int_{\Omega_0^\pm} T(v) : \nabla v dx - p'(\overline{\rho})\int_\Omega \theta\nabla\cdot v dx$$
$$+ \int_{\Gamma_0}[p'(\overline{\rho})\theta n - T(v)n]\cdot v dS = \int_\Omega f\cdot v dx,$$

which implies

$$\frac{1}{2}\frac{d}{dt}(\overline{\rho}\|v\|^2_{L_2(\Omega)} + \frac{p'(\overline{\rho})}{\overline{\rho}}\|\theta\|^2_{L_2(\Omega)}) + \sum_{\pm}(\frac{\mu^\pm}{2}\|S(v)\|^2_{L_2(\Omega_0^\pm)} + \mu_1^\pm\|\nabla\cdot v\|^2_{L_2(\Omega^\pm 0)})$$
$$= \int_\Omega (f\cdot v - \frac{p'(\overline{\rho})}{\overline{\rho}}h\theta)dx + \int_{\Gamma_0} b\cdot v dS. \tag{10.18}$$

Let $V(x,t)$, $x\in\Omega$, be a vector field such that

$$\begin{aligned}
&\nabla\cdot V(x,t) = -\theta(x,t), \quad x\in\Omega, \quad V(x,t) = 0, \quad x\in S,\\
&\|V\|_{W^1_2(\Omega)} \leqslant c\|\theta\|_{L_2(\Omega)},\\
&\|\mathscr{D}_t V\|_{W^1_2(\Omega)} \leqslant c\|\mathscr{D}_t\theta\|_{L_2(\Omega)} \leqslant c(\|v\|_{W^1_2(\Omega)} + \|h\|_{L_2(\Omega)})
\end{aligned} \tag{10.19}$$

(the problem (10.19) has been analyzed by many authors, see, e.g., [8–10]). We multiply the first equation in (10.11) by V and integrate over Ω, which leads to

$$\overline{\rho}(\frac{d}{dt}\int_\Omega v\cdot V dx - \int_\Omega v\cdot\mathscr{D}_t V dx) + \int_\Omega T(v) : \nabla V dx + p'(\overline{\rho})\|\theta\|^2_{L_2(\Omega)}$$
$$= \int_\Omega f\cdot V dx + \int_{\Gamma_0} b\cdot V dS. \tag{10.20}$$

Finally, we multiply (10.20) by a small positive γ and add to (10.18), which yields

$$\begin{aligned}&\frac{1}{2}\frac{dE_0(t)}{dt}+E_1(t)-\overline{\rho}\gamma\int_\Omega v\cdot\mathscr{D}_tV\,dx\\&=\int_\Omega(f\cdot(v+\gamma V)-\frac{p'(\overline{\rho})}{\overline{\rho}}h\theta)dx+\int_{\Gamma_0}b\cdot(v+\gamma V)dS,\end{aligned}\tag{10.21}$$

where

$$\begin{aligned}&E_0(t)=\overline{\rho}\|v\|^2_{L_2(\Omega)}+\frac{p'(\overline{\rho})}{\overline{\rho}}\|\theta\|^2_{L_2(\Omega)}+2\overline{\rho}\gamma\int_\Omega v\cdot V\,dx,\\&E_1(t)=\sum_\pm(\frac{\mu^\pm}{2}\|S(v)\|^2_{L_2(\Omega_0^\pm)}+\mu_1^\pm\|\nabla\cdot v\|^2_{L_2(\Omega_0^\pm)})\\&+\gamma p'(\overline{\rho})\|\theta\|^2_{L_2(\Omega)}+\gamma\sum_\pm\int_{\Omega_0^\pm}T(v):\nabla V\,dx.\end{aligned}\tag{10.22}$$

Due to (10.19) and the Korn inequality, we have

$$\begin{aligned}&c_1(\|v\|^2_{L_2(\Omega)}+\|\theta\|^2_{L_2(\Omega)})\leqslant E_0(t)\leqslant c_2(\|v\|^2_{L_2(\Omega)}+\|\theta\|^2_{L_2(\Omega)}),\\&E_1(t)\geqslant 2\beta E_0(t),\quad \beta>0,\\&E_1(t)\geqslant c_3(\|v\|^2_{W_2^1(\Omega)}+\|\theta\|^2_{L_2(\Omega)}),\end{aligned}\tag{10.23}$$

provided γ is sufficiently small. Hence

$$\frac{dE_0(t)}{dt}+2\beta E_0(t)+E_1(t)\leqslant I(t),$$

where $\beta>0$ and

$$I(t)=2\int_\Omega(f\cdot(v+\gamma V)-\frac{p'(\overline{\rho})}{\overline{\rho}}h\theta)dx+2\int_{\Gamma_0}b\cdot(v+\gamma V)dS+2\gamma\overline{\rho}\int_\Omega v\cdot\mathscr{D}_tV\,dx.$$

It follows that

$$e^{2\beta t}E_0(t)+\int_0^t e^{2\beta\tau}E_1(\tau)d\tau\leqslant E_0(0)+\int_0^t e^{2\beta\tau}|I(\tau)|d\tau,\tag{10.24}$$

which leads to (10.17) after easy calculations.

We pass to the local estimates of the higher order norms of v and θ. Let $\{\omega_k\}$ be the covering of Ω of the following form: ω_k, $k=1,\dots,m^--1$ are the balls $|x-x_k|\leqslant d$, $x_k\in S$, ω_k, $k=1+m^-,\dots,m^-+m^+-1$ are the balls $|x-x_k|\leqslant d$ with $x_k\in\Gamma_0$; ω_{m^-}, and $\omega_{m^-+m^+}$ are strictly interior subdomains of Ω^- and Ω^+, respectively. Let $\{\varphi_k\}$ be the partition of unity subordinate to this covering, and let ζ_k, $k=1,\dots,m^--1$, $k=m^-+1,\dots,m^-+m^+-1$, be the functions with $\mathrm{supp}\zeta_k\subset$

$B_k \subset \omega_k \cap \omega_{m^-}$, $k < m^-$, or $B_k \subset \omega_k \cap \omega_{m^-+m^+}$, $m^- < k < m^- + m^+$ such that $\int_{B_k} \varphi_k(x)dx = 1$, where B_k are balls with the radius αd, $\alpha < 1$. We set $v_k = v\varphi_k$, $\theta_k = \theta\varphi_k$, $\omega_k' = \omega_k \cap \Omega$,

$$w_k(x,t) = v_k(x,t) - \zeta_k(x) \int_{\omega_k'} v_k(y,t)dy, \quad \vartheta_k = \theta_k - \zeta_k \int_{\omega_k'} \theta_k(y,t)dy$$

for $k \neq m^-, m^- + m^+$, and

$$\widetilde{w}_{m^-}(x,t) = v_{m^-}(x,t) + \sum_{k=1}^{m^- -1} \zeta_k(x) \int_{\omega_k'} v_k(y,t)dy,$$

$$\widetilde{w}_{m^-+m^+} = v_{m^-+m^+} + \sum_{k=m^-+1}^{m^-+m^+-1} \zeta_k \int_{\omega_k'} v_k(y,t)dy,$$

$$\widetilde{\vartheta}_{m^-}(x,t) = \theta_{m^-}(x,t) + \sum_{k=1}^{m^- -1} \zeta_k(x) \int_{\omega_k'} \theta_k(y,t)dy,$$

$$\widetilde{\vartheta}_{m^-+m^+} = \theta_{m^-+m^+} + \sum_{k=m^-+1}^{m^-+m^+-1} \zeta_k \int_{\omega_k'} \theta_k(y,t)dy.$$

Let $\widetilde{\omega} = \Omega \setminus (\cup_{k=1}^{m^-} \omega_k \cup \omega_{m^-+m^+})$,

$$w_{m^-}(x,t) = \widetilde{w}_{m^-} - \widetilde{\zeta}(x) \int_{\omega_{m^-}} \widetilde{w}_{m^-}(y,t)dy,$$

$$w_{m^-+m^+}(x,t) = \widetilde{w}_{m^-+m^+} + \widetilde{\zeta}(x) \int_{\omega_{m^-}} \widetilde{w}_{m^-}(y,t)dy,$$

$$\vartheta_{m^-}(x,t) = \widetilde{\vartheta}_{m^-} - \widetilde{\zeta}(x) \int_{\omega_{m^-}} \widetilde{\vartheta}_{m^-}(y,t)dy,$$

$$\vartheta_{m^-+m^+}(x,t) = \widetilde{\vartheta}_{m^-+m^+} + \widetilde{\zeta}(x) \int_{\omega_{m^-}} \widetilde{\vartheta}_{m^-}(y,t)dy,$$

where $\widetilde{\zeta}(x)$ is a smooth function with $\mathrm{supp}\widetilde{\zeta} \subset \widetilde{\omega}$ and $\int_{\widetilde{\omega}} \widetilde{\zeta}(x)dx = 1$. It is clear that

$$\int_{\omega_k'} \vartheta_k(x,t)dx = 0, \quad k \neq m^-, m^- + m^+, \quad \int_{\omega_{m^-}} \widetilde{\vartheta}_{m^-}dx + \int_{\omega_{m^-+m^+}} \widetilde{\vartheta}_{m^-+m^+}dx = 0,$$

$$\int_{\widetilde{\omega}_{m^-}} \vartheta_{m^-}dx = 0, \quad \int_{\widetilde{\omega}_{m^-+m^+}} \vartheta_{m^-+m^+}dx = 0,$$

and

$$\sum_k \vartheta_k(x,t) = \theta(x,t), \quad \sum_k v_k(x,t) = v(x,t).$$

Moreover, for all $k = 1, ..., m^- + m^+$

$$\overline{\rho}\mathscr{D}_t w_k - \nabla \cdot T(w_k) + p'(\overline{\rho})\nabla\vartheta_k = f_k + \mathfrak{f}_k,$$
$$\mathscr{D}_t\vartheta_k + \overline{\rho}\nabla \cdot w_k = h_k + \mathfrak{h}_k, \tag{10.25}$$

where

$$f_k = f\varphi_k - \zeta_k \int_{\omega'_k} f\varphi_k dx, \quad h_k = h\varphi_k - \zeta_k \int_{\omega'_k} h\varphi_k dx, \quad k \neq m^-, m^- + m^+,$$

$$\widetilde{f}_{m^-} = f\varphi_{m^-} + \sum_{k=1}^{m^- -1} \zeta_k \int_{\omega'_k} f\varphi_k dx, \quad \widetilde{h}_{m^-} = h\varphi_{m^-} + \sum_{k=1}^{m^- -1} \zeta_k \int_{\omega'_k} h\varphi_k dx,$$

$$\widetilde{f}_{m^-+m^+} = f\varphi_{m^-+m^+} + \sum_{k=1+m^-}^{m^-+m^+-1} \zeta_k \int_{\omega'_k} f\varphi_k dx,$$

$$\widetilde{h}_{m^-+m^+} = h\varphi_{m^-+m^+} + \sum_{k=1+m^-}^{m^-+m^+-1} \zeta_k \int_{\omega'_k} h\varphi_k dx,$$

$$f_{m^-}(x,t) = \widetilde{f}_{m^-}(x,t) - \widetilde{\zeta}(x) \int_{\widetilde{\omega}_{m^-}} \widetilde{f}_{m^-} dy,$$

$$f_{m^-+m^+}(x,t) = \widetilde{f}_{m^-+m^+}(x,t) + \widetilde{\zeta}(x) \int_{\widetilde{\omega}_{m^-}} \widetilde{f}_{m^-} dy. \tag{10.26}$$

The expressions $\mathfrak{f}_k$, $\mathfrak{h}_k$ contain only lower order derivatives of v and θ.

In order to estimate (v, θ) in the interior of $\Omega_0^\pm$, we consider Eq. (10.25) for $k = m^-, k = m^- + m_+$ completed by initial conditions

$$w_k(x, 0) = w_{0k}(x), \quad \vartheta_k(x, 0) = \vartheta_{0k}, \quad k = m^-, m^- + m^+, \tag{10.27}$$

where

$$w_{0m^-}(x) = \widetilde{w}_{0m^-}(x) - \widetilde{\zeta}(x) \int_{\omega_{m^-}} \widetilde{w}_{0m^-}(y) dy,$$

$$w_{0m^-+m^+}(x) = \widetilde{w}_{0m^-+m^+} + \widetilde{\zeta}(x) \int_{\widetilde{\omega}_m} \widetilde{w}_{0m^-}(y) dy,$$

$$\vartheta_{0m^-}(x) = \widetilde{\vartheta}_{0m^-}(x) - \widetilde{\zeta}(x) \int_{\omega_{m^-}} \widetilde{\vartheta}_{0m^-}(y) dy,$$

$$\vartheta_{0m^-+m^+}(x) = \widetilde{\vartheta}_{0m^-+m^+} + \widetilde{\zeta}(x) \int_{\widetilde{\omega}_m} \widetilde{\vartheta}_{0m^-}(y) dy.$$

Since $\mathrm{supp} w_{m^-}, \mathrm{supp}\vartheta_{m^-} \subset \widetilde{\omega}_{m^-}$, $\mathrm{supp} w_{m^-+m^+}, \mathrm{supp}\vartheta_{m^-+m^+} \subset \widetilde{\omega}_{m^-+m^+}$, we can extend these functions by zero into larger domains $\mathfrak{Q}_k$: $|x_j - x_{jk}| \leqslant D$, $k = m^-$. $m^- + m^-$ with preservation of differential properties and apply Proposition 1 to the problems (10.25), (10.27). This gives

$$\int_0^T e^{2\beta t}(\|w_k(\cdot,t)\|^2_{W^1_2(\Omega_k)} + \|\vartheta_k(\cdot,t)\|^2_{L_2(\Omega_k)})dt$$
$$\leqslant c(\|w_k(\cdot,0)\|^2_{L_2(\Omega_k)} + \|\vartheta_k(\cdot,0)\|^2_{L_2(\Omega_k)} \tag{10.28}$$
$$+ \int_0^T e^{2\beta t}(\|f_k + \mathfrak{f}_k\|^2_{L_2(\Omega_k)} + \|h_k + \mathfrak{h}_k\|^2_{L_2(\Omega_k)})dt), \quad k = m^-, m^- + m^+.$$

To obtain the estimates of higher order norms of w_k and ϑ_k, we take finite differences $\Delta^q_j(z)u(x) = \sum_{k=0}^q (-1)^{q-k}\binom{q}{k}u(x+ke_jz)$ of (10.25) and (10.27) with $q > l+1$, which leads to

$$\begin{aligned}
&\overline{\rho}\Delta^q_j(z)\mathscr{D}_t w_k - \nabla\cdot T(\Delta^q_j(z)w_k) + p'(\overline{\rho})\nabla\Delta^q_j(z)\vartheta_k = \Delta^q_j(z)(f_k + \mathfrak{f}_k),\\
&\Delta^q_j(z)\mathscr{D}_t\vartheta_k + \overline{\rho}\nabla\cdot\Delta^q_j(z)w_k = \Delta^q_j(z)(h_k + \mathfrak{h}_k),\\
&\Delta^q_j(z)w_k(x,0) = \Delta^q_j(z)w_{0k}(x),\\
&\Delta^q_j(z)\varphi_k(x,0) = \Delta^q_j(z)\varphi_{0k},
\end{aligned} \tag{10.29}$$

$k = m^-, m^- + m^+$. We introduce the vector fields $W_k(x,t)$ such that

$$\begin{aligned}
&\nabla\cdot W_k(x,t) = -\Delta^q_j(z)\vartheta_k(x,t), \quad x \in \Omega_k,\\
&\|W_k\|_{W^1_2(\Omega_k)} \leqslant c\|\Delta^q_j(z)\vartheta_k\|_{L_2(\Omega_k)},\\
&\|\mathscr{D}_t W_k\|_{L_2(\Omega_k)} \leqslant c(\|\Delta^q_j(z)\nabla w_k\|_{L_2(\Omega_k)} + \|\Delta^q_j(z)(h_k + \mathfrak{h}_k)\|_{L_2(\Omega_k)}).
\end{aligned} \tag{10.30}$$

and in view of Proposition 1 applied to the system (10.29), we obtain

$$\begin{aligned}
&\int_0^T e^{2\beta t}(\|\Delta^q_j(z)w_k(\cdot,t)\|^2_{W^1_2(\Omega_k)} + \|\Delta^q_j(z)\vartheta_k(\cdot,t)\|^2_{L_2(\Omega_k)})dt\\
&\leqslant c\big(\|\Delta^q_j(z)w_{0k}\|^2_{L_2(\Omega_k)} + \|\Delta^q_j(z)\vartheta_{0k}\|^2_{L_2(\Omega_k)}\\
&+ 2\Big|\int_0^T e^{2\beta t}dt\int_{\Omega_k}(\Delta^q_j(z)(f_k + \mathfrak{f}_k)\cdot(\Delta^q_j(z)w_k + \gamma W_k(x,t))\\
&+ \Delta^q_j(z)(h_k + \mathfrak{h}_k)\Delta^q_j\vartheta_k)dx\Big| + 2\overline{\rho}\gamma\Big|\int_0^T e^{2\beta t}dt\int_{\Omega_k}\Delta^q_j(z)w_k\cdot\mathscr{D}_t W_k(x,t)dx\Big|\big).
\end{aligned} \tag{10.31}$$

We make use of

$$\begin{aligned}
&\int_0^T e^{2\beta t}dt\int_{\Omega_k}(\Delta^q_j(z)(f_k + \mathfrak{f}_k)\cdot(\Delta^q_j(z)w_k + \gamma W_k)dx\\
&= -\int_0^T e^{2\beta t}dt\int_{\Omega_k}(\Delta^{q-1}_j(z)(f_k + \mathfrak{f}_k)\cdot(\Delta^{q+1}_j w_k + \gamma\Delta^1_j(z)W_k)dx
\end{aligned} \tag{10.32}$$

and estimate the penultimate integral in (10.31) (denoted by J_k) as follows:

$$|J_k| \leqslant c\Big(\int_0^T e^{2\beta t}(\|\Delta_j^{q-1}(z)(f_k+\mathfrak{f}_k)\|_{L_2(\Omega_k)}(\|\Delta_j^{q+1}w_k\|_{L_2(\Omega_k)}+|z|\|\nabla W_k\|_{L_2(\Omega_k)})dt$$
$$+\int_0^T e^{2\beta t}(\|\Delta_j^q(z)(h_k+\mathfrak{h}_k)\|_{L_2(\Omega_k)}\|\Delta_j^q(z)\vartheta_k\|_{L_2(\Omega)}dt\Big)$$
$$\leqslant c\Big(\int_0^T e^{2\beta t}\|\Delta_j^{q-1}(z)(f_k+\mathfrak{f}_k)\|_{L_2\Omega)}(\|\Delta_j^{q+1}w_k\|_{L_2(\Omega_k)}+|z|\|\Delta_j^q(z)\vartheta_k\|_{L_2(\Omega_k)})dt$$
$$+\int_0^T e^{2\beta t}\|\Delta_j^q(z)(h_k+\mathfrak{h}_k)\|_{L_2(\Omega_k)}\|\Delta_j^q(z)\vartheta_k\|_{L_2(\Omega_k)}dt\Big). \qquad (10.33)$$

Next, we divide (10.31) by $z^{1+2(l+1)}$ and integrate over $z \in (0, z_0)$. Since the norms

$$\|u\|_{L_2(\Omega_k)}+\sum_{j=1}^3\Big(\int_0^{z_0}\|\Delta_j^s(z)u\|^2_{L_2(\Omega_k)}\frac{dz}{z^{1+2r}}\Big)^{1/2}$$

and $\|u\|_{W_2^r(\Omega_k)}$ are equivalent for all $s > r > 0$ [11], our calculations lead to

$$\|e^{\beta t}w_k\|^2_{W_2^{l+2,0}(\Omega_{kT})}+\|e^{\beta t}\vartheta_k\|^2_{W_2^{l+1,0}(\Omega_{kT})}\leqslant c\big(\|e^{\beta t}(f_k+\mathfrak{f}_k)\|^2_{W_2^{l,0}(\Omega_{kT})}$$
$$+\|e^{\beta t}(h_k+\mathfrak{h}_k)\|^2_{W_2^{l+1,0}(\Omega_{kT})}+\|w_k(\cdot,0)\|^2_{W_2^{l+1}(\Omega_k)}+\|\vartheta_k(\cdot,0)\|^2_{W_2^{l+1}(\Omega_k)}\big)\equiv A_k, \qquad (10.34)$$

where $\Omega_{kT}=\Omega_k\times(0,T)$. Moreover, we have

$$\|e^{\beta t}\mathscr{D}_t\vartheta_k\|^2_{W_2^{l+1,0}(\Omega_{kT})}\leqslant c(\|e^{\beta t}\nabla\cdot w_k\|^2_{W_2^{l+1,0}(\Omega_{kT})}+\|e^{\beta t}(h_k+\mathfrak{h}_k)\|^2_{W_2^{l+1,0}(\Omega_{kT})})\leqslant cA_k,$$
$$\|e^{\beta t}\mathscr{D}_t w_k\|^2_{W_2^{0,l/2}(\Omega_{kT})}\leqslant c\big(\|e^{\beta t}\mathscr{D}_x^2 w_k\|^2_{W_2^{0,l/2}(\Omega_{kT})}$$
$$+\|e^{\beta t}\nabla\vartheta_k\|^2_{W_2^{0,l/2}(\Omega_{kT})}+\|e^{\beta t}(f_k+\mathfrak{f}_k)\|^2_{W_2^{0,l/2}\Omega_{kT})}\big). \qquad (10.35)$$

Due to the interpolation inequalities, the right hand side in the last estimate is not greater than

$$\delta(\|e^{\beta t}\mathscr{D}_t w_k\|^2_{W_2^{0,l/2}(\Omega_{kT})}+\|e^{\beta t}\mathscr{D}_t\vartheta_k\|^2_{W_2^{l+1,0}(\Omega_{kT})})+c(\delta)(\|e^{\beta t}w_k\|^2_{W_2^{2+l,0}(\Omega_{kT})}$$
$$+\|e^{\beta t}\vartheta_k\|^2_{W_2^{l+1,0}(\Omega_{kT})})$$
$$+c(\|e^{\beta t}\mathscr{D}_t\vartheta_k\|^2_{W_2^{l+1,0}(\Omega_{kT})}+\|e^{\beta t}(f_k+\mathfrak{f}_k)\|^2_{W_2^{l,l/2}\Omega_{kT})}).$$

Collecting the estimates and choosing δ sufficiently small, we obtain the desired inequality

$$\|e^{\beta t} w_k\|^2_{W_2^{l+2,l/2+1}(\mathfrak{Q}_{kT})} + \|e^{\beta t}\vartheta_k\|^2_{W_2^{l+1,0}(\mathfrak{Q}_{kT})} + \|e^{\beta t}\mathscr{D}_t\vartheta_k\|^2_{W_2^{l+1,0}(\mathfrak{Q}_{kT})}$$
$$\leqslant c(A_k + \|e^{\beta t}(f_k + \mathfrak{f}_k)\|^2_{W_2^{0,l/2}\mathfrak{Q}_{kT})}), \quad k = m^-, m^- + m^+. \tag{10.36}$$

In order to estimate v and θ near S and Γ_0, we should consider Eq. (10.25) completed by appropriate initial and boundary conditions, for $k < m^-$ and $m^- < k < m^- + m^+$:

$$\begin{cases} \overline{\rho}\mathscr{D}_t w_k - \nabla \cdot T(w_k) + p'(\overline{\rho})\nabla\vartheta_k = f_k + \mathfrak{f}_k, \\ \mathscr{D}_t\vartheta_k + \overline{\rho}\nabla \cdot w_k = h_k + \mathfrak{h}_k, \quad x \in \omega'_k, \\ w_k(x,t) = 0, \quad x \in \overline{\omega}'_k \cap S, \\ w_k(x,0) = w_{0k}(x), \quad \vartheta_k(k,0) = \vartheta_{0k}(x), \quad x \in \omega'_k, \quad k < m^-, \end{cases} \tag{10.37}$$

$$\begin{cases} \overline{\rho}\mathscr{D}_t w_k - \nabla \cdot T(w_k) + p'(\overline{\rho})\nabla\vartheta_k = f_k + \mathfrak{f}_k, \\ \mathscr{D}_t\vartheta_k + \overline{\rho}\nabla \cdot w_k = h_k + \mathfrak{h}_k, \quad x \in \omega_k, \\ [w_k(x,t)] = 0, \quad [-p'(\overline{\rho})\vartheta_k n + T(w_k)n] = b_k(x,t) + \mathfrak{b}_k, \quad x \in \Gamma_0 \cap \omega_k, \\ w_k(x,0) = w_{0k}(x), \quad \vartheta_k(x,0) = \vartheta_{0k}(x), \quad x \in \omega_k, \quad m^- < k < m^- + m^+, \end{cases} \tag{10.38}$$

where f_k, h_k are defined in (10.26) and $b_k = b\varphi_k$. As above, $\mathfrak{f}_k$, $\mathfrak{h}_k$, $\mathfrak{b}_k$ contain lower order derivatives of v and θ. We also keep in mind that $\mathrm{supp} w_k$, $\mathrm{supp}\vartheta_k \subset \omega_k$.

We consider a more specific system (10.38). Without loss of generality we may assume that the point x_k coincides with the origin and the x_3-axis is directed along the interior normal $-n(x_k)$ with respect to Ω_0^+. Let $x_3 = \phi(x')$, $x' = (x_1, x_2) \in K$, be equation of Γ_0 near the origin; by K we mean the disc $|x| \leqslant d_1$; $d_1 > d$.

Assuming for simplicity that $\Gamma_0 \in W_2^{2+l}$, we make the change of variables

$$x_1 = y_1, \quad x_2 = y_2, \quad x_3 = y_3 + \phi(y'), \tag{10.39}$$

where $\phi \in W_2^{2+l}(K)$. The Jacobi matrix of this transformation is given by

$$\mathfrak{L} = \begin{pmatrix} 1 & 0 & 0 \\ 0 & 1 & 0 \\ \phi_{y_1} & \phi_{y_2} & 1 \end{pmatrix},$$

hence $det\mathfrak{L} = 1$, and $\mathfrak{L}^{-1}$ coincides with the co-factors matrix $\widehat{\mathfrak{L}}$. The transformation (10.39) converts $\omega_k \cap \Omega$ into a certain $\widetilde{\omega}_k \subset \mathbb{R}^3$. We extend w_k and ϑ_k by zero into larger domains

$$\mathfrak{Q}^\pm = \{0 < \pm y_3 < 2d_0, \quad y' \in \mathfrak{Q}' = \{|y_\alpha| < d_0\}, \quad \alpha = 1, 2,$$

where $d_0 > d$. In the new coordinates y, the system (10.38) takes the form

$$\begin{cases} \overline{\rho}\mathscr{D}_t w_k - \widetilde{\nabla}\cdot\widetilde{T}(w_k) + p'(\overline{\rho})\widetilde{\nabla}\vartheta_k = f_k + \mathfrak{f}_k, \\ \mathscr{D}_t\vartheta_k + \overline{\rho}\widetilde{\nabla}\cdot w_k = h_k + \mathfrak{h}_k, \quad y \in \mathfrak{Q}^{\pm}, \\ [w_k] = 0, \quad [-p'(\overline{\rho})\vartheta_k n + \widetilde{T}(w_k)n] = b_k(x,t) + \mathfrak{b}_k, \quad y_3 = 0, \\ w_k(y,0) = w_{0k}(y), \quad \vartheta_k(y,0) = \vartheta_{0k}(y), \quad y \in \mathfrak{Q}^{\pm}, \quad m^- < k < m^- + m^+, \end{cases} \tag{10.40}$$

where $\widetilde{\nabla} = \mathfrak{L}^{-T}\nabla$ is the transformed gradient ∇_x, $\widetilde{T}(u) = \mu\widetilde{S}(u) + \mu_1 I\widetilde{\nabla}\cdot u$ is the transformed stress tensor and $\widetilde{S}(u) = \widetilde{\nabla}u + (\widetilde{\nabla}u)^T$ is the transformed rate-of-strain tensor. We write (10.40) as

$$\begin{cases} \overline{\rho}\mathscr{D}_t w - \nabla\cdot T(w) + p'(\overline{\rho})\nabla\vartheta = f + \widetilde{\mathfrak{f}}, \\ \mathscr{D}_t\vartheta + \overline{\rho}\nabla\cdot w = h + \widetilde{\mathfrak{h}}, \quad y \in \mathfrak{Q}^{\pm}, \\ [w] = 0, \quad [\mu S_{\alpha 3}(w)] = b_\alpha + \widetilde{\mathfrak{b}}_\alpha, \quad \alpha = 1,2, \\ [-p'(\overline{\rho})\vartheta + T_{33}(w)] = b_3 + \widetilde{\mathfrak{b}}_3, \quad y_3 = 0, \quad y' \in \mathfrak{Q}', \\ w(y,t) = 0, \quad y_3 = \pm 2d_0, \quad y' \in \mathfrak{Q}', \\ w(y,0) = w_0(y), \quad \vartheta(y,0) = \vartheta_0(y), \quad y \in \widehat{\mathfrak{Q}}^{\pm} \end{cases} \tag{10.41}$$

(the index "k" is omitted). The expressions

$$\begin{aligned} \widetilde{\mathfrak{f}} &= \mathfrak{f} + \widetilde{\nabla}\cdot\widetilde{T}(w) - \nabla\cdot T(w) - p'(\overline{\rho})\widetilde{\nabla}\vartheta + p'(\overline{\rho})\nabla\vartheta, \\ \widetilde{\mathfrak{h}} &= \mathfrak{h} - \overline{\rho}\widetilde{\nabla}\cdot w + \overline{\rho}\nabla\cdot w, \\ \widetilde{\mathfrak{b}}_\alpha &= \mathfrak{b}_\alpha - [\mu(\widetilde{S}(w)n - n(n\cdot\widetilde{S}(w)n))_\alpha] + [\mu S_{\alpha 3}(w)], \\ \widetilde{\mathfrak{b}}_3 &= \mathfrak{b}_3 + [\mu(S_{33}(w) - n\cdot\widetilde{S}(w)n)], \end{aligned} \tag{10.42}$$

contain also higher order derivatives of w and ϑ with small coefficients proportional to ϕ or to the derivatives of ϕ. We note that also in the new coordinates

$$\int_{\mathfrak{Q}^+\cup\mathfrak{Q}^-} \vartheta(y,t)dy = 0. \tag{10.43}$$

Remark If $\phi \in W_2^{3/2+l}(K)$, then instead of (10.39) the transformation

$$x' = y', \quad x_3 = y_3 + \phi^*(y) \tag{10.44}$$

should be used, where ϕ^* is the extension of ϕ into $K \times (0, 2d_0)$ such that

$$\begin{aligned} &\phi^*(y)|_{y_3=0} = \phi(y'), \quad \frac{\partial\phi^*}{\partial y_3}\Big|_{y_3=0} = 0, \\ &\|\phi^*\|_{W_2^{l+2}(K\times(0,2d_0))} \leqslant c\|\phi\|_{W_2^{l+3/2}(K)}. \end{aligned} \tag{10.45}$$

Since $\phi^*(0) = 0, \quad \nabla\phi^*(0) = 0$, there holds

$$|\phi^*(y)| \leqslant c|y|^{1+\alpha}, \quad |\nabla\phi^*(y)| \leqslant c|y|^{\alpha}, \quad \alpha = l - 1/2. \tag{10.46}$$

The Jacobi matrix of the transformation (10.44) equals

$$\mathfrak{L} = \begin{pmatrix} 1 & 0 & 0 \\ 0 & 1 & 0 \\ \phi^*_{y_1} & \phi^*_{y_2} & 1 + \phi^*_{y_3} \end{pmatrix},$$

and $det\mathfrak{L} = 1 + \phi^*_{y_3}$. Consequently,

$$\int_{\Omega^+ \cup \Omega^-} \vartheta(y,t)(1 + \phi^*_{y_3})dy = 0.$$

In order to conserve the condition (10.43), we can introduce new function

$$\Theta(y,t) = \vartheta(y,t) + \Sigma(y) \int_{\Omega^+ \cup \Omega^-} \vartheta(y,t)\phi^*_{y_3} dy,$$

where $\Sigma(y)$ is smooth, $\text{supp}\Sigma \subset \Omega^+ \cup \Omega^-$ and $\int_{supp\Sigma} \Sigma(x)dx = 1$. Clearly, (w, Θ) satisfy the system of the type (10.41) with other $\widetilde{\mathfrak{f}}_k$ and $\widetilde{\mathfrak{h}}_k$, but possessing the same properties. Since the new system is treated in the same way as (10.41), we omit the details.

We proceed as above and apply Proposition 1 to the problem (10.41). Since finite differences can be taken this time only with respect to the tangential variables y_1, y_2, we obtain, instead of (10.36), the estimate

$$\begin{aligned} &\sum_{\pm} (\|e^{\beta t}\nabla w\|^2_{W^{l+1,0}_{2,tan}(\Omega^{\pm}_T)} + \|e^{\beta t} w\|^2_{W^{l+1,0}_{2,tan}(\Omega^{\pm}_T)} + \|e^{\beta t}\vartheta\|^2_{W^{l+1}_{2,tan}(\Omega^{\pm}_T)}) \\ &\leqslant c\Big(\sum_{\pm} \|e^{\beta t}(f + \widetilde{\mathfrak{f}})\|^2_{W^{l,l/2}_{2,tan}(\Omega^{\pm}_T)} + \|e^{\beta t}(h + \widetilde{\mathfrak{h}})\|^2_{W^{l+1,0}_{2,tan}(\Omega^{\pm}_T)} \\ &+ \|w(\cdot,0)\|^2_{W^{l+1}_{2,tan}(\Omega^{\pm})} + \|\vartheta(\cdot,0)\|^2_{W^{l+1}_{2,tan}(\Omega^{\pm})}) + \|e^{\beta t}(b + \widetilde{\mathfrak{b}})\|^2_{W^{l+1/2,0}_2(\Omega'_T)}\Big), \end{aligned} \tag{10.47}$$

where

$$W^r_{2,tan}(\Omega^{\pm}) = L_2(I^{\pm}, W^r_2(\Omega')), \quad I^{\pm} = \{0 < \pm y_3 < 2d_0\}.$$

To estimate the derivatives of w and ϑ with respect to t and y_3, we treat (10.38) as one-dimensional problems containing only these derivatives in the left hand side (other derivatives are included into $\mathfrak{f}'_i$, $\mathfrak{h}'$, $\mathfrak{b}'_j$):

$$\begin{cases} \overline{\rho}\mathscr{D}_t w_\alpha - \mu \dfrac{\partial^2 w_\alpha}{\partial y_3^2} = f_\alpha + \bar{\mathfrak{f}}_\alpha + \mathfrak{f}'_\alpha = g_\alpha, \\ y_3 \in I^\pm = \{\pm y_3 \in (0, 2d_0)\}, \quad t > 0, \quad \alpha = 1, 2, \\ [w_\alpha] = 0, \quad [\mu \dfrac{\partial w_\alpha}{\partial y_3}] = b_\alpha + \widetilde{\mathfrak{b}}_\alpha + \mathfrak{b}'_\alpha = a_\alpha(y', t), \quad y_3 = 0, \\ w_\alpha(y, t) = 0, \quad y_3 = \pm 2d_0, \\ w_\alpha(y, 0) = w_{\alpha 0}(y), \quad y_3 \in I^\pm, \end{cases} \tag{10.48}$$

$$\begin{cases} \overline{\rho}\mathscr{D}_t w_3 - (2\mu + \mu_1)\dfrac{\partial^2 w_3}{\partial y_3^2} + p'(\overline{\rho})\dfrac{\partial \vartheta}{\partial y_3} = f_3 + \bar{\mathfrak{f}}_3 + \mathfrak{f}'_3 = g_3, \\ \mathscr{D}_t \vartheta + \overline{\rho}\dfrac{\partial w_3}{\partial y_3} = h + \widetilde{\mathfrak{h}} + \mathfrak{h}' = e, \\ \pm y_3 \in (0, 2d_0), \quad t > 0, \\ [w_3] = 0, \quad [-p'(\overline{\rho})\vartheta + (2\mu + \mu_1)\dfrac{\partial w_3}{\partial y_3}] = b_3 + \widetilde{\mathfrak{b}}_3 + \mathfrak{b}'_3 = a_3, \quad y_3 = 0, \\ w_3 = 0, \quad y_3 = \pm 2d_0, \\ w_3(y, 0) = w_{30}(y), \quad \vartheta(y, 0) = \vartheta_0(y), \quad y_3 \in I^\pm. \end{cases} \tag{10.49}$$

We analyze only problem (10.49) as more complicated. We reduce it to a similar problem with homogeneous initial data. We set $w_3^{(1)\pm}(y_3, t) = u^\pm(y_3, t)\chi(t)$, where $\chi(t)$ is a smooth cut-off function equal to 1 for $t \in [0, 1]$, to zero for $t \geqslant 2$, and we require that

$$\begin{aligned} &u^\pm(y_3, 0) = w_{3,0}^\pm(y_3), \quad u^\pm(y_3, t)|_{y_3 = \pm 2d_0} = 0; \quad [u]|_{y_3=0} = 0, \\ &\|u^\pm\|_{W_2^{2+l, 1+l/2}(I^\pm \times (0,2))} \leqslant c\|w_{3,0}^\pm\|_{W_2^{1+l}(I^\pm)} \end{aligned} \tag{10.50}$$

for arbitrary fixed $y' \in \Omega'$; the functions $w_3^{(1)\pm}$ satisfy similar inequalities. The functions $\vartheta^{(1)\pm}$ we define by $\vartheta^{(1)\pm}(y, t) = \vartheta_0^\pm(y)\chi(t)$; it is clear that

$$\|\vartheta^{(1)\pm}\|_{W_2^{1+l,0}(I^\pm \times (0,2))} + \|\mathscr{D}_t \vartheta^{(1)\pm}\|_{W_2^{1+l,0}(I^\pm \times (0,2))} \leqslant c\|\vartheta_0^\pm\|_{W_2^{1+l}(I^\pm)}. \tag{10.51}$$

For $w_3^{(2)} = w_3 - w_3^{(1)}$, $\vartheta^{(2)} = \vartheta - \vartheta^{(1)}$ we obtain a problem with zero initial conditions

$$\begin{cases} \overline{\rho}\mathscr{D}_t w_3^{(2)} - (2\mu+\mu_1)\dfrac{\partial^2 w_3^{(2)}}{\partial y_3^2} + p'(\overline{\rho})\dfrac{\partial \vartheta^{(2)}}{\partial y_3} = g_3^{(1)}, \\ \mathscr{D}_t\vartheta^{(2)} + \overline{\rho}\dfrac{\partial w_3^{(2)}}{\partial y_3} = e^{(1)}, \quad y_3 \in I^{\pm}, \quad t>0, \\ [w_3^{(2)}]|_{y_3=0} = 0, \quad [-p'(\overline{\rho})\vartheta^{(2)} + (2\mu+\mu_1)\dfrac{\partial w_3^{(2)}}{\partial y_3}]|_{y_3=0} = a_3^{(1)}(s), \\ w_3^{(2)}|_{y_3=\pm 2d_0} = 0, \quad w_3^{(2)}|_{t=0} = 0, \quad \vartheta^{(2)}|_{t=0} = 0, \end{cases} \tag{10.52}$$

where

$$g_3^{(1)} = g_3 - \overline{\rho}\mathscr{D}_t w_3^{(1)} + (2\mu+\mu_1)\frac{\partial^2 w_3^{(1)}}{\partial y_3^2} - p'(\overline{\rho})\frac{\partial\vartheta^{(1)}}{\partial y_3},$$

$$e^{(1)} = e - \mathscr{D}_t\vartheta^{(1)} - \overline{\rho}\frac{\partial w_3^{(1)}}{\partial y_3}, \quad a_3^{(1)} = a_3 + [p'(\overline{\rho})\vartheta^{(1)} - (2\mu+\mu_1)\frac{\partial w_3^{(1)}}{\partial y_3}].$$

We extend all the functions in this problem by zero into the domain $t<0$ and apply the Laplace transformation defined by a standard formula

$$\widetilde{u}(s) = \int_0^\infty e^{-st}u(t)dt, \quad s = s_1 + is_2,$$

with a small negative $s_1 = \text{Re}s$. We also eliminate $\widetilde{\vartheta}^{(2)}$ by using the equation $s\widetilde{\vartheta}^{(2)} = \widetilde{e}^{(1)} - \overline{\rho}\frac{d\widetilde{w}_3^{(2)}}{dy_3}$, which leads to the following problem for $\widetilde{w}_3^{(2)}$:

$$\begin{cases} \overline{\rho}s\widetilde{w}_3^{(2)} - (2\mu+\mu_1(s))\dfrac{d^2\widetilde{w}_3^{(2)}}{dy_3^2} = \widetilde{g}_3^{(1)} - \dfrac{p'(\overline{\rho})\overline{\rho}}{s}\dfrac{d\widetilde{e}^{(1)}}{dy_3}, \quad y_3 \in I^{\pm}, \\ [\widetilde{w}_3^{(2)}] = 0, \quad [(2\mu+\mu_1(s))\dfrac{dw_3^{(2)}}{dy_3}] = \widetilde{a}_3^{(1)} + \dfrac{p'(\overline{\rho})\overline{\rho}}{s}\widetilde{e}^{(1)}(0), \quad y_3 = 0, \\ \widetilde{w}_3^{(2)}|_{y_3=\pm 2d_0} = 0, \end{cases} \tag{10.53}$$

where $\mu_1^{\pm}(s) = \mu_1^{\pm} + p'(\overline{\rho})\overline{\rho}/s$. It is equivalent to

$$\begin{cases} R(s)\widetilde{w}_3^{(2)} - \lambda(s)\dfrac{d^2}{dy_3^2}\widetilde{w}_3^{(2)} = \widetilde{G}, \quad y_3 \in I^{\pm}, \\ [\widetilde{w}_3^{(2)}] = 0, \quad [\lambda(s)\dfrac{d\widetilde{w}_3^{(2)}}{dy_3}] = \widetilde{E}, \quad y_3 = 0, \\ \widetilde{w}_3^{(2)}|_{y_3=\pm 2d_0} = 0 \end{cases} \tag{10.54}$$

with

$$R(s) = \frac{\overline{\rho} s^2}{a_0 s + b_0}, \quad \lambda^{\pm}(s) = \frac{a^{\pm} s + b}{a_0 s + b_0}, \quad a^{\pm} = 2\mu^{\pm} + \mu_1^{\pm}, \quad b = p^{'}(\overline{\rho})\overline{\rho},$$
$$\widetilde{G} = \frac{s\widetilde{g}_3^{(1)}}{a_0 s + b_0} - \frac{p^{'}(\overline{\rho})\overline{\rho}}{a_0 s + b_0}\frac{d\widetilde{e}^{(1)}}{dy_3}, \quad \widetilde{E} = \frac{s}{a_0 s + b_0}\widetilde{a}_3^{(1)} + \frac{p^{'}(\widetilde{\rho})\overline{\rho}}{a_0 s + b_0}\widetilde{e}^{(1)}$$

and $a_0, b_0 = const > 0$ (we assume that $a^{\pm} b_0 > a_0 b$).

Multiplying (10.54) by $\overline{\widetilde{w}}_3^{(2)}$ and $\bar{s}\overline{\widetilde{w}}_3^{(2)}$, integrating and taking the real part we obtain

$$\sum_{\pm}(\overline{\rho}\frac{s_1 a_0 |s|^2 + b_0(s_1^2 - s_2^2)}{|a_0 s + b_0|^2}\|\widetilde{w}_3^{(2)}\|^2_{L_2(I^{\pm})} + \mathrm{Re}\lambda^{\pm}(s)\|\frac{d\widetilde{w}_3^{(2)}}{dy_3}\|^2_{L_2(I^{\pm})})$$
$$= \mathrm{Re}\int_{-2d_0}^{2d_0} \widetilde{G}\overline{\widetilde{w}}_3^{(2)} dy_3 + \mathrm{Re}\widetilde{E}\overline{\widetilde{w}}_3^{(2)}|_{y_3=0}, \tag{10.55}$$
$$\sum_{\pm}(\frac{\overline{\rho}(a_0|s|^2 + b_0 s_1)}{|a_0 s + b_0|^2}\|s\widetilde{w}_3^{(2)}\|^2_{L_2(I^{\pm})} + \mathrm{Re}(\bar{s}\lambda^{\pm}(s))\|\frac{d\widetilde{w}_3^{(2)}}{dy_3}\|^2_{L_2(I^{\pm})})$$
$$= \mathrm{Re}\int_{-2d_0}^{2d_0} \widetilde{G}\bar{s}\overline{\widetilde{w}}_3^{(2)} dy_3 + \mathrm{Re}(\widetilde{E}\bar{s}\overline{\widetilde{w}}_3^{(2)}),$$

where $\bar{s}$ and $\overline{\widetilde{w}}_3$ are complex conjugate of s and $\widetilde{w}_3$, respectively. Since

$$\frac{s_1 a_0 |s|^2 + b_0(s_1^2 - s_2^2)}{|a_0 s + b_0|^2} \geqslant -c_1, \quad \mathrm{Re}\lambda^{\pm}(s) \geqslant c_2 > 0$$

with constants c_i independent of s_2, the first of Eq. (10.55) implies

$$\sum_{\pm}(\|\widetilde{w}_3^{(2)}\|^2_{L_2(I^{\pm})} + \|\widetilde{w}_3^{(2)}\|^2_{W_2^1(I^{\pm})}) \leqslant c(\sum_{\pm}\|\widetilde{G}\|^2_{L_2(I^{\pm})} + |\widetilde{E}|^2), \tag{10.56}$$

if d_0 is sufficiently small. Moreover, the condition $a^{\pm} b_0 > a_0 b$ implies $\mathrm{Re}\bar{s}\lambda^{\pm}(s) \geqslant c_3$ and $|s| \geqslant c\sqrt{|s_1|}$ implies

$$\frac{\overline{\rho}(a_0|s|^2 + b_0 s_1)}{|a_0 s + b_0|^2} \geqslant c_4 > 0,$$

hence by the second equation in (10.55),

$$\sum_{\pm}(\|s\widetilde{w}_3^{(2)}\|^2_{L_2(I^{\pm})}$$
$$\leqslant c(\sum_{\pm}\|\widetilde{G}\|^2_{L_2(I^{\pm})} + |\widetilde{E}|^2 + |\widetilde{e}^{(1)}(s)||\widetilde{w}_3^{(2)}(s,0)| + |\widetilde{a}_3^{(1)}||s||\widetilde{w}_3^{(2)}(s,0)|)|_{y_3=0})$$

(if $|s| \leqslant c_5\sqrt{|s_1|}$, then this inequality is also true, in view of (10.56)). Finally, by (10.54),

$$\sum_{\pm}(\|s\widetilde{w}_3^{(2)}\|^2_{L_2(I^\pm)} + \|\widetilde{w}_3^{(2)}\|^2_{W_2^2(I^\pm)}) \leqslant c(\sum_{\pm}\|\widetilde{G}\|^2_{L_2(I^\pm)} + |\widetilde{E}|^2 + |\widetilde{e}^{(1)}(s)||\widetilde{w}_3^{(2)}(s,0)| + |\widetilde{a}_3^{(1)}||s||\widetilde{w}_3^{(2)}(s,0)|)|_{y_3=0}). \tag{10.57}$$

We multiply (10.57) by $|s|^l$ and integrate over the line $\mathrm{Re} s = s_1$. This leads to the inequality equivalent to

$$\begin{aligned}&\sum_{\pm}(\|e^{\beta t}w_3^{(2)}\|^2_{W_2^{0,1+l/2}(I^\pm\times(0,\infty))} + \|e^{\beta t}w_3^{(2)}\|^2_{W_2^2(I^\pm,W_2^{l/2}(0,\infty))})\\ &\leqslant c\Big(\sum_{\pm}\|e^{\beta t}G\|^2_{W_2^{0,l/2}(I^\pm\times(0,\infty))}\\ &+\|e^{\beta t}E\|^2_{W_2^{l/2}(0,\infty)} + \|e^{\beta t}a_3^{(1)}\|_{W_2^{1/4+l/2}(0,\infty)}\|e^{\beta t}w_3^{(2)}(0)\|_{W_2^{3/4+l/2}(0,\infty)}\\ &+\|e^{\beta t}e^{(1)}(0)\|_{L_2(0,\infty)}\|e^{\beta t}w_3^{(2)}(0)\|_{W_2^l(0,\infty)}\Big),\quad \beta=-s_1.\end{aligned} \tag{10.58}$$

Using the Cauchy inequality and imbedding theorem we obtain

$$\sum_{\pm}(\|e^{\beta t}w_3^{(2)}\|^2_{W_2^{0,1+l/2}(I^\pm\times(0,\infty))} + \|e^{\beta t}w_3^{(2)}\|^2_{W_2^2(I^\pm;W_2^{l/2}(0,\infty))}) \leqslant c\Big(\sum_{\pm}(\|e^{\beta t}G\|^2_{W_2^{0,l/2}(I^\pm\times(0,\infty))} + \|e^{\beta t}e^{(1)}\|^2_{L_2(I^\pm\times(0,\infty))}) + \|e^{\beta t}a_3^{(1)}\|^2_{W_2^{l/2+1/4}(0,\infty)}\Big). \tag{10.59}$$

Finally, we estimate the $W_2^{2+l}(I^\pm)$- norms of $\widetilde{w}_3^{(2)}$ making use of (10.54). This yields

$$\sum_{\pm}\|\mathscr{D}^2_{y_3}\widetilde{w}_3^{(2)}\|_{W_2^l(I^\pm)} \leqslant c\sum_{\pm}(|R(s)|\|\widetilde{w}_3^{(2)}\|_{W_2^l(I^\pm)} + \|\widetilde{G}\|_{W_2^l(I^\pm)}).$$

From this and the preceding inequality we deduce

$$\begin{aligned}&\sum_{\pm}\|e^{\beta t}w_3^{(2)}\|^2_{W_2^{2+l,1+l/2}(I^\pm\times(0,\infty))} \leqslant c\Big(\sum_{\pm}(\|e^{\beta t}G\|^2_{W_2^{l,l/2}(I^\pm\times(0,\infty))}\\ &+\|e^{\beta t}e^{(1)}\|^2_{W_2^{1+l,0}(I^\pm\times(0,\infty))}) + \|e^{\beta t}a_3^{(1)}\|^2_{W_2^{l/2+1/4}(0,\infty)}\Big)\end{aligned} \tag{10.60}$$

and, in view of (10.50), (10.51),

$$\sum_{\pm}(\|e^{\beta t}w_3\|^2_{L_2(\Omega',W_2^{2+l,1+l/2}(I^\pm\times(0,\infty)))}+\|e^{\beta t}\vartheta\|^2_{L_2(\Omega',W_2^{1+l,0}(I^\pm\times(0,\infty)))}$$
$$+\|e^{\beta t}\mathscr{D}_t\vartheta\|^2_{L_2(\Omega',W_2^{1+l,0}(I^\pm\times(0,\infty)))})\leqslant c\Big(\sum_{\pm}(\|e^{\beta t}g_3\|^2_{L_2(\Omega';W_2^{l,l/2}(I^\pm\times(0,\infty)))} \tag{10.61}$$
$$+\|e^{\beta t}e\|^2_{L_2(\Omega',W_2^{l+1,0}(I^\pm\times(0,\infty)))})+\|e^{\beta t}a_3\|^2_{L_2(\Omega',W_2^{1/4+l/2}(0,\infty))}$$
$$+\sum_{\pm}(\|w_{30}\|^2_{L_2(\Omega',W_2^{1+l}(I^\pm))}+\|\vartheta_0\|_{L_2(\Omega',W_2^{l+1}(I^\pm))})\Big).$$

The functions w_α satisfying (10.48) can be estimated by similar (in fact, more elementary) calculations, so that we have

$$\sum_{\pm}(\|e^{\beta t}w\|^2_{L_2(\Omega',W_2^{2+l,1+l/2}(I^\pm\times(0,\infty)))}+\|e^{\beta t}\vartheta\|^2_{L_2(\Omega',W_2^{1+l,0}(I^\pm\times(0,\infty)))}$$
$$+\|e^{\beta t}\mathscr{D}_t\vartheta\|^2_{L_2(\Omega',W_2^{1+l,0}(I^\pm\times(0,\infty)))})\leqslant c\Big(\sum_{\pm}(\|e^{\beta t}g\|^2_{L_2(\Omega',W_2^{l,l/2}(I^\pm\times(0,\infty)))} \tag{10.62}$$
$$+\|e^{\beta t}e\|^2_{L_2(\Omega',W_2^{l+1,0}(I^\pm\times(0,\infty)))}+\|w_0\|^2_{L_2(\Omega',W_2^{l+1}(I^\pm))}+\|\vartheta_0\|^2_{L_2(\Omega',W_2^{l+1}(I^\pm))})$$
$$+\|e^{\beta t}a\|^2_{W_2^{0,l/2+1/4}(\Omega'\times(0,\infty))}\Big).$$

As in the case of parabolic problems (see [12]), the interval $(0,\infty)$ in (10.58)–(10.62) can be substituted by $(0,T)$, hence along with (10.62) there holds

$$\sum_{\pm}(\|e^{\beta t}w_k\|^2_{W_2^{l+2,1+l/2}(\Omega_T^\pm)}+\|e^{\beta t}\vartheta_k\|^2_{W_2^{l+1,0}(\Omega_T^\pm)}+\|e^{\beta t}\mathscr{D}_t\vartheta_k\|^2_{W_2^{l+1,0}(\Omega_T^\pm)})$$
$$\leqslant c\Big(\sum_{\pm}(\|e^{\beta t}g_k\|^2_{W_2^{l,l/2}(\Omega_T^\pm)}+\|e^{\beta t}e_k\|^2_{W_2^{l+1,0}(\Omega_T^\pm)}+\|w_k(\cdot,0)\|^2_{W_2^{l+1}(\Omega^\pm)}$$
$$+\|\vartheta_k(\cdot,0)\|^2_{W_2^{l+1}(\Omega^\pm)})+\|e^{\beta t}a_k\|^2_{W_2^{l+1/2,l/2+1/4}(\Omega'\times(0,T))}\Big),\quad m^-<k<m^-+m^+, \tag{10.63}$$

where
$$g_k=f_k+\widetilde{\mathfrak{f}}_k+\mathfrak{f}'_k,\quad e_k=h_k+\widetilde{\mathfrak{h}}_k+\mathfrak{h}'_k,\quad a_k=b_k+\widetilde{\mathfrak{b}}_k+\mathfrak{b}'_k.$$

We recall that the expressions $\mathfrak{f}'_k$, $\mathfrak{h}'_k$ $\mathfrak{b}'_k$ are linear combinations of the derivatives of w_k and ϑ_k, except $\mathscr{D}_t w_k$, $\mathscr{D}_t\vartheta_k$ $\mathscr{D}^2_{y_3}w_k$, $\mathscr{D}_{y_3}\vartheta_k$. They can be estimated by interpolation inequalities as follows:

$$\sum_{\pm}(\|e^{\beta t}\mathfrak{f}'_k\|^2_{L_2(\Omega',W_2^{l,l/2}(I^\pm\times(0,T)))}+\|e^{\beta t}\mathfrak{h}'_k\|^2_{L_2(\Omega',W_2^{l+1}(I^\pm\times(0,T)))})$$
$$\leqslant\delta\sum_{\pm}(\|e^{\beta t}w_k\|^2_{W_2^{l+2,1+l/2}(\Omega_T^\pm)}+\|e^{\beta t}\vartheta_k\|^2_{W_2^{l+1,0}(\Omega_T^\pm)}+\|e^{\beta t}\mathscr{D}_t\vartheta_k\|^2_{W_2^{l+1,0}(\Omega_T^\pm)})$$
$$+c(\delta)\sum_{\pm}(\|e^{\beta t}w_k\|^2_{W_{2,tan}^{l+2,0}(\Omega_T^\pm)}+\|e^{\beta t}\vartheta_k\|^2_{W_{2,tan}^{l+1,0}(\Omega_T^\pm)}+\|e^{\beta t}\mathscr{D}_t\vartheta_k\|^2_{W_{2,tan}^{l+1,0}(\Omega_T^\pm)}). \tag{10.64}$$

Hence, for small δ we have, in view of (10.47)

$$\sum_{\pm}(\|e^{\beta t}w_k\|^2_{W_2^{l+2,1+l/2}(\Omega_T^{\pm})}+\|e^{\beta t}\vartheta_k\|^2_{W_2^{l+1,0}(\Omega_T^{\pm})}+\|e^{\beta t}\mathscr{D}_t\vartheta_k\|^2_{W_2^{l+1,0}(\Omega_T^{\pm})})$$
$$\leqslant c\Big(\sum_{\pm}(\|e^{\beta t}(f_k+\widetilde{\mathfrak{f}}_k)\|^2_{W_2^{l,l/2}(\Omega_T^{\pm})}+\|e^{\beta t}(h_k+\widetilde{\mathfrak{h}}_k)\|^2_{W_2^{l+1,0}(\Omega_T^{\pm})}$$
$$+\|e^{\beta t}(b_k+\mathfrak{b}_k)\|_{W_2^{l+1/2,l/2+1/4}(\Omega'\times(0,T))})+\|w_k(\cdot,0)\|^2_{W_2^{l+1}(\Omega^+)}+\|\vartheta_k(\cdot,0)\|^2_{W_2^{l+1}(\Omega^+)}\Big), \tag{10.65}$$

$m^- < k < m^- + m^+$. Similar inequality, but without the norms of $b+\mathfrak{b}$, holds for the functions w_k and ϑ_k, $k < m^-$.

Since $v = \sum_{k=1}^m v\varphi_k$, the norms $\sum_\pm \|e^{\beta t}v\|^2_{W_2^{2+l,1+l/2}(Q_T^{\pm})}$ and

$$\sum_k\sum_{\pm}\|e^{\beta t}v_k\|^2_{W_2^{2+l,1+l/2}(\omega_k^{\pm}\times(0,T))}, \quad \omega_k^{\pm}=\Omega^{\pm}\cap\omega_k$$

are equivalent. The differences $w_k - v_k$ contain no higher order terms and can be estimated by interpolation inequalities, like $\mathfrak{f}_k$, $\mathfrak{h}_k$, $\mathfrak{b}_k$. The same is true for $\theta_k - \vartheta_k$.

Hence

$$\sum_{\pm}\sum_{k}(\|e^{\beta t}v_k\|^2_{W_2^{2+l,1+l/2}(\omega_k\times(0,T))}+\|e^{\beta t}\theta_k\|^2_{W_2^{1+l,0}(\omega_k\times(0,T))}+\|e^{\beta t}\mathscr{D}_t\theta_k\|^2_{W_2^{1+l,0}(\omega_k\times(0,T))})$$
$$\leqslant c\Big(\sum_{\pm}(\|v_0\|^2_{W_2^{l+1}(\Omega_0^{\pm})}+\|\theta_0\|^2_{W_2^{l+1}(\Omega_0^{\pm})}+\|e^{\beta t}f\|^2_{W_2^{l,l/2}(Q_T^{\pm})}+\|e^{\beta t}h\|^2_{W_2^{l+1,0}(Q_T^{\pm})})$$
$$+\|b\|_{W_2^{1/2+l,1/4+l/2}(\Gamma_0)})+\delta\sum_{\pm}(\|e^{\beta t}v\|^2_{W_2^{2+l,1+l/2}(Q_T^{\pm})}+\|e^{\beta t}\mathscr{D}_t\theta\|^2_{W_2^{1+l,0}(Q_T^{\pm})}$$
$$+\|e^{\beta t}\theta\|^2_{W_2^{1+l,0}(Q_T^{\pm})})+c(\delta)(\|e^{\beta t}v\|^2_{L_2(Q_T^{\pm})}+\|e^{\beta t}\theta\|^2_{L_2(Q_T^{\pm})})\Big). \tag{10.66}$$

The weighted L_2-norms of v and θ are already estimated in (10.17), hence the inequalities (10.17) and (10.66) imply (10.15), if δ is chosen small. This completes the proof of (10.15).

10.3 Nonlinear Problem

In this section, we outline the proof of Theorem 1. It is based on Theorem 2 and on the following estimates of nonlinear terms (10.7).

Proposition 2 *Let* $p \in C^2(\overline{\rho}/2, 3\overline{\rho}/2)$, $U(\xi,t)=\int_0^t u(\xi,\tau)d\tau$. *If*

$$\sup_{t<T}\|U(\cdot,t)\|_{W_2^{2+l}(\Omega_0^{\pm})}\leqslant\delta_1, \quad \delta_1>0,$$
$$\sup_{Q_T}|\theta(x,t)|\leqslant\overline{\rho}/2, \tag{10.67}$$

then

$$Z(T) \leqslant c(\delta_1 Y(T) + Y^2(T)), \tag{10.68}$$

where $T > 1$ *and*

$$\begin{aligned} Z(T) &\equiv Z(u,q) = \sum_{\pm} (\|l_1(u,\theta)\|_{W_2^{l,l/2}(Q_T^\pm)} + \|l_2(u,\theta)\|_{W_2^{l+1,0}(Q_T^\pm)}) \\ &+ \|l_3\|_{W_2^{l+1/2,l/2+1/4}(G_T)} + \|l_4\|_{W_2^{l+1/2,l/2+1/4}(G_T)}, \\ Y(T) &\equiv Y(u,q) = \sum_{\pm} (\|u\|_{W_2^{2+l,1+l/2}(Q_T^\pm)} + \|\theta\|_{W_2^{l+1,0}(Q_T^\pm)} + \|\mathscr{D}_t\theta\|_{W_2^{l+1,0}(Q_T^\pm)}), \\ l &\in (1/2, 1). \end{aligned} \tag{10.69}$$

Proof We make use of auxiliary inequalities

$$\|fg\|_{W_2^l(\Omega_0^\pm)} \leqslant c\|f\|_{W_2^l(\Omega_0^\pm)}\|g\|_{W_2^{l+1}(\Omega_0^\pm)}, \tag{10.70}$$

$$\|fg\|_{L_2(\Omega_0^\pm)} \leqslant c\|f\|_{L_p(\Omega_0^\pm)}\|g\|_{L_q(\Omega_0^\pm)} \leqslant c\|f\|_{W_2^l(\Omega_0^\pm)}\|g\|_{W_2^1(\Omega_0^\pm)}, \tag{10.71}$$

where $1/p + 1/q = 1/2$, $l - 3/2 + 3/p = 0$. The condition $l > 1/2$ implies $1 - 3/2 + 3/q > 0$. Setting

$$\|u\|^2_{W_2^\lambda(0,T)} = \|u\|^2_{L_2(0,T)} + \int_0^1 \frac{dh}{h^{1+2\lambda}} \int_h^T |\Delta_t(-h)u(t)|^2 dt, \quad 0 < \lambda < 1,$$

where $\Delta_t(-h)u = u(t-h) - u(t)$, we obtain

$$\begin{aligned} &\|\theta\mathscr{D}_t u\|_{W_2^{l,0}(Q_T^\pm)} \leqslant c \sup_{t<T} \|\theta(\cdot,t)\|_{W_2^{l+1}(\Omega_0^\pm)}\|\mathscr{D}_t u\|_{W_2^{l,0}(Q_T^\pm)}, \\ &\|\Delta_t(-h)(\theta\mathscr{D}_t u)\|_{L_2(\Omega_0^\pm)} \leqslant c(\sup_{\Omega_0^\pm} |\theta(x,t)|\|\Delta_t(-h)\mathscr{D}_t u_t\|_{L_2(\Omega_0^\pm)} \\ &+ \|\mathscr{D}_t u\|_{W_2^l(\Omega_0^\pm)} \int_0^h \|\mathscr{D}_t\theta(\cdot,t-\tau)\|_{W_2^{l+1}(\Omega_0^\pm)} d\tau,) \end{aligned} \tag{10.72}$$

which implies

$$\|\theta\mathscr{D}_t u\|_{W_2^{l,l/2}(Q_T^\pm)} \leqslant c\|\mathscr{D}_t u\|_{W_2^{l,l/2}(Q_T^\pm)}(\sup_{t<T}\|\theta\|_{W_2^{l+1}(\Omega_0^\pm)} + \|\mathscr{D}_t\theta\|_{W_2^{l+1,0}(Q_T^\pm)}). \tag{10.73}$$

Next, we consider the term $\nabla_u \cdot T_u(u) - \nabla \cdot T(u) = (\nabla_u - \nabla) \cdot T(u) + \nabla_u \cdot (T_u(u) - T(u))$. Since $\nabla_u - \nabla = (J_u^{-1}A - I)\nabla$ and

$$\|J_u^{-1} - 1\|_{W_2^{l+1}(\Omega_0^\pm)} + \|A - I\|_{W_2^{l+1}(\Omega_0^\pm)} \leqslant c\delta_1,$$

we have

$$\|(\nabla_u - \nabla) \cdot T(u)\|_{W_2^{l,0}(Q_T^\pm)} \leqslant c\delta_1 \|\mathscr{D}_x^2 u\|_{W_2^{l,0}(Q_T^\pm)}. \tag{10.74}$$

Moreover, from

$$\|\Delta_t(-h)(J_u^{-1}A - I)\|_{W_2^1(\Omega_0^\pm)} \leqslant c \int_0^h \|u\|_{W_2^2(\Omega_0^\pm)} dt \leqslant c\sqrt{h}\|u\|_{W_2^{2,0}(Q_T^\pm)},$$

it follows that

$$\|(\nabla_u - \nabla) \cdot T(u)\|_{W_2^{0,l/2}(Q_T^\pm)} \leqslant c(\delta_1 + \|u\|_{W_2^{2,0}(Q_T^\pm)})\|\mathscr{D}_x^2 u\|_{W_2^{l,l/2}(Q_T^\pm)}. \tag{10.75}$$

The expression $\nabla_u \cdot (T_u(u) - T(u))$ is estimated in a similar way.

Let us estimate the last term $-P(\theta)\nabla\theta + p(\overline{\rho} + \theta)(\nabla - \nabla_u)\theta$ in l_1 (see (10.7)), where $P(\theta) = p'(\overline{\rho} + \theta) - p'(\overline{\rho}))$. We have

$$\|P(\theta)\nabla\theta\|_{L_2(Q_T^\pm)} \leqslant c \sup_{Q_T^\pm} |\theta(x, t)| \|\nabla\theta\|_{L_2(Q_T^\pm)}. \tag{10.76}$$

Assuming that $\theta^\pm$ is extended into $\mathbb{R}^3$ with preservation of class we estimate the difference

$$\begin{aligned} &P(\theta(x + z, t))\nabla\theta(x + z, t) - P(\theta(x, t))\nabla\theta(x, t) \\ &= (P(\theta(x + z)) - P(\theta(x)))\nabla\theta(x + z) + P(\theta(x, t))(\nabla\theta(x + z) - \nabla\theta(x)). \end{aligned} \tag{10.77}$$

Since

$$\begin{aligned} &|P(\theta(x + z, t)) - P(\theta(x, t))| \leqslant c|\theta(x + z, t) - \theta(x, t)|, \\ &|P(\theta(x, t))| \leqslant c|\theta(x, t)|, \end{aligned}$$

the L_2-norm of the difference (10.77) is controlled by

$$\begin{aligned} &c(\|\theta(x + z, t) - \theta(x, t)\|_{L_q(\Omega_0^\pm)} \|\nabla\theta\|_{L_p(\Omega_0^\pm)} \\ &+ \sup_{\Omega_0^\pm} |\theta(x, t)| \|\nabla\theta(x + z, t) - \nabla\theta(x, t)\|_{L_2(\Omega_0^\pm)}) \\ &\leqslant c\|\theta(x + z, t) - \theta(x, t)\|_{W_2^1(\Omega_0^\pm)} \|\theta(\cdot, t)\|_{W_2^{l+1}(\Omega_0^\pm)}. \end{aligned} \tag{10.78}$$

Multiplying (10.77), (10.78) by $|z|^{-3-2l}$ and integrating with respect to z, we obtain

$$\|P(\theta)\nabla\theta\|_{W_2^{l,0}(Q_T^\pm)} \leqslant c \sup_{Q_T^\pm} \|\theta\|_{W_2^{l+1}(\Omega_0^\pm)} \|\theta\|_{W_2^{l+1,0}(Q_T^\pm)}. \tag{10.79}$$

The finite difference of $P(\theta)\nabla\theta$ with respect to t is estimated in the same way, and we obtain

$$\begin{aligned}
&\|P(\theta)\nabla\theta\|_{W_2^{0,l/2}(Q_T^\pm)} \\
&\leqslant c(\sup_{Q_T^\pm}|\theta(x,t)|\|\nabla\theta\|_{W_2^{0,l/2}(Q_T^\pm)} + \|\theta\|_{W_2^{l/2}(0,T),W_2^1(\Omega_0^\pm))}\sup_{t<T}\|\theta\|_{W_2^{l+1}(\Omega_0^\pm)}) \\
&\leqslant c\sup_{t<T}\|\theta\|_{W_2^{l+1}(\Omega_0^\pm)}(\|\mathscr{D}_t\theta\|_{W_2^{l+1,0}(Q_T^\pm)} + \|\theta\|_{W_2^{l+1,0}(Q_T^\pm)}),
\end{aligned} \tag{10.80}$$

hence

$$\|P(\theta)\nabla\theta\|_{W_2^{l,l/2}(Q_T^\pm)} \leqslant c\sup_{t<T}\|\theta\|_{W_2^{l+1}(\Omega_0^\pm)}(\|\theta\|_{W_2^{l+1,0}(Q_T^\pm)} + \|\mathscr{D}_t\theta\|_{W_2^{l+1,0}(Q_T^\pm)}). \tag{10.81}$$

It can be shown by similar arguments that the expression $p(\overline{\rho}+\theta)(\nabla_u - \nabla)\theta$ satisfies the same inequality with additional term

$$c\delta_1(\|\theta\|_{W_2^{l+1,0}(Q_T^\pm)} + \|\mathscr{D}_t\theta\|_{W_2^{l+1,0}(Q_T^\pm)}) + c\sup_{t<T}\|\theta\|_{W_2^{l+1}(\Omega_0^\pm)}\|u\|_{W_2^{2,0}(Q_T^\pm)}$$

on the right hand side.

As for $l_2 = -\theta\nabla_u\cdot u - \overline{\rho}(\nabla_u - \nabla)u$, we have

$$\|l_2(u,\theta)\|_{W_2^{l+1,0}(Q_T^\pm)} \leqslant c(\delta_1 + \sup_{t<T}\|\theta\|_{W_2^{l+1}(\Omega_0^\pm)})\|u\|_{W_2^{l+2,0}(Q_T^\pm)}. \tag{10.82}$$

It remains to estimate $l_3(u)$ and $l_4(u,\vartheta)$. At first we consider $S(u)n_0 - S_u(u)n$. Let $n_0^* \in W_2^{1+l}(\Omega_0^+)$ be the extension of n_0 into Ω_0^+ and $n^* = An_0^*/|An_0^*|$. We have

$$\begin{aligned}
&\|S(u)n_0 - S_u(u)n\|_{W_2^{1/2+l,0}(G_T)} \leqslant c\|S(u)n_0^* - S_u(u)n^*\|_{W_2^{1+l,0}(Q_T^+)} \\
&\leqslant c(\|(S(u)-S_u(u))n^*\|_{W_2^{1+l,0}(Q_T^+)} + \|S(u)(n_0^*-n^*)\|_{W_2^{1+l,0}(Q_T^+)}) \leqslant c\delta_1\|u\|_{W_2^{2+l,0}(Q_T^+)}.
\end{aligned} \tag{10.83}$$

Moreover, since

$$\begin{aligned}
&\|\Delta_t(-h)(S(u)-S_u(u))\|_{L_2(\Gamma_0)} \\
&\leqslant c(\sup_{t<T}|I-A|\|\Delta_t(-h)\nabla u\|_{L_2(\Gamma_0)} + h\sup_{Q_{t-h,t}^+}|\nabla u(\xi,\tau)|\|\nabla u\|_{L_2(\Gamma_0)}), \\
&|\Delta_t(-h)n| \leqslant c\sup_{t<T}|\Delta_t(-h)A| \leqslant ch\sup_{Q_{t-h,t}^+}|\nabla u(\xi,\tau)|,
\end{aligned}$$

there holds

$$\begin{aligned}
&\|S(u)n_0 - S_u(u)n\|_{W_2^{0,1/4+l/2}(G_T)} \leqslant \|(S(u)-S_u(u))n_0\|_{W_2^{0,1/4+l/2}(G_T)} \\
&+ \|S_u(u)(n_0-n)\|_{W_2^{0,1/4+l/2}(G_T)} \\
&\leqslant c(\delta_1\|\nabla u\|_{W_2^{0,l/2+1/4}(G_T)} + \sup_{Q_T^+}|\nabla u|\|\nabla u\|_{L_2(G_T)}).
\end{aligned} \tag{10.84}$$

Finally,

$$\begin{aligned} &\|[p(\overline{\rho}+\theta)-p(\overline{\rho})-p'(\overline{\rho})\theta]\|_{W_2^{1/2+l,1/4+l/2}(G_T)} \\ &\leqslant c\sum_{\pm}\|p(\overline{\rho}+\theta)-p(\overline{\rho})-p'(\overline{\rho})\theta\|_{W_2^{1+l,1/2+l/2}(Q_T^{\pm})} \\ &\leqslant c\sup_{t<T}\|\theta\|_{W_2^{l+1}(\Omega_0^{\pm})}(\|\theta\|_{W_2^{l+1,0}(Q_T^{\pm})}+\|\mathscr{D}_t\theta\|_{W_2^{l+1,0}(Q_T^{\pm})}) \end{aligned} \tag{10.85}$$

(see the proof of (10.81)).

From (10.83), (10.85) is easy to deduce

$$\begin{aligned} &\|l_3(u)\|_{W_2^{1/2+l,1/4+l/2}(G_T)}+\|l_4(u)\|_{W_2^{1/2+l,1/4+l/2}(G_T)} \\ &\leqslant c(\delta_1+\|u\|_{W_2^{2+l,1+l/2}(Q_T^1)})\|u\|_{W_2^{1+l,1/2}(Q_T^1)} \\ &+\sup_{t<T}\|\theta\|_{W_2^{l+1}(\Omega_0^{\pm})}(\|\theta\|_{W_2^{l+1,0}(Q_T^{\pm})}+\|\mathscr{D}_t\theta\|_{W_2^{l+1,0}(Q_T^{\pm})}). \end{aligned} \tag{10.86}$$

The estimate (10.68) is a consequence of (10.72)–(10.86). The proposition is proved.

We pass to the construction of solution of a nonlinear problem (10.6) in the time interval $t>0$. We recall that the solution of this problem satisfies the condition

$$\int_{\Omega}\theta(x,t)J_u(x,t)dx=0 \tag{10.87}$$

(in view of the mass conservation), while in the case of a linear problem we have

$$\int_{\Omega}\theta(x,t)dx=0. \tag{10.88}$$

Therefore we are constrained to construct the solution step by step, from the time interval $[(k-1)T,kT]$ to $[kT,(k+1)T]$ with a certain large enough but finite T, eliminating at every step the discrepancy between the conditions (10.87) and (10.88). This procedure was proposed in [13].

We set $u=u'+u''$, $\theta=\theta'+\theta''$, represent (u_0,θ_0) in the form $u_0=u_0'+u_0''$, $\theta_0=\theta_0'+\theta_0''$, where

$$u'=u_0,\quad u_0''=0,\quad \theta_0'=\theta_0,\quad \theta_0''=0, \tag{10.89}$$

and define (u',θ') and (u'',θ'') as solutions to the problems

$$\begin{cases} \overline{\rho}\mathscr{D}_t u' - \nabla \cdot T(u') + p'(\overline{\rho})\nabla\theta' = 0, \\ \mathscr{D}_t\theta' + \overline{\rho}\nabla \cdot u' = 0, \quad \xi \in \Omega_0^{\pm}, \\ [u'] = 0, \quad [-p'(\overline{\rho})\theta' n_0 + T(u')n_0] = 0, \quad \xi \in \Gamma_0, \\ u'(\xi, t) = 0, \quad \xi \in S, \\ u'(\xi, 0) = u_0(\xi), \quad \theta'(\xi, 0) = \theta_0(\xi), \quad \xi \in \Omega_0^{\pm}, \end{cases} \tag{10.90}$$

$$\begin{cases} \overline{\rho}\mathscr{D}_t u'' - \nabla \cdot T(u'') + p'(\overline{\rho})\nabla\theta'' = l_1(u' + u'', \theta' + \theta''), \\ \mathscr{D}_t\theta'' + \overline{\rho}\nabla \cdot u'' = l_2(u' + u'', \theta' + \theta''), \quad \xi \in \Omega_0^{\pm}, \\ [u''] = 0, [\Pi_0 T(u'')n_0] = l_3(u' + u''), \\ [-p'(\overline{\rho})\vartheta'' + n_0 \cdot T(u'')n_0] = l_4(u' + u'', \theta' + \theta''), \quad \xi \in \Gamma_0, \\ u''(\xi, t) = 0, \quad \xi \in S, \\ u''(\xi, 0) = 0, \quad \theta''(\xi, 0) = 0. \quad \xi \in \Omega_0^{\pm}, \end{cases} \tag{10.91}$$

Since $\int_{\Omega} \theta_0(x)dx = 0$ (because $X(\xi, 0) = I$ and $J_u(\xi, 0) = 1$), the problem (10.90) has a global solution satisfying (10.15) (with $f = 0, h = 0, b = 0$) for arbitrary $T > 0$, hence

$$e^{\beta T} N(u'(\cdot, T), \theta'(\cdot, T)) \leqslant c_1 N(u_0, \theta_0), \quad Y(u', \theta') \leqslant c_2 N(u_0, \theta_0), \tag{10.92}$$

where $N(u_0, \theta_0) = \sum_{\pm}(\|u_0\|_{W_2^{l+1}(\Omega_0^{\pm})} + \|\theta_0\|_{W_2^{l+1}(\Omega_0^{\pm})})$. We fix T by the condition

$$c_1 e^{-\beta T} \leqslant \frac{1}{32}. \tag{10.93}$$

According to Proposition 2, nonlinear terms in (10.91) satisfy inequality (10.68) with $\delta_1 = \sqrt{T}(Y(u', \theta') + Y(u'', \theta''))$, which yields

$$Z(u' + u'', \theta' + \theta'') \leqslant c(Y(u', \theta') + Y(u'', \theta''))^2 \leqslant c(Y^2(u'', \theta'') + \varepsilon N(u_0, \theta_0)).$$

In the case of small ε, the solution of (10.91) can be constructed in the time interval $(0, T)$ by successive approximations, and it can be shown that δ_1 is of order ε and

$$Y(u'', \theta'') \leqslant c_2 \varepsilon N_0(u_0, \theta_0). \tag{10.94}$$

It follows that

$$N(u(\cdot, T), \theta(\cdot, T)) \leqslant (\frac{1}{32} + c_2 \varepsilon N(u_0, \theta_0)) \leqslant \frac{1}{8} N_0(u_0, \theta_0). \tag{10.95}$$

Assume that the solution of (10.6) is found for $t < kT$ and

$$N_k \leqslant \frac{1}{8} N_{k-1} \leqslant \ldots \leqslant \frac{1}{8^k} N_0, \tag{10.96}$$

where $N_k = N(u(\cdot, kT), \theta(\cdot, kT))$. The function $\theta(\xi, kT)$ satisfies

$$\int_\Omega \theta(\xi, kT) J_u(\xi, kT) d\xi = 0,$$

and we set $\theta(\xi, kT) = \theta_k'(\xi) + \theta_k''(\xi)$, where

$$\begin{aligned} \theta_k'(\xi) &= \theta(\xi, kT) - |\Omega|^{-1} \int_\Omega \theta(\xi, kT) d\xi, \\ \theta_k''(\xi) &= |\Omega|^{-1} \int_\Omega \theta(\xi, kT)(1 - J_u(\xi, kT)) d\xi. \end{aligned} \tag{10.97}$$

We consider problems (10.90), (10.91) in the interval $t \in (kT, (k+1)T)$, taking as initial data $(u_k' = u(\xi, kT),\ \ \theta_k'(\xi))$ and $(u_k'' = 0,\ \ \theta_k''(\xi))$, respectively and assuming that (10.67) is satisfied with a certain small δ_1 for $t \in (0, kT)$. It is clear that

$$\begin{aligned} &N(u'(\cdot, (k+1)T), \theta'(\cdot, (k+1)T)) \\ &\leqslant c_1 e^{-\beta T} N(u'(\cdot, kT), \theta'(\cdot, kT)) \leqslant 2c_1 e^{-\beta T} N(u(\cdot, kT), \theta(\cdot, kT)), \end{aligned} \tag{10.98}$$

moreover, the nonlinear terms in (10.91) can be estimated for $t \in (kT, (k+1)T)$ as follows:

$$\begin{aligned} &Z_k(u' + u'', \theta' + \theta'') \leqslant c(\delta_1 (Y_k(u' + u'', \theta' + \theta'') + Y_k^2(u' + u'', \theta' + \theta'')^2) \\ &\leqslant c(\delta_1 Y_k(u'') + Y_k^2(u'') + \delta_1 Y_k(u') + Y_k^2(u') \\ &\leqslant c(\delta_1 Y_k(u'') + Y_k^2(u'', \theta'')) + c\delta_1 N(u(\cdot, kT, \theta(\cdot, kT)), \end{aligned}$$

where Y_k and Z_k stand for the norms (10.69) in the time interval $(kT, (k+1)T)$. The solution of (10.91) can be constructed in this interval by successive approximations. Since

$$N(u''(\cdot, kT, \theta''(\cdot, kT)) \leqslant c\delta_1 N(u(\cdot, kT, \theta(\cdot, kT)),$$

it can be shown that

$$Y_k(u'', \theta'') \leqslant c\delta_1 N(u(\cdot, kT), \theta(\cdot, kT)), \tag{10.99}$$

if δ_1 is small. Hence

$$N_0(u(\cdot, (k+1)T), \theta(\cdot, (k+1)T)) \leqslant c(e^{-\beta T} + \delta_1) N(u(\cdot, kT, \theta(\cdot, kT)),$$

and we require that $c(e^{\beta T} + \delta_1) \leqslant 1/8$, which can be achieved by taking small δ_1. Thus, (10.96) holds also for N_{k+1}.

Inequality (10.99) is satisfied for all intervals $t \in (jT, (j+1)T), j \leqslant k$, and if a is so small that $e^{aT} \leqslant 2$, then

$$\sum_{j=0}^{k} Y_j^2 e^{2ajT} \leqslant c \sum_{j=0}^{k} \frac{e^{2ajT}}{8^j} N_0^2 \leqslant cN_0^2,$$

which is equivalent to (10.10).

Verification of the condition (10.67) for U in the interval $(0, kT)$ reduces to

$$\sup_{t<kT} \|U(\cdot, t)\|_{W_2^{l+2}(\Omega_0^{\pm})} \leqslant c\Big(\int_0^{kT} e^{2at} \|u\|^2_{W_2^{l+2}(\Omega_0^{\pm})} dt\Big)^{1/2} \leqslant cN_0 \leqslant c\varepsilon,$$

hence we can set $\delta_1 = c\varepsilon$. The inequality $\|\theta\|_{W_2^{l+1}(\Omega_0^{\pm})} \leqslant \overline{p}/2$ is verified in a similar way.

More detailed arguments can be found in [14]. By the same method the solution of the non-homogeneous Navier-Stokes equation with mass forces $f(x, t)$ can be constructed, provided f decays exponentially as $t \to \infty$ (see [14, 15]).

Acknowledgements The research leading to these results has received funding from the People Programme (Marie Curie Actions) of the European Union's Seventh Framework Programme FP7/2007-2013/ under REA grant agreement n° 319012 and from the Funds for International Cooperation under Polish Ministry of Science and Higher Education grant agreement n° 2853/7.PR/2013/2. This research was partially supported by the model unit "Multiscale Analysis, Modeling and Simulation", Top Global University Project of Waseda University.

References

1. Solonnikov, V.A.: On the solvability of initial-boundary value problems for a viscous compressible fluid in an infinite time interval. Algebra Analiz **217**(3), 238–271
2. Enomoto, Y.: Shibata, Y: On the $\mathscr{R}$-sectoriality and the initial-boundary value problem for the viscous compressible fluid flow. Funkcialaj Equac. **56**(3), 441–505 (2013)
3. Kubo, T., Shibata, Y., Soga, K.: On the $\mathscr{R}$-boundedness for the two phase problem: compressible-compressible model problem, Preprint
4. I.V.Denisova, I.V.: Problem of motion of two compressible fluids separated by a closed free interface, Zap. Nauchn. Semin. POMI **243**, 61–86 (1997)
5. Padula, M.: On the exponential stability of the rest state of a viscous compressible fluid. J. Math. Fluid Mech. **1**, 62–77 (1999)
6. Padula, M.: Asymptotic stability of steady compressible fluids. Lect. Notes Math. **2024**, 1–229 (2011)
7. Denisova, I.V.: On energy inequality for the problem of the evolution of two fluids of different types without surface tension. J. Math. Fluid. Mech. **17**(1), 183–198 (2005)
8. Ladyzhenskaya, O.A., Solonnikov, V.A.: On some problems of vector analysis and on generalized formulations of boundary value problems for the Navier—Stokes equations. Zapiski Nauchn. Semin. POMI **59**, 81–116 (1976)
9. Bogovskii, M.E.: Solution of some problems of vector analysis related to the operators div and grad. Proc. Semin. S.L. Sobolev **1**, 5–40 (1980)

10. Galdi, J.P.: An introduction to the mathematical theory of the Navier-Stokes equations, vol. 1. Springer, Linear steady problems (1994)
11. Golovkin, K.K.: On the equivalent normalization of fractional spaces. Trudy Steklov Math. Inst. **66**, 364–383 (1962)
12. Agranovich, M.S., Vishik, M.I.: Elliptic parameter dependent problems and parabolic problems of general form. Uspehi Mat. Nauk **19**(3), 53–161 (1964)
13. Solonnikov, V.A.: On the problem of stability of equilibrium figures of uniformly rotating viscous incompressible liquid. In: Instability in Models Connected with Fluid Flows. II, pp. 189–254, International Mathematical Series (N.Y.), vol. 7. Springer, New York (2008)
14. Denisova, I.V.: Global L_2-solvability of a problem governing two-phase fluid motion without surface tension. Portugal. Math. **71**(1), 1–24 (2014)
15. Frolova, E.V.: Free boundary problem of magnetohydrodynamics. Zapiski Nauchn. Semin. POMI **425**, 149–178 (2014)

Chapter 11
Classical Solvability of the Two-Phase Radial Viscous Fingering Problem in a Hele-Shaw Cell

Atusi Tani and Hisasi Tani

Abstract We discuss two-phase radial viscous fingering problem in a Hele-Shaw cell, which is a nonlinear problem with a free boundary for elliptic equations. Unlike the Stefan problem for heat equations Hele-Shaw problem is of hydrodynamic type. In this paper the classical solvability of two-phase Hele-Shaw problem with radial geometry is established by applying the same method as for the Stefan problem and justifying the vanishing the coefficients of the derivative with respect to time in parabolic equations.

Keywords Radial viscous fingering · Two-phase Hele-Shaw problem · Classical solution

Mathematical Subject Classification (2010): 35R35 · 35Q35 · 76D27 · 35K55

11.1 Introduction

Hele-Shaw cell is a device which consists of two closely spaced parallel plates containing a thin layer of viscous fluid [8]. When a fluid of low viscosity displaces a fluid of higher viscosity, the interface between them becomes unstable and starts to deform. First rigorous result on it was due to Saffman and Taylor [19] in 1958. Since then a large number of physical experiments have appeared (see [3, 11, 26] and the literatures therein). Both experiment and theory focus on two basic flow geometries: (i) rectangular [19] and (ii) radial [17]. For both geometries the initial developments of the interface instability were discussed through the linear stability

A. Tani (✉)
Department of Mathematics, Keio University, 3-14-1 Hiyoshi,
Yokohama 223-8522, Japan
e-mail: tani@math.keio.ac.jp

H. Tani
Department of Mathematics, Meiji University, 1-1-1 Higashi-Mita,
Kawasaki 214-8571, Japan
e-mail: hisasitani@gmail.com

Y. Shibata and Y. Suzuki (eds.), *Mathematical Fluid Dynamics, Present and Future*, Springer Proceedings in Mathematics & Statistics 183,
DOI 10.1007/978-4-431-56457-7_11

theory in [3]. However, the significant differences have been reported between the theoretical and the experimental results, which call attention to the validity of Hele-Shaw approximation and the new formulation with various physically reasonable effects. Among them one is the wetting layer effect [9, 14, 16, 25] and another is viscous normal stress effect [12, 24].

Linear stability for the radial fingering phenomena was discussed in [14] with the former effect and in [12] with the latter effect. Weakly nonlinear analysis was done first by Miranda and Widom [15] without any effect, and recently by Tani in [25] with the wetting effect and in [24] with viscous normal stress effect.

Most of mathematical results have been established by applying the complex function theory (see [7, 10] and the literatures therein). There are very few results from the exact mathematical analysis: [5, 6] in the little Hölder spaces, [23] in the standard Hölder spaces and [18] in the Sobolev spaces for one-phase problem.

In this paper we discuss two-phase radial viscous fingering problem in a Hele-Shaw cell in the classical framework, which is in more realistic physical situation.

This paper is organized as follows. In Sect. 11.2 we formulate the problem discussed in this paper and describe the main theorems. In Sect. 11.3 the problem in Sect. 11.2 is reformulated. In Sect. 11.4 the linear problem of the reformulated one is analyzed by following the arguments due to Bazaliĭ [1, 2] and Bizhanova and Solonnikov [4]. The most importance is to get a uniform estimate of the solution with respect to a parameter as in [2, 21]. For that it necessitates to modify the discussion used in the Stefan problem since the boundary conditions on the interface for the Hele-Shaw problem cause further difficulty. In Sects. 11.5 and 11.6 the original nonlinear problem and the passing to the limit of the parameter are discussed.

11.2 Formulation of the Problem

We consider a slow quasi-stationary displacement of a fluid by another fluid in the Hele-Shaw cell. Both fluids are assumed to be immiscible and incompressible. The motion is quasi-two-dimensional and all characteristics of the flow are averaged over the cell thickness. This approximation, so-called Hele-Shaw approximation, is traditional for problems of this type.

The motion of such fluids is described by

$$\nabla \cdot \mathbf{v}_i = 0, \quad \mathbf{v}_i = -M_i \nabla p_i \quad \text{in } \Omega_i(t),\ t > 0 \quad (i = 1, 2). \tag{11.1}$$

Here the second equation in (11.1) means Darcy law ($M_i = b^2/12\mu_i$, mobility; μ_i, the fluid viscosity; b, the width of two plates), $\mathbf{v}_i$ is the velocity vector field in the fluid and p_i is the pressure ($i = 1$ or 2 for the displacing or the displaced fluid, respectively). Moreover,

$$\Omega_1(t) = \left\{ x \in \mathbb{R}^2 \mid R_* < |x| < R(t) + \zeta\left(\frac{x}{|x|}, t\right) \right\} \text{ is the displacing region,}$$
$$\Omega_2(t) = \left\{ x \in \mathbb{R}^2 \mid R(t) + \zeta\left(\frac{x}{|x|}, t\right) < |x| < R^* \right\} \text{ is the displaced region,}$$

where R_* is the radius of the hole through which the displacing fluid is injected at a flow rate $Q(t)$, R^* is the radius of the Hele-Shaw cell occupied by the displaced fluid, $R(t)$ is the time-dependent unperturbed radius satisfying

$$\pi R(t)^2 = \pi R_0^2 + \int_0^t Q(\tau)\, \mathrm{d}\tau, \quad R_0 \equiv R(0) > R_*$$

and ζ is the perturbed radius.

In addition, the following boundary conditions are imposed:

$$\begin{cases} \mathbf{v}_1 \cdot \mathbf{n} = \dfrac{Q(t)}{2\pi R_*} & \text{on } \Gamma_* = \{x \in \mathbb{R}^2 \mid |x| = R_*\},\ t > 0, \\ p_2 = p_e \quad \text{on } \Gamma^* = \{x \in \mathbb{R}^2 \mid |x| = R^*\},\ t > 0, \\ \mathbf{v}_1 \cdot \mathbf{n} = \mathbf{v}_2 \cdot \mathbf{n} = V_n, \quad p_1 = p_2 \quad \text{on } \Gamma(t),\ t > 0, \end{cases} \tag{11.2}$$

where

$$\Gamma(t) = \left\{ x \in \mathbb{R}^2 \mid |x| = R(t) + \zeta\left(\frac{x}{|x|}, t\right) \right\},$$

V_n is the normal velocity of the interface $\Gamma(t)$ and $\mathbf{n}$ is the unit normal vector on Γ_* or $\Gamma(t)$, and the initial conditions are

$$\begin{cases} \mathbf{v}_i|_{t=0} = \mathbf{v}_i^0, \quad p_i = p_i^0 \quad \text{on } \Omega_i(0) \equiv \Omega_i \quad (i = 1, 2), \\ \zeta|_{t=0} = \zeta^0 \quad \text{on } \Gamma(0) \equiv \Gamma \qquad (\zeta^0 \in (R_* - R_0, R^* - R_0)). \end{cases} \tag{11.3}$$

Our problem is to find $(\mathbf{v}_i, p_i)$ $(i = 1, 2)$ and ζ satisfying (11.1)–(11.3), which is reduced to find (p_1, p_2) and ζ satisfying

$$\begin{cases} \Delta p_i = 0 \quad \text{in } \Omega_i(t),\ t > 0\ (i = 1, 2), \\ -M_1 \nabla p_1 \cdot \mathbf{n} = \dfrac{Q(t)}{2\pi R_*} \quad \text{on } \Gamma_*,\ t > 0, \\ p_2 = p_e \quad \text{on } \Gamma^*,\ t > 0, \\ -M_1 \nabla p_1 \cdot \mathbf{n} = -M_2 \nabla p_2 \cdot \mathbf{n} = V_n, \quad p_1 = p_2 \quad \text{on } \Gamma(t),\ t > 0, \\ p_i \,\big|_{t=0} = p_i^0 \quad \text{on } \Omega_i\ (i = 1, 2), \quad \zeta \,\big|_{t=0} = \zeta^0 \quad \text{on } \Gamma. \end{cases} \tag{11.4}$$

As the compatibility conditions we assume that p_1^0 and p_2^0 satisfy

$$\begin{cases} \Delta p_i^0 = 0 \quad \text{in } \Omega_i \quad (i = 1, 2), \\ -M_1 \nabla p_1^0 \cdot \mathbf{n} = \dfrac{Q(0)}{2\pi R_*} \quad \text{on } \Gamma_*, \\ p_2^0 = p_e^0 \equiv p_e|_{t=0} \quad \text{on } \Gamma^*, \\ p_1^0 = p_2^0 \quad \text{on } \Gamma. \end{cases} \tag{11.5}$$

It is more convenient to rewrite (11.4) in polar coordinates (r, θ):

$$\begin{cases} \dfrac{1}{r}\dfrac{\partial}{\partial r}\left(r\dfrac{\partial p_1}{\partial r}\right) + \dfrac{1}{r^2}\dfrac{\partial^2 p_1}{\partial \theta^2} = 0 \\ \qquad\qquad (R_* < r < R(t) + \zeta(\theta, t), \ 0 \leqslant \theta < 2\pi, \ t > 0), \\ \dfrac{1}{r}\dfrac{\partial}{\partial r}\left(r\dfrac{\partial p_2}{\partial r}\right) + \dfrac{1}{r^2}\dfrac{\partial^2 p_2}{\partial \theta^2} = 0 \\ \qquad\qquad \big(R(t) + \zeta(\theta, t) < r < R^*, \ 0 \leqslant \theta < 2\pi, \ t > 0\big), \\ M_1 \dfrac{\partial p_1}{\partial r} = -\dfrac{Q(t)}{2\pi R_*} \qquad (r = R_*, \ 0 \leqslant \theta < 2\pi, \ t > 0), \\ p_2 = p_e \qquad \big(r = R^*, \ 0 \leqslant \theta < 2\pi, \ t > 0\big), \\ M_1 \left(\dfrac{\partial p_1}{\partial r} - \dfrac{1}{r^2}\dfrac{\partial \zeta}{\partial \theta}\dfrac{\partial p_1}{\partial \theta}\right) = M_2 \left(\dfrac{\partial p_2}{\partial r} - \dfrac{1}{r^2}\dfrac{\partial \zeta}{\partial \theta}\dfrac{\partial p_2}{\partial \theta}\right) \\ \qquad = -\dfrac{\partial}{\partial t}(R(t) + \zeta), \\ p_1 = p_2 \quad (r = R(t) + \zeta(\theta, t), \ 0 \leqslant \theta < 2\pi, \ t > 0), \\ p_1 \big|_{t=0} = p_1^0 \qquad \big(R_* < r < R_0 + \zeta^0(\theta), \ 0 \leqslant \theta < 2\pi\big), \\ p_2 \big|_{t=0} = p_2^0 \qquad \big(R_0 + \zeta^0(\theta) < r < R^*, \ 0 \leqslant \theta < 2\pi\big), \\ \zeta \big|_{t=0} = \zeta^0 \qquad (0 \leqslant \theta < 2\pi). \end{cases} \tag{11.6}$$

Besides the problem (11.6) we consider the following problem:

$$
\begin{cases}
\dfrac{1}{r}\dfrac{\partial}{\partial r}\left(r\dfrac{\partial p_1}{\partial r}\right)+\dfrac{1}{r^2}\dfrac{\partial^2 p_1}{\partial \theta^2}=\varepsilon\dfrac{\partial p_1}{\partial t}+\varepsilon f_1 \\
\qquad\qquad (R_* < r < R(t)+\zeta(\theta,t),\ 0\leqslant\theta<2\pi,\ t>0), \\
\dfrac{1}{r}\dfrac{\partial}{\partial r}\left(r\dfrac{\partial p_2}{\partial r}\right)+\dfrac{1}{r^2}\dfrac{\partial^2 p_2}{\partial \theta^2}=\varepsilon\dfrac{\partial p_2}{\partial t}+\varepsilon f_2 \\
\qquad\qquad \big(R(t)+\zeta(\theta,t) < r < R^*,\ 0\leqslant\theta<2\pi,\ t>0\big), \\
M_1\dfrac{\partial p_1}{\partial r}=-\dfrac{Q(t)}{2\pi R_*} \quad (r=R_*,\ 0\leqslant\theta<2\pi,\ t>0), \\
p_2=p_e \quad \big(r=R^*,\ 0\leqslant\theta<2\pi,\ t>0\big), \\
M_1\left(\dfrac{\partial p_1}{\partial r}-\dfrac{1}{r^2}\dfrac{\partial\zeta}{\partial\theta}\dfrac{\partial p_1}{\partial\theta}\right)=M_2\left(\dfrac{\partial p_2}{\partial r}-\dfrac{1}{r^2}\dfrac{\partial\zeta}{\partial\theta}\dfrac{\partial p_2}{\partial\theta}\right) \\
\quad =-\dfrac{\partial}{\partial t}\left(R(t)+\zeta\right), \\
p_1=p_2 \quad (r=R(t)+\zeta(\theta,t),\ 0\leqslant\theta<2\pi,\ t>0), \\
p_1\big|_{t=0}=p_1^0 \quad \big(R_*<r<R_0+\zeta^0(\theta),\ 0\leqslant\theta<2\pi\big), \\
p_2\big|_{t=0}=p_2^0 \quad \big(R_0+\zeta^0(\theta)<r<R^*,\ 0\leqslant\theta<2\pi\big), \\
\zeta\big|_{t=0}=\zeta^0 \quad (0\leqslant\theta<2\pi)
\end{cases}
\tag{11.7}
$$

with $\varepsilon > 0$ and f_i $(i = 1, 2)$ being given later.

Now let us transform the free boundary problem (11.7) into the problem on fixed domains. Introduce the transformation from

$$\Omega_1(t) = \{R_* < r < R(t)+\zeta(\theta,t),\ 0\leqslant\theta<2\pi\}$$

onto

$$\Omega_1 = \left\{R_* < r' < R_0+\zeta^0(\theta'),\ 0\leqslant\theta'<2\pi\right\}$$

by the change of the variables

$$r' = \frac{R_0+\zeta^0-R_*}{R+\zeta-R_*}(r-R_*)+R_*,\quad \theta'=\theta,\quad t'=t,$$

the transformation from

$$\Omega_2(t) = \left\{R(t)+\zeta(\theta,t) < r < R^*,\ 0\leqslant\theta<2\pi\right\}$$

onto

$$\Omega_2 = \left\{R_0+\zeta^0(\theta') < r' < R^*,\ 0\leqslant\theta'<2\pi\right\}$$

by the change of the variables

$$r' = \frac{R_0 + \zeta^0 - R^*}{R + \zeta - R^*}(r - R^*) + R^*, \quad \theta' = \theta, \quad t' = t,$$

and

$$p_i(r, \theta, t) = p_i'(r', \theta', t') \ (i = 1, 2), \quad \zeta(\theta, t) = \zeta'(\theta', t').$$

By omitting the *primes* for simplicity, problem (11.7) takes the form

$$\begin{cases} \varepsilon \dfrac{\partial p_1}{\partial t} = \mathscr{L}_\zeta^1 p_1 - \varepsilon f_1 \quad \text{in } \Omega_1, \ t > 0, \\ \varepsilon \dfrac{\partial p_2}{\partial t} = \mathscr{L}_\zeta^2 p_2 - \varepsilon f_2 \quad \text{in } \Omega_2, \ t > 0, \\ \dfrac{\partial p_1}{\partial r} = -\dfrac{Q(t)}{2\pi R_* M_1} \dfrac{R + \zeta - R_*}{R_0 + \zeta^0 - R_*} \\ \qquad\qquad\qquad \text{on } \Gamma_* \equiv \{r = R_*, \theta \in [0, 2\pi]\}, \ t > 0, \\ p_2 = p_e \quad \text{on } \Gamma^* \equiv \{r = R^*, \theta \in [0, 2\pi]\}, \ t > 0, \\ \dfrac{\partial \zeta}{\partial t} - b_2^1(\zeta) \dfrac{\partial p_1}{\partial r} - b_1^1(\zeta) \dfrac{\partial p_1}{\partial \theta} - b_2^2(\zeta) \dfrac{\partial p_2}{\partial r} - b_1^2(\zeta) \dfrac{\partial p_2}{\partial \theta} \\ \qquad = -\dfrac{Q(t)}{2\pi R}, \\ b_2^1(\zeta) \dfrac{\partial p_1}{\partial r} + b_1^1(\zeta) \dfrac{\partial p_1}{\partial \theta} = b_2^2(\zeta) \dfrac{\partial p_2}{\partial r} + b_1^2(\zeta) \dfrac{\partial p_2}{\partial \theta}, \\ p_1 = p_2 \quad \text{on } \Gamma \equiv \{r = R_0 + \zeta^0(\theta), \theta \in [0, 2\pi]\}, \ t > 0, \\ p_1 \big|_{t=0} = p_1^0 \quad \text{on } \Omega_1, \quad p_2 \big|_{t=0} = p_2^0 \quad \text{on } \Omega_2, \\ \zeta \big|_{t=0} = \zeta^0 \quad \text{on } [0, 2\pi]. \end{cases} \tag{11.8}$$

Here

$$\mathscr{L}_\zeta^1 \equiv \mathscr{L}_\zeta^1 \left(r, \theta; \frac{\partial}{\partial r}, \frac{\partial}{\partial \theta} \right) = \frac{1}{\left(R_* + \frac{R+\zeta-R_*}{R_0+\zeta^0-R_*}(r - R_*) \right)^2} \Bigg[\frac{\partial^2}{\partial \theta^2}$$

$$+2 \left(\frac{1}{R_0 + \zeta^0 - R_*} \frac{d\zeta^0}{d\theta} - \frac{1}{R + \zeta - R_*} \frac{\partial \zeta}{\partial \theta} \right) (r - R_*) \frac{\partial^2}{\partial r \partial \theta}$$

$$+ \left(\left(R_* + \frac{R + \zeta - R_*}{R_0 + \zeta^0 - R_*}(r - R_*) \right)^2 \left(\frac{R_0 + \zeta^0 - R_*}{R + \zeta - R_*} \right)^2 \right.$$

$$\left. + \left(\frac{1}{R_0 + \zeta^0 - R_*} \frac{d\zeta^0}{d\theta} - \frac{1}{R + \zeta - R_*} \frac{\partial \zeta}{\partial \theta} \right)^2 (r - R_*)^2 \right) \frac{\partial^2}{\partial r^2} \Bigg]$$

$$+\left\{\varepsilon\frac{r-R_*}{R+\zeta-R_*}\frac{\partial}{\partial t}(R+\zeta)+\frac{1}{R_*+\frac{R+\zeta-R_*}{R_0+\zeta^0-R_*}(r-R_*)}\frac{R_0+\zeta^0-R_*}{R+\zeta-R_*}\right.$$

$$+\frac{r-R_*}{\left(R_*+\frac{R+\zeta-R_*}{R_0+\zeta^0-R_*}(r-R_*)\right)^2}\left[\frac{\partial}{\partial\theta}\left(\frac{1}{R_0+\zeta^0-R_*}\frac{\mathrm{d}\zeta^0}{\mathrm{d}\theta}-\frac{1}{R+\zeta-R_*}\frac{\partial\zeta}{\partial\theta}\right)\right.$$

$$\left.\left.+\left(\frac{1}{R_0+\zeta^0-R_*}\frac{\mathrm{d}\zeta^0}{\mathrm{d}\theta}-\frac{1}{R+\zeta-R_*}\frac{\partial\zeta}{\partial\theta}\right)^2\right]\right\}\frac{\partial}{\partial r},$$

$$\mathscr{L}_\zeta^2\equiv\mathscr{L}_\zeta^2\left(r,\theta;\frac{\partial}{\partial r},\frac{\partial}{\partial\theta}\right)=\mathscr{L}_\zeta^1 \text{ with } R_* \text{ replaced by } R^*;$$

$$b_2^1(\zeta)=\frac{M_1}{2}\left[\frac{R_0+\zeta^0-R_*}{R+\zeta-R_*}\left(1+\frac{1}{(R_0+\zeta^0)^2}\left(\frac{\partial\zeta}{\partial\theta}\right)^2\right)-\frac{1}{(R_0+\zeta^0)^2}\frac{\partial\zeta}{\partial\theta}\frac{\mathrm{d}\zeta^0}{\mathrm{d}\theta}\right],$$

$$b_1^1(\zeta)=-\frac{M_1}{2}\frac{1}{(R_0+\zeta^0)^2}\frac{\partial\zeta}{\partial\theta},$$

$$b_2^2(\zeta)=\frac{M_2}{2}\left[\frac{R_0+\zeta^0-R^*}{R+\zeta-R^*}\left(1+\frac{1}{(R_0+\zeta^0)^2}\left(\frac{\partial\zeta}{\partial\theta}\right)^2\right)-\frac{1}{(R_0+\zeta^0)^2}\frac{\partial\zeta}{\partial\theta}\frac{\mathrm{d}\zeta^0}{\mathrm{d}\theta}\right],$$

$$b_1^2(\zeta)=-\frac{M_2}{2}\frac{1}{(R_0+\zeta^0)^2}\frac{\partial\zeta}{\partial\theta}.$$

Now we choose

$$f_1=\frac{r-R_*}{R_0+\zeta^0-R_*}\frac{\partial p_1^0}{\partial r}\frac{\partial}{\partial t}(R+\zeta)\Big|_{t=0}\quad\text{on }\Omega_1,$$

$$f_2=\frac{r-R^*}{R_0+\zeta^0-R^*}\frac{\partial p_2^0}{\partial r}\frac{\partial}{\partial t}(R+\zeta)\Big|_{t=0}\quad\text{on }\Omega_2.$$

Before describing the main result, we introduce function spaces. For a domain Ω in $\mathbb{R}^n$ ($n\in\mathbb{N}$) and any $T>0$ let $C^l(\bar{\Omega})$ and $C^{l,l/2}(\bar{Q}_T)$ ($\bar{Q}_T\equiv\bar{\Omega}\times[0,T]$) with $l=k+\alpha$, $k\in\mathbb{Z}$, $k\geqslant 0$, $\alpha\in(0,1)$ be the standard Hölder spaces constructed with the use of the following semi-norms:

$$\langle u\rangle_x^{(\alpha)}\equiv\sup_{x,y\in\bar{\Omega},\,t\in[0,T]}\frac{|u(x,t)-u(y,t)|}{|x-y|^\alpha},\quad\langle u\rangle_t^{(\alpha)}\equiv\sup_{x\in\bar{\Omega},\,t,t'\in[0,T]}\frac{|u(x,t)-u(x,t')|}{|t-t'|^\alpha},$$

$$\langle u\rangle^{(\alpha)}=\langle u\rangle_x^{(\alpha)}+\langle u\rangle_t^{(\alpha/2)}$$

(see [13]). We also use the semi-norm

$$[u]^{(\alpha,\beta)}\equiv\sup_{x,y\in\bar{\Omega}\,t,t'\in[0,T]}\frac{|u(x,t)-u(y,t)-u(x,t')+u(y,t')|}{|x-y|^\alpha|t-t'|^\beta},\quad\alpha,\beta\in(0,1),$$

and define the Banach spaces $E^{k+\alpha}(\bar{Q}_T)$ that are obtained as the completion of infinitely differential functions in respective norms

$$\|u\|_\alpha = \|u\|_{E^\alpha} \equiv E^{\alpha,\alpha/2}[u] = \sup_{(x,t)\in\bar{Q}_T} |u(x,t)| + \langle u\rangle^{(\alpha)} + [u]^{(\alpha,\alpha/2)},$$

$$\|u\|_{1+\alpha} = \|u\|_{E^{1+\alpha}} = E^{\alpha,\alpha/2}[\mathrm{D}_x^1 u] + D^{\alpha,\alpha}[u],$$

$$D^{\alpha,\alpha}[u] = \sup_{(x,t)\in\bar{Q}_T} |u(x,t)| + \langle u\rangle_x^{(\alpha)} + \langle u\rangle_t^{(\alpha)} + [u]^{(\alpha,\alpha)},$$

$$\|u\|_{2+\alpha} = \|u\|_{E^{2+\alpha}} = E^{\alpha,\alpha/2}[\mathrm{D}_x^2 u] + D^{\alpha,\alpha}[\mathrm{D}_x^1 u] + D^{\alpha,\alpha}[u]$$

$$\Big(\mathrm{D}_x^k = \sum_{|j|=k} \mathrm{D}_x^j \quad (j \text{ is a multi-index}), \quad k = 1, 2\Big),$$

and the space $P_\varepsilon^{2+\alpha}(\bar{Q}_T)$ with the norm

$$\|u\|_{P_\varepsilon^{2+\alpha}} = \|\varepsilon \mathrm{D}_t u\|_\alpha + \|u\|_{2+\alpha} \quad (\mathrm{D}_t = \partial/\partial t)\,.$$

We also introduce the spaces

$$\hat{E}^{2+\alpha}(\bar{Q}_T) = \big\{u \;\big|\; u \in E^{2+\alpha}(\bar{Q}_T),\ \mathrm{D}_t u \in E^{1+\alpha}(\bar{Q}_T)\big\}\,,$$
$$\check{E}^{2+\alpha}(\bar{Q}_T) = \big\{u \;\big|\; u \in E^{2+\alpha}(\bar{Q}_T),\ \mathrm{D}_t u \in E^{\alpha}(\bar{Q}_T)\big\}\,,$$

$$\|u\|_{\hat{E}^{2+\alpha}} = \|u\|_{2+\alpha} + \|\mathrm{D}_t u\|_{1+\alpha}, \quad \|u\|_{\check{E}^{2+\alpha}} = \|u\|_{2+\alpha} + \|\mathrm{D}_t u\|_\alpha.$$

Denote by $\underset{0}{C}^{l,l/2}(\bar{Q}_T)$, $\underset{0}{E}^{k+\alpha}(\bar{Q}_T)$, $\underset{0}{P_\varepsilon}^{2+\alpha}(\bar{Q}_T)$, $\underset{0}{\hat{E}}^{2+\alpha}(\bar{Q}_T)$ and $\underset{0}{\check{E}}^{2+\alpha}(\bar{Q}_T)$ the spaces of the corresponding spaces whose elements are equal to zero at $t = 0$ together with their admissible derivatives with respect to t.

For a smooth manifold Γ in $\mathbb{R}^n$ $C^l(\Gamma)$, $C^{l,l/2}(\Gamma_T)$, *etc.*, are defined with the help of partition of unity and of local maps.

The followings are our main results.

Theorem 11.2.1 *Let $T > 0$ and $\alpha \in (0,1)$. Assume that $(p_1^0, p_2^0, \zeta^0) \in C^{3+\alpha}(\bar{\Omega}_1) \times C^{3+\alpha}(\bar{\Omega}_2) \times C^{4+\alpha}([0, 2\pi])$ satisfy the compatibility conditions, $\partial p_1^0/\partial r - \partial p_2^0/\partial r > 0$ on Γ, $Q \in C^\alpha([0,T])$ and $p_e \in C^{3+\alpha,(3+\alpha)/2}(\Gamma_T^*)$. Then there exists $T_0 > 0$ depending on the data of the problem such that problem (11.8) with $\varepsilon = 0$ has a solution $(p_1, p_2, \zeta) \in E^{2+\alpha}(\bar{Q}_{1,T_0}) \times E^{2+\alpha}(\bar{Q}_{2,T_0}) \times \hat{E}^{2+\alpha}(\Gamma_{T_0})$.*

Theorem 11.2.1 is proved in several steps. First we consider problem (11.8) with $\varepsilon > 0$ in the classes of smooth functions, and then show that this solution has a convergent subsequence as $\varepsilon \to 0$. For that it is necessary to use the function classes where the uniform estimate in ε holds.

Theorem 11.2.2 *Let $\varepsilon_0 > 0$ and the corresponding assumptions in Theorem 11.2.1 hold. Then there exists $T_0 > 0$ depending on the data of the problem and ε_0 such that for any fixed $\varepsilon \in (0, \varepsilon_0]$ problem (11.8) has a unique solution $(p_1, p_2, \zeta) \in P_\varepsilon^{2+\alpha}(\bar{Q}_{1,T_0}) \times P_\varepsilon^{2+\alpha}(\bar{Q}_{2,T_0}) \times \hat{E}^{2+\alpha}(\Gamma_{T_0})$.*

11.3 Reformulation of the Problem

Let $\bar{\zeta} \in C^{4+\alpha,(4+\alpha)/2}([0, 2\pi] \times [0, T])$ be an extension of ζ^0 such that

$$\left(\bar{\zeta}, \frac{\partial \bar{\zeta}}{\partial t}, \frac{\partial^2 \bar{\zeta}}{\partial t^2}\right)\bigg|_{t=0} = \left(\zeta^0, \frac{\partial \zeta}{\partial t}, \frac{\partial^2 \zeta}{\partial t^2}\right)\bigg|_{t=0},$$

where $(\partial\zeta/\partial t, \partial^2\zeta/\partial t^2)\big|_{t=0}$ are obtained from the fifth equation in (11.8) and its derivative in t at $t = 0$.

We seek a solution of problem (11.8) in the form

$$\begin{cases} p_1 = p_1^* + p_1^0 + \dfrac{r - R_*}{R + \bar{\zeta} - R_*} \dfrac{\partial p_1^0}{\partial r} \zeta^*, \\ p_2 = p_2^* + p_2^0 + \dfrac{r - R^*}{R + \bar{\zeta} - R^*} \dfrac{\partial p_2^0}{\partial r} \zeta^*, \\ \zeta = \zeta^* + \bar{\zeta}. \end{cases} \tag{11.9}$$

Then (11.8) becomes

$$\begin{cases} \varepsilon \dfrac{\partial p_1^*}{\partial t} = \mathscr{L}_*^1 p_1^* + \Phi_1 \quad \text{in } \Omega_1,\ t > 0, \\ \varepsilon \dfrac{\partial p_2^*}{\partial t} = \mathscr{L}_*^2 p_2^* + \Phi_2 \quad \text{in } \Omega_2,\ t > 0, \\ \dfrac{\partial p_1^*}{\partial r} = \Psi_* \quad \text{on } \Gamma_*,\ t > 0, \\ p_2^* = \Psi^* \quad \text{on } \Gamma^*,\ t > 0, \\ \dfrac{\partial \zeta^*}{\partial t} - b_2^1(\bar{\zeta}) \dfrac{\partial p_1^*}{\partial r} - b_1^1(\bar{\zeta}) \dfrac{\partial p_1^*}{\partial \theta} - b_2^2(\bar{\zeta}) \dfrac{\partial p_2^*}{\partial r} - b_1^2(\bar{\zeta}) \dfrac{\partial p_2^*}{\partial \theta} = \Psi_1 + \Psi_2, \\ b_2^1(\bar{\zeta}) \dfrac{\partial p_1^*}{\partial r} + b_1^1(\bar{\zeta}) \dfrac{\partial p_1^*}{\partial \theta} - b_2^2(\bar{\zeta}) \dfrac{\partial p_2^*}{\partial r} - b_1^2(\bar{\zeta}) \dfrac{\partial p_2^*}{\partial \theta} = -\Psi_1 + \Psi_2, \\ p_1^* - p_2^* + d(\bar{\zeta})\zeta^* = \Psi_3 \quad \text{on } \Gamma,\ t > 0, \\ p_1^* \big|_{t=0} = 0 \quad \text{on } \Omega_1, \qquad p_2^* \big|_{t=0} = 0 \quad \text{on } \Omega_2, \\ \zeta^* \big|_{t=0} = 0 \quad \text{on } [0, 2\pi). \end{cases} \tag{11.10}$$

Here

$$\mathscr{L}_*^1 \equiv \mathscr{L}_*^1\left(r, \theta; \frac{\partial}{\partial r}, \frac{\partial}{\partial \theta}\right) = \frac{1}{\left(R_* + \frac{R+\bar{\zeta}-R_*}{R_0+\zeta^0-R_*}(r-R_*)\right)^2}\left[\frac{\partial^2}{\partial \theta^2}\right.$$

$$+2\left(\frac{1}{R_0+\zeta^0-R_*}\frac{\mathrm{d}\zeta^0}{\mathrm{d}\theta} - \frac{1}{R+\bar{\zeta}-R_*}\frac{\partial \bar{\zeta}}{\partial \theta}\right)(r-R_*)\frac{\partial^2}{\partial r \partial \theta}$$

$$+\left(\left(R_* + \frac{R+\bar{\zeta}-R_*}{R_0+\zeta^0-R_*}(r-R_*)\right)^2\left(\frac{R_0+\zeta^0-R_*}{R+\bar{\zeta}-R_*}\right)^2\right.$$

$$\left.\left.+\left(\frac{1}{R_0+\zeta^0-R_*}\frac{\mathrm{d}\zeta^0}{\mathrm{d}\theta} - \frac{1}{R+\bar{\zeta}-R_*}\frac{\partial \bar{\zeta}}{\partial \theta}\right)^2 (r-R_*)^2\right)\frac{\partial^2}{\partial r^2}\right],$$

$$\mathscr{L}_*^2 \equiv \mathscr{L}_*^2\left(r, \theta; \frac{\partial}{\partial r}, \frac{\partial}{\partial \theta}\right) = \mathscr{L}_*^1 \text{ with } R_* \text{ replaced by } R^*;$$

$$\Phi_1 = \Phi_1(p_1^*, \zeta^*) = \mathscr{L}_\zeta^1 p_1 - \mathscr{L}_*^1 p_1^* - \varepsilon\,(r-R_*)\,\frac{\partial p_1^0}{\partial r}\,\frac{\partial}{\partial t}\left(\frac{\zeta^*}{R+\bar{\zeta}-R_*}\right) - \varepsilon f_1,$$

$$\Phi_2 - \Phi_2(p_2^*, \zeta^*) - \mathscr{L}_\zeta^2 p_2 - \mathscr{L}_*^2 p_2^* - \varepsilon\left(r-R^*\right)\frac{\partial p_2^0}{\partial r}\,\frac{\partial}{\partial t}\left(\frac{\zeta^*}{R+\zeta-R^*}\right) - \varepsilon f_2;$$

$$\Psi_* = \Psi_*(\zeta^*) = -\frac{\partial}{\partial r}\left(p_1^0 + \frac{r-R_*}{R+\bar{\zeta}-R_*}\frac{\partial p_1^0}{\partial r}\right)\zeta^* - \frac{R+\zeta-R_*}{R_0+\zeta^0-R_*}\frac{Q(t)}{2\pi R_* M_1},$$

$$\Psi^* = p_e - p_2^0,$$

$$\Psi_j = \Psi_j\left(p_1^*, p_2^*, \zeta^*\right) = b_2^j(\zeta)\frac{\partial p_j}{\partial r} + b_1^j(\zeta)\frac{\partial p_j}{\partial \theta} - b_2^j(\bar{\zeta})\frac{\partial p_j^*}{\partial r} - b_1^j(\bar{\zeta})\frac{\partial p_j^*}{\partial \theta}$$

$$-\frac{Q}{4\pi R} - \frac{1}{2}\frac{\partial \bar{\zeta}}{\partial t} \quad (j=1,2),$$

$$\Psi_3 = p_2^0 - p_1^0;$$

$$d(\bar{\zeta}) = \frac{R_0+\zeta^0-R_*}{R+\bar{\zeta}-R_*}\frac{\partial p_1^0}{\partial r} - \frac{R_0+\zeta^0-R^*}{R+\bar{\zeta}-R^*}\frac{\partial p_2^0}{\partial r},$$

with (11.9).

11.4 Linear Problem

In this section we consider the following linear problem:

$$\begin{cases} \varepsilon \dfrac{\partial u_1}{\partial t} - \mathscr{L}_*^1 u_1 = \phi_1 \quad \text{in } \Omega_1,\ t > 0, \\ \varepsilon \dfrac{\partial u_2}{\partial t} - \mathscr{L}_*^2 u_2 = \phi_2 \quad \text{in } \Omega_2,\ t > 0, \\ \dfrac{\partial u_1}{\partial r} = \psi_* \quad \text{on } \Gamma_*,\ t > 0, \quad u_2 = \psi^* \quad \text{on } \Gamma^*,\ t > 0, \\ \dfrac{\partial \rho}{\partial t} - b_2^1(\bar{\zeta}) \dfrac{\partial u_1}{\partial r} - b_1^1(\bar{\zeta}) \dfrac{\partial u_1}{\partial \theta} - b_2^2(\bar{\zeta}) \dfrac{\partial u_2}{\partial r} - b_1^2(\bar{\zeta}) \dfrac{\partial u_2}{\partial \theta} = \psi_1 + \psi_2, \\ b_2^1(\bar{\zeta}) \dfrac{\partial u_1}{\partial r} + b_1^1(\bar{\zeta}) \dfrac{\partial u_1}{\partial \theta} - b_2^2(\bar{\zeta}) \dfrac{\partial u_2}{\partial r} - b_1^2(\bar{\zeta}) \dfrac{\partial u_2}{\partial \theta} = -\psi_1 + \psi_2, \\ u_1 - u_2 + d(\bar{\zeta})\rho = \psi_3 \quad \text{on } \Gamma,\ t > 0, \\ (u_1, u_2, \rho) \big|_{t=0} = 0 \end{cases} \tag{11.11}$$

for given $\phi_1,\ \phi_2,\ \psi_*,\ \psi^*,\ \psi_1,\ \psi_2,\ \psi_3$ under the conditions $b_2^1 > 0,\ b_2^2 > 0,\ d > 0$.

First we study four model problems in the whole- and half-spaces:

$$\varepsilon \frac{\partial u}{\partial t} - \mathscr{L} u = f \ \left((x_1, x_2) \in \mathbb{R}^2,\ t > 0\right), \quad u \big|_{t=0} = 0; \tag{11.12}$$

$$\begin{cases} \varepsilon \dfrac{\partial u}{\partial t} - \mathscr{L} u = f \ \left((x_1, x_2) \in \mathbb{R}_+^2,\ t > 0\right), \\ u \big|_{x_2=0} = 0, \quad u \big|_{t=0} = 0; \end{cases} \tag{11.13}$$

$$\begin{cases} \varepsilon \dfrac{\partial u}{\partial t} - \mathscr{L} u = f \ \left((x_1, x_2) \in \mathbb{R}_+^2,\ t > 0\right), \\ \dfrac{\partial u}{\partial x_2} \Big|_{x_2=0} = 0, \quad u \big|_{t=0} = 0; \end{cases} \tag{11.14}$$

$$\begin{cases} \varepsilon \dfrac{\partial u^+}{\partial t} - \mathscr{L} u^+ = 0 \quad \left((x_1, x_2) \in \mathbb{R}_+^2,\ t > 0\right), \\ \varepsilon \dfrac{\partial u^-}{\partial t} - \mathscr{L} u^- = 0 \quad \left((x_1, x_2) \in \mathbb{R}_-^2,\ t > 0\right), \\ \dfrac{\partial \rho}{\partial t} - b^+ \dfrac{\partial u^+}{\partial x_2} - b^- \dfrac{\partial u^-}{\partial x_2} \Big|_{x_2=0} = g_1, \\ -b^+ \dfrac{\partial u^+}{\partial x_2} + b^- \dfrac{\partial u^-}{\partial x_2} \Big|_{x_2=0} = g_2, \quad -u^+ + u^- + d\rho \big|_{x_2=0} = g_3, \\ (u^+, u^-, \rho)\big|_{t=0} = 0 \end{cases} \tag{11.15}$$

$$\left(\mathbb{R}^2_\pm \equiv \left\{(x_1, x_2) \in \mathbb{R}^2 \mid \pm x_2 > 0\right\}\right).$$

In the above $\mathscr{L} = \sum_{j,k=1}^{2} a_{jk}\partial^2/\partial x_j \partial x_k$ is the second order partial differential operator with positive real coefficients a_{jk} which constitute positive definite symmetric matrix, and $b^\pm$ and d are positive constants.

For problems (11.12)–(11.15) we can assume without loss of generality $a_{jk} = \delta_{jk}$ by changing the independent variables (cf. [13]). Then the solutions to the problems (11.12)–(11.14) for $\mathscr{L} = \Delta$ are given by

$$u(x,t) = \int_0^t \mathrm{d}\tau \int_{\mathbb{R}^2} \Gamma_\varepsilon(x-y, t-\tau) f(y,\tau)\,\mathrm{d}y, \tag{11.16}$$

$$u(x,t) = \int_0^t \mathrm{d}\tau \int_{\mathbb{R}^2} G_\varepsilon(x-y, t-\tau) f(y,\tau)\,\mathrm{d}y \tag{11.17}$$

and

$$u(x,t) = \int_0^t \mathrm{d}\tau \int_{\mathbb{R}^2} N_\varepsilon(x-y, t-\tau) f(y,\tau)\,\mathrm{d}y, \tag{11.18}$$

respectively, where

$$\begin{aligned} \Gamma_\varepsilon(x,t) &= \varepsilon^{n/2-1}(4\pi t)^{-n/2} \exp\left[-\frac{\varepsilon|x|^2}{4t}\right], \\ G_\varepsilon(x,t) &= \Gamma_\varepsilon(x', x_n, t) - \Gamma_\varepsilon(x', -x_n, t), \quad x = (x', x_n), \\ N_\varepsilon(x,t) &= \Gamma_\varepsilon(x', x_n, t) + \Gamma_\varepsilon(x', -x_n, t), \quad n = 2. \end{aligned}$$

Applying the same arguments as in the estimation of the volume potential for the heat equation to the integrals in (11.16)–(11.18), we have a uniform estimate with respect to ε for the solutions of the problems (11.12)–(11.14) (see, [2, 13]):

$$\varepsilon \left\| \frac{\partial u}{\partial t} \right\|_\alpha + \|u\|_{2+\alpha} \leqslant C_1 \|f\|_\alpha. \tag{11.19}$$

For the problem (11.15) by making use of the Fourier-Laplace transformation

$$\mathscr{F}\mathscr{L}[u](\xi, x_2, s) \equiv \tilde{u}(\xi, x_2, s) = \int_0^\infty \mathrm{e}^{-st}\,\mathrm{d}t \int_{-\infty}^\infty \mathrm{e}^{-\mathrm{i}\xi x_1} u(x_1, x_2, t)\,\mathrm{d}x_1$$

as in [1, 2, 4], the parabolic equations are reduced to the ordinary differential equations in x_2, so that we get

$$\begin{cases} \tilde{u}^+(\xi, x_2, s) = \tilde{v}^+(\xi, s)\, \mathrm{e}^{-r_\varepsilon x_2} \quad (x_2 > 0), \\ \tilde{u}^-(\xi, x_2, s) = \tilde{v}^-(\xi, s)\, \mathrm{e}^{r_\varepsilon x_2} \quad (x_2 < 0), \\ r_\varepsilon = r_\varepsilon(s, \xi) = \sqrt{\varepsilon s + \xi^2}, \quad \operatorname{Re} r_\varepsilon > 0. \end{cases} \tag{11.20}$$

From the boundary conditions in (11.15) it follows that

$$\begin{cases} b^+ r_\varepsilon \tilde{v}^+ - b^- r_\varepsilon \tilde{v}^- + s\tilde{\rho} = \tilde{g}_1, \\ b^+ r_\varepsilon \tilde{v}^+ + b^- r_\varepsilon \tilde{v}^- = \tilde{g}_2, \\ \tilde{v}^+ - \tilde{v}^- - d\tilde{\rho} = -\tilde{g}_3. \end{cases}$$

Solving these, we obtain

$$\begin{cases} \tilde{v}^+ = \dfrac{1}{s + \Phi_\varepsilon}\left(\dfrac{db^-}{b^+ + b^-}\tilde{g}_1 - \dfrac{b - db^-}{b^+ + b^-}\tilde{g}_2 + \dfrac{bb^-}{b^+ + b^-} r_\varepsilon \tilde{g}_3\right) \\ \qquad + \dfrac{1}{(b^+ + b^-)\, r_\varepsilon}\tilde{g}_2 - \dfrac{b^-}{b^+ + b^-}\tilde{g}_3, \\ \tilde{v}^- = \dfrac{1}{s + \Phi_\varepsilon}\left(-\dfrac{db^+}{b^+ + b^-}\tilde{g}_1 - \dfrac{b - db^+}{b^+ + b^-}\tilde{g}_2 - \dfrac{bb^+}{b^+ + b^-} r_\varepsilon \tilde{g}_3\right) \\ \qquad + \dfrac{1}{(b^+ + b^-)\, r_\varepsilon}\tilde{g}_2 + \dfrac{b^+}{b^+ + b^-}\tilde{g}_3, \\ \tilde{\rho} = \dfrac{1}{s + \Phi_\varepsilon}\left(\tilde{g}_1 - \dfrac{b^+ - b^-}{b^+ + b^-}\tilde{g}_2 + \dfrac{2b^+ b^-}{b^+ + b^-} r_\varepsilon \tilde{g}_3\right), \\ \Phi_\varepsilon = \dfrac{2db^+ b^-}{b^+ + b^-} r_\varepsilon \equiv b\, r_\varepsilon. \end{cases} \tag{11.21}$$

The solution of problem (11.15) is given through the inverse Fourier-Laplace transformation

$$v(x_1, t) \equiv (\mathscr{F}\mathscr{L})^{-1}[\tilde{v}] = \frac{1}{2\pi \mathrm{i}} \int_{\mathbb{R}} \mathrm{e}^{\mathrm{i} x_1 \xi}\, \mathrm{d}\xi \int_{\operatorname{Re} s = a > 0} \mathrm{e}^{st}\, \tilde{v}(\xi, s)\, \mathrm{d}s$$

as follows.

$$\begin{cases} v^+(x_1,t) = (\mathscr{F}\mathscr{L})^{-1}\left[\dfrac{1}{s+\Phi_\varepsilon}\right] * \left(\dfrac{db^-}{b^++b^-}\,g_1 - \dfrac{b-db^-}{b^++b^-}\,g_2 \right. \\ \qquad\qquad \left. + \dfrac{bb^-}{b^++b^-}\,(\mathscr{F}\mathscr{L})^{-1}\left[r_\varepsilon\tilde{g}_3\right]\right) \\ \qquad\qquad + \dfrac{1}{b^++b^-}\,(\mathscr{F}\mathscr{L})^{-1}\left[\dfrac{1}{r_\varepsilon}\right] * g_2 - \dfrac{b^-}{b^++b^-}\,g_3, \\ v^-(x_1,t) = (\mathscr{F}\mathscr{L})^{-1}\left[\dfrac{1}{s+\Phi_\varepsilon}\right] * \left(-\dfrac{db^+}{b^++b^-}\,g_1 - \dfrac{b-db^+}{b^++b^-}\,g_2 \right. \\ \qquad\qquad \left. - \dfrac{bb^-}{b^++b^-}\,(\mathscr{F}\mathscr{L})^{-1}\left[r_\varepsilon\tilde{g}_3\right]\right) \\ \qquad\qquad + \dfrac{1}{b^++b^-}\,(\mathscr{F}\mathscr{L})^{-1}\left[\dfrac{1}{r_\varepsilon}\right] * g_2 + \dfrac{b^+}{b^++b^-}\,g_3, \\ \rho(x_1,t) = (\mathscr{F}\mathscr{L})^{-1}\left[\dfrac{1}{s+\Phi_\varepsilon}\right] * \left(g_1 - \dfrac{b^+-b^-}{b^++b^-}\,g_2 \right. \\ \qquad\qquad \left. + \dfrac{2b^+b^-}{b^++b^-}\,(\mathscr{F}\mathscr{L})^{-1}\left[r_\varepsilon\tilde{g}_3\right]\right), \end{cases} \tag{11.22}$$

where $*$ means a convolution with respect to x_1 and t.

By following the arguments in [1] let us derive the explicit representation of

$$Z_\varepsilon(x_1,t) \equiv (\mathscr{F}\mathscr{L})^{-1}\left[\frac{1}{s+\Phi_\varepsilon(s,\xi)}\right], \quad t>0. \tag{11.23}$$

For that it is sufficient to consider the case $b=1$. Since

$$\left(s+\sqrt{\varepsilon s+\xi^2}\right)^{-1} = \int_0^\infty \exp\left[-\tau\left(s+\sqrt{\varepsilon s+\xi^2}\right)\right]\mathrm{d}\tau,$$

$$\begin{aligned} &\frac{1}{2\pi \mathrm{i}}\int_{a-\mathrm{i}\infty}^{a+\mathrm{i}\infty} \exp\left[-\tau\left(s+\sqrt{\varepsilon s+\xi^2}\right)+st\right]\mathrm{d}s \\ &= \frac{1}{2\pi \mathrm{i}}\int_{a-\mathrm{i}\infty}^{a+\mathrm{i}\infty} \exp[-\tau s+st]\,\mathrm{d}s \underset{t}{*} \frac{1}{2\pi \mathrm{i}}\int_{a-\mathrm{i}\infty}^{a+\mathrm{i}\infty} \exp\left[-\tau\sqrt{\varepsilon s+\xi^2}+st\right]\mathrm{d}s \\ &= \delta(t-\tau) \underset{t}{*} \frac{\sqrt{\varepsilon}}{2\sqrt{\pi}}\,\tau t^{-3/2}\exp\left[-\frac{\xi^2 t}{\varepsilon}-\frac{\varepsilon\tau^2}{4t}\right] \quad (a>0), \end{aligned}$$

where $\underset{t}{*}$ means a convolution with respect to t, so that

$$\begin{aligned}
\tilde{Z}_\varepsilon(\xi,t) &= \int_0^\infty \delta(t-\tau) \underset{t}{*} \frac{\sqrt{\varepsilon}}{2\sqrt{\pi}}\, \tau t^{-3/2} \exp\left[-\frac{\xi^2 t}{\varepsilon} - \frac{\varepsilon\tau^2}{4t}\right] \mathrm{d}\tau \\
&= \frac{\sqrt{\varepsilon}}{2\sqrt{\pi}} \int_0^t \tau (t-\tau)^{-3/2} \exp\left[-\frac{\xi^2}{\varepsilon}(t-\tau) - \frac{\varepsilon\tau^2}{4(t-\tau)}\right] \mathrm{d}\tau \\
&= \frac{1}{2\sqrt{\pi}} \int_0^{t/\varepsilon} \frac{t-\varepsilon z}{z^{3/2}} \exp\left[-\xi^2 z - \frac{(t-\varepsilon z)^2}{4z}\right] \mathrm{d}z.
\end{aligned}$$

Therefore, we have

$$\begin{aligned}
Z_\varepsilon(x_1,t) &= \frac{1}{2\pi} \int_0^\infty \tilde{Z}_\varepsilon(\xi,t)\cos(\xi x_1)\,\mathrm{d}\xi \\
&= \frac{1}{4\pi} \int_0^{t/\varepsilon} \frac{t-\varepsilon z}{z^2} \exp\left[-\frac{x_1^2}{4z} - \frac{(t-\varepsilon z)^2}{4z}\right] \mathrm{d}z.
\end{aligned} \tag{11.24}$$

From (11.24) it easily follows

Lemma 11.4.1 *Following inequalities hold with a constant C_2 independent of ε:*

$$\begin{cases}
|Z_\varepsilon(x_1,t)| \leqslant C_2\, t\, \dfrac{1+\sqrt{\varepsilon t}}{x_1^2+t^2}, \\[2ex]
\left|\dfrac{\partial}{\partial t} Z_\varepsilon(x_1,t)\right| + \left|\dfrac{\partial}{\partial x_1} Z_\varepsilon(x_1,t)\right| \leqslant C_2\, \dfrac{1+\sqrt{\varepsilon t}}{x_1^2+t^2}, \\[2ex]
\left|\dfrac{\partial^2}{\partial t\,\partial x_1} Z_\varepsilon(x_1,t)\right| + \left|\dfrac{\partial^2}{\partial x_1^2} Z_\varepsilon(x_1,t)\right| \leqslant C_2\, \dfrac{1+\varepsilon t}{(x_1^2+t^2)^{3/2}}.
\end{cases} \tag{11.25}$$

Lemma 11.4.1 implies that for any bounded continuous function $f(x_1)$

$$\lim_{t\to 0} \int_{-\infty}^{\infty} Z_\varepsilon(x_1-\xi,t) f(\xi)\,\mathrm{d}\xi = f(x_1). \tag{11.26}$$

Introduce the notation

$$\begin{aligned}
w(x_1,t) &= (Z_\varepsilon * g)(x_1,t) = \int_0^t \mathrm{d}\tau \int_{-\infty}^{\infty} Z_\varepsilon(x_1-y,t-\tau)\, g(y,\tau)\,\mathrm{d}y, \\
w_h(x_1,t) &= \int_0^{t-h} \mathrm{d}\tau \int_{-\infty}^{\infty} Z_\varepsilon(x_1-y,t-\tau)\, g(y,\tau)\,\mathrm{d}y \quad (h>0).
\end{aligned}$$

For w_h it is clear to hold

$$\frac{\partial}{\partial t} w_h(x_1,t) = \int_0^{t-h} \mathrm{d}\tau \int_{-\infty}^{\infty} \frac{\partial}{\partial t} Z_\varepsilon(x_1-y,t-\tau)\,(g(y,\tau)-g(x_1,\tau))\,\mathrm{d}y$$
$$+\int_0^{t-h} g(x_1,\tau)\,\mathrm{d}\tau \int_{-\infty}^{\infty} \frac{\partial}{\partial t} Z_\varepsilon(x_1-y,t-\tau)\,\mathrm{d}y + \int_{-\infty}^{\infty} Z_\varepsilon(x_1-y,h)\,g(y,t-h)\,\mathrm{d}y.$$

Making use of (11.26) and the formula

$$\frac{\partial}{\partial t}\tilde{Z}_\varepsilon(\xi,t) = -\frac{1}{\sqrt{\pi}}\int_0^{t/\varepsilon} \frac{\varepsilon(t-\varepsilon z)}{z^{1/2}(t+\varepsilon z)^2} \exp\left[-\xi^2 z - \frac{(t-\varepsilon z)^2}{4z}\right] \mathrm{d}z$$
$$-\frac{1}{\sqrt{\pi}}\int_0^{t/\varepsilon} \frac{\xi^2(t-\varepsilon z)}{z^{1/2}(t+\varepsilon z)} \exp\left[-\xi^2 z - \frac{(t-\varepsilon z)^2}{4z}\right] \mathrm{d}z \qquad (11.27)$$

together with the estimates in Lemma 11.4.1, we have after passing to the limit $h \to 0$

$$\frac{\partial}{\partial t} w(x_1,t) = \int_0^{t} \mathrm{d}\tau \int_{-\infty}^{\infty} \frac{\partial}{\partial t} Z_\varepsilon(x_1-y,t-\tau)\,(g(y,\tau)-g(x_1,\tau))\,\mathrm{d}y$$
$$+\int_0^{t} g(x_1,\tau)\,\mathrm{d}\tau \int_{-\infty}^{\infty} \frac{\partial}{\partial t} Z_\varepsilon(x_1-y,t-\tau)\,\mathrm{d}y + g(x_1,t). \qquad (11.28)$$

Analogously we have

$$\frac{\partial^2}{\partial x_1^2} w(x_1,t)$$
$$= \int_0^{t} \mathrm{d}\tau \int_{-\infty}^{\infty} \frac{\partial}{\partial x_1} Z_\varepsilon(x_1-y,t-\tau)\left(\frac{\partial}{\partial y} g(y,\tau) - \frac{\partial}{\partial x_1} g(x_1,\tau)\right)\mathrm{d}y. \qquad (11.29)$$

We begin by estimating the first term denoted by $w^\dagger$ in the right hand side of (11.28). Following the arguments in [13], we get

$$w^\dagger(x_1,t) - w^\dagger(x_1',t)$$
$$= \int_0^{t} \mathrm{d}\tau \int_{|x_1-y|\leqslant 2|x_1-x_1'|} \frac{\partial}{\partial t} Z_\varepsilon(x_1-y,t-\tau)\,(g(y,\tau)-g(x_1,\tau))\,\mathrm{d}y$$
$$-\int_0^{t} \mathrm{d}\tau \int_{|x_1-y|\leqslant 2|x_1-x_1'|} \frac{\partial}{\partial t} Z_\varepsilon(x_1'-y,t-\tau)\,(g(y,\tau)-g(x_1',\tau))\,\mathrm{d}y$$
$$+\int_0^{t} \mathrm{d}\tau \int_{|x_1-y|\geqslant 2|x_1-x_1'|} \left(\frac{\partial}{\partial t} Z_\varepsilon(x_1-y,t-\tau) - \frac{\partial}{\partial t} Z_\varepsilon(x_1'-y,t-\tau)\right)$$
$$\times (g(y,\tau)-g(x_1,\tau))\,\mathrm{d}y$$

$$
\begin{aligned}
&+\int_0^t \big(g(x_1', \tau) - g(x_1, \tau)\big)\,\mathrm{d}\tau \int_{|x_1-y|\geqslant 2|x_1-x_1'|} \frac{\partial}{\partial t} Z_\varepsilon(x_1' - y, t-\tau)\,\mathrm{d}y \\
&\equiv \sum_{j=1}^{4} I_j. \qquad (11.30)
\end{aligned}
$$

Applying Lemma 11.4.1, we have

$$
\begin{aligned}
|I_1| &\leqslant C_3''\,\langle g\rangle_x^{(\alpha)} \int_{|x_1-y|\leqslant 2|x_1-x_1'|} |x_1-y|^\alpha\,\mathrm{d}y \int_0^t \frac{1+\sqrt{\varepsilon(t-\tau)}}{|x_1-y|^2+(t-\tau)^2}\,\mathrm{d}\tau \\
&\leqslant C_3'\,(1+\sqrt{\varepsilon t})\,\langle g\rangle_x^{(\alpha)} \int_{|x_1-y|\leqslant 2|x_1-x_1'|} |x_1-y|^{\alpha-1}\,\mathrm{d}\tau \\
&\leqslant C_3(\varepsilon_0, T)\,\langle g\rangle_x^{(\alpha)}\,|x_1-x_1'|^\alpha.
\end{aligned}
$$

I_2 is estimated by just the same way as I_1. For I_3 we do in the same way with the help of the mean value theorem. Finally for I_4 we get

$$
\begin{aligned}
|I_4| &\leqslant C_4'\,\langle g\rangle_x^{(\alpha)}\,|x_1-x_1'|^\alpha \int_0^t \mathrm{d}\tau \left| \int_{|x_1-y|\geqslant 2|x_1-x_1'|} \frac{\partial}{\partial t} Z_\varepsilon(x_1'-y, \tau)\,\mathrm{d}y \right| \\
&\leqslant C_4(\varepsilon_0, T)\,\langle g\rangle_x^{(\alpha)}\,|x_1-x_1'|^\alpha
\end{aligned}
$$

due to (11.27) and Lemma 11.4.1. Second term in the right hand side of (11.28) is estimated similarly. Hence we obtain

$$
\left\langle \frac{\partial w}{\partial t} \right\rangle_x^{(\alpha)} \leqslant C_5(\varepsilon_0, T)\,\langle g\rangle_x^{(\alpha)}. \qquad (11.31)
$$

Next the difference with respect to t of $w^\dagger$ is expressed for $t' \leqslant t$:

$$
\begin{aligned}
&w^\dagger(x_1, t) - w^\dagger(x_1, t') \\
&= \int_{2t'-t}^{t} \mathrm{d}\tau \int_{-\infty}^{\infty} \frac{\partial}{\partial t} Z_\varepsilon(x_1-y, t-\tau)\,(g(y,\tau)-g(x_1,\tau))\,\mathrm{d}y \\
&\quad - \int_{2t'-t}^{t'} \mathrm{d}\tau \int_{-\infty}^{\infty} \frac{\partial}{\partial t'} Z_\varepsilon(x_1-y, t'-\tau)\,(g(y,\tau)-g(x_1,\tau))\,\mathrm{d}y \\
&\quad + \int_{-\infty}^{2t'-t} \mathrm{d}\tau \int_{-\infty}^{\infty} \left(\frac{\partial}{\partial t} Z_\varepsilon(x_1-y, t-\tau) - \frac{\partial}{\partial t'} Z_\varepsilon(x_1-y, t'-\tau) \right) \\
&\qquad\qquad \times (g(y,\tau)-g(x_1,\tau))\,\mathrm{d}y \\
&\equiv \sum_{j=1}^{3} I_j'.
\end{aligned}
$$

Lemma 11.4.1 yields

$$
\begin{aligned}
|I_1'| &\leqslant C_6'' \langle g\rangle_x^{(\alpha)} \int_{2t'-t}^{t} \mathrm{d}\tau \int_{-\infty}^{\infty} \frac{|x_1-y|^{\alpha}(1+\sqrt{\varepsilon(t-\tau)})}{(x_1-y)^2+(t-\tau)^2}\,\mathrm{d}y \\
&\leqslant C_6' \langle g\rangle_x^{(\alpha)} \int_{2t'-t}^{t} \frac{1+\sqrt{\varepsilon(t-\tau)}}{(t-\tau)^{1-\alpha}}\,\mathrm{d}\tau \\
&\leqslant C_6(\varepsilon_0, T)\, \langle g\rangle_x^{(\alpha)}\, |t-t'|^{\alpha}.
\end{aligned}
$$

I_2' is estimated by just the same way as I_1'. For I_3' we do in the same way with the help of the mean value theorem and the estimate of $\partial^2 Z_\varepsilon(x_1,t)/\partial t^2$:

$$
\left|\frac{\partial^2}{\partial t^2} Z_\varepsilon(x_1,t)\right| \leqslant C_7 \frac{\varepsilon}{t^2} \exp\left[-\frac{\varepsilon x_1^2}{4t}\right] + C_7' \int_0^{t/\varepsilon} z^{-5/2} \exp\left[-\frac{x_1^2}{4z} - \frac{(t-\varepsilon z)^2}{4z}\right] \mathrm{d}z.
$$

After some lengthy calculations we get

$$
\langle w^{\dagger}\rangle_t^{(\alpha)} \leqslant C_8(\varepsilon_0, T)\, \langle g\rangle_x^{(\alpha)}.
$$

Second term in the right hand side of (11.28) is estimated similarly. Hence we obtain

$$
\left\langle \frac{\partial w}{\partial t}\right\rangle_t^{(\alpha)} \leqslant C_9(\varepsilon_0, T) \left(\langle g\rangle_x^{(\alpha)} + \langle g\rangle_t^{(\alpha)}\right). \tag{11.32}
$$

From (11.28) we can derive

$$
\begin{aligned}
\frac{\partial^2}{\partial t \partial x_1} w(x_1,t) &= \int_0^t \mathrm{d}\tau \int_{-\infty}^{\infty} \frac{\partial}{\partial t} Z_\varepsilon(x_1-y, t-\tau) \left(\frac{\partial}{\partial y} g(y,\tau) - \frac{\partial}{\partial x_1} g(x_1,\tau)\right) \mathrm{d}y \\
&\quad + \int_0^t \frac{\partial}{\partial x_1} g(x_1,\tau)\,\mathrm{d}\tau \int_{-\infty}^{\infty} \frac{\partial}{\partial t} Z_\varepsilon(x_1-y,t-\tau)\,\mathrm{d}y + \frac{\partial}{\partial x_1} g(x_1,t).
\end{aligned}
$$

Then repeating the above arguments, we obtain the estimates

$$
\begin{cases}
\left\langle \dfrac{\partial^2 w}{\partial t \partial x_1}\right\rangle_x^{(\alpha)} \leqslant C_{10}(\varepsilon_0, T) \left\langle \dfrac{\partial g}{\partial x_1}\right\rangle_x^{(\alpha)}, \\[2ex]
\left\langle \dfrac{\partial^2 w}{\partial t \partial x_1}\right\rangle_t^{(\alpha/2)} \leqslant C_{10}(\varepsilon_0, T)\, \|g\|_{1+\alpha}.
\end{cases} \tag{11.33}
$$

We come to the estimate $[\partial w/\partial t]^{(\alpha,\alpha)}$. Since $w^{\dagger}$ can be written as

$$
w^{\dagger}(x_1,t) = \int_{-\infty}^{\infty} \mathrm{d}\tau \int_{-\infty}^{\infty} \frac{\partial}{\partial t} Z_\varepsilon(x_1-y,\tau)\, (g(y,t-\tau) - g(x_1,t-\tau))\,\mathrm{d}y,
$$

$$
\begin{aligned}
\frac{w^{\dagger}(x_1,t)-w^{\dagger}(x_1,t-\Delta t)}{(\Delta t)^{\alpha}} &\equiv W^{\dagger}(x_1,t)\\
&= \int_{-\infty}^{\infty} \mathrm{d}\tau \int_{-\infty}^{\infty} \frac{\partial}{\partial t} Z_{\varepsilon}(x_1-y,\tau)\,(\varphi(y,t-\tau)-\varphi(x_1,t-\tau))\,\mathrm{d}y\\
&= \int_{0}^{t} \mathrm{d}\tau \int_{-\infty}^{\infty} \frac{\partial}{\partial t} Z_{\varepsilon}(x_1-y,t-\tau)\,(\varphi(y,\tau)-\varphi(x_1,\tau))\,\mathrm{d}y
\end{aligned}
$$

for $\Delta t > 0$, where

$$
\varphi(x_1,t) = \frac{g(x_1,t)-g(x_1,t-\Delta t)}{(\Delta t)^{\alpha}}.
$$

Again the above argument implies

$$
\begin{aligned}
\langle W^{\dagger}\rangle_x^{(\alpha)} &\leqslant C_{11}(\varepsilon_0,T)\,\langle\varphi\rangle_x^{(\alpha)} = C_{11}(\varepsilon_0,T)\sup_{x_1,x_1',t}\frac{|\varphi(x_1,t)-\varphi(x_1',t)|}{|x_1-x_1'|^{\alpha}}\\
&= C_{11}(\varepsilon_0,T)\sup_{x_1,x_1',t}\frac{|g(x_1,t)-g(x_1',t)-g(x_1,t-\Delta t)+g(x_1',t-\Delta t)|}{|x_1-x_1'|^{\alpha}(\Delta t)^{\alpha}},
\end{aligned}
$$

and hence

$$
[w^{\dagger}]^{(\alpha,\alpha)} \leqslant C_{11}(\varepsilon_0,T)\,[g]^{(\alpha,\alpha)}.
$$

Second term in the right hand side of (11.28) is estimated similarly. Thus we obtain

$$
\left[\frac{\partial w}{\partial t}\right]^{(\alpha,\alpha)} \leqslant C_{11}'(\varepsilon_0,T)\,[g]^{(\alpha,\alpha)}. \tag{11.34}
$$

Similarly we have

$$
\left[\frac{\partial^2 w}{\partial t\,\partial x_1}\right]^{(\alpha,\alpha/2)} \leqslant C_{12}(\varepsilon_0,T)\left[\frac{\partial g}{\partial x_1}\right]^{(\alpha,\alpha/2)}. \tag{11.35}
$$

Now it is necessary to get the representation of $(\mathscr{F}\mathscr{L})^{-1}[r_{\varepsilon}\tilde{g}_3]$:

$$
\begin{aligned}
(\mathscr{F}\mathscr{L})^{-1}[r_{\varepsilon}\tilde{g}_3] &= 2\frac{\partial^2}{\partial x_2^2}\int_0^t \mathrm{d}\tau\int_{-\infty}^{\infty}\Gamma_{\varepsilon}(x_1-y,x_2,t-\tau)\,g_3(y,\tau)\,\mathrm{d}y\Big|_{x_2=0}\\
&= 2\frac{\partial^2}{\partial x_2^2}\int_0^{t'} \mathrm{d}\tau\int_{-\infty}^{\infty}\Gamma_{1}(x_1-y,x_2,t'-\tau)\,G_3(y,\tau)\,\mathrm{d}y\Big|_{x_2=0},
\end{aligned}
$$

where $t' = t/\varepsilon,\ G_3(x_1, t') = g_3(x_1, t),\ \Gamma_1 = \Gamma_\varepsilon|_{\varepsilon=1}$ (see [21]). From this we have

$$\begin{aligned}\langle(\mathscr{F}\mathscr{L})^{-1}[r_\varepsilon \tilde{g}_3]\rangle^{(\alpha,\alpha/2)} &\leqslant C_{13}\langle G_3\rangle^{(1+\alpha,(1+\alpha)/2)}\\ &= C_{13}\left(\langle g_3\rangle_x^{(1+\alpha)} + \varepsilon^{(1+\alpha)/2}\langle g_3\rangle_t^{((1+\alpha)/2)}\right).\end{aligned}$$

Equation (11.22) and estimates (11.31)–(11.35) lead to

$$\left\|\frac{\partial\rho}{\partial t}\right\|_{1+\alpha} \leqslant C_{14}(\varepsilon_0, T)\ (\|g_1\|_{1+\alpha} + \|g_2\|_{1+\alpha} + \|g_3\|_{2+\alpha})\,. \tag{11.36}$$

The same arguments can be applied to (11.29) with the help of Lemma 11.4.1, so that similar estimates to (11.31), (11.32) and (11.34) hold for $\partial^2 w/\partial x_1^2$:

$$\left\|\frac{\partial^2 w}{\partial x_1^2}\right\|_\alpha \leqslant C_{15}(\varepsilon_0, T)\left\|\frac{\partial g}{\partial x_1}\right\|_\alpha ;$$

moreover, we have

$$\left\langle\frac{\partial^2 w}{\partial x_1^2}\right\rangle_t^{(\alpha)} \leqslant C_{16}(\varepsilon_0, T)\left\langle\frac{\partial g}{\partial x_1}\right\rangle_x^{(\alpha)}.$$

Lower order derivatives of w are easily estimated.

Therefore, from (11.22) we get

$$\begin{cases}\|\rho\|_{2+\alpha} \leqslant C_{17}(\varepsilon_0, T)\ (\|g_1\|_{1+\alpha} + \|g_2\|_{1+\alpha} + \|g_3\|_{2+\alpha})\,,\\ \left\langle\dfrac{\partial^2\rho}{\partial x_1^2}\right\rangle_t^{(\alpha)} \leqslant C_{18}(\varepsilon_0, T)\ \left(\left\langle\dfrac{\partial g_1}{\partial x_1}\right\rangle_x^{(\alpha)} + \left\langle\dfrac{\partial g_2}{\partial x_1}\right\rangle_x^{(\alpha)}\right.\\ \qquad\qquad\left. + \left\langle\dfrac{\partial g_3}{\partial x_1}\right\rangle^{(1+\alpha,(1+\alpha)/2)}\right).\end{cases} \tag{11.37}$$

Functions $v^\pm(x_1, t)$ in (11.22) are easily estimated. Indeed, since

$$(\mathscr{F}\mathscr{L})^{-1}\left[\frac{1}{r_\varepsilon}\right] * g(x_1, t) = \int_0^t \mathrm{d}\tau \int_{-\infty}^{\infty} \Gamma_\varepsilon(x_1 - y, 0, t-\tau)\, g(y, \tau)\, \mathrm{d}y,$$

we have a similar estimate to (11.36) and (11.37):

$$\left\|\frac{\partial v^\pm}{\partial t}\right\|_\alpha + \|v^\pm\|_{2+\alpha} \leqslant C_{19}\left(\|g_1\|_{1+\alpha} + \|g_2\|_{1+\alpha} + \|g_3\|_{\check{E}^{2+\alpha}}\right). \tag{11.38}$$

For $u^{\pm}(x,t)$ we prepare two representations:

$$u^{\pm}(x,t) = -2\int_0^t \mathrm{d}\tau \int_{-\infty}^{\infty} \frac{\partial}{\partial x_2}\Gamma_\varepsilon(x_1 - y, x_2, t-\tau)\, v^{\pm}(y,\tau)\,\mathrm{d}y, \tag{11.39}$$

and

$$u^{\pm}(x,t) = -2\int_0^t \mathrm{d}\tau \int_{-\infty}^{\infty} \Gamma_\varepsilon(x_1 - y, x_2, t-\tau)\, g^{\pm}(y,\tau)\,\mathrm{d}y, \tag{11.40}$$

where

$$g^{+} = -\frac{1}{2b^{+}}\left(g_1 + g_2 - \frac{\partial\rho}{\partial t}\right), \quad g^{-} = \frac{1}{2b^{-}}\left(-g_1 + g_2 + \frac{\partial\rho}{\partial t}\right),$$

since $\tilde{v}^{\pm}$ are represented as

$$\tilde{v}^{+} = \frac{1}{2b^{+}r_\varepsilon}\left(\tilde{g}_1 + \tilde{g}_2 - s\tilde{\rho}\right), \quad \tilde{v}^{-} = \frac{1}{2b^{-}r_\varepsilon}\left(-\tilde{g}_1 + \tilde{g}_2 + s\tilde{\rho}\right).$$

Here we use (11.39) for the estimates of $\partial^2 u^{\pm}/\partial x_1^2$, $\partial^2 u^{\pm}/\partial x_2^2$, $\partial u^{\pm}/\partial x_1$, and (11.40) for the estimates of $\partial^2 u^{\pm}/\partial x_1 \partial x_2$, $\partial u^{\pm}/\partial x_2$.

Introduce the notation with $t' = t/\varepsilon$

$$u^{\pm}(x,\varepsilon t') = U^{\pm}(x,t'), \quad v^{\pm}(x_1,\varepsilon t') = V^{\pm}(x_1,t'), \quad g^{\pm}(x_1,\varepsilon t') = G^{\pm}(x_1,t').$$

Then, (11.39) and (11.40) are represented as

$$U^{\pm}(x,t') = -2\frac{\partial\Gamma_1}{\partial x_2} * V^{\pm} \quad \text{and} \quad U^{\pm}(x,t') = -2\Gamma_1 * G^{\pm},$$

respectively. We can find the following estimates (see [13]):

$$\left\langle \frac{\partial\Gamma_1}{\partial x_2} * V^{\pm}\right\rangle_{t'}^{(\alpha/2)} \leqslant C_{20}\, \langle V^{\pm}\rangle_{t'}^{(\alpha/2)}, \quad \left\langle \frac{\partial\Gamma_1}{\partial x_2} * V^{\pm}\right\rangle_{x}^{(\alpha)} \leqslant C_{20}\left(\langle V^{\pm}\rangle_{x}^{(\alpha)} + \langle V^{\pm}\rangle_{t'}^{(\alpha/2)}\right).$$

Simultaneously we have

$$\left\langle \frac{\partial^2 U^{\pm}}{\partial x_1^2}\right\rangle_{t'}^{(\alpha/2)} + \left\langle \frac{\partial^2 U^{\pm}}{\partial x_2^2}\right\rangle_{t'}^{(\alpha/2)} \leqslant C_{21}\left(\left\langle \frac{\partial^2 V^{\pm}}{\partial x_1^2}\right\rangle_{t'}^{(\alpha/2)} + \left\langle \frac{\partial V^{\pm}}{\partial t'}\right\rangle_{t'}^{(\alpha/2)}\right),$$

$$\left\langle \frac{\partial^2 U^{\pm}}{\partial x_1^2}\right\rangle_{x}^{(\alpha)} + \left\langle \frac{\partial^2 U^{\pm}}{\partial x_2^2}\right\rangle_{x}^{(\alpha)}$$
$$\leqslant C_{21}\left(\left\langle \frac{\partial^2 V^{\pm}}{\partial x_1^2}\right\rangle_{x}^{(\alpha)} + \left\langle \frac{\partial^2 V^{\pm}}{\partial x_1^2}\right\rangle_{t'}^{(\alpha/2)} + \left\langle \frac{\partial V^{\pm}}{\partial t'}\right\rangle_{x}^{(\alpha)} + \left\langle \frac{\partial V^{\pm}}{\partial t'}\right\rangle_{t'}^{(\alpha/2)}\right).$$

These inequalities yield

$$\begin{cases} \left\langle \dfrac{\partial^2 u^\pm}{\partial x_1^2} \right\rangle_t^{(\alpha/2)} + \left\langle \dfrac{\partial^2 u^\pm}{\partial x_2^2} \right\rangle_t^{(\alpha/2)} \leqslant C_{21} \left(\left\langle \dfrac{\partial^2 v^\pm}{\partial x_1^2} \right\rangle_t^{(\alpha/2)} + \varepsilon \left\langle \dfrac{\partial v^\pm}{\partial t} \right\rangle_t^{(\alpha/2)} \right), \\ \left\langle \dfrac{\partial^2 u^\pm}{\partial x_1^2} \right\rangle_x^{(\alpha)} + \left\langle \dfrac{\partial^2 u^\pm}{\partial x_2^2} \right\rangle_x^{(\alpha)} \leqslant C_{21} \left(\left\langle \dfrac{\partial^2 v^\pm}{\partial x_1^2} \right\rangle_x^{(\alpha)} \right. \\ \qquad \left. + \varepsilon^{\alpha/2} \left\langle \dfrac{\partial^2 v^\pm}{\partial x_1^2} \right\rangle_t^{(\alpha/2)} + \varepsilon \left\langle \dfrac{\partial v^\pm}{\partial t} \right\rangle_x^{(\alpha)} + \varepsilon^{1+\alpha/2} \left\langle \dfrac{\partial v^\pm}{\partial t} \right\rangle_t^{(\alpha/2)} \right). \end{cases} \tag{11.41}$$

Furthermore, by the same method as that for $w^\dagger$ we have

$$\begin{aligned} \left[\frac{\partial^2 U^\pm}{\partial x_1^2} \right]^{(\alpha,\alpha/2)} &\leqslant C'_{22} \left(\left[\frac{\partial^2 V^\pm}{\partial x_1^2} \right]^{(\alpha,\alpha/2)} + \sup_{x_1,t',s,\Delta t'} \frac{1}{|\Delta t'|^{\alpha/2}|s|^{\alpha/2}} \left| \frac{\partial^2 V^\pm}{\partial x_1^2}(x_1,t'-s) \right. \right. \\ &\qquad \left. \left. - \frac{\partial^2 V^\pm}{\partial x_1^2}(x_1,t') - \frac{\partial^2 V^\pm}{\partial x_1^2}(x_1,t'+\Delta t'-s) + \frac{\partial^2 V^\pm}{\partial x_1^2}(x_1,t'+\Delta t') \right| \right) \\ &\leqslant C_{22} \left(\left[\frac{\partial^2 V^\pm}{\partial x_1^2} \right]^{(\alpha,\alpha/2)} + \left\langle \frac{\partial^2 V^\pm}{\partial x_1^2} \right\rangle_{t'}^{(\alpha)} \right), \end{aligned}$$

and hence

$$\left[\frac{\partial^2 u^\pm}{\partial x_1^2} \right]^{(\alpha,\alpha/2)} \leqslant C_{23} \left(\left[\frac{\partial^2 v^\pm}{\partial x_1^2} \right]^{(\alpha,\alpha/2)} + \varepsilon^{\alpha/2} \left\langle \frac{\partial^2 v^\pm}{\partial x_1^2} \right\rangle_t^{(\alpha)} \right). \tag{11.42}$$

Analogously we obtain the estimate for $[\partial^2 u^\pm / \partial x_2^2]^{(\alpha,\alpha/2)}$.

By virtue of the expression (11.40) we get

$$\frac{\partial U^\pm}{\partial x_2} = -2 \frac{\partial \Gamma_1}{\partial x_2} * G^\pm, \qquad \frac{\partial^2 U^\pm}{\partial x_2 \partial x_1} = -2 \frac{\partial \Gamma_1}{\partial x_2} * \frac{\partial G^\pm}{\partial x_1},$$

from which it follows

$$\left\langle \frac{\partial U^\pm}{\partial x_2} \right\rangle_{t'}^{(\alpha)} \leqslant C_{24} \langle G^\pm \rangle_{t'}^{(\alpha)}, \qquad \left\langle \frac{\partial^2 U^\pm}{\partial x_1 \partial x_2} \right\rangle_{t'}^{(\alpha/2)} \leqslant C_{24} \left\langle \frac{\partial G^\pm}{\partial x_1} \right\rangle_{t'}^{(\alpha/2)},$$

and hence

$$\left\langle \frac{\partial u^\pm}{\partial x_2} \right\rangle_t^{(\alpha)} \leqslant C_{25} \langle g^\pm \rangle_t^{(\alpha)}, \qquad \left\langle \frac{\partial^2 u^\pm}{\partial x_1 \partial x_2} \right\rangle_t^{(\alpha/2)} \leqslant C_{25} \left\langle \frac{\partial g^\pm}{\partial x_1} \right\rangle_t^{(\alpha/2)}. \tag{11.43}$$

Each term in the right hand side of (11.41)–(11.43) is estimated by virtue of (11.36)–(11.38).

Similar arguments are applicable to other terms appearing in the norm $\|u^{\pm}\|_{2+\alpha}$. And finally from the first and second equations in (11.15) we derive the uniform estimates $\varepsilon\|\partial u^{\pm}/\partial t\|_{\alpha}$ with respect to ε.

Lemma 11.4.2 *Let ε_0 be a fixed positive number and b^+, b^- and d be positive constants. Assume that $g_1,\ g_2 \in \underset{0}{E}^{1+\alpha}(\mathbb{R}_T)$ and $g_3 \in \underset{0}{\check{E}}^{2+\alpha}(\mathbb{R}_T)$ with $\alpha \in (0, 1)$ and $T > 0$. Then problem (11.15) with any $\varepsilon \in (0, \varepsilon_0]$ has a unique solution*

$$u^+ \in \underset{0}{P_\varepsilon}^{2+\alpha}(\mathbb{R}^2_{+,T}),\quad u^- \in \underset{0}{P_\varepsilon}^{2+\alpha}(\mathbb{R}^2_{-,T}),\quad \rho \in \underset{0}{\hat{E}}^{2+\alpha}(\mathbb{R}_T)$$
$$\big(\mathbb{R}^2_{\pm,T} \equiv \mathbb{R}^2_{\pm} \times [0, T],\quad \mathbb{R}_T \equiv \mathbb{R} \times [0, T]\big)$$

satisfying the inequality

$$\varepsilon\left\|\frac{\partial u^+}{\partial t}\right\|_{\alpha} + \varepsilon\left\|\frac{\partial u^-}{\partial t}\right\|_{\alpha} + \|u^+\|_{2+\alpha} + \|u^-\|_{2+\alpha} + \|\rho\|_{2+\alpha} + \left\|\frac{\partial \rho}{\partial t}\right\|_{1+\alpha}$$
$$\leqslant C_{26}\left(\|g_1\|_{1+\alpha} + \|g_2\|_{1+\alpha} + \|g_3\|_{2+\alpha} + \left\|\frac{\partial g_3}{\partial t}\right\|_{\alpha}\right), \qquad (11.44)$$

where C_{26} is a positive constant depending on ε_0, but not on ε.

Now we shall solve problem (11.15) by the regularizer method.

In the followings (θ, r) corresponds to the above (x_1, x_2). For a suitably small positive number λ we construct two systems of coverings $\{\omega^k\}$ and $\{\Omega^k\}$ of $\overline{\Omega_1 \cup \Omega_2} \equiv \bar{\Omega}$ as follows (cf. [20, 22]):

Let k be in $\mathcal{M}_1$ if two dimensional squares ω^k and Ω^k included completely in Ω_1 are mapped by Π_1 from those with common center (r_k, θ_k) and with the length of their edges, in parallel directions of axes, equal to $\lambda/2$ and λ, respectively;

$\mathcal{M}_2$ and Π_2 for Ω_2 are the same as $\mathcal{M}_1$ and Π_1 for Ω_1;

For $k \in \mathcal{M}_3$ satisfying $\omega^k \cap \Gamma \neq \emptyset$, ω^k and $\Omega^k \subset \overline{\Omega_1}$ are defined in the local coordinates (z_1, z_2) with an origin at $(r_k, \theta_k) \in \Gamma$ by

$$\omega^k = \Pi_3\left\{|z_2| \leqslant \lambda/2,\ -\lambda \leqslant z_1 - F_\Gamma(z_2) \leqslant 0\right\},$$
$$\Omega^k = \Pi_3\left\{|z_2| \leqslant \lambda,\ -2\lambda \leqslant z_1 - F_\Gamma(z_2) \leqslant 0\right\},$$

where equation $z_1 = F_\Gamma(z_2)$ represents Γ around $(r_k, \theta_k) \in \Gamma$ and Π_3 is a transformation from (z_1, z_2) to $(r, \theta) \in \Omega^k$;

$\mathscr{M}_4$ is the same set as $\mathscr{M}_3$, but to use in distinction of the coverings in $\overline{\Omega_2}$:

$$\omega^k = \Pi_4 \{|z_2| \leqslant \lambda/2,\ 0 \leqslant z_1 - F_\Gamma(z_2) \leqslant \lambda\},$$
$$\Omega^k = \Pi_4 \{|z_2| \leqslant \lambda,\ 0 \leqslant z_1 - F_\Gamma(z_2) \leqslant 2\lambda\};$$

By $\{\omega^k\}$ and $\{\Omega^k\}$, $k \in \mathscr{M}_5$, we denote the coverings of Γ_* in $\overline{\Omega_1}$ and by Π_5 the transformation from z to $(r, \theta) \in \Omega^k$;
$\mathscr{M}_6$ and Π_6 for Γ^* are the same as $\mathscr{M}_5$ and Π_5 for Γ_*, respectively.

Now we introduce partitions of unity $\{\eta_k\}$ and $\{\eta_k^*\}$ subordinated to $\{\omega^k\}$ and $\{\Omega^k\}$ such that

$$\eta_k(r,\theta) = \begin{cases} 1 & \text{for } (r,\theta) \in \omega^k, \\ 0 & \text{for } (r,\theta) \in \bar{\Omega} \setminus \Omega^k, \end{cases} \qquad 0 \le \eta_k(r,\theta) \le 1, \quad \eta_k \in C_0^\infty(\bar{\Omega}),$$

$$\left|\mathrm{D}^j_{r,\theta}\eta_k(r,\theta)\right| \leqslant C_{27}\lambda^{-|j|}, \quad \eta_k^*(r,\theta) \equiv \frac{\eta_k(r,\theta)}{\sum_k(\eta_k(r,\theta))^2}.$$

Obviously, $\{\eta_k^*(r,\theta)\}$ have properties

$$\eta_k^*(r,\theta) = 0 \text{ if } (r,\theta) \in \bar{\Omega} \setminus \Omega^k, \quad \sum_k \eta_k(r,\theta)\eta_k^*(r,\theta) = 1,$$

$$\left|\mathrm{D}^j_{r,\theta}\eta_k^*(r,\theta)\right| \leqslant C_{28}\lambda^{-|j|}.$$

First let $T_* = \gamma\lambda^2$, where $\gamma \in (0,1)$ will be specified later. Then the regularizer $\mathscr{R}$ is defined by

$$\mathscr{R}H = \sum_{m=1,3,5}\sum_{k\in\mathscr{M}_m} \eta_k^* u_1^k + \sum_{m=2,4,6}\sum_{k\in\mathscr{M}_m} \eta_k^* u_2^k + \sum_{k\in\mathscr{M}_3} \eta_k^* \rho^k,$$

$$H = (\phi_1, \phi_2, \psi_*, \psi^*, \psi_1, \psi_2, \psi_3).$$

Here $u_1^k = \Pi_1 \bar{u}_1^k$, where $\bar{u}_1^k$ is a solution of problem

$$\begin{cases} \dfrac{\partial \bar{u}_1^k}{\partial t} - \Pi_1^{-1}\left(\mathscr{L}^1_{*,k}\right)\bar{u}_1^k = \Pi_1^{-1}\eta_k\phi_1 & \text{in } \mathbb{R}^2 \times (0,T), \\ \bar{u}_1^k\big|_{t=0} = 0 & \text{on } \mathbb{R}^2, \end{cases} \tag{11.45}$$

$$\left(\mathscr{L}^j_{*,k} \equiv \mathscr{L}^j_*\left(r_k, \theta_k; \partial/\partial r, \partial/\partial\theta\right)\Big|_{t=0} \quad (j = 1, 2)\right);$$

$u_2^k = \Pi_2 \bar{u}_2^k$, where $\bar{u}_2^k$ is a solution of problem

$$\begin{cases} \dfrac{\partial \bar{u}_2^k}{\partial t} - \Pi_2^{-1}\left(\mathscr{L}_{*,k}^2\right)\bar{u}_2^k = \Pi_2^{-1}\eta_k\phi_2 \quad \text{in} \quad \mathbb{R}^2 \times (0,T), \\ \bar{u}_2^k\big|_{t=0} = 0 \quad \text{on} \quad \mathbb{R}^2; \end{cases} \tag{11.46}$$

$u_1^k = \Pi_5 \bar{u}_1^k$, where $\bar{u}_1^k$ is a solution of problem

$$\begin{cases} \dfrac{\partial \bar{u}_1^k}{\partial t} - \Pi_5^{-1}\left(\mathscr{L}_{*,k}^1\right)\bar{u}_1^k = \Pi_5^{-1}\eta_k\phi_1 \quad \text{in} \quad \mathbb{R}_+^2 \times (0,T), \\ \bar{u}_1^k\big|_{t=0} = 0 \quad \text{on} \quad \mathbb{R}_+^2, \\ \left.\dfrac{\partial \bar{u}_1^k}{\partial z_1}\right|_{z_1=0} = \Pi_5^{-1}\eta_k\psi_*; \end{cases} \tag{11.47}$$

$u_2^k = \Pi_6 \bar{u}_2^k$, where $\bar{u}_2^k$ is a solution of problem

$$\begin{cases} \dfrac{\partial \bar{u}_2^k}{\partial t} - \Pi_6^{-1}\left(\mathscr{L}_{*,k}^2\right)\bar{u}_2^k = \Pi_6^{-1}\eta_k\phi_2 \quad \text{in} \quad \mathbb{R}_+^2 \times (0,T), \\ \bar{u}_2^k\big|_{t=0} = 0 \quad \text{on} \quad \mathbb{R}_+^2, \\ \bar{u}_2^k\big|_{z_1=0} = \Pi_6^{-1}\eta_k\psi^*; \end{cases} \tag{11.48}$$

$u_1^k = \Pi_3 \bar{u}_1^k,\ u_2^k = \Pi_4 \bar{u}_2^k,\ \rho^k = \Pi_3 \bar{\rho}^k$, where $\left(\bar{u}_1^k, \bar{u}_2^k, \bar{\rho}^k\right)$ is a solution of problem

$$\begin{cases} \dfrac{\partial \bar{u}_1^k}{\partial t} - \Pi_3^{-1}\left(\mathscr{L}_{*,k}^1\right)\bar{u}_1^k = \Pi_3^{-1}\eta_k\phi_1 \quad \text{in} \quad \mathbb{R}_-^2 \times (0,T), \\ \dfrac{\partial \bar{u}_2^k}{\partial t} - \Pi_4^{-1}\left(\mathscr{L}_{*,k}^2\right)\bar{u}_2^k = \Pi_4^{-1}\eta_k\phi_2 \quad \text{in} \quad \mathbb{R}_+^2 \times (0,T), \\ \left(\bar{u}_1^k, \bar{u}_2^k, \bar{\rho}^k\right)\big|_{t=0} = 0, \\ \left.\dfrac{\partial \bar{\rho}^k}{\partial t} - b_k^+ \dfrac{\partial \bar{u}_2^k}{\partial z_1} - b_k^- \dfrac{\partial \bar{u}_1^k}{\partial z_1}\right|_{z_1=0} = \Pi_3^{-1}\eta_k(\psi_1 + \psi_2), \\ \left.-b_k^+ \dfrac{\partial \bar{u}_2^k}{\partial z_1} + b_k^- \dfrac{\partial \bar{u}_1^k}{\partial z_1}\right|_{z_1=0} = \Pi_3^{-1}\eta_k(-\psi_1 + \psi_2), \\ -\bar{u}_2^k + \bar{u}_1^k + d_k\bar{\rho}^k\big|_{z_1=0} = \Pi_3^{-1}\eta_k\psi_3 \end{cases} \tag{11.49}$$

with $(b_k^+, b_k^-)\,\mathbf{e}_{z_1} = \Pi_3\left((b_2^2, b_1^2)(\bar{\zeta}), (b_2^1, b_1^1)(\bar{\zeta})\right)\big|_{\theta=\theta_k, t=0}$, $d_k = d(\bar{\zeta}))\big|_{\theta=\theta_k, t=0}$ and $\mathbf{e}_{z_1} = (1,0)$, where $\mathbb{R}_\pm^2 \equiv \{(z_1, z_2) \in \mathbb{R}^2 \mid \pm z_1 > 0\}$.

Then it is not difficult to see that $\mathscr{R}H = (u_1', u_2', \rho')$ satisfies

$$\begin{cases} \varepsilon \dfrac{\partial u_1'}{\partial t} - \mathscr{L}_*^1 u_1' = \phi_1 - \mathscr{T}_1 H \quad \text{in } \Omega_1,\ t > 0, \\ \varepsilon \dfrac{\partial u_2'}{\partial t} - \mathscr{L}_*^2 u_2' = \phi_2 - \mathscr{T}_2 H \quad \text{in } \Omega_2,\ t > 0, \\ \dfrac{\partial u_1'}{\partial r} = \psi_* - \mathscr{T}_3 H \quad \text{on } \Gamma_*,\ t > 0, \\ u_2' = \psi^* - \mathscr{T}_4 H \quad \text{on } \Gamma^*,\ t > 0, \\ \dfrac{\partial \rho'}{\partial t} - b_2^1(\bar{\zeta}) \dfrac{\partial u_1'}{\partial r} - b_1^1(\bar{\zeta}) \dfrac{\partial u_1'}{\partial \theta} - b_2^2(\bar{\zeta}) \dfrac{\partial u_2'}{\partial r} - b_1^2(\bar{\zeta}) \dfrac{\partial u_2'}{\partial \theta} \\ \qquad\qquad\qquad = \psi_1 + \psi_2 - \mathscr{T}_5 H, \\ b_2^1(\bar{\zeta}) \dfrac{\partial u_1'}{\partial r} + b_1^1(\bar{\zeta}) \dfrac{\partial u_1'}{\partial \theta} - b_2^2(\bar{\zeta}) \dfrac{\partial u_2'}{\partial r} - b_1^2(\bar{\zeta}) \dfrac{\partial u_2'}{\partial \theta} \\ \qquad\qquad\qquad = -\psi_1 + \psi_2 - \mathscr{T}_6 H, \\ u_1' - u_2' + d(\bar{\zeta})\rho' = \psi_3 - \mathscr{T}_7 H \quad \text{on } \Gamma,\ t > 0, \\ \left(\bar{u}_1', \bar{u}_2', \bar{\rho}'\right)\big|_{t=0} = 0. \end{cases} \tag{11.50}$$

Here the operator $\mathscr{T} = (\mathscr{T}_1, \mathscr{T}_2, \mathscr{T}_3, \mathscr{T}_4, \mathscr{T}_5, \mathscr{T}_6, \mathscr{T}_7)$ is defined on

$$\begin{aligned} \mathscr{H}_T = \underset{0}{E}^{\alpha}(\bar{Q}_{1,T}) \times \underset{0}{E}^{\alpha}(\bar{Q}_{2,T}) \times \underset{0}{E}^{1+\alpha}(\Gamma_{*,T}) \times \underset{0}{E}^{2+\alpha}(\Gamma_T^*) \\ \times \underset{0}{E}^{1+\alpha}(\Gamma_T) \times \underset{0}{E}^{1+\alpha}(\Gamma_T) \times \underset{0}{\check{E}}^{2+\alpha}(\Gamma_T) \end{aligned}$$

by the formulae

$$\begin{aligned} \mathscr{T}_1 H &= \mathscr{L}_*^1 u_1' - \sum_{j=1,3,5} \sum_{k \in \mathscr{M}_j} \eta_k^* \Pi_j \mathscr{L}_{*,k}^1 \bar{u}_1^k, \\ \mathscr{T}_2 H &= \mathscr{L}_*^2 u_2' - \sum_{j=2,4,6} \sum_{k \in \mathscr{M}_j} \eta_k^* \Pi_j \mathscr{L}_{*,k}^2 \bar{u}_2^k, \\ \mathscr{T}_3 H &= \frac{\partial u_1'}{\partial r} - \sum_{k \in \mathscr{M}_5} \eta_k^* \Pi_5 \frac{\partial \bar{u}_1^k}{\partial z_1}, \\ \mathscr{T}_4 H &= 0, \\ \mathscr{T}_5 H &= b_2^1(\bar{\zeta}) \frac{\partial u_1'}{\partial r} + b_1^1(\bar{\zeta}) \frac{\partial u_1'}{\partial \theta} - \sum_{k \in \mathscr{M}_3} \eta_k^* \Pi_3 b_k^- \frac{\partial \bar{u}_1^k}{\partial z_1} \\ &\quad + b_2^2(\bar{\zeta}) \frac{\partial u_2'}{\partial r} + b_1^2(\bar{\zeta}) \frac{\partial u_2'}{\partial \theta} - \sum_{k \in \mathscr{M}_4} \eta_k^* \Pi_4 b_k^+ \frac{\partial \bar{u}_2^k}{\partial z_1}, \end{aligned}$$

$$\mathscr{T}_6 H = -b_2^1(\bar{\zeta})\frac{\partial u_1'}{\partial r} - b_1^1(\bar{\zeta})\frac{\partial u_1'}{\partial \theta} + \sum_{k\in\mathscr{M}_3} \eta_k^* \Pi_3 b_k^- \frac{\partial \bar{u}_1^k}{\partial z_1}$$
$$+b_2^2(\bar{\zeta})\frac{\partial u_2'}{\partial r} + b_1^2(\bar{\zeta})\frac{\partial u_2'}{\partial \theta} - \sum_{k\in\mathscr{M}_4} \eta_k^* \Pi_4 b_k^+ \frac{\partial \bar{u}_2^k}{\partial z_1},$$
$$\mathscr{T}_7 H = -d(\bar{\zeta})\rho' + \sum_{k\in\mathscr{M}_3} \eta_k^* \Pi_3 d_k \bar{\rho}^k.$$

One can find the solution of problem (11.11) in the form

$$(u_1, u_2, \rho) = \mathscr{R}(\mathscr{I} + \mathscr{T} + \mathscr{T}^2 + \ldots)H = \mathscr{R}(\mathscr{I} - \mathscr{T})^{-1}H$$

($\mathscr{I}$ is an identity operator),

for which it is necessary to show that the operator $\mathscr{T}$ is a contraction on $\mathscr{H}_{T_*}$. We first note that $\Gamma \in C^{2+\alpha}$ implies $F_\Gamma \in C^{2+\alpha}(\bar{B}_\delta)$ ($B_\delta \equiv \{z_2 \in \mathbb{R} \,|\, |z_2| < \delta\}$) satisfying $F_\Gamma(0) = 0$, $\partial F_\Gamma/\partial z_2\,(0) = 0$, $\|F_\Gamma\|_{C^{2+\alpha}(\bar{B}_\delta)} \leqslant C$ with some constants δ and C being independent of z_2. We take λ small enough to satisfy $\lambda \leqslant \delta/2$. Clearly,

$$|F_\Gamma(z_2)| \leqslant C|z_2|^{1+\alpha}, \quad \left|\frac{\partial F_\Gamma(z_2)}{\partial z_2}\right| \leqslant C|z_2|^\alpha. \tag{11.51}$$

We estimate each term in $\mathscr{T}_1 H, \mathscr{T}_2 H, \ldots, \mathscr{T}_7 H$ in such a way that for the lower order terms the interpolation inequalities, for example,

$$\sup_{x\in\Omega} |\nabla u(x)| \leqslant c\left(\langle u\rangle_x^{(\alpha)}\right)^{\frac{1+\alpha}{2}} \left(\langle u\rangle_x^{(2+\alpha)}\right)^{\frac{1-\alpha}{2}}, \quad \langle \nabla^2 u\rangle_t^{(\alpha/2)} \leqslant c\lambda^\alpha [\nabla^2 u]^{(\alpha,\alpha/2)}$$

for $u \in \underset{0}{E}^{2+\alpha}(\bar{Q}_T)$ are used, while for the highest order terms the smallness of their coefficients like (11.51) derived from the smallness of λ and γ are used. Then we get the following estimate after some lengthy, but straightforward calculations:

$$\|\mathscr{T}H\|_{\mathscr{H}_t} \equiv \|\mathscr{T}_1 H\|_{E^\alpha(\bar{Q}_{1,t})} + \|\mathscr{T}_2 H\|_{E^\alpha(\bar{Q}_{2,t})} + \|\mathscr{T}_3 H\|_{E^{1+\alpha}(\Gamma_{*,t})}$$
$$+\|\mathscr{T}_4 H\|_{E^{2+\alpha}(\Gamma_t^*)} + \|\mathscr{T}_5 H\|_{E^{1+\alpha}(\Gamma_t)} + \|\mathscr{T}_6 H\|_{E^{1+\alpha}(\Gamma_t)} + \|\mathscr{T}_7 H\|_{\check{E}^{2+\alpha}(\Gamma_t)}$$
$$\leqslant C_{29}(\lambda,\gamma)\left(\sum_{j=1,3,5}\sum_{k\in\mathscr{M}_j} \|u_1^k\|_{E^{2+\alpha}(\bar{Q}_{1,t}^k)} + \sum_{j=2,4,6}\sum_{k\in\mathscr{M}_j} \|u_2^k\|_{E^{2+\alpha}(\bar{Q}_{2,t}^k)}\right.$$
$$\left. + \sum_{k\in\mathscr{M}_3} \|\rho^k\|_{\hat{E}^{2+\alpha}(\Gamma_t^k)}\right) \tag{11.52}$$

for any $t \in (0, T_*)$, where $Q_t^k = \Omega^k \times (0, t)$, $\Gamma_t^k = \Gamma_t \cap Q_t^k$ and $C_{29}(\lambda, \gamma) = C_{29,1}(\lambda) + C_{29,2}(\gamma) + C_{29,3}(\lambda)C_{29,4}(\gamma)$ is a positive constant with $C_{29,j}(\cdot)$ being dependent on their argument non-decreasingly and $C_{29,j}(0) = 0\,(j = 1, 2, 3, 4)$. By virtue of Lemma 11.4.2 and (11.19) we see that $\mathscr{T}$ is the contraction operator on $\mathscr{H}_{T_*}$ for suitably small λ and γ.

Repeating the above argument finite times, we come to

Lemma 11.4.3 *For any $T > 0$ and any $\varepsilon_0 > 0$ the problem (11.11) with a fixed $\varepsilon \in (0, \varepsilon_0]$ has a unique solution $(u_1, u_2, \rho) \in \underset{0}{P_\varepsilon}{}^{2+\alpha}(\bar{Q}_{1,T}) \times \underset{0}{P_\varepsilon}{}^{2+\alpha}(\bar{Q}_{2,T}) \times \underset{0}{\hat{E}}{}^{2+\alpha}(\Gamma_T)$ satisfying*

$$\|u_1\|_{P_\varepsilon^{2+\alpha}(\bar{Q}_{1,T})} + \|u_2\|_{P_\varepsilon^{2+\alpha}(\bar{Q}_{2,T})} + \|\rho\|_{\hat{E}^{2+\alpha}(\Gamma_T)} \leqslant C_{30}\|H\|_{\mathscr{H}_T}, \tag{11.53}$$

where C_{30} is a positive constant depending on ε_0, but not on ε.

11.5 Nonlinear Problem: Proof of Theorem 11.2.2

In this section we construct the solution to problem (11.10) by the successive approximation method.

Let $(p_{1,n}^*, p_{2,n}^*, \zeta_n^*)$ $(n = 1, 2, 3, \dots)$ be a solution of problem (11.11) with

$$\begin{aligned} &\phi_1 = \Phi_1(p_{1,n-1}^*, \zeta_{n-1}^*), \quad \phi_2 = \Phi_2(p_{2,n-1}^*, \zeta_{n-1}^*), \quad \psi_* = \Psi_*(\zeta_{n-1}^*), \quad \psi^* = \Psi^*, \\ &\psi_1 = \Psi_1(p_{1,n-1}^*, p_{2,n-1}^*, \zeta_{n-1}^*), \quad \psi_2 = \Psi_2(p_{1,n-1}^*, p_{2,n-1}^*, \zeta_{n-1}^*), \quad \psi_3 = \Psi_3 \end{aligned}$$

for a given $(p_{1,n-1}^*, p_{2,n-1}^*, \zeta_{n-1}^*) \in \underset{0}{P_\varepsilon}{}^{2+\alpha}(\bar{Q}_{1,T}) \times \underset{0}{P_\varepsilon}{}^{2+\alpha}(\bar{Q}_{2,T}) \times \underset{0}{\hat{E}}{}^{2+\alpha}(\Gamma_T)$ and

$$(p_{1,0}^*, p_{2,0}^*, \zeta_0^*) = (0, 0, 0).$$

It is easily seen that there exists a constant $M > 0$ dependent on ε_0 and T, but not on ε such that

$$\big\|\big(\Phi_1(0,0), \Phi_2(0,0), \Psi_*(0), \Psi^*, \Psi_1(0,0,0), \Psi_2(0,0,0), \Psi_3\big)\big\|_{\mathscr{H}_T} \leqslant M. \tag{11.54}$$

From the same argument as in (11.52) we derive the following inequality with the help of (11.54):

$$\begin{aligned} &\big\|\big(\Phi_1(p_{1,n-1}^*, \zeta_{n-1}^*), \Phi_2(p_{2,n-1}^*, \zeta_{n-1}^*), \Psi_*(\zeta_{n-1}^*), \Psi^*, \Psi_1(p_{1,n-1}^*, p_{2,n-1}^*, \zeta_{n-1}^*), \\ &\qquad\qquad \Psi_2(p_{1,n-1}^*, p_{2,n-1}^*, \zeta_{n-1}^*), \Psi_3\big)\big\|_{\mathscr{H}_t} \\ &\leqslant \big\|\big(\Phi_1(0,0), \Phi_2(0,0), \Psi_*(0), \Psi^*, \Psi_1(0,0,0), \Psi_2(0,0,0), \Psi_3\big)\big\|_{\mathscr{H}_t} \\ &\quad + \big\|\big(\Phi_1(p_{1,n-1}^*, \zeta_{n-1}^*) - \Phi_1(0,0), \Phi_2(p_{2,n-1}^*, \zeta_{n-1}^*) - \Phi_2(0,0), \end{aligned}$$

$$\begin{aligned}
&\Psi_*(\zeta_{n-1}^*) - \Psi_*(0), 0, \Psi_1(p_{1,n-1}^*, p_{2,n-1}^*, \zeta_{n-1}^*) - \Psi_1(0,0,0), \\
&\qquad \Psi_2(p_{1,n-1}^*, p_{2,n-1}^*, \zeta_{n-1}^*) - \Psi_2(0,0,0), 0\big) \big\|_{\mathscr{H}_t} \\
&\leqslant M + \beta \left(\|p_{1,n-1}^*\|_{E^{2+\alpha}(\bar{Q}_{1,t})} + \|p_{2,n-1}^*\|_{E^{2+\alpha}(\bar{Q}_{2,t})} + \|\zeta_{n-1}^*\|_{\hat{E}^{2+\alpha}(\Gamma_t)} \right) \\
&\quad + C_\beta t^\chi F \left(\|p_{1,n-1}^*\|_{E^{2+\alpha}(\bar{Q}_{1,t})} + \|p_{2,n-1}^*\|_{E^{2+\alpha}(\bar{Q}_{2,t})} + \|\zeta_{n-1}^*\|_{\hat{E}^{2+\alpha}(\Gamma_t)} \right) \qquad (11.55)
\end{aligned}$$

for any $t \in (0, T)$ and any $\beta > 0$, where C_β is a positive constant depending on β non-increasingly, $\chi > 0$ is a constant depending on α (possibly) and $F(\cdot)$ is a polynomial in its argument. Thus, Lemma 11.4.3 yields

$$\begin{aligned}
&\|p_{1,n}^*\|_{P_\varepsilon^{2+\alpha}(\bar{Q}_{1,t})} + \|p_{2,n}^*\|_{P_\varepsilon^{2+\alpha}(\bar{Q}_{2,t})} + \|\zeta_n^*\|_{\hat{E}^{2+\alpha}(\Gamma_t)} \\
&\leqslant C_{30} \Big(M + \beta \left(\|p_{1,n-1}^*\|_{E^{2+\alpha}(\bar{Q}_{1,t})} + \|p_{2,n-1}^*\|_{E^{2+\alpha}(\bar{Q}_{2,t})} + \|\zeta_{n-1}^*\|_{\hat{E}^{2+\alpha}(\Gamma_t)} \right) \\
&\quad + C_\beta t^\chi F \left(\|p_{1,n-1}^*\|_{E^{2+\alpha}(\bar{Q}_{1,t})} + \|p_{2,n-1}^*\|_{E^{2+\alpha}(\bar{Q}_{2,t})} + \|\zeta_{n-1}^*\|_{\hat{E}^{2+\alpha}(\Gamma_t)} \right) \Big) \qquad (11.56)
\end{aligned}$$

for any $t \in (0, T)$. Now we take first $\beta = 1/(4C_{30})$, and then

$$T' = \min \left\{ T, \left(\frac{M}{2C_\beta F(2C_{30}M)} \right)^{1/\chi} \right\}.$$

Therefore, (11.56) yields

$$\|p_{1,n}^*\|_{P_\varepsilon^{2+\alpha}(\bar{Q}_{1,t})} + \|p_{2,n}^*\|_{P_\varepsilon^{2+\alpha}(\bar{Q}_{2,t})} + \|\zeta_n^*\|_{\hat{E}^{2+\alpha}(\Gamma_t)} \leqslant 2C_{30}M$$

for any $t \in (0, T')$ if

$$\|p_{1,n-1}^*\|_{P_\varepsilon^{2+\alpha}(\bar{Q}_{1,T})} + \|p_{2,n-1}^*\|_{P_\varepsilon^{2+\alpha}(\bar{Q}_{2,T})} + \|\zeta_{n-1}^*\|_{\hat{E}^{2+\alpha}(\Gamma_T)} \leqslant 2C_{30}M,$$

that is, for $n = 0, 1, 2, \ldots$

$$\begin{cases}
(p_{1,n}^*, p_{2,n}^*, \zeta_n^*) \in \underset{0}{P_\varepsilon}^{2+\alpha}(\bar{Q}_{1,T'}) \times \underset{0}{P_\varepsilon}^{2+\alpha}(\bar{Q}_{2,T'}) \times \underset{0}{\hat{E}}^{2+\alpha}(\Gamma_{T'}), \\
\|p_{1,n}^*\|_{P_\varepsilon^{2+\alpha}(\bar{Q}_{1,T'})} + \|p_{2,n}^*\|_{P_\varepsilon^{2+\alpha}(\bar{Q}_{2,T'})} + \|\zeta_n^*\|_{\hat{E}^{2+\alpha}(\Gamma_{T'})} \leqslant 2C_{30}M.
\end{cases} \qquad (11.57)$$

In order to prove the convergence of the approximate sequence we consider the difference $(p_{1,n}^* - p_{1,n-1}^*, p_{2,n}^* - p_{2,n-1}^*, \zeta_n^* - \zeta_{n-1}^*)$, which satisfies (11.11) with

$$\begin{cases} u_1 = p^*_{1,n} - p^*_{1,n-1}, \quad u_2 = p^*_{2,n} - p^*_{2,n-1}, \quad \rho = \zeta^*_n - \zeta^*_{n-1}, \\ \phi_1 = \Phi_1(p^*_{1,n-1}, \zeta^*_{n-1}) - \Phi_1(p^*_{1,n-2}, \zeta^*_{n-2}), \\ \phi_2 = \Phi_2(p^*_{2,n-1}, \zeta^*_{n-1}) - \Phi_2(p^*_{2,n-2}, \zeta^*_{n-1}), \\ \psi_* = \Psi_*(\zeta^*_{n-1}) - \Psi_*(\zeta^*_{n-2}), \quad \psi^* = 0, \\ \psi_1 = \Psi_1(p^*_{1,n-1}, p^*_{2,n-1}, \zeta^*_{n-1}) - \Psi_1(p^*_{1,n-2}, p^*_{2,n-2}, \zeta^*_{n-2}), \\ \psi_2 = \Psi_2(p^*_{1,n-1}, p^*_{2,n-1}, \zeta^*_{n-1}) - \Psi_2(p^*_{1,n-2}, p^*_{2,n-2}, \zeta^*_{n-2}), \quad \psi_3 = 0. \end{cases}$$

Similarly as above, with the help of the interpolation inequalities we can obtain a similar estimate as (11.55):

$$\begin{aligned} &\big\| \big(\Phi_1(p^*_{1,n-1}, \zeta^*_{n-1}) - \Phi_1(p^*_{1,n-2}, \zeta^*_{n-2}), \Phi_2(p^*_{2,n-1}, \zeta^*_{n-1}) - \Phi_2(p^*_{2,n-2}, \zeta^*_{n-2}), \\ &\quad \Psi_*(\zeta^*_{n-1}) - \Psi_*(\zeta^*_{n-2}), 0, \Psi_1(p^*_{1,n-1}, p^*_{2,n-1}, \zeta^*_{n-1}) - \Psi_1(p^*_{1,n-2}, p^*_{2,n-2}, \zeta^*_{n-2}), \\ &\qquad \Psi_2(p^*_{1,n-1}, p^*_{2,n-1}, \zeta^*_{n-1}) - \Psi_2(p^*_{1,n-2}, p^*_{2,n-2}, \zeta^*_{n-2}), 0\big) \big\|_{\mathscr{H}_t} \\ &\leqslant \Big(\beta' + C_{\beta'} t^{\chi'} F'\Big(\|p^*_{1,n-1}\|_{E^{2+\alpha}(\bar{Q}_{1,t})} + \|p^*_{1,n-2}\|_{E^{2+\alpha}(\bar{Q}_{1,t})} + \|p^*_{2,n-1}\|_{E^{2+\alpha}(\bar{Q}_{2,t})} \\ &\qquad + \|p^*_{2,n-2}\|_{E^{2+\alpha}(\bar{Q}_{2,t})} + \|\zeta^*_{n-1}\|_{\hat{E}^{2+\alpha}(\Gamma_t)} + \|\zeta^*_{n-2}\|_{\hat{E}^{2+\alpha}(\Gamma_t)}\Big)\Big) \\ &\quad \times \Big(\|p^*_{1,n-1} - p^*_{1,n-2}\|_{E^{2+\alpha}(\bar{Q}_{1,t})} + \|p^*_{2,n-1} - p^*_{2,n-2}\|_{E^{2+\alpha}(\bar{Q}_{2,t})} \\ &\qquad + \|\zeta^*_{n-1} - \zeta^*_{n-2}\|_{\hat{E}^{2+\alpha}(\Gamma_t)}\Big) \end{aligned}$$

for any $t \in (0, T')$ and any $\beta' > 0$, where $C_{\beta'}$ is a positive constant depending on β' non-increasingly, $\chi' > 0$ is a constant depending on α (possibly) and $F'(\cdot)$ is a polynomial in its argument. Therefore, Lemma 11.4.3 and (11.57) yield

$$\begin{aligned} &\|p^*_{1,n} - p^*_{1,n-1}\|_{P^{2+\alpha}_{\varepsilon}(\bar{Q}_{1,t})} + \|p^*_{2,n} - p^*_{2,n-1}\|_{P^{2+\alpha}_{\varepsilon}(\bar{Q}_{2,t})} + \|\zeta^*_n - \zeta^*_{n-1}\|_{\hat{E}^{2+\alpha}(\Gamma_t)} \\ &\quad \leqslant C_{30}\Big(\beta' + C_{\beta'} t^{\chi'} F'(4C_{30}M)\Big) \Big(\|p^*_{1,n-1} - p^*_{1,n-2}\|_{E^{2+\alpha}(\bar{Q}_{1,t})} \\ &\qquad + \|p^*_{2,n-1} - p^*_{2,n-2}\|_{E^{2+\alpha}(\bar{Q}_{2,t})} + \|\zeta^*_{n-1} - \zeta^*_{n-2}\|_{\hat{E}^{2+\alpha}(\Gamma_t)}\Big) \end{aligned}$$

for any $t \in (0, T')$. Again taking first $\beta' = 1/(4C_{30})$, and then

$$T_0 = \min\left\{ T', \left(\frac{1}{4C_{30} C_{\beta'} F'(4C_{30}M)} \right)^{1/\chi'} \right\},$$

we obtain

$$
\begin{aligned}
&\|p^*_{1,n} - p^*_{1,n-1}\|_{P_\varepsilon^{2+\alpha}(\bar{Q}_{1,t})} + \|p^*_{2,n} - p^*_{2,n-1}\|_{P_\varepsilon^{2+\alpha}(\bar{Q}_{2,t})} + \|\zeta^*_n - \zeta^*_{n-1}\|_{\hat{E}^{2+\alpha}(\Gamma_t)} \\
&\quad \leqslant \frac{1}{2}\Big(\|p^*_{1,n-1} - p^*_{1,n-2}\|_{E^{2+\alpha}(\bar{Q}_{1,t})} + \|p^*_{2,n-1} - p^*_{2,n-2}\|_{E^{2+\alpha}(\bar{Q}_{2,t})} \\
&\qquad + \|\zeta^*_{n-1} - \zeta^*_{n-2}\|_{\hat{E}^{2+\alpha}(\Gamma_t)}\Big) \quad \text{for } t \in (0, T_0). \qquad (11.58)
\end{aligned}
$$

Consequently, $(p^*_{1,n}, p^*_{2,n}, \zeta^*_n)$ converges to (p^*_1, p^*_2, ζ^*) as $n \to \infty$ uniformly in $\underset{0}{P_\varepsilon}^{2+\alpha}(\bar{Q}_{1,T_0}) \times \underset{0}{P_\varepsilon}^{2+\alpha}(\bar{Q}_{2,T_0}) \times \underset{0}{\hat{E}}^{2+\alpha}(\Gamma_{T_0})$, which satisfies

$$
\|p^*_1\|_{P_\varepsilon^{2+\alpha}(\bar{Q}_{1,T_0})} + \|p^*_2\|_{P_\varepsilon^{2+\alpha}(\bar{Q}_{2,T_0})} + \|\zeta^*\|_{\hat{E}^{2+\alpha}(\Gamma_{T_0})} \leqslant 2C_{30}M.
$$

The uniqueness of such a solution follows from the similar inequality as (11.58) for two solutions to problem (11.10).

11.6 Passing to the Limit $\varepsilon \to 0$: Proof of Theorem 11.2.1

Theorem 11.2.2 means that for any $\varepsilon \in (0, \varepsilon_0]$ the problem (11.10) admits a smooth solution $(p^*_{1,\varepsilon}, p^*_{2,\varepsilon}, \zeta^*_\varepsilon)$, and hence the following integral identity holds for any sufficiently smooth test function $\varphi(r, \theta, t)$ with $\varphi(r, \theta, T^*) = 0$ $(0 < T^* \leqslant T_0)$:

$$
\begin{aligned}
&-\int_{\Omega_1} \varepsilon p^*_{1,\varepsilon} \frac{\partial \varphi}{\partial t}\, \mathrm{d}r\, \mathrm{d}\theta - \int_{\Omega_2} \varepsilon p^*_{2,\varepsilon} \frac{\partial \varphi}{\partial t}\, \mathrm{d}r\, \mathrm{d}\theta - \int_{Q_{1,T^*}} \mathcal{L}^1_* p^*_{1,\varepsilon}\, \varphi\, \mathrm{d}r\, \mathrm{d}\theta\, \mathrm{d}t \\
&\quad - \int_{Q_{2,T^*}} \mathcal{L}^2_* p^*_{2,\varepsilon}\, \varphi\, \mathrm{d}r\, \mathrm{d}\theta\, \mathrm{d}t \\
&= \int_{Q_{1,T^*}} \Phi_1(p^*_{1,\varepsilon}, \zeta^*_\varepsilon)\, \varphi\, \mathrm{d}r\, \mathrm{d}\theta\, \mathrm{d}t + \int_{Q_{2,T^*}} \Phi_2(p^*_{2,\varepsilon}, \zeta^*_\varepsilon)\, \varphi\, \mathrm{d}r\, \mathrm{d}\theta\, \mathrm{d}t. \qquad (11.59)
\end{aligned}
$$

Since the set of $(p^*_{1,\varepsilon}, p^*_{2,\varepsilon}, \zeta^*_\varepsilon)$ belongs to the compact subset

$$
\left\{(p^*_{1,\varepsilon}, p^*_{2,\varepsilon}, \zeta^*_\varepsilon) \;\middle|\; \|p^*_{1,\varepsilon}\|_{E^{2+\alpha}(\bar{Q}_{1,T_0})} + \|p^*_{1,\varepsilon}\|_{E^{2+\alpha}(\bar{Q}_{2,T_0})} + \|\zeta^*_\varepsilon\|_{\hat{E}^{2+\alpha}(\Gamma_{T_0})} \leqslant 2C_{30}M\right\},
$$

one can select a convergent subsequence of it such that

$$
(p^*_{1,\varepsilon_k}, p^*_{2,\varepsilon_k}, \zeta^*_{\varepsilon_k}) \;\to\; (p^*_1, p^*_2, \zeta^*) \quad \text{as } \varepsilon_k \;\to\; 0
$$

in $E^{2+\alpha}(\bar{Q}_{1,T_0}) \times E^{2+\alpha}(\bar{Q}_{2,T_0}) \times \hat{E}^{2+\alpha}(\Gamma_{T_0})$. Passing to the limit in (11.59) along the selected subsequence, we establish that the function (p^*_1, p^*_2, ζ^*) is a solution of (11.10) with $\varepsilon = 0$, which proves Theorem 11.2.1.

References

1. Bazaliĭ, B.V.: On estimates for the solution of a model conjugation problem in the theory of problems with a free boundary. Differ. Uravn. **33**, 1374–1381 (1997). (Russian)
2. Bazaliĭ, B.V.: On a proof of the classical solvability of the Hele-Shaw problem with a free surface. Ukr. Mat. Zh. **50**, 1452–1462 (1998). (Russian)
3. Bensimon, D., Kadanoff, L.P., Liang, S., Shraiman, B.I., Tang, C.: Viscous flow in two dimensions. Rev. Mod. Phys. **58**, 977–999 (1986)
4. Bizhanova, G.I., Solonnikov, V.A.: On some model problems for second order parabolic equations with a time derivative in the boundary conditions. Algebra and Analysis **6**, 30–50 (1994). (Russian)
5. Escher, J., Simonett, G.: Classical solutions of multidimensional Hele-Shaw model. SIAM J. Math. Anal. **28**, 1028–1047 (1997)
6. Escher, J., Simonett, G.: Classical solutions for Hele-Shaw models with surface tension. Adv. Differ. Equ. **2**, 619–642 (1997)
7. Gustafsson, B., Vasil'ev, A.: Conformal and potential analysis in Hele-Shaw cells. Birkhaüser (2000)
8. Hele-Shaw, H.S.: The flow of water. Nature **58**, 33–36 (1898)
9. Homsy, G.: Viscous fingering in porous media. Annu. Rev. Fluid Mech. **19**, 271–311 (1987)
10. Howison, S.D.: Complex variable methods in Hele-Shaw moving boundary problems. Euro. J. Appl. Math. **3**, 209–224 (1992)
11. Kessler, D.A., Koplik, J., Levine, H.: Pattern selection in fingering growth phenomena. Adv. Phys. **37**, 255–339 (1988)
12. Kim, H., Funada, T., Joseph, D.D., Homsy, G.: Viscous potential flow analysis of radial fingering in a Hele-Shaw cell. Phys. Fluids **21**, 074106-1–074106-9 (2009)
13. Ladyzhenskaya, O.A., Solonnikov V.A., Ural'tseva, N.N.: Linear and Quasi-linear Equations of Parabolic Type. Nauka (1967) (Russian)
14. Martyushev, L.M., Birzina, A.I.: Specific feature of the loss of stability during radial displacement of fluid in the Hele-Shaw cell. J. Phys.: Condens. Matter **20**, 045201+08 (2008)
15. Miranda, J.A., Widom, M.: Radial fingering in a Hele-Shaw cell: a weakly nonlinear analysis. Phys. D **120**, 315–328 (1998)
16. Park, C.-W., Homsy, G.: Two-phase displacement in Hele-Shaw cell: theory. J. Fluid Mech. **139**, 291–308 (1984)
17. Paterson, L.: Radial fingering in a Hele-Shaw cell. J. Fluid Mech. **113**, 513–529 (1981)
18. Prokert, G.: Existence results for Hele-Shaw flow driven by surface tension. Euro. J. Appl. Math. **9**, 195–221 (1998)
19. Saffman, P.G., Taylor, G.I.: The penetration of a fluid into a porous medium or Hele-Shaw cell containing a more viscous liquid. Proc. Royal Soc. London **A245**, 312–329 (1958)
20. Solonnikov, V.A.: On boundary value problems for linear parabolic systems of differential equations of general form. Trudy Mat. Inst. Steklov **83**, 3–163 (1965). (Russian)
21. Solonnikov, V.A.: On the justification of the quasistationary approximation in the problem of motion of a viscous capillary drop. Interfaces Free Bound. **1**, 125–173 (1999)
22. Tani, A.: Two-phase free boundary problem for compressible viscous fluid motion. J. Math. Kyoto Univ. **24**, 243–267 (1984)
23. Tani, A., Tani, H.: Classical solvability of the radial viscous fingering problem in a Hele-Shaw cell, to be submitted
24. Tani, H.: Weakly nonlinear analysis on radial growing interface with the effect of viscous normal stress. J. Mol. Liquids **200**, 38–41 (2014)
25. Tani, H., Tani, A.: Effect of the wetting layer on the fingering pattern in a Hele-Shaw cell. J. Phys. Soc. Japan **83**, 034401 (2014)
26. Tanveer, S.: Surprises in viscous fingering. J. Fluid Mech. **409**, 273–308 (2000)

Chapter 12
Investigation of Bubble Clouds in a Cavitating Jet

Katsuhiro Yamamoto

Abstract It is possible for cloud cavitation to severely damage a material's surface. In this study, the unsteady behavior of cloud cavitation in a high-speed water jet is investigated by experimental observation and numerical simulations. Using a high-speed video camera with a frame rate of approximately 5×10^5 fps, it is found that high-pressure pulses are formed by collapsing bubble clouds, and that those pulses rise at a few microseconds before the cloud collapses. An erosion test is carried out by the injection of the water jet into the aluminum specimen. This test shows that the mass loss curve has two peaks and that the mass loss at the second peak located some distance below the nozzle outlet comes from the erosive property of the cloud cavitation. To explain these experimental results, two cavitation models are employed. The first is a simplified continuum model of a homogeneous two-phase flow, and the other is a spherical cloud model filled with the cavitation bubbles. The intermittent generation of the cavitating jets is simulated numerically by the first model, and the focusing effect of a spherical wave is computed by the second model. The second model reproduces the large impulsive pressure and the time lag between the pressure pulse and the cloud collapse. Some problems in the computational models are also identified by comparing them with the experimental results.

Keywords Cavitation cloud · Impulsive pressure · High-speed photography · Bubble dynamics · Numerical simulation

12.1 Introduction

Rayleigh [16] and Plesset [13] analyzed the collapse of a single spherical bubble in a liquid under hydrostatic pressure to explain the cavitation damage in solid materials. They regarded the liquid surrounding a bubble as an incompressible fluid. The liquid compressibility in the Rayleigh-Plesset model was considered by Gilmore [7],

---

K. Yamamoto (✉)
Waseda University, 3-4-1 Ohkubo Shinjuku-ku, Tokyo, Japan
e-mail: yamamott@waseda.jp

Y. Shibata and Y. Suzuki (eds.), *Mathematical Fluid Dynamics, Present and Future*, Springer Proceedings in Mathematics & Statistics 183,
DOI 10.1007/978-4-431-56457-7_12

Keller [9] and Prosperetti [15], among others. One of the most important problems in cavitation is determining the magnitude of the maximum pressure during the bubble collapse. However, the maximum pressure generated by a single bubble collapse is much smaller than the yield stress of common-use metals. Thus, the collapse of a single bubble cannot be considered as a predominant cause of cavitation erosion.

By the way, some researchers have predicted that a micro-jet may be formed, if a bubble collapse is asymmetric. Such micro-jet formation was shown by the numerical simulations of Plesset and Chapman in the case of a bubble collapse near a solid surface [14]. In addition, such asymmetric collapses and high-speed micro-jets have been observed by Kling [11], Lauterborn [12], and Tomita [22], among others. Today, the micro-jet may be considered as a dominant cause of cavitation erosion rather than a spherical bubble collapse.

It has been known over the last two decades that impulsive pressures due to the collapse of cavitation clouds cause cavitation noise and hard material damage rather than the impulsive pressures arising from individual cavitation bubbles [1, 3, 17]. Mathematical models of such cavitation clouds were firstly introduced by Brennen [6, 24] and Matsumoto [19]. Both models come from the two-phase flow model of Wijngaarden [23], that is, they considered the flow in the cloud as a homogeneous two-phase flow containing many cavitation bubbles as represented by the Rayleigh-Plesset equation. The main assumptions in the models are that the flow in the cloud is spherically symmetric, the small bubble radius is much smaller than the distance between each bubble and the distance is much smaller than the cloud radius. However, the validity of their models has still not been verified by experimental results. Recently, the bubble clouds were numerically analyzed by ETH Zurich and the IBM team [18]. They conducted three-dimensional direct simulation of the liquid flow containing 15,000 small bubbles using high-performance computing. Their two-phase flow model is almost free of the assumptions of Wijngaarden's model. The model shows the generation of micro-jets and impulsive pressure during the collapse of the bubble cloud. However, the results have also not been compared with experimental results.

Water jet technology is well known as a technique to use the damage potential of cavitation. Particularly, a high-speed submerged water jet is composed of a large number of cavitation clouds with high erosive properties [27]. Thus, the purpose of this study is to investigate the properties of cloud cavitation in a high-speed water jet by experimental observation and numerical simulation.

12.2 Experimental Observations

12.2.1 Experimental Apparatus

Figure 12.1 shows the experimental apparatus of the high-speed submerged water jet. The tap water was pressurized by a plunger pump with a maximum delivery pressure

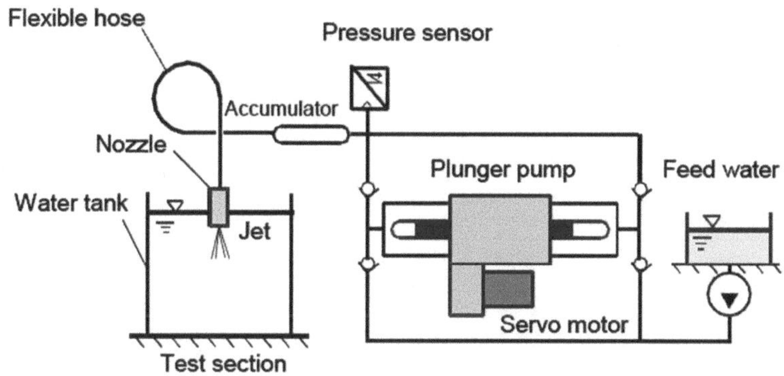

Fig. 12.1 Layout of high-speed submerged water jet system

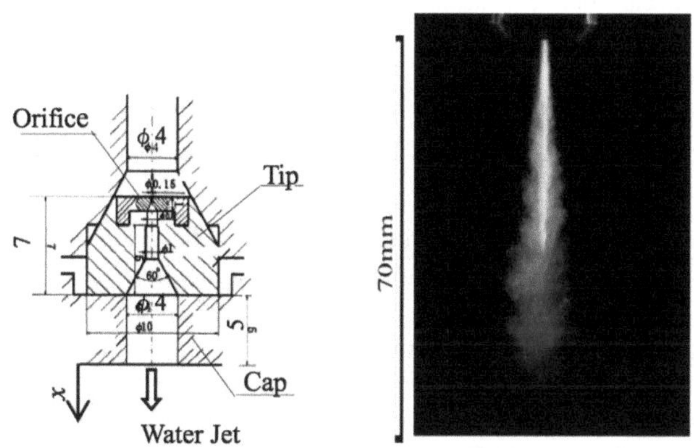

(a) Detail view of orifice nozzle ($d_n = 0.15$ mm)

(b) Snap shot of cavitating jet ($P_u = 100$ MPa, $P_d = 101.3$ kPa, and shutter speed of 1 ms)

Fig. 12.2 High-speed cavitating jet from orifice nozzle

of 200 MPa and a flow rate of 0.5 L/min. The pressurized water was injected vertically into the still water at atmospheric pressure through an orifice nozzle with an internal diameter d_n of 0.15 mm. The observation of the submerged jets and the erosion test of the aluminum specimen were carried out in the water tank. The upstream injection pressure P_u was varied from 30 to 200 MPa and the downstream pressure P_d was kept constant at atmospheric pressure.

Figure 12.2 shows the details of the orifice nozzle and a snapshot of the high-speed cavitating jet, which is the target of this study, taken with a usual shutter speed of 1 ms.

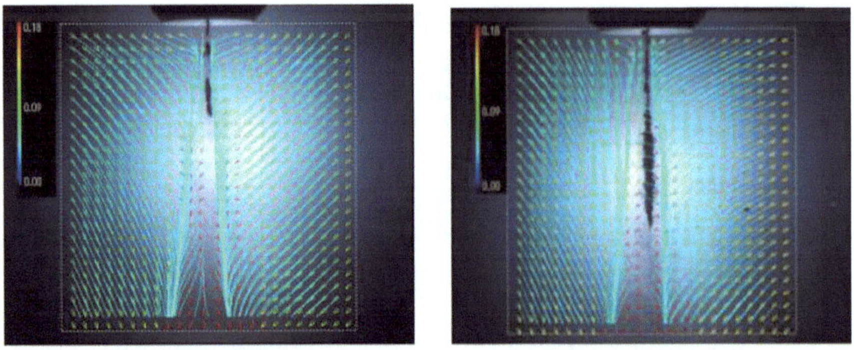

(a) $P_u = 50$ MPa (b) $P_u = 100$ MPa

Fig. 12.3 Flow structure of free cavitating jet ($P_d = 101.3$ kPa)

12.2.2 *Instantaneous Photograph of Cavitating Jet and Visualization of Flow Around the Jet*

To realize the structure of the cavitating jets, the flow field was observed using an instantaneous photograph and the particle image velocimetry (PIV) method. The instantaneous photograph of the cavitating jet was taken using a charge-coupled device (CCD) camera with an exposure time of 150 ns. Figure 12.3 illustrates the averaged flow field of the streamline and the velocity vector around the free cavitating jet over a period of 5 s. These were measured using the PIV method with injection pressures of $P_u = 50$ and 100 MPa.

The flow field is separated into two regions. The outer region has a low speed of 0.2 m/s maximum, whereas the inner region is the high-speed cavitating jet with a velocity of a few hundred meters per second. The outer region has an almost axisymmetric steady flow, whereas the inner region has a non-symmetric and intermittent jet flow.

12.2.3 *Observation of Unsteady Behavior of Bubble Cloud Using Shadowgraph Technique and Schlieren Method with High-Speed Video Camera*

To measure the unsteady behavior of the cavitating jets, a shadowgraph and a Schlieren photograph of the jets were taken for the distance $x = 30$–60 mm below the nozzle outlet and the injection pressure of $P_u = 50$, and 100 MPa. A high-speed video camera of a complementary metal oxide semiconductor (CMOS) was used. It had a maximum frame rate of 1.3 million frames second, but in this study, the rate was used at approximately a half million frames per a second to maintain the resolution

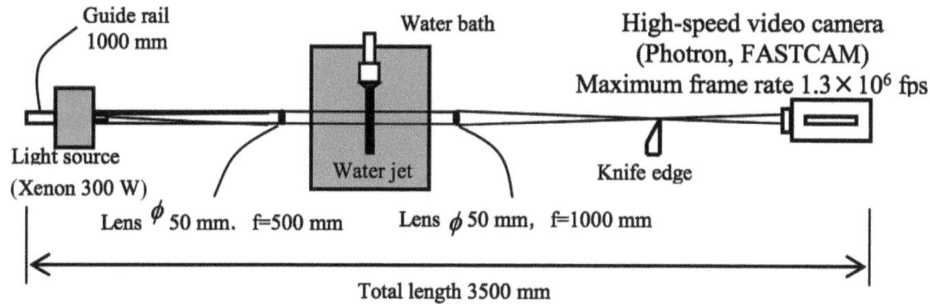

Fig. 12.4 Optical arrangement for a shadowgraph or Schlieren photograph of cavitation jet using high-speed video camera

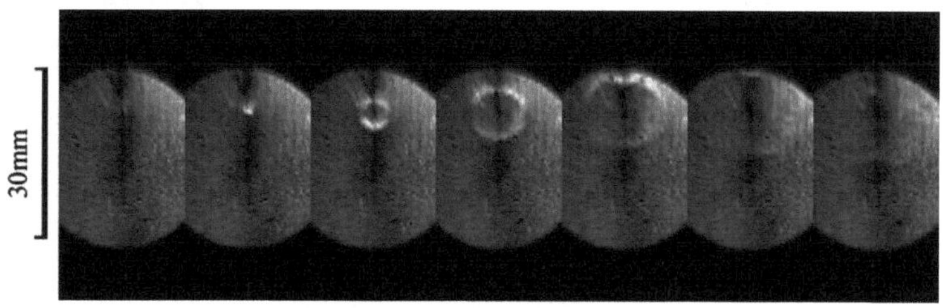

Fig. 12.5 Schlieren photograph of pressure pulse from the cavitation cloud (at frame rate of 4.65×10^5 fps, $x = 30 - 60$ mm, $P_u = 100$ MPa, $P_d = 101.3$ kPa)

of the picture (the time interval between the frames was approximately 2.2 μs). The optical arrangement for the Schlieren photograph is shown in Fig. 12.4.

The collapse of the bubble cloud and the severe pressure pulses were observed for both injection pressures. From the binary image of the bubble cloud picture, I calculated the spherical radius equivalent to the volume of the bubble cloud. Figure 12.5 shows a typical example of the Schlieren photograph of the pressure pulse for the injection pressure of 100 MPa. It shows the bright ring of the pressure wave spreading after the collapse of the bubble cloud. The Schlieren photograph is clearer than the shadowgraph and can be used for more precise measurements of the propagation speeds of the pressure pulses.

The time histories of the spherical waves and the radius change of the bubble cloud are shown in Fig. 12.6 for the injection pressure P_u = 50–200 MPa. It can be observed from Fig. 12.6b that the pressure pulses are generated 1–6 μs before the volume of the cloud reaches a minimum value.

The propagation speed of the first pressure pulses is in the range of 1000±200 m/s as shown in Fig. 12.7a. From this figure, it can be reconfirmed that the starting times are approximately 1–6 μs before the cloud collapse. Figure 12.7b indicates that the propagation speed decreases rapidly as the number of collapsing times increases. The void fraction increases after collapsing, due to small bubbles in the cavitation clouds dispersing into the water.

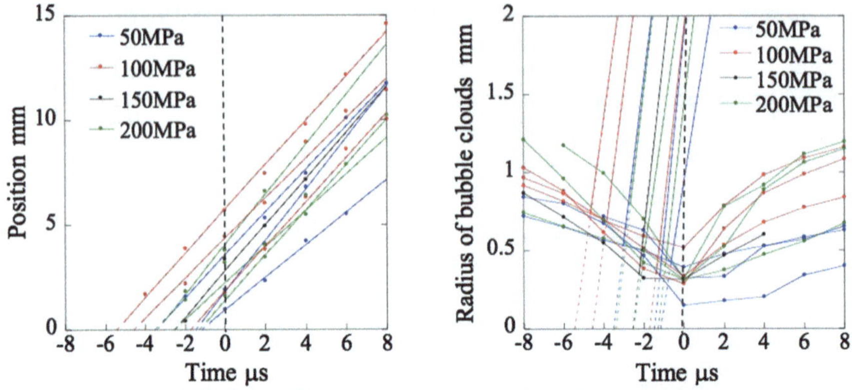

(a) Propagation of pressure pulse (b) Origin of pressure pulse

Fig. 12.6 Origin and propagation of pressure pulse with the first collapse of the cavitation clouds ($P_d = 101.3$ kPa)

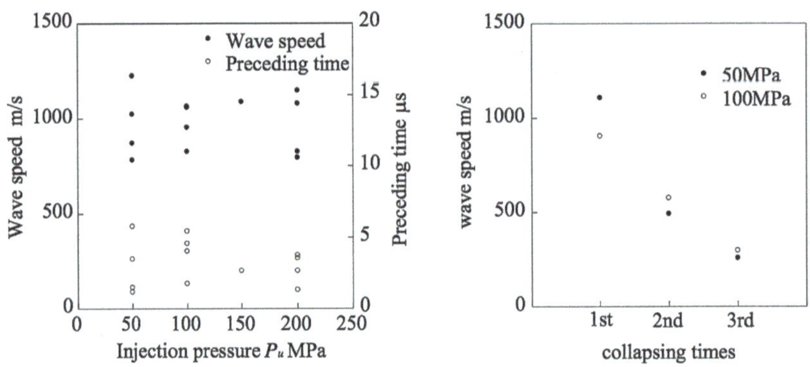

(a) First collapse and rebound (b) Decay of wave speed with collapsing times

Fig. 12.7 Propagation speed of pressure pulse and its preceding time before cloud collapse ($P_d = 101.3$ kPa)

Figure 12.8 presents the typical behaviors of the bubble cloud, which is separated from the cavitating jet. The time histories of the leading edge of the continuous cavitating jet, its velocity, the equivalent radius of the bubble cloud are shown in this figure. The equivalent radius was estimated from a binary image of the cloud which could be seen in the shadowgraph. From this measurement, we can see that the velocities of the leading edge of the cavitating jet are 100–150 m/s. The initial radii of the cavitation cloud are 1–3 mm, and the rebounding periods of the clouds are approximately 100–150 µs.

From the experimental observation until this section, the unsteady behavior of the free cavitation jets can be summarized as illustrated in Fig. 12.9.

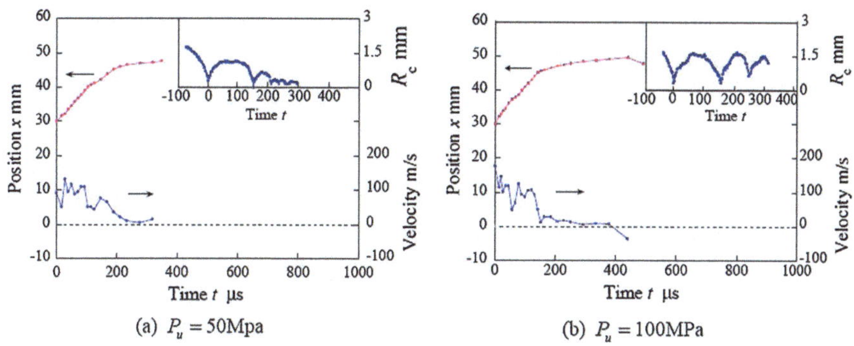

Fig. 12.8 Examples of time histories of jet length and cloud radius in cavitating jet (P_d = 101.3 kPa)

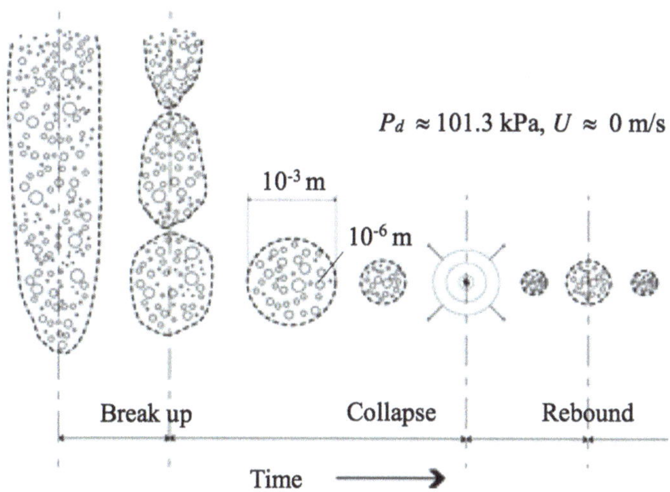

Fig. 12.9 Schematic diagram of unsteady behavior of free cavitating jet

Phase 1: A continuous cavitating jet from the orifice nozzle breaks up some bubble clouds of radii in the order of a few millimeters.

Phase 2: The bubble clouds are separated from the cavitating jet and collapsed by the downstream pressure. Then the impulsive pressure generated during the collapse of the cloud. Such pressure rises at a few micro seconds before the cloud reaches the minimum volume.

Phase 3: Each cloud disappears after a few rebounds.

12.2.4 Incidence Frequencies of Pressure Pulses Generated by Cavitating Jet

To measure the incident frequencies of the pressure pulses generated by the cavitating jet, a hydrophone (with a maximum frequency of 250 kHz) was positioned at $x = 40\,\text{mm}$ below the nozzle outlet and $y = 30\,\text{mm}$ from the jet axis as shown in Fig. 12.10a. Figure 12.10b represents a sample of the output signal of the hydrophone and it shows the strong impulsive pressure described in Sect. 12.2.3. The Fourier components of such signals were analyzed by fast Fourier transformation (FFT).

Figure 12.11 presents an example of the pressure pulse detected by the hydrophone. The average Fourier spectra analyzed by FFT for the changes of 125 samples are shown in Fig. 12.12. The peak frequency of the pressure pulses is 2.36 kHz for an

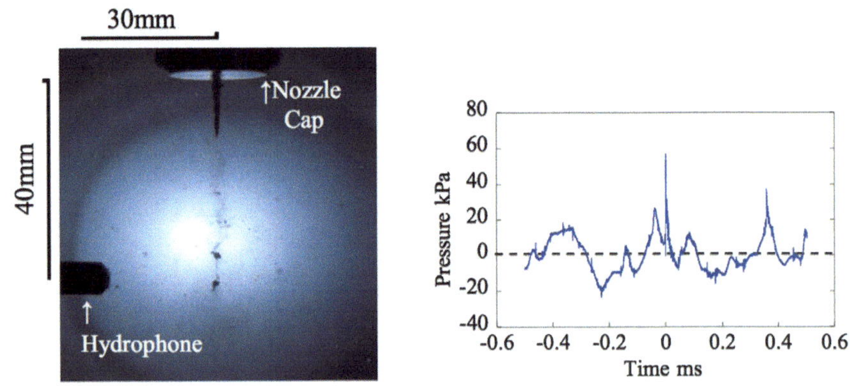

(a) Instantaneous photo of jet synchronized with pressure impulse

(b) Output signal of hydrophone

Fig. 12.10 Pressure pulses generated by collapse of cavitation clouds ($P_u = 50$ MPa, $P_d = 101.3$ kPa)

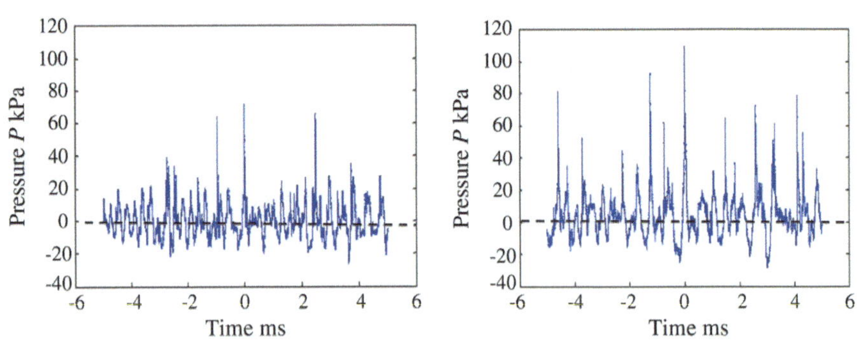

(a) $P_u = 50$ MPa

(b) $P_u = 100$ MPa

Fig. 12.11 Pressure pulse generated by collapse of cavitation clouds at $x = 40\,\text{mm}$, $y = 30\,\text{mm}$

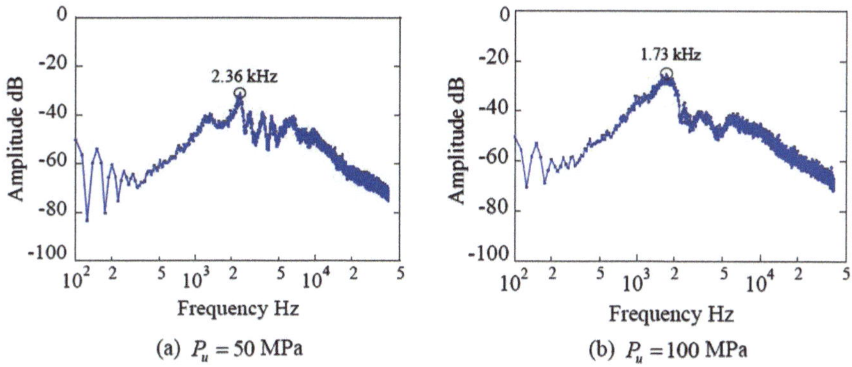

Fig. 12.12 Fourier spectra of pressure pulse from cavitating jet (average of 125 samples, $x =$ 40 mm, $y = 30$ mm)

injection pressure of 50 MPa, and 1.73 kHz for an injection pressure of 100 MPa. These incident frequencies of the pressure pulses are in the audible range of 1–10 kHz. Pressure pulses of this type generate cavitation noise. The peak frequency decreases as the injection pressure increases because the sound speed of the fluid around the jets decreases rapidly as the void fraction increases with the injection pressure.

12.2.5 Erosion Test of Aluminum Specimen Using Cavitating Jet

To examine the erosive properties of the cavitating jet, the erosion test was carried out using pure aluminum specimens. The water jet was directed perpendicularly at each specimen for 1200 s, and the mass loss and diameter of the damaged area were measured. The stand-off distance of the specimens x below the nozzle outlet varied between 0 and 70 mm. Typical damage patterns on the specimen surfaces for each stand-off distance when P_u = 100 MPa are shown in Fig. 12.13. In Fig. 12.13a, c, the mass losses of the specimens have maximum values, whereas in Fig. 12.13b, the damaged area reaches maximum, as described in the following.

Figure 12.14 shows the mass loss curves and the damaged area of the aluminum specimens at the defined stand-off distance (S.O.D.) with the injection pressure varying between 50 and 100 MPa. Figure 12.14a indicates that the mass loss curves have two peaks. The first peak occurs immediately below the nozzle outlet and the mass loss increases remarkably with the injection pressure. However, the stand-off distance of the second peak increases linearly with the injection pressure and the mass loss is slightly dependent on the injection pressure. This likely means that the damage for the second peak is significantly affected by cloud cavitation. Figure 12.15 shows the contour map of the damage characteristics with the injection pressure $P_u = 0–150$

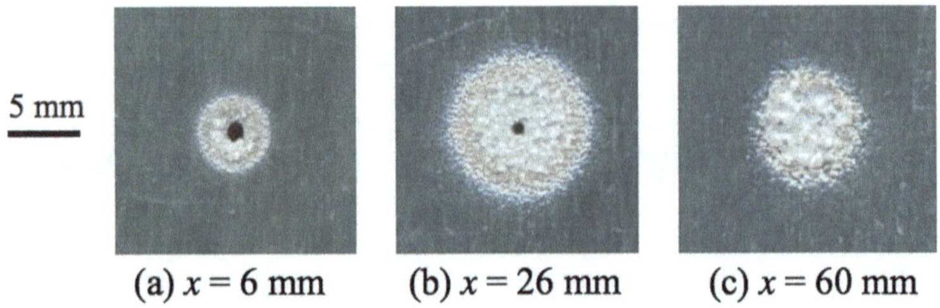

Fig. 12.13 Eroded surface of aluminum specimen (P_u = 100 MPa, P_d = 101.3 kPa)

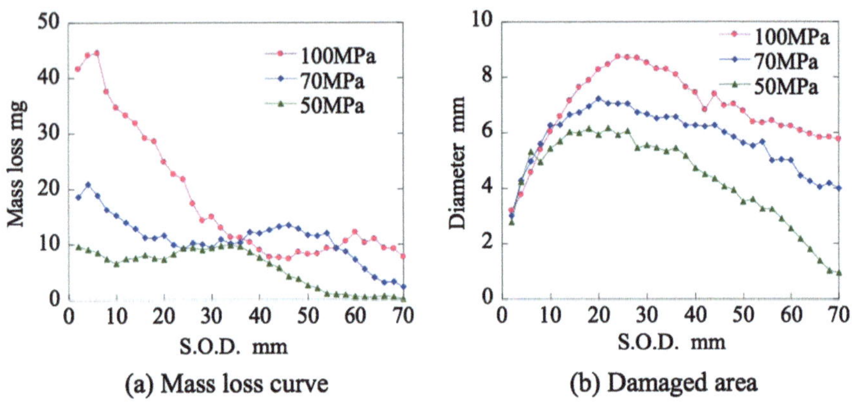

Fig. 12.14 Erosive properties of cavitating jet for aluminum specimen (S.O.D.: stand-off distance, P_d = 101.3 kPa)

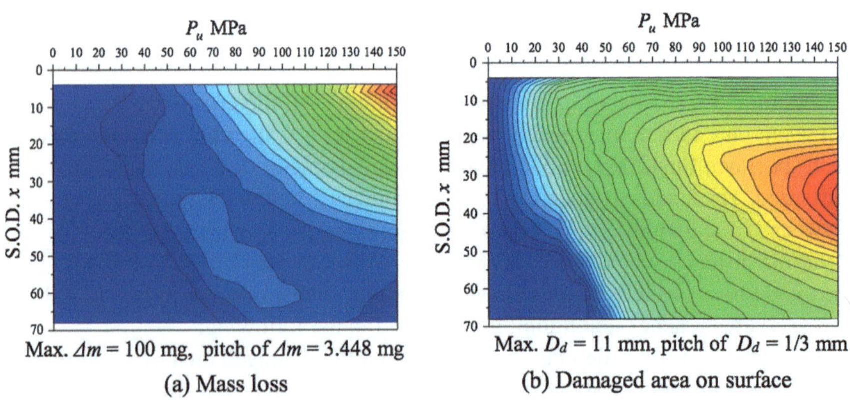

Fig. 12.15 Contour map of the damaged characteristics with injection pressure P_u and S.O.D. x (S.O.D.: stand-off distance, test peace: Pure aluminum, P_d = 101.3 kPa)

MPa and the stand-off distance $x = 0$–70 mm. The 2nd peak region in the mass loss curves can be seen in Fig. 12.15a. The mass loss in this region is mostly caused by the cloud cavitation as described before.

12.3 Numerical Analysis

12.3.1 Continuum Model of Homogeneous Two-Phase Flow of Cavitating Jet

The orifice nozzle in Fig. 12.2 has an internal diameter $d_n = 0.15$ mm, and a tip length of 7.0 mm. The internal diameter of the cap is 4.0 mm, and a length of the straight section is 5.0 mm. High-pressure water was injected from this nozzle into still water at atmospheric pressure to produce a high-speed submerged water jet. Thus the computational object should be a three-dimensional flow of the jet. To examine the separation of the cavitation cloud from the submerged water jet, a continuum model of the homogeneous two-phase (gas-liquid) flow was initially employed.

The governing equations of the flow are the compressible Navier-Stokes equations, written in the non-conservative form to apply the constrained interpolation profile (CIP) scheme [21, 26].

$$\frac{\partial \rho}{\partial t} + (\mathbf{u} \cdot \nabla)\rho = -\rho(\nabla \cdot \mathbf{u}) \tag{12.1}$$

$$\frac{\partial \mathbf{u}}{\partial t} + (\mathbf{u} \cdot \nabla)\mathbf{u} = -\frac{1}{\rho}\nabla p + \nu \left\{\nabla^2 \mathbf{u} + \frac{1}{3}\nabla(\nabla \cdot \mathbf{u})\right\} \tag{12.2}$$

where t is the time; $\mathbf{x} = (x, y, z)$ is the position; $\rho(\mathbf{x}, t)$ is the fluid density; $\mathbf{u}(\mathbf{x}, t) = (u, v, w)^T$ is the velocity; $p(\mathbf{x}, t)$ is the static pressure; ν is the kinematic viscosity; and ∇ is the nabla operator.

With the assumption that water is a barotoropic fluid, the modified Tait equation for water in the high-pressure region and the continuum mixture of the cavitating flow in the low-pressure region can be expressed as

$$p > P_v : \rho = \rho_0 \left(\frac{p + B}{P_0 + B}\right)^{1/N} \tag{12.3}$$

$$p = P_v : \rho = (1 - \alpha)\rho_{vl} + \alpha\rho_{vg} \approx (1 - \alpha)\rho_{vl} \tag{12.4}$$

where $B = 293.1$ MPa; $N = 7.15$; P_0 is the atmospheric pressure; ρ_0 is the density of water at atmospheric pressure; P_v is the vapor pressure of water at room temperature; ρ_{vl} is the density of liquid water at $p = P_v$; ρ_{vg} is the density of vaporous water at $p = P_v$; and α is the void fraction, which is determined by Eqs. (12.1) and (12.4).

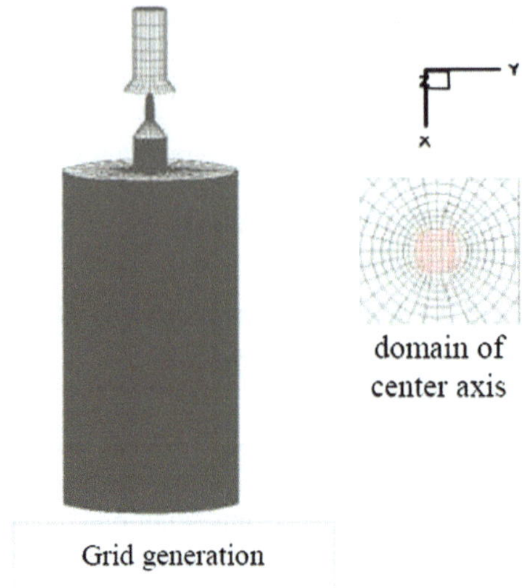

Fig. 12.16 Computational domain

Introducing the sound speed of water $a = \sqrt{dp/d\rho}$, Eq. (12.1) becomes

$$\frac{\partial p}{\partial t} + (\mathbf{u} \cdot \nabla)p = -\rho a^2(\nabla \cdot \mathbf{u}). \tag{12.5}$$

The initial conditions are approximated by the steady one-dimensional flow; that is, at $t = 0$:

$$\mathbf{u}(\mathbf{x}, 0) = (u(x), 0, 0)^T, \quad \text{and} \quad p(\mathbf{x}, 0) = p(x),$$

where $u(x)$ and $p(x)$ are the one-dimensional pipe flow in the steady state.

It is assumed that the metal specimen at the downstream end is a rigid wall, the boundary conditions are written as follows:

At $\mathbf{x} = \mathbf{x}_{\text{in}}$ (inflow side): $p(\mathbf{x}, t) = P_u$ (injection pressure), $v = 0$ and $w = 0$.

At $\mathbf{x} = \mathbf{x}_{\text{solid}}$ (solid surface of the nozzle and specimen): $\mathbf{u}(\mathbf{x}, t) = \mathbf{0}$.

At $\mathbf{x} = \mathbf{x}_{\text{out}}$ (outflow side): $p(\mathbf{x}, t) = P_d$.

Figure 12.16 shows a grid generation method for the computational domain including the nozzle. The number of grids is 3×10^5. The CIP scheme of the finite difference method [21, 26] was used to maintain the numerical stability for a steep change in the two-phase flow. The time increment of the calculation was $\Delta t \approx 2.0$ ns, and the Courant number was $C_n \approx 0.1$.

12.3.2 Computational Results of Cavitating Jet

When the injection pressure upstream of the nozzle $P_u = 50$ MPa, the downstream pressure $P_d = P_0 = 101.3$ kPa (atmospheric pressure), and the minimum pressure $P_v = 2.3$ kPa (vapor pressure of water), the computed density distribution on the vertical cross-section, including the center axis of the flow field, is as illustrated in Fig. 12.17. In this figure, the cavitation cloud can be regarded as a low-density fluid mass with void fraction $\alpha = 1 - (\rho/\rho_{vl})$, which is greater than $1.0 \times 10^{-3} (0.1\,\%)$. From this figure, we can confirm that the cavitating jet is a non-symmetric three-dimensional flow and that the cavitation clouds are intermittently generated by the submerged water jet.

The fluctuation of the flow rate at the nozzle throat is shown in Fig. 12.18, where $C_d = (Q/A_n)/\sqrt{2(P_u - P_d)/\rho_l}$ is the flow coefficient. The measured value of the flow coefficient for steady flow is 0.66. Thus the computational result is proper. The frequency of the fluctuation is 32 kHz, but this value should not be compared with the experimental results of the pressure pulses described in Sect. 12.2.4, because the flow fluctuation is different from the collapse of cavitation clouds.

The pressure and velocity fluctuations of the free jet and the pressure of the impinging jet on the rigid wall, as obtained from the numerical simulation, are shown in Fig. 12.19. The characteristics of the cavitating jet are as follows: At $x = 20$ mm and $y = 0$ mm, the pressure change is between $p = 2.3$ kPa and 1000 kPa, the frequency range is between $f_p = 10$ kHz and 30 kHz, and the velocity range is between $u = 150$ m/s and 250 m/s. In addition, the pressure range of the impinging jet is between p

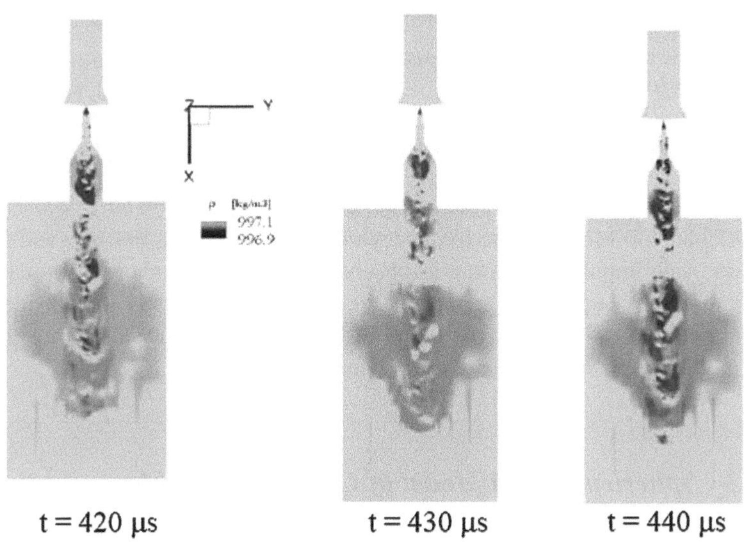

Fig. 12.17 Distribution of void fraction in non-symmetric three-dimensional non-symmetric flow of cavitating jet ($P_u = 50$ MPa, $P_d = 101.3$ kPa and $P_v = 2.3$ kPa)

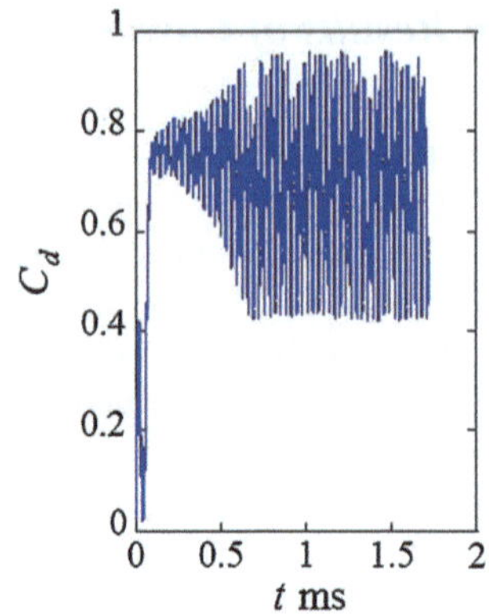

Fig. 12.18 Fluctuation of flow rate at the nozzle throat

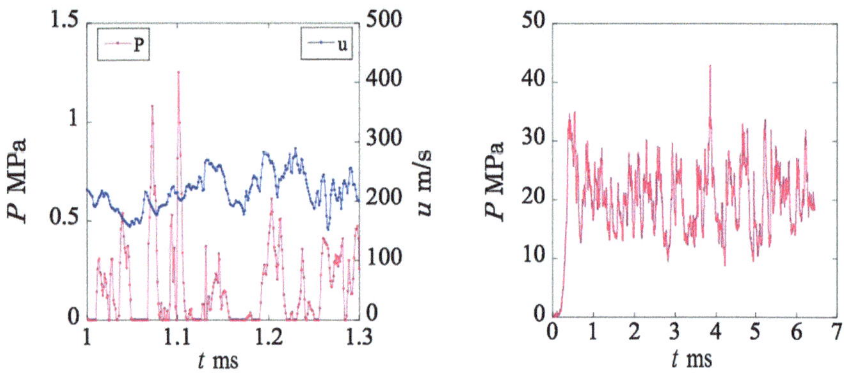

(a) Pressure and velocity at $x = 20$ mm and $y = 0$ mm

(b) Impinging pressure on rigid wall at $x = 35$ mm and $y = 0$ mm

Fig. 12.19 Computational results of unsteady free jet and impinging jet on rigid wall

= 10 MPa and 30 MPa, and the frequency range is between f_p = 1 kHz and 5 kHz. However, such impulsive pressure as observed in Fig. 12.5 cannot be simulated by this computation, because the motion of the cavitation bubbles is not considered in the continuum model of the cavitating flow. I therefore introduce a spherical bubble cloud model for the impulsive pressure in the next section.

12.3.3 Spherical Cloud Model of Cavitation Bubbles

To examine the impulsive pressure produced by the collapse of the cavitation cloud described in Sects. 12.2.3 and 12.2.4, I hereby introduce the simplified cavitation

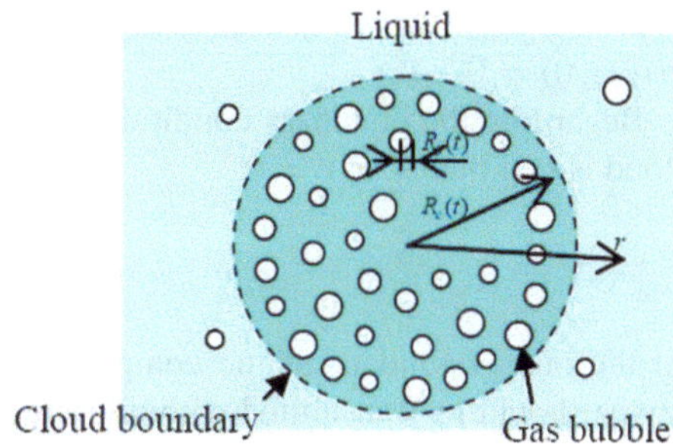

Fig. 12.20 Spherical cloud model of cavitation bubbles

model, which is a spherical cloud model of the cavitation bubble that takes into account the many small bubble motions as shown in Fig. 12.20 [6, 19, 20, 24, 25].

Assuming that the flow field around the cloud is spherical in symmetry and that the fluid is compressible and inviscid, the density ρ of the continuum mixture for a void fraction α is given by

$$\rho = (1-\alpha)\rho_l + \alpha\rho_g \approx (1-\alpha)\rho_l \tag{12.6}$$

where the density ρ_l of water is given by the modified Tait equation in the high-pressure region.

The conservation laws of the spherical flow are expressed by the non-conservative form [4, 5], and the CIP scheme is used to obtain the numerical solution.

$$\frac{\partial \rho}{\partial t} + u\frac{\partial \rho}{\partial r} = -\rho\left(\frac{\partial u}{\partial r} + \frac{2}{r}u\right) \tag{12.7}$$

$$\frac{\partial u}{\partial t} + u\frac{\partial u}{\partial r} = -\frac{1}{\rho}\frac{\partial p}{\partial r} \tag{12.8}$$

where t is the time, r is the distance from the center of the bubble cloud; $u(r,t)$ is the radial velocity of the flow field; and $p(r,t)$ is the pressure of the flow field. Eliminating ρ and ρ_l from Eqs. (12.3), (12.6), and (12.7), the unknown variables are $u(r,t)$, $p(r,t)$, and $\alpha(r,t)$. An equation relating these variables is therefore required, which would be the constitutive equation.

In Fig. 12.20, the bubble cloud of radius R_{c0} and void fraction α_0 at a low pressure P_{l0} is abruptly released into still water with void fraction α_1 at atmospheric pressure P_0 (=101.3 kPa). If the pressure discontinuity $P_0 - P_{l0}$ at the boundary $r = r_{end}$ is removed instantaneously, the spherical compression wave begins to enter into the cloud and a rarefaction wave simultaneously spreads to the outer region. This situation is represented by the following initial conditions at $t = 0$:

At $0 \leqslant r \leqslant r_{end}$ (in the computational domain): $u(r,0) = 0$.

At $0 \leqslant r < R_{c0}$ (inside the cloud): $p(r,0) = P_{l0}$ and $\alpha(r,0) = \alpha_0$.

At $R_{c0} \leqslant r < r_{end}$ (outside the cloud): $p(r,0) = P_{l0}$ and $\alpha(r,0) = \alpha_1(< \alpha_0)$.

At $r = r_{end}$ (at the boundary of the computational domain): $p(r_{end}, 0) = P_0$ and $\alpha(r_{end}, 0) = 0$.

Regarding the boundary conditions ($t > 0$), the fluid at the center of the bubble cloud is still that is, at $r = 0$:

$$u(0, t) = 0 \tag{12.9}$$

At the outer boundary of the computational domain, $r = r_{end}$, all the expanding waves should be transmitted without reflection. The Rayleigh-Plesset-Keller equation [9, 13, 16] may be used to represent this condition, because it is an approximate high-order integration between infinity and any concentric spherical surface of a compressible inviscid fluid. Because the movement of a fluid particle at $r = r_{end}$ is very small, thus we can assume that $r_{end} \approx R_{end}$, and $u(r_{end}, t) \approx u(R_{end}, t) = dR_{end}/dt = u_{end}(t)$. This specifies a constraint condition between the unknown variables u_{end} and p_{end} at $r = r_{end}$;

$$\begin{aligned} &r_{end}\left(1 - \frac{u_{end}}{c_l}\right)\frac{du_{end}}{dt} + \frac{3}{2}\left(1 - \frac{u_{end}}{3c_l}\right)u_{end}^2 \\ &= \frac{1}{\rho_l}\left(1 + \frac{u_{end}}{c_l}\right)(P_{end} - P_0) + \frac{r_{end}}{\rho_l c_l}\frac{dp_{end}}{dt} \end{aligned} \tag{12.10}$$

In this model, the bubble cloud is distinguished from the surrounding fluid, and the fluid particle in the flow has either a high or low void fraction in the initial condition. The relation between p and α is determined by the oscillation of the bubbles.

I assume that the frequencies of the bubbles are sufficiently high compared to the fluctuation of the flow field. The surface tension at the liquid/gas interface, and the viscosity of the liquid are considered. Moreover, the bubble radius is very small. Using the small bubble radius $R_b(r, t)$, and the velocity $U_b = DR_b/Dt$ in which $D/Dt = \partial/\partial t + u\partial/\partial r$ is the Lagrangian derivative and taking into account the compressibility of the surrounding liquid, the Rayleigh-Plesset-Keller equation [9, 15] can be used again to determine the motion of individual bubble.

$$\begin{aligned} &R_b\left(1 - \frac{U_b}{c_l}\right)\frac{DU_b}{Dt} + \frac{3}{2}\left(1 - \frac{U_b}{3c_l}\right)U_b^2 \\ &= \frac{1}{\rho_l}\left(1 + \frac{U_b}{c_l} + \frac{R_b}{c_l}\frac{D}{Dt}\right)\left(p_B - p_\infty - \frac{4\mu U_b}{R_b} - \frac{2\sigma_T}{R_b}\right) \end{aligned} \tag{12.11}$$

We suppose that the distance between any two bubbles is very large compared with their radii, although there are many bubbles in the computational cell of the homogeneous two-phase flow. The pressure at infinity can therefore be considered equal to the pressure of the two-phase flow namely, $p_\infty = p(r, t)$. Moreover, the pressure p_B inside a small bubble is the sum of the pressure p_g of the non-condensable gas and the vapor pressure P_v of water. The non-condensable gas in each bubble has a constant mass and its pressure change is isothermal; that is,

$$p_B = p_g + P_v, \quad p_g = p_{g0}\left(\frac{R_{b0}}{R_b}\right)^3 \tag{12.12}$$

The small bubbles are frozen to the flowing fluid, and if the density number of the bubbles is represented by n_b m^{-3}, the conservation of the bubble number can be expressed as

$$\frac{\partial n_b}{\partial t} + u\frac{\partial n_b}{\partial r} = -n_b\left(\frac{\partial u}{\partial r} + \frac{2}{r}u\right) \tag{12.13}$$

In this model, the void fraction α of the continuum mixture can be expressed in terms of the distribution of the small bubble nuclei size [20, 25], using the radius of the bubble $R_{b,k}$ and the density number of the bubble $n_{b,k}$

$$\alpha = \sum_k n_{b,k}\frac{4}{3}\pi R_{b,k}^3 \tag{12.14}$$

where k = 1–5. It can be seen that the set of Eqs. (12.11)–(12.14) is a differential constitutive relation between p and α for the spherical cloud model. In Eqs. (12.12) and (12.14), p_g, R_b, and α have no time lag, but in Eq. (12.11), a time lag arises between $p(r, t)$ and p_g and then between $p(r, t)$ and $\alpha(r, t)$. These time lags are induced by the mass of the liquid surrounding the small bubble, which is known as the added mass effect.

Equation (12.11) is integrated by the fourth-order Runge-Kutta scheme, and the system of Eqs. (12.7) and (12.8) is solved by the CIP scheme as explained previously [21, 26].

12.3.4 Using Parameters for Flow Simulation of Cloud Cavitation

Table 12.1 lists the numerical data of the physical property, the initial and boundary conditions, and the computational constants for the flow simulation of cloud cavitation. In this table, the physical properties of the fluids are standard ones. As for the constants of the initial and boundary conditions, the initial radius of the bubble cloud R_{C0} is determined by considering the measured values described in Sect. 12.2.3. The initial pressures in the computational domain and in the small bubbles $P_l = p_{g0}$ is low pressure in the cavitating flow. The initial void fraction $\alpha_0(0 \leqslant r \leqslant R_{C0})$ and $\alpha_1(R_{C0} \leqslant r_{end})$ and the initial radii of the small bubbles R_{b0} are determined by the references [20, 25]. Then the initial bubble density numbers is between $n_{b,k} = 5 \times 10^7$ and 1×10^8m^{-3}. The mesh sizes of space increment Δr and time increment $\Delta t = C_n(\Delta r/c_{l0})$ are used to solve Eqs. (12.7) and (12.8). The

Table 12.1 Numerical data for flow simulation of cavitation cloud

Quantities	Numerical values
Density of water: ρ_{l0}	997.1 kg/m^3
Sound speed of water: c_{l0}	1496 m/s
Viscosity of water: μ_l	891 μPa $\cdot$ s
Surface tension of water/air: σ_T	72.74×10^{-3} Nm
Vapor pressure of water: P_v	2.3 kPa
Atmospheric pressure: P_0	101.3 kPa
Downstream pressure: $P_d (= P_0)$	101.3 kPa
Initial pressure: $P_{l0} = p_{g0}(0 \leqslant r \leqslant r_{end})$	10 kPa
Initial radius of the bubble cloud: R_{C0}	3 mm
Initial void fraction: $\alpha_0(0 \leqslant r < R_{C0})$	1.0×10^{-3}
$\alpha_1(R_{C0} \leqslant r < r_{end})$	5.5×10^{-4}
Initial radii of small bubbles: R_{b0}	60, 70, 80, 90 and 100 μm
Radius of computational domain: r_{end}	6 mm
Space increment: Δr	3 μm
Time increment: Δt	$\approx$2 ns ($C_n = 0.5$)

parameter Cn is Courant number. To solve Eq. (12.11), the time increment is set to 1/10 of Δt.

12.3.5 Results and Discussion on Pressure Pulse in Cavitation Cloud

First, it must be emphasized that in my model, the impulsive pressure is generated by the focusing effect of the spherical wave even in the case of pure water without bubbles. Figure 12.21 presents a computational result in the case of water that does not contain any small bubbles in the domain. The spherical pressure wave converges to the center $r = 0$ at $t = 4\mu$s and the atmospheric pressure becomes a pressure impulse of approximately 10 MPa at $r = 3\Delta r = 9\mu$m near the center. In addition, it is confirmed that there is no flection wave at the outer boundary $r = r_{end}$ from Eq. (12.9).

In the case of the cavitation cloud which is specified in Table 12.1, the computational result is shown in Fig. 12.22. This figure shows that the spherical wave converges to the center $r = 0$ at $t = 31.4$ μs and that the atmospheric pressure reaches a pressure impulse of approximately 176 MPa at $r = 3\Delta r = 9$ μm.

Figure 12.23 shows the propagation speeds of the pressure waves in the computational domain. The speed of the inward pressure wave outside the cloud ($R_C < r < r_{end}$) is reduced by the elastic effect of the small bubbles. However, the bubbles cause not only a time lag of the changes between the pressure and the void fraction, but also very large pressure owing to the effect of added mass of the water surrounding a bubble. The speed of the outward pressure wave inside the cloud($0 < r < R_C$) becomes $c \approx 1400$ m/s because the mixture is at a high pressure and almost of the same order as pure water. Outside the cloud the speed becomes $c \approx 1100$ m/s, which coincides with the experimental value, as shown in Fig. 12.7a.

The propagation of the spherical waves in the cavitation cloud is illustrated in Fig. 12.24. The left side is a void fraction, and the right side is a pressure distribution. The parameters are those given in Table 12.1.

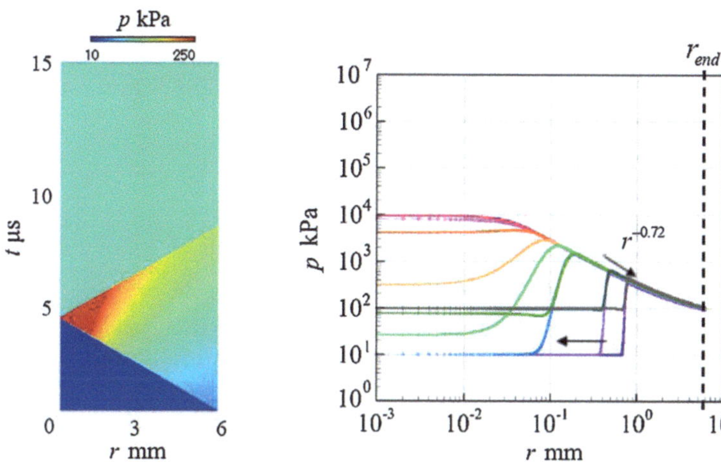

(a) Wave diagram on r,t-plane (b) Pressure profile

Fig. 12.21 Generation of impulsive pressure in pure water ($P_d = 101.3$ kPa, $P_{l0} = 10$ kPa, $\alpha_0 = 0$)

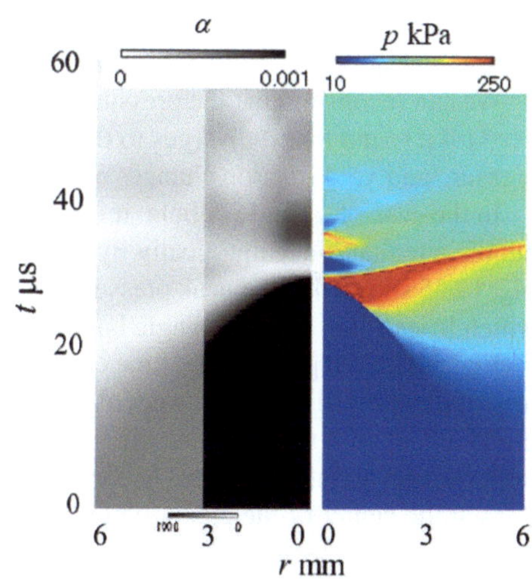

Fig. 12.22 Wave diagram on r, t—plane (*Parameters* see Table 12.1)

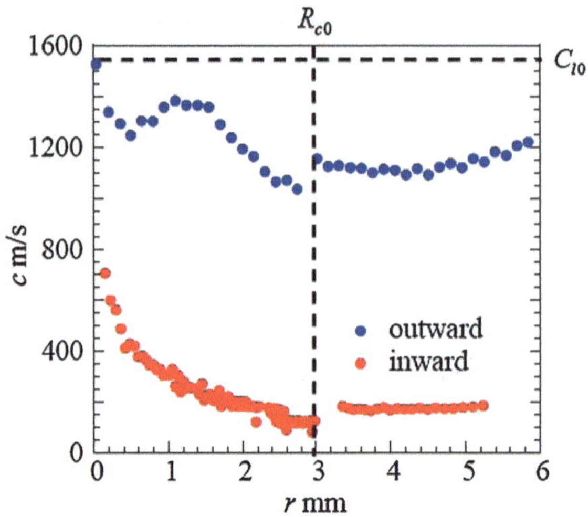

Fig. 12.23 Propagation speed of spherical wave in bubbly flow

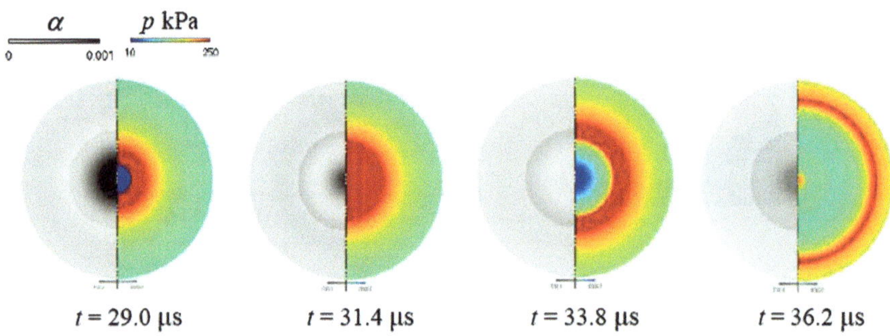

Fig. 12.24 Computational results of cavitation cloud (parameters are given in Table 12.1)

We can reconfirm that the pressure wave converges to the center of cloud at time t = 31.4 μs, and that it diverges to the outside of the cloud. A time lag between the pressure and void fraction changes can be seen at the center of the cloud.

In the case of a single bubble, it is known that the peak pressure surrounding the bubble attenuates asymptotically by r^{-1} [2, 8]. In the case of the cavitation cloud, Fig. 12.25 shows that the peak pressure attenuates at a rate between $r^{-0.97}$ and $r^{-0.63}$, which is less than that of a single bubble.

Figure 12.26 represents the time changes of the pressure p and the void fraction α at $r = 9$ μm which is near the center of the cloud. As described previously, the spherical cloud model with the small bubbles causes the time lag between two changes of the void fraction and the pressure. As shown in Fig. 12.26, the time lag is approximately 2 μs at the center of the cloud, which is within the experimentally measured range of 1–6 μs, as presented in Fig. 12.7a.

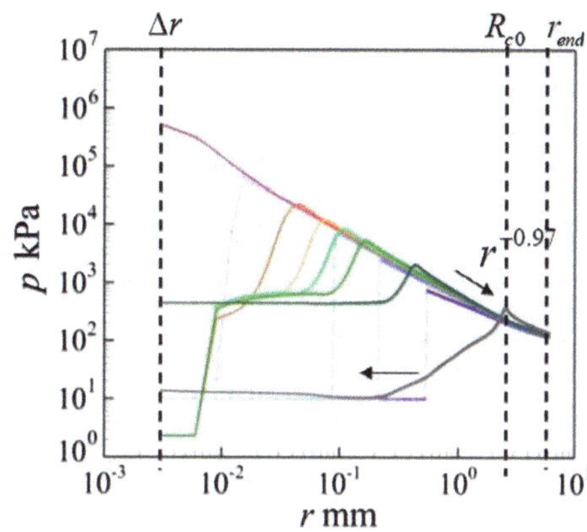

Fig. 12.25 Changes in pressure profile in the cavitation cloud

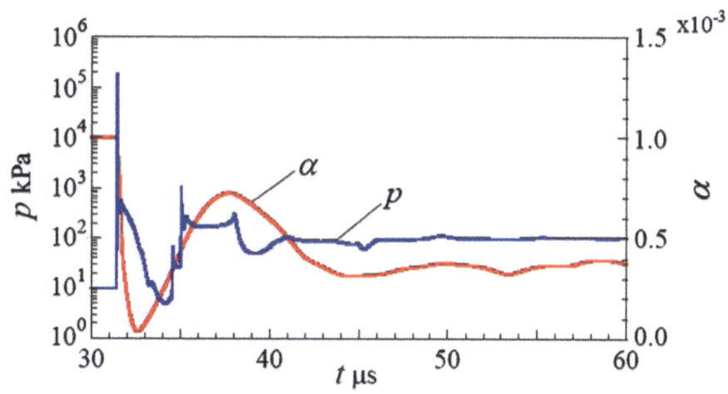

Fig. 12.26 Time histories of pressure and void fraction near the center of the cloud

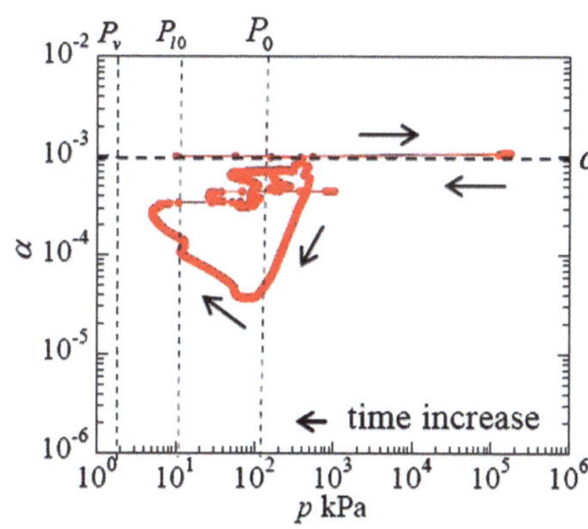

Fig. 12.27 Void fraction versus pressure diagram near the center of the cloud

Figure 12.27 displays the $p - \alpha$ curve at $r = 9\mu\text{m}$. It can be observed that the pressure reaches the maximum value before the void fraction decreases to the minimum value. This time lag is caused by the acceleration terms in Eq. (12.11) and is the so-called added mass effect.

The time changes of the pressure and radius of the small bubble at the center of the cloud are shown in Fig. 12.28. The period of the bubble oscillation is approximately

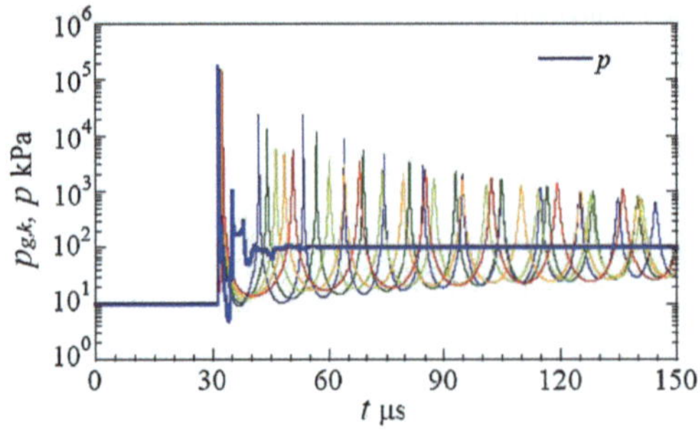

(a) Pressures in bubbles and flow field

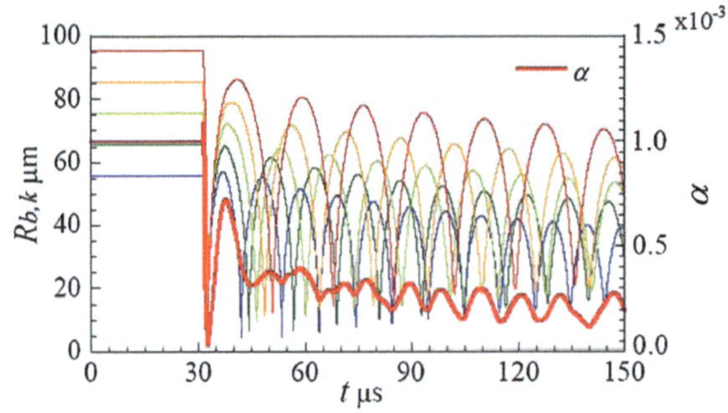

(b) Bubble radius and void fraction

Fig. 12.28 Time histories of pressure of small bubbles and bubble radiuses at the center of the cloud

Table 12.2 Comparison of calculated pressure pulse with measured values

y mm	P_{cal}kPa	P_{exp}kPa
3	250	---
6	120	---
30	24	60
90	8	25

10–18 μs, which is almost equal to the natural oscillation of each bubble under atmospheric pressure. It can be seen from Fig. 12.28b that the period of rebounding of the cloud is approximately 12 μs, which is nearly equal to that of the natural oscillation described previously. However, this rebounding period is too short compared with the measured value of approximately 100–150 μs. Therefore it is necessary to revise the spherical cloud model under consideration to simulate the rebounding of the bubble cloud.

Table 12.2 gives the peak value of the outward pressure pulse from the cloud. In this table, P_{cal} denotes the calculated peak pressures and P_{exp} denotes the measured values. The calculated pressures P_{cal} at $y = 30$ and 90 mm are extrapolated from the inside of the cloud by the approximate theory of Kirkwood-Bethe [10], which is $r(h + u^2/2) =$ constant along an outward wave $dr/dt = u + c$, where $h = \int_{p_0}^{p} (1/\rho)dp$.

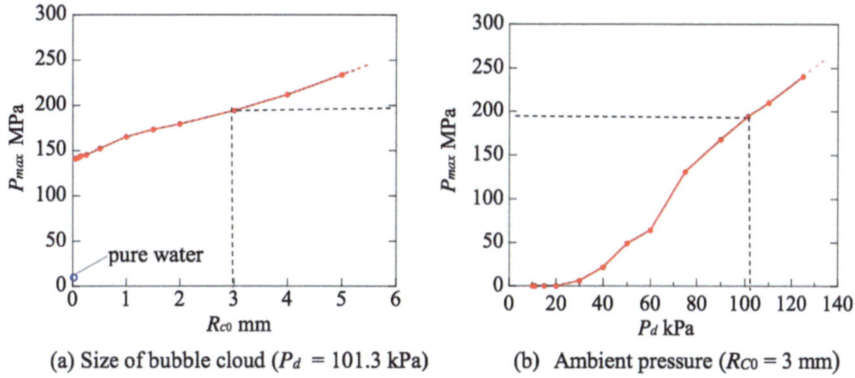

Fig. 12.29 Influence of the cloud size and downstream pressure on the impulsive pressure ($P_{l0} = 10$ kPa, $\alpha_0 = 1.0 \times 10^{-3}$)

This table indicates that the calculated pressure pulse P_{cal} is much lower than the measured values P_{exp}. Therefore the cloud model should be revised again.

12.3.6 Influence of Cloud Size and Downstream Pressure on the Impulsive Pressure

Up to this section, the behavior of a bubble cloud has been simulated for a cloud of one size only. However, there are many sizes of cavitation clouds in practice. Figure 12.29a shows that the impulsive pressure increases as the cloud size becomes large. In an application with cavitating jets, it is very important to know the influence of the downstream pressure on the impulsive pressure. Figure 12.29b shows that the impulsive pressure increases as the downstream pressure increases. The model of the bubble cloud does not explicitly include injection pressure. The collapse of the cavitation cloud is a phenomenon that is strongly dependent on the downstream pressure more than the injection pressure.

12.3.7 Concluding Remarks and Further Outlook

Finally, this work can be concluded with a few remarks and the outlook for the future.

1. It should be stressed that a high-speed video camera with a grade of a million fps at a flame rate is a very powerful tool for the study of the unsteady behavior of bubble clouds in a cavitating jet.
2. I simulated numerically the shedding of bubble clouds from a cavitating jet and the large impulsive pressure in a bubble cloud by using two computational models.

The results show that it is possible for cavitation clouds to severely damage a material's surface.

3. Some problems of the two mathematical models have also been identified by my experiments. The calculated rebounding period of the cloud is too short compared with the measured values, and the pressure pulse is much lower than the measured values.

Consequently to obtain more realistic numerical results, some breakthrough improvements will be required for the current mathematical models of cavitation.

Acknowledgements This research was partially supported by the JSP Grant-in-Aid 23560269, JST-CREST, and the Waseda University Grant for SR 2010A-606. The author wishes to thank the Professors Emeritus of Waseda University, Eisuke Outa and the late Kiyohiro Tajima, for their steadfast encouragements of this work. Most of the research was completed with help from Oka, H., Kitajima, H., Simomura, K., Yoshida, A., Yoshida, T., and Iida, H., former graduate students at Waseda University. The author thanks them for their experimental and computational contributions.

References

1. Brennen, C.E.: Cavitation and bubble dynamics, pp. 82–91. Oxford University Press (1995)
2. Brennen, C.E.: Ibid, pp. 67–68
3. Brujan, E.A., Ikeda, T., Yoshinaka, K., Matsumoto, Y.: The final stage of the collapse of a cloud of bubbles close to a rigid boundary. Ultrason. Sonochemistry **8**, 59–64 (2011)
4. Cole, R.H.: Underwater explosions, pp.27–28. Dover Pub. Inc. (1948)
5. Courant, R., Friedrichs, K.O.: Supresonic flow and shock waves, pp. 416–418. Interscience Pub. Inc. (1948)
6. Gilmore, F.R.: The collapse and growth of a spherical bubble in a viscous compressible liquid. California Institute of Technology. Hydrodynamics Lab. Report. No. 26-4(1952)
7. D'agosina, L., Brennen, C.E.: On the acoustical dynamics of bubble clouds. ASME Cavitation and Multiphase Flow Forum, 72–75 (1983)
8. Hickling, R., Plesset, M.S.: Collapse and rebound of a spherical bubble oscillations. Phys. Fluids **7**, 7–14 (1964)
9. Keller, J.B., Kolodner, I.I.: Damping of underwater explosion bubble oscillations. J. Appl. Phys. **27**, 1152–1161 (1956)
10. Kirkwood, J.G., Bathe, H.A.: The pressure wave produced by an underwater explosion. O.S.R.D. Report 588 (1942)
11. Kling, C.L., Hammitt, F.G.: A photographic study of spark-induced cavitation bubble collapse, J. Basic Eng. Trans. ASME Ser. D, **94**-4, 825–833 (1972)
12. Lauterborn, W., Bolle, H.: Experimental investigations of cavitation-bubble collapse in the neighbourhood of a solid boundary. J. Fluid Mech. **72**, 391–399 (1975)
13. Plesset, M.S.: The dynamics of cavitation bubbles. J. Appl. Mech. Trans. ASME **16**, 228–231 (1949)
14. Plesset, M.S., Chapman, R.B.: Collapse of an initially spherical vapour cavity in the neighbourhood of a solid boundary. J. Fluid Mech. **47**, 283–290 (1971)
15. Prosperetti, A.: The equation of bubble dynamics in a compressible liquid. Phys. Fluids **30**, 3626–3628 (1987)
16. Rayleigh, L.: On the pressure developed in a liquid during the collapse of a spherical cavity. Phil. Mag. **34**, 94–98 (1917)
17. Reisman, G.E., Wang, Y.C., Brennen, C.E.: Observations of shock waves in cloud cavitation. J. Fluid Mech. **355**, 255–283 (1998)

18. Rossinelli, D. et al.: 11PFLOP/s simulations of cloud cavitation collapse. In: Proceedings SC13: International Conference for High Performance Computing, Networking, Storage and Analysis, Article No. 3 (2013)
19. Shimada, M., Matsumoto, Y., Kobayashi, T.: Dynamics of the cloud cavitation and cavitation erosion. Trans. JSME (B) **65**(634), 1934–1941 (1999, in Japanese)
20. Shimada, M., Matsumoto, Y., Kobayashi, T.: Influence of the nuclei size distribution on the collapsing behavior of the cloud cavitation. Trans. JSME (B) **65**(637), 3002–3007 (1999, in Japanese)
21. Takewaki, H., Nishiguti, A., Yabe, T.: The cubic-interpolated pseudo-particle (CIP) method for solving hyperbolic-type equations. J. Comput. Phys. **61**, 261–268 (1985)
22. Tomita, Y., Shima, A.: On the behavior of a spherical bubble and the impulse pressure in a viscous compressible liquid. Bull. JSME **20**, 1453–1460 (1977)
23. van Wijngaarden, L.: On the equations of motion for mixtures of liquid and gas bubbles. J. Fluid Mech. **33**, 465–474 (1698)
24. Wan, Y.C., Brennen, C.E.: Numerical computation of shock waves in a spherical cloud of cavitation bubbles. Trans. ASME J. Fluid Eng. **121**, 872–880 (1999)
25. Wan, Y.C.: Effect of nuclei size distribution on the dynamics of a spherical cloud of cavitation bubbles. Trans. ASME J. Fluid Eng. **121**, 881–886 (1999)
26. Yabe, T., Xiao, F., Utumi, T.: Constrained interpolation profile method for multiphase analysis. J. Comput. Phys. **169**, 556–593 (2001)
27. Yoshida, T., Iida, H., Yoshida, A., Yamamoto, K.: High speed observation of unsteady behavior of bubble clouds split from cavitating jet. In: Proceedings 21st International Conference on Water Jetting, BHRG, 207–317 (2012)

Part II
Other Related Topics

Chapter 13
Weak Solutions to Problems Involving Inviscid Fluids

Eduard Feireisl

Abstract We consider an abstract functional-differential equation derived from the pressureless Euler system with variable coefficients that includes several systems of partial differential equations arising in the fluid mechanics. Using the method of convex integration we show the existence of infinitely many weak solutions for prescribed initial data and kinetic energy.

Keywords Euler system · Weak solution · Convex integration

13.1 Introduction

The concept of *weak solution* is indispensable in the mathematical theory of inviscid fluids, where solutions of the underlying non-linear systems of partial differential equations are known to develop singularities in a finite lap of time no matter how smooth the initial data might be. The weak solutions are being used even in the analysis of certain viscous fluids like the standard Navier-Stokes system, where a rigorous theory in the classical framework represents one of the major open problems of modern mathematics. In the absence of a sufficiently strong dissipative mechanism, solutions of non-linear systems of conservation laws may develop fast oscillations and/or concentrations that inevitably give rise to singularities of various types. As shown in the nowadays classical work of Tartar [19], oscillations are involved in many problems, in particular in those arising in the context of inviscid fluids.

The well know deficiency of weak solutions is that they may not be uniquely determined in terms of the data and suitable admissibility criteria must be imposed in order to pick up the physically relevant ones, cf. Dafermos [9]. Although most of the admissibility constraints are derived from fundamental physical principles as the Second law of thermodynamics, their efficiency in eliminating the nonphysical

E. Feireisl (✉)
Institute of Mathematics of the Academy of Sciences of the Czech Republic,
Žitná 25, 115 67 Praha 1, Czech Republic
e-mail: feireisl@math.cas.cz

Y. Shibata and Y. Suzuki (eds.), *Mathematical Fluid Dynamics, Present and Future*, Springer Proceedings in Mathematics & Statistics 183,
DOI 10.1007/978-4-431-56457-7_13

solutions is still dubious, cf. Dafermos [10]. Recently, DeLellis and Székelyhidi [11] developed the method previously known as *convex integration* in the context of fluid mechanics, in particular for the Euler system. Among other interesting results, they show the existence of infinitely many solutions to the incompressible Euler system violating many of the standard admissibility criteria. Later, the method was adapted to the compressible case by Chiodaroli [7].

In this note, we introduce an abstract functional-differential equation that may be viewed as the pressureless Euler system with variable (functionally solution dependent) coefficients. We present an abstract version of the so-called oscillatory lemma and use it in order to show the existence of infinitely many solutions adapting the method of [11]. Various specific systems arising in fluid dynamics will be then identified as special cases of the abstract problem.

The paper is organized as follows. In Sect. 13.2, we introduce the abstract problem and formulate our main result proved in the remaining part of the paper. To this end, we adapt the apparatus of convex integration including the concept of *subsolution* in Sect. 13.3. In Sect. 13.4, we present the oscillatory lemma and show the existence of infinitely many solutions. Several specific examples are discussed in Sect. 13.5. Finally, Sect. 13.6 addresses the problem of strong continuity of the weak solutions at the initial time.

13.2 Abstract Problem, Main Result

The symbol $R^{N\times N}_{\rm sym}$ will denote the space of $N \times N$ symmetric matrices over the Euclidean space R^N, $N = 2, 3$, $R^{N\times N}_{\rm sym,0}$ is its subspace of those with zero trace. For two vectors $\mathbf{v}, \mathbf{w} \in R^N$, we denote

$$\mathbf{v}\otimes\mathbf{w} \in R^{N\times N}_{\rm sym},\ [\mathbf{v}\otimes\mathbf{w}]_{i,j} = v_i v_j,\ \text{and } \mathbf{v}\odot\mathbf{w} \in R^{N\times N}_{\rm sym,0},\ \mathbf{v}\odot\mathbf{w} = \mathbf{v}\otimes\mathbf{w} - \frac{1}{N}\mathbf{v}\cdot\mathbf{w}\mathbb{I}.$$

For the sake of simplicity, we suppose the physical space to be the "flat" torus

$$\Omega = \left([-1,1]\Big|_{\{-1,1\}} \right)^N,$$

meaning, the functions of $x \in \Omega$ are (2-)periodic in R^N.

13.2.1 Abstract Problem

We consider the following problem:

Find a vector field $\mathbf{u} \in C_{\mathrm{weak}}([0, T]; L^2(\Omega; R^N))$ satisfying

$$\partial_t \mathbf{u} + \mathrm{div}_x \left(\frac{(\mathbf{u} + \mathbf{h}[\mathbf{u}]) \odot (\mathbf{u} + \mathbf{h}[\mathbf{u}])}{r[\mathbf{u}]} + \mathbb{H}[\mathbf{u}] \right) = 0, \ \ \mathrm{div}_x \mathbf{u} = 0, \tag{13.1}$$

in $\mathcal{D}'((0, T) \times \Omega; R^N)$,

$$\frac{1}{2} \frac{|\mathbf{u} + \mathbf{h}[\mathbf{u}]|^2}{r[\mathbf{u}]}(t, x) = e[\mathbf{u}](t, x) \text{ for a.a. } (t, x) \in (0, T) \times \Omega, \tag{13.2}$$

$$\mathbf{u}(0, \cdot) = \mathbf{u}_0, \ \ \mathbf{u}(T, \cdot) = \mathbf{u}_T, \tag{13.3}$$

where $\mathbf{h}[\mathbf{u}]$, $r[\mathbf{u}]$, $\mathbb{H}[\mathbf{u}]$, and $e[\mathbf{u}]$ are given (nonlinear) operators.

Remark 13.2.1 The problem (13.1)–(13.3) is seemingly overdetermined as both the initial and the end state are prescribed. Moreover, the associated "kinetic energy" is constrained by (13.2). Specific applications will be given in Sect. 13.5.

Remark 13.2.2 The choice

$$\mathbf{h} = 0, r = 1, \mathbb{H} = 0, e = e(t)$$

gives rise to the incompressible Euler system

$$\partial_t \mathbf{u} + \mathrm{div}_x (\mathbf{u} \otimes \mathbf{u}) + \nabla_x \Pi = 0, \ \ \mathrm{div}_x \mathbf{u} = 0$$

with the prescribed kinetic energy

$$\frac{1}{2} |\mathbf{u}|^2 = e(t)$$

studied by Chiodaroli [7] and DeLellis, Székelyhidi [11].

Remark 13.2.3 Note that a "more complex" problem

$$\partial_t \mathbf{u} + \mathrm{div}_x \left(\frac{(\mathbf{u} + \mathbf{h}[\mathbf{u}]) \otimes (\mathbf{u} + \mathbf{h}[\mathbf{u}])}{r[\mathbf{u}]} + \mathbb{H}[\mathbf{u}] \right) + \nabla_x \Pi[\mathbf{u}] = 0, \ \ \mathrm{div}_x \mathbf{u} = 0 \tag{13.4}$$

can be converted to (13.1) and (13.2), with

$$e[\mathbf{u}] = Z[\mathbf{u}](t) - \frac{N}{2} \Pi[\mathbf{u}],$$

where Z is an arbitrary spatially homogeneous function.

Remark 13.2.4 The “pressure” Π in (13.4) can be incorporated in $\mathbb{H}$ by solving the problem

$$\mathrm{div}_x \mathbb{H}_\Pi = \nabla_x \Pi \text{ in } \Omega, \ \mathbb{H}_\Pi(x) \in R^{N\times N}_{\mathrm{sym},0}, \ x \in \Omega.$$

We can take, for instance, the solution of the Lamé system

$$\mathbb{H}_\Pi = \nabla_x \mathbf{U} + \nabla_x \mathbf{U}^T - \frac{2}{N}\mathrm{div}_x \mathbf{U}\mathbb{I}.$$

As observed by Desvillettes and Villani [12, Sect. 4.1, Proposition 11], the vector field $\mathbf{U}$ is uniquely determined up to an additive constant. Of course, in order to preserve certain continuity of $\mathbb{H}_\Pi$, more regularity of Π is needed.

The quantities $\mathbf{h}$, r, $\mathbb{H}$, and e are operators depending on the solution $\mathbf{u}$. In order to specify their properties, we introduce the following definition:

Definition 13.2.1 Let $Q \subset (0,T) \times \Omega$ be an open set such that

$$|Q| = |(0,T) \times \Omega|.$$

An operator

$$b : C_{\mathrm{weak}}([0,T]; L^2(\Omega; R^N)) \cap L^\infty((0,T) \times \Omega; R^N) \to C_b(Q, R^M)$$

is *Q-continuous* if:

- b maps bounded sets in $L^\infty((0,T) \times \Omega; R^N)$ on bounded sets in $C_b(Q, R^M)$;
- b is continuous, specifically,

$$b[\mathbf{v}_n] \to b[\mathbf{v}] \text{ in } C_b(Q; R^M) \text{ (uniformly for } (t,x) \in Q \text{)}$$

whenever

$$\mathbf{v}_n \to \mathbf{v} \text{ in } C_{\mathrm{weak}}([0,T]; L^2(\Omega; R^N)) \text{ and weakly-(*) in } L^\infty((0,T) \times \Omega; R^N); \tag{13.5}$$

- b is causal (non-anticipative), meaning

$$\mathbf{v}(t,\cdot) = \mathbf{w}(t,\cdot) \text{ for } 0 \le t \le \tau \le T \text{ implies } b[\mathbf{v}] = b[\mathbf{w}] \text{ in } [(0,\tau] \times \Omega] \cap Q. \tag{13.6}$$

In this paper, we suppose

$$\begin{aligned} \mathbf{h} = \mathbf{h}[\mathbf{u}] &: C_{\text{weak}}([0,T]; L^2(\Omega; R^N)) \to C_b(Q; R^N), \\ r = r[\mathbf{u}] &: C_{\text{weak}}([0,T]; L^2(\Omega; R^N)) \to C_b(Q; R),\ r > 0, \\ e = e[\mathbf{u}] &: C_{\text{weak}}([0,T]; L^2(\Omega; R^N)) \to C_b(Q; R),\ e \geqslant 0, \\ \mathbb{H} = \mathbb{H}[\mathbf{u}] &: C_{\text{weak}}([0,T]; L^2(\Omega; R^N)) \to C_b(Q; R^{N\times N}_{\text{sym},0}) \end{aligned} \tag{13.7}$$

are given Q-continuous operators for a certain open set Q.

13.2.2 Subsolutions

Before stating our main result concerning solvability of problem (13.1)–(13.3), it is convenient to introduce the set of *subsolutions*. Let $\lambda_{\max}[\mathbb{A}]$ denote the maximal eigenvalue of a matrix $\mathbb{A} \in R^{N\times N}_{\text{sym}}$. Similarly to DeLellis and Székelyhidi [11], we introduce the set of *subsolutions*:

$$X_0 = \Bigg\{ \mathbf{v} \ \Bigg|\ \mathbf{v} \in C_{\text{weak}}([0,T]; L^2(\Omega; R^N)) \cap L^\infty((0,T)\times\Omega; R^N), \tag{13.8}$$

$$\mathbf{v}(0,\cdot) = \mathbf{u}_0,\ \mathbf{v}(T,\cdot) = \mathbf{u}_T,$$

$$\partial_t \mathbf{v} + \operatorname{div}_x \mathbb{F} = 0,\ \operatorname{div}_x \mathbf{v} = 0 \text{ in } \mathrm{d}'((0,T)\times\Omega; R^N),$$

$$\text{for some } \mathbb{F} \in L^\infty((0,T)\times\Omega; R^{N\times N}_{\text{sym},0}) \cap C(Q; R^{N\times N}_{\text{sym},0}),\ \mathbf{v} \in C(Q; R^N),$$

$$\sup_{(t,x)\in Q, t>\tau} \frac{N}{2}\lambda_{\max}\left[\frac{(\mathbf{v}+\mathbf{h}[\mathbf{v}])\otimes(\mathbf{v}+\mathbf{h}[\mathbf{v}])}{r[\mathbf{v}]} - \mathbb{F} + \mathbb{H}[\mathbf{v}]\right] - e[\mathbf{v}] < 0$$

$$\text{for any } 0 < \tau < T \Bigg\}.$$

Remark 13.2.5 Note that, in contrast with [11], the inequality

$$\frac{N}{2}\lambda_{\max}\left[\frac{(\mathbf{v}+h[\mathbf{v}])\otimes(\mathbf{v}+h[\mathbf{v}])}{r[\mathbf{v}]} - \mathbb{F} + \mathbb{H}[\mathbf{v}]\right] < e[\mathbf{v}]$$

is satisfied only on the open set Q, where all quantities are continuous. Moreover, the inequality is strict on any open time interval (τ, T), $0 < \tau < T$.

13.2.3 Main Result

We are ready to state our main result.

Theorem 13.2.1 *Let the operators* $\mathbf{h}$, r, $\mathbb{H}$, *and* e *given by (13.7) be* Q*-continuous, where* $Q \subset [(0,T) \times \Omega]$ *is an open set,*

$$|Q| = |(0,T) \times \Omega|.$$

In addition, suppose that $r[\mathbf{v}] > 0$ *and that the mapping* $\mathbf{v} \mapsto 1/r[\mathbf{v}]$ *is continuous in the sense specified in (13.5). Finally, assume that the set of subsolutions* X_0 *is* non-empty *and* bounded *in* $L^\infty((0,T) \times \Omega; R^N)$.

Then problem (13.1)–(13.3) admits infinitely many solutions.

The next two sections will be devoted to the proof of Theorem 13.2.1. For the set of subsolutions to be non-empty, the energy e must be chosen large enough. For instance, taking $\mathbf{u}_0 = \mathbf{u}_T \in C(\Omega; R^N)$, $\mathrm{div}_x \mathbf{u}_0 = 0$ we check easily that X_0 is non-empty, specifically $\mathbf{u}_0 \in X_0$, as soon as

$$\frac{N}{2} \lambda_{\max} \left[\frac{(\mathbf{u}_0 + h[\mathbf{u}_0]) \otimes (\mathbf{u}_0 + h[\mathbf{u}_0])}{r[\mathbf{u}_0]} + \mathbb{H}[\mathbf{u}_0] \right] < e[\mathbf{u}_0]. \tag{13.9}$$

Recalling the purely algebraic inequality (cf. [11])

$$\frac{1}{2} \frac{|\tilde{\mathbf{h}}|^2}{\tilde{r}} \leq \frac{N}{2} \lambda_{\max} \left[\frac{\tilde{\mathbf{h}} \otimes \tilde{\mathbf{h}}}{\tilde{r}} - \tilde{H} \right], \tag{13.10}$$

where the equality holds only if

$$\tilde{H} = \frac{\tilde{\mathbf{h}} \otimes \tilde{\mathbf{h}}}{\tilde{r}} - \frac{1}{N} \frac{|\tilde{\mathbf{h}}|^2}{\tilde{r}} \mathbb{I}, \tag{13.11}$$

we get from (13.9) that

$$\frac{1}{2} \frac{|\mathbf{u}_0 + h[\mathbf{u}_0]|^2}{r[\mathbf{u}_0]} < e[\mathbf{u}_0],$$

meaning the relation (13.2) is *violated* at the initial time. This is the undesirable initial "energy jump" characteristic for the weak solutions obtained by the method of convex integration. A possible remedy for this problem will be discussed in Sect. 13.6.

13.3 Convex Integration

As the set X_0 is bounded, there exists a positive constant $\overline{e}$ such that

$$e[\mathbf{v}] \leq \overline{e} \text{ for any } \mathbf{v} \in X_0. \tag{13.12}$$

Under the hypotheses of Theorem 13.2.1 we may define a topological space $\overline{X}_0$ as the closure of the space of subsolutions X_0 with respect to the (metrizable) topology of the space $C_{\text{weak}}([0,T]; L^2(\Omega; R^N))$. Accordingly, $\overline{X}_0$ is a (non-empty) complete metric space with the distance of two functions $\mathbf{v}$, $\mathbf{w}$ given by

$$\sup_{t \in [0,T]} d[\mathbf{v}(t,\cdot); \mathbf{w}(t,\cdot)],$$

where d is the metrics induced by the weak topology on bounded sets of the Hilbert space $L^2(\Omega; R^N)$. Note that, in view of (13.10) and (13.12) and boundedness of all operators involved in the definition of X_0, the associated fluxes $\mathbb{F}$ are bounded in L^∞, in particular,

$$\partial_t \mathbf{v} + \mathrm{div}_x \mathbb{F} = 0, \ \mathrm{div}_x \mathbf{v} = 0 \text{ in } \mathrm{d}'((0,T) \times \Omega; R^N), \tag{13.13}$$

for any $\mathbf{v} \in \overline{X}_0$, where the flux $\mathbb{F} \in L^\infty((0,T) \times \Omega; R^{N \times N}_{\mathrm{sym},0})$ can be obtained as a weak limit of fluxes in X_0. Moreover, by convexity of the function

$$\frac{N}{2} \lambda_{\max} \left[\frac{(\mathbf{v} + \mathbf{h}) \otimes (\mathbf{v} + \mathbf{h})}{r} - \mathbb{F} + \mathbb{H} \right]$$

in $\mathbf{v}$ and $\mathbb{F}$, we get

$$\frac{N}{2} \lambda_{\max} \left[\frac{(\mathbf{v} + \mathbf{h}[\mathbf{v}]) \otimes (\mathbf{v} + \mathbf{h}[\mathbf{v}])}{r[\mathbf{v}]} - \mathbb{F} + \mathbb{H}[\mathbf{v}] \right] \leq e[\mathbf{v}] \text{ a.a. in } (0,T) \times \Omega. \tag{13.14}$$

Next, we introduce a countable family of functionals

$$I_n[\mathbf{v}] = \int_{1/n}^T \int_\Omega \left[\frac{1}{2} \frac{|\mathbf{v} + \mathbf{h}[\mathbf{v}]|^2}{r[\mathbf{v}]} - e[\mathbf{v}] \right] \mathrm{d}x : \overline{X}_0 \to (-\infty, 0].$$

In accordance with the hypotheses (13.7), each I_n can be seen as a lower semi-continuous functional on $\overline{X}_0$. In particular, by means of Baire's category argument, the set

$$\mathscr{S} = \cap_{n>0} \left\{ \mathbf{v} \in \overline{X}_0 \;\middle|\; \mathbf{v} \text{ is a point of continuity of } I_n \right\}$$

has infinite cardinality.

In the next section, we show that

$$\text{if } \mathbf{v} \text{ is a point of continuity of } I_n \text{ in } \overline{X}_0, \text{ then } I_n[\mathbf{v}] = 0. \tag{13.15}$$

In accordance with (13.10) and (13.11), combined with the previous observations stated in (13.13) and (13.14), this implies that $\mathscr{S}$ consists of weak solutions to problem (13.1)–(13.3). Consequently, the proof of Theorem 13.2.1 reduces to showing (13.15).

13.4 Oscillatory Lemma, Infinitely Many Solutions

In accordance with the previous discussion, the final step in the proof of Theorem 13.2.1 is to show (13.15). The main tool we shall use is the following variant of the oscillatory lemma (cf. De Lellis and Székelyhidi [11, Proposition 3], Chiodaroli [7, Sect. 6, formula (6.9)]) proved in [13, Lemma 3.1]:

Lemma 13.4.1 *Let $U \subset R \times R^N$, $N = 2, 3$ be a bounded open set. Suppose that*

$$\tilde{\mathbf{h}} \in C(U; R^N),\ \tilde{\mathbb{H}} \in C(U; R^{N\times N}_{\mathrm{sym},0}),\ \tilde{e},\ \tilde{r} \in C(U),\ \tilde{r} > 0,\ \tilde{e} \le \overline{e} \text{ in } U$$

are given such that

$$\frac{N}{2}\lambda_{\max}\left[\frac{\tilde{\mathbf{h}} \otimes \tilde{\mathbf{h}}}{\tilde{r}} - \tilde{\mathbb{H}}\right] < \tilde{e} \text{ in } U. \tag{13.16}$$

Then there exist sequences

$$\mathbf{w}_n \in C^\infty_c(U; R^N),\ \mathbb{G}_n \in C^\infty_c(U; R^{N\times N}_{\mathrm{sym},0}),\ n = 0, 1, \dots$$

such that

$$\partial_t \mathbf{w}_n + \mathrm{div}_x \mathbb{G}_n = 0,\ \mathrm{div}_x \mathbf{w}_n = 0 \text{ in } R^N,$$

$$\frac{N}{2}\lambda_{\max}\left[\frac{(\tilde{\mathbf{h}} + \mathbf{w}_n) \otimes (\tilde{\mathbf{h}} + \mathbf{w}_n)}{\tilde{r}} - (\tilde{\mathbb{H}} + \mathbb{G}_n)\right] < \tilde{e} \text{ in } U, \tag{13.17}$$

and

$$\mathbf{w}_n \to 0 \text{ weakly in } L^2(U; R^N),\ \liminf_{n\to\infty} \int_U \frac{|\mathbf{w}_n|^2}{\tilde{r}}\, \mathrm{d}x\, \mathrm{d}t \geqslant \Lambda(\overline{e}) \int_U \left(\tilde{e} - \frac{1}{2}\frac{|\tilde{\mathbf{h}}|^2}{\tilde{r}}\right)^2 \mathrm{d}x\, \mathrm{d}t \tag{13.18}$$

for a certain $\Lambda(\overline{e}) > 0$ depending only on the energy upper bound $\overline{e}$.

Remark 13.4.1 Note that Lemma 13.4.1 applies to continuous, not necessarily bounded, functions on the open set U.

With Lemma 13.4.1 at hand, we may show the following result that contains (13.15) as a particular case.

Lemma 13.4.2 *Let*

$$I_D = \int_D \left[\frac{1}{2} \frac{|\mathbf{v} + \mathbf{h}[\mathbf{v}]|^2}{r[\mathbf{v}]} - e[\mathbf{v}] \right] \mathrm{d}x \, \mathrm{d}t : \overline{X}_0 \to (-\infty, 0]$$

be a functional defined on an open set $D \subset [(\tau, T) \times \Omega] \cap Q$, $0 < \tau < T$.

Then I_D *vanishes at any of its points of continuity.*

Proof Arguing by contradiction we assume that $\mathbf{v} \in \overline{X}_0$ is a point of continuity of I_D such that

$$I_D[\mathbf{v}] < 0.$$

Since I_D is continuous at $\mathbf{v}$, there is a sequence $\{\mathbf{v}_m\}_{m=1}^{\infty} \subset X_0$ (with the associated fluxes $\mathbb{F}_m$) such that

$$\mathbf{v}_m \to \mathbf{v} \text{ in } C_{\mathrm{weak}}([0, T]; L^2(\Omega; R^N)), \ I_D[\mathbf{v}_m] \to I_D[\mathbf{v}] \text{ as } m \to \infty.$$

As $[\mathbf{v}_m, \mathbb{F}_m]$ are subsolutions and $\tau > 0$, we get, thanks to (13.8),

$$\frac{N}{2} \lambda_{\max} \left[\frac{(\mathbf{v}_m + \mathbf{h}[\mathbf{v}_m]) \otimes (\mathbf{v}_m + \mathbf{h}[\mathbf{v}_m])}{r[\mathbf{v}_m]} - \mathbb{F}_m + \mathbb{H}[\mathbf{v}_m] \right]$$

$$< e[\mathbf{v}_m] - \delta_m \text{ in } [[\tau, T) \times \Omega)] \cap Q \text{ for some } \delta_m \searrow 0.$$

Now, fixing m for a while, we apply Lemma 13.4.1 with

$$N = 2, 3, \ U = D, \ \tilde{r} = r[\mathbf{v}_m], \ \tilde{\mathbf{h}} = \mathbf{v}_m + \mathbf{h}[\mathbf{v}_m], \ \tilde{\mathbb{H}} = \mathbb{F}_m - \mathbb{H}[\mathbf{v}_m], \ \tilde{e} = e[\mathbf{v}_m] - \delta_m.$$

For $\left\{ [\mathbf{w}_{m,n}, \mathbb{G}_{m,n}] \right\}_{n=1}^{\infty}$ the quantities resulting from the conclusion of Lemma 13.4.1, we set

$$\mathbf{v}_{m,n} = \mathbf{v}_m + \mathbf{w}_{m,n}, \ \mathbb{F}_{m,n} = \mathbb{F}_m + \mathbb{G}_{m,n}.$$

Obviously,

$$\partial_t \mathbf{v}_{m,n} + \mathrm{div}_x \mathbb{F}_{m,n} = 0, \ \mathrm{div}_x \mathbf{v}_{m,n} = 0 \text{ in } \mathscr{D}'((0, T) \times \Omega),]$$

$$vcv_{m,n}(0, \cdot) = \mathbf{v}_0, \ \mathbf{v}_{m,n}(T, \cdot) = \mathbf{v}_T.$$

Moreover, in accordance with (13.17) and the fact that $\mathbf{w}_n$, $\mathbb{G}_{m,n}$ vanish outside D,

$$\frac{N}{2}\lambda_{\max}\left[\frac{(\mathbf{v}_{m,n}+\mathbf{h}[\mathbf{v}_m])\otimes(\mathbf{v}_{m,n}+\mathbf{h}[\mathbf{v}_m])}{r[\mathbf{v}_m]}-\mathbb{F}_{m,n}+\mathbb{H}[\mathbf{v}_m]\right]$$

$$< e[\mathbf{v}_m]-\delta_m \text{ in } [\tau,T)\times\Omega\cap Q,$$

and, by virtue of the causality property (13.6),

$$\sup_{(t,x)\in Q, s<t\leq\tau}\frac{N}{2}\lambda_{\max}\left[\frac{(\mathbf{v}_{m,n}+\mathbf{h}[\mathbf{v}_{m,n}])\otimes(\mathbf{v}_{m,n}+\mathbf{h}[\mathbf{v}_{m,n}])}{r[\mathbf{v}_{m,n}]}-\mathbb{F}_{m,n}+\mathbb{H}[\mathbf{v}_{m,n}]\right]-e[\mathbf{v}_{m,n}]$$

$$<0$$

for any $0<s<\tau$. Consequently, in view of continuity of the operators $\mathbf{v}\mapsto\mathbf{h}[\mathbf{v}], r[\mathbf{v}], e[\mathbf{v}], \mathbb{H}[\mathbf{v}]$ specified in (13.5), we may infer that for each m there exists $n=n(m)$ such that

$$\mathbf{v}_{m,n(m)}\in X_0,\ m=1,2,\ldots$$

Moreover, by virtue of (13.18), we may suppose that

$$\mathbf{v}_{m,n(m)}\to\mathbf{v} \text{ in } C_{\text{weak}}([0,T];L^2(\Omega;R^2))$$

in particular,

$$I_D[\mathbf{v}_{m,n(m)}]\to I_D[\mathbf{v}] \tag{13.19}$$

as $m\to\infty$.

Finally, using again the conclusion of Lemma 13.4.1 combined with Jensen's inequality, we observe that the sequence $\mathbf{v}_{m,n(m)}$ can be taken in such a way that

$$\liminf_{m\to\infty} I_D[\mathbf{v}_{m,n(m)}]$$

$$=\liminf_{m\to\infty}\int_D\left(\frac{1}{2}\frac{|\mathbf{v}_m+\mathbf{w}_{m,n(m)}+\mathbf{h}[\mathbf{v}_m+\mathbf{w}_{m,n(m)}]|^2}{r[\mathbf{v}_m+\mathbf{w}_{m,n(m)}]}-e[\mathbf{v}_m+\mathbf{w}_{m,n(m)}]\right)\mathrm{d}x\,\mathrm{d}t$$

$$=\lim_{m\to\infty}\int_D\left(\frac{1}{2}\frac{|\mathbf{v}_m+\mathbf{h}[\mathbf{v}_m+\mathbf{w}_{m,n(m)}]|^2}{r[\mathbf{v}_m+\mathbf{w}_{m,n(m)}]}-e[\mathbf{v}_m+\mathbf{w}_{m,n(m)}]\right)\mathrm{d}x\,\mathrm{d}t$$

$$+\liminf_{m\to\infty}\int_D\frac{1}{2}\frac{|\mathbf{w}_{m,n(m)}|^2}{r[\mathbf{v}_m+\mathbf{w}_{m,n(m)}]}\mathrm{d}x\,\mathrm{d}t$$

$$\geqslant I_D[\mathbf{v}]+\frac{\Lambda(\overline{e})}{2}\liminf_{m\to\infty}\int_D\left(e([\mathbf{v}_m])-\delta_m-\frac{1}{2}\frac{|\mathbf{v}_m+\mathbf{h}[\mathbf{v}_m]|^2}{r[\mathbf{v}_m]}\right)^2\mathrm{d}x\,\mathrm{d}t$$

$$\geqslant I_D[\mathbf{v}] + \frac{\Lambda(\overline{e})}{2|G|} \liminf_{m\to\infty} \left(\int_D \left(e([\mathbf{v}_m]) - \delta_m - \frac{1}{2} \frac{|\mathbf{v}_m + \mathbf{h}[\mathbf{v}_m]|^2}{r[\mathbf{v}_m]} \right) \mathrm{d}x \ \mathrm{d}t \right)^2$$

$$= I_D[\mathbf{v}] + \frac{\Lambda(\overline{e})}{2|G|} (I_D[\mathbf{v}])^2,$$

which is compatible with (13.19) only if $I_D[\mathbf{v}] = 0$. □

We have shown (13.15); whence Theorem 13.2.1.

13.5 Examples

There are many systems arising in mathematical fluid dynamics that can be written in the abstract form (13.1)–(13.3). We review some of them already studied in the available literature.

13.5.1 Euler-Fourier System

The Euler-Fourier system describes the time evolutions of the mass density $\rho = \rho(t,x)$, the velocity $\mathbf{u} = \mathbf{u}(t,x)$, and the (absolute) temperature $\vartheta = \vartheta(t,x)$:

$$\partial_t \rho + \mathrm{div}_x(\rho \mathbf{u}) = 0, \tag{13.20}$$

$$\partial_t(\rho \mathbf{u}) + \mathrm{div}_x(\rho \mathbf{u} \otimes \mathbf{u}) + \nabla_x(\rho \vartheta) = 0, \tag{13.21}$$

$$\frac{3}{2}\Big(\partial_t(\rho\vartheta) + \mathrm{div}_x(\rho\vartheta\mathbf{u})\Big) - \Delta\vartheta = -\rho\vartheta\,\mathrm{div}_x\mathbf{u}. \tag{13.22}$$

Following [8] we first write the momentum $\rho\mathbf{u}$ as its Helmholtz decomposition

$$\rho\mathbf{u} = \mathbf{v} + \nabla_x \Phi, \ \mathrm{div}_x \mathbf{v} = 0.$$

Accordingly, we may *fix* the density ρ and the acoustic potential Φ so that

$$\partial_t \rho + \Delta\Phi = 0 \text{ holds,}$$

meaning equation (13.20) is satisfies as $\mathrm{div}_x\mathbf{v} = 0$.

With ρ, Φ given we may determine the temperature field $\vartheta = \vartheta[\mathbf{v}]$ as the (unique solution) of (13.22), specifically

$$\frac{3}{2}\Big(\rho\partial_t\vartheta + (\mathbf{v} + \nabla_x\Phi)\cdot\nabla_x\vartheta\Big) - \Delta\vartheta = -\rho\vartheta\,\mathrm{div}_x\left(\frac{1}{\rho}(\mathbf{v} + \nabla_x\Phi)\right)$$

endowed with appropriate initial data.

Finally, we rewrite (13.21) in the form

$$\partial_t \mathbf{v} + \mathrm{div}_x \left(\frac{(\mathbf{v} + \nabla_x \Phi) \otimes (\mathbf{v} + \nabla_x \Phi)}{\rho} \right) + \nabla_x (\partial_t \nabla_x \Phi + \rho \vartheta [\mathbf{v}]) = 0. \quad (13.23)$$

Fixing the "energy" so that

$$\frac{1}{2} \frac{|\mathbf{v} + \nabla_x \Phi|^2}{r} = e[\mathbf{v}] \equiv Z - \frac{N}{2} (\partial_t \nabla_x \Phi - \rho \vartheta [\mathbf{v}]), \quad (13.24)$$

where $Z = Z(t)$ is a suitable spatially homogeneous function, we reduce (13.23) to

$$\partial_t \mathbf{v} + \mathrm{div}_x \left(\frac{(\mathbf{v} + \nabla_x \Phi) \odot (\mathbf{v} + \nabla_x \Phi)}{\rho} \right) = 0, \ \mathrm{div}_x \mathbf{v} = 0, \quad (13.25)$$

which is an equation in the form (13.1).

With certain effort, it is possible to show that the hypotheses of Theorem 13.2.1 are satisfied for $Q = (0, T) \times \Omega$, and we obtain the following result, see [8, Theorem 3.1]:

Theorem 13.5.1 *Let $T > 0$ be given, along with the initial data*

$$\left\{ \begin{array}{c} \rho(0, \cdot) = \rho_0 \in C^3(\Omega), \ \rho_0 > 0, \ \vartheta(0, \cdot) = \vartheta_0 \in C^2(\Omega), \ \vartheta_0 > 0, \\ \mathbf{u}(0, \cdot) = \mathbf{u}_0 \in C^3(\Omega; R^N) \end{array} \right\}, \quad (13.26)$$

$$\Omega = \left([-1; 1]_{\{-1;1\}} \right)^N, \ N = 2, 3.$$

Then the Euler-Fourier system (13.20)–(13.22) admits infinitely many weak solutions in $(0, T) \times \Omega$ emanating from the same initial data (13.26).

As already pointed out, the solutions obtained in Theorem 13.5.1 may be non-physical in the sense they violate the principle of energy conservation. However, this drawback can be removed at least for certain initial data. We will discuss this issue in Sect. 13.6.1.

13.5.2 Quantum Fluids

The Euler-Korteweg-Poisson system describes the time evolution of the density $\rho = \rho(t, x)$ and the momentum $\mathbf{J} = \mathbf{J}(t, x)$ of an inviscid fluid:

$$\partial_t \rho + \mathrm{div}_x \mathbf{J} = 0, \quad (13.27)$$

$$\partial_t \mathbf{J} + \operatorname{div}_x\left(\frac{\mathbf{J}\times\mathbf{J}}{\rho}\right) + \nabla_x p(\rho) = -\alpha\mathbf{J} + \rho\nabla_x\left(K(\rho)\Delta_x\rho + \frac{1}{2}K'(\rho)|\nabla_x\rho|^2\right) + \rho\nabla_x V, \tag{13.28}$$

$$\Delta_x V = \rho - \overline{\rho}, \tag{13.29}$$

where $K : (0,\infty) \to (0,\infty)$ is a given function, see Audiard [3], Benzoni-Gavage et al. [4, 5]. The choice $K = \overline{K} > 0$ yields the standard equations of an inviscid capillary fluid (see Bresch et al. [6], Kotchote [16, 17]), while $K(\rho) = \frac{\hbar}{4\rho}$ gives rise to the quantum fluid system (see for instance Antonelli and Marcati [1, 2], Jüngel [15, Chapter 14] and the references therein).

For

$$\chi(\rho) = \rho K(\rho),$$

it can be shown that system (13.27)–(13.29) can be recast in the form

$$\partial_t \mathbf{v} + \operatorname{div}_x\left(\frac{(\mathbf{v}+\mathbf{h})\otimes(\mathbf{v}+\mathbf{h})}{r} + \mathbb{H}\right) + \nabla_x \Pi = 0, \tag{13.30}$$

with

$$r = e^t\rho,\ \mathbf{h} = e^t\nabla_x M,$$

$$\mathbb{H}(t,x) = 4e^t\Big(\chi(\rho)\nabla_x\sqrt{\rho}\otimes\nabla_x\sqrt{\rho} - \frac{1}{3}\chi(\rho)|\nabla_x\sqrt{\rho}|^2\mathbb{I} - \frac{1}{4}\nabla_x V\otimes\nabla_x V$$
$$+\frac{1}{12}|\nabla_x V|^2\mathbb{I}\Big),$$

and

$$\Pi(t,x) = e^t\Big(p(\rho) + \partial_t M + M - \chi(\rho)\Delta_x\rho - \frac{1}{2}\chi'(\rho)|\nabla_x\rho|^2$$
$$+\frac{4}{3}\chi(\rho)|\nabla_x\sqrt{\rho}|^2 - \overline{\rho}V + \frac{1}{6}|\nabla_x V|^2\Big),$$

where ρ and M are suitably chosen functions, see [13].

Now, Theorem 13.2.1 can be applied to obtain the following result, see [13, Theorem 2.1] and the proof therein.

Theorem 13.5.2 *Let $T > 0$ be given. Suppose that p and χ satisfy*

$$p \in C^1[0,\infty)\cap C^2(0,\infty),\ p(0) = 0,\ \chi \in C^2[0,\infty),\ \chi > 0 \textit{ in } (0,\infty).$$

Let the initial data be given such that

$$\rho(0,\cdot) = \rho_0 = r_0^2,\ r_0 \in C^2(\Omega),\ \text{meas}\Big\{x\in\Omega\ \Big|\ r_0(x) = 0\Big\} = 0, \tag{13.31}$$

$$\mathbf{J}(0,\cdot) = \mathbf{J}_0 = \rho_0 \mathbf{U}_0, \ \mathbf{U}_0 \in C^3(\Omega; R^3). \tag{13.32}$$

Then the initial value problem (13.27)–(13.29), (13.31), (13.32) admits infinitely many weak solutions in $(0, T) \times \Omega$.

In the situation described in Theorem 13.5.2, the set Q must be taken

$$Q = (0, T) \times \Omega \setminus \left\{ (t, x) \;\middle|\; \rho(t, x) = 0 \right\}.$$

13.5.3 Binary Mixtures of Compressible Fluids

We consider a physically motivated regularization of the Euler equations proposed in the seminal paper by Lowengrub and Truskinovsky [18]. The model describes the motion of a mixture of two immiscible compressible fluids in terms of the density $\rho = \rho(t, x)$, the velocity $\mathbf{u} = \mathbf{u}(t, x)$, and the concentration difference $c = c(t, x)$. The fluid is described by means of the standard Euler system coupled with the Cahn-Hilliard equation describing the evolution of c:

$$\partial_t \rho + \mathrm{div}_x(\rho \mathbf{u}) = 0, \tag{13.33}$$

$$\partial_t(\rho \mathbf{u}) + \mathrm{div}_x(\rho \mathbf{u} \otimes \mathbf{u}) + \nabla_x p_0(\rho, c) = \mathrm{div}_x\left(\rho \nabla_x c \otimes \nabla_x c - \frac{\rho}{2}|\nabla_x c|^2 \mathbb{I}\right), \tag{13.34}$$

$$\partial_t(\rho c) + \mathrm{div}_x(\rho c \mathbf{u}) = \Delta\left(\mu_0(\rho, c) - \frac{1}{\rho}\mathrm{div}_x(\rho \nabla_x c)\right), \tag{13.35}$$

where

$$p_0(\rho, c) = \rho^2 \frac{\partial f_0(\rho, c)}{\partial \rho}, \ \mu_0(\rho, c) = \frac{\partial f_0(\rho, c)}{\partial c} \tag{13.36}$$

for a given free energy function f_0. The system is neither purely hyperbolic nor parabolic as the dissipation mechanism acts in a very subtle way through the coupling of the Euler and the Cahn-Hilliard systems.

The machinery of convex integration can be applied, first fixing ρ and Φ, similarly to Sect. 13.5.1, to solve

$$\partial_t \rho + \Delta \Phi = 0,$$

then taking $c = c[\mathbf{v}]$, $\mathrm{div}_x \mathbf{v} = 0$ to be the unique solution of the equation

$$\partial_t(\rho c) + \mathrm{div}_x(\rho c(\mathbf{v} + \nabla_x \Phi)) = \Delta\left(\mu_0(\rho, c) - \frac{1}{\rho}\mathrm{div}_x(\rho \nabla_x c)\right).$$

Accordingly, we obtain

$$\mathbf{v}(0,\cdot) = \mathbf{v}_0, \ \mathrm{div}_x \mathbf{v} = 0, \tag{13.37}$$

$$\partial_t \mathbf{v} + \mathrm{div}_x \left(\frac{(\mathbf{v} + \nabla_x \Phi) \odot (\mathbf{v} + \nabla_x \Phi)}{\rho} - \rho(\nabla_x c[\mathbf{v}] \odot \nabla_x c[\mathbf{v}]) \right) = 0, \tag{13.38}$$

$$\frac{1}{2} \frac{|\mathbf{v} + \nabla_x \Phi|^2}{\rho} = e[\mathbf{v}] \equiv Z(t) - \frac{N}{2} \left(\frac{1}{6} |\nabla_x c[\mathbf{v}]|^2 + p_0(\rho, c[\mathbf{v}]) + \partial_t \nabla_x \Phi \right), \tag{13.39}$$

where Z is a spatially homogeneous function.

Theorem 13.2.1 yields the following result, see [14] for details:

Theorem 13.5.3 *Let the potential* $f_0 = f_0(\rho, c)$ *satisfy*

$$f_0(\rho, c) = H(c) + \log(\rho) \left(\alpha_1 \frac{1-c}{2} + \alpha_2 \frac{1+c}{2} \right),$$

$$H \in C^2(R), \ |H''(c)| \le \overline{H} \textit{ for all } c \in R^1.$$

Then for any choice of initial conditions

$$\rho(0,\cdot) = \rho_0 \in C^3(\Omega), \ \inf_{\Omega} \rho_0 > 0, \ \mathbf{u}(0,\cdot) = \mathbf{u}_0 \in C^3(\Omega; R^3),$$

$$c(0,\cdot) = c_0 \in C^2(\Omega),$$

the problem (13.33–13.35) admits infinitely many weak solutions in $(0,T) \times \Omega$.

13.6 Continuity at the Initial Time, Admissible Solutions

The major drawback of the construction delineated in the previous part of the paper and the main reason why the weak solution obtained via convex integration can be eliminated as physically unacceptable is the energy jump at the initial time discussed in Sect. 13.2.3. On the other hand, however, once a subsolution $\mathbf{v}$ along with the associated energy $e[\mathbf{v}]$ are obtained, it is possible to show the existence of another subsolution defined on a possibly shorter time interval for which the initial energy is attained. Such a subsolution can be then used in the process of convex integration to produce weak solutions that are *strongly continuous* at the initial time and dissipate energy.

We first state the result for the abstract system and then shortly comment on possible applications. Modifying slightly the procedure used in the proof of Theorem 13.2.1 we can show the following assertion:

Theorem 13.6.1 *In addition to the hypotheses of Theorem 13.2.1, suppose that*

$$\left|\left\{x \in \Omega \,\middle|\, (t,x) \in Q\right\}\right| = |\Omega| \text{ for any } 0 < t < T. \tag{13.40}$$

Then there exists a set of times $\mathscr{R} \subset (0,T)$ dense in $(0,T)$ such that for any $\tau \in \mathscr{R}$ there is $\mathbf{v} \in \overline{X}_0$ with the following properties:

-
$$\left\{ \begin{array}{c} \mathbf{v} \in C_b([(0,\tau) \cup (\tau,T) \times \Omega] \cap Q; R^N) \cap C_{\rm weak}([0,T]; L^2(\Omega; R^N)), \\ \mathbf{v}(0,\cdot) = \mathbf{u}_0, \ \mathbf{v}(T,\cdot) = \mathbf{u}_T \end{array} \right\}; \tag{13.41}$$

-
$$\partial_t \mathbf{v} + {\rm div}_x \mathbb{F} = 0, \ {\rm div}_x \mathbf{v} = 0 \text{ in } \mathrm{d}'((0,T) \times \Omega; R^N) \tag{13.42}$$

  for some $\mathbb{F} \in C_b([(0,\tau) \cup (\tau,T) \times \Omega] \cap Q; R^{3\times 3}_{\rm sym,0})$;

-
$$\frac{N}{2} \lambda_{\max} \left[\frac{(\mathbf{v} + h[\mathbf{v}]) \otimes (\mathbf{v} + h[\mathbf{v}])}{r[\mathbf{v}]} - \mathbb{F} + \mathbb{H}[\mathbf{v}] \right] < e[\mathbf{v}] \text{ in } [(0,\tau) \times \Omega] \cap Q, \tag{13.43}$$

-
$$\sup_{(t,x) \in Q, t > \tau + s} \frac{N}{2} \lambda_{\max} \left[\frac{(\mathbf{v} + h[\mathbf{v}]) \otimes (\mathbf{v} + h[\mathbf{v}])}{r[\mathbf{v}]} - \mathbb{F} + \mathbb{H}[\mathbf{v}] \right] - e[\mathbf{v}] < 0 \tag{13.44}$$

  for any $0 < s < T - \tau$,

-
$$\frac{1}{2} \int_\Omega \frac{|\mathbf{v} + \mathbf{h}[\mathbf{v}]|^2}{r[\mathbf{v}]}(\tau,\cdot) \,\mathrm{d}x = \int_\Omega e[\mathbf{v}](\tau,\cdot)\,\mathrm{d}x. \tag{13.45}$$

Remark 13.6.1 Unlike the subsolutions considered in the proof of Theorem 13.2.1, the function $\mathbf{v}$ satisfies (13.45) and is therefore *strongly* continuous at the point τ attaining the desired energy $e[\mathbf{v}](\tau,\cdot)$ in the integral sense. There is no energy jump at the time $t = \tau$! Moreover, the set of such times is dense in $(0,T)$.

Remark 13.6.2 In view of hypothesis (13.40) we have

$$C_b(Q; R^M) \subset C_{\rm loc}(0,T; L^q(\Omega; R^M)) \text{ for any } 1 \le q < \infty. \tag{13.46}$$

This observation justifies (13.45) and will be frequently used in the proof below.

Proof The function $\mathbf{v}$ will be constructed recursively as a limit

$$\mathbf{v}_k \to \mathbf{v} \text{ in } C_{\rm weak}([0,T]; L^2(\Omega; R^N)) \text{ for suitable } \mathbf{v}_k \in X_0. \tag{13.47}$$

We start by fixing the open interval $(a_0, b_0) \subset (0, T)$ in which the time τ is to be localized. As the space X_0 of subsolutions is non-empty, we take

$$\mathbf{v}_0 \in X_0, \text{ along with the associated flux } \mathbb{F}_0.$$

Next, we construct a sequence of functions $\mathbf{v}_k$, open intervals $(a_k, b_k) \subset (0, T)$, times $\tau_k \in (a_k, b_k)$, and a decreasing sequence of positive numbers $\delta_k \searrow 0$ such that:

- $$\partial_t \mathbf{v}_k + \mathrm{div}_x \mathbb{F}_k = 0,\ \mathrm{div}_x \mathbf{v}_k = 0 \text{ in } \mathrm{d}'((0,T) \times \Omega),\ \mathbf{v}(0) = \mathbf{u}_0,\ \mathbf{v}(T) = \mathbf{u}_T, \tag{13.48}$$

  for a certain field $\mathbb{F}_k \in C(Q; R^{N \times N}_{\mathrm{sym},0})$,

  $$\mathbf{v}_k - \mathbf{v}_{k-1} \in C^\infty_c(Q; R^N),\ \mathrm{supp}[\mathbf{v}_k - \mathbf{v}_{k-1}] \subset [(a_k, b_k) \times \Omega] \cap Q,$$

  $$\mathbb{F}_k - \mathbb{F}_{k-1} \in C^\infty_c(Q; R^{N \times N}_{\mathrm{sym},0}),\ \mathrm{supp}[\mathbb{F}_k - \mathbb{F}_{k-1}] \subset [(a_k, b_k) \times \Omega] \cap Q, \tag{13.49}$$

  where

  $$0 < a_{k-1} < a_k < b_k < b_{k-1},\ \varepsilon_k = b_k - a_k \to 0 \text{ for } k \to \infty;$$

- $$\sup_{t \in [0,T]} d(\mathbf{v}_k(t), \mathbf{v}_{k-1}(t)) < \frac{1}{2^k}; \tag{13.50}$$

- $$\sup_{t \in (0,T)} \left| \int_\Omega \frac{1}{r[\mathbf{v}_j]} (\mathbf{v}_k - \mathbf{v}_{k-1}) \cdot \mathbf{v}_j \, \mathrm{d}x \right| < \frac{1}{2^k} \tag{13.51}$$

  for all $j = 0, \ldots, k-1$;
- there exists $\tau_k \in (a_k, b_k)$ and a positive constant λ *independent* of k such that

  $$\frac{1}{2} \int_\Omega \frac{|\mathbf{v}_k + \mathbf{h}[\mathbf{v}_k]|^2}{r[\mathbf{v}_k]} (\tau_k, \cdot) \, \mathrm{d}x \geqslant \frac{1}{2} \int_\Omega \frac{|\mathbf{v}_{k-1} + \mathbf{h}[\mathbf{v}_{k-1}]|^2}{r[\mathbf{v}_{k-1}]} (t, \cdot) \, \mathrm{d}x + \lambda \frac{1}{\varepsilon_k^2} \alpha_k^2 \tag{13.52}$$

  $$> \frac{1}{2} \int_\Omega \frac{|\mathbf{v}_{k-1} + \mathbf{h}[\mathbf{v}_{k-1}]|^2}{r[\mathbf{v}_{k-1}]} (\tau_{k-1}, \cdot) \, \mathrm{d}x + \frac{1}{2} \lambda \frac{1}{\varepsilon_k^2} \alpha_k^2 \text{ for all } t \in (a_k, b_k),$$

  where we have set

  $$\alpha_k = \int_{a_k}^{b_k} \int_\Omega \left(e[\mathbf{v}_{k-1}] - \frac{1}{2} \frac{|\mathbf{v}_{k-1} + \mathbf{h}[\mathbf{v}_{k-1}]|^2}{r[\mathbf{v}_{k-1}]} \right) \mathrm{d}x\mathrm{d}t > 0;$$

•

$$\frac{N}{2}\lambda_{\max}\left[\frac{(\mathbf{v}_k + h[\mathbf{v}_k]) \otimes (\mathbf{v}_k + h[\mathbf{v}_k])}{r[\mathbf{v}_k]} - \mathbb{F}_k + \mathbb{H}[\mathbf{v}_k]\right] < e[\mathbf{v}_k] \text{ in } [(0, a_k] \times \Omega] \cap Q, \tag{13.53}$$

$$\frac{N}{2}\lambda_{\max}\left[\frac{(\mathbf{v}_k + h[\mathbf{v}_k]) \otimes (\mathbf{v}_k + h[\mathbf{v}_k])}{r[\mathbf{v}_k]} - \mathbb{F}_k + \mathbb{H}[\mathbf{v}_k]\right] < e[\mathbf{v}_k] - \delta_k\left(1 + \frac{1}{2^k}\right) \tag{13.54}$$

$$\text{in } [(a_k, b_k) \times \Omega] \cap Q,$$

$$\frac{N}{2}\lambda_{\max}\left[\frac{(\mathbf{v}_k + h[\mathbf{v}_k]) \otimes (\mathbf{v}_k + h[\mathbf{v}_k])}{r[\mathbf{v}_k]} - \mathbb{F}_k + \mathbb{H}[\mathbf{v}_k]\right] < e[\mathbf{v}_k] - \delta_{j-1}\left(1 + \frac{1}{2^k}\right) \tag{13.55}$$

$$\text{in } [[b_j, b_{j-1}) \times \Omega] \cap Q, \; j = 0, 1, \ldots, k, \; b_{-1} \equiv T.$$

Step 1

It follows from the properties of X_0 that $\mathbf{v}_0$ satisfies (13.48), along with the bounds (13.53)–(13.55) for a certain

$$\delta_0 = \delta_{-1} > 0.$$

Step 2

Suppose we have already constructed the functions $\mathbf{v}_j$, along with intervals (a_j, b_j), the times τ_j, and the constants δ_j, for $j = 0, 1, \ldots k - 1$ enjoying the properties (13.48)–(13.55). Our goal is to find $\mathbf{v}_k$, (a_k, b_k), τ_k, and δ_k.

First, we fix the interval (a_k, b_k). To this end, compute

$$\alpha_k = \int_{a_k}^{b_k} \int_{\Omega} \left(e[\mathbf{v}_{k-1}] - \frac{1}{2} \frac{|\mathbf{v}_{k-1} + \mathbf{h}[\mathbf{v}_{k-1}]|^2}{r[\mathbf{v}_{k-1}]} \right) \mathrm{d}x \mathrm{d}t.$$

As a consequence of (13.46), the integrand is a continuous function of time continuous function of time; whence

$$\frac{\alpha_k}{\varepsilon_k} = \frac{1}{\varepsilon_k} \int_{a_k}^{b_k} \int_{\Omega} \left(e[\mathbf{v}_{k-1}] - \frac{1}{2} \frac{|\mathbf{v}_{k-1} + \mathbf{h}[\mathbf{v}_{k-1}]|^2}{r[\mathbf{v}_{k-1}]} \right) \mathrm{d}x \mathrm{d}t$$

$$\to$$

$$\int_{\Omega} \left(e[\mathbf{v}_{k-1}] - \frac{1}{2} \frac{|\mathbf{v}_{k-1} + \mathbf{h}[\mathbf{v}_{k-1}]|^2}{r[\mathbf{v}_{k-1}]} \right) (\tau_{k-1}) \, \mathrm{d}x \text{ as } \varepsilon_k = b_k - a_k \to 0.$$

Consequently, keeping in mind that $\alpha_k > 0$ and repeating the same continuity argument, we may choose $a_{k-1} < a_k < b_k < b_{k-1}$ and ε_k so small that

$$\frac{1}{\varepsilon_k}\int_{a_k}^{b_k}\int_\Omega \frac{1}{2}\frac{|\mathbf{v}_{k-1}+\mathbf{h}[\mathbf{v}_{k-1}]|^2}{r[\mathbf{v}_{k-1}]}\,\mathrm{d}x\mathrm{d}t + \Lambda(\overline{e})\frac{\alpha_k^2}{\varepsilon_k^2} \tag{13.56}$$

$$\geqslant \int_\Omega \frac{1}{2}\frac{|\mathbf{v}_{k-1}+\mathbf{h}[\mathbf{v}_{k-1}]|^2}{r[\mathbf{v}_{k-1}]}(t,\cdot)\,\mathrm{d}x + \Lambda(\overline{e})\frac{\alpha_k^2}{2\varepsilon_k^2}$$

$$\geqslant \int_\Omega \frac{1}{2}\frac{|\mathbf{v}_{k-1}+\mathbf{h}[\mathbf{v}_{k-1}]|^2}{r[\mathbf{v}_{k-1}]}(\tau_{k-1},\cdot)\,\mathrm{d}x + \Lambda(\overline{e})\frac{\alpha_k^2}{4\varepsilon_k^2} \text{ for all } t\in(a_k,b_k),$$

where $\Lambda(\overline{e})$ is the universal constant introduced in Lemma 13.4.1.

At this stage, we apply Lemma 13.4.1 for

$$U=[(a_k,b_k)\times\Omega]\cap Q,\ \tilde{\mathbf{h}}=\mathbf{v}_{k-1}+\mathbf{h}[\mathbf{v}_{k-1}],\ \tilde{r}=r[\mathbf{v}_{k-1}],\ \tilde{\mathbb{H}}=\mathbb{F}_{k-1}-\mathbb{H}[\mathbf{v}_{k-1}],$$

and

$$\tilde{e}=e[\mathbf{v}_{k-1}]-\delta_k\left(1+\frac{1}{2^{k-1}}\right),$$

where $\delta_k>0$ is chosen small enough so that (13.16) may hold.

Now we claim that it is possible to take

$$\mathbf{v}_k=\mathbf{v}_{k-1}+\mathbf{w}_n,\ \mathbb{F}_k=\mathbb{F}_{k-1}+\mathbb{G}_n,\ n \text{ large enough,} \tag{13.57}$$

where $\mathbf{w}_n$, $\mathbb{G}_n$ are the quantities constructed in Lemma 13.4.1. Obviously, the functions $\mathbf{v}_k$ satisfy (13.48)–(13.51) provided n is large enough. Indeed we observe that n can be chosen so large for (13.51) to be satisfied. To see this we realize that, by virtue of (13.46), the image

$$\cup_{t\in[a_k,b_k]}\frac{\mathbf{v}_j}{r[\mathbf{v}_j]}(t,\cdot) \text{ is compact in } L^2(\Omega;R^N),\ j=0,\dots,k-1.$$

Next, we use continuity of the operators $\mathbf{h}$, r specified in (13.46) to compute

$$\int_{a_k}^{b_k}\int_\Omega \frac{1}{2}\frac{|\mathbf{v}_k+\mathbf{h}[\mathbf{v}_k]|^2}{r[\mathbf{v}_k]}\,\mathrm{d}x\,\mathrm{d}t = \int_{a_k}^{b_k}\int_\Omega \frac{1}{2}\frac{|\mathbf{v}_{k-1}+\mathbf{h}[\mathbf{v}_{k-1}+\mathbf{w}_n]|^2}{r[\mathbf{v}_{k-1}+\mathbf{w}_n]}\,\mathrm{d}x\,\mathrm{d}t$$

$$+\int_{a_k}^{b_k}\int_\Omega \frac{1}{2}\frac{|\mathbf{w}_n|^2}{r[\mathbf{v}_{k-1}+\mathbf{w}_n]}\,\mathrm{d}x\,\mathrm{d}t + 2\int_{a_k}^{b_k}\int_\Omega \frac{\mathbf{w}_n\cdot(\mathbf{v}_{k-1}+\mathbf{h}[\mathbf{v}_{k-1}+\mathbf{w}_n])}{r[\mathbf{v}_{k-1}+\mathbf{w}_n]}\,\mathrm{d}x\,\mathrm{d}t$$

$$\geqslant \int_{a_k}^{b_k}\int_\Omega \frac{1}{2}\frac{|\mathbf{v}_{k-1}+\mathbf{h}[\mathbf{v}_{k-1}]|^2}{r[\mathbf{v}_{k-1}]}\,\mathrm{d}x\,\mathrm{d}t + \frac{1}{2}\int_{a_k}^{b_k}\int_\Omega \frac{1}{2}\frac{|\mathbf{w}_n|^2}{r[\mathbf{v}_{k-1}]}\,\mathrm{d}x\,\mathrm{d}t + e_n$$

provided n is large enough, where $e_n\to 0$ as $n\to\infty$ for any fixed k. Consequently, (13.18) gives rise to

$$\int_{a_k}^{b_k}\int_{\Omega}\frac{1}{2}\frac{|\mathbf{v}_k+\mathbf{h}[\mathbf{v}_k]|^2}{r[\mathbf{v}_k]}\,\mathrm{d}x\,\mathrm{d}t+e_n \geqslant \int_{a_k}^{b_k}\int_{\Omega}\frac{1}{2}\frac{|\mathbf{v}_{k-1}+\mathbf{h}[\mathbf{v}_{k-1}]|^2}{r[\mathbf{v}_{k-1}]}\,\mathrm{d}x\,\mathrm{d}t \quad (13.58)$$

$$+\frac{\Lambda(\overline{e})}{4}\int_{a_k}^{b_k}\int_{\Omega}\left(e[\mathbf{v}_{k-1}]-\frac{1}{2}\frac{|\mathbf{v}_{k-1}+\mathbf{h}[\mathbf{v}_{k-1}]|^2}{r[\mathbf{v}_{k-1}]}\right)^2\mathrm{d}x\,\mathrm{d}t$$

$$\geqslant \int_{a_k}^{b_k}\int_{\Omega}\frac{1}{2}\frac{|\mathbf{v}_{k-1}+\mathbf{h}[\mathbf{v}_{k-1}]|^2}{r[\mathbf{v}_{k-1}]}\,\mathrm{d}x\,\mathrm{d}t+\frac{\Lambda(\overline{e})}{4}\frac{1}{|\Omega|}\frac{\alpha_k^2}{\varepsilon_k},$$

where the last line follows from Jensen's inequality. Thus, using (13.56), (13.58), we may find n large enough and $\tau_k \in (a_k, b_k)$ such that (13.52) holds with some λ that can be determined in terms of $\Lambda(\overline{e})$ and $|\Omega|$.

Finally, our goal is to check that $\mathbf{v}_k$, $\mathbb{F}_k$ satisfy (13.53)–(13.55). First we claim that (13.53) is a direct consequence of the causality property (13.6). Next, Lemma 13.4.1, specifically (13.17), yields

$$\frac{N}{2}\lambda_{\max}\left[\frac{(\mathbf{v}_k+\mathbf{h}[\mathbf{v}_{k-1}])\otimes(\mathbf{v}_k+\mathbf{h}[\mathbf{v}_{k-1}])}{r[\mathbf{v}_{k-1}]}-\mathbb{F}_k+\mathbb{H}[\mathbf{v}_{k-1}]\right]$$

$$< e[\mathbf{v}_{k-1}]-\delta_k\left(1+\frac{1}{2^{k-1}}\right)$$

in $[(a_k, b_k)\times\Omega]\cap Q$; whence (13.54) follows from uniform continuity of $\mathbf{h}$, r, $\mathbb{H}$ and e provided n is large enough. To see (13.55), we have to realize that $\mathbf{v}_k=\mathbf{v}_{k-1}$ and $\mathbb{F}_k=\mathbb{F}_{k-1}$ in $[(b_k, T)\times\Omega]\cap Q$, and, similarly to the above, relation (13.55) follows from continuity of $\mathbf{h}$, r, $\mathbb{H}$ and e as son as n is chosen large enough.

Step 3

Our ultimate goal is to observe that $\mathbf{v}$, determined by the limit (13.47), enjoys the desired properties claimed in Theorem 13.6.1. We set $\tau=\lim_{k\to\infty}\tau_k$. Since the functions $\mathbf{v}_k$, $\mathbb{F}_k$ coincide with $\mathbf{v}_{k-1}$, $\mathbb{F}_{k-1}$ on the time intervals $(0, a_k)$, (b_k, T), the properties (13.41)–(13.44) follow by taking the limit in (13.48), (13.53)–(13.55) for $k\to\infty$.

To see (13.45), we first observe that, by virtue of (13.52),

$$\frac{1}{2}\int_{\Omega}\frac{|\mathbf{v}_k+\mathbf{h}[\mathbf{v}_k]|^2}{r[\mathbf{v}_k]}(\tau_k)\,\mathrm{d}x \nearrow Y \text{ as } k\to\infty,$$

and the convergence is uniform on the time intervals (a_k, b_k); whence

$$\frac{\alpha_k}{\varepsilon_k}=\frac{1}{\varepsilon_k}\int_{a_k}^{b_k}\int_{\Omega}\left(e[\mathbf{v}_{k-1}]-\frac{1}{2}\frac{|\mathbf{v}_{k-1}+\mathbf{h}[\mathbf{v}_{k-1}]|^2}{r[\mathbf{v}_{k-1}]}\right)\mathrm{d}x\mathrm{d}t\to 0,$$

which in turn implies

$$\frac{1}{2}\int_{\Omega} \frac{|\mathbf{v}_k + \mathbf{h}[\mathbf{v}_k]|^2}{r[\mathbf{v}_k]}(t)\,\mathrm{d}x \to \int_{\Omega} e[\mathbf{v}](\tau)\,\mathrm{d}x \text{ as } k \to \infty \text{ uniformly for } t \in (a_k, b_k). \tag{13.59}$$

We show that (13.59) yields

$$\mathbf{w}_k(\tau, \cdot) \to \mathbf{w}(\tau, \cdot)$$

which completes the proof of Theorem 13.6.1. Indeed we may write

$$\int_{\Omega} \frac{|\mathbf{v}_m - \mathbf{v}_n|^2}{r[\mathbf{v}_n]}\,\mathrm{d}x = \int_{\Omega} \frac{|\mathbf{v}_m|^2}{r[\mathbf{v}_n]}\,\mathrm{d}x - \int_{\Omega} \frac{|\mathbf{v}_n|^2}{r[\mathbf{v}_n]}\,\mathrm{d}x - 2\int_{\Omega} \frac{(\mathbf{v}_m - \mathbf{v}_n)\cdot\mathbf{v}_n}{r[\mathbf{v}_n]}\,\mathrm{d}x,\ m > n,$$

where the difference of the first two integrals vanishes as $n \to \infty$ uniformly for $t \in (0, T)$; whereas

$$\left|\int_{\Omega} \frac{(\mathbf{v}_m - \mathbf{v}_n)\cdot\mathbf{v}_n}{r[\mathbf{v}_n]}\,\mathrm{d}x\right| = \left|\sum_{k=n}^{m-1}\int_{\Omega} \frac{(\mathbf{v}_{k+1} - \mathbf{v}_k)\cdot\mathbf{v}_n}{r[\mathbf{v}_n]}\,\mathrm{d}x\right| \le \frac{1}{2^{n-1}} \text{ uniformly in } (0, T)$$

in view of (13.51). □

Now, we can define a set of subsolutions on the time interval (τ, T), with

$$\mathbf{u}_0 = \mathbf{v}(\tau),\ Q = Q_\tau = Q \cap [(\tau, T) \times \Omega],$$

and the operators $\mathbf{h}_\tau, r_\tau, e_\tau, \mathbb{H}_\tau$ defined as

$$\mathbf{h}_\tau[\mathbf{w}] = \mathbf{h}[\tilde{\mathbf{w}}]|_{(\tau,T)},\ \text{where } \tilde{\mathbf{w}} = \begin{cases} \mathbf{v} \text{ in } [0, \tau] \\ \mathbf{w} \in [\tau, T] \end{cases},$$

where $\mathbf{v}$ is the function constructed in Theorem 13.6.1. In accordance with (13.45), we have $\mathbf{v}|_{[\tau,T]}$ is a subsolution, and

$$\frac{1}{2}\int_{\Omega} \frac{|\mathbf{u}_0 + \mathbf{h}_\tau[\mathbf{u}_0]|^2}{r_\tau[\mathbf{u}_0]}\,\mathrm{d}x = \int_{\Omega} e_\tau[\mathbf{u}_0]\,\mathrm{d}x.$$

Finally we note that in this case the weak solutions $\mathbf{u}$ constructed via Theorem 13.2.1 will satisfy

$$\frac{1}{2}\frac{|\mathbf{u} + \mathbf{h}[\mathbf{u}]|^2}{r[\mathbf{u}]}(t, x) = e[\mathbf{u}](t, x)$$

for a.a. $(t, x) \in (0, T) \times \Omega$ and *including* the initial time $t = 0$.

13.6.1 Example, Dissipative Solutions to the Euler-Fourier System

Revisiting the Euler-Fourier system introduced in Sect. 13.5.1, we say that $\rho, \vartheta, \mathbf{u}$ is a *dissipative solution* of (13.20)–(13.22), if, in addition, the energy balance

$$E(t) \equiv \int_\Omega \left(\frac{1}{2}\rho|\mathbf{u}|^2 + \frac{3}{2}\rho\vartheta \right)(t,\cdot)\,\mathrm{d}x = \int_\Omega \left(\frac{1}{2}\rho_0|\mathbf{u}_0|^2 + \frac{3}{2}\rho_0\vartheta_0 \right)\mathrm{d}x \tag{13.60}$$

holds for a.a. $t \in (0,T)$.

As a possible application of Theorem 13.6.1, one can show the following result, see [8, Theorem 4.2]:

Theorem 13.6.2 *Under the hypotheses of Theorem 13.5.1, let $T > 0$ and the data*

$$\rho_0,\ \vartheta_0 \in C^2(\Omega),\ \rho_0, \vartheta_0 > 0$$

be given.

Then there exists $\mathbf{u}_0 \in L^\infty(\Omega; R^N)$ such that the Euler-Fourier system (13.20)–(13.22), with the initial conditions (13.26), admits infinitely many dissipative *solutions in $(0,T) \times \Omega$.*

Acknowledgements The research of E.F. leading to these results has received funding from the European Research Council under the European Union's Seventh Framework Programme (FP7/2007-2013)/ ERC Grant Agreement 320078.

References

1. Antonelli, P., Marcati, P.: On the finite energy weak solutions to a system in quantum fluid dynamics. Commun. Math. Phys. **287**, 657–686 (2009)
2. Antonelli, P., Marcati, P.: The quantum hydrodynamics system in two space dimensions. Arch. Rational Mech. Anal. **203**, 499–527 (2012)
3. Audiard, C.: Dispersive smoothing for the Euler-Korteweg model. SIAM J. Math. Anal. **44**(4), 3018–3040 (2012)
4. Benzoni-Gavage, S.: Spectral transverse instability of solitary waves in Korteweg fluids. J. Math. Anal. Appl. **361**(2), 338–357 (2010)
5. Benzoni-Gavage, S.: Planar traveling waves in capillary fluids. Differ. Integral Equ. **26**(3–4), 439–485 (2013)
6. Bresch, D., Desjardins, B., Ducomet, B.: Quasi-neutral limit for a viscous capillary model of plasma. Ann. Inst. Poincare **22**, 1–9 (2005)
7. Chiodaroli, E.: A counterexample to well-posedness of entropy solutions to the compressible Euler system. J. Hyperbolic Differ. Equ. **11**(3), 493–519 (2014)
8. Chiodaroli, E., Feireisl, E., Kreml, O.: On the weak solutions to the equations of a compressible heat conducting gas. Annal. Inst. Poincaré, Anal. Nonlinear **32**, 225–243 (2015)
9. Dafermos, C.M.: Hyperbolic Conservation Laws in Continuum Physics. Springer, Berlin (2000)
10. Dafermos, C.M.: The second law of thermodynamics and stability. Arch. Rational Mech. Anal. **70**, 167–179 (1979)

11. De Lellis, C., Székelyhidi Jr., L.: On admissibility criteria for weak solutions of the Euler equations. Arch. Ration. Mech. Anal. **195**(1), 225–260 (2010)
12. Desvillettes, L., Villani, C.: On the trend to global equilibrium for spatially inhomogeneous kinetic systems: the Boltzmann equation. Invent. Math. **159**(2), 245–316 (2005)
13. Donatelli, D., Feireisl, E., Marcati, P.: Well/ill posedness for the Euler-Korteweg-poisson system and related problems. Commun. Partial Differ. Equ. **40**(7), 1314–1335 (2015)
14. Feireisl, E.: On weak solutions to a diffuse interface model of a binary mixture of compressible fluids. Discret. Contin. Dyn. S.—Ser. S. **9**(1), 173–183 (2014)
15. Jüngel, A.: Transport Equations for Semiconductors. Lecture Notes in Physics, vol. 773. Springer, Berlin (2009)
16. Kotschote, M.: Strong well-posedness for a Korteweg-type model for the dynamics of a compressible non-isothermal fluid. J. Math. Fluid Mech. **12**(4), 473–484 (2010)
17. Kotschote, M.: Dynamics of compressible non-isothermal fluids of non-Newtonian Korteweg type. SIAM J. Math. Anal. **44**(1), 74–101 (2012)
18. Lowengrub, J., Truskinovsky, L.: Quasi-incompressible Cahn-Hilliard fluids and topological transitions. R. Soc. Lond. Proc. Ser. A Math. Phys. Eng. Sci. **454**(1978), 2617–2654 (1998)
19. Tartar, L.: Compensated compactness and applications to partial differential equations. In: Knopps, L.J. (ed.) Nonlinear Analysis and Mechanics, Heriot-Watt Symposium. Research Notes in Math, vol. 39, pp. 136–211. Pitman, Boston (1975)

Chapter 14
Enstrophy Variations in the Incompressible 2D Euler Flows and α Point Vortex System

Takeshi Gotoda and Takashi Sakajo

Abstract The dissipation of the enstrophy, which is the L^2 norm of the vorticity, in the zero-viscous limit gives rise to the emergence of inertial range in the ensemble average of the energy density spectrum in 2D fluid turbulence. However, it is mathematically known that not only smooth solutions but also weak solutions in $L^p(\mathbb{R}^2)$, $p > 2$ to the 2D Euler equations never dissipates the enstrophy [7]. This indicates that weak solutions for initial vorticity distributions belonging to weaker function spaces such as the space of Radon measure on $\mathbb{R}^2$ should be constructed to obtain such singular solutions with the enstrophy dissipation, but no existence result in this function space has not yet been established. We here consider the 2D Euler-α equations, which is a dispersive regularization of the Euler equations with a scaling parameter α, for the initial vorticity distributions whose support consists of a set of N points, called α-point vortices. We shall construct singular weak solutions to the Euler equations from those of the evolution equations of the α point vortices by taking their $\alpha \to 0$ limit. We then numerically demonstrate that the self-similar collapse of the α point vortices gives rise to the anomalous enstrophy dissipation in the distributional sense and it is a robust mechanism of the enstrophy dissipation observed for a wide range of initial configurations of α point vortices.

Keywords Hamiltonian dynamics · Point Vortex collapse · 2D turbulence · Euler-alpha model · Onsager's conjecture

14.1 Introduction

The isotropic turbulence is a model of fluid turbulence, in which the flow field is assumed to be statistically steady, homogeneous and isotropic. Kolmogorov success-

T. Gotoda · T. Sakajo (✉)
Department of Mathematics, Kyoto University,
Kitashirakawa Oiwake-cho, Sakyo-ku, Kyoto, Japan
e-mail: sakajo@math.kyoto-u.ac.jp

T. Gotoda
e-mail: gotoda@math.kyoto-u.ac.jp

Y. Shibata and Y. Suzuki (eds.), *Mathematical Fluid Dynamics, Present and Future*, Springer Proceedings in Mathematics & Statistics 183,
DOI 10.1007/978-4-431-56457-7_14

fully revealed some fundamental properties of 3D fluid turbulence with this isotropic model [13, 14]. Numerical investigation of the incompressible Navier-Stokes equations with high Reynolds number in a 3D periodic box has also supported the statistical properties. See, e.g., [11]. In spite of these great achievements, it is still far from a complete theoretical understanding how a set of solutions of the nonlinear fluid equations such as the Navier-Stokes equations gives rise to their statistical properties of the isotropic turbulence.

One of the characteristic properties of the isotropic turbulence is the emergence of a region of wavenumbers $\boldsymbol{k}$, called the *inertial range*, in the ensemble average of the energy density spectrum $\langle E(k)\rangle$ with $k = |\boldsymbol{k}|$, where the energy dissipates self-similarly at a constant rate as $\langle E(k)\rangle \sim k^{-\frac{5}{3}}$ known as the *energy cascade*. In the Kolmogorov's theory, it is derived under the assumption that the energy dissipation rate of the flow field converges to a strictly positive constant in the zero-viscous limit. However, naively speaking, this assumption contradicts the fact that smooth solutions to the incompressible Euler equations describing non-viscous flow evolutions conserve the energy. To this problem, Onsager [26] has conjectured that non-smooth solutions to the Euler equations with Hölder continuity of exponent greater than $1/3$ can conserve the energy, which is known as *Onsager's conjecture*. This assertion has been justified mathematically by Constantin et al. [3] and Duchon and Robert [6] under the assumption that weak solutions to the 3D incompressible Euler equations exist. Onsager's conjecture, in other words, also claims that the Euler flows with less than $1/3$-Hölder continuity may dissipate the energy. Regarding the existence of weak solutions with a Hölder continuity, a great progress has been made recently by Buckmaster et al. [2], in which weak solutions to the 3D Euler equation exist in $L^1([0,T); C^{\frac{1}{3}}(\mathbb{T}^3))$ with *any* prescribed energy profile. This supports the occurrence of the energy dissipation in the weak Euler flows. In spite of the recent developments, it is uncertain, from a viewpoint of fluid dynamics, what kind of physical mechanism of those singular Euler flows triggers the energy dissipation, which is our concern here. However, it is difficult to consider this problem in 3D flows, since the global well-posedness of the 3D Navier-Stokes equations as well as the 3D Euler equations has not yet been established and we are thus unable to define the evolution of solutions as a dynamical system. So, for simplicity, we reformulate the same problem for 2D turbulent flows in this paper.

In 2D fluid turbulence, it is pointed out that there also appears an inertial range in the energy density spectra caused by the dissipation of not the energy but the enstrophy, which is the L^2 norm of the vorticity, for sufficiently small viscosity [1, 15, 18]. Since the enstrophy is conserved by smooth solutions to the 2D Euler equations, the 2D turbulence is characterized by non-smooth singular weak solutions to the 2D Euler equations that dissipate the enstrophy, which is a 2D analogue of the energy in Onsager's conjecture for the 3D isotropic turbulence. Hence, we expect that the theoretical description of the singular enstrophy dissipation in terms of 2D fluid dynamics shall shed some lights upon the theory of 2D fluid turbulence.

The 2D Euler equations have been well-investigated mathematically. See the book by Marchioro and Pulvirenti [22] for a list of references. The global existence of the unique solution has been established for the initial vorticity data in $L^1(\mathbb{R}^2) \cap L^\infty(\mathbb{R}^2)$ [29] and in $L^p(\mathbb{R}^2)$ with $1 < p < \infty$ [5, 8]. In regards to the enstrophy dissipation, Eyink [7] has shown that weak solutions of the 2D Euler equations in $L^p(\mathbb{R}^2)$, $p > 2$ can not dissipate the enstrophy in a weak sense. This indicates that it is necessary to deal with the initial vorticity data in a weaker space, the space of Radon measure $\mathscr{M}(\mathbb{R}^2)$ on $\mathbb{R}^2$ for instance, to obtain such weak solutions with the singular enstrophy dissipation. The global existence of (non-unique) weak solutions has been established for the initial vorticity $\omega_0 \in \mathscr{M}(\mathbb{R}^2)$ and its induced velocity $\boldsymbol{v}_0 \in L^2_{loc}(\mathbb{R}^2)$ [4, 21], but no existence has not yet been proven only for $\omega_0 \in \mathscr{M}(\mathbb{R}^2)$.

In the meantime, when the initial vorticity distribution consists of a finite number of δ measures belonging to $\mathscr{M}(\mathbb{R}^2)$, one can formally reduce the Euler equations to the system of ordinary differential equations describing the evolution of the δ-singularities, called the *point vortex (PV) system*. The PV system has been used as inviscid models of 2D flows [23], but solutions of the PV system do not define a weak solutions to the 2D Euler equations, since the velocity field induced by point vortices does not belong to $L^2_{loc}(\mathbb{R}^2)$. Accordingly, it is impossible to investigate the evolution of weak solutions to the 2D Euler equations in terms of the evolution of point vortices.

In order to construct such a singular weak solution to the 2D Euler equations in $\mathscr{M}(\mathbb{R}^2)$, we consider weak solutions to a regularized the Euler equations, called the 2D Euler-α equations; For incompressible velocity field $\boldsymbol{u}^\alpha(\boldsymbol{x}, t)$ in space $\boldsymbol{x} \in \mathbb{R}^2$ and time $t \in \mathbb{R}$, they are given by

$$(1 - \alpha^2 \Delta)\partial_t \boldsymbol{u}^\alpha + \boldsymbol{u}^\alpha \cdot \nabla(1 - \alpha^2 \Delta)\boldsymbol{u}^\alpha - \alpha^2 (\nabla \boldsymbol{u}^\alpha)^T \cdot \Delta \boldsymbol{u}^\alpha = -\nabla p, \quad (14.1)$$

$$\nabla \cdot \boldsymbol{u}^\alpha = 0, \quad \boldsymbol{u}^\alpha(\boldsymbol{x}, 0) = \boldsymbol{u}_0^\alpha(\boldsymbol{x}), \quad (14.2)$$

where $p(\boldsymbol{x}, t)$ denotes the pressure and α is a regularization parameter [9, 10]. Note that (14.1) and (14.2) are equivalent to the Euler equations when $\alpha = 0$. The momentum equation (14.1) is a dispersive regularization, since it is derived by the least action principle of the regularized energy,

$$\mathcal{E} = \int_0^T \int_{\mathbb{R}^2} |\boldsymbol{u}(\boldsymbol{x}, t)|^2 + \alpha^2 |\nabla \boldsymbol{u}(\boldsymbol{x}, t)|^2 \, d\boldsymbol{x} dt.$$

Physically, the momentum equation (14.1) is obtained by taking an average of the spatial flow information below the small α. It has also been numerically confirmed in [20] that the Navier-Stokes-α equations acquire the inertial ranges in the energy density spectrum corresponding to the backward energy cascade and the forward enstrophy cascade for small viscosity and small scale α, which is a common property with 2D fluid turbulence. Hence, the Euler-α equations are regarded as a model of 2D fluid turbulence in the nonviscous limit. Mathematically, Linshiz and Titi [19] have shown that there exists a unique global solution of (14.1) and (14.2) for the initial velocity field in the Sobolev space $H^m(\mathbb{R}^2)$, $m > 2$ and that the solution converges to that of the Euler equations in $L^\infty([0, \infty); H^m)$. Moreover, the 2D Euler-α equations still have the unique global weak solution for the initial vorticity distributions in $\mathscr{M}(\mathbb{R}^2)$ [25]. Hence, the initial pointwise discrete vorticity distribution gives rise to a weak solution to the Euler-α equations. This is a significant difference from the relation between the Euler equations and the PV system. Consequently, we are able to describe a singular weak solution to the Euler equations that dissipates the enstrophy as $\alpha \to 0$ in terms of the evolution of α point vortices.

With the scalar α-vorticity, $q = (1 - \alpha^2 \Delta)\nabla^{\perp} \boldsymbol{u}^{\alpha}$, the 2D Euler-$\alpha$ equations are reduced to the following transport equation for q:

$$\partial_t q(\boldsymbol{x}, t) + (\boldsymbol{u}^{\alpha}(\boldsymbol{x}, t) \cdot \nabla) q(\boldsymbol{x}, t) = 0, \qquad \boldsymbol{u}^{\alpha}(\boldsymbol{x}, t) = \int_{\mathbb{R}^2} K^{\alpha}(\boldsymbol{x}, y) q(\boldsymbol{y}, t) d\boldsymbol{y}, \tag{14.3}$$

$$q(\boldsymbol{x}, 0) = q_0(\boldsymbol{x}) = (1 - \alpha^2 \Delta)\nabla^{\perp} u_0(\boldsymbol{x}), \tag{14.4}$$

where

$$K^{\alpha} = \nabla^{\perp} G^{\alpha}, \quad -\Delta G^{\alpha}(\boldsymbol{x} - \boldsymbol{y}) = \frac{1}{2\pi\alpha^2} K_0\left(\frac{|\boldsymbol{x} - \boldsymbol{y}|}{\alpha}\right).$$

Here $K_0(x)$ denotes the modified Bessel function of the second kind [30]. We define the Lagrangian flow map induced by $\boldsymbol{u}^{\alpha}(\boldsymbol{x}, t)$ as follows.

$$\frac{d}{dt}\boldsymbol{\eta}^{\alpha}(\boldsymbol{x}, t) = \boldsymbol{u}^{\alpha}(\boldsymbol{\eta}^{\alpha}(\boldsymbol{x}, t), t), \qquad \boldsymbol{\eta}^{\alpha}(\boldsymbol{x}, 0) = \boldsymbol{x}. \tag{14.5}$$

We note that if $q_0 \in L^1(\mathbb{R}^2) \cap L^{\infty}(\mathbb{R}^2)$, then there exists a unique global weak solution $q \in C(\mathbb{R}; L^1(\mathbb{R}^2) \cap L^{\infty}(\mathbb{R}^2))$ and a unique flow map $\boldsymbol{\eta}^{\alpha} \in C^1(\mathbb{R}; \mathcal{G})$, where $\mathcal{G}$ denotes the group of all homeomorphisms on $\mathbb{R}^2$, which is easily shown with the same argument for the 2D Euler equations [22, 29].

Suppose that the initial vorticity field is represented by α-point vortices:

$$q_0(\boldsymbol{x}) = \sum_{m=1}^{N} \Gamma_m \delta(\boldsymbol{x} - \boldsymbol{x}_m^0), \tag{14.6}$$

where $\boldsymbol{x}_m^0$, $m = 1, \ldots, N$ are the support of point singularities and Γ_m denotes the strength of the α-point vortices. Substituting (14.6) into (14.5) and setting $\boldsymbol{\eta}^{\alpha}(\boldsymbol{x}_n^0, t) = \boldsymbol{x}_n^{\alpha}(t)$, we obtain the evolution of the α-point vortices.

$$\frac{d}{dt}\boldsymbol{x}_m^{\alpha} = \frac{1}{2\pi} \sum_{n \neq m}^{N} \Gamma_n \frac{(\boldsymbol{x}_m^{\alpha} - \boldsymbol{x}_n^{\alpha})^{\perp}}{(l_{mn}^{\alpha})^2} B_K\left(\frac{l_{mn}^{\alpha}}{\alpha}\right), \qquad \boldsymbol{x}_m^{\alpha}(0) = \boldsymbol{x}_m^0, \qquad m = 1, \ldots, N, \tag{14.7}$$

where $l_{mn}^{\alpha}(t) = |\boldsymbol{x}_m^{\alpha}(t) - \boldsymbol{x}_n^{\alpha}(t)|$ and $B_K(x) = 1 - xK_1(x)$. Here, $K_1(x)$ denotes the modified Bessel function of the second kind [30]. We call the Eq. (14.7) the α-point vortex (αPV) system. Let us note that the Eq. (14.7) is equivalent to the PV system when we set $\alpha = 0$:

$$\frac{d}{dt}\boldsymbol{x}_m = \frac{1}{2\pi}\sum_{n\neq m}^{N}\Gamma_n\frac{(\boldsymbol{x}_m - \boldsymbol{x}_n)^{\perp}}{(l_{mn})^2}, \qquad \boldsymbol{x}_m(0) = \boldsymbol{x}_m^0, \qquad m = 1, \ldots, N, \tag{14.8}$$

where $l_{mn}(t) = |\boldsymbol{x}_m(t) - \boldsymbol{x}_n(t)|$.

In this paper, we investigate singular weak solutions to the 2D Euler equations in terms of the $\alpha \to 0$ limit solutions of the αPV system. In Sect. 14.2, we consider the relation between solutions to the 2D Euler-α equations and those of the αPV system and show that the αPV system gives rise to a weak solution of the 2D Euler-α equations. In Sect. 14.3, we consider the evolution of three or four α point vortices for initial configurations leading to a self-similar collapse in the PV system. In particular, we pay close attention to how the $\alpha \to 0$ limit evolution of the αPV system and numerically demonstrate that this gives rise to an anomalous enstrophy dissipation in a weak sense. Section 14.4 gives a summary and discussion for the future direction.

14.2 The 2D Euler-α Equations and the αPV System

We consider the *vortex model* of the Euler-α Eqs. (14.1) and (14.2), where the initial distribution of the α-vorticity is concentrated in a set of small domains with constant vorticity containing $\boldsymbol{x}_m \in \mathbb{R}^2$ for $m = 1, \ldots, N$. When the size of these domains tends to zero with keeping the circulations unchanged, one can expect that the evolution of these domains is approximated by the αPV system (14.7). The vortex model for the Euler equations was originally introduced by Marchioro and Pulvirenti [22] to compare the Euler equations and the PV system (14.8). We follow the same argument as theirs to figure out the similarity and the difference between the PV system and the αPV system.

The first step is considering the evolution of an isolated single small vorticity domain in the presence of a smooth divergence-free external force.

Theorem 14.2.1 *Let $\boldsymbol{F}^\alpha$ be a divergence-free and uniformly bounded time-dependent vector field satisfying the global Lipschitz condition*

$$|\boldsymbol{F}^\alpha(\boldsymbol{x}, t) - \boldsymbol{F}^\alpha(\boldsymbol{y}, t)| \leq L|\boldsymbol{x} - \boldsymbol{y}|$$

for some $L > 0$ independent of α. We consider the weak form of the α-vorticity Eqs. (14.3) and (14.4) as follows.

$$\begin{cases} \dfrac{d}{dt}\displaystyle\int_{\mathbb{R}^2} q_\varepsilon(\boldsymbol{x}, t) f(\boldsymbol{x}) d\boldsymbol{x} = \int_{\mathbb{R}^2} q_\varepsilon(\boldsymbol{x}, t)(\boldsymbol{u}_\varepsilon^\alpha(\boldsymbol{x}, t) + \boldsymbol{F}^\alpha(\boldsymbol{x}, t)) \cdot \nabla f(\boldsymbol{x}) d\boldsymbol{x}, \\ \boldsymbol{u}_\varepsilon^\alpha(\boldsymbol{x}, t) = \displaystyle\int_{\mathbb{R}^2} K^\alpha(\boldsymbol{x}, \boldsymbol{y}) q_\varepsilon(\boldsymbol{y}, t) d\boldsymbol{y}, \end{cases}$$

with the initial vorticity profile $q_\varepsilon(\boldsymbol{x},0)=\varepsilon^{-2}\chi_{\Lambda_\varepsilon}(\boldsymbol{x})$, *where* $\chi_{\Lambda_\varepsilon}$ *denotes the characteristic function for a family of open sets* $\{\Lambda_\varepsilon\}_{\varepsilon>0}$ *satisfying*

$$\text{meas.}\,\Lambda_\varepsilon=\varepsilon^2,\quad \Lambda_\varepsilon\subset B(\boldsymbol{x}^*,\beta\varepsilon).$$

Here, $B(\boldsymbol{x},r)$ *denotes a ball centered at* $\boldsymbol{x}$ *with radius* r *and* $\beta>0$ *is a constant. Then, the solution of the Euler-*α *equations is given by* $q_\varepsilon(x,t)=\varepsilon^{-2}\chi_{\Lambda_\varepsilon(t)}(x)$. *Moreover, for an arbitrary fixed* $T>0$, *we have the following properties.*

(i)

$$\lim_{\varepsilon\to0}\boldsymbol{G}_\varepsilon(t)=\boldsymbol{G}(t),$$

where $\boldsymbol{G}_\varepsilon(t)$ *represents the center of vorticity,*

$$\boldsymbol{G}_\varepsilon(t)=\int_{\mathbb{R}^2}\boldsymbol{x}q_\varepsilon(\boldsymbol{x},t)d\boldsymbol{x},$$

and $\boldsymbol{G}(t)$ *is the solution of the initial value problem*

$$\frac{d}{dt}\boldsymbol{G}(t)=\boldsymbol{F}^\alpha(\boldsymbol{G}(t),t),\qquad \boldsymbol{G}(0)=\boldsymbol{x}^*.$$

(ii) For any $f\in C_0^\infty(\mathbb{R}^2)$, *we have*

$$\lim_{\varepsilon\to0}\int_{\mathbb{R}^2}q_\varepsilon(\boldsymbol{x},t)f(\boldsymbol{x})d\boldsymbol{x}=f(\boldsymbol{G}(t)),\quad \forall t\in[0,T].$$

(iii) For all $d>0$, *we choose* $\varepsilon_0=\varepsilon_0(d,T,\|\boldsymbol{F}^\alpha\|_{L^\infty})>0$ *such that, if* $\varepsilon<\varepsilon_0$, *then*

$$\Lambda_\varepsilon(t)\subset B(\boldsymbol{G}_\varepsilon(t),d),\quad \forall t\in[0,T].$$

Proof Since $q_\varepsilon(\cdot,0)$ belongs to $L^1(\mathbb{R}^2)\cap L^\infty(\mathbb{R}^2)$, there exist a unique weak solution $q_\varepsilon(\cdot,t)\in L^1(\mathbb{R}^2)\cap L^\infty(\mathbb{R}^2)$ and the flow map $\boldsymbol{\eta}^\alpha(\boldsymbol{x},t)$ globally in time as we note in the introduction. Moreover, it is easy to see that $q_\varepsilon(\boldsymbol{x},t)=\varepsilon^{-2}\chi_{\Lambda_\varepsilon(t)}(\boldsymbol{x})$, where $\Lambda_\varepsilon(t)\equiv\{\boldsymbol{\eta}^\alpha(\boldsymbol{x},t)|\boldsymbol{x}\in\Lambda_\varepsilon\}=\{\boldsymbol{x}|\boldsymbol{\eta}^\alpha(\boldsymbol{x},-t)\in\Lambda_\varepsilon\}$, since $q_\varepsilon(\boldsymbol{x},t)=q_\varepsilon(\boldsymbol{\eta}^\alpha(\boldsymbol{x},-t),0)=\varepsilon^{-2}\chi_{\Lambda_\varepsilon}(\boldsymbol{\eta}^\alpha(\boldsymbol{x},-t))=\varepsilon^{-2}\chi_{\Lambda_\varepsilon(t)}(\boldsymbol{x})$.

First, we prove that the moment of inertia around $\boldsymbol{G}_\varepsilon(t)$,

$$I_\varepsilon(t)=\int_{\mathbb{R}^2}q_\varepsilon(\boldsymbol{x},t)|\boldsymbol{x}-\boldsymbol{G}_\varepsilon(t)|^2d\boldsymbol{x},$$

is vanishing in the limit of $\varepsilon\to0$. By taking the time derivative of $\boldsymbol{G}_\varepsilon(t)$, we have

$$\frac{d}{dt}\boldsymbol{G}_\varepsilon(t)=\int_{\mathbb{R}^2}q_\varepsilon(\boldsymbol{x},t)(\boldsymbol{u}^\alpha(\boldsymbol{x},t)+\boldsymbol{F}^\alpha(\boldsymbol{x},t))d\boldsymbol{x}=\int_{\mathbb{R}^2}q_\varepsilon(\boldsymbol{x},t)\boldsymbol{F}^\alpha(\boldsymbol{x},t)d\boldsymbol{x}. \tag{14.9}$$

Similarly, we have

$$\begin{aligned}\frac{d}{dt}I_\varepsilon(t) &= 2\int_{\mathbb{R}^2} q_\varepsilon(\boldsymbol{x},t)(\boldsymbol{u}^\alpha(\boldsymbol{x},t)+\boldsymbol{F}^\alpha(\boldsymbol{x},t))\cdot(\boldsymbol{x}-\boldsymbol{G}_\varepsilon(t))d\boldsymbol{x}\\ &\quad +2\frac{d}{dt}\boldsymbol{G}_\varepsilon(t)\cdot\int_{\mathbb{R}^2} q_\varepsilon(\boldsymbol{x},t)(\boldsymbol{x}-\boldsymbol{G}_\varepsilon(t))d\boldsymbol{x}\\ &= 2\int_{\mathbb{R}^2} q_\varepsilon(\boldsymbol{x},t)\boldsymbol{F}^\alpha(\boldsymbol{x},t)\cdot(\boldsymbol{x}-\boldsymbol{G}_\varepsilon(t))d\boldsymbol{x}\\ &= 2\int_{\mathbb{R}^2} q_\varepsilon(\boldsymbol{x},t)(\boldsymbol{F}^\alpha(\boldsymbol{x},t)-\boldsymbol{F}^\alpha(\boldsymbol{G}_\varepsilon(t),t))\cdot(\boldsymbol{x}-\boldsymbol{G}_\varepsilon(t))d\boldsymbol{x},\end{aligned}$$

from which

$$\left|\frac{d}{dt}I_\varepsilon(t)\right| \le 2LI_\varepsilon(t).$$

Hence, $I_\varepsilon(t)\le I_\varepsilon(0)\exp(2LT)$ and we get the convergence of $I_\varepsilon(t)$, since

$$I_\varepsilon(0)=\int_{\mathbb{R}^2}\frac{1}{\varepsilon^2}\chi_{\Lambda_\varepsilon}|\boldsymbol{x}-\boldsymbol{x}^*|^2d\boldsymbol{x}\le\int_{\mathbb{R}^2}\frac{1}{\varepsilon^2}\chi_{\Lambda_\varepsilon}(\beta\varepsilon)^2d\boldsymbol{x}=(\beta\varepsilon)^2\to 0,$$

as $\varepsilon\to 0$. Here, we recall that $|\boldsymbol{x}-\boldsymbol{x}^*|\le\beta\varepsilon$, if $\boldsymbol{x}\in\chi_{\Lambda_\varepsilon}$.

Next, we prove (i) and (ii) which are straightforward. We have

$$\begin{aligned}|\boldsymbol{G}(t)-\boldsymbol{G}_\varepsilon(t)| &\le \left|\boldsymbol{x}^*-\boldsymbol{G}_\varepsilon(0)\right|+\int_0^t|\boldsymbol{F}^\alpha(\boldsymbol{G}(s),s))-\boldsymbol{F}^\alpha(\boldsymbol{G}_\varepsilon(s),s)|\,ds\\ &\quad+\int_0^t\left|\boldsymbol{F}^\alpha(\boldsymbol{G}_\varepsilon(s),s)-\int_{\mathbb{R}^2}q_\varepsilon(\boldsymbol{x},s)\boldsymbol{F}^\alpha(\boldsymbol{x},s)d\boldsymbol{x}\right|ds\\ &\le \left|\boldsymbol{x}^*-\boldsymbol{G}_\varepsilon(0)\right|+L\int_0^t|\boldsymbol{G}(s)-\boldsymbol{G}_\varepsilon(s)|\,ds+TL\,(I_\varepsilon(t))^{1/2}\\ &\le \left(\left|\boldsymbol{x}^*-\boldsymbol{G}_\varepsilon(0)\right|+TL\,(I_\varepsilon(0))^{1/2}\exp(LT)\right)\exp(Lt).\end{aligned}$$

By Gronwall's inequality, we finish the proof of (i). Regarding (ii), we use the mean value theorem.

$$\left|\int_{\mathbb{R}^2}q_\varepsilon(\boldsymbol{x},t)f(\boldsymbol{x})d\boldsymbol{x}-f(\boldsymbol{G}_\varepsilon(t))\right|=\int_{\mathbb{R}^2}|q_\varepsilon(\boldsymbol{x},t)|\,|f(\boldsymbol{x})-f(\boldsymbol{G}_\varepsilon(s))|\,d\boldsymbol{x}\le\|\nabla f\|_{L^\infty}\,(I_\varepsilon(t))^{\frac{1}{2}}.$$

Therefore, we have

$$\left|\int_{\mathbb{R}^2}q_\varepsilon(\boldsymbol{x},t)f(\boldsymbol{x})d\boldsymbol{x}-f(\boldsymbol{G}(t))\right|\le\|\nabla f\|_{L^\infty}\,(I_\varepsilon(t))^{1/2}+\|\nabla f\|_{L^\infty}\,|\boldsymbol{G}(t)-\boldsymbol{G}_\varepsilon(t)|\,,$$

and (ii) follows by using the above results.

The last step is proving the localization property (iii). To control the vorticity flux, we introduce the following function $W_R \in C_0^\infty(\mathbb{R}^2)$, depending only on $|\boldsymbol{x}|$, such that:

$$W_R(\boldsymbol{x}) = \begin{cases} 1 \ , |\boldsymbol{x}| \le R, \\ 0 \ , |\boldsymbol{x}| > 2R, \end{cases}$$

such that, for some $C > 0$,

$$|\nabla W_R(\boldsymbol{x})| \le \frac{C}{R}, \qquad |\nabla W_R(\boldsymbol{x}) - \nabla W_R(\boldsymbol{y})| \le \frac{C}{R^2}|\boldsymbol{x} - \boldsymbol{y}|.$$

Define the quantity $\mu_R(t)$ by

$$\begin{aligned}\mu_R(t) &= 1 - \varepsilon^{-2}\int_{\Lambda_\varepsilon} W_R(\boldsymbol{G}_\varepsilon(t) - \boldsymbol{\eta}_\varepsilon^\alpha(\boldsymbol{x},t))d\boldsymbol{x} \\ &= 1 - \int_{\mathbb{R}^2} q_\varepsilon(\boldsymbol{x},0) W_R(\boldsymbol{G}_\varepsilon(t) - \boldsymbol{\eta}_\varepsilon^\alpha(\boldsymbol{x},t))d\boldsymbol{x} \\ &= 1 - \int_{\mathbb{R}^2} q_\varepsilon(\boldsymbol{x},t) W_R(\boldsymbol{G}_\varepsilon(t) - \boldsymbol{x})d\boldsymbol{x},\end{aligned}$$

where $\boldsymbol{\eta}_\varepsilon^\alpha(\boldsymbol{x},t)$ is the solution of the initial value problem

$$\frac{d}{dt}\boldsymbol{\eta}_\varepsilon^\alpha(\boldsymbol{x},t) = \boldsymbol{u}(\boldsymbol{\eta}_\varepsilon^\alpha(\boldsymbol{x},t),t) + \boldsymbol{F}^\alpha(\boldsymbol{\eta}_\varepsilon^\alpha(\boldsymbol{x},t),t), \qquad \boldsymbol{\eta}_\varepsilon^\alpha(\boldsymbol{x},0) = \boldsymbol{x} \in \Lambda_\varepsilon.$$

Moreover, setting $m_t(R) = \varepsilon^{-2}|\Lambda_\varepsilon(t) \cap B(\boldsymbol{G}_\varepsilon(t), R)^C|$, we have

$$m_t(R) \le \frac{1}{R^2} I_\varepsilon(t) \le \frac{\exp(2LT)}{R^2} I_\varepsilon(0), \qquad m_t(R) \le \mu_{R/2}(t).$$

We estimate the time derivative

$$\begin{aligned}\frac{d}{dt}\mu_R(t) &= \varepsilon^{-4}\int_{\Lambda_\varepsilon(t)} (\nabla W_R)(\boldsymbol{G}_\varepsilon(t) - \boldsymbol{x}) \cdot \int_{\Lambda_\varepsilon(t)} K^\alpha(\boldsymbol{x} - \boldsymbol{y}) d\boldsymbol{y} d\boldsymbol{x} \\ &\quad + \varepsilon^{-4}\int_{\Lambda_\varepsilon(t)} (\nabla W_R)(\boldsymbol{G}_\varepsilon(t) - \boldsymbol{x}) \cdot \int_{\Lambda_\varepsilon(t)} (\boldsymbol{F}^\alpha(\boldsymbol{x},t) - \boldsymbol{F}^\alpha(\boldsymbol{y},t)) d\boldsymbol{y} d\boldsymbol{x} \\ &= J_1 + J_2.\end{aligned}$$

To estimate the first term in the right-hand side, we use the estimate of the kernel K^α,

$$|J_1| \le \varepsilon^{-4}\frac{C}{R}\int_{\Lambda_\varepsilon(t) \cap \{R < |\boldsymbol{G}_\varepsilon(t) - \boldsymbol{x}| < 2R\}} \int_{\Lambda_\varepsilon(t)} |K^\alpha(\boldsymbol{x} - \boldsymbol{y})| d\boldsymbol{y} d\boldsymbol{x} \le \frac{C}{\alpha R} m_t(R).$$

In the second term, we use the properties of $\boldsymbol{F}^\alpha$,

$$|J_2| \le \varepsilon^{-4}\frac{C}{R}\int_{\Lambda_\varepsilon(t)\cap\{R<|\boldsymbol{G}_\varepsilon(t)-\boldsymbol{x}|<2R\}}\int_{\Lambda_\varepsilon(t)\cap B(\boldsymbol{G}_\varepsilon(t),R)^C}|\boldsymbol{F}^\alpha(\boldsymbol{x},t)-\boldsymbol{F}^\alpha(\boldsymbol{y},t)|d\boldsymbol{y}d\boldsymbol{x}$$
$$+\varepsilon^{-4}\frac{C}{R}\int_{\Lambda_\varepsilon(t)\cap\{R<|\boldsymbol{G}_\varepsilon(t)-\boldsymbol{x}|<2R\}}\int_{\Lambda_\varepsilon(t)\cap B(\boldsymbol{G}_\varepsilon(t),R)}|\boldsymbol{F}^\alpha(\boldsymbol{x},t)-\boldsymbol{F}^\alpha(\boldsymbol{y},t)|d\boldsymbol{y}d\boldsymbol{x}$$
$$\le \frac{C}{R}\|\boldsymbol{F}^\alpha\|_{L^\infty}m_t(R)^2 + CLm_t(R).$$

Therefore, we achieve the following estimate,

$$\frac{d}{dt}\mu_R(t) \le \frac{C}{\alpha R}m_t(R) + \frac{C}{R}\|\boldsymbol{F}^\alpha\|_{L^\infty}m_t(R)^2 + CLm_t(R)$$
$$\le CL\mu_{R/2}(t) + \frac{C_F\exp(2LT)}{R^5}I_\varepsilon(0)\left(\exp(2LT)I_\varepsilon(0)+\frac{R^2}{\alpha}\right)$$
$$= CL\mu_{R/2}(t) + A(R,\varepsilon).$$

Here, we choose $R>0$ and $k\in\mathbb{N}$ sufficiently small but large enough that $2^{-k}R>\beta\varepsilon$, then, by iterating the inequality, we can obtain

$$\mu_R(t) \le TA(R,\varepsilon) + CL\int_0^t \mu_{R/2}(\tau)d\tau \le T\sum_{s=0}^{k-1}(TL)^s A(R2^{-s},\varepsilon) + \frac{(Lt)^k}{k!}.$$

Let $R=\varepsilon^{1/n}$, $k=[-D\log\varepsilon]$ and ε, D be small enough, then

$$\mu_R(t) \le C_{F,T,L}\left(\varepsilon^{2-2/n}+\frac{1}{\alpha}\right)\varepsilon^{2-3/n-DC_{T,L}} = c_0\varepsilon^l.$$

The above estimates conclude the proof. Consider the disk $B_1 = B(\boldsymbol{G}_\varepsilon(t),\varepsilon^{1/n})$ and $B_2 = B(\boldsymbol{G}_\varepsilon(t),\varepsilon^{1/m})$ for $m>n$. Let $n = \frac{\boldsymbol{x}-\boldsymbol{G}_\varepsilon(t)}{|\boldsymbol{x}-\boldsymbol{G}_\varepsilon(t)|}$ be the normal vector. We now evaluate the velocity field $\boldsymbol{u}_1$ generated by vorticity inside of the disk B_1 and $\boldsymbol{u}_2$ generated by vorticity outside of the disk B_1.

$$|\boldsymbol{u}_1(\boldsymbol{x})\cdot\boldsymbol{n}| = \left|n\cdot\varepsilon^{-2}\int_{B_1\cap\Lambda_\varepsilon(t)}(K^\alpha(\boldsymbol{x}-\boldsymbol{y})-K^\alpha(\boldsymbol{x}-\boldsymbol{G}_\varepsilon(t)))\,d\boldsymbol{y}\right|$$
$$\le \varepsilon^{-2}\int_{B_1\cap\Lambda_\varepsilon(t)}|\boldsymbol{y}-\boldsymbol{G}_\varepsilon(t)|\int_0^1|\nabla K^\alpha(\boldsymbol{x}-\boldsymbol{y}+\tau(\boldsymbol{y}-\boldsymbol{G}_\varepsilon(t)))|\,d\tau d\boldsymbol{y}$$
$$\le C\frac{\varepsilon^{1/n}}{\alpha^2}\left(-\log\frac{\varepsilon^{1/n}}{\alpha}+1\right).$$

Moreover,

$$|\boldsymbol{u}_2(x)| = \left| \varepsilon^{-2} \int_{B_1^C \cap \Lambda_\varepsilon(t)} K^\alpha(\boldsymbol{x}-\boldsymbol{y}) d\boldsymbol{y} \right| \leq C \frac{c_0}{\alpha} \varepsilon^l.$$

Therefore, the time derivative of the distance from $\boldsymbol{G}_\varepsilon(t)$ to the particle $\boldsymbol{\eta}_\varepsilon^\alpha(\boldsymbol{x}, t)$ is given by

$$\begin{aligned}
\frac{d}{dt} \left|\boldsymbol{\eta}_\varepsilon^\alpha(\boldsymbol{x}, t) - \boldsymbol{G}_\varepsilon(t)\right|^2 \leq\ & 2 \left|\boldsymbol{\eta}_\varepsilon^\alpha(\boldsymbol{x}, t) - \boldsymbol{G}_\varepsilon(t)\right| \Big\{ |\boldsymbol{u}_1(\boldsymbol{\eta}_\varepsilon^\alpha(\boldsymbol{x}, t))| + |\boldsymbol{u}_2(\boldsymbol{\eta}_\varepsilon^\alpha(\boldsymbol{x}, t))| \\
& + \varepsilon^{-2} \int_{\Lambda_\varepsilon(t)} \left|\boldsymbol{F}^\alpha(\boldsymbol{\eta}_\varepsilon^\alpha(\boldsymbol{x}, t), t) - \boldsymbol{F}^\alpha(\boldsymbol{y}, t)\right| d\boldsymbol{y} \Big\} \\
\leq\ & C \frac{\varepsilon^{1/n}}{\alpha^2} (1 + \|\boldsymbol{F}^\alpha\|_{L^\infty}) \left|\boldsymbol{\eta}_\varepsilon^\alpha(\boldsymbol{x}, t) - \boldsymbol{G}_\varepsilon(t)\right| \\
& + 2L \left|\boldsymbol{\eta}_\varepsilon^\alpha(\boldsymbol{x}, t) - \boldsymbol{G}_\varepsilon(t)\right|^2.
\end{aligned}$$

On the other hand, if there is some $\boldsymbol{x}_0 \in \Lambda_\varepsilon$ such that $\boldsymbol{\eta}_\varepsilon^\alpha(\boldsymbol{x}_0, t) \in B(\boldsymbol{G}_\varepsilon(t), d)^C$, then we have

$$\left|\boldsymbol{\eta}_\varepsilon^\alpha(\boldsymbol{x}_0, t) - \boldsymbol{G}_\varepsilon(t)\right| \leq |\boldsymbol{x}_0 - \boldsymbol{G}_\varepsilon(0)| \exp\left\{ C \frac{T\varepsilon^{1/n}}{d\alpha^2} (1 + \|\boldsymbol{F}^\alpha\|_{L^\infty}) + LT \right\}.$$

This implies that $\Lambda_\varepsilon(t)$ must be contained in a disk $B(\boldsymbol{G}_\varepsilon(t), d)$, for an arbitrary d, provided that ε is sufficiently small. This fact contradicts the choice of $\boldsymbol{x}_0$. □

We now extend Theorem 14.2.1 to many vorticity distributions. Let $\{\Lambda_\varepsilon^n\}_{\varepsilon>0}$ for $n = 1, \ldots, N$ be a family of open sets satisfying the conditions

$$\text{meas. } \Lambda_\varepsilon^n = \varepsilon^2, \quad \Lambda_\varepsilon^n \subset B(\boldsymbol{x}_n^*, \beta\varepsilon)$$

and we consider the following initial distributions of vorticity

$$q_\varepsilon(\boldsymbol{x}, 0) = \sum_{n=1}^N q_0^n(\boldsymbol{x}),$$

where $q_0^n(\boldsymbol{x}) = \varepsilon^{-2} \chi_{\Lambda_\varepsilon^n}(\boldsymbol{x})$. Suppose that, for any fixed T, we can choose $\varepsilon(T)$ sufficiently small such that there is some positive constant $d(T)$ satisfying

$$\min_{0 \leq t \leq T} \min_{n \neq m} \inf_{\boldsymbol{x}_n \in \Lambda_\varepsilon^n(t), \boldsymbol{x}_m \in \Lambda_\varepsilon^m(t)} |\boldsymbol{x}_n - \boldsymbol{x}_m| > d(T) > 0,$$

for $\varepsilon < \varepsilon(T)$. Then, in the same way as in the proof of Theorem 14.2.1, for each $\boldsymbol{x}_n^*$ one can define $\Lambda_\varepsilon^n(t)$, with which the unique solution to the Euler-α equations is expressed by

$$q_\varepsilon(\boldsymbol{x},t) = \sum_{n=1}^{N} q_\varepsilon^n(\boldsymbol{x},t) = \sum_{n=1}^{N} \varepsilon^{-2}\chi_{\Lambda_\varepsilon^n(t)}(\boldsymbol{x}),$$

and $q_\varepsilon^n(\boldsymbol{x},t)$ satisfies following equations:

$$\frac{d}{dt}\int_{\mathbb{R}^2} q_\varepsilon^n(\boldsymbol{x},t) f(\boldsymbol{x}) d\boldsymbol{x} = \int_{\mathbb{R}^2} q_\varepsilon^n(\boldsymbol{x},t)(\boldsymbol{u}_\varepsilon^n(\boldsymbol{x},t) + \boldsymbol{F}^\alpha(\boldsymbol{x},t)) \cdot \nabla f(\boldsymbol{x}) d\boldsymbol{x},$$
$$\boldsymbol{u}_\varepsilon^n(\boldsymbol{x},t) = \int_{\mathbb{R}^2} K^\alpha(\boldsymbol{x},\boldsymbol{y}) q_\varepsilon^n(\boldsymbol{y},t) d\boldsymbol{y},$$

where $\boldsymbol{F}^\alpha(\boldsymbol{x},t)$ represents the velocity field induced by the other vorticity distributions:

$$\boldsymbol{F}^\alpha(\boldsymbol{x},t) = \sum_{m\neq n}^{N} \int_{\mathbb{R}^2} K^\alpha(\boldsymbol{x}, \boldsymbol{\eta}^\alpha(z,t)) q_0^m(z) dz.$$

We just confirm that $\boldsymbol{F}^\alpha$ satisfies the Lipschitz condition with the Lipschitz constant independent of α. Let us set $r = |\boldsymbol{x} - \boldsymbol{y}|$. Then we have

$$\begin{aligned}
\left|\boldsymbol{F}^\alpha(\boldsymbol{x},t) - \boldsymbol{F}^\alpha(\boldsymbol{y},t)\right| \leq & \sum \int_{|\boldsymbol{x}-\boldsymbol{\eta}^\alpha(z)|\leq 2r} \left|K^\alpha(\boldsymbol{x},\boldsymbol{\eta}^\alpha(z)) - K^\alpha(\boldsymbol{y},\boldsymbol{\eta}^\alpha(z))\right| \left|q_0^m(z)\right| dz \\
& + \sum \int_{|\boldsymbol{x}-\boldsymbol{\eta}^\alpha(z)|>2r} \left|K^\alpha(\boldsymbol{x},\boldsymbol{\eta}^\alpha(z)) - K^\alpha(\boldsymbol{y},\boldsymbol{\eta}^\alpha(z))\right| \left|q_0^m(z)\right| dz \\
\leq & \sum \int_{|\boldsymbol{x}-\boldsymbol{\eta}^\alpha(z)|\leq 3r} \left|K(\boldsymbol{x},\boldsymbol{\eta}^\alpha(z))\right| + \frac{1}{\alpha} K_1\left(\frac{|\boldsymbol{x}-\boldsymbol{\eta}^\alpha(z)|}{\alpha}\right) \left|q_0^m(z)\right| dz \\
& + C\sum \int_{|\boldsymbol{x}-\boldsymbol{\eta}^\alpha(z)|>2r} \frac{1}{|\boldsymbol{x}-\boldsymbol{\eta}^\alpha(z)|^2} + \frac{1}{\alpha^2} K_0\left(\frac{|\boldsymbol{x}-\boldsymbol{\eta}^\alpha(z)|}{2\alpha}\right) \\
& + \frac{1}{\alpha|\boldsymbol{x}-\boldsymbol{\eta}^\alpha(z)|} K_1\left(\frac{|\boldsymbol{y}-\boldsymbol{\eta}^\alpha(z)|}{2\alpha}\right) \left|q_0^m(z)\right| dz \\
\leq & C\frac{r}{d(T)^2} \sum \|q_0^m\|_{L^1}.
\end{aligned}$$

Moreover, we have

$$\|\boldsymbol{F}^\alpha\|_{L^\infty} \leq \frac{C}{\alpha} \sum \|q_0^m\|_{L^1}.$$

Therefore, we can apply Theorem 14.2.1 and Corollary 14.2.2 to $q_\varepsilon^n(\boldsymbol{x},t)$ and get the same result as we had for the vortex model for the Euler equations.

The theorem is the same as that for the Euler equations shown by Marchioro and Pulvirenti [22]. The idea of the proof is similar to theirs, while, in the Euler-α equations, the proof becomes much simpler, since the kernel is regularized. On the other hand, we can show that the orbit of any point in Λ_ε converges to that of the center of vorticity for the Euler-α equations. This fact does not hold true for the vortex model for the Euler equation. See Remark 1 of Theorem 4.1 in [22].

Corollary 14.2.2 *Under the same conditions and notations with Theorem 14.2.1, we have*

$$\lim_{\varepsilon\to 0}\sup_{0\le t\le T}\sup_{x_0\in\Lambda_\varepsilon}\left|\boldsymbol{\eta}_\varepsilon^\alpha(\boldsymbol{x}_0,t)-\boldsymbol{G}(t)\right|=0.$$

Proof We have

$$\begin{aligned}
|\boldsymbol{\eta}_\varepsilon^\alpha(\boldsymbol{x}_0,t)-\boldsymbol{G}(t)| &\le |\boldsymbol{x}_0-\boldsymbol{x}^*|+\int_0^t|\boldsymbol{u}^\alpha(\boldsymbol{\eta}_\varepsilon^\alpha(\boldsymbol{x}_0,s))|ds\\
&\quad+\int_0^t\left|\boldsymbol{F}^\alpha(\boldsymbol{\eta}_\varepsilon^\alpha(\boldsymbol{x}_0,s),s)-\boldsymbol{F}^\alpha(\boldsymbol{G}(s),s)\right|ds\\
&\le |\boldsymbol{x}_0-\boldsymbol{x}^*|+L\int_0^t\left|\boldsymbol{\eta}_\varepsilon^\alpha(\boldsymbol{x}_0,s)-\boldsymbol{G}(s)\right|ds\\
&\quad+\varepsilon^{-2}\int_0^t\int_{\Lambda_\varepsilon}|K^\alpha(\boldsymbol{\eta}_\varepsilon^\alpha(\boldsymbol{x}_0,s)-\boldsymbol{\eta}_\varepsilon^\alpha(\boldsymbol{z},s))|d\boldsymbol{z}ds\\
&\le |\boldsymbol{x}_0-\boldsymbol{x}^*|+L\int_0^t\left|\boldsymbol{\eta}_\varepsilon^\alpha(\boldsymbol{x}_0,s)-\boldsymbol{G}(s)\right|ds\\
&\quad+\frac{C}{\alpha}\varepsilon^{-2}\int_0^t\int_{\Lambda_\varepsilon}\phi\left(\frac{|\boldsymbol{\eta}_\varepsilon^\alpha(\boldsymbol{x}_0,s)-\boldsymbol{\eta}_\varepsilon^\alpha(\boldsymbol{z},s)|}{\alpha}\right)d\boldsymbol{z}ds,
\end{aligned}$$

where ϕ is defined by

$$\phi(\boldsymbol{x})=\begin{cases}|\boldsymbol{x}|(1-\log|\boldsymbol{x}|)\ , & |\boldsymbol{x}|<1,\\ 1 & ,\ |\boldsymbol{x}|\ge 1.\end{cases}$$

We use $K^\alpha(r)\sim r\log\, r+\mathcal{O}(r)$ as $r\to 0$ and, by Theorem 2.1, $|\boldsymbol{\eta}_\varepsilon^\alpha(\boldsymbol{x}_0)-\boldsymbol{\eta}_\varepsilon^\alpha(\boldsymbol{z})|$ is small enough. To obtain explicit bounds, we set

$$\rho(t)=\frac{1}{\alpha}\sup_{x\in\Lambda_\varepsilon}|\boldsymbol{\eta}_\varepsilon^\alpha(\boldsymbol{x},t)-\boldsymbol{G}(t)|,$$

then it follows that ρ satisfies the integral inequality:

$$\rho(t)\le\rho(0)+L\int_0^t\rho(s)+\frac{C}{\alpha^2}\phi\left(\rho(s)\right)ds\le\rho(0)+c_0\int_0^t\phi\left(\rho(s)\right)ds,$$

where $c_0=L+\frac{C}{\alpha^2}$. By using Gronwall's inequality, we obtain

$$\rho(t)\le\rho(0)^{\exp(-c_0t)}e^{1-\exp(-c_0t)}.$$

That is

$$\sup_{x\in\Lambda_\varepsilon(t)} |\boldsymbol{x} - \boldsymbol{G}(t)| \leq \alpha e \left(\frac{\beta\varepsilon}{\alpha e}\right)^{\exp\left(L+\frac{C}{\alpha^2}\right)t}.$$

□

The vortex model where the vorticity is localized in small regions is approximated by the evolution of the 2D Euler-α equations as well as the 2D Euler equations. On the other hand, since the PV system is not a weak solution of the Euler equations, we have a stronger result for the Euler-α equations.

Suppose now that the α-vorticity belongs to $\mathscr{M}(\mathbb{R}^2)$ at the initial moment. Then, Oliver and Shkoller [25] have established the global well-posedness of the Euler-α equations.

Theorem 14.2.3 *(Oliver and Shkoller [25]) For initial data $q_0 \in \mathcal{M}(\mathbb{R}^2)$, there exists a unique global weak solution to (14.3), (14.4) and (14.5) with*

$$\boldsymbol{\eta}^\alpha \in C^1(\mathbb{R}; \mathcal{G}), \quad \boldsymbol{u}^\alpha \in C(\mathbb{R}; C(\mathbb{R}^2, \mathbb{R}^2)), \quad q \in C(\mathbb{R}; \mathcal{M}(\mathbb{R}^2)),$$

where $\mathcal{G}$ denotes the group of all homeomorphisms of $\mathbb{R}^2$ that preserves the Lebesgue measure.

We further assume that the initial vorticity field is represented by discrete point distributions as (14.6). Then, the evolution of the α-point vortices gives rise to a weak solution of the Euler-α equations.

Proposition 14.2.4 *The solution to (14.3) and (14.4) with the initial data (14.6) is expressed by*

$$q(\boldsymbol{x}, t) = \sum_{m=1}^{N} \Gamma_m \delta(\boldsymbol{x} - \eta^\alpha(\boldsymbol{x}_m^0, t)). \tag{14.10}$$

Moreover, the α-point vortices never collapse.

Proof We know the global existence and uniqueness of the Euler-α flow from Theorem 1.1. Thus,

$$q(\boldsymbol{x}, t) = q_0(\eta^\alpha(\boldsymbol{x}, -t)) = \sum_{m=1}^{N} \Gamma_m \delta(\eta^\alpha(\boldsymbol{x}, -t) - \boldsymbol{x}_m^0).$$

If $\boldsymbol{\eta}^\alpha(\boldsymbol{x}, -t) = \boldsymbol{x}_m^0$ then $\boldsymbol{x} = \boldsymbol{\eta}^\alpha(\boldsymbol{x}_m^0, t)$ else $\boldsymbol{x} \neq \boldsymbol{\eta}^\alpha(\boldsymbol{x}_m^0, t)$. This implies

$$\delta(\eta^\alpha(\boldsymbol{x}, -t) - \boldsymbol{x}_m^0) = \delta(\boldsymbol{x} - \eta^\alpha(\boldsymbol{x}_m^0, t)).$$

Moreover, by the uniqueness of the flow, it follows that if $n \neq m$ then $\boldsymbol{\eta}^\alpha(\boldsymbol{x}_m^0, t) \neq \boldsymbol{\eta}^\alpha(\boldsymbol{x}_n^0, t)$ for an arbitrary $t > 0$, namely, there is no collapse. □

Proposition 14.2.4 shows a significant difference between the point vortex approximation of the Euler equations and the α-point vortex approximation of the Euler-α

equations. Namely, the evolution of α-point vortices is a weak solution of the Euler-α equations, whereas that of point vortices is no longer a weak solution of the Euler equations. In addition, the α-point vortices never collapse in finite time, which is another difference from the PV system where the self-similar collapse occurs.

14.3 Enstrophy Variations in the αPV System

14.3.1 The αPV System

Let $\boldsymbol{x}_m^\alpha(t) = (x_m^\alpha(t), y_m^\alpha(t))$ for $m = 1, \ldots, N$. Then the αPV system (14.7) is described as a Hamiltonian dynamical system:

$$\frac{dx_m^\alpha}{dt} = \left\{x_m^\alpha, H^\alpha\right\}, \quad \frac{dy_m^\alpha}{dt} = \left\{y_m^\alpha, H^\alpha\right\}, \quad x_m^\alpha(0) = x_m^0, \quad y_m^\alpha(0) = y_m^0, \tag{14.11}$$

where the Hamiltonian H^α and the Poisson bracket between functions $f, g : \mathbb{R}^{2N} \to \mathbb{R}$ are defined by

$$H^\alpha = -\frac{1}{2\pi}\sum_{n=1}^{N}\sum_{m=n+1}^{N} \Gamma_n\Gamma_m \log l_{nm}^\alpha - \frac{1}{2\pi}\sum_{n=1}^{N}\sum_{m=n+1}^{N} \Gamma_n\Gamma_m K_0\left(\frac{l_{nm}^\alpha}{\alpha}\right), \tag{14.12}$$

$$\{f, g\} = \sum_{m=1}^{N} \frac{1}{\Gamma_m}\left(\frac{\partial f}{\partial x_m}\frac{\partial g}{\partial y_m} - \frac{\partial g}{\partial x_m}\frac{\partial f}{\partial y_m}\right).$$

From (14.7), we can derive the following evolution equation for the distance $l_{mn}^\alpha(t)$:

$$\frac{d}{dt}(l_{mn}^\alpha)^2 = \frac{2}{\pi}\sum_{k\neq m\neq n}^{N} \Gamma_k\sigma_{mnk}A_{mnk}^\alpha\left(\frac{1}{(l_{nk}^\alpha)^2}B_K\left(\frac{l_{nk}^\alpha}{\alpha}\right) - \frac{1}{(l_{km}^\alpha)^2}B_K\left(\frac{l_{km}^\alpha}{\alpha}\right)\right), \tag{14.13}$$

$$l_{mn}^\alpha(0) = |\boldsymbol{x}_m^\alpha - \boldsymbol{x}_n^\alpha|,$$

where σ_{mnk}^α and A_{mnk}^α denote the sign of the arrangement and the area of the triangle formed by the three α point vortices, respectively. That is to say, $\sigma_{mnk} = 1$ if the three point vortices at $\boldsymbol{x}_m^\alpha$, $\boldsymbol{x}_n^\alpha$ and $\boldsymbol{x}_k^\alpha$ are arranged in the counterclockwise direction, and $\sigma_{mnk} = -1$ otherwise. The area A_{mnk}^α is computed from Heron's formula:

$$A_{mnk}^\alpha = \frac{1}{2}\left[2(l_{mn}^\alpha)^2(l_{nk}^\alpha)^2 + 2(l_{nk}^\alpha)^2(l_{km}^\alpha)^2 + 2(l_{km}^\alpha)^2(l_{mn}^\alpha)^2 - (l_{mn}^\alpha)^4 - (l_{nk}^\alpha)^4 - (l_{km}^\alpha)^4\right]^{\frac{1}{2}}. \tag{14.14}$$

The quantities

$$Q = \sum_{m=1}^{N} \Gamma_m x_m^\alpha, \qquad P = \sum_{m=1}^{N} \Gamma_m y_m^\alpha, \qquad I = \sum_{m=1}^{N} \Gamma_m \left[(x_m^\alpha)^2 + (y_m^\alpha)^2 \right],$$

are invariant owing to $\{Q, H\} = \{P, H\} = \{I, H\} = 0$. It follows from $\{P^2 + Q^2, H\} = 0$ that I, $P^2 + Q^2$ and H are in involution and the three αPV system (14.11) is thus integrable for all values of Γ_m. Let us note that, for $N = 4$, it is integrable when the strengths of the four α-point vortices satisfy $\sum_{m=1}^{N} \Gamma_m = 0$. This indicates that the integrability of the αPV system is the same as that of the PV system [23, 28].

In this paper, we are concerned with the variations of the energy and the enstrophy associated with the evolution of the αPV system. According to [27], the energy variation $E^\alpha(t)$ and the enstrophy variation $\mathscr{Z}^\alpha(t)$ in the sense of Novikov [24] are given by

$$E^\alpha(t) = -\frac{1}{2\pi} \sum_{n=1}^{n} \sum_{m=n+1}^{N} \Gamma_n \Gamma_m \left[\log l_{mn}^\alpha + K_0\left(\frac{l_{mn}^\alpha}{\alpha}\right) + \frac{l_{mn}^\alpha}{2\alpha} K_1\left(\frac{l_{mn}^\alpha}{\alpha}\right) \right], \tag{14.15}$$

$$\mathscr{Z}^\alpha(t) = \frac{1}{4\pi\alpha^2} \sum_{n=1}^{N} \sum_{m=n+1}^{N} \Gamma_n \Gamma_m \frac{l_{mn}^\alpha}{\alpha} K_1\left(\frac{l_{mn}^\alpha}{\alpha}\right). \tag{14.16}$$

We are also interested in the limits of $\boldsymbol{x}_m^\alpha(t)$ and $l_{mn}^\alpha(t)$ as $\alpha \to 0$. In order to take these limits, introducing the scaled variables

$$\boldsymbol{X}_m(t) = \frac{1}{\alpha} \boldsymbol{x}_m^\alpha(\alpha^2 t), \qquad L_{mn}(t) = |\boldsymbol{X}_m(t) - \boldsymbol{X}_n(t)| = \frac{1}{\alpha} l_{mn}^\alpha(\alpha^2 t),$$

we consider the following equations for $\boldsymbol{X}_m(t)$, $m = 1, \ldots, N$:

$$\frac{d\boldsymbol{X}_m}{dt} = \frac{1}{2\pi} \sum_{n \neq m}^{N} \Gamma_n \frac{(\boldsymbol{X}_m - \boldsymbol{X}_n)^\perp}{L_{mn}^2} B_K(L_{mn}), \qquad \boldsymbol{X}_m(0) = \frac{\boldsymbol{x}_m^0}{\alpha}. \tag{14.17}$$

The equation for the distance $L_{mn}(t)$ is given by

$$\frac{d}{dt} L_{mn}^2 = \frac{2}{\pi} \sum_{k \neq m \neq n}^{N} \Gamma_k \sigma_{mnk} A_{mnk} \left(\frac{B_K(L_{nk})}{L_{nk}^2} - \frac{B_K(L_{km})}{L_{km}^2} \right), \quad L_{mn}(0) = \frac{l_{mn}^\alpha(0)}{\alpha}, \tag{14.18}$$

where A_{mnk} denotes the area of the triangle formed by the three points at $\boldsymbol{X}_m$, $\boldsymbol{X}_n$ and $\boldsymbol{X}_k$.

$$A_{mnk}^2 = \frac{1}{4} \left[2L_{mn}^2 L_{nk}^2 + 2L_{nk}^2 L_{km}^2 + 2L_{km}^2 L_{mn}^2 - L_{mn}^4 - L_{nk}^4 - L_{km}^4 \right]. \tag{14.19}$$

Since the Eq. (14.17) is equivalent to (14.7) with $\alpha = 1$, the Eq. (14.17) also defines a Hamiltonian dynamical system with the Hamiltonian,

$$H = -\frac{1}{2\pi}\sum_{n=1}^{N}\sum_{m=n+1}^{N} \Gamma_n\Gamma_m \log L_{nm} - \frac{1}{2\pi}\sum_{n=1}^{N}\sum_{m=n+1}^{N} \Gamma_n\Gamma_m K_0\left(L_{nm}\right),$$

but the initial configurations are obtained by magnifying the original initial data (14.7) by the factor α^{-1}. We then recover the solutions of (14.7) and (14.13) from those of the scaled systems (14.17) and (14.18) via

$$\boldsymbol{x}_m^\alpha(t) = \alpha \boldsymbol{X}_m\left(\frac{t}{\alpha^2}\right), \qquad l_{nm}^\alpha(t) = \alpha L_{nm}\left(\frac{t}{\alpha^2}\right). \tag{14.20}$$

Before solving the Eqs. (14.17) and (14.18), we consider a necessary condition for the occurrence of the energy and enstrophy variations in the $\alpha \to 0$ limit of the αPV system.

Proposition 14.3.1 *Suppose that there exist a time interval I and functions $C_{mn}(t) \neq 0$ for all $t \in I$ such that*

$$\lim_{\alpha\to 0}\sup_{t\in I}\left|l_{mn}^\alpha(t) - C_{mn}(t)\right| = 0, \qquad m \neq n = 1, \ldots, N. \tag{14.21}$$

Then the energy converges to the Hamiltonian energy of the PV system and the enstrophy converges zero. Namely, we have

$$\lim_{\alpha\to 0}\sup_{t\in I}|E^\alpha(t) - H| = 0, \qquad \lim_{\alpha\to 0}\sup_{t\in I}|\mathscr{Z}^\alpha(t)| = 0,$$

where H denotes the Hamiltonian of the PV system.

Proof Let us remark that the evolution of point vortices (14.8) is described by the Hamiltonian dynamical system with

$$H(t) = -\frac{1}{2\pi}\sum_{n=1}^{N}\sum_{m=n+1}^{N} \Gamma_n\Gamma_m \log l_{nm}(t).$$

First, we prove the convergence of the energy. Owing to (14.12), we have

$$|E^\alpha - H| \leq |H^\alpha - H| + \frac{1}{2\pi}\sum\sum|\Gamma_n\Gamma_m|\frac{l_{mn}^\alpha}{2\alpha}K_1\left(\frac{l_{mn}^\alpha}{\alpha}\right). \tag{14.22}$$

Since Hamiltonian $H^\alpha(t)$ is invariant in time, we find

$$-\frac{1}{2\pi}\sum\sum\Gamma_n\Gamma_m\log l^\alpha_{nm}(t)-\frac{1}{2\pi}\sum\sum\Gamma_n\Gamma_m K_0\left(\frac{l^\alpha_{nm}(t)}{\alpha}\right)$$
$$=-\frac{1}{2\pi}\sum\sum\Gamma_n\Gamma_m\log|\boldsymbol{x}^0_n-\boldsymbol{x}^0_m|-\frac{1}{2\pi}\sum\sum\Gamma_n\Gamma_m K_0\left(\frac{|\boldsymbol{x}^0_n-\boldsymbol{x}^0_m|}{\alpha}\right).$$

By the assumption (14.21), in the limit of $\alpha\to 0$, we have

$$-\frac{1}{2\pi}\sum\sum\Gamma_n\Gamma_m\log C_{mn}(t)=-\frac{1}{2\pi}\sum\sum\Gamma_n\Gamma_m\log|\boldsymbol{x}^0_n-\boldsymbol{x}^0_m|,$$

for $t\in I$. On the other hand, from the time invariance of $H(t)$, we have

$$-\frac{1}{2\pi}\sum\sum\Gamma_n\Gamma_m\log l_{mn}(t)=-\frac{1}{2\pi}\sum\sum\Gamma_n\Gamma_m\log|\boldsymbol{x}^0_n-\boldsymbol{x}^0_m|.$$

Hence, we obtain

$$-\frac{1}{2\pi}\sum\sum\Gamma_n\Gamma_m\log C_{mn}(t)=-\frac{1}{2\pi}\sum\sum\Gamma_n\Gamma_m\log l_{mn}(t).$$

This shows that the first term of the right hand side in (14.22) vanishes as α approaches zero. Regarding the second term, it follows from $K_0(x),\ K_1(x)\sim e^{-x}$ as $x\to\infty$ that it tends to zero in the limit of α.

The proof of the convergence of the enstrophy variation is straightforward. By the assumption, we have

$$\frac{l^\alpha_{mn}}{\alpha}\to\infty\quad\text{as}\quad\alpha\to 0.$$

Thus, owing to the decay rate of K_1, we have the conclusion. □

This proposition indicates that the evolution of α-point vortices is equivalent to that of point vortices with the Hamiltonian H as $\alpha\to 0$ as long as their distance function $l^\alpha_{mn}(t)$ converges a non-vanishing function uniformly in time. Furthermore, we have no enstrophy variation in the same time period $t\in I$. In other words, taking its contraposition, we have the following corollary on the necessary condition for the enstrophy variation.

Corollary 14.3.2 *Suppose that, in the limit of $\alpha\to 0$, the enstrophy $\mathscr{Z}^\alpha(t_c)\neq 0$ at time t_c. Then $l^\alpha_{mn}(t_c)\to 0$ as $\alpha\to 0$.*

Let us remember that the αPV system has no collapse, i.e. $l^\alpha_{mn}(t)\neq 0$ for all $t\in\mathbb{R}$ and $\alpha\neq 0$, as shown in Proposition 14.2.4. Nevertheless, the distance function could vanish at some time t_c in the limit of $\alpha\to 0$, at which we might observe the enstrophy variation. The limit evolution of $\boldsymbol{x}^\alpha_m(t)$ or $l^\alpha_{mn}(t)$ as $\alpha\to 0$ is obtained by solving the canonical equation (14.17) or (14.18) and taking $\lim_{\alpha\to 0}\alpha\boldsymbol{X}_m(t/\alpha)$ or

$\lim_{\alpha\to 0} \alpha L_{mn}(t/\alpha)$ owing to (14.20). According to Corollary 14.3.2, the enstrophy variation occurs at the time when the $\alpha \to 0$ limit of $l^{\alpha}_{mn}(t)$ vanishes. We must note that, even if Proposition 14.2.4 shows that the α-point vortices never collapse in finite time, there is a possibility that the distance l^{α}_{mn} vanishes at a certain time as $\alpha \to 0$.

It is straightforward to see that the enstrophy variation never occurs for the evolution of two α point vortices. That is to say, it follows from $dL_{12}^2/dt = 0$ owing to the distance equation (14.18) that we have $L_{12}(t) = L_{12}(0) = l_{12}(0)/\alpha$ for all $t \in \mathbb{R}$ and thus $\lim_{\alpha\to 0} \alpha L_{12}(t/\alpha^2) = l_{12}(0) \neq 0$. Hence, in order to obtain the enstrophy variation in the evolution of the $\alpha \to 0$ limit of the evolution of α point vortices, we need to consider the αPV system (14.17) with $N \geq 3$.

14.3.2 Collapse of α-Point Vortices and the Enstrophy Dissipation

14.3.2.1 Enstrophy Dissipation via the Triple Collapse

It is well known that, under a certain circumstance, the PV system (14.8) admits a singular solution, where N point vortices collide self-similarly in finite time. Kimura [12] has actually constructed such a singular evolution of three point vortices, in which there exists an initial configuration of point vortices and a positive constant C_{mn} such that the distance of point vortices is represented by $l_{mn}(t) = C_{mn}\sqrt{|t|}$. This means that the three point vortices are approaching self-similarly and collide at the origin as $t \to 0-$, which is *a self-similar collapse*, and they emerge abruptly at the origin at $t = 0$ and then diverge self-similarly to infinity for $t > 0$, which we call *a self-similar expansion*. Let us note that these collapsing and diverging self-similar solutions are not connected as one solution of the PV system. Since the distances of the three point vortices vanish at the critical time for the triple collapse, one can easily expect that distance of three α point vortices may vanish in the $\alpha \to 0$ limit under the same conditions as Kimura's, which are

$$\Gamma_1\Gamma_2 + \Gamma_2\Gamma_3 + \Gamma_3\Gamma_1 = 0, \qquad M = \Gamma_1\Gamma_2 L_{12}^2 + \Gamma_2\Gamma_3 L_{23}^2 + \Gamma_3\Gamma_1 L_{31}^2 = 0. \tag{14.23}$$

This problem has been investigated in [27, 28]. Based on these results, we explain how to obtain $\lim_{\alpha\to 0} l^{\alpha}_{mn}$ from the solution of the canonical equation (14.17) and (14.18). Let us first notice that $M = \sum_{n=1}^{N}\sum_{m=n+1}^{N} \Gamma_n\Gamma_m L_{nm}^2$ is an invariant quantity in the αPV system owing to $M = \left(\sum_{m=1}^{N} \Gamma_m\right) I - Q^2 - P^2$. As discussed in Sect. 14.3.1, the three αPV system is integrable for any strengths and initial configurations. Thus we can investigate the behaviour of solutions of the canonical equations (14.17) for $N = 3$ with the same technique as used in the analysis of the integrable PV system [23].

The canonical equations (14.17) are reduced to the equations (14.18) for the distances $L_{12}(t)$, $L_{23}(t)$, $L_{31}(t)$ and the one for the area $A(t) := A_{123}(t)$. Since $A(t)$ is obtained from the distances $L_{mn}(t)$ via (14.19), we have only to consider the Eq. (14.18). Introducing the new variables,

$$b_1 = \frac{L_{23}^2}{\Gamma_1}, \qquad b_2 = \frac{L_{31}^2}{\Gamma_2}, \qquad b_3 = \frac{L_{12}^2}{\Gamma_3},$$

we rewrite the condition $M = 0$ as $b_1 + b_2 + b_3 = 0$, from which we can reduce (14.18) to a two-dimensional dynamical system in the phase space (b_1, b_2). Furthermore, since the canonical system conserves the Hamiltonian,

$$\begin{aligned} H = &-\frac{1}{4\pi}\left[\Gamma_2\Gamma_3 \log \Gamma_1 b_1 + \Gamma_3\Gamma_1 \log \Gamma_2 b_2 + \Gamma_1\Gamma_2 \log \Gamma_3 b_3\right] \\ &-\frac{1}{2\pi}\left[\Gamma_2\Gamma_3 K_0\left(\sqrt{\Gamma_1 b_1}\right) + \Gamma_3\Gamma_1 K_0\left(\sqrt{\Gamma_2 b_2}\right) + \Gamma_1\Gamma_2 K_0\left(\sqrt{\Gamma_3 b_3}\right)\right], \end{aligned} \tag{14.24}$$

each level curve of the Hamiltonian in the phase space (b_1, b_2) corresponds to the orbit of a solution of the canonical system.

Since the area of the triangle $A(t)$ must be positive for all t, it follows from $|A(t)| \geq 0$ and $M = 0$ that there exist two reals k_1 and k_2 with $0 < k_1 < 1 < k_2$ such that $k_1 \leq L_{23}^2/L_{31}^2 \leq k_2$. This gives rise to a wedge region $\Gamma_2/\Gamma_1 k_1 \leq b_1/b_2 \leq \Gamma_2/\Gamma_1 k_2$ in the (b_1, b_2)-space, in which solutions of the canonical system exist and thus level curves of the Hamiltonian are plotted. Let us note that every point at the boundary $b_1/b_2 = \Gamma_2/\Gamma_1 k_1$ and $\Gamma_2/\Gamma_1 k_2$ of the wedge region corresponds to a collinear configuration, where three points are aligned along a straight line. Figure 14.1a shows the level curves of the Hamiltonian (14.24) for $\Gamma_1 = \Gamma_2 = 1$. The steady solutions of (14.18) with (14.23) are similar equilateral triangles, i.e., $L_{12} = L_{23} = L_{31}$, which correspond to the diagonal line connecting to the origin. The other level curves except the diagonal line are connected to the boundary of the wedge region and go to infinity. Hence, when we choose the initial configuration at a point off the diagonal line in the phase pace, the solution of the canonical equation moves along the level curve going through this point. Hence, without loss of generality, we choose a collinear configuration at the boundary of the wedge region as an initial data. Then, the solution of (14.18) with $A(0) = 0$ satisfies $A(t) = -A(-t)$ and $L_{mn}(t) = L_{mn}(-t)$ for $t \in \mathbb{R}$.

Let us also note that that each point in the phase space represents two mirror symmetric configurations, $(L_{12}, L_{23}, L_{31}, A)$ and $(L_{12}, L_{23}, L_{31}, -A)$. It thus goes to infinity along the level curves as $t \to \pm\infty$. As $b_1, b_2 \to \infty$, the contour lines of the Hamiltonian tend asymptotically to $\Gamma_1 b_1/\Gamma_2 b_2 = L_{23}^2/L_{31}^2 = k$ for any $k \in (k_1, k_2)$ with $k \neq 1$, which correspond to the self-similar evolutions. Therefore, as $t \to \pm\infty$, the solution of the canonical system for any collinear initial configuration behaves like

$$\lim_{t\to\pm\infty} L_{mn}(t) \sim C_{mn}^{(k)}\sqrt{|t|}, \tag{14.25}$$

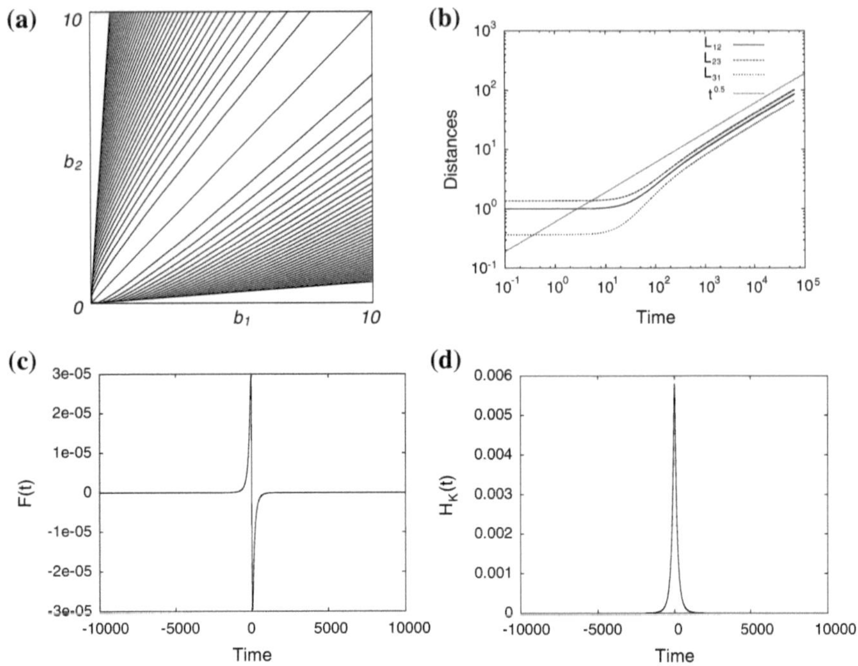

Fig. 14.1 Analysis of the canonical system (14.17) for $\Gamma_1 = \Gamma_2 = 1$. **a** Level curves of the Hamiltonian (14.24). **b** Evolution of the distances $L_{mn}(t)$ for a collinear initial configuration. **c** Plot of $F(t)$. **d** Plot of $H_K(t)$

for some constant $C^{(k)}_{mn} > 0$ and the ratios L_{12}/L_{31} and L_{23}/L_{31} converge to the constants $C^{(k)}_{12}/C^{(k)}_{31}$ and $C^{(k)}_{23}/C^{(k)}_{31}$ as $t\pm \to \infty$. Figure 14.1b shows the evolutions of the distances $L_{mn}(t)$ starting from a collinear initial configuration for $\Gamma_1 = \Gamma_2 = 1$, which clearly demonstrates the asymptotic behavior (14.25) as $t \to \infty$.

We now consider the behaviour of solutions for the sequence of initial configurations $L_{mn}(0) = l_{mn}(0)/\alpha$ and observe their $\alpha \to 0$ limit. Since the initial data satisfies $L_{23}/L_{31} = l_{23}/l_{31} := \widetilde{k}$ regardless of α, the initial configurations are represented as the straight line $b_1/b_2 = \Gamma_2/\Gamma_1\widetilde{k}$ in the phase space (b_1, b_2). As $\alpha \to 0$, the initial configuration diverges along $L_{23}/L_{31} = \widetilde{k}$. Then, the evolution starting from the infinitely large initial configuration corresponds to the solution (14.25) with $k = \widetilde{k}$. Hence, noting that $t/\alpha^2 \to \infty$ for $t \neq 0$ as $\alpha \to 0$, we have the convergence of the solution of the αPV system (14.13) owing to (14.20) and (14.25) as follows.

$$l^{\alpha}_{mn}(t) = \alpha L_{mn}\left(\frac{t}{\alpha^2}\right) \longrightarrow C^{(\widetilde{k})}_{mn}\sqrt{|t|}, \qquad \alpha \to 0, \tag{14.26}$$

which indicates that the solution of αPV system converges to the self-similar triple collapse for $t < 0$ and to the self-similar triple expansion for $t > 0$.

It follows from (14.26) and Proposition 14.3.1 that we have $|E^\alpha - H| \to 0$ and $|\mathscr{Z}^\alpha(t)| \to 0$ for $t \neq 0$ as $\alpha \to 0$. On the contrary, since $\alpha L_{mn}(t/\alpha^2) \to 0$ at $t = 0$, the energy and the enstrophy variations could occur at this time according to Corollary 14.3.2, which is examined in what follows. Since the energy variation in the αPV system is represented by

$$E^\alpha(t) = H^\alpha + \frac{1}{4\pi} H_K\left(\frac{t}{\alpha^2}\right),$$

owing to (14.12) and (14.15), where

$$H_K(t) = -\frac{1}{4\pi}\sum_{m=1}^{N}\sum_{n=m+1}^{N} \Gamma_m\Gamma_n L_{mn}(t) K_1(L_{mn}(t)),$$

we have the energy variation at $t = 0$ as follows.

$$\lim_{\alpha\to 0} E^\alpha(0) - H = H_K(0) \neq 0.$$

This indicates the energy variation acquires a finite jump discontinuity at $t = 0$. Moreover, since the Hamiltonian H^α is invariant, the energy dissipation rate $\mathscr{D}_E^\alpha(t)$ is given by

$$\begin{aligned}
\mathscr{D}_E^\alpha(t) &= \frac{1}{4\pi}\frac{d}{dt} H_K(t) \\
&= \frac{1}{4\pi\alpha^2}\sum_{m=1}^{N}\sum_{n=m+1}^{N} \Gamma_m\Gamma_n \frac{dL_{mn}}{dt}\left(\frac{t}{\alpha^2}\right) L_{mn}\left(\frac{t}{\alpha^2}\right) K_0\left(L_{mn}\left(\frac{t}{\alpha^2}\right)\right) \\
&:= \frac{1}{\alpha^2} F\left(\frac{t}{\alpha^2}\right).
\end{aligned}$$

Figure 14.1c shows the plot of the function $F(t)$ for a collinear initial configuration. As $\alpha \to 0$, the energy dissipation rate $\mathscr{D}_E^\alpha(t)$ acquires an infinite discontinuity at $t = 0$ and thus it is no longer a function. On the other hand, since $F(t)$ is odd, i.e. $F(t) + F(-t) = 0$, for any compactly supported smooth function $\varphi(t)$, we have

$$\langle \mathscr{D}_E^\alpha, \varphi\rangle = \int_{-\infty}^{\infty} \frac{1}{\alpha^2} F\left(\frac{t}{\alpha^2}\right)\varphi(t)dt = \int_{-\infty}^{\infty} F(s)\varphi(\alpha^2 s)ds \to \varphi(0)\int_{-\infty}^{\infty} F(s)ds = 0.$$

Hence, the energy dissipation rate becomes zero in the sense of distributions.

Figure 14.1d is the plot of $H_K(t)$, which shows that the function is even, positive and rapidly decreasing. Since the enstrophy variation (14.16) is represented by

$$\mathscr{Z}^\alpha(t) = -\frac{1}{\alpha^2} H_K\left(\frac{t}{\alpha^2}\right),$$

it also diverges at $t = 0$ as $\alpha \to 0$. On the other hand, for any compactly supported smooth function $\varphi(t)$, we have

$$\langle \mathscr{Z}^{\alpha}, \varphi \rangle = -\int_{-\infty}^{\infty} \frac{1}{\alpha^2} H_K\left(\frac{t}{\alpha^2}\right) \varphi(t)dt = -\int_{-\infty}^{\infty} H_K(s)\varphi(\alpha^2 s)ds \to -z_0\varphi(0),$$

in which, in view of Fig. 14.1d,

$$z_0 = \int_{-\infty}^{\infty} H_K(s)ds > 0.$$

We thus obtain the convergence of the enstrophy variation in the distributional sense as follows.

$$\lim_{\alpha \to 0} \mathscr{Z}^{\alpha} = -z_0\delta_0,$$

where δ_0 is Dirac's δ function with singularity at $t = 0$. By integrating it in the sense of distributions, we have the convergence of the total enstrophy variation:

$$\int_{-\infty}^{T} \mathscr{Z}^{\alpha}(t)dt \to -z_0 H(T),$$

where $H(T)$ denotes the Heaviside function with a finite jump discontinuity at $T = 0$. This means that the total enstrophy remains zero, until the three point vortices collapse. It suddenly drops to $-z_0$ at $t = 0$, and then it remains at the same level for $t > 0$, where the three point vortices exhibit the self-similar expansion. This indicates that the enstrophy dissipation occurs at the event of the collapse of the three point vortices. As shown in [28], the solution of the canonical system has the same behavior for any collinear initial configuration, its corresponding $H_K(t)$ is similarly even, positive and rapidly decreasing. We thus observe the enstrophy dissipation for any initial configurations in the limit of $\alpha \to 0$ for $\Gamma_1 = \Gamma_2 = 1$. We refer to this distributional loss of the enstrophy as *an anomalous enstrophy dissipation* via the triple collapse.

Figure 14.2a, b are the level curves of the Hamiltonian for $(\Gamma_1, \Gamma_2) = (2, 1)$ and $(4, 3)$, respectively, which also indicate that the same observation as above still holds true. The plots of $F(t)$ and $H_K(t)$ for various Γ_1 and Γ_2 in Fig. 14.2c, d show that $F(t)$ is odd and $H_K(t)$ is even, positive and rapidly decreasing. Accordingly, the anomalous enstrophy dissipation via the triple collapse occurs for the other strengths Γ_1 and Γ_2 as long as initial collinear configurations satisfy (14.23). As a matter of fact, even if the initial configuration does not satisfy the second condition of (14.23), namely

$$M = \Gamma_1\Gamma_2 L_{12}^2 + \Gamma_2\Gamma_3 L_{23}^2 + \Gamma_3\Gamma_1 L_{31}^2 \neq 0, \tag{14.27}$$

the anomalous enstrophy variation has also been confirmed by numerical means [28]. In this case, we need to use another two-dimensional phase space representation to

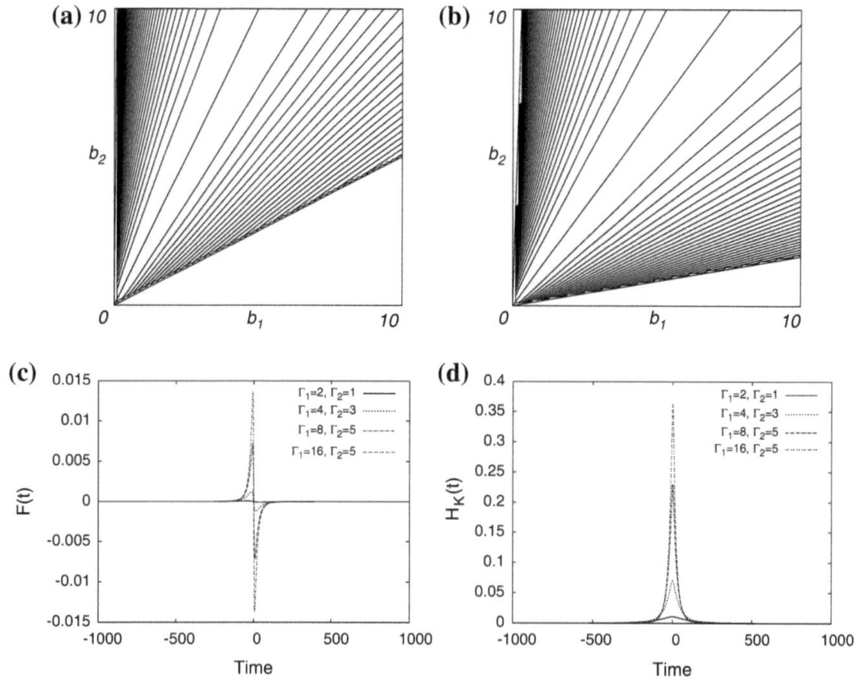

Fig. 14.2 **a** Level curves of the Hamiltonian (14.24) for $\Gamma_1 = 2$ and $\Gamma_2 = 1$. **b** Level curves of the Hamiltonian (14.24) for $\Gamma_1 = 4$ and $\Gamma_2 = 3$. **c** Plots of $F(t)$ for various Γ_1 and Γ_2. **d** Plots of $H_K(t)$ for various Γ_1 and Γ_2

observe the contour plot of the Hamiltonian, since (14.27) is reduced to $b_1 + b_2 + b_3 = 3$, where

$$b_1 = \frac{L_{23}^2}{\Gamma_1 \widetilde{M}}, \qquad b_2 = \frac{L_{31}^2}{\Gamma_2 \widetilde{M}}, \qquad b_3 = \frac{L_{12}^2}{\Gamma_3 \widetilde{M}},$$

with $\widetilde{M} = \frac{M}{3\Gamma_1\Gamma_2\Gamma_3}$. This change of variables gives rise to a two-dimensional phase space (x, y), called *the trilinear coordinate*,

$$(x, y) = \left(\sqrt{3} - \frac{1}{\sqrt{3}} b_1 - \frac{2}{\sqrt{3}} b_2, b_1\right),$$

as shown in Fig. 14.3a. Figure 14.3b shows the contour plot of the Hamiltonian in the trilinear coordinates for $\Gamma_1 = \Gamma_2 = 1$ and $M = 0.02$. The boundary $A = 0$ becomes a hyperbola and each contour line is also a hyperbola approaching asymptotically to two straight lines on which the ratio b_1/b_2 becomes a constant [28]. Accordingly, with the same argument as we have done in the case of $M = 0$, we can conclude that the evolution of the canonical system for the sequence of initial configurations $l_{mn}(0)/\alpha$ also behaves similarly like (14.25) as α tends to zero. Figure 14.3c, d show

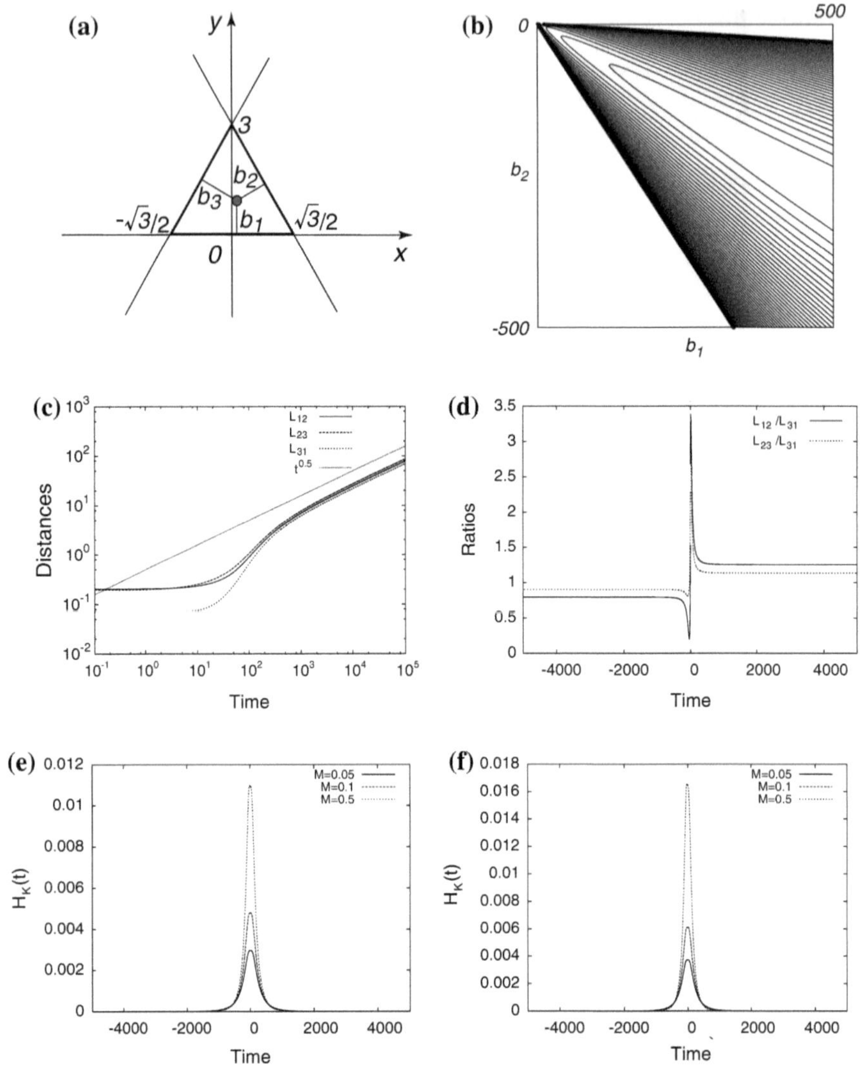

Fig. 14.3 **a** Trilinear coordinates to plot the level curves of the Hamiltonian with $M \neq 0$. **b** Level curves of the Hamiltonian in the trilinear coordinates for $M = 0.02$. **c** Evolution of the distances $L_{mn}(t)$ between the three α point vortices for $\Gamma_1 = \Gamma_2 = 1$. The invariant quantity for the initial condition is $M = 0.02$. **d** Evolution of the ratios between these distances. **e** Plots of $H_K(t)$ for various $M \neq 0$ when $\Gamma_1 = 1$ and $\Gamma_2 = 1$. **f** Plots of $H_K(t)$ for various $M \neq 0$ when $\Gamma_1 = 2$ and $\Gamma_2 = 1$

the evolution of the distances L_{mn} and the ratios between them for $\Gamma_1 = \Gamma_2 = 1$ and $M = 0.02$. They indicate that the distances certainly tend asymptotically to $\sqrt{|t|}$ and their ratios converge to different constants asymptotically as $t \to \pm\infty$. Hence, the evolution of the αPV system converges to a self-similar collapse for $t < 0$ and a self-similar expansion for $t > 0$, while the shape of the vortex triangle for $t < 0$ is different from that for $t > 0$ in this case.

Computing the evolution of the canonical system for $M = 0.05, 0.1$ and 0.5 when $(\Gamma_1, \Gamma_2) = (1, 1)$ and $(2, 1)$, we give the plots of $H_K(t)$ for them in Fig. 14.3e and f, respectively. They show that $H_K(t)$ is even, positive and rapidly decreasing. Hence, the enstrophy variation converges to the δ_0 function with a negative weight in the sense of distributions for $M > 0$ as well as $M = 0$. We must note that the collapse of three point vortices never occurs if $M \neq 0$, which seems to be a contradiction in terms of the continuity of the evolution of αPV system in the $\alpha \to 0$ limit and the PV system with $\alpha = 0$. But this is not the case, since, owing to (14.18) and (14.27) at the initial moment, we have

$$\begin{aligned}\alpha M &= \Gamma_1\Gamma_2 l_{12}^2(0) + \Gamma_2\Gamma_3 l_{23}^2(0) + \Gamma_3\Gamma_1 l_{31}^2(0),\\ &= \Gamma_1\Gamma_2(l_{12}^{\alpha})^2(0) + \Gamma_2\Gamma_3(l_{23}^{\alpha}(0))^2 + \Gamma_3\Gamma_1(l_{31}^{\alpha}(0))^2,\end{aligned}$$

and thus, if the evolution l_{mn}^{α} converges as $\alpha \to 0$, the limit evolution is subject to

$$0 = \Gamma_1\Gamma_2(l_{12})^2 + \Gamma_2\Gamma_3(l_{23})^2 + \Gamma_3\Gamma_1(l_{31})^2,$$

which is equivalent to the condition of the initial configuration for the triple collapse. Therefore, we conclude that the anomalous enstrophy dissipation is a robust phenomenon observed for a wide range of initial configurations of the three α point vortices.

14.3.2.2 Collapse of Four α-Point Vortices and Enstrophy Variations

In this section, we consider the evolution of four α point vortices under the same conditions as the self-similar collapse of N point vortices. According to Kimura [12], the necessary condition for the self-similar collapse in terms of the vortex strengths is given by

$$\sum_{m=1}^{N}\sum_{n=m+1}^{N} \Gamma_m\Gamma_n = 0,$$

which is equivalent to the first condition of (14.23). Since the canonical system with this condition is no longer integrable for $N \geq 4$, it is not easy to solve the canonical equation nor to consider the $\alpha \to 0$ limit of the evolution analytically. Thus, we use an implicit fourth order symplectic Runge-Kutta method [17] with time step size $\Delta t = 0.001$ to obtain long-time evolutions of (14.17) numerically. Initial configurations of four α point vortices are chosen so that point vortices collapse self-similarly in finite time. Kudela [16] has been shown that N point vortices collapse self-similarly in finite time, if their initial configuration satisfies the following $2N$ equations:

$$\sum_{m=1}^{N} \Gamma_m |z_m|^2 = 0, \quad \sum_{m=1}^{N} \Gamma_m z_m = 0, \tag{14.28}$$

$$v_1 z_m = v_m z_1, \quad m = 1, \ldots, N-2, \tag{14.29}$$

$$\mathrm{Re}\left(\sum_{m=1}^{N} \Gamma_m v_m \right) = 0, \tag{14.30}$$

in which $z_m = x_m + i y_m$ is the location of the mth point vortex in the complex plane, and v_m represents the complex velocity at z_m induced by the other $N-1$ point vortices given as follows.

$$v_m = -\frac{1}{2\pi i} \sum_{n \neq m}^{N} \frac{\Gamma_n}{\bar{z}_m - \bar{z}_n}.$$

As shown in [16], the algebraic equations (14.28)–(14.30) for $N = 4$ are solved numerically with high accuracy using Mathematica. Figure 14.4 shows an example of the self-similar evolution of four point vortices with $\Gamma_1 = 2$, $\Gamma_2 = 1$, $\Gamma_3 = 1$ and $\Gamma_4 = -\frac{5}{4}$ for an initial configuration obtained by solving (14.28)–(14.30).

Figure 14.5a shows the long-time evolution of four α point vortices with $\Gamma_1 = 2$, $\Gamma_2 = 1$, $\Gamma_3 = 1$ and $\Gamma_4 = -\frac{5}{4}$ for the same initial configuration as in Fig. 14.4. We solve the canonical system numerically forward as well as backward in time. After the four point vortices are getting closer with each other without collapse, they separate away. In order to see the evolution more clearly, we plot the evolution of the distances $L_{mn}(t)$ for $\alpha = 1$ in Fig. 14.5b. All distances take their minimum values at a certain time, say $t_c^\alpha < 0$, and each of them behaves as $C_{mn}\sqrt{|t - t_c^\alpha|}$ for a constant $C_{mn} > 0$. Figure 14.5c shows the close-up of the evolution of the distance

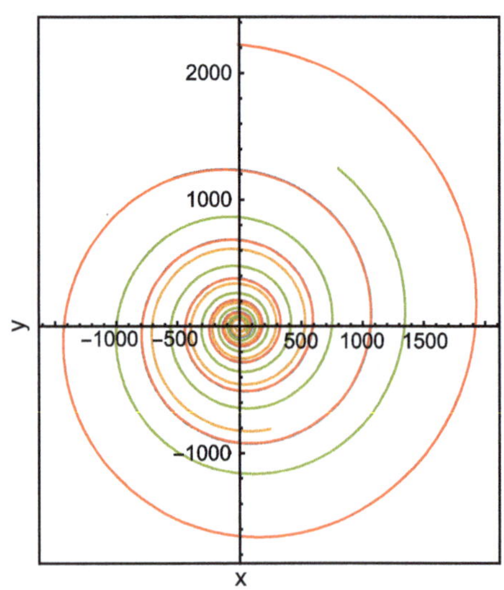

Fig. 14.4 Self-similar evolution of four point vortices with the strengths $\Gamma_1 = 2$, $\Gamma_2 = 1$, $\Gamma_3 = 1$ and $\Gamma_4 = -\frac{5}{4}$. The initial configuration is obtained by solving (14.28)–(14.30) numerically

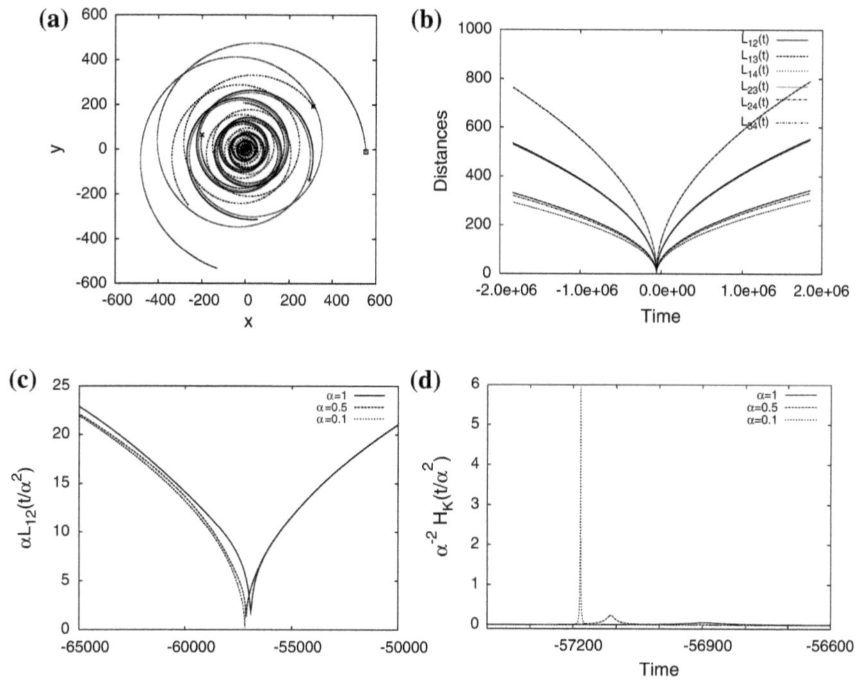

Fig. 14.5 **a** Evolution of the four α point vortices with $\alpha = 1$ for the same initial configurations as in Fig. 14.4. We can solve the equation numerically forward and backward in time, since there is no collapse in the αPV system. **b** Evolution of the distances $L_{mn}(t)$ between the four α point vortices corresponding to the evolution. **c** Convergence of th distance $\alpha L_{12}(t/\alpha^2)$ as $\alpha \to 0$. **d** Convergence of the enstrophy variation $1/\alpha^2 H_K(t/\alpha^2)$ as $\alpha \to 0$

$\alpha L_{12}(t/\alpha^2)$, which indicates that the minimum distance converges to zero and we observe $\lim_{\alpha\to 0} t_c^\alpha = t_c^0$. This numerical solution suggests that, as α tends to zero, the distance $\alpha L_{12}(t/\alpha^2)$ converges to $C_{mn}\sqrt{|t - t_c^0|}$. Consequently, the solution of the canonical system converges to the self-similar collapse for $t < t_c^0$ and to the self-similar expansion for $t > t_c^0$ as we have observed in the three vortex problem. Furthermore, the plots of the enstrophy variation $\frac{1}{\alpha^2} H_K(t/\alpha^2)$ in Fig. 14.5d illustrate the convergence to the delta function $\delta\left(t - t_c^0\right)$ as $\alpha \to 0$. Hence, we conclude that the anomalous enstrophy dissipation occurs via the quadruple collapse as $\alpha \to 0$.

According to [16], one can obtain the other initial configurations of α point vortices with different Hamiltonian value by a continuation. That is to say, we first compute the value of Hamiltonian, say H_0, for the initial configuration obtained from (14.28)–(14.30). We then solve the Eqs. (14.28), (14.29) and $\widetilde{H} = H_0 + \epsilon$ for a given small ϵ instead of (14.30) to obtain the initial configuration with this Hamiltonian value. Figure 14.6a, b show the evolution of the distances $L_{mn}(t)$ for another initial configuration and the plot of $H_K(t)$ corresponding to the evolution, respectively. They also indicate that the anomalous enstrophy dissipation via the quadruple collapse is observed as $\alpha \to 0$.

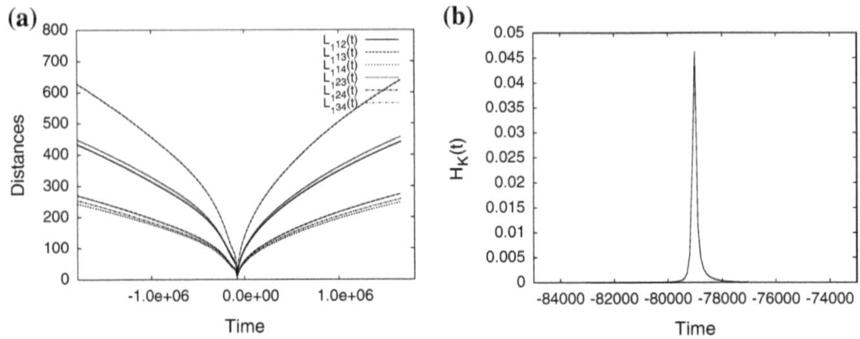

Fig. 14.6 **a** Evolution of the distances $L_{mn}(t)$ between the four α point vortices with the strengths $\Gamma_1 = 2$, $\Gamma_2 = 1$, $\Gamma_3 = 1$ and $\Gamma_4 = -\frac{5}{4}$ for the initial configuration satisfying (14.28) and (14.29) with a different value of Hamiltonian. **b** Plot of the enstrophy variation $H_K(t)$ in the neighborhood of the critical time

A different evolution of four α point vortices is observed when the vortex strengths are chosen as $\Gamma_1 = 1$, $\Gamma_2 = 1$, $\Gamma_3 = 1$ and $\Gamma_4 = -1$. Figure 14.7a shows the orbits of $\boldsymbol{X}_m(t)$ for the initial configuration satisfying (14.28)–(14.30). After the four α point vortices approach sufficiently close to each other, an interaction caused by the α regularization begins, and we then observe that the two α point vortices with $\Gamma_1 = 1$ and $\Gamma_4 = -1$ are moving towards infinity. Figure 14.7b shows the evolution of the distances $L_{mn}(t)$. There exists a critical time $\widetilde{t_c^\alpha}$ such that $L_{mn}(t) \sim \sqrt{|t - \widetilde{t_c^\alpha}|}$ for $t < \widetilde{t_c^\alpha}$, which shows that the four α point vortices approach self-similarly. For $t > \widetilde{t_c^\alpha}$, $L_{14}(t)$ and $L_{23}(t)$ become constant functions, while the other distances are increasing linearly. In order to see more closely how they evolve after $t > \widetilde{t_c^\alpha}$, we plot the evolutions of each component of $\boldsymbol{X}_m(t) = (X_m(t), Y_m(t))$ in Fig. 14.7c, which indicates that the four α point vortices are divided into two pairs. The pair of the α point vortices with $\Gamma_1 = 1$ and $\Gamma_4 = -1$ move together with a constant speed to infinity as a vortex dipole. The other α-vortex pair with $\Gamma_2 = 1$ and $\Gamma_3 = 1$ is co-rotating and stays in the neighborhood of the origin. The two vortex pairs are separated away linearly, since the vortex dipole moves with a constant speed. We can expect that, as $\alpha \to 0$, $\alpha L_{mn}(t/\alpha^2)(t)$ behaves like $\sqrt{|t - \widetilde{t_c^0}|}$ for $t < \widetilde{t_c^0} = \lim_{\alpha \to 0} \widetilde{t_c^\alpha}$, which means that the evolution converges to a self-similar quadruple collapse. On the other hand, after $t > \widetilde{t_c^0}$, the distances $\alpha L_{14}(t/\alpha^2)$ and $\alpha L_{23}(t/\alpha^2)$ converge to zero, but the other distances diverge as $\alpha \to 0$. Hence, the limit evolution consists of a singular vortex dipole going infinity with infinite speed and a singular pair of identical point vortices co-rotating with infinite angular speed after $\widetilde{t_c^0}$. Finally, we plot the function $H_K(t)$ for the evolution in Fig. 14.7d. Since $\alpha L_{14}(t/\alpha^2)$ and $\alpha L_{23}(t/\alpha^2)$ converge to zero for $t > \widetilde{t_c^0}$ as $\alpha \to 0$, the enstrophy variation $\mathscr{Z}^\alpha(t)$ never converges to zero after $\widetilde{t_c^0}$ owing to Corollary 14.3.2. As a matter of fact, since $H_K(t)$ becomes a non-zero constant for $t > \widetilde{t_c^0}$, the enstrophy diverges as $\alpha \to 0$ for $t > 0$, which suggests that the infinite enstrophy jump occurs at the event of the self-similar quadruple collapse.

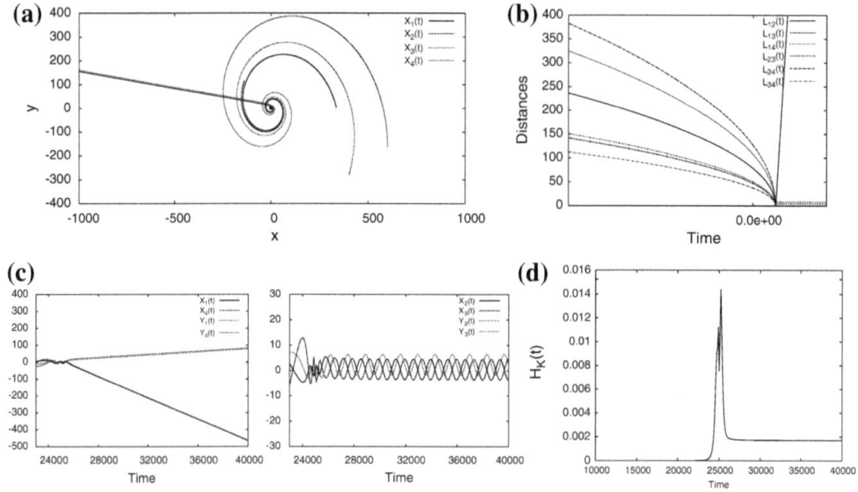

Fig. 14.7 **a** Trajectories $X_m(t)$ for $m = 1, 2, 3, 4$ of the four α point vortices with $\alpha = 1$. The vortex strengths are given by $\Gamma_1 = 1$, $\Gamma_2 = 1$, $\Gamma_3 = 1$ and $\Gamma_4 = -1$ and their initial configuration is computed by solving (14.28)–(14.30). **b** Evolution of the distances $L_{mn}(t)$ between the four α point vortices corresponding to the evolution. **c** Evolutions of $X_m(t)$ and $Y_m(t)$ for $m = 1, 2, 3, 4$ after $\tilde{t}_c^0$. **d** Evolution of $H_K(t)$ corresponding to the evolution

14.4 Summary and Discussion

We have investigated the $\alpha \to 0$ limit of weak solutions to the 2D Euler-α equations for the initial vorticity distribution in $\mathscr{M}(\mathbb{R}^2)$ whose support consists of a set of N discrete points, in order to understand how singular solutions to the Euler equations dissipate the enstrophy anomalously. The evolution of the N points, called α point vortices, is described by a Hamiltonian dynamical system with N degrees of freedom, whose solution exists globally in time and gives rise to a weak solution to the Euler-α equations. The enstrophy variation, which is defined as a function of the distances between α point vortices, vanishes as $\alpha \to 0$, unless the distances converges to zero in the limit, i.e. α point vortices collapse at a certain time. As long as $\alpha \neq 0$, the collapse of α point vortices never occurs, but we have found that it occurs in the $\alpha \to 0$ limit when we consider the three α point vortices with the vortex strengths and the initial configurations satisfying (14.23). The evolution of the three α point vortices converges to the self-similar collapse for $t < 0$ and the self-similar expansion for $t > 0$ in the limit. The enstrophy variation converges to the δ-measure at the critical time $t = 0$ with a negative weight, which indicates that the enstrophy dissipates at the event of the collapse. We numerically observe this phenomenon, even if the initial configuration does not satisfy (14.23). Moreover, we have also confirmed numerically that the anomalous enstrophy dissipation occurs for a collapse of four α point vortices in the $\alpha \to 0$ limit. Hence, the collapse of α point vortices is a robust mechanism that induces the anomalous enstrophy dissipation. Let us note that the collapse of α-point vortices with the anomalous enstrophy variation in the $\alpha \to 0$ limit is just confirmed by numerical means. Hence, it is important to prove this fact with a mathematical rigor, which will be reported in near future.

By considering the limit solutions of the αPV system, we successfully describe the physical mechanism of the anomalous enstrophy dissipation induced by singular evolutions of incompressible fluid flows in terms of vortex dynamics. Since the enstrophy dissipation is one of the characteristic properties observed in 2D high-Reynolds number turbulence, the αPV system works as a model of singular weak solutions to the 2D Euler equations. However, on the other hand, there are many questions to be considered in understanding of 2D turbulence, part of which are proposed as follows. (i) Mathematically, it is uncertain whether or not the $\alpha \to 0$ limit solution of the αPV system defines a weak solution to the 2D Euler equations. (ii) The αPV system suggests that the enstrophy variation could occur when the coherent vortex structures approach closely with each other. We need to check computationally if this phenomenon is observed in 2D Navier-Stokes-α flows for sufficiently small viscosity and α. (iii) We have considered a special class of singular solutions in the PV system, the self-similar evolutions, as a reference in order to construct singular evolution of the αPV system as $\alpha \to 0$. It is interesting to investigate the case of non-integrable αPV system with $N \geq 4$, in which we observe how complex (chaotic) evolutions and their corresponding enstrophy variation converges as $\alpha \to 0$.

References

1. Batchelor, G.K.: Computation of the energy spectrum in homogeneous two-dimensional turbulence. Phys. Fluids Supppl. **II**(12), 233–239 (1969)
2. Buckmaster, T., De Lellis, C., Székelyhidi Jr., L.: Dissipative Euler flows with Onsager-critical spatioal regularity. arXiv:1404.6915 (2014)
3. Constantin, P.: E, W. and Titi, E. S.: Onsager's Conjecture on the Energy Conservation for Solutions of Euler's Equation. Comm. Math. Phys. **165**, 207–209 (1994)
4. Delrort, J.-M.: Existence de nappe de tourbillion en dimension deux. J. Am. Math. Soc. **4**, 553–586 (1991)
5. Diperna, R.J., Majda, A.J.: Concentrations in regularizations for 2-D incompressible flow. Comm. Pure Appl. Math. **40**, 301–345 (1987)
6. Duchon, J., Robert, R.: Inertial energy dissipation for weak solutions of incompressible Euler and Navier-Stokes equations. Nonlinearity. **13**, 249–255 (2000)
7. Eyink, G.L.: Dissipation in turbulence solutions of 2D Euler equations. Nonlinearity. **14**, 787–802 (2001)
8. Giga, Y., Miyakawa, T., Osada, H.: Two-dimensional Navier-Stokes flow with measures as initial vorticity and Morrey spaces. Arch. Rat. Mech. Anal. **104**, 223–250 (1988)
9. Holm, D.D., Marsden, J.E., Ratiu, T.S.: Euler-Poincaré models of ideal fluids with nonlinear dispersion. Phys. Rev. Lett. **80**, 4173–4176 (1998)
10. Holm, D.D.: Variational principles for Lagrangian-averaged fluid dynamics. J. Phys. A **35**, 679–688 (2002)
11. Kaneda, Y., Ishihara, T., Yokokawa, M., Itakura, K., Uno, A.: Energy dissipation rate and energy spectrum in high resolution direct numerical simulations of turbulence in a periodic box. Phys. Fluids **15**, L21–L24 (2003)

12. Kimura, Y.: Similarity solution of two-dimensional point vortices. J. Phys. Soc. Jpn. **56**, 2024–2030 (1987)
13. Kolmogorov, A.N.: The local structure of turbulence in incompressible viscous fluid for very large Reynolds number. Dokl. Akad. Nauk SSSR **30**, 301–305 (1941) [reprinted in Proc. R. Soc. Lond. A **434**, 9–13 (1991).]
14. Kolmogorov, A.N.: A refinement of previous hypothesis concerning the local structure of turbulence in a viscous incompressible fluid at high Reynolds number. J. Fluid Mech. **13**, 82–85 (1962)
15. Kraichnan, R.H.: Inertial ranges in two-dimensional turbulence. Phys. Fluids. **10**, 1417–1423 (1967)
16. Kudela, H.: Collapse of n-point vortices in self-similar motion. Fluid Dyn. Res. **46**, 031414 (2014). doi:10.1088/0169-5983/46/3/031414
17. Leimkuhler, B., Reich, S.: Simulating Hamiltonian Dynamics. Cambridge University Press, Cambridge (2005)
18. Leith, C.E.: Diffusion approximation for two-dimensional turbulence. Phys. Fluids. **11**, 671–672 (1968)
19. Linshiz, J.S., Titi, E.S.: On the convergence rate of the Euler-α, an inviscid second-grade complex fluid, model to the Euler equations. J. Stat. Phys. **138**, 305–332 (2010). doi:10.1007/s10955-009-9916-9
20. Lunasin, E., Kurien, S., Taylor, M.A., Titi, E.S.: A study of the Navier-Stokes-α model for two-dimensional turbulence. J. Turbul. **8**, 1–21 (2007). doi:10.1080/14685240701439403
21. Majda, A.J.: Remarks on weak solutions for vortex sheets with a distinguished sign. Indiana Univ. Math. J. **42**, 921–939 (1993)
22. Marchioro, C., Pulvirenti, M.: Mathematical Theory of Incompressible Nonviscous Fluids. Applied Mathematical Sciences, vol. 96. Springer, New York (1994)
23. Newton, P.K.: The N-vortex Problem. Analytical Techniques. Springer, New York (2001)
24. Novikov, E.A.: Dynamics and statistic of a system of vortices. Sov. Phys. JETP. **41**, 937–943 (1976)
25. Oliver, M., Shkoller, S.: The vortex blob method as a second-grade non-Newtonian fluid. Comm. Partial Differ. Equ. **26**, 295–314 (2001)
26. Onsager, L.: Statistical hydrodynamics. Nouvo Cimento Suppl. **6**, 287–297 (1949)
27. Sakajo, T.: Instantaneous energy and enstrophy variations in Euler-alpha point vortices via triple collapse. J. Fluid Mech. **702**, 188–214 (2012)
28. Sakajo, T.: Anomalous enstrophy dissipation via the self-similar triple collapse of the Euler-α point vortices. In: Chen, H., et al. (eds.) Emerging Topics on Differential Equations and Their Applications. Nankai Series in Pure, Applied Mathematics and Theoretical Physics, vol. 10, pp. 155–169 (2013)
29. Yudovich, V.I.: Nonstationary motion of an ideal incompressible liquied. USSR Comput. Math. Phys. **3**, 1407–1456 (1963)
30. Watson, G.N.: A Treatise on the Theory of Bessel Functions. Merchant Books (2008)

Chapter 15
Thermodynamical Consistent Modeling and Analysis of Nematic Liquid Crystal Flows

Matthias Hieber and Jan Prüss

Abstract The general Ericksen-Leslie model for the flow of nematic liquid crystals is reconsidered in the non-isothermal case aiming for thermodynamically consistent models. The non-isothermal simplified model is then investigated analytically. A fairly complete dynamic theory is developed by analyzing these systems as quasilinear parabolic evolution equations in an $L_p - L_q$-setting. First, the existence of a unique, local strong solution is proved. It is then shown that this solution extends to a global strong solution provided the initial data are close to an equilibrium or the solution is eventually bounded in the natural norm of the underlying state space. In these cases the solution converges exponentially to an equilibrium in the natural state manifold.

Keywords Nematic liquid crystals · Quasilinear parabolic evolution equations · Regularity · Global solutions · Convergence to equilibria

15.1 Introduction

The continuum theory of liquid crystals was developed by Ericksen and Leslie during the 1960s in their pioneering work [9, 22]. This theory models nematic liquid crystal flow from a hydrodynamical point of view and reduces to the Oseen-Frank theory in the static case, see [15, 31]. It describes the evolution of the complete system under the influence of the velocity u of the fluid and the orientation configuration d of rod-like liquid crystals. Hence, $d = d(t, x)$ is a unit vector in $\mathbb{R}^3$. The original derivation [9, 22] is based on the conservation laws for mass and linear as well as

M. Hieber
Technische Universität Darmstadt, Fachbereich Mathematik,
Schlossgarten-Strasse 7, 64289 Darmstadt, Germany
e-mail: hieber@mathematik.tu-darmstadt.de

J. Prüss (✉)
Institut Für Mathematik, Martin-Luther-Universität Halle-Wittenberg,
Theodor-Lieser-Strasse 5, 06120 Halle, Germany
e-mail: jan.pruess@mathematik.uni-halle.de

Y. Shibata and Y. Suzuki (eds.), *Mathematical Fluid Dynamics, Present and Future*, Springer Proceedings in Mathematics & Statistics 183,
DOI 10.1007/978-4-431-56457-7_15

angular momentums. General liquid crystal materials are described by the Landau-de Gennes theory [5] from a unified point of view.

The Ericksen-Leslie theory is nowadays widely used as a model for the flow of nematic liquid crystals, see for example the works of Ericksen and Kinderlehrer [10], Chandrasekhar [2], DeGennes and Prost [6] as well as Virga [37].

Note that these models are mostly formulated in an isothermal environment and are, in general, neither thermodynamically consistent nor thermodynamically stable. To the best of our knowledge, only very few articles are dealing so far with the *thermodynamical consistency* of these models. Concerning the Ericksen-Leslie model, for a physically rigorous derivation we refer to the work of Müller [30] concentrating on the modeling aspect, and for recent analytical work to Feireisl et al. [11], Feireisl et al. [12] and Li and Xin [23]. Non-isothermal Landau-De Gennes nematic liquid crystal flows were investigated in the recent articles [13, 14].

The aim of this paper is twofold: first, we reconsider the Ericksen-Leslie approach from the perspective of thermodynamical consistency and stability. Following arguments from thermodynamics and employing entropy principles, we derive consistent models in a mathematically efficient way, even in the case of compressible fluids. Let us emphasize that, in the end, our model contains the classical Ericksen-Leslie model in its general form as a special case.

Secondly, we investigate our model analytically. Restricting ourselves to the case of constant density and not taking into account so called stretching, we develop a rather complete dynamic theory for the equations representing these models. More precisely, we first prove the existence of a unique, local strong solution to this system. We further show that this solution extends to a global, strong solution, provided the initial data are close to an equilibrium or the solution is eventually bounded in the natural norm of the underlying state space. In this case the solution converges exponentially to an equilibrium in the natural state manifold. The results obtained thus parallel those proved recently by Hieber et al. [17] dealing with the isothermal situation. For results concerning the asymptotic behaviour of solutions in the situation of the whole space $\mathbb{R}^3$, we refer to the work of Dai and Schonbek [4].

The nowadays called simplified Ericksen-Leslie model in the isothermal situation was introduced and investigated first by Lin [24, 25]. Lin and Liu [26, 27] studied the situation, where the nonlinearity in the equation for the director d is replaced by a Ginzburg-Landau energy functional. The existence of global weak solutions to this system in dimension 2 or 3 was proved under suitable assumptions on the intial data. For related results see [18, 28]. Wang proved in [38] global well-posedness for the simplified system for initial data being small in $BMO^{-1} \times BMO$ in the case of a whole space by combining techniques of Koch and Tataru with methods from harmonic maps to certain Riemannian manifolds.

The general Ericksen-Leslie model (in the isothermal situation) is based on the Oseen-Frank energy density functional which takes into account stretching as well as rotational effects for the director field. In the special case of homogeneous isotropic elasticity the equation for d reads as

$$\partial_t d + u \cdot \nabla d - Vd + \frac{\lambda_2}{\lambda_1} Dd = -\frac{1}{\lambda_1}(\Delta d + |\nabla d|_2^2 d) + \frac{\lambda_2}{\lambda_1}(Dd \cdot d)d \quad \text{in } (0,T) \times \Omega. \tag{15.1}$$

Here $|\cdot|_2$ means the l_2-norm, $D = \frac{1}{2}(\nabla u + [\nabla u]^T)$ denotes the symmetric, $V = \frac{1}{2}(\nabla u - [\nabla u]^\mathsf{T})$ the anti-symmetric part of the deformation tensor and $\lambda_1, \lambda_2 \in \mathbb{R}\backslash\{0\}$ are material coefficients. Modifications of this model were considered by Coutard and Shkoller [3] in which the above Eq. (15.1) for d is replaced by a Ginzburg-Landau type approximation.

$$\gamma(\partial_t d + u \cdot \nabla d + d \cdot \nabla u) = \Delta d - \frac{1}{\varepsilon^2}(|d|_2^2 - 1)d \quad \text{in } (0,T) \times \Omega. \tag{15.2}$$

They proved local wellposedness for this system as well as a global existence result for small data within this setting. Note, however, that in this case the presence of the stretching term $d \cdot \nabla u$ causes loss of total energy balance and, moreover, the condition $|d|_2 = 1$ in $(0, T) \times \Omega$, is not preserved anymore. For recent results on the general Ericksen-Leslie model with d satisfying (15.1), we refer to the articles [19, 28, 29, 39], which contain well-posedness criteria for the general system under various assumptions on the Leslie coefficients. For results on non-isothermal models including the above mentioned stretching term, see [12] and references therein.

Let us stress at this point that an important novelty of our approach lies in the fact that the complete model described in Sect. 2.7 is rigoroulsy proven to be thermodynamically consistent and stable. Specialising to the isothermal situation, we rediscover in particular the classical general Ericksen-Leslie system. It is interesting to compare our approach with the approach of Müller [30], and with the energy variational approach developed by Liu and coworkers [29] and by Virga [37].

The plan for this contribution is as follows: Sect. 15.2 is devoted to the modeling of liquid crystals. In particular, based on the entropy principle, we derive a model of Ericksen-Leslie type which is thermodynamically consistent and stable. In Sect. 15.3 the equilibria of the system are identified—which are zero velocities and constant temperature and director—and it is proved that these are thermodynamically stable. The negative total entropy is shown to be a strict Ljapunov functional, in particular the model is thermodynamically consistent. In Sect. 15.4 we prove local well-posedness of the non-isothermal simplified model and construct the resulting local semiflow in the natural state manifold of the system. We show that each solution which does not develop singularities in a sense to be specified converges to a unique equilibrium. These results are proved by means of techniques involving maximal L_p-regularity and results on quasilinear paprabolic evolution equations. For these methods, we refer to the booklet by Denk et al. [7] and to the work of Prüss and Simonett [34], Köhne et al. [20], and LeCrone et al. [21].

By means of these techniques we are also able to prove analogous results for the full model, which, however, due to limitation of space will be presented elsewhere.

15.2 Thermodynamical Consistent Modeling

In this section we aim to give a self-contained presentation of a *thermodynamically consistent modeling of liquid crystals*. Like this we are able to refrain from refering to the orginal papers [9, 22], which are not easily accessible to a mathematical audience. Our approach does not only extend the classical Ericksen-Leslie model to the non-isothermal situation in a thermodynamical consistent and stable way but it also allows to exhibit the physical and mathematical beauty of this model.

In this section, $\Omega \subset \mathbb{R}^n$ always denotes a domain with C^1-boundary.

1. First Principles

We begin with the balance laws of mass, momentum, and energy. They read as

$$\begin{aligned} \partial_t \rho + \operatorname{div}(\rho u) &= 0 && \text{in } \Omega, \\ \rho(\partial_t + u \cdot \nabla)u + \nabla \pi &= \operatorname{div} S && \text{in } \Omega, \\ \rho(\partial_t + u \cdot \nabla)\varepsilon + \operatorname{div} q &= S : \nabla u - \pi \operatorname{div} u && \text{in } \Omega, \\ u = 0, \quad q \cdot \nu &= 0 && \text{on } \partial\Omega. \end{aligned} \tag{15.3}$$

Here ρ means density, u velocity, π pressure, ε internal energy, S extra stress and q heat flux. This immediately gives conservation of the total energy. In fact, we have

$$\rho(\partial_t + u \cdot \nabla)e + \operatorname{div}(q + \pi u - Su) = 0 \quad \text{in } \Omega,$$

where $e := |u|^2/2 + \varepsilon$ means the total mass specific energy density (kinetic and internal). The energy flux Φ_e is given by $\Phi_e := q + \pi u - Su$. Integrating over Ω yields

$$\partial_t \mathsf{E}(t) = 0, \quad \mathsf{E}(t) = \mathsf{E}_{kin}(t) + \mathsf{E}_{int}(t) = \int_\Omega \rho(t,x)e(t,x)dx,$$

provided

$$q \cdot \nu = u = 0 \quad \text{on } \partial\Omega. \tag{15.4}$$

Hence, if (15.4) holds, total energy is preserved, independent of the particular choice of S and q.

2. Thermodynamics

Assume a given free energy ψ of the form $\psi = \psi(\rho, \theta, \tau)$, where θ denotes the (absolute) temperature and τ will be specified later. We then have the following thermodynamical relations:

$$\begin{aligned} \varepsilon &= \psi + \theta\eta && \text{internal energy}, \\ \eta &= -\partial_\theta \psi && \text{entropy}, \\ \kappa &= \partial_\theta \varepsilon = -\theta \partial_\theta^2 \psi && \text{heat capacity}. \end{aligned} \tag{15.5}$$

Later on, for well-posedness of the heat problem, we require $\kappa > 0$, i.e. ψ to be strictly concave with respect to $\theta \in (0, \infty)$.

In the classical case, where ψ depends only on ρ and θ, we have the *Clausius-Duhem equation*

$$\rho(\partial_t + u \cdot \nabla)\eta + \operatorname{div}(q/\theta) = S : \nabla u/\theta - q \cdot \nabla\theta/\theta^2 + (\rho^2\partial_\rho\psi - \pi)(\operatorname{div} u)/\theta \quad \text{in } \Omega.$$

Hence, in this case the entropy flux Φ_η is given by $\Phi_\eta := q/\theta$ and the entropy production by

$$\theta r := S : \nabla u - q \cdot \nabla\theta/\theta + (\rho^2\partial_\rho\psi - \pi)(\operatorname{div} u).$$

Employing the boundary conditions (15.4), we obtain for the total entropy N by integration over Ω

$$\partial_t\mathsf{N}(t) = \int_\Omega r(t,x)dx \geq 0, \quad \mathsf{N}(t) = \int_\Omega \rho(t,x)\eta(t,x)dx,$$

provided $r \geq 0$ in Ω. As $\operatorname{div} u$ has no sign we require

$$\pi = \rho^2\partial_\rho\psi, \tag{15.6}$$

which is the famous *Maxwell relation*. Further, as S and q are independent, this requirement leads to the classical conditions

$$S : \nabla u \geq 0 \quad \text{and} \quad q \cdot \nabla\theta \leq 0. \tag{15.7}$$

Summarizing, we see that whatever one chooses for S and q, one always has conservation of energy and the total entropy is non-decreasing provided (15.7), (15.6) and (15.4) are satisfied. Thus, these conditions ensure the thermodynamic consistency of the model.

As an example for S and q consider the classical laws due to Newton and Fourier which are given by

$$S := S_N := 2\mu_s D + \mu_b \operatorname{div} u\, I, \quad 2D = (\nabla u + [\nabla u]^\mathsf{T}), \quad q = -\alpha_0\nabla\theta.$$

In this case, (15.7) is satisfied as soon as $\mu_s \geq 0$, $2\mu_s + n\mu_b \geq 0$ and $\alpha_0 \geq 0$ hold. Note that it does not matter at all whether μ_s, μ_b, α_0 are constants or whether they depend on ρ, θ, or on other variables.

3. Nematic Liquid Crystals

For isotropic nematic liquid crystals we assume a free energy density ψ of the form

$$\psi = \psi(\rho, \theta, \tau), \quad \text{with } \tau = \frac{1}{2}|\nabla d|_2^2.$$

Here d means the orientation vector, also called the *director*, which should satisfy the condition

$$|d|_2^2 := \sum_{j=1}^{n} d_j^2 = 1.$$

Note that $2\tau = \mathrm{tr}(\nabla d[\nabla d]^\mathsf{T})$ is the first invariant of the matrix $\nabla d[\nabla d]^\mathsf{T}$, and for its last invariant it holds $\det(\nabla d[\nabla d]^\mathsf{T}) = (\det\nabla d)^2 = 0$, as $\nabla dd = 0$ by $|d|_2 = 1$.

We neglect spin energy below but take into account transport of energy due to couple stress. This means that the energy flux is replaced by

$$\Phi_e := q + \pi u - Su - \Pi \mathscr{D}_t d, \quad \mathscr{D}_t = \partial_t + u \cdot \nabla,$$

where Π has to be modeled.

As constitutive laws we will employ

$$S = S_N + S_E + S_L, \quad S_E = -\lambda \nabla d[\nabla d]^\mathsf{T}, \quad q = -\alpha_0 \nabla\theta - \alpha_1 (d \cdot \nabla\theta) d. \quad (15.8)$$

S_N means the *Newton stress* introduced above, S_E the *Ericksen stress*, and S_L the *Leslie stress* which will be defined later. Assuming these two constitutive laws we derive in the following the balance of entropy, i.e. the *Clausius-Duhem equation.* A short computation gives

$$\rho(\partial_t + u \cdot \nabla)\eta + \mathrm{div}\, \Phi_\eta = r, \quad (15.9)$$

with $\Phi_\eta = q/\theta$, and

$$\begin{aligned}\theta r = &-q \cdot \nabla\theta/\theta + 2\mu_s |D|_2^2 + \mu_b |\mathrm{div}\, u|^2 + (\rho^2 \partial_\rho \psi - \pi)\mathrm{div}\, u \\ &+ (\rho \partial_\tau \psi - \lambda)\nabla d[\nabla d]^\mathsf{T} : \nabla u + (\Pi - \rho \partial_\tau \psi \nabla d) : \nabla \mathscr{D}_t d \\ &+ S_L : \nabla u + (\mathrm{div}\Pi + \beta d) \cdot \mathscr{D}_t d.\end{aligned}$$

for some scalar function β. Note that $d \cdot \mathscr{D}_t d = 0$ as $|d|_2 = 1$, hence $\beta \in \mathbb{R}$ can be chosen arbitrarily.

For the entropy production r to be nonnegative, we require

$$\mu_s \geq 0, \quad 2\mu_s + n\mu_b \geq 0, \quad \alpha_0 \geq 0, \quad \alpha_0 + \alpha_1 \geq 0.$$

Except for the last one, these conditions are the well-known conditions from fluid dynamics, see Sect. 2.2. The subsequent terms in the definition of r have no sign, hence we require them to vanish, which yields the relations

$$\pi = \rho^2 \partial_\rho \psi, \quad \lambda = \rho \partial_\tau \psi \quad \Pi = \rho \partial_\tau \psi \nabla d. \quad (15.10)$$

Finally, to obtain nonnegativity of the last two terms, in the simplest case, we may assume that the Leslie stress S_L vanishes, and

$$\gamma \mathscr{D}_t d = \mathrm{div}[(\rho\partial_\tau\psi)\nabla]d + \beta d,$$

for some $\gamma = \gamma(\rho, \theta, \tau) \geq 0$. The condition $|d|_2 = 1$ then requires $\beta = \lambda|\nabla d|_2^2$, which leads to the equation

$$\gamma(\partial_t + u \cdot \nabla)d = \mathrm{div}[\lambda\nabla]d + \lambda|\nabla d|_2^2 d, \tag{15.11}$$

a nonlinear convection-diffusion equation for d. This is the basic equation governing the evolution of the director field d. With these assumptions the entropy production reads as

$$\theta r = -q \cdot \nabla\theta/\theta + 2\mu_s|D|_2^2 + \mu_b|\mathrm{div}\, u|^2 + \frac{1}{\gamma}|\mathsf{a}|_2^2,$$

where

$$\mathsf{a} = \mathrm{div}[\lambda\nabla]d + \lambda|\nabla d|_2^2 d = \gamma\mathscr{D}_t d.$$

At the boundary $\partial\Omega$, energy should be preserved, which means $\Phi_e \cdot \nu = 0$. As $q \cdot \nu = 0$ and $u = 0$ this yields

$$\lambda\partial_\nu d \cdot \partial_t d = 0.$$

This is clearly valid if d satisfies the Neumann condition $\partial_\nu d = 0$, which is physically reasonable.

4. Stretching and Vorticity

Observe that the equation (15.11) for d admits the solutions $d = const$, no matter how the velocity field and the temperature field are defined. In this case the director field is not at all affected by the fluid dynamics. This seems to be physically unrealistic and so the model should be adapted.

This can be done by introducing a so-called *stretching* stress. To introduce this stress we follow Leslie. Define $P_d = I - d \otimes d$ the orthogonal projection onto $E_d := \{d\}^\perp$, the vorticity V according to $2V = \nabla u - [\nabla u]^\mathsf{T}$, and set

$$\mathsf{n} = \mu_V V d + \mu_D P_d D d - \gamma\mathscr{D}_t d,$$

where μ_V, μ_D, γ are scalar functions of ρ, θ, τ and $\gamma > 0$. Note that $\mathsf{n} \cdot d = 0$. For brevity we use the notation

$$\mathsf{a} := P_d \mathrm{div}(\lambda\nabla)d = \mathrm{div}(\lambda\nabla)d + \lambda|\nabla d|_2^2 d.$$

Now we define the stretch tensor

$$S_L^{stretch} = \frac{\mu_D + \mu_V}{2\gamma}\mathsf{n} \otimes d + \frac{\mu_D - \mu_V}{2\gamma} d \otimes \mathsf{n}. \tag{15.12}$$

This modification of the model does not change the entropy flux $\Phi_\eta = q/\theta$, and the relevant entropy production becomes,

$$S_L^{stretch} : \nabla u + \mathscr{D}_t d \cdot \mathsf{a} = \frac{1}{\gamma}(|\mathsf{n}|^2 + \gamma \mathscr{D}_t d \cdot \mathsf{n})) + \mathsf{a} \cdot \mathscr{D}_t d$$
$$= \frac{1}{\gamma}(|\mathsf{a}|^2 + (\mathsf{n} + \mathsf{a}) \cdot (\mu_V V d + \mu_D P_d D d - \mathsf{a}).$$

If we want to keep the total entropy production at the same level as in the previous section, the simplest way to achieve this is to set $\mathsf{n} + \mathsf{a} = 0$, which yields the equation

$$\gamma(\partial_t d + u \cdot \nabla d) = \operatorname{div}(\lambda \nabla) d + \lambda |\nabla d|_2^2 d + \mu_V V d + \mu_D P_d D d. \tag{15.13}$$

This is the stretched equation for d. Note that it preserves the constraint $|d|_2 = 1$. The entropy production is the same as before, we have

$$\theta r = [\alpha_0 |\nabla \theta|_2^2 + \alpha_1 (d|\nabla \theta)^2]/\theta + 2\mu_s |D|_2^2 + \mu_b |\operatorname{div} u|^2 + \frac{1}{\gamma} |P_d \operatorname{div}(\lambda \nabla) d|_2^2.$$

In particular, N satisfies $\partial_t \mathsf{N}(t) = \int_\Omega r(t,x)dx$, and so $-\mathsf{N}$ will be shown below to be a strict Lyapunov functional for the system, as soon as

$$\mu_s > 0, \quad 2\mu_s + n\mu_b > 0, \quad \alpha_0 > 0, \quad \alpha_0 + \alpha_1 > 0, \quad \gamma > 0, \tag{15.14}$$

and

$$\kappa > 0, \quad \lambda > 0, \quad \partial_\rho \pi > 0. \tag{15.15}$$

Note that no conditions on the new parameter functions μ_D, μ_V are needed, so far.

5. Additional Dissipation

We may add additional dissipative terms in the stress tensor of the form

$$S_L^{diss} = \frac{\mu_P}{\gamma}(\mathsf{n} \otimes d + d \otimes \mathsf{n}) + \frac{\gamma \mu_L + \mu_P^2}{2\gamma}(P_d D d \otimes d + d \otimes P_d D d) + \mu_0 (D d|d) d \otimes d, \tag{15.16}$$

where as before $2D = \nabla u + [\nabla u]^{\mathsf{T}}$ and $2V = \nabla u - [\nabla u]^{\mathsf{T}}$ are the symmetric and antisymmetric parts of the rate of strain tensor ∇u. Note that the tensor S_L^{diss} is symmetric. Adding these terms to the stress tensor will be thermodynamically consistent provided their contribution to the entropy production ensures that the total entropy production remains nonnegative. By a simple calculation we obtain

$$S_L^{diss} : \nabla u = S_L^{diss} : D = 2\frac{\mu_P}{\gamma}(\mathsf{n}|P_d D d) + (\mu_L + \frac{\mu_P^2}{\gamma})|P_d D d|_2^2 + \mu_0 (D d|d)^2. \tag{15.17}$$

So with $\mathsf{n} = -\mathsf{a}$, the total relevant dissipation amounts to

$$\frac{1}{\gamma}(|\mathsf{a}|_2^2 + 2\mu_P(\mathsf{n}|P_d Dd) + \mu_P^2|P_d Dd|_2^2) + \mu_L|P_d Dd|_2^2 + \mu_0(Dd|d)^2$$
$$= \frac{1}{\gamma}|\mathsf{a} - \mu_P P_d Dd|_2^2 + \mu_L|P_d Dd|_2^2 + \mu_0(Dd|d)^2,$$

hence the total entropy production becomes

$$\theta r = [\alpha_0|\nabla\theta|_2^2 + \alpha_1(d|\nabla\theta)^2]/\theta + 2\mu_s|D|_2^2 + \mu_b|\mathrm{div}\, u|^2$$
$$+ \frac{1}{\gamma}|P_d \mathrm{div}(\lambda\nabla)d - \mu_P P_d Dd|_2^2 + \mu_L|P_d Dd|_2^2 + \mu_0(Dd|d)^2.$$

Note that so far the parameter functions $\mu_j, j = 0, s, b, V, D, P, L, \alpha_0, \alpha_1$, and γ for thermodynamical consistency are only subject to the requirements

$$\alpha_0, \alpha_0 + \alpha_1 \geq 0, \quad \mu_s, 2\mu_s + n\mu_b \geq 0, \quad \mu_0, \mu_L \geq 0, \quad \gamma > 0. \tag{15.18}$$

Recall that all parameters functions are allowed to be functions of ρ, θ, τ.

Remark (i) A more refined algebra shows that it is enough to require

$$2\mu_s + \mu_L \geq 0, \quad 2\mu_s + \mu_0 \geq 0$$

in the incompressible case, and additionally

$$\frac{\mu_0^2}{n^2} \leq (2\mu_s + \mu_0)(\frac{2\mu_s}{n} + \mu_b + \frac{\mu_0}{n^2})$$

in the compressible case.

(ii) We want to stress that in case $\mu_V = \gamma$, our parameters $\mu_s, \mu_0, \mu_V, \mu_D, \mu_P, \mu_L$ are in one-to-one correspondence to the famous Leslie parameters $\alpha_1, \ldots, \alpha_6$. This shows that our model contains the isotropic Ericksen-Leslie model as a special case.

6. Conservation of Angular Momentum in 3D

We briefly discuss conservation of momentum in the physical important three-dimensional case. Recall that the mass specific density m of angular momentum is defined by

$$\mathsf{m} = x \times u.$$

Balance of angular momentum reads as follows

$$\mathscr{D}_t(\rho\mathsf{m}) + \mathrm{div}(\rho u \times \mathsf{m} - x \times T) = -e_i \times T_i,$$

where we use Einstein's sum convention and T_i denotes the i-th row of the stress tensor T. Thus the flux of angular momentum Φ_m is given by

$$\Phi_m = \rho u \times \mathsf{m} - x \times T.$$

It is well-known that $e_i \times T_i = 0$ in case T is symmetric. Therefore we may concentrate on the non-symmetric part of T which is given by

$$T^{as} = \frac{\mu_V}{2\gamma}(\mathsf{n} \otimes d - d \otimes \mathsf{n}).$$

This implies

$$\begin{aligned} e_i \times T_i^{as} &= \frac{\mu_V}{2\gamma}(\mathsf{n} \times d - d \times \mathsf{n}) = -\frac{\mu_V}{\gamma} d \times \mathsf{n} \\ &= \frac{\mu_V}{\gamma} d \times \operatorname{div}(\lambda\nabla) d \\ &= \partial_i(\frac{\mu_V}{\gamma}\lambda d \times \partial_i d) - \partial_i(\frac{\mu_V}{\gamma})\lambda d \times \partial_i d - \frac{\mu_V \lambda}{\gamma}\partial_i d \times \partial_i d \\ &= \partial_i(\frac{\mu_V}{\gamma}\lambda d \times \partial_i d) - \partial_i(\frac{\mu_V}{\gamma})\lambda d \times \partial_i d. \end{aligned}$$

This shows that $e_i \times T_i^{as}$ is a divergence provided $\nabla_x \frac{\mu_V}{\gamma} = 0$, i.e. if

$$\mu_V = c_0 \gamma, \quad \text{for some constant } c_0 \in \mathbb{R}.$$

Then the flux of angular momentum becomes

$$\Phi_m = \rho u \times \mathsf{m} - x \times T + c_0 \lambda d \times \nabla d.$$

We mention that if we also require so-called *objectivity* of the model, then $c_0 = 1$, which means $\mu_V = \gamma$.

7. The Complete Model: Non-isothermal, Compressible Fluid, Isotropic Elasticity

Summarizing, the complete model may be represented as

$$\begin{aligned} \partial_t \rho + \operatorname{div}(\rho u) &= 0 && \text{in } \Omega, \\ \rho(\partial_t + u \cdot \nabla)u + \nabla\pi &= \operatorname{div} S && \text{in } \Omega, \\ \rho(\partial_t + u \cdot \nabla)\varepsilon + \operatorname{div} q - \operatorname{div}(\lambda\nabla d \mathscr{D}_t d) &= S : \nabla u - \pi \operatorname{div} u && \text{in } \Omega, \\ \gamma(\partial_t + u \cdot \nabla)d - \mu_V V d - \mu_D P_d D d - \operatorname{div}[\lambda\nabla]d &= \lambda|\nabla d|_2^2 d, && \text{in } \Omega, \\ u = 0, \quad q \cdot \nu = 0, \quad \partial_\nu d &= 0 && \text{on } \partial\Omega. \\ \rho(0) = \rho_0, \quad u(0) = u_0, \quad \theta(0) = \theta_0, \quad d(0) &= d_0 && \text{in } \Omega \end{aligned} \tag{15.19}$$

These equations have to be supplemented by the thermodynamical laws

$$\begin{aligned} &\varepsilon = \psi + \theta\eta, \quad \eta = -\partial_\theta \psi, \quad \kappa = \partial_\theta \varepsilon, \\ &\pi = \rho^2 \partial_\rho \psi, \quad \lambda = \rho \partial_\tau \psi, \end{aligned} \tag{15.20}$$

and by the constitutive laws

$$\begin{aligned} S &= S_N + S_E + S_L^{stretch} + S_L^{diss}, \\ S_N &= 2\mu_s D + \mu_b \mathrm{div}\, u\, I, \quad S_E = -\lambda \nabla d [\nabla d]^\mathsf{T}, \\ S_L^{stretch} &= \frac{\mu_D + \mu_V}{2\gamma} \mathsf{n} \otimes d + \frac{\mu_D - \mu_V}{2\gamma} d \otimes \mathsf{n}, \quad \mathsf{n} = \mu_V V d + \mu_D P_d D d - \gamma \mathscr{D}_t d, \\ S_L^{diss} &= \frac{\mu_P}{\gamma} (\mathsf{n} \otimes d + d \otimes \mathsf{n}) + \frac{\gamma \mu_L + \mu_P^2}{2\gamma} (P_d D d \otimes d + d \otimes P_d D d) + \mu_0 (Dd|d) d \otimes d, \\ q &= -\alpha_0 \nabla \theta - \alpha_1 (d|\nabla\theta) d. \end{aligned} \tag{15.21}$$

Here all coefficients μ_j, α_j and γ are functions of ρ, θ, τ. For thermodynamic consistency we require

$$\mu_s \geq 0, \quad 2\mu_s + n\mu_b \geq 0, \quad \alpha_0 \geq 0, \quad \alpha_0 + \alpha_1 \geq 0, \quad \mu_0, \mu_L \geq 0, \quad \gamma > 0, \tag{15.22}$$

Finally, we will use in addition the following conditions

$$\begin{aligned} &\mu_s > 0, \quad 2\mu_s + n\mu_b > 0, \quad \alpha_0 > 0, \quad \alpha_0 + \alpha_1 > 0, \quad \gamma > 0, \\ &\kappa > 0, \quad \lambda > 0, \quad \partial_\rho \pi > 0, \end{aligned} \tag{15.23}$$

to identify the equilibria and to investigate their thermodynamic stability in Sect. 15.3.

8. The Complete Model: Non-isothermal, Compressible Fluid, Non-isotropic Elasticity

For the sake of completeness, we comment briefly on the non-isotropic case. Then $\psi = \psi(\rho, \theta, d, \nabla d)$, and the Ericksen stress tensor becomes $S_E = -\rho \frac{\partial \psi}{\partial \nabla d} [\nabla d]^\mathsf{T}$. Following the derivation in Sects. 2.3 and 2.4, here the energy and entropy fluxes read again as

$$\Phi_e := q + \pi u - Su - \Pi \mathscr{D}_t d, \quad \Phi_\eta = q/\theta,$$

and the equation for d becomes

$$\gamma \mathscr{D}_t d = P_d \mathsf{a}, \quad \mathsf{a} = \partial_i (\rho \nabla_{\partial_i d} \psi) - \rho \nabla_d \psi,$$

in the case without stretching, and

$$\gamma \mathscr{D}_t d = P_d \mathsf{a} + \mu_V V d + \mu_D P_d D d$$

in the stretched case. The couple stress here is $\Pi = \rho\partial_{\nabla d}\psi$, and the entropy production now reads as

$$\theta r = [\alpha_0|\nabla\theta|_2^2 + \alpha_1(d|\nabla\theta)^2]/\theta + 2\mu_s|D|_2^2 + \mu_b|\mathrm{div}\, u|^2 \\ + \frac{1}{\gamma}|P_d(\mathsf{a} - \mu_P Dd)|_2^2 + \mu_L|P_d Dd|^2 + \mu_0(Dd|d)^2.$$

Summarizing, the complete model in the case of non-isotropic elasticity becomes

$$\begin{aligned} \partial_t\rho + \mathrm{div}(\rho u) &= 0 && \text{in } \Omega, \\ \rho(\partial_t + u\cdot\nabla)u + \nabla\pi &= \mathrm{div}\, S && \text{in } \Omega, \\ \rho(\partial_t + u\cdot\nabla)\varepsilon + \mathrm{div}\, q - \mathrm{div}(\rho\partial_{\nabla d}\psi\mathscr{D}_t d) - S : \nabla u + \pi\,\mathrm{div}\, u &= 0 && \text{in } \Omega, \\ \gamma(\partial_t + u\cdot\nabla)d - \mu_V V d - \mu_D P_d D d - P_d\big(\mathrm{div}(\rho\frac{\partial\psi}{\partial\nabla d}) - \rho\nabla_d\psi\big) &= 0, && \text{in } \Omega, \\ u = 0, \quad q\cdot\nu = 0, \quad \partial_\nu d &= 0 && \text{on } \partial\Omega. \\ \rho(0) = \rho_0, \quad u(0) = u_0, \quad \theta(0) = \theta_0, \quad d(0) &= d_0 && \text{in } \Omega \end{aligned} \tag{15.24}$$

These equations have to be supplemented by the thermodynamical laws

$$\varepsilon = \psi + \theta\eta, \quad \eta = -\partial_\theta\psi, \quad \kappa = \partial_\theta\varepsilon, \quad \pi = \rho^2\partial_\rho\psi, \tag{15.25}$$

and by the constitutive laws

$$\begin{aligned} S &= S_N + S_E + S_L^{stretch} + S_L^{diss}, \\ S_N &= 2\mu_s D + \mu_b \mathrm{div}\, u I, \quad S_E = -\rho\frac{\partial\psi}{\partial\nabla d}[\nabla d]^{\mathsf{T}}, \\ S_L^{stretch} &= \frac{\mu_D + \mu_V}{2\gamma}\mathsf{n}\otimes d + \frac{\mu_D - \mu_V}{2\gamma} d\otimes\mathsf{n}, \quad \mathsf{n} = \mu_V V d + \mu_D P_d D d - \gamma\mathscr{D}_t d, \\ S_L^{diss} &= \frac{\mu_P}{\gamma}(\mathsf{n}\otimes d + d\otimes\mathsf{n}) + \frac{\gamma\mu_L + \mu_P^2}{2\gamma}(P_d Dd\otimes d + d\otimes P_d Dd) + \mu_0(Dd|d)d\otimes d, \\ q &= -\alpha_0\nabla\theta - \alpha_1(d|\nabla\theta)d. \end{aligned} \tag{15.26}$$

Here all coefficients μ_j, α_j and γ are functions of $\rho, \theta, \nabla d$. For thermodynamic consistency we require as before only (15.22). We also note that the natural boundary condition at $\partial\Omega$ here becomes

$$\nu_i\nabla_{\partial_i d}\psi = 0.$$

Observe that this condition is fully nonlinear, in general, in contrast to the isotropic case.

Concluding, we mention as an example the classical *Oseen-Frank* free energy density for the isothermal incompressible case, which is given by

$$\psi^{FO} = k_1(\mathrm{div}\, d)^2 + k_2|d\times(\nabla\times d)|_2^2 + k_3|d\cdot(\nabla\times d)|^2 + (k_2+k_4)[\mathrm{tr}(\nabla d)^2 - (\mathrm{div}\, d)^2],$$

where k_i are given constants.

15.3 Thermodynamical Consistency and Stability

In this section we determine the equilibria set of the complete system described above in Sect. 2.7, show that the critical points of the entropy functional coincide with these equilibria and prove that they are thermodynamically stable. We begin investigating the set of equilibria.

1. Equilibria

Suppose that in some time interval $t \in (t_1, t_2)$ we have $\partial_t \mathsf{N}(t) = 0$. Then $r \geq 0$ implies $r(t, x) = 0$ in Ω. This yields $\nabla\theta(t, x) = 0$ in Ω as $\alpha_0 > 0$ and $\alpha_0 + \alpha_1 > 0$. Hence, $\theta = \theta_*$ is constant in Ω.

Next, by $\mu_s > 0, 2\mu_s + n\mu_b > 0$, we also have $D = 0$ in Ω. By Korn's inequality and the no-slip boundary condition for u, we hence obtain $u = u_* = 0$ in Ω, $t \in (t_1, t_2)$. Therefore $\partial_t\rho = \partial_t u = 0$, which implies $\nabla\pi = 0$.

Finally, $\gamma > 0$ yields $\mathscr{D}_t d = 0$ in Ω, which implies that d satisfies the nonlinear eigenvalue problem

$$\begin{aligned} \operatorname{div}(a(x)\nabla)d + a(x)|\nabla d|_2^2 d &= 0 \ \text{ in } \Omega, \\ |d|_2 &= 1 \ \text{ in } \Omega, \\ \partial_\nu d &= 0 \ \text{ on } \partial\Omega, \end{aligned} \tag{15.27}$$

where $a(x) = \lambda(\rho(t, x), \theta(t), \tau(t, x))$, for each fixed $t \in (t_1, t_2)$. But, as the next lemma shows, this implies $\nabla d = 0$ in Ω. Hence, $d = d_*$ is constant.

Lemma 1 *Let $q > n$, $a \in H^1_q(\Omega)$, $a > 0$ and suppose that $d \in H^2_q(\Omega; \mathbb{R}^n)$ satisfies (15.27). Then d is constant in Ω.*

Proof The idea is to reduce inductively the dimension $N = n$ of the vector d. This can be achieved by introducing polar coordinates according to

$$d_1 = c_1 \cos\varphi, \ d_2 = c_1 \sin\varphi, \ d_j = c_{j-1}, \quad j \geq 3.$$

Simple computations yield

$$1 = |d|_2^2 = |c|_2^2, \quad |\nabla d|_2^2 = |\nabla c|_2^2 + c_1^2|\nabla\varphi|_2^2,$$

and

$$\operatorname{div}(a\nabla)c_j + a[|\nabla c|_2^2 + c_1^2|\nabla\varphi|_2^2]c_j = 0 \quad \text{in } \Omega,$$

as well as $\partial_\nu c_j = 0$ on $\partial\Omega$ for $j = 2, \dots, n-1$. Moreover, by some more calculations we further obtain

$$-\operatorname{div}(a\nabla)c_1 + ac_1|\nabla\varphi|_2^2 = a[|\nabla c|_2^2 + c_1^2|\nabla\varphi|_2^2]c_1 \quad \text{in } \Omega,$$

and

$$c_1 \mathrm{div}(a\nabla)\varphi + 2a\nabla c_1 \cdot \nabla\varphi = 0 \quad \text{in } \Omega,$$

as well as

$$\partial_\nu c_1 = c_1 \partial_\nu \varphi = 0 \quad \text{on } \partial\Omega.$$

Multiplying the second of the last equations by $c_1\varphi$ and integrating over Ω we deduce

$$0 = \int_\Omega [c_1 \mathrm{div}(a\nabla)\varphi + 2a\nabla c_1 \cdot \nabla\varphi] c_1\varphi dx = \int_\Omega \mathrm{div}[c_1^2 a\nabla\varphi]\varphi dx = -\int_\Omega c_1^2 a|\nabla\varphi|^2 dx.$$

Hence, $c_1\nabla\varphi = 0$ as $a > 0$ by assumption. This implies that c satisfies equation (15.27), where the vector c has dimension $N-1$. Inductively, we arrive at dimension $N = 1$ and if d is a solution of (15.27) with dimension 1, then $d = 1$ or $d = -1$ by the connectedness of Ω. □

Knowing that θ and d are constant in Ω, and $\nabla\pi = 0$, we see that $\pi = \rho^2\partial_\rho\psi(\rho, \theta, 0)$ is constant, hence $\rho = \rho_*$ is constant, provided the function $\rho \mapsto \pi(\rho, \theta, 0)$ is strictly increasing. This shows that we are at an equilibrium $(\rho_*, u_*, \theta_*, d_*) \in \mathcal{E}$ with

$$\mathcal{E} = \{(\rho_*, u_*, \theta_*, d_*) \in (0,\infty) \times \{0\} \times (0,\infty) \times \mathbb{R}^n : |d_*|_2 = 1\},$$

the set of physical equilibria. In particular, the functional $-\mathsf{N}$ is a strict Lyapunov functional.

Observe that $\mathcal{E}$ forms an $n+1$-dimensional manifold. If we take into account conservation of mass and energy,

$$\mathsf{M}_0 := \int_\Omega \rho dx = \rho_*|\Omega|, \quad \mathsf{E}_0 := \int_\Omega (\rho|u|_2^2/2 + \rho\varepsilon) dx = \rho_*\varepsilon_*|\Omega|,$$

at an equilibrium, then the values of ρ_* and θ_* are uniquely determined by

$$\rho_* = \mathsf{M}_0/|\Omega|, \quad \varepsilon_* := \varepsilon(\rho_*, \theta_*, 0) = \mathsf{E}_0/\mathsf{M}_0,$$

whenever $\theta \mapsto \varepsilon(\rho, \theta, 0)$ is strictly increasing, i.e. whenever $\kappa > 0$.

2. Critical Points of Total Entropy

(a) Consider the entropy functional N with constraints of prescribed mass $\mathsf{M} = \mathsf{M}_0$ and energy $\mathsf{E} = \mathsf{E}_0$, as well as $G(d) := (|d|_2^2 - 1)/2 = 0$. Suppose we have a sufficiently smooth critical point (ρ, u, θ, d) of N with $\rho, \theta > 0$, subject to the constraints. Then the method of Lagrange multipliers yields $\kappa_M, \kappa_E \in \mathbb{R}$ and $\kappa_G \in L_2(\Omega)$ such that

$$\langle \mathsf{N}' + \kappa_M \mathsf{M}' + \kappa_E \mathsf{E}' + \kappa_G G' | z \rangle = 0,$$

where $z = (\sigma, v, \vartheta, \delta)$. We have

$$\langle \mathsf{M}'|z\rangle = \int_\Omega \sigma dx, \quad \langle \kappa_G G'|z\rangle = \int_\Omega \kappa_G d \cdot \delta dx,$$

and

$$\langle \mathsf{N}'|z\rangle = \int_\Omega [(\partial_\rho(\rho\eta))\sigma + \rho\partial_\theta\eta\vartheta + \rho\partial_\tau\eta\nabla d : \nabla\delta]dx,$$

as well as

$$\langle \mathsf{E}'|z\rangle = \int_\Omega [\rho u \cdot v(\partial_\rho(\rho\varepsilon))\sigma + \rho\partial_\theta\varepsilon\vartheta + \rho\partial_\tau\varepsilon\nabla d : \nabla\delta]dx.$$

This yields the relation

$$\begin{aligned} 0 = &\int_\Omega \{[\partial_\rho(\rho\eta) + \kappa_M + \kappa_E(\frac{1}{2}|u|_2^2 + \partial_\rho(\rho\varepsilon))]\sigma + [\rho\partial_\theta\eta + \kappa_E\rho\partial_\theta\varepsilon]\vartheta\}dx \\ &+ \int_\Omega \{\kappa_E\rho u \cdot v + [\rho\partial_\tau\eta + \kappa_E\rho\partial_\tau\varepsilon]\nabla d : \nabla\delta + \kappa_G d \cdot \delta\}dx. \end{aligned}$$

We first vary ϑ to obtain $\rho(\partial_\theta\eta + \kappa_E\partial_\theta\varepsilon) = 0$, which by $\rho > 0$ and by the definition of η, ε and $\kappa > 0$ yields $\kappa_E = -1/\theta$. Hence, θ is constant and $\kappa_E < 0$. Next, varying v we obtain $u = 0$, as κ_E and ρ are not zero. Next we vary δ, which after an integration by parts, employing the boundary condition $\partial_\nu d = 0$, implies

$$\mathrm{div}(\lambda\nabla)d + \kappa_G d = 0 \quad \text{in } \Omega.$$

But then $|d|_2 = 1$ implies $\kappa_G = \lambda|\nabla d|_2^2$, and d is a solution of the problem (15.27), which by Lemma 1 shows that d is constant. Finally, we vary σ to the result that $\partial_\rho(\rho\psi) = \theta\kappa_M$ is constant. As $\pi = \rho^2\partial_\rho\psi$ is strictly increasing in the variable ρ, this shows that ρ is constant in Ω as well. Therefore, the critical points of the entropy functional are precisely the equilibria of the problem.

(b) Let

$$H := \mathsf{N}'' + \kappa_E\mathsf{E}''$$

denote the second variation of N. Note that $\mathsf{M}'' = 0$ and $\kappa_G = \lambda|\nabla d|_2^2 = 0$. The identities

$$\begin{aligned} \rho(\partial_\tau\eta - \frac{1}{\theta}\partial_\tau\varepsilon) = -\lambda, \quad \partial_\rho\partial_\theta(\rho\eta) - \frac{1}{\theta}\partial_\rho\partial_\theta(\rho\varepsilon) = 0, \\ \partial_\rho^2(\rho\eta) - \frac{1}{\theta}\partial_\rho^2(\rho\varepsilon) = -\frac{\partial_\rho\pi}{\rho\theta}, \quad \partial_\theta^2\eta - \frac{1}{\theta}\partial_\theta^2\varepsilon = -\frac{\kappa}{\theta^2}, \end{aligned}$$

imply

$$-\langle Hz|z\rangle = \int_\Omega \Big[\frac{\partial_\rho \pi}{\rho\theta}\sigma^2 + \frac{\kappa}{\theta^2}\vartheta^2 + \lambda|\nabla\delta|_2^2\Big]dx \geq 0,$$

by κ, λ, $\partial_\rho\pi \geq 0$. This shows that the second variation of N at an equilibrium is negative semi-definite, which means that the equilibria are *thermodynamically stable*.

(c) Summarizing we have the following basic result

Theorem 1 *The complete model has the following properties.*
(i) Along smooth solutions total mass M *and energy* E *are preserved.*
(ii) Along smooth solutions the total entropy N *is non-decreasing.*
(iii) The negative total entropy is a strict Lyapunov functional.
(iv) The condition $|d|_2 = 1$ *is preserved along smooth solutions.*
(v) The equilibria are given by the set of constants

$$\mathcal{E} = \{(\rho_*, 0, \theta_*, d_*) : \rho_*, \theta_* \in (0, \infty),\ d_* \in \mathbb{R}^n,\ |d_*|_2 = 1\}.$$

Here ρ_*, θ_* *are uniquely determined by the identities*

$$\rho_* = \mathsf{M}_0/|\Omega|, \quad \varepsilon(\rho_*, \theta_*, 0) = \mathsf{E}_0/\mathsf{M}_0.$$

(vi) The equilibria are precisely the critical points of the total entropy with prescribed mass and energy.
(vii) The second variation of N *with given mass and energy at equilibrium is negative semidefinite.*

In particular, the model is **thermodynamically consistent** *and it is also* **thermodynamically stable**.

3. The Isothermal Case

In the isothermal case we set $\theta = const$ and ignore the equation for the energy. In this case, instead of the total mass specific energy $e = |u|_2^2/2 + \varepsilon$, we employ the *available energy* e_a which is defined by $e_a = |u|_2^2/2 + \psi$. We have the following balance of e_a which is a direct consequence of balance of total energy and entropy

$$\rho(\partial_t + u\cdot\nabla)e_a + \operatorname{div}(\Phi_e - \theta\Phi_\eta) = -\theta r - \rho\eta\mathcal{D}_t\theta - \Phi_\eta\cdot\nabla\theta.$$

In the case where θ is constant this reduces to

$$\rho(\partial_t + u\cdot\nabla)e_a + \operatorname{div}(\Phi_e - \theta\Phi_\eta) = -r_a,$$

with

$$r_a = 2\mu_s|D|_2^2 + \mu_b|\operatorname{div} u|^2 + \mu_0(Dd|d)^2 + \mu_L|P_d Dd|_2^2 + |P_d(\operatorname{div}(\lambda\nabla)d - \mu_{PDd})|_2^2/\gamma.$$

Therefore, in the isothermal case, the total available energy E_a is a strict Ljapunov functional for the system, i.e.

$$\partial_t \mathsf{E}_a(t) = -\int_\Omega r_a(t,x)dx, \quad \mathsf{E}_a(t) = \int_\Omega \rho(t,x)e_a(t,x)dx.$$

As a consequence, the equilibrium set is the same as in the non-isothermal case, dropping temperature, hence is a manifold of dimension n, and when we incorporate preserved mass it is isomorphic to the unit sphere in $\mathbb{R}^n$. In this case the equations read

$$\begin{aligned} \partial_t\rho + \operatorname{div}(\rho u) &= 0 && \text{in } \Omega, \\ \rho(\partial_t + u\cdot\nabla)u + \nabla\pi &= \operatorname{div} S && \text{in } \Omega, \\ \gamma(\partial_t + u\cdot\nabla)d - \mu_V V d - \operatorname{div}(\lambda\nabla)d &= \lambda|\nabla d|_2^2 d + \mu_D P_d D d && \text{in } \Omega \end{aligned} \tag{15.28}$$

where $\psi = \psi(\rho,\tau)$, $\pi = \rho^2\partial_\rho\psi$, $\lambda = \rho\partial_\tau\psi/\theta$, and

$$\begin{aligned} S &= 2\mu_s D + \mu_b \operatorname{div} u\, I - \theta\lambda\nabla d[\nabla d]^{\mathsf{T}} + S_L^{stretch} + S_L^{diss}, \\ S_L^{stretch} &= \frac{\mu_D + \mu_V}{2\gamma}\mathsf{n}\otimes d + \frac{\mu_D - \mu_V}{2\gamma} d\otimes\mathsf{n}, \quad \mathsf{n} = \mu_V V d + \mu_D P_d D d - \gamma\mathscr{D}_t d. \\ S_L^{diss} &= \frac{\mu_P}{\gamma}(\mathsf{n}\otimes d + d\otimes\mathsf{n}) + \frac{\gamma\mu_L + \mu_P^2}{2\gamma}(P_d D d\otimes d + d\otimes P_d D d) + \mu_0 (Dd|d) d\otimes d, \end{aligned} \tag{15.29}$$

If one further restricts to the incompressible case $\rho = const > 0$, λ constant, $\mu_0 = \mu_D = \mu_V = \mu_P = \mu_L = 0$, with $\mu = \mu_s$ one obtains the so-called *isothermal simplified Ericksen-Leslie model*

$$\begin{aligned} \rho(\partial_t + u\cdot\nabla)u + \nabla\pi &= \mu_s\Delta u - \lambda\operatorname{div}(\nabla d[\nabla d]^{\mathsf{T}}) && \text{in } \Omega, \\ |d|_2 = 1, \quad \operatorname{div} u &= 0 && \text{in } \Omega, \\ \gamma(\partial_t + u\cdot\nabla)d - \lambda\Delta d &= \lambda|\nabla d|^2 d && \text{in } \Omega. \end{aligned} \tag{15.30}$$

Of course, in all cases we have to add initial conditions as well as boundary conditions $u = \partial_\nu d = 0$ on $\partial\Omega$. Problem (15.30) subject to the condition $|d|_2 = 1$ in Ω has been analyzed in a fairly complete manner in the recent article [17] by Hieber, Nesensohn, Prüss and Schade.

15.4 Analysis of the Non-isothermal Simplified Model

In this section we consider the *incompressible case* $\rho = const$ and we let $\mu = \mu_s$. Hence, the pressure π is no longer determined by Maxwell's relation; it is now a free variable, a Lagrangian multiplier to cover the constraint $\operatorname{div} u = 0$. Furthermore, in the following we neglect stretching, i.e. we assume $\mu_D = \mu_V = \mu_P = \mu_L =$

$\mu_0 = 0$. For simplicity we also set $\alpha_1 = 0$ and $\alpha = \alpha_0$. Then the resulting model—which we call the *non-isothermal simplified Ericksen-Leslie model*—reads as follows.

$$\begin{aligned}
\rho\mathscr{D}_t u - 2\mathrm{div}(\mu D) + \nabla\pi &= -\mathrm{div}(\lambda\nabla d[\nabla d]^{\mathsf{T}}) && \text{in } \Omega,\\
|d|_2 = 1, \quad \mathrm{div}\, u &= 0 && \text{in } \Omega,\\
\rho\kappa\mathscr{D}_t\theta - \mathrm{div}(\alpha\nabla\theta) &= 2\mu|D|_2^2 - \lambda\nabla d[\nabla d]^{\mathsf{T}} : D\\
&\quad - \rho\partial_\tau\varepsilon\nabla d : \mathscr{D}_t\nabla d + \mathrm{div}(\lambda\nabla d\mathscr{D}_t d) && \text{in } \Omega,\\
\gamma\mathscr{D}_t d - \mathrm{div}(\lambda\nabla)d &= \lambda|\nabla d|_2^2 d && \text{in } \Omega,\\
u = \partial_\nu\theta = \partial_\nu d &= 0 && \text{on } \partial\Omega,\\
u(0) = u_0,\ \theta(0) = \theta_0,\ d(0) &= d_0 && \text{in } \Omega.
\end{aligned} \tag{15.31}$$

Recall that $\rho > 0$ is constant and α, γ, μ as well as $\lambda = \rho\partial_\tau\psi$, $\kappa = \partial_\theta\varepsilon = -\theta\partial_\theta^2\psi$ are functions of $\theta > 0$ and $\tau \geq 0$.

1. Regularity Assumptions (R)

The parameter functions should have the following minimal regularity properties:

$$\mu, \alpha, \gamma \in C^2((0,\infty) \times [0,\infty)), \quad \psi \in C^4((0,\infty) \times [0,\infty));$$

We also require the positivity conditions

$$\mu > 0, \quad \alpha > 0, \quad \kappa > 0, \quad \gamma > 0, \quad \lambda > 0,$$

which have been mentioned before, but for well-posedness of the problem for d we need in addition to require $\lambda + 2\tau\partial_\tau\lambda > 0$. We assume that $\Omega \subset \mathbb{R}^n$ is a bounded domain with C^{3-}-boundary.

2. Maximal L_p-Regularity of the Principal Linearization

The equation for u will turn out to be only weakly coupled, so we first concentrate on the system for $w := [\theta, d]^{\mathsf{T}}$. The *principal part of the linearization* becomes

$$\begin{aligned}
\partial_t w + \mathscr{A}(w_0, \nabla)w &= f && \text{in } \Omega,\\
\partial_\nu w &= 0 && \text{on } \partial\Omega,\\
w(0) &= w_0 && \text{in } \Omega.
\end{aligned} \tag{15.32}$$

The matrix $\mathscr{A} = \mathscr{A}(w_0, \nabla)$ reads as

$$\mathscr{A} = \begin{bmatrix} -a_0\Delta - a_1\nabla d_0[\nabla d_0]^{\mathsf{T}} : \nabla^2, & b_0\nabla d_0 : (\lambda_0\Delta + \partial_\tau\lambda_0[\nabla d_0]^{\mathsf{T}}\nabla d_0 : \nabla^2)\nabla \\ b_1[\nabla d_0]^{\mathsf{T}}\nabla, & -\gamma_0^{-1}(\lambda_0\Delta + \partial_\tau\lambda_0[\nabla d_0]^{\mathsf{T}} \otimes \nabla d_0 : \nabla^2). \end{bmatrix}.$$

Here $\kappa_0 = \kappa(\theta_0, \tau_0)$ etc., and we used the abbreviations

$$a_0 = \frac{\alpha_0}{\rho\kappa_0}, \quad a_1 = \frac{\rho\theta_0[\partial_\tau\eta_0]^2}{\gamma_0\kappa_0}, \quad b_0 = \frac{\theta_0\partial_\tau\eta_0}{\gamma_0\kappa_0}, \quad b_1 = \frac{\rho\partial_\tau\eta_0}{\gamma_0}.$$

Note that $\mathscr{A}(w_0, \nabla)$ is second order in the diagonal, but third and first order off-diagonal! This is a mixed-order problem subject to Neumann boundary conditions and subject to variable, non-smooth coefficients. For resolvent estimates within the L^p-setting for various mixed-order systems we refer to the work of Grubb [16]. Regarding the *maximal L^p-regularity* of this nonstandard problem, we do not know of any general theory covering the above situation. However, we note for the whole space case there is the theory of Denk and Kaip [8] available. Nevertheless, we prove maximal L_p-regularity for this problem in the following.

To this end, fix $q \in (1, \infty)$ and choose as a base space

$$Y_0 := L_q(\Omega) \times H^1_q(\Omega; \mathbb{R}^n),$$

and as a regularity space

$$Y_1 := \{w = (\theta, d) \in H^2_q(\Omega) \times H^3_q(\Omega; \mathbb{R}^n) : \partial_\nu\theta = \partial_\nu d = 0 \text{ on } \partial\Omega\},$$

equipped with their natural norms. We will also employ the *time-weighted spaces* defined by

$$\mathsf{y} \in H^m_{p,\mu}(J; Y) \quad \Leftrightarrow \quad t^{1-\mu}\mathsf{y} \in H^m_p(J; Y), \quad m \in \mathbb{N}_0, \ \mu \in (1/p, 1].$$

The time trace space $Y_{\gamma,\mu}$ is then given by

$$Y_{\gamma,\mu} = \{(\theta, d) \in B^{2(\mu-1/p)}_{qp}(\Omega) \times B^{2(\mu-1/p)+1}_{qp}(\Omega; \mathbb{R}^n) : \partial_\nu d = 0 \text{ on } \partial\Omega\},$$

provided

$$\frac{1}{p} < \mu < \frac{1}{2} + \frac{1}{p} + \frac{1}{2q};$$

otherwise one has to add $\partial_\nu\theta = 0$ in the definition of $Y_{\gamma,\mu}$. In order to profit from the embedding

$$Y_{\gamma,\mu} \hookrightarrow C(\overline{\Omega}) \times C^1(\overline{\Omega}; \mathbb{R}^n), \tag{15.33}$$

we will always assume

$$1 \geq \mu > \frac{1}{p} + \frac{n}{2q}.$$

Then by means of the assumptions stated before we obtain the following result.

Theorem 2 *Assume (R), $1/p + n/2q < \mu \le 1$, and suppose that $w_0 \in Y_{\gamma,\mu}$. Then the differential operator $A_2(w_0)$ defined by $A_2(w_0)w := \mathscr{A}(w_0, \nabla)w$ with domain $\mathsf{D}(A_2(w_0)) := Y_1$ has maximal L_p-regularity in Y_0 and thus also maximal $L_{p,\mu}$-regularity in Y_0.*

Proof The proof is based on the results and techniques developed by Denk et al. [7] for the case $\mu = 1$. By the results due to Prüss and Simonett [34], these results extend to general $\mu \in (1/p + n/2q, 1]$, as the coefficients have enough regularity by the embedding (15.33).

(a) The case $\Omega = \mathbb{R}^n$ with constant coefficients
In the sequel, we denote the covariable for t by z and that for x by ξ. The symbol $\mathscr{A}(\xi)$ of $\mathscr{A}(w_0, \nabla)$ reads as

$$\mathscr{A}(\xi) = \begin{bmatrix} a_0|\xi|^2 + a_1|c(\xi)|^2 & -ib_0(\lambda_0|\xi|^2 + \partial_\tau\lambda_0|c(\xi)|^2)c(\xi)^{\mathsf{T}} \\ ib_1 c(\xi) & \frac{\lambda_0}{\gamma_0}|\xi|^2 + \frac{\partial_\tau\lambda_0}{\gamma_0}c(\xi)\otimes c(\xi) \end{bmatrix},$$

where $c(\xi) = \xi \cdot \nabla d_0$. It is convenient to reduce this symbol for the variable $w_{red} = [\theta, d_{red}]^{\mathsf{T}}$ where $d_{red} = c(\xi)\cdot d$. The reduced symbol $\mathscr{A}_{red}(\xi)$ becomes

$$\mathscr{A}_{red}(\xi) = \begin{bmatrix} a_0|\xi|^2 + a_1|c(\xi)|^2 & -ib_0(\lambda_0|\xi|^2 + \partial_\tau\lambda_0|c(\xi)|^2) \\ ib_1|c(\xi)|^2 & \frac{\lambda_0}{\gamma_0}|\xi|^2 + \frac{\partial_\tau\lambda_0}{\gamma_0}|c(\xi)|^2 \end{bmatrix}.$$

This symbol is homogeneous of second order and *not* strongly elliptic. However, it is *normally elliptic* in the sense of [7] as its spectrum satisfies $\sigma(\mathscr{A}_{red}(\xi)) \subset (0, \infty)$ for each $\xi \neq 0$. The latter can be seen by considering

$$\begin{aligned}\det(z + \mathscr{A}_{red}(\xi)) &= (z + a_0|\xi|^2 + a_1|c(\xi)|^2)(z + \frac{\lambda_0}{\gamma_0}|\xi|^2 + \frac{\partial_\tau\lambda_0}{\gamma_0}|c(\xi)|^2) \\ &\quad - b_0 b_1|c(\xi)|^2(\lambda_0|\xi|^2 + \partial_\tau\lambda_0|c(\xi)|^2), \quad \xi \neq 0,\end{aligned}$$

which has two negative zeros, as $\alpha, \gamma, \kappa > 0$ and $\lambda + 2\tau\partial_\tau\lambda > 0$. Therefore, $\mathscr{A}_{red}$ is normally elliptic, and by Sect. 6 of [7], the L_p-realization A_{red} of $\mathscr{A}_{red}$ has maximal L_p-regularity. This shows that whenever $f_\theta \in L_p(J; L_q(\mathbb{R}^n))$ and $f_d \in L_p(J; H^1_q(\Omega; \mathbb{R}^n))$ are given, there is a unique solution

$$w_{red} = [\theta, \bar{d}]^{\mathsf{T}} \in {}_0H^1_p(J; L_q(\mathbb{R}^n; \mathbb{R}^2)) \cap L_p(J; H^2_q(\mathbb{R}^n; \mathbb{R}^2))$$

of

$$\partial_t w_{red} + A_{red}w_{red} = f_{red}, \quad t > 0, \; w_{red}(0) = 0,$$

where $f_{red} = [f_\theta, -ic(\nabla)\cdot f_d]^{\mathsf{T}}$. To obtain d, it remains to solve the problem

$$\partial_t d - \frac{\lambda_0}{\gamma_0}\Delta d = f^1_d := f_d + i\frac{\partial_\tau\lambda_0}{\gamma_0}c(\nabla)d_{red} - b_1 c(\nabla)\theta, \quad t > 0, \; d(0) = 0,$$

with maximal L_p-regularity of $-\Delta$ to obtain a unique solution

$$d \in {}_0H^1_p(J; H^1_q(\mathbb{R}^n; \mathbb{R}^n)) \cap L_p(J; H^3_q(\mathbb{R}^n; \mathbb{R}^n)),$$

as $\lambda, \gamma > 0$ and $f^1_d \in L_p(J; H^1_q(\mathbb{R}^n; \mathbb{R}^n))$. This proves Theorem 2 in the case $\Omega = \mathbb{R}^n$ with constant coefficients. As detailed in Sect. 6 of [7], this assertion extends by perturbation and localization to variable coefficients, still in the case $\Omega = \mathbb{R}^n$.

(b) The case $\Omega = \mathbb{R}^n_+$ with constant coefficients
It is convenient to replace $x \in \mathbb{R}^n_+$ by $(x, y) \in \mathbb{R}^{n-1} \times \mathbb{R}_+$. On the symbolic level we have to replace ξ by $\xi - i\nu\partial_y$, where ν denotes the outer normal at a boundary point of Ω, and $\xi \cdot \nu = 0$. Then $c(\xi)$ becomes

$$c(\xi - i\nu\partial_y) = (\xi - i\nu\partial_y) \cdot \nabla d_0 = \xi \cdot \nabla d_0 - i\nu \cdot \nabla d_0 \partial_y = \xi \cdot \nabla d_0,$$

as $\partial_\nu d_0 = 0$. Therefore the symbol $\mathscr{A}(\xi)$ from Step (a) is replaced by

$$-E\partial_y^2 + z + \mathscr{A}(\xi),$$

where

$$E = \begin{bmatrix} a_0 & -ib_0\lambda_0 c(\xi)^{\mathsf{T}} \\ 0 & (\lambda_0/\gamma_0)I \end{bmatrix}.$$

Considering again the reduced variables $w_{red} = (\theta, d_{red})$ with $d_{red} = c(\xi) \cdot d$, the reduced symbol becomes

$$-E_{red}\partial_y^2 + z + \mathscr{A}_{red}(\xi),$$

with

$$E_{red} = \begin{bmatrix} a_0 & -ib_0\lambda_0 \\ 0 & \lambda_0/\gamma_0 \end{bmatrix}.$$

To apply the half-space theory for normally elliptic operators in Sect. 7 of [7], we need to verify the corresponding Lopatinskii-Shapiro condition (LS) which states the following.

If $w \in C_0(\mathbb{R}_+; \mathbb{R}^2)$ *satisfies*

$$\text{(LS)} \quad -E_{red}\partial_y^2 w(y) + (z + \mathscr{A}_{red}(\xi))w(y) = 0, \quad y > 0, \quad \partial_y w(0) = 0,$$

then $w = 0$.

Condition (LS) can be proved without much pain. In fact, we observe that $E_{red}^{-1}(z + \mathscr{A}_{red}(\xi))$ has no eigenvalues $-\omega^2 \le 0$; otherwise $z \in \mathbb{C} \setminus (-\infty, 0]$ would be a solution of

$$0 = \det(z + \mathscr{A}_{red}(\xi) - \omega^2 E_{red}) = \det(z + \mathscr{A}_{red}(\xi - i\omega\nu)),$$

which by Step (a) is impossible. Therefore $B = (E_{red}^{-1}(z + \mathscr{A}_{red}(\xi)))^{1/2}$ is well-defined and has spectrum in $\mathbb{C}_+$. Thus, $w_{red}(y) := e^{-By}w_b$ is the unique stable solution of

$$-E_{red}\partial_y^2 w_{red}(y) + (z + \mathscr{A}_{red}(\xi))w_{red}(y) = 0, \quad y > 0, \quad w_{red}(0) = w_b.$$

The Neumann condition implies

$$0 = \partial_y w_{red}(0) = -Bw_b,$$

hence $w_b = 0$, as B is invertible, for all $(z, \xi) \neq (0, 0)$, $z \in \mathbb{C} \setminus (-\infty, 0]$, $\xi \in \mathbb{R}^n$, $\xi \cdot \nu = 0$. Therefore, the techniques of Sect. 7 in [7] apply and show that the reduced problem has maximal L_p-regularity in $\mathbb{R}^n_+$.

As a result, given $f_\theta \in L_p(J; L_q(\mathbb{R}^n_+))$, $f_d \in L_p(J; H_q^1(\mathbb{R}^n_+; \mathbb{R}^n))$, with $f_{red} = [f_\theta, -ic(\nabla_x) \cdot f_d]^\mathsf{T}$, we find a unique solution $w_{red} = [\theta, d_{red}]^\mathsf{T}$ of the problem

$$\partial_t w_{red} + A_{red}w_{red} = f_{red} \quad \text{in } \mathbb{R}^n_+, \quad w_{red}(0) = 0,$$

within the class

$$w_{red} \in {}_0H_p^1(J; L_q(\mathbb{R}^n_+; \mathbb{R}^2)) \cap L_p(J; H_q^2(\mathbb{R}^n_+; \mathbb{R}^2)),$$

where A_{red} denotes the realization of $\mathscr{A}_{red}$ in $L_q(\mathbb{R}^n_+; \mathbb{R}^2)$ with Neumann boundary condition.

Next, we solve the remaining problem for d

$$\partial_t d - \frac{\lambda_0}{\gamma_0}\Delta d = f_d^1 \quad \text{in } \mathbb{R}^n_+, \ d(0) = 0,$$
$$\partial_\nu d = 0 \quad \text{on } \partial\mathbb{R}^n_+$$

with

$$f_d^1 := f_d + i\frac{\partial_\tau \lambda_0}{\gamma_0}c(\nabla)d_{red} - b_1 c(\nabla)\theta,$$

in a similar way as in Step (a) by employing maximal L_p-regularity for $-\Delta$. This yields a unique solution

$$d \in {}_0H_p^1(J; L_q(\mathbb{R}^n_+; \mathbb{R}^n)) \cap L_p(J; H_q^2(\mathbb{R}^n_+; \mathbb{R}^n)).$$

As $f_d^1 \in L_p(J; H_q^1(\mathbb{R}^n_+; \mathbb{R}^n))$ we may differentiate the equation for d tangentially to obtain also

$$\nabla_x d \in {}_0H_p^1(J; L_q(\mathbb{R}^n_+; \mathbb{R}^{(n-1)\times n})) \cap L_p(J; H_q^2(\mathbb{R}^n_+; \mathbb{R}^{(n-1)\times n})).$$

On the other hand, we may also take the derivative with respect to the normal variable y in order to obtain a problem with Dirichlet boundary conditions for $v := \partial_y d$. We solve this with maximal L_p-regularity to obtain

$$\partial_y d \in {}_0H^1_p(J; L_q(\mathbb{R}^n_+; \mathbb{R}^n)) \cap L_p(J; H^2_q(\mathbb{R}^n_+; \mathbb{R}^n)).$$

This proves Theorem 2 for the case $\Omega = \mathbb{R}^n_+$ with constant coefficients. As described in [7], Sect. 7, this assertion extends by perturbation and localization to variable coefficients, still in the case $\Omega = \mathbb{R}^n_+$.

(c) General domains and variable coefficients
Here we follow the line given in [7], Sect. 8. We may use a perturbation argument to extend the result for the half-space to a bent half-space and then employ the localization method to prove Theorem 2 for general domains with C^{3-}-boundary.

3. Local-Wellposedess
We rewrite the above problem as an abstract quasilinear evolution equation of the form

$$\dot{z} + A(z)z = F(z), \quad t > 0, \ z(0) = z_0. \tag{15.34}$$

Here $z = (u, w) = (u, \theta, d)$ and we apply the *Helmholtz projection* $\mathbb{P}$ to the equation for u. The base space will be $X_0 := L_{q,\sigma}(\Omega) \times Y_0$, where the subscript σ means *solenoidal*. Then with the *generalized Stokes operator* $A_1(w) = -\mathbb{P}\mu(\theta, \tau)\Delta$, we define the regularity space by

$$X_1 := \mathsf{D}(A_1) \times Y_1, \quad \mathsf{D}(A_1) = \{u \in H^2_q(\Omega; \mathbb{R}^n) \cap L_{q,\sigma}(\Omega) : u = 0 \text{ on } \partial\Omega\}.$$

The operator $A(z)$ is defined by $A(z) = \operatorname{diag}(A_1(w), A_2(w))$, and $F(z)$ collects all lower order terms.

In order to prove local well-posedness of (15.34), we may now resort to abstract theory, e.g. to the results by Köhne et al. [20] and by LeCrone et al. [21].

Then, by Theorem 2 and by the maximal regularity of the generalized Stokes operator, see e.g. Bothe and Prüss [1], $A(z)$ has maximal L_p-regularity. For an interval $J = [0, a]$, the *solution space* $\mathbb{E}_\mu(J)$ will be

$$\mathbb{E}_\mu(J) = H^1_{p,\mu}(J; X_0) \cap L_{p,\mu}(J; X_1).$$

The time-trace space $X_{\gamma,\mu}$ of $\mathbb{E}_\mu(J)$ is given by

$$X_{\gamma,\mu} = \{u \in B^{2(\mu-1/p)}_{qp}(\Omega)^n \cap L_{q,\sigma}(\Omega) : u_{|\partial\Omega} = 0\} \times Y_{\gamma,\mu};$$

it satisfies

$$X_{\gamma,\mu} \hookrightarrow B^{2(\mu-1/p)}_{qp}(\Omega)^{n+1} \times B^{1+2(\mu-1/p)}_{qp}(\Omega)^n \hookrightarrow C(\overline{\Omega})^{n+1} \times C^1(\overline{\Omega})^n,$$

provided

$$\frac{1}{p}+\frac{n}{2q}<\mu\leq 1. \tag{15.35}$$

Here B^s_{pq} denote as usual the Besov spaces; see e.g. Triebel [36]. Then A, F satisfy the requirements in the paper by LeCrone et al. [21], and so we have local well-posedness. If $\frac{1}{p}+\frac{n}{2q}+\frac{1}{2}<\mu\leq 1$, the conditions of Köhne et al. [20] also hold. In particular, defining the *state manifold* of (15.34) by

$$\mathcal{SM}=\{(u,\theta,d)\in X_\gamma : \theta>0,\ |d|_2=1\},\quad X_\gamma:=X_{\gamma,1},$$

then $\mathcal{SM}$ is *locally positive invariant* for the semi-flow, total energy E is preserved, and the negative total entropy $-\mathsf{N}$ is a strict Lyapunov functional for the semi-flow on $\mathcal{SM}$. Summarizing we have the following result

Theorem 3 *Assume (R), let p, q, μ be subject to (15.35), and let $z_0\in X_{\gamma,\mu}$. Then for some $a=a(z_0)>0$, there is a unique solution*

$$z\in H^1_{p,\mu}(J,X_0)\cap L_{p,\mu}(J;X_1),\quad J=[0,a],$$

of (15.34), i.e. (15.31) on J. Moreover,

$$z\in C([0,a];X_{\gamma,\mu})\cap C((0,a];X_\gamma),$$

i.e. the solution regularizes instantly in time. It depends continuously on z_0 and exists on a maximal time interval $J(z_0)=[0,t^+(z_0))$. Moreover,

$$t[\frac{d}{dt}]z\in H^1_{p,\mu}(J;X_0)\cap L_{p,\mu}(J;X_1),$$

and $|d(t,x)|_2\equiv 1$, $\mathsf{E}(t)\equiv \mathsf{E}_0$, and $-N$ is a strict Lyapunov functional. Furthermore, the problem (15.34) generates a local semi-flow in its natural state manifold $\mathcal{SM}$.

4. The Generalized Principle of Linearized Stability

Consider the autonomous quasilinear problem

$$\dot{z}(t)+A(z(t))z(t)=F(z(t)),\quad t>0,\quad z(0)=z_0. \tag{15.36}$$

Here we assume

$$(A,F)\in C^1(V,\mathcal{B}(X_1,X_0)\times X_0), \tag{15.37}$$

where $V\subset X_\gamma$ is open. Let $\mathcal{E}\subset V\cap X_1$ denote the set of equilibrium solutions of (15.36), which means that

$$z\in\mathcal{E}\quad\text{if and only if}\quad z\in V\cap X_1,\ A(z)z=F(z).$$

Given an element $z_* \in \mathcal{E}$, we assume that z_* is contained in an m-dimensional manifold of equilibria. This means that there is an open subset $U \subset \mathbb{R}^m$, $0 \in U$, and a C^1-function $\psi : U \to X_1$, such that

$$\psi(U) \subset \mathcal{E}, \quad \psi(0) = z_*, \quad A(\psi(\zeta))\psi(\zeta) = F(\psi(\zeta)), \quad \zeta \in U. \tag{15.38}$$

and the rank of $\psi'(0)$ equals m.

Let A_0 denote the linearization of $A(z)z - F(z)$ at z_*, i.e.

$$A_0 h = A(z_*)h + [A'(z_*)h]z_* - F'(z_*)h.$$

We call $z_* \in \mathcal{E}$ *normally stable* if the following conditions hold.

(i) near z_* the set $\mathcal{E}$ is a C^1-manifold in X_1, $\dim \mathcal{E} = m \in \mathbb{N}_0$,
(ii) the tangent space for $\mathcal{E}$ at z_* is isomorphic to $\mathsf{N}(A_0)$,
(iii) 0 is a semi-simple eigenvalue of A_0, i.e. $\mathsf{N}(A_0) \oplus \mathsf{R}(A_0) = X_0$,
(iv) $\sigma(A_0) \setminus \{0\} \subset \mathbb{C}_+ = \{\zeta \in \mathbb{C} : \operatorname{Re} \zeta > 0\}$.

The following result is due to Prüss et al. [35].

Theorem 4 *Let $1 < p < \infty$. Suppose $z_* \in V \cap X_1$ is an equilibrium of (15.36) and that (A, F) satisfy (15.37) and that $A(z_*)$ has the property of maximal L_p-regularity. Assume further that z_* is normally stable.*

Then z_* is stable in X_γ, and there exists $\delta > 0$ such that the unique solution $z(t)$ of (15.36) with initial value $z_0 \in X_\gamma$ satisfying $|z_0 - z_*|_\gamma < \delta$ exists on $\mathbb{R}_+$ and converges at an exponential rate in X_γ to some $z_\infty \in \mathcal{E}$ as $t \to \infty$.

It is worthwhile to note that in case $m = 0$, z_* is necessarily isolated by (i). Then Theorem 4 reduces to the usual *principal of linearized stability*, as (ii), (iii), (iv) are equivalent to $\sigma(A_0) \subset \mathbb{C}_+$.

5. Linear Stability of Equilibria

The linearization of (15.34), i.e. of (15.31) at an equilibrium $z_* = (0, \theta_*, d_*)$ is given by the operator

$$A_* = -\operatorname{diag}\big((\mu_*/\rho)\mathbb{P}\Delta, (\alpha_*/\rho\kappa_*)\Delta, (\lambda_*/\gamma_*)\Delta\big)$$

in the base space X_0 with domain $\mathsf{D}(A_*) = X_1$. This operator has maximal L_p-regularity, it is the negative generator of a compact analytic C_0-semigroup, and it has compact resolvent. So its spectrum consists only of countably many eigenvalues of finite multiplicity, which are all positive, hence stable, except for 0. The eigenvalue 0 is semi-simple, its eigenspace is given by

$$\mathsf{N}(A_*) = \{(0, \vartheta, \mathsf{d}) : \vartheta \in \mathbb{R}, \mathsf{d} \in \mathbb{R}^n\},$$

hence it coincides with the set of constant equilibria $\bar{\mathcal{E}}$, when ignoring the constraint $|d|_2 = 1$ and conservation of energy. Therefore each such equilibrium is normally stable.

6. Nonlinear Stability
We have *stability with asymptotic phase* for the equilibria of (15.34).

Theorem 5 *Assume (R). Then any equilibrium $z_* \in \bar{\mathscr{E}}$ of (15.34) is stable in X_γ. Moreover, for each $z_* \in \bar{\mathscr{E}}$ there is $\varepsilon > 0$ such that if $|z_0 - z_*|_{X_{\gamma,\mu}} \leq \varepsilon$, then the solution z of (15.34) with initial value z_0 exists globally in time and converges at an exponential rate in X_γ to some $z_\infty \in \bar{\mathscr{E}}$.*

This result is proved by means of the *generalized principle of linearized stability*, Theorem 4. In fact, by the previous section we know that each equilibrium $z_* = (0, \theta_*, d_*)$ is normally stable.

7. Long-Time Behaviour
We conclude this paper with a result on the convergence of solutions to equilibria in the topology of the state manifold $\mathscr{SM}$.

Theorem 6 *Assume (R) and let z be the solution of (15.34), i.e. of (15.31), with initial value $z_0 \in \mathscr{SM}$. Then the following assertions hold.*
(a) If we suppose

$$\sup_{t \in (0, t^+(z_0))} [|z(t)|_{X_{\gamma,\mu}} + |1/\theta(t)|_{L_\infty}] < \infty,$$

then $t^+(z_0) = \infty$ and z is a global solution.
(b) If z is a global solution, bounded in $X_{\gamma,\mu}$ and with $1/\theta$ bounded, then z converges exponentially in $\mathscr{SM}$ to an equilibrium $z_\infty \in \mathscr{E}$ of (15.34), as $t \to \infty$.

This result follows from abstract dynamical system arguments involving the strict Lyapunov functional $-\mathsf{N}$, as well as the nonlinear stability result; see Köhne et al. [20]. Note that, by a compactness argument, the converse of (b) is also valid.

References

1. Bothe, D., Prüss, J.: L^p-theory for a class of non-Newtonian fluids. SIAM J. Math. Anal. **39**, 379–421 (2007)
2. Chandrasekhar, S.: Liquid Crystals. Cambridge University Press (1992)
3. Coutand, D., Shkoller, S.: Well-posedness of the full Ericksen-Leslie model of nematic liquid crystals. C. R. Acad. Sci. Paris Sér. I Math. **333**, 919–924 (2001)
4. Dai, M., Schonbek, M.: Asymptotic behaviour of solutions to the liquid crystal system in $H^m(\mathbb{R}^3)$. SIAM J. Math. Anal. **46**, 3131–3150 (2014)
5. DeGennes, P.G.: The Physics of Liquid Crystals. Oxford University Press (1974)
6. DeGennes, P.G.,Prost, J.: The Physics of Liquid Crystals. Oxford University Press (1995)
7. Denk, R., Hieber, M., Prüss, J.: $\mathbb{R}$-boundedness, Fourier multipliers and problems of elliptic and parabolic type. Mem. Am. Math. Soc. **166** (2003)
8. Denk, R., Kaip, M.: General Parabolic Mixed Order Systems in L_p and Applications. Operator Theory: Advances and Applications. Birkhäuser, Basel (2013)

9. Ericksen, J.L.: Hydrostatic theory of liquid crystals. Arch. Ration. Mech. Anal. **9**, 371–378 (1962)
10. Ericksen, J.L., Kinderlehrer, D. (eds.): Theory and Applications of Liquid Crystals, The IMA Volumes in Mathematics and its Applications, vol. 5. Springer, New York (1987)
11. Feireisl, E., Rocca, E., Schimperna, G.: On a non-isothermal model for nematic liquid crystals. Nonlinearity **24**, 243–257 (2011)
12. Feireisl, E., Frémond, M., Rocca, E., Schimperna, G.: A new approach to non-isothermal models for nematic liquid crystals. Arch. Ration. Mech. Anal. **205**, 651–672 (2012)
13. Feireisl, E., Rocca, E., Schimperna, G., Zarnescu, A.: Evolution of non-isothermal Landau-de Gemmes nematic liquid crystal flows with singular potential. arXiv:1207.1643
14. Feireisl, E., Rocca, E., Schimperna, G., Zarnescu, A.: Nonisothermal nematic liquid crystal flows with Ball-Majumdar free energy, arXiv:1310.8474
15. Frank, F.C.: On the theory of liquid crystals. Discuss. Faraday Soc. **25**, 19–28 (1958)
16. Grubb, G.: Boundary problems for systems of partial differential operators of mixed order. J. Funct. Anal. **26**, 131–165 (1977)
17. Hieber, M., Nesensohn, M., Prüss, J., Schade, K.: Dynamics of nematic lquid crystals: the quasilinear approach. Ann. Inst. H. Poincaré Anal. Non Linéaire, online 2014 (2015) (to appear)
18. Hu, X., Wang, D.: Global solutions to the three-dimensional incompressible flow of liquid crystals. Comm. Math. Phys. **296**, 861–880 (2010)
19. Huang, J., Lin, F., Wang, C.: Regularity and existence of global solutions to the Ericksen-Leslie system in $\mathbb{R}^2$, Comm. Math. Phys. (to appear)
20. Köhne, M., Prüss, J., Wilke, M.: On quasilinear parabolic evolution equations in l_p-spaces. J. Evol. Equ. **10**, 443–463 (2010)
21. LeCrone, J., Prüss, J., Wilke, M.: On quasilinear parabolic evolution equations in weighted L^p-spaces II. J. Evol. Equ. **14**, 509–533 (2014)
22. Leslie, F.M.: Some constitutive equations for liquid crystals. Arch. Ration. Mech. Anal. **28**, 265–283 (1968)
23. Li, J., Xin, Z.: Global weak solutions to non-isothermal neamtic liquid crystal in 2D. arXiv:1307.2065
24. Lin, F.: Nonlinear theory of defects in nematic liquid crystals: phase transition and flow phenomena. Comm. Pure Appl. Math. **42**, 789–814 (1989)
25. Lin, F.: On nematic liquid crystals with variable degree of freedom. Comm. Pure Appl. Math. **44**, 453–468 (1991)
26. Lin, F., Liu, Ch.: Nonparabolic dissipative systems modeling the flow of liquid crystals. Comm. Pure Appl. Math. **48**, 501–537 (1995)
27. Lin, F., Liu, Ch.: Existence of solutions for the Ericksen-Leslie system. Arch. Ration. Mech. Anal. **154**, 135–156 (2000)
28. Lin, F., Lin, J., Wang, C.: Liquid crystal flows in two dimensions. Arch. Ration. Mech. Anal. **197**, 297–336 (2010)
29. Liu, C., Wu, H., Xu, X.: On the general Ericksen-Leslie system: Parodi's relation, well-posedness and stability. Arch. Ration. Mech. Anal. **208**, 59–107 (2013)
30. Müller, I.: Thermodynamics.Interaction of Mechanics and Mathematics, Pitman (1985)
31. Oseen, C.W.: The theory of liquid crystals. Trans. Faraday Soc. **29**, 883–899 (1933)
32. Parodi, O.: Stress tensor for a nematic liquid crystal. J. Phys. **31**, 581–584 (1970)
33. Prüss, J.: Maximal regularity for evolution equations in L_p-spaces. Conf. Semin. Mat. Univ. Bari **2002**(285), 1–39 (2003)
34. Prüss, J., Simonett, G.: Maximal regularity for evolution equations in weighted l_p-spaces. Arch. Math. (Basel) **82**, 415–431 (2004)
35. Prüss, J., Simonett, G., Zacher, R.: On convergence of solutions to equilibria for quasilinear parabolic problems. J. Diff. Eqns. **246**, 3902–3931 (2009)
36. Triebel, H.: Theory of Function Spaces II. Birkhäuser, Basel (1992)
37. Virga, E.G.: Variational Theories for Liquid Crystals. Chapman-Hall, London (1994)
38. Wang, C.: Well-posedness for the heat flow of harmonic maps and the liquid crystal flow with rough initial data. Arch. Ration. Mech. Anal. **200**, 1–19 (2011)
39. Wang, W., Zhang, P., Zhang, Z.: Well-posedness of the Ericksen-Leslie system. Arch. Ration. Mech. Anal. **210**, 837–855 (2013)

Chapter 16
Statistical Mechanics of Quasi-geostrophic Vortices

Takeshi Miyazaki, Yuichi Shimoda and Keisei Saga

Abstract The statistical mechanics of quasi-geostrophic vortices is investigated numerically and theoretically. Direct numerical simulations of a point vortex system of mixed sign under periodic boundary conditions are performed using a fast special-purpose computer for molecular dynamics (GRAPE9). Clustering of point vortices of like sign is observed and a columnar dipole structure appears as an equilibrium state. These numerical results are explained from the viewpoint of the classical statistical mechanics. A three-dimensional mean field equation is derived based on the maximum entropy theory. The numerically obtained end states are shown to be the two-dimensional sn-sn dipole solutions of the mean field equation (i.e., the sinh-Poisson equation). We present other branches of two- and three-dimensional solution of the mean field equation. The entropy of these solution branches is found to be smaller than that of the two-dimensional sn-sn dipole branch. The stability of the maximum entropy states is studied theoretically and numerically. The two-dimensional (sn-sn dipole and zonal) solutions are stable against disturbances of finite amplitude, whereas the three-dimensional solutions are shown to be unstable. These findings explain the reason why only the two-dimensional sn-sn dipole states are found in the numerical simulations of point vortices. The influence of the aspect ratio of periodic unit box on the maximum entropy states and their stability is investigated. When the horizontal aspect ratio (L_y/L_x) is less than unity, the entropy of the zonal flow solution becomes larger than that of the dipole solution if the energy is less than a certain critical value. This critical energy increases as the aspect ratio is decreased. In contrast, the dipole solution in a box of square cross section ($L_y/L_x = 1$) has the largest entropy, even if L_z/L_x is changed.

T. Miyazaki (✉) · Y. Shimoda · K. Saga
University of Electro-Communications, 1-5-1 Chofugaoka,
Chofu, Tokyo 182-8585, Japan
e-mail: miyazaki@mce.uec.ac.jp

Y. Shimoda
e-mail: Shimodai@miyazaki.mce.uec.ac.jp

K. Saga
e-mail: keisei@miyazaki.mce.uec.ac.jp

Y. Shibata and Y. Suzuki (eds.), *Mathematical Fluid Dynamics, Present and Future*, Springer Proceedings in Mathematics & Statistics 183,
DOI 10.1007/978-4-431-56457-7_16

Keywords Quasi-geostrophic vortices · Statistical mechanics · Maximum entropy theory · Mean field equation

16.1 Introduction

Statistical mechanics of two-dimensional turbulence has been investigated extensively, for applications in plasma physics and geophysical fluid dynamics as well as for its own intrinsic interest. Geophysical flows are subject to strong influence of the stable density stratification and the Coriolis force. Both effects suppress vertical motion and the geophysical flows are considered to be two-dimensional at the lowest order of approximation. In two-dimensional turbulence, many coherent vortices appear spontaneously and their interactions dominate the turbulence dynamics. The investigation on the statistical mechanics of two-dimensional vortex system has a long history, starting from Onsager [14] who was the first to illustrate the existence of negative temperature states. Joyce and Montgomery [7] and Montgomery and Joyce [12] derived the mean field equation (so called the sinh-Poisson equation) for a system of point vortices of mixed sign, based on the maximum entropy theory. The sinh-Poisson equation is known to be integrable and Gurarie and Chow [4] presented doubly periodic exact solutions, including the sn-sn dipole solutions. There have been also many numerical works on two-dimensional decaying turbulence, in which the relaxation process towards a maximum entropy state is investigated (see [13]). These classical studies are succeeded by more sophisticated researches focusing on the delicate characteristics of the statistical mechanics of systems with long-range interactions (see e.g. [8, 18] and references therein).

In actual geophysical flows, the fluid motion is almost confined within a horizontal plane, but different flow patterns are realized on different horizontal planes. This three-dimensionality is incorporated in the 'quasi-geostrophic approximation', which takes the first order terms of the Rossby number expansions into account (see, e.g., [15, 16]). This approximation yields a very simple governing equation quite similar to the Euler equation. The numerical simulations by McWilliams [9] of decaying quasi-geostrophic turbulence indicated that the vorticity field developed coherent vortex structures spontaneously. These vortices kept their identity for a long time, dominating the dynamics of geostrophic turbulence. A following vortex-based statistical analysis showed that there was a period of self-similar dissipative temporal evolution, which terminated as the number of vortices decreased due to merger and alignment of like-sign vortices. The end state was consisted of a pair of tall columnar vortices of different sign [10]. In order to understand the turbulence dynamics fully, it will be of importance to investigate a simpler system, i.e., the statistical mechanics of inviscid quasi-geostrophic point vortices.

The statistical properties of mono-disperse quasi-geostrophic point vortices in an infinite space, were studied numerically and theoretically (see, e.g., [6, 11]). Large direct numerical simulations of point vortices were performed using a fast special-purpose computer for molecular dynamics (GRAPE-2, -3 and -DR), and axisymmetric equilibrium states were found to form. The radial vorticity distributions of the equilibrium states changed in accord with the vertical distribution of vortices $P(z)$ and the total energy of the vortex system E. Here, the vertical vorticity distribution is invariant, because each vortex moves in the horizontal plane on which it was located initially. It is known that the angular momentum I is conserved, besides the energy E, under the axisymmetric boundary condition. At a certain energy level, the radial vorticity distribution becomes the Gaussian distribution at any vertical height. When the energy is lower than this critical value (in the positive temperature region), the radial vorticity distribution of the center region becomes flatter. In contrast, if the energy is higher (in the negative temperature region), the vortices in the center region concentrate tighter near the axis of symmetry. These numerical equilibrium states were shown to be actually the maximum Shannon entropy state, under the constraints of the fixed vertical distribution $P(z)$, the energy E and the angular momentum I, of which the last one was found to be crucial for the existence of negative temperature states. These theoretical studies based on the statistical mechanics provided the most probable vorticity distributions, which agreed with the numerically obtained equilibrium states.

The statistical properties of bi-disperse quasi-geostrophic point vortices under periodic boundary conditions were studied by Funakoshi et al. [5]. They derived a mean field equation based on the maximum entropy theory. This equation was a natural extension of the sinh-Poisson equation derived for two-dimensional point vortices. They solved the mean field equation utilizing an iteration procedure proposed by Turkington and Whitaker [17]. Although two-dimensional solutions were obtained in a wide energy range, they could present three-dimensional solutions only at very low energy level, where the mean field equation was approximated by a linear Poisson equation. They also carried out direct numerical computations of quasi-geostrophic point vortices, to find the equilibrium states were two-dimensional at any energy level considered, i.e., the sn-sn dipole solutions of the sinh-Poisson equation. They conjectured that the three-dimensional states with high energy might be unstable and that only two-dimensional states with the largest entropy could be found in the numerical simulations.

The aim of the present paper is to extend the previous work by Funakoshi et al. [5]. In order to provide more quantitative discussions, we carry out numerical simulations with more point vortices, and also we determine the two- and three-dimensional maximum entropy states at higher energy level, theoretically. It is interesting to investigate the influence of the aspect ratio of the periodic box. We present the solutions of the mean field equation in a unit box of various size. The stability of these maximum entropy states is studied using Arnold's method and by direct numerical simulations of the continuous quasi-geostrophic equation. It is shown that some of the two-dimensional maximum entropy states are Lyapunov stable. In contrast, we are not able to determine the stability characteristics of the three-dimensional states by Arnold's method alone. The numerical simulations indicate that the three-dimensional states are weakly unstable against small disturbances and that they switches to a two-dimensional structure resembling the sn-sn dipole, in the later stage of the simulations. The influence of the aspect ratio of unit box on the stability is also studied.

The paper is organized as follows. In the next section, we will make a few comments on the quasi-geostrophic approximation as well as on the equations of motion of quasi-geostrophic point vortices. We review in Sect. 16.3 the derivation of the mean field equation, based on the maximum entropy theory. In Sect. 16.4, the results from the numerical simulations of the point vortex system are illustrated in some detail. The most probable states with maximum entropy are given theoretically, and compared with the numerical equilibrium states in Sect. 16.5, where the influence of the aspect ratio of the periodic box on the maximum entropy states is discussed. In Sect. 16.6, the stability of the maximum entropy states is investigated theoretically and numerically. The last section is devoted to a brief summary.

16.2 Quasi-geostrophic Approximation and the Equations of Motion of Point Vortices

We consider geophysical flows in a uniformly rotating (f-plane), continuously stratified fluid with uniform vertical density gradient (i.e., constant Brunt-Vaisala frequency). In the quasi-geostrophic approximation, the fluid motions occur on each horizontal plane, whereas they are assumed to be different on different horizontal planes. Then, we are able to introduce a stream function $\psi(r, t), r = (x, y, z)$ by,

$$u = \frac{\partial \psi}{\partial y}, \quad v = -\frac{\partial \psi}{\partial x}. \tag{16.1}$$

In this paper, we follow the sign convention in the fluid dynamical society. It is noted that $\psi(r, t)$ and $u(r, t), v(r, t)$ depend on the vertical coordinate z, because different flows are realized at different vertical heights.

The quasi-geostrophic equations of motion represents the conservation of the potential vorticity:

$$\left(\frac{\partial}{\partial t} + \frac{\partial \psi}{\partial y}\frac{\partial}{\partial x} - \frac{\partial \psi}{\partial x}\frac{\partial}{\partial y}\right) q(r, t) = 0. \tag{16.2}$$

Here, the potential vorticity is defined as,

$$q(r, t) = -\Delta\psi(r, t). \tag{16.3}$$

The vertical coordinate z is rescaled appropriately (see, e.g., [15, 16]) to make the equation isotropic (so called Charney isotropy). These equations are quite similar to the two-dimensional Euler equation. The differences appear only in the z-dependence of the stream function $\psi(r, t)$ and in the three-dimensional Δ in the definition of the potential vorticity $q(r, t)$.

As a model of coherent vortices, we consider the simplest one, i.e., point vortices with infinitesimal size. The potential vorticity of point vortices is concentrated δ-

function like at the location of vortices $R_i = (X_i, Y_i, Z_i)$:

$$q(r) = \sum_{i=1}^{N_+ + N_-} \hat{\Gamma}_i \delta(r - R_i). \tag{16.4}$$

Here, N_+ and N_- denote the numbers of vortices with positive and negative sings, respectively. $\hat{\Gamma}_i$ is the strength of the i-th vortex. We consider a symmetric bi-disperse case, taking $N_+ = N_- = N$ and $|\hat{\Gamma}_i| = \hat{\Gamma}_0$. The point vortices move by mutual advection, and the motion of the i-th vortex is governed by the canonical equations:

$$\frac{dX_i}{dt} = \frac{1}{\hat{\Gamma}_i}\frac{\partial H}{\partial Y_i}, \quad \frac{dY_i}{dt} = -\frac{1}{\hat{\Gamma}_i}\frac{\partial H}{\partial X_i}. \tag{16.5}$$

These equations look similar to those describing the motion of two-dimensional point vortices, in which the interaction energy is proportional to the logarithm of the distance between two vortices. It should, however, be noted that the interaction energy between two quasi-geostrophic point vortices is proportional to the inverse of the distance between them, corresponding to the fact that Δ in the definition of the potential vorticity is the three-dimensional Laplace operator. The vertical coordinate Z_i of the ith-vortex remains constant, i.e., the vortices move on the horizontal plane on which they are initially located, under the quasi-geostrophic approximation.

The Hamiltonian under periodic boundary conditions is given by evaluating the Ewald sum. Several efficient methods have been proposed for molecular dynamics simulations dealing with the Coulomb interactions between molecules (see e.g., [1]). We can utilize them with some modifications, because the Green function of the present problem is the same to that of the Coulomb potential. We will show the results of direct numerical simulations in Sect. 16.4. There is no dissipative term in the canonical equations and the energy H is conserved. In contrast, the angular momentum is not conserved, unlike in an infinite fluid domain, because the axial symmetry is lost under the periodic boundary conditions.

16.3 Maximum Entropy Theory and the Three-Dimensional Mean Field Equation

We will review the formulation of the maximum entropy theory, whose details were provided by Funakoshi et al. [5]. In this section, we consider the statistical mechanics of bi-disperse vortices ($\hat{\Gamma}_+ = -\hat{\Gamma}_- = \hat{\Gamma}_0$) located inside a unit box of size L_x, L_y and L_z, assuming the periodicities in the $x-$, $y-$ and $z-$directions. The probability distributions of positive and negative vortices are denoted by $F_+(x, y, z)$ and $F_-(x, y, z)$, respectively. We saw in the previous section that the energy of the vortex system is conserved. In addition, the vertical distribution of the vortices is unchanged, because each point vortex moves on the same horizontal plane where it is initially. This means that the vertical distributions

$$P_{\pm}(z) = \iint F_{\pm}(x, y, z)dxdy \tag{16.6}$$

are unchanged. Note that they satisfy the following normalization constraints:

$$\int_{-L_z/2}^{L_z/2} P_{\pm}(z)dz = 1. \tag{16.7}$$

The Shannon entropy of this system is defined as,

$$S = -\iiint [F_+(r)\log F_+(r) + F_-(r)\log F_-(r)]d^3r. \tag{16.8}$$

The normalized energy $E = H/(\hat{\Gamma}_0 N)^2$ is computed to be

$$E = \frac{1}{2}\iiiint\!\!\iint G(r, r')[F_+(r)F_+(r') + F_-(r)F_-(r') - 2F_+(r)F_-(r')]d^3rd^3r'. \tag{16.9}$$

Here, $G(r, r')$ denotes the Green function under the periodic boundary conditions, which is provided basically by evaluating the Ewald sum of the Coulomb potential. In the following iteration procedure, we expand the probability distribution functions in the form of Fourier series expansions (normally 64^3 and 128^3 for higher energy cases) and the inversion from the potential vorticity to the stream function is performed in the Fourier space. Then, the energy is computed in the real space by integrating the product of the potential vorticity and the stream function (see Funakoshi et al. [5] for the details).

The maximum Shannon entropy state under the constraints of the energy E and the vertical vorticity distributions $P_{\pm}(z)$, is determined by the method of Lagrange's multipliers, i.e., making the following functional $\bar{S}$ stationary.

$$\begin{aligned}\bar{S} = S &- \int \alpha_+(z)P_+(z)dz - \int \alpha_-(z)P_-(z)dz \\ &- \frac{1}{2}\beta \iiiint\!\!\iint G(r, r')[F_+(r)F_+(r') + F_-(r)F_-(r') - 2F_+(r)F_-(r')]d^3rd^3r'.\end{aligned} \tag{16.10}$$

Here, $\alpha_{\pm}(z)$ are the Lagrange's multipliers corresponding to the invariants $P_{\pm}(z)$ and β is that corresponding to the energy E. β is called the inverse temperature in the literature of the statistical mechanics. We will give solutions for negative values of β, in Sect. 16.5.

We obtain the following equations by taking the variation with respect to $F_{\pm}(r)$:

$$-1 - \log F_+(r) - \alpha_+(z) - \beta \iiint G(r, r')[F_+(r') - F_-(r')]d^3r' = 0, \tag{16.11}$$

$$-1 - \log F_-(r) - \alpha_-(z) + \beta \iiint G(r, r')[F_+(r') - F_-(r')]d^3r' = 0. \tag{16.12}$$

These equations with the constraints of the vertical distributions $P_\pm(z)$ and the energy E, are solved to determine the maximum entropy states.

If we focus on a symmetric case in which $P_+(z) = P_-(z) = P(z)$, we can assume the symmetry of $\alpha_+(z) = \alpha_-(z) = \alpha(z)$. Then, the variations with respect to $F_\pm(r)$ yield the following equations, respectively:

$$F_+(r) = \exp[-\beta\psi(r) - \alpha(z) - 1], \tag{16.13}$$

$$F_-(r) = \exp[\beta\psi(r) - \alpha(z) - 1]. \tag{16.14}$$

The stream function $\psi(r)$ and the potential vorticity $q(r)$ are related to $F_\pm(r)$ as

$$\psi(r) = \iiint G(r, r')[F_+(r') - F_-(r')]d^3r', \tag{16.15}$$

$$q(r) = F_+(r) - F_-(r). \tag{16.16}$$

Then, taking the difference between $F_\pm(r)$, we have

$$\begin{aligned} q(r) &= e^{-\alpha(z)-1}[e^{-\beta\psi(r)} - e^{\beta\psi(r)}] \\ &= -2e^{-\alpha(z)-1} \sinh \beta\psi(r). \end{aligned} \tag{16.17}$$

Finally, we obtain the mean field equation, after recalling the relation between the potential vorticity and the stream function $q(r) = -\Delta\psi(r)$:

$$\Delta\bar{\psi}(r) + \lambda^2(z) \sinh \bar{\psi}(r) = 0. \tag{16.18}$$

Here, we have introduced new variables $\bar{\psi}(r) = -\beta\psi(r)$ and $\lambda^2(z) = -2\beta e^{-\alpha(z)-1}$, for later convenience. Coherent vortex structures appear only in the negative temperature region with $\beta < 0$. This equation is a natural extension of the sinh-Poisson equation derived for the two-dimensional point vortices [5].

16.4 Direct Numerical Simulation of the Point Vortex System

We will illustrate the results of numerical simulation for a point vortex system in a cubic box of size $L_x = L_y = L_z = 2\pi$, in this section. We integrate the equations of motion for $N = N_\pm = 2400$ and 4000 point vortices of strength $\hat{\Gamma}_0 = 0.103$ and 0.062, respectively. The time is normalized by the mean potential vorticity $N\hat{\Gamma}_0/(2\pi)^3 = 1.00$, which is taken to be unity. We are to make sure that the same equilibrium states are formed irrespective of the number of point vortices. The energy is conserved within the accuracy of five significant figures. The vertical distribution $P_{\pm}(z) = L_z^{-1} = 1/2\pi$ is assumed to be uniform. This simple vertical distribution was studied repeatedly in the dissipative numerical simulations by McWilliams et al. [9] (spectral code).

Figures 16.1a and 16.2a show the initial distributions of bigger computations with $N = 4000$. Two energy levels of $E = 1.0 \times 10^{-2}$ (Case A) and $E = 8.1 \times 10^{-3}$ (Case B) are considered. Red points represent positive vortices and blue points do negative ones. The initial distributions have somewhat characteristic structures, i.e., Case A has a three-dimensional dipole structure, whereas Case B with the slightly lower energy is characterized by a two-dimensional zonal structure.

We show, in Fig. 16.1b–d, the time evolution of Case A with higher energy. The number of vortices N is 2400. In all figures, 20 snapshots during a time interval of $\Delta t = 2$ are superposed. The initial three-dimensional dipole structure becomes unstable soon (see Fig. 16.1b), but the initial clusters of vortices of like sign are not destroyed completely. After about $t = 37.5$, we find that a tall dipole-pair of columnar vortices is formed (Fig. 16.1c), as a result of vertical alignment. This almost two-dimensional vortex pair, however, moves around the whole computational domain until $t = 390$, at which time we stopped the numerical computation. The top view of the final state (not yet in the equilibrium) is shown in Fig. 16.1d. The positive columnar vortex (red) is tilted slightly even at this time. The system of point vortices seems to approach to a purely two-dimensional dipole state very slowly. The corresponding time evolution of the bigger system with $N = 4000$ vortices is shown in Fig. 16.1e–g. We selected the times so that similar vortex structures are observed. A dipole structure is almost formed by $t = 50$, which is later than in the smaller computation. The dipole moves around slowly and randomly, again. Then, the point vortex system cannot reach the true equilibrium until $t = 500$, at which time we stopped the numerical computation.

Figure 16.2b, c illustrate the time evolution of Case B using $N = 2400$ point vortices, and Fig. 16.2d–e show that using $N = 4000$ vortices. We show only the top views of the vortex structures, since the flow field remains almost two-dimensional in the course of time. It takes much longer for the two-dimensional distribution to reach the equilibrium, probably because the two dimensional zonal flow is more stable and robust than the three-dimensional vorticity distribution. In the smaller computation (b–c), the initial zonal distribution becomes deformed significantly by $t = 345$, whereas similar deformation cannot be observed until $t = 630$ in the bigger computation. Two columnar clusters of positive and negative vortices are formed at about $t = 430$ (Fig. 16.2c) and this two-dimensional dipole structure lasts for a quite long time, telling that the system has settled down into an equilibrium state. However, in the bigger computation, it takes much longer for the latter point vortex system to be equilibrium. In fact, the flow pattern at $t = 990$ (Fig. 16.2e) is still evolving. Generally speaking, the simulation using more point vortices is a better approximation of the continuous quasi-geostrophic equation, which has an infinite number of conserved quantity. Then the influence of the initial condition may last longer as the number of point vortices is increased.

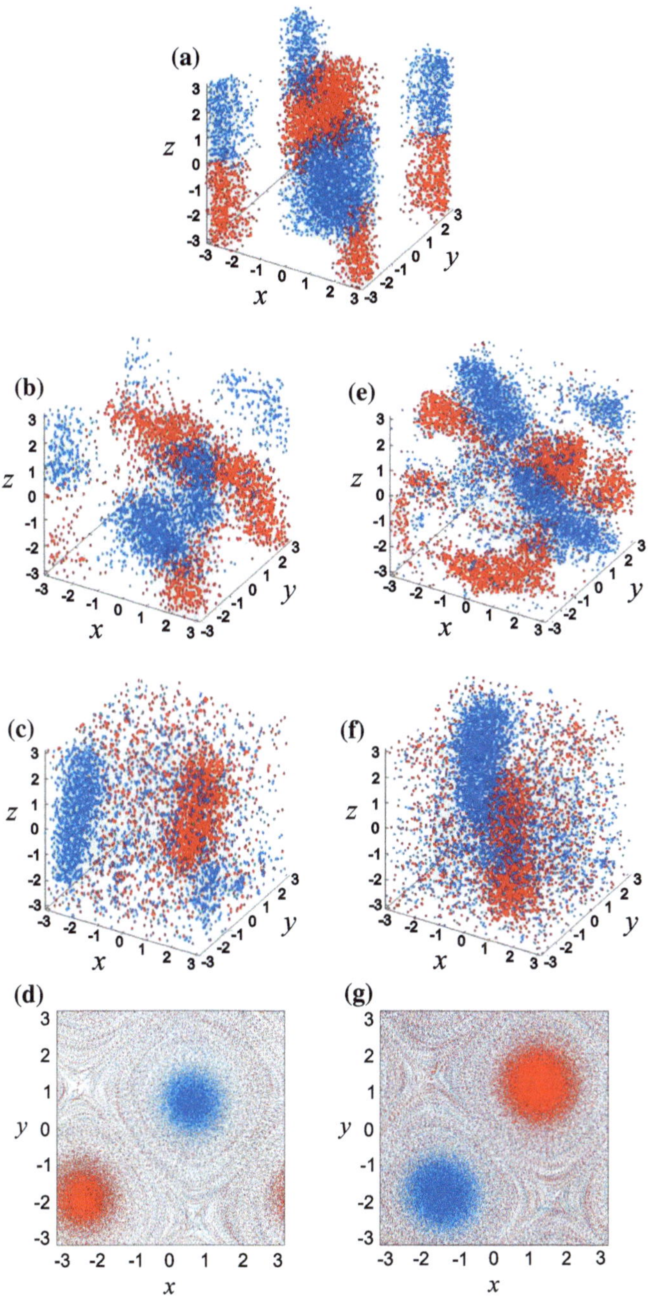

Fig. 16.1 Distributions of point vortices for $E = 1.0 \times 10^{-2}$. *Red* and *blue points* denote positive and negative vortices, respectively. **a** the initial condition ($t = 0$, $N = 4000$), **b** seemingly random phase ($t = 37.5$, $N = 2400$), **c** the end state ($t = 390$, $N = 2400$), **d** *top view* of the end state ($t = 390$, $N = 2400$), **e** seemingly random phase ($t = 50$, $N = 4000$), **f** the end state ($t = 500$, $N = 4000$), **g** *top view* of the end state ($t = 500$, $N = 4000$)

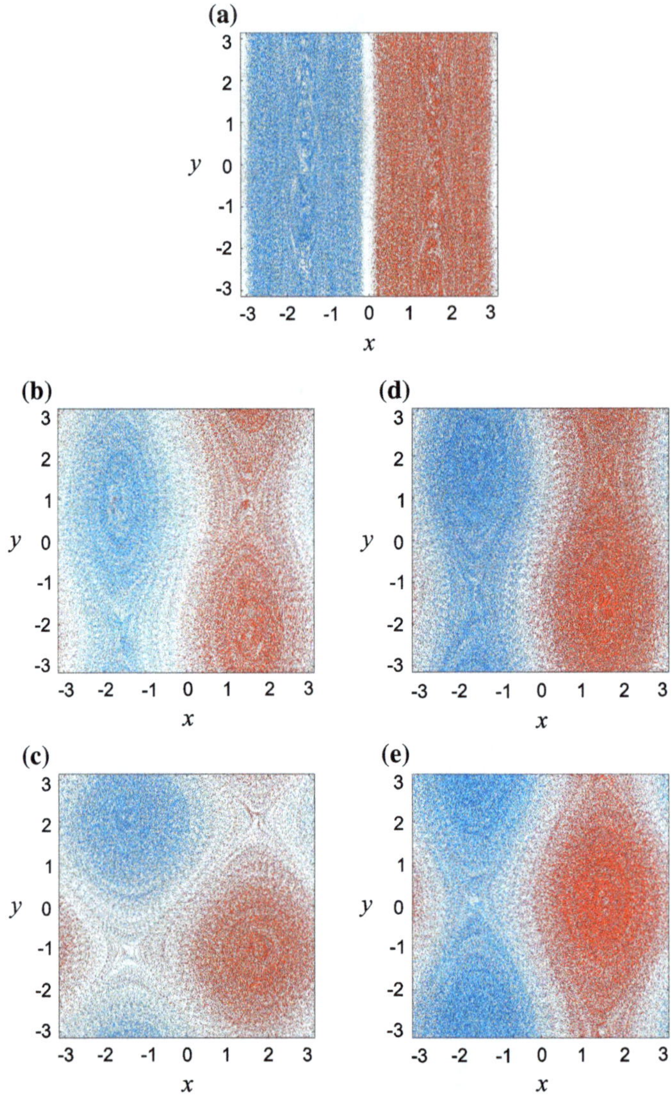

Fig. 16.2 Distributions of point vortices for $E = 8.1 \times 10^{-3}$. *Red* and *blue points* denote positive and negative vortices, respectively. **a** the initial condition ($t = 0$, $N = 4000$), **b** wavy zonal structure ($t = 345$, $N = 2400$), **c** the end state ($t = 990$, $N = 2400$), **d** wavy zonal structure ($t = 630$, $N = 4000$), **f** the end state ($t = 990$, $N = 4000$)

In these simulations, we have observed the transitions from both three- and two-dimensional initial vorticity distributions to the sn-sn dipole states. Note that it took much longer for the latter transition (from the two-dimensional zonal flow) to be completed. As we will make clear in Sect. 16.6, this deference is attributed to the deference of the stability of initial flows. Any way, we may conclude both the clustering of like-sign vortices and the formation of tall columnar vortex structures are purely inviscid phenomena. The next task is to explain the underlining mechanism, whose first step is to determine the maximum entropy states at various energy level.

16.5 Maximum Entropy States

In the previous section, we have investigated the time evolution of a point vortex system at relatively high energy, numerically. A uniform vertical distribution $P_{\pm}(z) = 1/2\pi$ is assumed in order to study the basic statistical properties of the system. We observed clustering of vortices of like sign and the formation of columnar dipole structures.

Funakoshi et al. [5] argued that the equilibrium state should be the maximum entropy state with the initially specified energy. They also pointed out that the solutions corresponding to the numerical equilibrium states belong to a class of soliton solutions of the sinh-Poisson equation, given by Gurarie and Chow [4]. Since the analytical form is presented using the Jacobian elliptic sn-function, they named this solution branch the 'sn-sn dipole'. There are a number of other solution branches, including the zonal flow and quadrupole branches.

In this section, we first compare the numerically obtained end states with the two-dimensional maximum entropy states, which were the solutions of the classical sinh-Poisson equation with a constant $\lambda(z)^2 = \lambda^2$. Figure 16.3 shows three solution branches in the E–S (energy-entropy) plane, where the horizontal and ver-

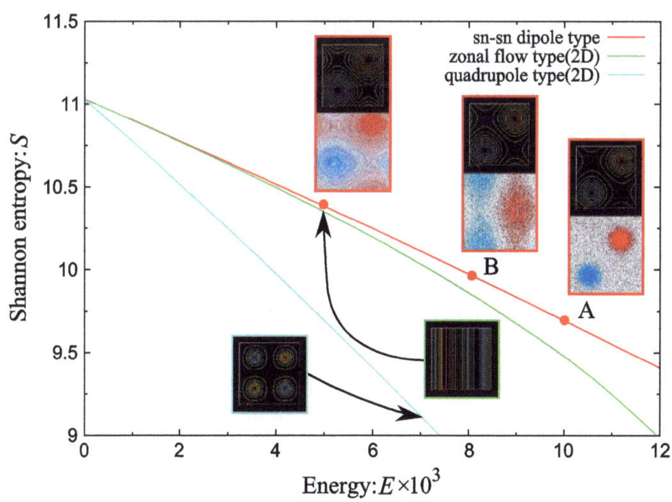

Fig. 16.3 Two-dimensional solution branches on the E–S plane: the inserted figures show the potential vorticity distribution of each branch

tical axes denote the energy and entropy, respectively. The inserted figures illustrate the contours of the potential vorticity of each branch. All branches emanate from the point $E = 0$, $S = 6\log(2\pi) \approx 11.027$, i.e., the uniformly random distributions $F_{\pm} = 1/(2\pi)^3$. As can be seen in Fig. 16.3, the sn-sn dipole branch has the largest entropy. The entropy difference between the zonal- and dipole-solution branches is rather small, although their vorticity distributions are quite different. We see that the entropy of the quadrupole branch is much smaller. Funakoshi et al. [5] conjectured that only the sn-sn dipole branch with the largest entropy, could be found in the numerical simulation of point vortices. We show the end states of Cases A and B computed in the previous section, by two points on the sn-sn dipole branch. The numerical end state of Case B is elongated in the $y-$direction slightly and it is not the true equilibrium. The dipole distribution at a still lower energy ($E = 4.9 \times 10^{-3}$) is shown for comparison. It can be seen that the distribution of the potential vorticity shrinks as the energy increases.

In order to see how close the numerical distributions are to the maximum entropy states, we have computed the entropy of the numerically obtained end states in the computations using 2400 and 4000 point vortices. Since the number of vortices is not large enough, 1400 snapshots during a time interval of $\Delta t = 140$ are superposed by adjusting the coordinate origin, so that the dipole patterns coincide as close as possible. The volume integration of the entropy is carried out using 64^3 grid points. The obtained entropy value in Case A is 9.58 for bigger computation with $N = 4000$, which is smaller than the theoretical value 9.69 in Fig. 16.3. This difference is partially attributed to the inaccurate shift of the coordinate origin and mainly to the fact that the numerical end state hasn't reached the true final equilibrium. In contrast, the numerical value 9.76 in the smaller computation ($N = 2400$) slightly exceeds the theoretical value 9.69, probably because the inaccurate shit of the origin contaminated the data. As for Case B, the theoretical entropy value is 9.96 and the numerical values are 9.88 (bigger computation with $N = 4000$) and 9.93 (smaller computation with $N = 2400$). The numerically obtained entropy is a little smaller than the theoretical value, because the numerical end states don't reach the true equilibrium yet. The approaching speed to the equilibrium is much slower for Case B, in which case the initial entropy value is closer to the equilibrium value, and the initial state is more stable. It can be reasonably concluded that the numerically found end states are very close to the maximum entropy states.

Funakoshi et al. [5] also searched for three-dimensional solutions, but they were unsuccessful except at very low energy level, because of poor convergence of their iteration procedure. In the next subsection, we first reconsider the same problem in a cubic box of $(2\pi)^3$, and determine the three-dimensional solution branches extending the energy range. After that, we investigate the influence of the box aspect ratio on the maximum entropy states, by changing the shape of the unit box.

16.5.1 *Three-Dimensional Maximum Entropy States in a Cubic Box*

We give the three-dimensional maximum entropy states in the cubic box of size $(2\pi)^3$, in this subsection. The method proposed by Turkington and Whitaker [17] is utilized again, but with several refinements. The main modification is to introduce an appropriate damping factor between the succeeding iteration steps, which has made the iteration procedure more robust. Also, the initial guess has been improved by using a data-fitting technique and an appropriate extrapolation method, as the energy value is changed.

We have traced three branches, corresponding to the three-dimensional dipole, zonal and quadrupole solutions. They have larger entropy than other three-dimensional solutions with more complicated structure. We show three solution branches on the E–S plane, in Fig. 16.4. The horizontal and vertical axes denote the energy and entropy, respectively. From top to bottom, three branches representing the dipole (purple), zonal flow (blue) and quadrupole (black) are drawn. All branches emanate from the point $E = 0$, $S = 6\log(2\pi) \approx 11.027$, corresponding the uniform distributions $F_{\pm} = 1/(2\pi)^3$, again. It is interesting to note that the entropy of the three-dimensional dipole branch is slightly larger than the two-dimensional quadrupole branch (see Fig. 16.3). We are successful in extending the energy range considerably. The potential vorticity distributions of three-dimensional dipole solution branch at $E = 1.30 \times 10^{-3}$ and $E = 6.40 \times 10^{-3}$ are inserted. The potential vorticity distribution around the center of vortex structure shrinks as the energy increases.

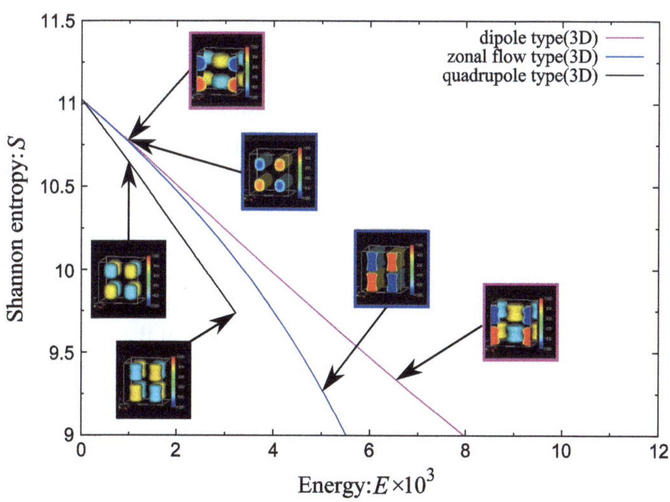

Fig. 16.4 Three-dimensional maximum entropy states on the E–S plane. Inserted figures are three-dimensional potential vorticity distributions of the three-dimensional dipole, zonal and quadrupole solution branches ($q \times 10^4$)

The same trend was reported by Miyazaki et al. [11], in which the maximum entropy states in an infinite domain was investigated. The three-dimensional zonal type solutions at $E = 1.30 \times 10^{-3}$ and $E = 6.40 \times 10^{-3}$ are shown in the inserted figures. It was rather difficult to have three-dimensional quadrupole solutions at a high energy level. So, we insert two figures demonstrating the potential vorticity distributions at relatively low energy levels, i.e., $E = 1.01 \times 10^{-3}$ and $E = 3.23 \times 10^{-3}$. The center region of each vortex structure becomes narrower with the energy, again. As the value of $\lambda^2(z)$ increases, it becomes more difficult to have a well converged solution.

16.5.2 Influence of the Aspect Ratio of the Periodic Domain

We will determine two- and three-dimensional maximum entropy states in various unit box, by changing the length L_y or L_z, while keeping $L_x = 2\pi$ unchanged. The solution procedure is the same as in the case of the cubic box of $(2\pi)^3$.

Before showing the obtained solution branches, we explain an important difference that appears when the symmetry between the $x-$ and $y-$ coordinates is lost. We take the zonal flow branches for the case of $L_x = 2\pi$, $L_y = 1.8\pi$, $L_z = 2\pi$, as an example. As it can be seen in Fig. 16.5, there are presented two zonal flow branches of different entropy, i.e., the x—independent solution branches (yellow and gray lines) and the $y-$independent solution branches (green and blue lines). It is seen that the latter branches have larger entropy. Generally speaking, solutions with smaller spacial gradient have larger entropy, because they are closer to the uniform distribution with the largest entropy. Hereafter, we present only solution branches with larger entropy, although there are always accompanied solution branches with smaller entropy.

When the size of periodic box is $L_x = 2\pi$, $L_y = 1.8\pi$, $L_z = 2\pi$, we observe a similar but slightly different solution branch diagram in the $E - S$ plane, compared with the purely cubic case (Fig. 16.6). All branches emanate from $E = 0$, $S = 2\log(7.2\pi^3)$, corresponding to the uniformly random distribution without any structure. This entropy value differs slightly from that of the cubic box. The more important is the fact that the dipole type branch has less entropy than the zonal flow, at very low energy level. As the energy level is increased, the entropy of the dipole solution becomes larger than that of the zonal flow. This change occurs at about $E = 5.5 \times 10^{-3}$. The difference in the entropy value is very small at lower energy level, and then the difference between two distributions of the potential vorticity is very small, too. We see the similar exchange between the three-dimensional zonal flow and the three-dimensional dipole solution branches, which occurs at about $E = 1.6 \times 10^{-3}$ (Fig. 16.6).

The two- and three-dimensional solution branches for the case $L_x = 2\pi$, $L_y = 1.6\pi$, $L_z = 2\pi$ are illustrated in Fig. 16.7, where the aspect ratio is decreased further to $L_y/L_x = 0.8$. Similar exchange of the branch with maximum entropy, is observed at still higher energy level (at about $E = 9.6 \times 10^{-3}$ for the two-dimensional solutions). At lower energy level, the zonal flow (green line) has the largest entropy, but the difference in the entropy value is quite small (see the inserted close up). Then it

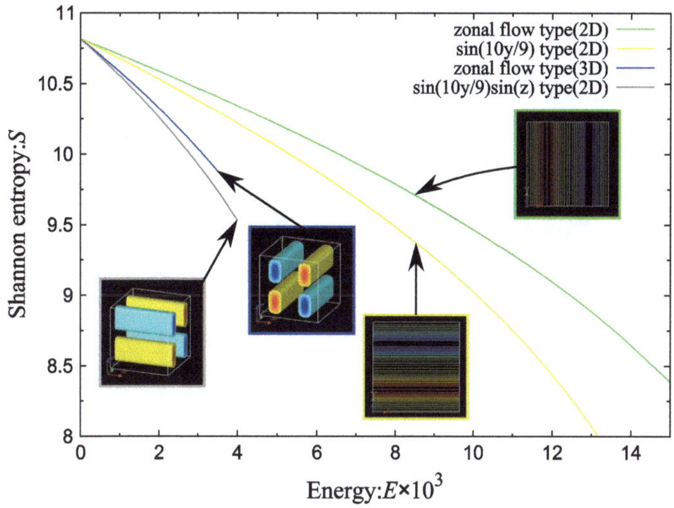

Fig. 16.5 Breakdown of symmetry of zonal flow branches on the E–S plane in the unit box of $L_x = 2\pi$, $L_y = 1.8\pi$, $L_z = 2\pi$. Inserted figures are two- and three-dimensional potential vorticity distributions of the zonal branches ($q \times 10^4$). The y axis is elongated by a factor 10/9, and the unit box is represented by a cubic box

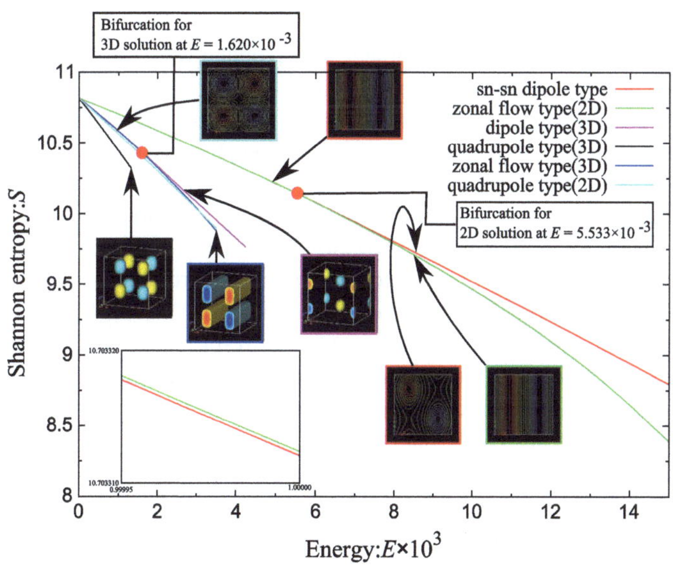

Fig. 16.6 Three-dimensional extremum states on the E–S plane in the unit box of $L_x = 2\pi$, $L_y = 1.8\pi$, $L_z = 2\pi$. Inserted figures are two- and three-dimensional potential vorticity distributions of the dipole, zonal flow and quadrupole solution branches ($q \times 10^4$). The y axis is elongated by a factor 10/9, and the unit box is represented by a cubic box. The close up at low energy level of two-dimensional dipole and zonal flow branches

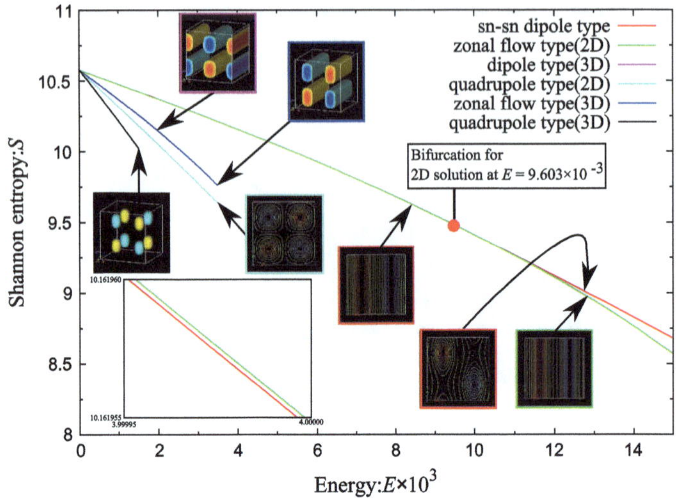

Fig. 16.7 Maximum entropy states on the E–S plane in the unit box of $L_x = 2\pi$, $L_y = 1.6\pi$, $L_z = 2\pi$. Inserted figures are two- and three-dimensional potential vorticity distributions of the dipole, zonal flow and quadrupole solution branches ($q \times 10^4$). The y axis is elongated by a factor 5/4, and the unit box is represented by a cubic box

may be very difficult to identify the branch type (zonal flow or dipole) of the end state realized in the numerical simulations of point vortices. As the energy is increased more and more, the dipole pattern (red line) will be formed clearly as an equilibrium state, because it has significantly larger entropy than the zonal flow solution even in a box of the rectangular cross section $L_y/L_x = 0.8$. It is difficult to distinguish the three-dimensional zonal flow branch (blue line) from the three-dimensional dipole branch (purple line) in the present figure, for the corresponding two lines overlap.

For the case of $L_y = \pi$, we see the zonal flow branch (green line) has larger entropy than the sn-sn dipole branch (red line) up to $E = 1.5 \times 10^{-2}$ (Fig. 16.8), although the difference remains very small (see the inserted close-up). We can hardly distinguish the vorticity distributions of the two solution branches (compare the inserted potential vorticity distributions). The fact that the entropy values of two-dimensional zonal flow and dipole solutions are larger than those of three-dimensional solutions remains unchanged. Then, we can conclude columnar structures will form irrespective of the aspect ratio L_y/L_x of the horizontal cross section. It becomes much harder to distinguish the three-dimensional zonal flow (blue line) from the three-dimensional dipole (purple line), and the purple line is masked with the blue line completely.

Next, we consider the influence of the vertical box size. The solution branches in the E–S plane for a shallower box with $L_x = L_y = 2\pi$, $L_z = \pi$ are shown in Fig. 16.9. A qualitative difference from the cubic case is found, i.e., the two-dimensional quadrupole solution branch has larger entropy than any three-dimensional solution branch. When L_z is decreased, the entropy of three-dimensional solutions decreases quite rapidly as the energy increases. This decay rate is much

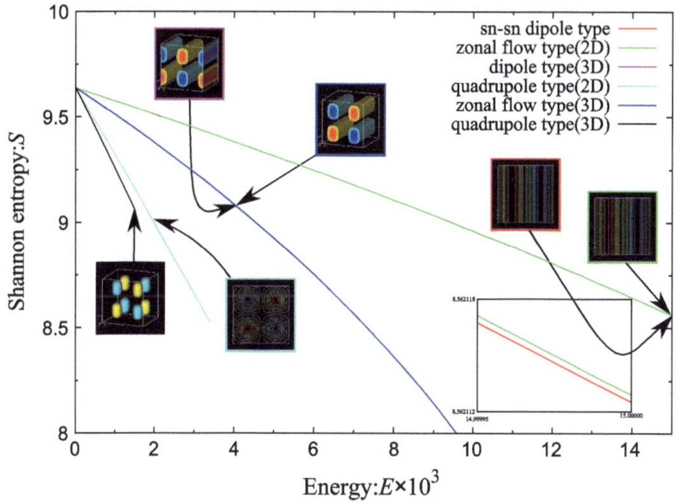

Fig. 16.8 Maximum entropy states on the E–S plane in the unit box of $L_x = 2\pi$, $L_y = \pi$, $L_z = 2\pi$. Inserted figures are two- and three-dimensional potential vorticity distributions of the dipole, zonal flow and quadrupole solution branches ($q \times 10^4$). The y axis is elongated by a factor 2, and the unit box is represented by a cubic box

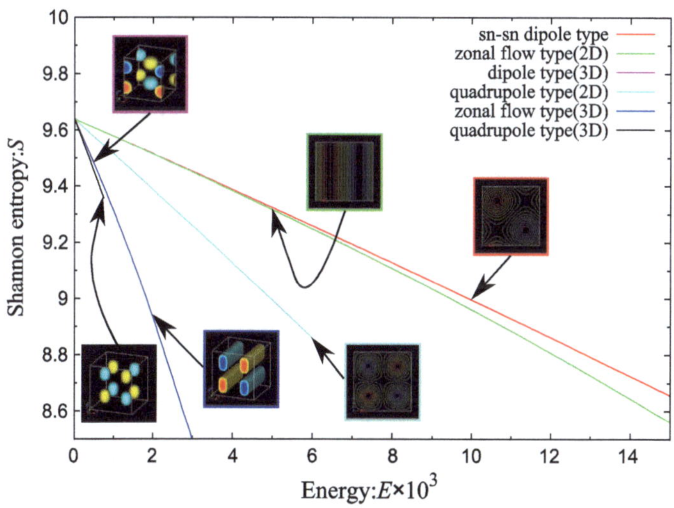

Fig. 16.9 Maximum entropy states on the E–S plane in the unit box of $L_x = L_y = 2\pi$, $L_z = \pi$. Inserted figures are two- and three-dimensional potential vorticity distributions of the dipole, zonal flow and quadrupole solution branches ($q \times 10^4$). The z axis is elongated by a factor 2, and the unit box is represented by a cubic box

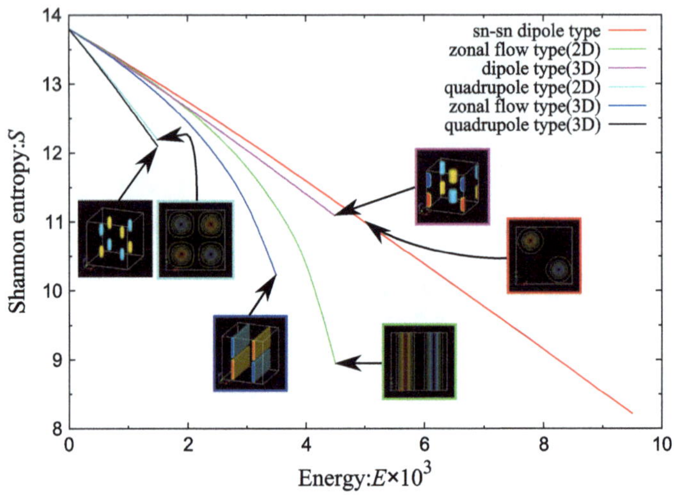

Fig. 16.10 Maximum entropy states on the E–S plane in the unit box of $L_x = L_y = 2\pi$, $L_z = 8\pi$. Inserted figures are two- and three-dimensional potential vorticity distributions of the dipole, zonal flow and quadrupole solution branches ($q \times 10^4$). The z axis is shrunk by a factor 4, and the unit box is represented by a cubic box

larger than those of two-dimensional solutions. Then, the two-dimensional sn-sn dipole solution will become more dominant equilibrium state in a shallower box.

In contrast, the entropy of three-dimensional solutions increases in a deeper box with increased L_z. Figure 16.10 show the solution branches in the E–S plane for the unit box of $L_x = L_y = 2\pi$, $L_z = 8\pi$. The entropy difference between two- and three-dimensional solutions becomes less significant. In fact, the three-dimensional dipole branch has larger entropy than the two-dimensional zonal flow. The difference between the entropy values of two- and three-dimensional dipole solutions becomes much smaller, although the two-dimensional dipole branch still has the largest entropy.

We have obtained three-dimensional maximum entropy states in addition to two-dimensional states in a periodic box of various size. It is noted that the equilibrium state with the largest entropy can be two-dimensional zonal flow when the $x - y$ symmetry in the horizontal plane is lost. This interesting fact may be attributed to the inverse energy cascade mechanism in geostrophic turbulence, that leads to the formation of larger and larger vortex structures in the course of time. Then the final states are constrained strongly by the boundary conditions at the largest scale, here the shape of the periodic box. It makes a sharp contrast to the energy cascade observed in the three-dimensional Navier-Stokes turbulence, in which the energy is transfered to the smallest scale and dissipated there. When we try to construct a sub-grid turbulence model of geophysical flow, we should keep these findings in mind.

Although various two- and three-dimensional solution branches of the mean field equations are found under the periodic boundary conditions, only the two-dimensional sn-sn dipole states are realized in the numerical simulations of point vortices in a cubic box. A possible explanation is that the state with the largest entropy at a specified energy level is selected among other maximum entropy states of the same energy. This conjecture is simple and attracting, but we should be more careful because the entropy is merely a measure of geometric arrangement, in which no dynamical information is included. In order to understand the dynamical aspects behind vortex clustering and the formation of columnar vortices fully, the stability of the solutions of the mean field equation should be investigated, which is the subject in the following section.

16.6 Stability of the Maximum Entropy States

We have obtained several branches of two- and three-dimensional maximum entropy states in various unit box. Funakoshi et al. [5] argued that these solutions might be stable, because they had larger entropy than nearby states. It is a suspicious conjecture and a more quantitative investigation is required to explore their stability characteristics.

In this section, we present theoretical and numerical studies about the stability of the maximum entropy states. First, we apply the method developed by Arnold [2, 3], yielding exact theoretical conjectures for some of two-dimensional solutions. Unfortunately, this method alone doesn't provide decisive results for other solution branches. Secondly, we study the time evolution of small perturbations imposed on the solutions of the mean field equation (16.18), by integrating the quasi-geostrophic equation (16.2) numerically. In the following subsection, we will explain the application of Arnold's method for the cases in the $(2\pi)^3$ box in some detail, while only the results will be presented briefly for other cases.

16.6.1 Arnold's Method

The mean field equation (16.18) can be derived in a framework of different variational problem. It is well known that there are infinite number of conserved quantity of the quasi-geostrophic equation (16.2). The volume integral of an arbitrary function of the potential vorticity q and z, is invariant under the the quasi-geostrophic approximation:

$$C = \iiint F(q, z) dx dy dz. \tag{16.19}$$

These conserved quantities are called Casimir invariants. By selecting an appropriate function for $F(q, z)$ and adding the associated Casimir C to the energy, we can formulate another variational problem, that leads to the same mean field equation:

$$H_C = H + C. \tag{16.20}$$

The mean field equation (16.18) can be re-derived by taking the first variation of (16.20) with respect to ψ:

$$\begin{aligned}\delta H_C &= \iiint q_0 \delta\psi dxdydz - \iiint \Delta\left(\frac{\partial F}{\partial q_0}\right)\delta\psi dxdydz\\ &= \iiint \left[q_0 - \Delta\left(\frac{\partial F}{\partial q_0}\right)\right]\delta\psi dxdydz.\end{aligned} \tag{16.21}$$

Here, q_0 denotes the potential vorticity in the equilibrium. In order to be consistent with the mean field equation (16.18), $F(q, z)$ should satisfy the following relation:

$$\frac{\partial F}{\partial q_0} = -\sinh^{-1}\left(\frac{q_0}{\lambda^2(z)}\right). \tag{16.22}$$

Integrating and differentiating this relation with respect to q_0, we have

$$\begin{aligned}F &= -q_0 \sinh^{-1}\left(\frac{q_0}{\lambda^2(z)}\right) + \sqrt{\lambda^4(z) + q_0^2},\\ \frac{\partial^2 F}{\partial q_0^2} &= -\frac{1}{\sqrt{\lambda^4(z) + q_0^2}}.\end{aligned} \tag{16.23}$$

Next, we calculate the second variation with respect to ψ:

$$\begin{aligned}\delta^2 H_C &= \frac{1}{2}\iiint [-\Delta(\delta\psi)]\delta\psi dxdydz + \frac{1}{2}\iiint \frac{1}{\sqrt{\lambda^4(z) + q_0^2}}(\Delta(\delta\psi))^2 dxdydz\\ &= \frac{1}{2}\iiint \left[|\mathrm{grad}\delta\psi|^2 - \frac{(\Delta(\delta\psi))^2}{\sqrt{\lambda^4(z) + q_0^2}}\right] dxdydz.\end{aligned} \tag{16.24}$$

Here, the second equation of (16.23) is used. The sign of the second variation (16.24) seems indeterminate, because the first term is positive definite whereas the second is negative definite. This situation is similar to that of the second theorem by Arnold [2, 3]. Under the periodic boundary conditions, we notice that the volume integral of $(\Delta\delta\psi)^2$ is always larger than (or equal to) that of $|\mathrm{grad}\delta\psi|^2$. Then, the second variation (16.24) is known to be negative definite, if the factor of the second term is larger than unity everywhere in the cubic box $(2\pi)^3$. In that case, the Lyapunov stability of the equilibrium solution can be assured.

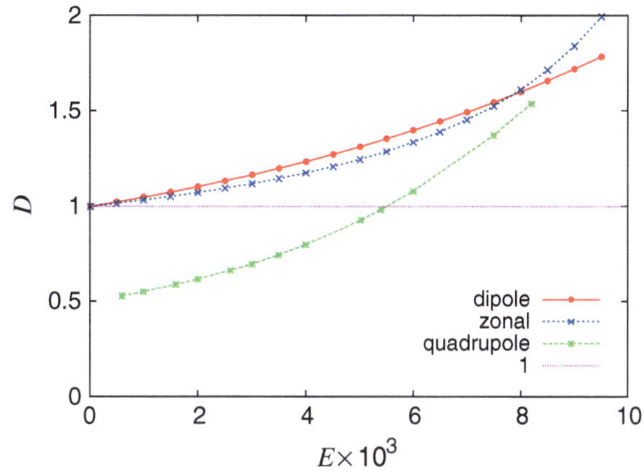

Fig. 16.11 The minimum values of the factor D as functions of the energy: Two-dimensional branches

Let us denote by D the minimum values of the factor in front of the second term $(\Delta\delta\psi)^2$ in the x–y plane:

$$D(z) = \min_{(x,y)\in[-\pi,\pi]\times[-\pi,\pi]} \frac{1}{\sqrt{\lambda^4 + q_0^2}}. \tag{16.25}$$

We plot, in Fig. 16.11, the values of D for the sn-sn dipole (red circle), the two-dimensional zonal flow (blue cross) and the quadrupole (green asterisk) solutions, as functions of the energy. The horizontal axis is the energy and the vertical axis shows the minimum value of D, which is independent of z for two-dimensional solutions. We see that D both for the two-dimensional zonal flow (blue cross) branch and for the sn-sn dipole (red circle) branch, increases with the energy and it is always greater than unity. These solution-branches are known to be stable in the Lyapunov sense. This finding is consistent with the results of numerical simulations of the two-dimensional Euler equation carried out by Gurarie and Chow [4], in which the sn-sn dipole solutions were shown to be stable against two-dimensional disturbances. In contrast, D of the two-dimensional quadrupole branch is less than unity at low energy level, and we cannot prove its stability.

The situation becomes more complex for the three-dimensional solutions, because the maximum value in the horizontal plane $D(z)$ depends on the vertical coordinate z. Generally $D(z)$ for any three-dimensional solutions is smaller than unity in some interval of z at any energy level, although we will not show the details. Thus, we are not able to judge the stability of these solution branches based on the method by Arnold.

We give quickly the results of theoretical stability analysis of maximum entropy states in various periodic box. When L_y (or L_z) is decreased from 2π, the method of Arnold provides completely the same results. The two-dimensional dipole and zonal flow solutions are shown to be stable in the Lyapunov's sense, whereas we cannot know the stability characteristics of other solutions from the theoretical analysis of

Arnold's type. In contrast, if L_z is increased, the minimum of the factor $D(z)$ should be larger than $(L_z/2\pi)^2$ to assure the stability. As we can see in Fig. 16.11, this is not the case at lower energy level. Thus, we are not able to judge the stability of any solution branch based on the method by Arnold in a taller unit box.

16.6.2 Numerical Simulation of Quasi-geostrophic Equation

There are two remaining methods to study stability. The first is a linear instability analysis, and the second is a full numerical simulation of the continuous quasi-geostrophic equation (16.2). Since a linear modal instability analysis of the three-dimensional flow field, treating an extremely big eigenvalue problem, is not an easy task. So, in this subsection, we will present some preliminary results of the spectral simulations with 64^3 Fourier modes. We study the time evolution of perturbations imposed on the solutions of the mean field equation (16.18), by integrating the quasi-geostrophic equation (16.2) numerically, just as Gurarie and Chow [4] carried out direct numerical simulations of the two-dimensional Euler equation. We use the fourth order Runge-Kutta method for the time-marching.

We add small perturbations to the three-dimensional dipole and zonal flow solutions with the energy $E = 6.4 \times 10^{-3}$, and to the two-dimensional quadrupole solution with the energy $E = 1.5 \times 10^{-3}$ then we integrate numerically the continuous quasi-geostrophic equation (16.2). The superposed disturbance consists of the three-dimensional dipole and zonal distributions of very small amplitude. It is produced by multiplying a factor 10^{-3} to both basic solutions and adding them up. In order to stabilize the numerical time marching, we introduce weak dissipation in the form of a usual viscous term $\nu \Delta q$ with $\nu = 10^{-6}$. Therefore the energy decreases gradually, and the vortex structures deviate slightly from their initial structures, even if they are stable. Figure 16.12 illustrates the time evolution of the volume integral of the difference $|q - q_0|^2$ of the potential vorticity. Here, q_0 denotes the potential vorticity field of the initial maximum entropy solutions (without perturbations). The horizontal axis denotes the time and the vertical axis represents the deviation measure denoted by $||q - q_0||^2$.

As for the three-dimensional zonal flow, the deviation measure grows quickly (blue broken line), suggesting the occurrence of strong instability. After $t = 10$, the zonal structure becomes wavy and an imperfect quadrupole structure develops by $t = 35$. In a later stage of simulation, a two-dimensional pattern appears, which resembles the sn-sn dipole solution (see the inserted potential vorticity fields). The 3D dipole state keeps its structure (green solid line) until about $t = 40$, telling that it is weakly unstable. A drastic increase of the deviation is observed after $t = 40$, where the 3D dipole structure relaxes and the coherent vortex blobs begin to move around. By the final stage of the numerical simulation ($t = 160$), a two-dimensional dipole structure seems to be forming (see the inserted potential vorticity distribution). In contrast, the two-dimensional quadrupole solution (red solid line) seems to be stable,

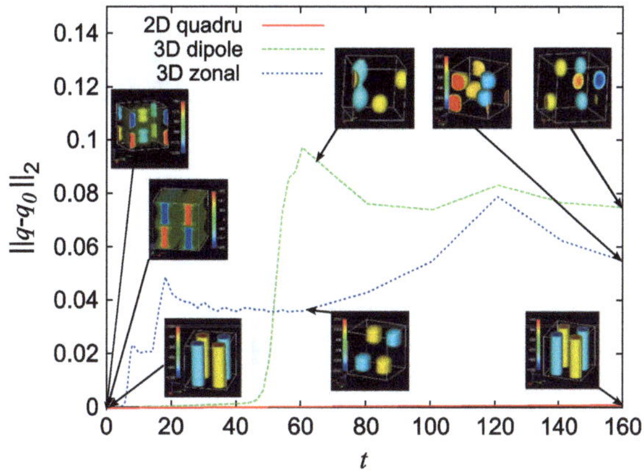

Fig. 16.12 Time evolution of the three-dimensional perturbed zonal flow and dipole solutions, and the two-dimensional perturbed quadrupole solution (continuous model). Inserted figures show the potential vorticity distributions

because the deviation norm remains very small and the quadrupole structure keeps its initial structure up to $t = 160$.

Finally, we investigate the stability of the two-dimensional solutions in a tall unit box $L_x = L_y = 2\pi$, $L_z = 8\pi$ numerically, because we can't provide any information by the method of Arnold. The sn-sn dipole solution with $E = 3.5 \times 10^{-3}$, the zonal flow solution with $E = 4.5 \times 10^{-3}$ and the quadrupole solution with $E = 1.5 \times 10^{-3}$ are taken as basic flows. In addition, the three-dimensional dipole solution with $E = 3.5 \times 10^{-3}$ is considered. The results are shown in Fig. 16.13, where only the two-dimensional dipole structure (red solid line) is stable. Other solutions are unstable, of which the most unstable one is the two-dimensional zonal flow (green broken line). This may be linked with the fact that its entropy is less than that of the three-dimensional dipole solution. At the final stage of computation ($t = 160$), a seemingly two-dimensional dipole structure is formed. As for the two-dimensional quadrapole solution (denoted by blue broken line), it keeps its initial structure until about $t = 45$, and after that a three-dimensional instability mode grows and transient three-dimensional structures appear. Finally, the flow settles down to the two-dimensional dipole as shown by the inserted figure. The three-dimensional dipole solution (purple dotted line) is weakly unstable, keeping its initial structure until about $t = 75$. Then, it is broken up into small fragments and three-dimensionality remains until the final stage ($t = 160$) of the computation, although the two-dimensional dipole structure may form gradually. We have carried out similar computations about the three-dimensional zonal flow and quadrupole solutions to find out that they are unstable. It will be worth mentioning that the instability of the three-dimensional dipole solution is relatively weak. In fact, a seemingly three-dimensional dipole structure appears even in the time evolution starting from the three-dimensional zonal flow and keep the structure for a while, before it changes to the two-dimensional dipole structure finally. Anyway, only the two-dimensional dipole solution is stable in a tall box, and therefore it seems to be a common final structure starting from any initial flow field.

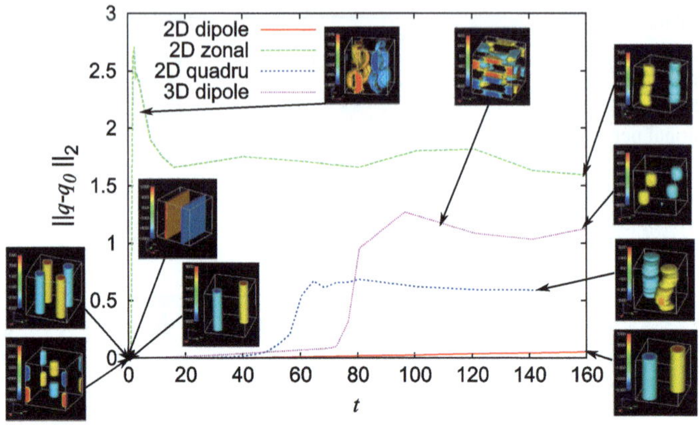

Fig. 16.13 Time evolution of the two-dimensional perturbed dipole, zonal flow and quadrupole solutions, and the three-dimensional perturbed dipole solution (continuous model). Inserted figures show the potential vorticity distributions. The z axis is shrunk by a factor 4, and the unit box is represented by a cubic box

The stability analysis presented here, is quite insufficient and much more extensive investigation is required to understand the stability of each solution branch and to complete the bifurcation diagram in the E–S plane. Instead of the highly elaborate normal mode analysis, a linear instability analysis using the Krylov subspace method might be helpful, which is left for future work.

The above observations are consistent with the results of the previous dissipative simulations of geostrophic turbulence, as well as with the results of our inviscid simulations of point vortices. We may conclude that the three-dimensional solution branches are unstable. As the depth of the box is increased, the two-dimensional zonal flow and quadrupole solution branches may become unstable. It is commonly observed that a two-dimensional dipole structure appears at a later stage of numerical simulations starting from unstable initial conditions. It took longer for the continuous system to reach the two-dimensional end state than for point vortices. The dynamical system of point vortices is less constrained compared with the system governed by the continuous quasi-geostrophic equation. In fact, the latter system has an infinite number of conserved quantities in the inviscid limit, whereas the energy is a single conserved quantity of point vortices. Then, the system of point vortices is subject to strong fluctuations, and it reaches the maximum entropy state sooner.

16.7 Summary

We have investigated the statistical mechanics of quasi-geostrophic point vortices of mixed sign under periodic boundary conditions. In the direct numerical simulations of the point vortices, two-dimensional dipole structures are found to form as end states.

The mean field equation, which was derived as an extension of the sinh-Poisson equation, is solved to obtain the most probable states for the case with the uniform vertical distributions $P_{\pm}(z) = 1/2\pi$. The two- and three-dimensional solutions in a periodic unit box of various size are determined in a considerably wide energy range. The two-dimensional sn-sn dipole solution has the largest Shannon entropy, among all solutions in a cubic box, whereas the two-dimensional zonal flow has larger entropy at lower energy level, if the aspect ratio L_y/L_x is less than unity. When L_z is decreased while keeping $L_x = L_y = 2\pi$, the entropy of three-dimensional solutions decreases in comparison with that of two-dimensional solutions, while it increases if L_z is increased. The stability of these solutions is investigated theoretically and numerically. The two-dimensional sn-sn dipole and zonal flow solutions are shown to be stable in a cubic (and smaller) box, but their stability cannot be assured theoretically in a taller box. Numerical simulations of the continuous quasi-geostrophic equation suggest that the two-dimensional quadrupole solution is stable in a cubic box, but the three-dimensional solutions are unstable against small disturbances. In a tall box with $L_z/L_x = 4$, even the two-dimensional zonal flow and quadrupole solutions are found to be unstable. Unstable solutions change their structure rather abruptly and transitions to two-dimensional dipole structures take place, which is in accord with the observation in the direct numerical simulations of the quasi-geostrophic point vortices, i.e., the equilibrium states are on the sn-sn dipole solution branch.

Acknowledgements This work is supported by a Grant-in-Aid for Scientific Research (c) (No. 25400462) from the Ministry of Education, Culture, Sports, Science and Technology, Japan. We are grateful for the support of Mr. Fukushige (K & F Computing Research) throughout this work, and we thank Mr. Ishihara for preparing figures.

References

1. Allen, M.P., Tildesley, D.J.: Computer Simulation of Liquids, p. 385. Oxford University Press, Oxford (1987)
2. Arnold, V.I.: On an a priori estimate in the theory of hydrodynamical stability Izv. Uchebn Zaved Mat. **54**, 3–5 (1966)
3. Arnold, V.I.: Mathematical Methods of Classical Mechanics, p. 508. Springer, Berlin (1978)
4. Gurarie, D., Chow, K.W.: Vortex array for sinh-Poisson equation of two-dimensional fluids: equilibria and stability. Phys. Fluids **16**, 3296–3305 (2004)
5. Funakoshi, S., Sato, T., Miyazaki, T.: Maximum entropy state of the quasi-geostrophic bi-disperse point vortex system: bifurcation phenomena under periodic boundary conditions. Fluid Dyn. Res. **44**, 031407 (13 pp.) (2012)
6. Hoshi, S., Miyazaki, T.: Statistics of Quasi-geostrophic point vortices. Fluid Dyn. Res. **40**, 662–678 (2008)
7. Joyce, G., Montgomery, D.: Negative temperature states for the two-dimensional guiding center plasma. J. Plasma Phys. **10**, 107–121 (1973)
8. Majda, A.J., Wang, X.: Nonlinear Dynamics and Statistical Theories for Basic Geophysical Flows, p. 551. Cambridge University Press, Cambridge (2006)
9. McWilliams, J.C.: Anisotropy and coherent vortex structures in planetary turbulence. Science **264**, 410–413 (1994)

10. McWilliams, J.C., Weiss, J.B., Yavneh, I.: The vortices of homogeneous geostrophic turbulence. J. Fluid Mech. **401**, 1–26 (1999)
11. Miyazaki, T., Sato, T., Takahashi, N.: Maximum entropy states of quasi-geostrophic point vortices. Phys. Fluids **24**, 056601 (15 pp.) (2012)
12. Montgomery, D., Joyce, G.: Statistical mechanics of "negative temperature" states. Phys. Fluids **17**, 1139–1145 (1974)
13. Montgomery, D., Stribling, W.T., Martinez, D., Oughton, S.: Relaxation in two-dimensions and the "sinh-Poisson" equation. Phys. Fluids A **4**, 3–6 (1992)
14. Onsager, L.: Statistical hydrodynamics. Nuovo Cimento Suppl. **6**, 279–287 (1949)
15. Pedlosky, J.: *Geophysical Fluid Dynamics*, p. 624. Springer, New York (1979)
16. Salmon, R.: *Lectures on Geophysical Fluid Dynamics*, p. 378. Oxford University Press, New York (1998)
17. Turkington, B., Whitaker, N.: Statistical equilibrium computations of coherent structures in turbulent shear layers. SIAM J. Sci. Comput. **17**, 1414–1433 (1996)
18. Venaille, A., Bouchet, F.: Solvable phase diagrams and ensemble inequivalence for two-dimensional and geophysical turbulent flows. J. Stat. Phys. **143**, 346–380 (2011)

Chapter 17
Heat Convection of Compressible Viscous Fluids. III

Takaaki Nishida, Mariarosaria Padula and Yoshiaki Teramoto

Abstract Heat convection problems of the compressible viscous fluids are considered. The initial and boundary value problem of the linearized system is solved globally in time and is examined about the comparison with its incompressible Oberbeck-Boussinesq limit.

Keywords Heat convection · Compressible viscous and heat-conducting fluids · Incompressible Oberbeck-Boussinesq model

17.1 Introduction

This article is the third of a series of papers on mathematical aspects of perturbations to an equilibrium configuration S_e of a heat conducting compressible viscous fluid in a horizontal rigid layer. S_e is also known as Benard rest state. In our perspective the rest state is considered a parameter-dependent physical system, where there are several parameters: the Rayleigh and the Prandtl numbers, a typical length scale L and many other variables. In the first part we have proved the existence of steady motions around S_e, in the presence of small steady external forces by taking as independent variables the pressure, the velocity and the temperature. These variables have been chosen because they are more suitable to make a comparison with the more known

---

Mariarosaria Padula is deceased.

T. Nishida (✉)
Kyoto University, Yoshida Honmachi, Sakyo-ku, Kyoto, Japan
e-mail: tkknish@acs.i.kyoto-u.ac.jp

M. Padula
Depatment of Mathematics, Faculty of Engineering, University of Ferrara,
via Machiavelli 35, 44124 Ferrara, Italy
e-mail: pad@unife.it

Y. Teramoto
Setsunan University, Neyagawa, Osaka-prefecture, Japan
e-mail: teramoto@mpg.setsunan.ac.jp

Y. Shibata and Y. Suzuki (eds.), *Mathematical Fluid Dynamics, Present and Future*, Springer Proceedings in Mathematics & Statistics 183,
DOI 10.1007/978-4-431-56457-7_17

results for the simpler incompressible Oberbeck-Boussinesq model. The existence has been proved for Rayleigh number smaller than a critical value $\mathscr{R}_c(L)$. In the second part we have studied the stability and the bifurcation of flows close to S_e. Our aim has been the study of elementary local stationary bifurcation from $\mathscr{R}_c(L)$ with comparison to that of the well known Oberbeck-Boussinesq model.

In this paper we consider the initial boundary value problem for the linearized system with fixing the Prandtl number and all other variables confining our attention to study the non-stationary motions for Rayleigh number less than the critical Rayleigh number $\mathscr{R}_c(L)$, and letting the length scale L go to infinity.

The main results consist in the proof of existence of global in time linearized un-steady flows around S_e, and in the analysis of the limit flow as L goes to infinity. In particular it is proved that more thin the layer becomes more close to the Oberbeck-Boussinesq solution the non-stationary solutions become with a small correction term.

17.2 Formulation

Following Spiegel's dimensionalization [2] and [7] with the vertical axis (e_3) pointing downward, we consider the system in the horizontal domain $z_0 < z < z_0 + 1$, where

$$\mathscr{R}^2 = \frac{P^2 \beta R_* c_p (m+1)^3 d^{2m+3}}{g^2 \mu \kappa}, \quad z_0 = \frac{T_u}{\beta_0 d}, \quad \beta_0 = \frac{T_l - T_u}{d},$$

d is the width of the layer and $\mathscr{P}_r$ is the Prandtl number. The non-dimensional system for the heat convection problem of the compressible viscous and heat conducting fluids is the following [4, 5].

The mass conservation:

$$\frac{\partial \rho}{\partial t} + \mathscr{R} \nabla \cdot (\rho u) = 0,$$

The momentum conservation:

$$\frac{1}{\mathscr{P}_r} \rho \frac{\partial u}{\partial t} - \Delta u - \frac{1}{3} \nabla(\nabla \cdot u) + \frac{\mathscr{R}}{b\gamma(m+1)} \nabla p + \frac{\mathscr{R}}{\mathscr{P}_r} \rho u \cdot \nabla u = \frac{\mathscr{R}}{b\gamma} \rho e_3,$$

The energy conservation:

$$\rho \frac{\partial T}{\partial t} - \Delta T + \mathscr{R}(\gamma - 1) p \nabla \cdot u + \mathscr{R} \rho u \cdot \nabla T = \frac{2 g b \gamma}{\beta_0 c_v} \{ D : D - \frac{1}{3} (\nabla \cdot u)^2 \}.$$

The equilibrium solution in the horizontal strip domain $z_0 \leqslant z \leqslant z_0 + 1$ is the heat conduction state:

$$\overline{u} = 0\,, \quad \overline{\rho} = z^m\,, \quad \overline{T} = z\,, \quad \overline{p} = z^{m+1}\,,$$

where and hereafter we use the state equation of fluid $p = \rho\, T$.

We consider the non-stationary problem in $[\,0,\, +\infty\,) \times \Omega$, where

$$\Omega = \{0 \leqslant x \leqslant 2\pi/a\,,\ 0 \leqslant y \leqslant 2\pi/b\,,\ z_0 \leqslant z \leqslant z_0 + 1\,\}$$

with the periodic boundary condition with respect to x, y and the boundary conditions.

$$u\,(z_0) = u\,(z_0 + 1) = 0\,, \quad T\,(z_0) = z_0\,, \quad T\,(z_0 + 1) = z_0 + 1\,,$$

We use the unknown variables p_*, $u_* = (u,\, v,\, w)$, T_* for the above system and rewrite the system for the perturbation p, $u = (u,\, v,\, w)$, θ from the equilibrium state by

$$u_* \to u\,, \quad p_* \to z^{m+1} + p\,, \quad T_* \to z + \theta\,, \quad \rho_* \to z^m + \rho\,,$$

where $p = (z + \theta)\,\rho + z^m\,\theta$, i.e., $\rho = (\,p - z^m\,\theta)\,/\,(\,z + \theta\,)$. Then u and θ vanish on the horizontal boundaries and

$$\int \rho\; d\,\Omega = 0\,.$$

We notice that nonlinearities depend on the choice of the variables.

The mass conservation law has the form:

$$\frac{\partial}{\partial t}\left(\frac{p - z^m\,\theta}{z + \theta}\right) + \mathscr{R}\,\nabla \cdot \left(\frac{z^{m+1} + p}{z + \theta}\,u\right) = 0\,. \tag{17.1}$$

We rewrite it as follows:

$$\frac{1}{z + \theta}\,\frac{\partial\, p}{\partial\, t} - \frac{z^{m+1} + p}{(z + \theta)^2}\,\frac{\partial\, \theta}{\partial\, t} + \mathscr{R}\,\nabla \cdot (\,z^m\,u\,)$$
$$+ \mathscr{R}\,\nabla \cdot \left(\frac{p}{z + \theta}\,u\right) - \mathscr{R}\,\nabla \cdot \left(\frac{z^m\,\theta}{z + \theta}\,u\right) = 0\,. \tag{17.2}$$

If we multiply (17.2) by $(\,z + \theta\,)$, we have

$$\frac{\partial p}{\partial t} - \frac{z^{m+1} + p}{z + \theta} \frac{\partial \theta}{\partial t} + \mathscr{R} \nabla \cdot (z^{m+1} u) + \mathscr{R} \nabla \cdot (p u) - \mathscr{R} \left(\frac{z^{m+1} + p}{z + \theta} u \cdot \nabla (z + \theta) \right) = 0 . \quad (17.3)$$

The momentum equation is the following:

$$\frac{1}{\mathscr{P}_r} \frac{z^{m+1} + p}{z + \theta} \frac{\partial u}{\partial t} - \Delta u - \frac{1}{3} \nabla (\nabla \cdot u) + \frac{\mathscr{R}}{b\gamma(m+1)} \nabla p - \frac{\mathscr{R}}{b\gamma} \frac{p - z^m \theta}{z} e_3 + \frac{\mathscr{R}}{\mathscr{P}_r} \frac{z^{m+1} + p}{z + \theta} u \cdot \nabla u = 0 . \quad (17.4)$$

The energy equation is the following:

$$\frac{z^{m+1} + p}{z + \theta} \frac{\partial \theta}{\partial t} - \Delta \theta + \mathscr{R}(\gamma - 1)(z^{m+1} + p) \nabla \cdot u + \mathscr{R} \frac{z^{m+1} + p}{z + \theta} u \cdot \nabla (z + \theta) = \frac{2gb\gamma}{\beta_0 c_v} \left(D : D - \frac{1}{3} (\nabla \cdot u)^2 \right) .$$

The energy equation can be rewritten by the mass conservation (17.3) in the following form:

$$\gamma \left(\frac{z^{m+1} + p}{z + \theta} \right) \frac{\partial \theta}{\partial t} - (\gamma - 1) \frac{\partial p}{\partial t} - \Delta \theta + \mathscr{R} (1 - m(\gamma - 1)) z^m w - \mathscr{R} (\gamma - 1) u \cdot \nabla p = - \mathscr{R} \gamma \left(\frac{z^{m+1} + p}{z + \theta} \right) u \cdot \nabla \theta - \mathscr{R} \gamma \frac{p - z^m \theta}{z + \theta} w + \frac{2gb\gamma}{\beta_0 c_v} \left(D : D - \frac{1}{3} (\nabla \cdot u)^2 \right) \quad (17.5)$$

We are considering the system in the horizontal domain $z_0 \leqslant z \leqslant z_0 + 1$, and we use the following scale for the time, velocity and pressure, which is the same as [4] and [5], where $L = z_0 + \frac{1}{2}$.

$$t = L^m \tau , \quad u = \frac{\tilde{u}}{\sqrt{L}} , \quad p = L^{m-1} \tilde{p} , \quad \theta = \tilde{\theta} .$$

Then we have the following system in the horizontal strip $L - \frac{1}{2} \leqslant z \leqslant L + \frac{1}{2}$:

$$\frac{1}{L(z + \theta)} \frac{\partial \tilde{p}}{\partial \tau} - \left(\frac{1}{L} \left(\frac{z}{L} \right)^{m-1} \frac{z^2}{(z + \theta)^2} + \frac{\tilde{p}}{L(z + \theta)^2} \right) \frac{\partial \theta}{\partial \tau} + \mathscr{R}_m \nabla \cdot \left(\left(\frac{z}{L} \right)^m \tilde{u} \right) + \mathscr{R}_m \nabla \cdot \left(\frac{\tilde{p}}{L(z + \theta)} \tilde{u} \right) - \mathscr{R}_m \nabla \cdot \left(\frac{\theta}{z + \theta} \left(\frac{z}{L} \right)^m \tilde{u} \right) = 0,$$

$$\frac{1}{\mathscr{P}_r}\left((\frac{z}{L})^m\frac{z}{z+\theta}+\frac{\tilde{p}}{L(z+\theta)}\right)\frac{\partial\tilde{u}}{\partial\tau}-\Delta\tilde{u}$$
$$-\frac{1}{3}\nabla(\nabla\cdot\tilde{u})+\frac{\mathscr{R}_m}{b\gamma(m+1)}\nabla\tilde{p}-\frac{\mathscr{R}_m}{b\gamma}\frac{\tilde{p}}{z}e_3+\frac{\mathscr{R}_m}{b\gamma}(\frac{z}{L})^{m-1}\theta\,e_3$$
$$=-\frac{\mathscr{R}_m}{\mathscr{P}_r}\left((\frac{z}{L})^m\frac{z}{z+\theta}+\frac{\tilde{p}}{L(z+\theta)}\right)\tilde{u}\cdot\nabla\tilde{u}\,,$$
$$\frac{\gamma}{\gamma-1}\left((\frac{z}{L})^m\frac{z}{z+\theta}+\frac{\tilde{p}}{L(z+\theta)}\right)\frac{\partial\theta}{\partial\tau}-\frac{1}{L}\frac{\partial\tilde{p}}{\partial\tau}$$
$$-\frac{1}{\gamma-1}\Delta\theta+\mathscr{R}_m\frac{(1-m(\gamma-1))}{\gamma-1}(\frac{z}{L})^m\tilde{w}$$
$$=-\mathscr{R}_m\frac{\gamma}{\gamma-1}\left((\frac{z}{L})^m\frac{z}{z+\theta}+\frac{\tilde{p}}{L(z+\theta)}\right)\tilde{u}\cdot\nabla\theta$$
$$-\mathscr{R}_m\frac{\gamma}{\gamma-1}\left(\frac{\tilde{p}}{L(z+\theta)}-(\frac{z}{L})^m\frac{\theta}{z+\theta}\right)\tilde{w}$$
$$-\frac{\mathscr{R}_m}{L}\tilde{u}\cdot\nabla\tilde{p}+\frac{2gb\gamma}{L\beta_0c_v(\gamma-1)}\left(\tilde{D}:\tilde{D}-\frac{1}{3}(\nabla\cdot\tilde{u})^2\right)\,,$$

where we are using the Rayleigh number which was pointed out by [7]

$$\mathscr{R}_m=\mathscr{R}\,L^{m-\frac{1}{2}}\,.$$

If we neglect those terms of $\frac{1}{L}\times$ nonliner, we have the following system.

$$\frac{1}{Lz}\frac{\partial\tilde{p}}{\partial\tau}-\frac{1}{L}(\frac{z}{L})^{m-1}\frac{\partial\theta}{\partial\tau}+\mathscr{R}_m\nabla\cdot\left((\frac{z}{L})^m\tilde{u}\right)=0\,,$$

$$\frac{1}{\mathscr{P}_r}(\frac{z}{L})^m\frac{\partial\tilde{u}}{\partial\tau}-\Delta\tilde{u}-\frac{1}{3}\nabla(\nabla\cdot\tilde{u})+\frac{\mathscr{R}_m}{b\gamma(m+1)}\nabla\tilde{p}$$
$$-\frac{\mathscr{R}_m}{b\gamma}\frac{\tilde{p}}{z}e_3+\frac{\mathscr{R}_m}{b\gamma}(\frac{z}{L})^{m-1}\theta\,e_3=-\frac{\mathscr{R}_m}{\mathscr{P}_r}(\frac{z}{L})^m\tilde{u}\cdot\nabla\tilde{u}\,,$$

$$\frac{\gamma}{\gamma-1}(\frac{z}{L})^m\frac{\partial\theta}{\partial\tau}-\frac{1}{L}\frac{\partial\tilde{p}}{\partial\tau}-\frac{1}{\gamma-1}\Delta\theta$$
$$+\mathscr{R}_m\frac{(1-m(\gamma-1))}{\gamma-1}(\frac{z}{L})^m\tilde{w}=-\mathscr{R}_m\frac{\gamma}{\gamma-1}(\frac{z}{L})^m\tilde{u}\cdot\nabla\theta.$$

This system has the same nonlinear terms as those of the Oberbeck-Boussinesq system as $L\to+\infty$.

17.3 Linearized System

Thus we have the following linearized system that we want to solve here, where tilde is omitted and τ is replaced by t.

$$\frac{1}{L z}\frac{\partial p}{\partial t} - \frac{1}{L}(\frac{z}{L})^{m-1}\frac{\partial \theta}{\partial t} + \mathscr{R}_m \nabla \cdot \Big((\frac{z}{L})^m u\Big) = 0, \qquad (17.6)$$

$$\frac{b\gamma(m+1)}{\mathscr{P}_r}(\frac{z}{L})^m \frac{\partial u}{\partial t} - b\gamma(m+1)\Big(\Delta u + \frac{1}{3}\nabla(\nabla\cdot u)\Big)$$
$$+ \mathscr{R}_m \nabla p - \mathscr{R}_m (m+1)\frac{p}{z} e_3 + \mathscr{R}_m (m+1)(\frac{z}{L})^{m-1}\theta e_3 = 0, \quad (17.7)$$

$$\frac{\gamma}{\gamma-1}(\frac{z}{L})^m \frac{\partial \theta}{\partial t} - \frac{1}{L}\frac{\partial p}{\partial t} - \frac{1}{\gamma-1}\Delta\theta$$
$$+ \mathscr{R}_m \frac{1-m(\gamma-1)}{\gamma-1}(\frac{z}{L})^m w = 0. \qquad (17.8)$$

It follows from the mass conservation (17.6) that the constraint of mass conservation $\int \rho \, d\Omega = 0$ for the linearized system is the following.

$$\int \frac{p}{z} d\Omega - \int (\frac{z}{L})^{m-1}\theta \, d\Omega = 0 \quad \text{for any } t \geqslant 0. \qquad (17.9)$$

We are going to consider the case that L tends to $+\infty$ and z/L tends to 1, we use the following rescale for the unknown variables:

$$(\frac{L}{z})^{m+1} p = \tilde{p}, \quad (\frac{z}{L})^m u = \tilde{u}, \quad (\frac{z}{L})^{m-1}\theta = \tilde{\theta}. \qquad (17.10)$$

If we notice the following identity in the Eq. (17.7)

$$\nabla p - (m+1)\frac{p}{z} e_3 = (\frac{z}{L})^{m+1}\nabla\big((\frac{L}{z})^{m+1} p\big),$$

the system can be rewritten in the following form, where we omit the tilde.

$$\frac{1}{L^2}(\frac{z}{L})^m \frac{\partial p}{\partial t} - \frac{1}{L}\frac{\partial \theta}{\partial t} + \mathscr{R}_m \nabla \cdot u = 0, \qquad (17.11)$$

$$\frac{b\,\gamma\,(m+1)}{\mathscr{P}_r}\,(\frac{L}{z})^{m+1}\,\frac{\partial\,u}{\partial\,t} \;-\; b\,\gamma\,(m+1)\,(\frac{L}{z})^{2m+1}\,\Big(\Delta\,u \;+\; \frac{1}{3}\,\nabla\,(\nabla\cdot u)$$
$$-\;\frac{2\,m}{z}\,\frac{\partial\,u}{\partial\,z} \;-\; \frac{m}{3\,z}\,(\nabla\cdot u)\,e_3 \;-\; \frac{m}{3\,z}\,\nabla\,w \;+\; \frac{m(m+1)}{z^2}\,(u \;+\; \frac{w}{3}\,e_3)\Big)$$
$$+\;\mathscr{R}_m\,\nabla\,p \;+\; \mathscr{R}_m\,(m+1)\,(\frac{L}{z})^{m+1}\,\theta\,e_3 \;=\; 0\,, \qquad (17.12)$$

$$\frac{\gamma}{\gamma-1}\,(\frac{L}{z})^{m}\,\frac{\partial\,\theta}{\partial\,t} \;-\; \frac{1}{L}\,\frac{\partial\,p}{\partial\,t} \;-\; \frac{1}{(\gamma-1)}\,(\frac{L}{z})^{2m}\,\Big(\Delta\,\theta \;-\; \frac{2(m-1)}{z}\,\frac{\partial\,\theta}{\partial\,z}$$
$$+\;\frac{(m-1)m}{z^2}\,\theta\Big) \;+\; \mathscr{R}_m\,\frac{1-m(\gamma-1)}{\gamma-1}\,(\frac{L}{z})^{m+1}\,w \;=\; 0\,. \qquad (17.13)$$

The constraint of mass conservation $\int\,\rho\,d\,\Omega \;=\; 0$ for this linearized system is the following.

$$\int\,\frac{1}{L}\,(\frac{z}{L})^{m}\,p\,d\,\Omega \;-\; \int\,\theta\,\,d\,\Omega \;=\; 0 \quad \text{for any } t \,\geqslant\, 0\,. \qquad (17.14)$$

This system with $\gamma \,>\, 1$ and $L \,\geqslant\, 1$ in the domain $L - \frac{1}{2} \,<\, z \,<\, L + \frac{1}{2}$ is similar to the symmetric hyperbolic-parabolic systems which are considered in Kawashima and Shizuta [3] and in Galdi and Padula [1].

Lemma 1 *We have the following energy estimates uniformly for* $L \,\geqslant\, L_0$ *and* $\mathscr{R}_m \,\leqslant\, \mathscr{R}_0$.

$$\frac{\partial}{\partial t}\int\Big(\frac{1}{L^2}\,(\frac{z}{L})^{m}\,\frac{p^2}{2} \;-\; \frac{p\,\theta}{L} \;+\; \frac{\gamma}{\gamma-1}\,(\frac{L}{z})^{m}\,\frac{\theta^2}{2} \;+\; \frac{b\,\gamma\,(m+1)}{\mathscr{P}_r}\,(\frac{L}{z})^{m+1}\,\frac{|\,u\,|^2}{2}\Big)\,d\,\Omega$$
$$+\;C_0\int\Big(b\,\gamma\,(m+1)\,(\frac{L}{z})^{2m+1}\,(\,|\,\nabla u\,|^2 \;+\; \frac{1}{3}\,|\,\nabla\cdot u\,|^2\,) \;+\; \frac{1}{\gamma-1}\,(\frac{L}{z})^{2m}\,|\,\nabla\,\theta\,|^2\Big)\,d\,\Omega$$
$$\leqslant\; 0\,. \qquad (17.15)$$

Proof Multiply (17.11) by p, (17.12) by u and (17.13) by θ, add them and integrate it in Ω:

$$\frac{\partial}{\partial t}\int\Big(\frac{1}{L^2}\,(\frac{z}{L})^{m}\,\frac{p^2}{2} \;-\; \frac{p\,\theta}{L} \;+\; \frac{\gamma}{\gamma-1}\,(\frac{L}{z})^{m}\,\frac{\theta^2}{2} \;+\; \frac{b\,\gamma\,(m+1)}{\mathscr{P}_r}\,(\frac{L}{z})^{m+1}\,\frac{|\,u\,|^2}{2}\Big)\,d\,\Omega$$
$$+\;\int\Big(b\,\gamma\,(m+1)\,(\frac{L}{z})^{2m+1}\,(\,|\,\nabla u\,|^2 \;+\; \frac{1}{3}|\,\nabla\cdot u\,|^2\,) \;+\; \frac{1}{\gamma-1}\,(\frac{L}{z})^{2m}\,|\,\nabla\,\theta\,|^2\Big)\,d\,\Omega$$
$$+\;\mathscr{R}_m\,\frac{\gamma}{\gamma-1}\int(\frac{L}{z})^{m+1}\,\theta\,w\,d\,\Omega \;-\; b\,\gamma\,(m+1)\int(\frac{L}{z})^{2m+1}\Big(\frac{1}{z}\,u\cdot\frac{\partial\,u}{\partial\,z}$$
$$+\;\frac{m+1}{3\,z}\,w\,\nabla\cdot u \;-\; \frac{m}{3\,z}\,u\cdot\nabla\,w \;+\; \frac{m(m+1)}{z^2}\,(|u|^2 \;+\; \frac{w^2}{3}\,)\Big)\,d\,\Omega$$
$$-\;\frac{1}{\gamma-1}\int(\frac{L}{z})^{2m}\,\Big(\frac{2}{z}\,\theta\,\frac{\partial\,\theta}{\partial\,z} \;-\; \frac{(m-1)\,m}{z^2}\,\theta^2\Big)\,d\,\Omega\,. \; (17.16)$$

Here we know that the integrand of the first term is positive definite with respect to p, θ, u for $\gamma \,>\, 1$. Since the second integral term is positive definite with respect to

$\nabla\theta,\ \nabla u$, the last three integrals can be absorbed in the second integral term for $L \geqslant L_1$ and for $\mathscr{R}_m \leqslant \mathscr{R}_1$. Time derivatives have the similar estimates.

The dissipation for the pressure can be obtained by considering Padula's auxiliary vector function Ψ and Bogovskii lemma.

$$\nabla\cdot\Psi = p - \int p\,d\Omega\,, \qquad \Psi|_{z=L\pm 0.5} = 0\,, \quad \Psi \text{ is periodic in } x,\ y.$$

$$||\nabla\Psi|| \leqslant C_1\,||p - \int p\,d\Omega\,||\,, \qquad ||\Psi|| \leqslant C_0\,||p - \int p\,d\Omega\,||\,. \tag{17.17}$$

Multiply the momentum equation (17.12) by Ψ and integrate it by integration by parts, then we have

Lemma 2 *We have the following energy estimates uniformly for* $L \geqslant L_1$ *and* $\mathscr{R}_m \leqslant \mathscr{R}_1$.

$$\begin{aligned}&\mathscr{R}_m\,||\,p - \int p\,d\Omega\,||\\ &\leqslant C\left(||\frac{\partial u}{\partial t}|| + ||\nabla u|| + \frac{1}{3}||\nabla\cdot u|| + \mathscr{R}_m\,||\theta||\right).\end{aligned} \tag{17.18}$$

If we combine the estimates of Lemmas 1 and 2 and those estimates for time derivatives, we can obtain the uniform estimate for $L \geqslant L_2$ and $\mathscr{R}_m \leqslant \mathscr{R}_2$ in $0 \leqslant t < \infty$.

$$\begin{aligned}\int_0^\infty\int\Big(&C_1\mathscr{R}_m\,(p - \int p\,d\Omega\,)^2 + b\gamma\,(m+1)\,(\frac{L}{z})^{2m+1}\,(\,|\nabla u|^2 + \frac{1}{3}\,|\nabla\cdot u|^2\,)\\ &+ \frac{1}{\gamma-1}\,(\frac{L}{z})^{2m}\,|\nabla\theta|^2\Big)\,d\Omega\,dt \leqslant C_2\,.\end{aligned} \tag{17.19}$$

Time derivatives have the similar estimates.

If we use this estimate for the dissipation and those estimates for time derivatives, we can take the limit of $L \to \infty$ in the system (17.11), (17.12), (17.13) and get the following theorem for the limit system.

Theorem

$$\mathscr{R}_m\,\nabla\cdot u = 0\,, \tag{17.20}$$

$$\frac{1}{\mathscr{P}_r}\frac{\partial u}{\partial t} - \Delta u - \frac{1}{3}\nabla(\nabla\cdot u) + \frac{\mathscr{R}_m}{b\gamma\,(m+1)}\nabla p + \frac{\mathscr{R}_m}{b\gamma}\,\theta\,e_3 = 0\,, \tag{17.21}$$

$$\gamma\,\frac{\partial\theta}{\partial t} - (\gamma-1)\,\frac{\partial}{\partial t}\int\theta\,d\Omega - \Delta\theta + \mathscr{R}_m\,(1 - m(\gamma-1))\,w = 0\,. \tag{17.22}$$

Proof The first and second terms in (17.11) decay on $0 < t < \infty$ as $L \to \infty$, because of the estimate (17.19) and those for the time derivatives. Thus we have the equation (17.20). The second equations (17.21) are obtained from (17.12) by the uniform estimates (17.19) for u. The second term of the third equation (17.22) comes from the second term of (17.13) and by the constraint for the mass conservation from (17.14) as follows.

$$\frac{1}{L}\frac{\partial p}{\partial t} = \frac{1}{L}\frac{\partial}{\partial t}\Big(p - \int p\, d\Omega\Big) + \frac{1}{L}\frac{\partial}{\partial t}\int p\, d\Omega$$

$$= \frac{1}{L}\frac{\partial}{\partial t}\Big(p - \int p\, d\Omega\Big) + \int \frac{\partial}{\partial t}\theta\, d\Omega + \frac{1}{L}\frac{\partial}{\partial t}\int \left(1 - \left(\frac{z}{L}\right)^m\right) p\, d\Omega\ .$$

The energy equation (17.22) can be rewritten also in the form:

$$\gamma\frac{\partial\theta}{\partial t} - \Delta\theta + \mathscr{R}_m(1 - m(\gamma-1))w$$
$$-(\gamma-1)\left(\int \Delta\theta - \mathscr{R}_m(1 - m(\gamma-1))\int w\right) = 0\ . \quad (17.23)$$

This gives a correction to the Oberbeck-Boussinesq system from the system of the compressible viscous and heat-conductive fluids. The pressure effect remains in the energy equation (17.22) or (17.23) as $L \to +\infty$. It is remarked by the asymptotic expansion in [6].

Acknowledgements T. Nishida is supported in part by JSPS Grant-in-Aid for Scientific Research (26400163).

References

1. Galdi, G.P., Padula, M.: A new approach to energy theory in the stability of fluid motion. Arch. Ration. Mech. Anal. **110**(3), 187–286 (1990)
2. Gough, D.O., Moore, D.R., Spiegel, E.A., Weiss, N.O.: Convective instability in a compressible atmosphere. II. Astrophys. J. **206**, 536–542 (1976)
3. Kawashima, S., Shizuta, Y.: On the normal form of the symmetrei hyperbolic-parabolic systems associated with the conservation laws. Tohoku Math. J. Second Ser. **40**(3), 449–464 (1988)
4. Nishida, T., Padula, M., Teramoto, Y.: Heat convection of compressible viscous fluids: I. J. Math. Fluid Mech. **15**, 525–536 (2012)
5. Nishida, T., Padula, M., Teramoto, Y.: Heat convection of compressible viscous fluids. II. J. Math. Fluid Mech. 689–700 (2013)
6. Sone, Y.: Equations derived from the compressible Navier-Stokes set when the Mach number and the temperature variation are small, Section 3.1.3 in Supplement (Version 5-02) to Kinetic Theory and Fluid Dynamics (2002, Birkhaeuser) and Sections 3.1.1, 3.2.4 in Supplement (Version 12-00) to Molecular Gas Dynamics (2007, Birkhaeuser), Kyoto University Research Information Repository (2013)
7. Spiegel, E.A.: Convective instability in a compressible atomosphere. I. Astrophys. J. **141** (1965)

Chapter 18
Error Estimates of a Stabilized Lagrange–Galerkin Scheme of Second-Order in Time for the Navier–Stokes Equations

Hirofumi Notsu and Masahisa Tabata

Abstract Error estimates with optimal convergence orders are proved for a stabilized Lagrange–Galerkin scheme of second-order in time for the Navier–Stokes equations. The scheme is a combination of Lagrange–Galerkin method and Brezzi–Pitkäranta's stabilization method. It maintains the advantages of both methods; (i) It is robust for convection-dominated problems and the system of linear equations to be solved is symmetric. (ii) Since the P1 finite element is employed for both velocity and pressure, the number of degrees of freedom is much smaller than that of other typical elements for the equations, e.g., P2/P1. Therefore, the scheme is efficient especially for three-dimensional problems. The second-order accuracy in time is realized by Adams-Bashforth's (two-step) method for the discretization of the material derivative along the trajectory of fluid particles. The theoretical convergence orders are recognized by two- and three-dimensional numerical results.

Keywords Error estimates · Stabilized Lagrange-Galerkin scheme · Second-order scheme · Navier-Strokes equations

18.1 Introduction

In this paper, a stabilized Lagrange–Galerkin scheme of second-order in time for the Navier–Stokes equations is presented and its stability and convergence with optimal error estimates are proved. It is a higher-order scheme in time of a stabilized

H. Notsu (✉)
Faculty of Mathematics and Physics, Kanazawa University,
Kakuma, Kanazawa 920-1192, Japan
e-mail: notsu@se.kanazawa-u.ac.jp

H. Notsu
Japan Science and Technology Agency (JST), PRESTO,
Kawaguchi, Saitama, Japan

M. Tabata
Department of Mathematics, Waseda University, 3-4-1, Ohkubo, Shinjuku,
Tokyo 169-8555, Japan
e-mail: tabata@waseda.jp

Y. Shibata and Y. Suzuki (eds.), *Mathematical Fluid Dynamics, Present and Future*, Springer Proceedings in Mathematics & Statistics 183,
DOI 10.1007/978-4-431-56457-7_18

Lagrange–Galerkin scheme of first-order in time for the Navier–Stokes equations proposed and analyzed in [14, 15, 17], and is a combination of a Lagrange–Galerkin (LG) method and Brezzi–Pitkäranta's stabilization method [5]. The second-order accuracy in time is realized by Adams-Bashforth's (two-step) method, which has been introduced for convection-diffusion equations in [9] and applied to the Navier–Stokes equations in [3].

The LG method is a combined finite element method with the method of characteristics, and has such common advantages that robustness for convection-dominated problems and symmetry of the resulting matrix. The well-known LG schemes for the Navier–Stokes equations have been mathematically studied in [1, 3, 18, 23], where error estimates have been proved in [18, 23] for the first-order schemes (in time), in [3] for the second-order scheme and in [1] for a first-order projection scheme. The schemes in these literature require a stable element, e.g., P2/P1 finite element [11], which leads to a large number of degrees of freedom (DOF).

A stabilized LG scheme of first-order in time for the Navier–Stokes equations has been proposed in [14, 15] in order to reduce the number of DOF. It is one of the earliest works of LG schemes combined with the so-called stabilization method. The conditional stability and convergence with optimal error estimates have been proved for the scheme in [17], where such a condition is not required for a corresponding scheme for the Oseen equations [16].

The scheme to be proposed and analyzed in this paper has second-order accuracy in time realized by Adams-Bashforth's method in addition to the advantages of the first-order stabilized LG scheme, i.e., robustness for convection-dominated problems, symmetry of the resulting matrix and the small number of DOF. The stability and convergence with optimal error estimates are proved for the velocity in the H^1-norm and the pressure in the L^2-norm (Theorem 1) and for the velocity in the L^2-norm (Theorem 2) under a condition of $\Delta t = O(h^{d/6})$, which is the same form as in [3] for a stable LG scheme of second-order in time for the Navier–Stokes equations.

This paper is organized as follows. Our stabilized LG scheme of second-order in time for the Navier–Stokes equations is presented in Sect. 18.2. The main results on the stability and convergence with optimal error estimates are stated in Sect. 18.3, and they are proved in Sect. 18.4. The theoretical convergence orders are recognized numerically by two- and three-dimensional computations in Sect. 18.5. The conclusions are given in Sect. 18.6. In the Appendix three lemmas used in Sect. 18.4 are proved.

18.2 A Lagrange–Galerkin Scheme of Second-Order in Time

We prepare the function spaces and the notation to be used throughout the paper. Let Ω be a bounded domain in $\mathbb{R}^d (d = 2, 3)$, $\Gamma \equiv \partial\Omega$ the boundary of Ω, and T a positive constant. For an integer $m \geqslant 0$ and a number $p \in [1, \infty]$ we use the Sobolev spaces $W^{m,p}(\Omega)$, $W_0^{1,\infty}(\Omega)$, $H^m(\Omega)(= W^{m,2}(\Omega))$, $H_0^1(\Omega)$ and $H^{-1}(\Omega)$. For any normed space X with norm $\|\cdot\|_X$, we define function spaces $C^m([0, T]; X)$ and

$H^m(0,T;X)$ consisting of X-valued functions in $C^m([0,T])$ and $H^m(0,T)$, respectively, and the superscript "0" is often omitted from $C^0([0,T];X)$. We use the same notation $(\cdot,\cdot)$ to represent the $L^2(\Omega)$ inner product for scalar-, vector- and matrix-valued functions. The dual pairing between X and the dual space X' is denoted by $\langle\cdot,\cdot\rangle$. The norms on $W^{m,p}(\Omega)^d$ and $H^m(\Omega)^d$ are simply denoted as

$$\|\cdot\|_{m,p} \equiv \|\cdot\|_{W^{m,p}(\Omega)^d}, \quad \|\cdot\|_m \equiv \|\cdot\|_{H^m(\Omega)^d} \ (= \|\cdot\|_{m,2})$$

and the notation $\|\cdot\|_m$ is employed not only for vector-valued functions but also for scalar-valued ones. We also denote the norm on $H^{-1}(\Omega)^d$ by $\|\cdot\|_{-1}$. $L_0^2(\Omega)$ is a subspace of $L^2(\Omega)$ defined by

$$L_0^2(\Omega) \equiv \{q \in L^2(\Omega);\ (q,1)=0\}.$$

We often omit $[0,T]$, Ω and/or d if there is no confusion, e.g., we shall write $C(L^\infty)$ in place of $C([0,T];L^\infty(\Omega)^d)$. For t_0 and $t_1 \in \mathbb{R}$ we introduce the function space

$$Z^m(t_0,t_1) \equiv \{v \in H^j(t_0,t_1;H^{m-j}(\Omega)^d);\ j=0,\dots,m,\ \|v\|_{Z^m(t_0,t_1)} < \infty\}$$

with the norm

$$\|v\|_{Z^m(t_0,t_1)} \equiv \left\{\sum_{j=0}^m \|v\|^2_{H^j(t_0,t_1;H^{m-j}(\Omega)^d)}\right\}^{1/2},$$

and set $Z^m \equiv Z^m(0,T)$. The abbreviation LHS means left-hand side.

We consider the Navier–Stokes problem; find $(u,p):\Omega\times(0,T)\to\mathbb{R}^d\times\mathbb{R}$ such that

$$\frac{Du}{Dt} - \nabla\cdot\left[2\nu D(u)\right] + \nabla p = f \quad \text{in } \Omega\times(0,T), \tag{18.1a}$$

$$\nabla\cdot u = 0 \quad \text{in } \Omega\times(0,T), \tag{18.1b}$$

$$u = 0 \quad \text{on } \Gamma\times(0,T), \tag{18.1c}$$

$$u = u^0 \quad \text{in } \Omega,\ \text{at } t=0, \tag{18.1d}$$

where u is the velocity, p is the pressure, $f:\ \Omega\times(0,T)\to\mathbb{R}^d$ is a given external force, $u^0:\ \Omega\to\mathbb{R}^d$ is a given initial velocity, $\nu>0$ is a viscosity, $D(u)$ is the strain-rate tensor defined by

$$D_{ij}(u) \equiv \frac{1}{2}\left(\frac{\partial u_i}{\partial x_j} + \frac{\partial u_j}{\partial x_i}\right), \quad i,j=1,\dots,d,$$

and D/Dt is the material derivative defined by

$$\frac{D}{Dt} \equiv \frac{\partial}{\partial t} + u\cdot\nabla.$$

Letting $V \equiv H_0^1(\Omega)^d$ and $Q \equiv L_0^2(\Omega)$, we define the bilinear forms a on $V \times V$, b on $V \times Q$ and $\mathscr{A}$ on $(V \times Q) \times (V \times Q)$ by

$$a(u, v) \equiv 2\nu\big(D(u), D(v)\big), \quad b(v, q) \equiv -(\nabla \cdot v, q),$$
$$\mathscr{A}\big((u, p), (v, q)\big) \equiv a(u, v) + b(v, p) + b(u, q),$$

respectively. Then, we can write the weak formulation of (18.1) as follows; find $(u, p) : (0, T) \to V \times Q$ such that, for $t \in (0, T)$,

$$\Big(\frac{Du}{Dt}(t), v\Big) + \mathscr{A}\big((u, p)(t), (v, q)\big) = (f(t), v), \quad \forall (v, q) \in V \times Q, \tag{18.2}$$

with $u(0) = u^0$.

Let Δt be a time increment and $t^n \equiv n\Delta t$ for $n \in \mathbb{N} \cup \{0\}$. For a function g defined in $\Omega \times (0, T)$ we denote generally $g(\cdot, t^n)$ by g^n. Let $g^{(n-1)*}$ be a second-order approximate function of g^n defined by

$$g^{(n-1)*} \equiv 2g^{n-1} - g^{n-2}.$$

Let $X : (0, T) \to \mathbb{R}^d$ be a solution of the system of ordinary differential equations,

$$\frac{dX}{dt} = u(X, t). \tag{18.3}$$

Then, it holds that

$$\frac{Du}{Dt}(X(t), t) = \frac{d}{dt}u\big(X(t), t\big),$$

when u is smooth. Let $X(\cdot; x, t^n)$ be the solution of (18.3) subject to an initial condition $X(t^n) = x$. For a velocity $w : \Omega \to \mathbb{R}^d$ let $X_1(w, \Delta t) : \Omega \to \mathbb{R}^d$ be a mapping defined by

$$X_1(w, \Delta t)(x) \equiv x - w(x)\Delta t. \tag{18.4}$$

Since the positions $X_1(u^{(n-1)*}, \Delta t)(x)$ and $X_1(u^{(n-1)*}, 2\Delta t)(x)$ are approximations of $X(t^{n-1}; x, t^n)$ and $X(t^{n-2}; x, t^n)$ for $n \geqslant 2$, respectively, we can consider a second order approximation of the material derivative at (x, t^n),

$$\begin{aligned}\frac{Du}{Dt}(x, t^n) &= \frac{d}{dt}u\big(X(t; x, t^n), t\big)\Big|_{t=t^n}\\ &= \frac{3u^n\big(X(t^n; x, t^n)\big) - 4u^{n-1}\big(X(t^{n-1}; x, t^n)\big) + u^{n-2}\big(X(t^{n-2}; x, t^n)\big)}{2\Delta t} + O(\Delta t^2)\\ &= \frac{3u^n - 4u^{n-1} \circ X_1(u^{(n-1)*}, \Delta t) + u^{n-2} \circ X_1(u^{(n-1)*}, 2\Delta t)}{2\Delta t}(x) + O(\Delta t^2),\end{aligned}$$

where the symbol $\circ$ stands for the composition of functions,

$$(v \circ w)(x) \equiv v\big(w(x)\big),$$

for $v : \Omega \to \mathbb{R}^d$ and $w : \Omega \to \Omega$. $X_1(w, \Delta t)(x)$ is called an upwind point of x with respect to the velocity w and the time increment Δt. The next proposition gives a sufficient condition to guarantee that all upwind points by $X_1(w, \Delta t)$ are in Ω.

Proposition 1 *([20, Proposition 1]) Let $w \in W_0^{1,\infty}(\Omega)^d$ be a given function, and assume that*

$$\Delta t \|w\|_{1,\infty} < 1.$$

Then, it holds that

$$X_1(w, \Delta t)(\Omega) = \Omega.$$

For the sake of simplicity we assume that Ω is a polygonal ($d = 2$) or polyhedral ($d = 3$) domain. Let $\mathscr{T}_h = \{K\}$ be a triangulation of $\bar{\Omega}$ $(= \bigcup_{K \in \mathscr{T}_h} K)$, h_K a diameter of $K \in \mathscr{T}_h$, and $h \equiv \max_{K \in \mathscr{T}_h} h_K$ the maximum element size. Throughout this paper we consider a regular family of triangulations $\{\mathscr{T}_h\}_{h \downarrow 0}$ satisfying the inverse assumption [6], i.e., there exists a positive constant α_0 independent of h such that

$$\frac{h}{h_K} \leqslant \alpha_0, \quad \forall K \in \mathscr{T}_h, \quad \forall h. \tag{18.5}$$

We define the function spaces X_h, M_h, V_h and Q_h by

$$X_h \equiv \{v_h \in C(\bar{\Omega})^d;\ v_{h|K} \in P_1(K)^d,\ \forall K \in \mathscr{T}_h\},$$
$$M_h \equiv \{q_h \in C(\bar{\Omega});\ q_{h|K} \in P_1(K),\ \forall K \in \mathscr{T}_h\},$$

$V_h \equiv X_h \cap V$ and $Q_h \equiv M_h \cap Q$, respectively, where $P_1(K)$ is the space of linear functions on $K \in \mathscr{T}_h$. Let $N_T \equiv \lfloor T/\Delta t \rfloor$ be a total number of time steps, δ_0 a small positive constant fixed arbitrarily and $(\cdot, \cdot)_K$ the $L^2(K)^d$ inner product. We define the bilinear forms $\mathscr{C}_h$ on $H^1(\Omega) \times H^1(\Omega)$ and $\mathscr{A}_h$ on $(V \times H^1(\Omega)) \times (V \times H^1(\Omega))$ by

$$\mathscr{C}_h(p, q) \equiv \delta_0 \sum_{K \in \mathscr{T}_h} h_K^2 (\nabla p, \nabla q)_K,$$
$$\mathscr{A}_h\big((u, p), (v, q)\big) \equiv a(u, v) + b(v, p) + b(u, q) - \mathscr{C}_h(p, q). \tag{18.6}$$

The bilinear form $\mathscr{C}_h$ corresponds to Brezzi–Pitkäranta's pressure-stabilization [5].

Suppose $f \in C([0, T]; L^2(\Omega)^d)$ and that approximate functions $u_h^i \in V_h$ of $u^i \in V$, $i = 0, 1$, are given. Our stabilized LG scheme for (18.1) is to find $\{(u_h^n, p_h^n)\}_{n=2}^{N_T} \subset V_h \times Q_h$ such that, for $n = 2, \ldots, N_T$,

$$\frac{1}{2\Delta t}\big(3u_h^n - 4u_h^{n-1}\circ X_1(u_h^{(n-1)*},\Delta t) + u_h^{n-2}\circ X_1(u_h^{(n-1)*},2\Delta t), v_h\big)$$
$$+\mathscr{A}_h\big((u_h^n,p_h^n),(v_h,q_h)\big) = (f^n, v_h), \quad \forall (v_h,q_h)\in V_h\times Q_h. \quad (18.7)$$

Remark 1 (i) For the P1/P1 finite element, while the conventional inf-sup condition [11],

$$\inf_{q_h\in Q_h}\sup_{v_h\in V_h}\frac{b(v_h,q_h)}{\|v_h\|_1\|q_h\|_0}\geqslant \beta^*,$$

does not hold, the bilinear form $\mathscr{A}_h$ satisfies a generalized version [10],

$$\inf_{(u_h,p_h)\in V_h\times Q_h}\ \sup_{(v_h,q_h)\in V_h\times Q_h}\frac{\mathscr{A}_h((u_h,p_h),(v_h,q_h))}{\|(u_h,p_h)\|_{V\times Q}\|(v_h,q_h)\|_{V\times Q}}\geqslant \gamma^*,$$

thanks to Brezzi–Pitkäranta's stabilization, where β^* and γ^* are positive constants independent of h.
(ii) If the inequality $2\Delta t\|u_h^{(n-1)*}\|_{1,\infty} < 1$ is satisfied for given u_h^{n-1} and $u_h^{n-2}\in V_h$, we have $X_1(u_h^{(n-1)*},\Delta t)(\Omega) = X_1(u_h^{(n-1)*},2\Delta t)(\Omega) = \Omega$ by Proposition 1. Then, there exists a unique solution $(u_h^n,p_h^n)\in V_h\times Q_h$ of (18.7), since the resulting matrix is invertible. The invertibility is obtained from the fact that $(u_h^n,p_h^n) = (0,0)$ when $u_h^{n-1} = u_h^{n-2} = f^n = 0$ since we have

$$\frac{3}{2\Delta t}\|u_h^n\|_0^2 + 2\nu\|D(u_h^n)\|_0^2 + \delta_0\sum_{K\in\mathscr{T}_h} h_K^2\|\nabla p_h^n\|_{L^2(K)^d}^2 = 0 \quad (18.8)$$

by substituting $(u_h^n, -p_h^n)\in V_h\times Q_h$ into (v_h,q_h) in (18.7).

18.3 Main Results

In this section we state the main results, conditional stability and optimal error estimates for the scheme (18.7), which are proved in Sect. 18.4.

We use the following norms and a seminorm, $\|\cdot\|_{V_h}\equiv\|\cdot\|_V\equiv\|\cdot\|_1$, $\|\cdot\|_{Q_h}\equiv\|\cdot\|_Q\equiv\|\cdot\|_0$,

$$\|u\|_{l^\infty(X)}\equiv\max_{n=0,\ldots,N_T}\|u^n\|_X,\quad \|u\|_{l^2(X)}\equiv\Big\{\Delta t\sum_{n=1}^{N_T}\|u^n\|_X^2\Big\}^{1/2},$$

$$|p|_h\equiv\Big\{\sum_{K\in\mathscr{T}_h}h_K^2(\nabla p,\nabla p)_K\Big\}^{1/2},$$

for $X = L^\infty(\Omega), L^2(\Omega)$ and $H^1(\Omega)$. Let $\overline{D}_{\Delta t}$ and $\overline{D}_{\Delta t}^{(2)}$ be backward difference operators defined by

$$\overline{D}_{\Delta t} u^n \equiv \frac{u^n - u^{n-1}}{\Delta t}, \quad \overline{D}_{\Delta t}^{(2)} u^n \equiv \frac{3u^n - 4u^{n-1} + u^{n-2}}{2\Delta t} = \frac{3}{2}\overline{D}_{\Delta t} u^n - \frac{1}{2}\overline{D}_{\Delta t} u^{n-1},$$

which correspond to the first-order backward difference formula (BDF1) and the second-order backward difference formula (BDF2), respectively.

Definition 1 (Stokes projection) For $(w, r) \in V \times Q$ we define the Stokes projection $(\hat{w}_h, \hat{r}_h) \in V_h \times Q_h$ of (w, r) by

$$\mathscr{A}_h\big((\hat{w}_h, \hat{r}_h), (v_h, q_h)\big) = \mathscr{A}\big((w, r), (v_h, q_h)\big), \quad \forall (v_h, q_h) \in V_h \times Q_h. \tag{18.9}$$

Remark 2 The Stokes projection is well-defined, since when $(w, r) = (0, 0)$, we have an equation corresponding to (18.8), $2\nu \|D(\hat{w}_h)\|_0^2 + \delta_0 \sum_{K \in \mathscr{T}_h} h_K^2 \|\nabla \hat{r}_h\|_{L^2(K)^d}^2 = 0$, which derives $(\hat{w}_h, \hat{r}_h) = (0, 0)$.

Hypothesis 1 The solution (u, p) of (18.2) satisfies $u \in C([0, T]; V \cap W^{1,\infty}(\Omega)^d) \cap Z^3$ and $p \in H^1(0, T; Q \cap H^1(\Omega))$.

Remark 3 (i) It is known that the regularity assumption such as Hypothesis 1 leads to the "nonlocal compatibility condition" on the given data [12, 13]. Here we do not touch on the behavior of the solution close to the initial time and consider only the ideal case.

(ii) The regularity assumptions for the velocity, $u \in C(W^{1,\infty})$ and $u \in Z^3$, are used in the condition (18.26) and the estimate (18.29a) below, respectively, which imply that the mappings $X_1(u^{(n-1)*}, \Delta t)$ and $X_1(u^{(n-1)*}, 2\Delta t)$ are bijective and that scheme (18.7) has second-order accuracy in time for the material derivative.

Let $(\hat{u}_h, \hat{p}_h)(t) \in V_h \times Q_h$ be the Stokes projection of $(u, p)(t)$ by (18.9) for $t \in [0, T]$, and let

$$e_h^n \equiv u_h^n - \hat{u}_h^n, \quad \varepsilon_h^n \equiv p_h^n - \hat{p}_h^n, \quad \eta(t) \equiv (u - \hat{u}_h)(t).$$

It is known that there exists $\delta_1 \in (0, 1)$ such that

$$J(x) \equiv \det \frac{\partial X_1(a, \Delta t)}{\partial x}(x) \geqslant \frac{1}{2}, \quad \forall x \in \Omega, \tag{18.10}$$

for any Δt and $a \in W_0^{1,\infty}(\Omega)^d$ satisfying $\Delta t \|a\|_{1,\infty} \leqslant \delta_1$, since $J_{ij} = \delta_{ij} - \Delta t \partial a_i / \partial x_j$, $i, j = 1, \ldots, d$. Hereafter, let δ_1 be the constant above. In [24, Lemma 5.7] the inequality (18.10) is proved under the condition $\Delta t |a|_{1,\infty} \leqslant 1/4$, which implies that δ_1 can be taken as $\delta_1 = 1/4$.

Hypothesis 2 $(u_h^i, p_h^i) \in V_h \times Q_h$, $i = 0, 1$, satisfy

$$\|u_h^0\|_{0,\infty},\ \|u_h^1\|_{0,\infty},\ \|u_h^{1*}\|_{0,\infty} \leqslant \|u\|_{C(L^\infty)} + 1, \qquad (18.11a)$$

$$\Delta t\|u_h^0\|_{1,\infty},\ \Delta t\|u_h^1\|_{1,\infty},\ 2\Delta t\|u_h^{1*}\|_{1,\infty} \leqslant \delta_1, \qquad (18.11b)$$

$$b(u_h^1, q_h) - \mathscr{C}_h(p_h^1, q_h) = 0, \qquad \forall q_h \in Q_h, \quad (18.11c)$$

$$\sum_{i=0,1}\Big(\sqrt{\nu}\|D(e_h^i)\|_0 + \sqrt{\frac{\delta_0}{2}}|\varepsilon_h^i|_h\Big) + \sqrt{\frac{\Delta t}{8}}\|\overline{D}_{\Delta t}e_h^1\|_0 + \sqrt{\Delta t}\|\varepsilon_h^1\|_0 \leqslant c_I(\Delta t^2 + h), \qquad (18.11d)$$

where c_I is a positive constant independent of h and Δt.

Remark 4 Hypothesis 2 is satisfied by, e.g., (u_h^0, p_h^0) and $(u_h^1, p_h^1) \in V_h \times Q_h$ prepared by the Stokes projection (18.9) of $(u^0, 0)$ and the stabilized LG scheme of first-order in time, cf. (18.47), respectively.

Theorem 1 *Suppose Hypotheses 1 and 2 hold. Then, there exist positive constants h_0 and c_0 such that for any pair $(h, \Delta t)$ with*

$$h \in (0, h_0], \quad \Delta t \leqslant c_0 h^{d/6}, \qquad (18.12)$$

the following hold.

(i) Scheme (18.7) *has a unique solution $(u_h, p_h) = \{(u_h^n, p_h^n)\}_{n=2}^{N_T} \subset V_h \times Q_h$.*

(ii) There exists a positive constant $\bar{c}$ independent of h and Δt such that

$$\|u_h - u\|_{l^\infty(H^1)},\ \|p_h - p\|_{l^2(L^2)} \leqslant \bar{c}(\Delta t^2 + h). \qquad (18.13)$$

Hypothesis 3 The Stokes problem is regular, i.e., for any $g \in L^2(\Omega)^d$ the solution $(w, r) \in V \times Q$ of the Stokes problem,

$$\mathscr{A}\big((w, r), (v, q)\big) = (g, v), \quad \forall (v, q) \in V \times Q,$$

belongs to $H^2(\Omega)^d \times H^1(\Omega)$ and the estimate

$$\|w\|_2 + \|r\|_1 \leqslant c_R\|g\|_0$$

holds, where c_R is a positive constant independent of g, w and r.

Remark 5 Hypothesis 3 holds, e.g., if Ω is convex in $\mathbb{R}^2$, cf. [11].

Hypothesis 4 $u_h^i \in V_h$, $i = 0, 1$, satisfy

$$\sum_{i=0,1} \|e_h^i\|_0 \leqslant \tilde{c}_I(\Delta t^2 + h^2), \tag{18.14}$$

where $\tilde{c}_I$ is a positive constant independent of h and Δt.

Theorem 2 *Suppose Hypotheses 1–4 hold. Then, there exists a positive constant* $\tilde{c}_0$ *such that for any pair* $(h, \Delta t)$ *with*

$$h \in (0, h_0], \quad \Delta t \leqslant \tilde{c}_0 h^{d/6}, \tag{18.15}$$

the estimate

$$\|u_h - u\|_{l^\infty(L^2)} \leqslant \tilde{c}(\Delta t^2 + h^2), \tag{18.16}$$

holds, where h_0 *is the constant in Theorem 1,* $u_h = \{u_h^n\}_{n=2}^{N_T} \subset V_h$ *is the first component of the solution of scheme* (18.7), *and* $\tilde{c}$ *is a positive constant independent of* h *and* Δt.

18.4 Proofs of Theorems 1 and 2

We use c, c_u and $c_{(u,p)}$ to represent the generic positive constants independent of the discretization parameters h and Δt. c_u and $c_{(u,p)}$ are constants depending on u and (u, p), respectively. The symbol "′ (prime)" is sometimes used in order to distinguish between two constants, e.g., c_u and c_u'.

18.4.1 Preparations

We recall some lemmas and a proposition, which are directly used in our proofs. The next lemma is derived from Korn's inequality [8].

Lemma 1 *Let* Ω *be a bounded domain with a Lipschitz-continuous boundary. Then, there exists a positive constant* α_1 *and the following inequalities hold.*

$$\|D(v)\|_0 \leqslant \|v\|_1 \leqslant \alpha_1 \|D(v)\|_0, \qquad \forall v \in H_0^1(\Omega)^d. \tag{18.17}$$

We use inverse inequalities and interpolation properties.

Lemma 2 *([6]) There exist positive constants α_{2i}, $i = 0, \ldots, 4$, independent of h and the following inequalities hold.*

$$
\begin{aligned}
|q_h|_h &\leqslant \alpha_{20}\|q_h\|_0, & \forall q_h &\in Q_h, && (18.18a)\\
\|v_h\|_{0,\infty} &\leqslant \alpha_{21}h^{-d/6}\|v_h\|_1, & \forall v_h &\in V_h, && (18.18b)\\
\|v_h\|_{1,\infty} &\leqslant \alpha_{22}h^{-d/2}\|v_h\|_1, & \forall v_h &\in V_h, && (18.18c)\\
\|\Pi_h v\|_{0,\infty} &\leqslant \|v\|_{0,\infty}, & \forall v &\in C(\bar{\Omega})^d, && (18.18d)\\
\|\Pi_h v\|_{1,\infty} &\leqslant \alpha_{23}\|v\|_{1,\infty}, & \forall v &\in W^{1,\infty}(\Omega)^d, && (18.18e)\\
\|\Pi_h v - v\|_1 &\leqslant \alpha_{24}h\|v\|_2, & \forall v &\in H^2(\Omega)^d, && (18.18f)
\end{aligned}
$$

where $\Pi_h : C(\bar{\Omega})^d \to X_h$ is the Lagrange *interpolation operator.*

Remark 6 (i) Although (18.18b) is not optimal for $d = 2$, it is sufficient in this paper. (ii) The inequality $\alpha_{23} \geqslant 1$ holds.

Lemma 3 *([10, Lemma 3.2]) There exists a positive constant α_{30} independent of h such that for any h*

$$\inf_{(w_h,r_h)\in V_h\times Q_h} \sup_{(v_h,q_h)\in V_h\times Q_h} \frac{\mathscr{A}_h\big((w_h, r_h), (v_h, q_h)\big)}{\|(w_h, r_h)\|_{V\times Q}\|(v_h, q_h)\|_{V\times Q}} \geqslant \alpha_{30}. \tag{18.19}$$

Remark 7 The stability inequality (18.19) holds for the P1/P1 finite element spaces, while we do not have the conventional inf-sup condition [11] for the spaces.

Proposition 2 *([4]) (i) Suppose $(w, r) \in (V \cap H^2(\Omega)^d) \times (Q \cap H^1(\Omega))$. Then, there exists a positive constant α_{31} independent of h such that for any h the* Stokes *projection $(\hat{w}_h, \hat{r}_h)$ of (w, r) by (18.9) satisfies*

$$\|\hat{w}_h - w\|_1,\ \ \|\hat{r}_h - r\|_0,\ \ |\hat{r}_h - r|_h \leqslant \alpha_{31}h\|(w, r)\|_{H^2\times H^1}. \tag{18.20a}$$

(ii) Suppose Hypothesis 3 additionally holds. Then, there exists a positive constant α_{32} independent of h such that for any h

$$\|\hat{w}_h - w\|_0 \leqslant \alpha_{32}h^2\|(w, r)\|_{H^2\times H^1}. \tag{18.20b}$$

For the evaluation of composite functions we recall Lemma 4 of [17], which is mainly due to Lemma 4.5 in [1] and Lemma 1 in [7]. In the next lemma a and b are any functions in $W_0^{1,\infty}(\Omega)^d$ satisfying

$$\Delta t\|a\|_{1,\infty},\ \ \Delta t\|b\|_{1,\infty} \leqslant \delta_1.$$

Lemma 4 *([17, Lemma 4]) There exist positive constants α_{4i}, $i = 0, \ldots, 3$, independent of Δt such that the following inequalities hold.*

$$\|g - g \circ X_1(a, \Delta t)\|_0 \leqslant \alpha_{40} \Delta t \|a\|_{0,\infty} \|g\|_1, \qquad \forall g \in H^1(\Omega)^d, \tag{18.21a}$$

$$\|g - g \circ X_1(a, \Delta t)\|_{-1} \leqslant \alpha_{41} \Delta t \|a\|_{1,\infty} \|g\|_0, \qquad \forall g \in L^2(\Omega)^d, \tag{18.21b}$$

$$\|g \circ X_1(b, \Delta t) - g \circ X_1(a, \Delta t)\|_0 \leqslant \alpha_{42} \Delta t \|b - a\|_0 \|g\|_{1,\infty}, \quad \forall g \in W^{1,\infty}(\Omega)^d, \tag{18.21c}$$

$$\|g \circ X_1(b, \Delta t) - g \circ X_1(a, \Delta t)\|_{0,1} \leqslant \alpha_{43} \Delta t \|b - a\|_0 \|g\|_1, \qquad \forall g \in H^1(\Omega)^d. \tag{18.21d}$$

At the end of this subsection we prepare a lemma on a discrete Gronwall's inequality, which is proved in Appendix.

Lemma 5 *Let a_0 be a non-negative number, and $\Delta t \in (0, 1/(2a_0)]$ a number. Let $\{x_n\}_{n \geqslant 0}$, $\{y_n\}_{n \geqslant 1}$, $\{z_n\}_{n \geqslant 2}$ and $\{b_n\}_{n \geqslant 2}$ be non-negative sequences. Suppose that*

$$\frac{1}{\Delta t}\Big(\frac{3}{2}x_n - 2x_{n-1} + \frac{1}{2}x_{n-2} + y_n - y_{n-1}\Big) + z_n \leqslant a_0(x_{n-1} + x_{n-2}) + b_n, \quad \forall n \geqslant 2, \tag{18.22}$$

holds. Then, it holds that

$$x_n + \frac{2}{3}y_n + \frac{2}{3}\Delta t \sum_{i=2}^{n} z_i \leqslant \frac{3}{2}\exp(2a_0 n \Delta t)\Big(x_1 + \frac{2}{3}y_1 + \frac{2}{3}\Delta t \sum_{i=2}^{n} b_i\Big), \quad \forall n \geqslant 2. \tag{18.23}$$

18.4.2 An Estimate at Each Time Step

We have that for $n \geqslant 2$

$$\big(\overline{D}_{\Delta t}^{(2)} e_h^n, v_h\big) + \mathscr{A}_h\big((e_h^n, \varepsilon_h^n), (v_h, q_h)\big) = \langle R_h^n, v_h\rangle, \quad \forall (v_h, q_h) \in V_h \times Q_h, \tag{18.24}$$

where

$$R_h^n \equiv \sum_{i=1}^{4} R_{hi}^n,$$

$$R_{h1}^n \equiv \frac{Du^n}{Dt} - \frac{1}{2\Delta t}\{3u^n - 4u^{n-1} \circ X_1(u^{(n-1)*}, \Delta t) + u^{n-2} \circ X_1(u^{(n-1)*}, 2\Delta t)\},$$

$$R_{h2}^n \equiv \frac{1}{2\Delta t}\Big[\big\{-4u^{n-1}\circ X_1(u^{(n-1)*},\Delta t)+u^{n-2}\circ X_1(u^{(n-1)*},2\Delta t)\big\}$$
$$-\big\{-4u^{n-1}\circ X_1(u_h^{(n-1)*},\Delta t)+u^{n-2}\circ X_1(u_h^{(n-1)*},2\Delta t)\big\}\Big],$$
$$R_{h3}^n \equiv \frac{1}{2\Delta t}\big\{3\eta^n-4\eta^{n-1}\circ X_1(u_h^{(n-1)*},\Delta t)+\eta^{n-2}\circ X_1(u_h^{(n-1)*},2\Delta t)\big\},$$
$$R_{h4}^n \equiv \frac{1}{2\Delta t}\Big[\big\{-4e_h^{n-1}+e_h^{n-2}\big\}-\big\{-4e_h^{n-1}\circ X_1(u_h^{(n-1)*},\Delta t)+e_h^{n-2}\circ X_1(u_h^{(n-1)*},2\Delta t)\big\}\Big].$$

(18.24) is derived from (18.7), (18.9) and (18.2).

Proposition 3 *(i) Let $n \in \{2,\ldots,N_T\}$ be a fixed number and let u_h^{n-1} and $u_h^{n-2} \in V_h$ be known. Suppose the inequality*

$$2\Delta t\|u_h^{(n-1)*}\|_{1,\infty} \leqslant \delta_1 \tag{18.25$_n$}$$

holds. Then, there exists a unique solution $(u_h^n, p_h^n) \in V_h \times Q_h$ of (18.7).

(ii) Furthermore, suppose Hypothesis 1 and the inequality

$$6\Delta t\|u\|_{C(W^{1,\infty})} \leqslant \delta_1 \tag{18.26}$$

hold. Let $p_h^{n-1} \in Q_h$ be known and suppose the equation

$$b(u_h^{n-1}, q_h) - \mathscr{C}_h(p_h^{n-1}, q_h) = 0, \quad \forall q_h \in Q_h, \tag{18.27$_n$}$$

holds. Then, it holds that

$$\overline{D}_{\Delta t}\Big(\nu\|D(e_h^n)\|_0^2 + \frac{\delta_0}{2}|\varepsilon_h^n|_h^2\Big) + \frac{1}{2}\|\overline{D}_{\Delta t}e_h^n\|_0^2$$
$$\leqslant A_1(\|u_h^{(n-1)*}\|_{0,\infty})\nu\big(\|D(e_h^{n-1})\|_0^2 + \|D(e_h^{n-2})\|_0^2\big) + \frac{1}{8}\|\overline{D}_{\Delta t}e_h^{n-1}\|_0^2$$
$$+ A_2(\|u_h^{(n-1)*}\|_{0,\infty})\Big[\Delta t^3\|u\|_{Z^3(t^{n-2},t^n)}^2 + h^2\Big(\frac{1}{\Delta t}\|(u,p)\|_{H^1(t^{n-2},t^n;H^2\times H^1)}^2 + 1\Big)\Big], \tag{18.28$_n$}$$

where A_i, $i = 1, 2$, are functions defined by

$$A_i(\xi) \equiv c_i(\xi^2 + 1)$$

and c_i, $i = 1, 2$, are positive constants independent of h and Δt. They are defined by (18.35) *below.*

For the proof we use the next lemma, which is proved in Appendix.

Lemma 6 *Suppose Hypothesis 1 holds. Let $n \in \{2, \dots, N_T\}$ be a fixed number and let u_h^{n-1} and $u_h^{n-2} \in V_h$ be known. Then, under the conditions (18.25)$_n$ and* (18.26) *it holds that*

$$\|R_{h1}^n\|_0 \leqslant c_u \Delta t^{3/2} \|u\|_{Z^3(t^{n-2},t^n)}, \tag{18.29a}$$

$$\|R_{h2}^n\|_0 \leqslant c_{(u,p)}(\|e_h^{n-1}\|_0 + \|e_h^{n-2}\|_0 + h), \tag{18.29b}$$

$$\|R_{h3}^n\|_0 \leqslant \frac{ch}{\sqrt{\Delta t}}(\|u_h^{(n-1)*}\|_{0,\infty} + 1)\|(u,p)\|_{H^1(t^{n-2},t^n;H^2\times H^1)}, \tag{18.29c}$$

$$\|R_{h4}^n\|_0 \leqslant c\|u_h^{(n-1)*}\|_{0,\infty}(\|e_h^{n-1}\|_1 + \|e_h^{n-2}\|_1). \tag{18.29d}$$

Proof of Proposition 3. (i) is obtained from (18.25)$_n$ and Remark 1-(ii).

We prove (ii). Substituting $(\overline{D}_{\Delta t} e_h^n, 0)$ into (v_h, q_h) in (18.24) and using the inequalities $(ab \leqslant a^{2/4} + b^2$ and

$$\begin{aligned}
(\overline{D}_{\Delta t}^{(2)} e_h^n, \overline{D}_{\Delta t} e_h^n) &= \Big(\frac{3}{2}\overline{D}_{\Delta t} e_h^n - \frac{1}{2}\overline{D}_{\Delta t} e_h^{n-1}, \overline{D}_{\Delta t} e_h^n\Big) \\
&= \frac{3}{2}\|\overline{D}_{\Delta t} e_h^n\|_0^2 - \frac{1}{2}(\overline{D}_{\Delta t} e_h^{n-1}, \overline{D}_{\Delta t} e_h^n) \\
&\geqslant \frac{3}{2}\|\overline{D}_{\Delta t} e_h^n\|_0^2 - \frac{1}{2}\Big(\frac{1}{4}\|\overline{D}_{\Delta t} e_h^{n-1}\|_0^2 + \|\overline{D}_{\Delta t} e_h^n\|_0^2\Big) \\
&= \|\overline{D}_{\Delta t} e_h^n\|_0^2 - \frac{1}{8}\|\overline{D}_{\Delta t} e_h^{n-1}\|_0^2,
\end{aligned}$$

we have

$$\|\overline{D}_{\Delta t} e_h^n\|_0^2 - \frac{1}{8}\|\overline{D}_{\Delta t} e_h^{n-1}\|_0^2 + \overline{D}_{\Delta t}\big(\nu\|D(e_h^n)\|_0^2\big) + b(\overline{D}_{\Delta t} e_h^n, \varepsilon_h^n) \leqslant \sum_{i=1}^4 \langle R_{hi}^n, \overline{D}_{\Delta t} e_h^n\rangle. \tag{18.30}$$

Here, we have noted that both $X_1(u^{(n-1)*}, \Delta t)$ and $X_1(u^{(n-1)*}, 2\Delta t)$ in R_{hi}^n, $i = 1, 2$, map Ω onto Ω, since we have

$$\begin{aligned}
2\Delta t\|u^{(n-1)*}\|_{1,\infty} = 2\Delta t\|2u^{n-1} - u^{n-2}\|_{1,\infty} &\leqslant 2\Delta t(2\|u^{n-1}\|_{1,\infty} + \|u^{n-2}\|_{1,\infty}) \\
&\leqslant 6\Delta t\|u\|_{C(W^{1,\infty})} \leqslant \delta_1 \quad \text{(by (18.26))}.
\end{aligned}$$

From (18.27)$_n$ and (18.7) with $v_h = 0 \in V_h$ we have

$$b(u_h^k, q_h) - \mathscr{C}_h(p_h^k, q_h) = 0, \quad \forall q_h \in Q_h, \tag{18.31}$$

for $k = n-1$ and n. Since $(\hat{u}_h^n, \hat{p}_h^n)$ is the Stokes projection of (u^n, p^n) by (18.9), we have

$$b(\hat{u}_h^k, q_h) - \mathscr{C}_h(\hat{p}_h^k, q_h) = b(u^k, q_h) = 0, \quad \forall q_h \in Q_h, \tag{18.32}$$

for $k = n-1$ and n. The equalities (18.31) and (18.32) imply

$$b(\overline{D}_{\Delta t} e_h^n, q_h) - \mathscr{C}_h(\overline{D}_{\Delta t} \varepsilon_h^n, q_h) = 0, \quad \forall q_h \in Q_h,$$

which leads to

$$-b(\overline{D}_{\Delta t} e_h^n, \varepsilon_h^n) + \mathscr{C}_h(\overline{D}_{\Delta t} \varepsilon_h^n, \varepsilon_h^n) = 0 \tag{18.33}$$

by putting $q_h = -\varepsilon_h^n \in Q_h$. Adding (18.33) to (18.30) and using Lemma 6 and the inequality $ab \leqslant \beta a^2/2 + b^2/(2\beta)$ $(\beta > 0)$, we have

$$\begin{aligned}
&\|\overline{D}_{\Delta t} e_h^n\|_0^2 + \overline{D}_{\Delta t}\Big(\nu\|D(e_h^n)\|_0^2 + \frac{\delta_0}{2}|\varepsilon_h^n|_h^2\Big) \leqslant \sum_{i=1}^4 \langle R_{hi}^n, \overline{D}_{\Delta t} e_h^n\rangle + \frac{1}{8}\|\overline{D}_{\Delta t} e_h^{n-1}\|_0^2 \\
&\leqslant \Big(\sum_{i=1}^4 \beta_i\Big)\|\overline{D}_{\Delta t} e_h^n\|_0^2 + c_{(u,p)}\frac{\alpha_1^2}{\nu}\Big(\frac{1}{\beta_2} + \frac{\|u_h^{(n-1)*}\|_{0,\infty}^2}{\beta_4}\Big)\nu\big(\|D(e_h^{n-1})\|_0^2 + \|D(e_h^{n-2})\|_0^2\big) \\
&\quad + \frac{1}{8}\|\overline{D}_{\Delta t} e_h^{n-1}\|_0^2 + c'_{(u,p)}\Big[\frac{\Delta t^3}{\beta_1}\|u\|_{Z^3(t^{n-2},t^n)}^2 \\
&\qquad + h^2\Big(\frac{1}{\beta_2} + \frac{\|u_h^{(n-1)*}\|_{0,\infty}^2 + 1}{\beta_3 \Delta t}\|(u,p)\|_{H^1(t^{n-2},t^n;H^2\times H^1)}^2\Big)\Big]
\end{aligned} \tag{18.34}$$

for any positive numbers β_i, $i = 1, \ldots, 4$, where the inequality $\|e_h^{n-1}\|_0 \leqslant \|e_h^{n-1}\|_1$ has been used. By setting $\beta_i = 1/8$ for $i = 1, \ldots, 4$ in (18.34) we have

$$\begin{aligned}
&\overline{D}_{\Delta t}\Big(\nu\|D(e_h^n)\|_0^2 + \frac{\delta_0}{2}|\varepsilon_h^n|_h^2\Big) + \frac{1}{2}\|\overline{D}_{\Delta t} e_h^n\|_0^2 \\
&\leqslant \frac{c_{(u,p)}}{\nu}(\|u_h^{(n-1)*}\|_{0,\infty}^2 + 1)\nu\big(\|D(e_h^{n-1})\|_0^2 + \|D(e_h^{n-2})\|_0^2\big) + \frac{1}{8}\|\overline{D}_{\Delta t} e_h^{n-1}\|_0^2 \\
&\quad + c'_{(u,p)}(\|u_h^{(n-1)*}\|_{0,\infty}^2 + 1)\Big[\Delta t^3\|u\|_{Z^3(t^{n-2},t^n)}^2 + h^2\Big(\frac{1}{\Delta t}\|(u,p)\|_{H^1(t^{n-2},t^n;H^2\times H^1)}^2 + 1\Big)\Big].
\end{aligned}$$

Putting

$$c_1 \equiv c_{(u,p)}/\nu, \qquad c_2 \equiv c'_{(u,p)}, \tag{18.35}$$

we obtain $(18.28)_n$. □

18.4.3 Proof of Theorem 1

The proof is performed through three steps.

Step 1 (Setting c_0 and h_0): Let c_I and A_i, $i = 1, 2$, be the constant in Hypothesis 2 and the functions in Proposition 3, respectively. Let a_1, a_2 and c_* be constants defined by

$$
\begin{aligned}
a_1 &\equiv A_1(3\|u\|_{C(L^\infty)} + 1), \quad a_2 \equiv A_2(3\|u\|_{C(L^\infty)} + 1),\\
c_* &\equiv \frac{\alpha_1}{\sqrt{\nu}} \exp(a_1 T)\\
&\quad \times \max\Big\{\big[2(c_I^2 + a_2\|u\|_{Z^3}^2)\big]^{1/2}, \big[2c_I^2 + a_2(2\|(u,p)\|_{H^1(H^2\times H^1)}^2 + T)\big]^{1/2}\Big\}.
\end{aligned}
$$

We can choose sufficiently small positive constants c_0 and h_0 such that

$$
\begin{aligned}
&3\alpha_{21}\big\{c_*(c_0^2 h_0^{d/6} + h_0^{1-d/6}) + (\alpha_{24} + \alpha_{31})h_0^{1-d/6}\|(u,p)\|_{C(H^2\times H^1)}\big\}\\
&\qquad\qquad + c_0^2 h_0^{d/3}\|u\|_{C^2(L^\infty)} \leqslant 1, \qquad (18.36\text{a})\\
&6c_0\big[\alpha_{22}\big\{c_*(c_0^2 + h_0^{1-d/3}) + (\alpha_{24} + \alpha_{31})h_0^{1-d/3}\|(u,p)\|_{C(H^2\times H^1)}\big\}\\
&\qquad\qquad + \alpha_{23}h_0^{d/6}\|u\|_{C(W^{1,\infty})}\big] \leqslant \delta_1, \qquad (18.36\text{b})
\end{aligned}
$$

since all the powers of h_0 are positive in (18.36a) and non-negative in (18.36b). In the following we consider a pair $(h, \Delta t)$ satisfying (18.12) with c_0 and h_0 above.

For the induction in Step 2 we define a property P(n), $n \in \{2, \ldots, N_T\}$, by

$$
\begin{aligned}
\mathrm{P}(n):\ &\nu\|D(e_h^n)\|_0^2 + \frac{\delta_0}{2}|\varepsilon_h^n|_h^2 + \frac{3}{8}\Delta t\sum_{i=2}^{n-1}\|\overline{D}_{\Delta t}e_h^i\|_0^2 + \frac{1}{2}\Delta t\|\overline{D}_{\Delta t}e_h^n\|_0^2\\
&\leqslant \exp\{a_1(2n-3)\Delta t\}\Big[\sum_{i=0,1}\Big(\nu\|D(e_h^i)\|_0^2 + \frac{\delta_0}{2}|\varepsilon_h^i|_h^2\Big) + \frac{1}{8}\Delta t\|\overline{D}_{\Delta t}e_h^1\|_0^2\\
&+ a_2\Big\{\Delta t^4\sum_{i=2}^{n}\|u\|_{Z^3(t^{i-2},t^i)}^2 + h^2\Big(\sum_{i=2}^{n}\|(u,p)\|_{H^1(t^{i-2},t^i;H^2\times H^1)}^2 + (n-1)\Delta t\Big)\Big\}\Big],
\end{aligned}
$$

which can be rewritten as

$$
x_n + \frac{3}{4}\Delta t\sum_{i=2}^{n-1}y_i + \Delta t y_n \leqslant \exp\{a_1(2n-3)\Delta t\}\Big(X_0 + \Delta t\sum_{i=2}^{n}b_i\Big), \qquad (18.37)_n
$$

where

$$
\begin{aligned}
&x_n \equiv \nu\|D(e_h^n)\|_0^2 + \frac{\delta_0}{2}|\varepsilon_h^n|_h^2, \quad y_i \equiv \frac{1}{2}\|\overline{D}_{\Delta t}e_h^i\|_0^2, \quad X_0 \equiv x_0 + x_1 + \frac{1}{4}\Delta t y_1,\\
&b_i \equiv a_2\Big[\Delta t^3\|u\|_{Z^3(t^{i-2},t^i)}^2 + h^2\Big(\frac{1}{\Delta t}\|(u,p)\|_{H^1(t^{i-2},t^i;H^2\times H^1)}^2 + 1\Big)\Big].
\end{aligned}
$$

Let us note the following four facts.

(F1): The condition (18.26) is satisfied.

(F2): For any $k \in \{2, \ldots, N_T\}$, P(k) implies

$$\|e_h^k\|_1 \leqslant c_*(\Delta t^2 + h). \tag{18.38$_k$}$$

(F3): For any $k \in \{2, \ldots, N_T - 1\}$, (18.38)$_k$ and (18.38)$_{k-1}$ imply

$$\left\|u_h^{k*}\right\|_{0,\infty} \leqslant 3\|u\|_{C(L^\infty)} + 1, \qquad 2\Delta t \left\|u_h^{k*}\right\|_{1,\infty} \leqslant \delta_1, \tag{18.39$_k$}$$

and there exists a unique solution (u_h^{k+1}, p_h^{k+1}) of Eq. (18.7) with $n = k + 1$, where the latter inequality of (18.39)$_k$ is nothing but (18.25)$_{k+1}$.

(F4): For any $k \in \{2, \ldots, N_T\}$, under $\|u_h^{(k-1)*}\|_{0,\infty} \leqslant 3\|u\|_{C(L^\infty)} + 1$, the inequality (18.28)$_k$ implies

$$x_k + \Delta t y_k \leqslant (1 + a_1\Delta t)x_{k-1} + a_1\Delta t x_{k-2} + \frac{1}{4}\Delta t y_{k-1} + \Delta t b_k. \tag{18.40$_k$}$$

(F1) is derived from the estimate,

$$6\Delta t\|u\|_{C(W^{1,\infty})} \leqslant 6c_0 h_0^{d/6}\|u\|_{C(W^{1,\infty})} \leqslant 6c_0\alpha_{23}h_0^{d/6}\|u\|_{C(W^{1,\infty})} \leqslant \delta_1,$$

where (18.12), Remark 6-(ii) and (18.36b) are used.

(F2) is obtained from the estimate,

$$\begin{aligned}
&\nu\|D(e_h^k)\|_0^2 + \frac{\delta_0}{2}|\varepsilon_h^k|_h^2 + \frac{3}{8}\Delta t\sum_{i=2}^{k-1}\|\overline{D}_{\Delta t}e_h^i\|_0^2 + \frac{1}{2}\Delta t\|\overline{D}_{\Delta t}e_h^k\|_0^2 \\
&\leqslant \exp\bigl\{a_1(2k-3)\Delta t\bigr\}\Bigl[c_I^2(\Delta t^2 + h)^2 \\
&\quad + a_2\Bigl\{\Delta t^4\sum_{i=2}^{k}\|u\|_{Z^3(t^{i-2},t^i)}^2 + h^2\Bigl(\sum_{i=2}^{k}\|(u,p)\|_{H^1(t^{i-2},t^i;H^2\times H^1)}^2 + (k-1)\Delta t\Bigr)\Bigr\}\Bigr] \\
&\hspace{20em}\text{(by P}(k)\text{and Hypothesis 2)} \\
&\leqslant \exp(2a_1T)\Bigl[c_I^2(\Delta t^2 + h)^2 + a_2\Bigl\{2\Delta t^4\|u\|_{Z^3}^2 + h^2\Bigl(2\|(u,p)\|_{H^1(H^2\times H^1)}^2 + T\Bigr)\Bigr\}\Bigr] \\
&\leqslant \exp(2a_1T)\Bigl[2\Delta t^4(c_I^2 + a_2\|u\|_{Z^3}^2) + h^2\Bigl\{2c_I^2 + a_2\bigl(2\|(u,p)\|_{H^1(H^2\times H^1)}^2 + T\bigr)\Bigr\}\Bigr] \\
&\leqslant \frac{\nu}{\alpha_1^2}\bigl\{c_*(\Delta t^2 + h)\bigr\}^2,
\end{aligned} \tag{18.41}$$

which implies (18.38)$_k$ as $\|e_h^k\|_1 \leqslant \alpha_1\|D(e_h^k)\|_0 \leqslant c_*(\Delta t^2 + h)$.

We show (F3). Let us remind that Π_h is the Lagrange interpolation operator stated in Lemma 2. Noting that $\|u^{k*}\|_X \leqslant 2\|u^k\|_X + \|u^{k-1}\|_X \leqslant 3\|u\|_{C(X)}$ for $X = L^\infty(\Omega)^d$ and $W^{1,\infty}(\Omega)^d$, we have

$$
\begin{aligned}
\|u_h^{k*}\|_{0,\infty} &\leqslant \|u_h^{k*} - \Pi_h u^{k*}\|_{0,\infty} + \|\Pi_h u^{k*}\|_{0,\infty} \\
&\leqslant 2\|u_h^k - \Pi_h u^k\|_{0,\infty} + \|u_h^{k-1} - \Pi_h u^{k-1}\|_{0,\infty} + \|\Pi_h u^{k*}\|_{0,\infty} \\
&\leqslant \alpha_{21} h^{-d/6}(2\|u_h^k - \Pi_h u^k\|_1 + \|u_h^{k-1} - \Pi_h u^{k-1}\|_1) + \|\Pi_h u^{k*}\|_{0,\infty} \\
&\leqslant \alpha_{21} h^{-d/6}\{2(\|u_h^k - \hat{u}_h^k\|_1 + \|\hat{u}_h^k - u^k\|_1 + \|u^k - \Pi_h u^k\|_1) \\
&\quad + (\|u_h^{k-1} - \hat{u}_h^{k-1}\|_1 + \|\hat{u}_h^{k-1} - u^{k-1}\|_1 + \|u^{k-1} - \Pi_h u^{k-1}\|_1)\} + \|u^{k*}\|_{0,\infty} \\
&\leqslant \alpha_{21} h^{-d/6}\big[2\{c_*(\Delta t^2 + h) + \alpha_{31} h \|(u,p)^k\|_{H^2\times H^1} + \alpha_{24} h \|u^k\|_2\} \\
&\quad + \{c_*(\Delta t^2 + h) + \alpha_{31} h \|(u,p)^{k-1}\|_{H^2\times H^1} + \alpha_{24} h \|u^{k-1}\|_2\}\big] + 3\|u\|_{C(L^\infty)} \\
&\qquad\qquad \text{(by (18.38)}_k \text{ and (18.38)}_{k-1}) \\
&\leqslant 3\alpha_{21} h^{-d/6}\{c_*(\Delta t^2 + h) + (\alpha_{24} + \alpha_{31}) h \|(u,p)\|_{C(H^2\times H^1)}\} + 3\|u\|_{C(L^\infty)} \\
&\leqslant 3\alpha_{21}\{c_*(c_0^2 h_0^{d/6} + h_0^{1-d/6}) + (\alpha_{24} + \alpha_{31}) h_0^{1-d/6} \|(u,p)\|_{C(H^2\times H^1)}\} + 3\|u\|_{C(L^\infty)} \\
&\qquad\qquad \text{(by (18.12))} \\
&\leqslant 3\|u\|_{C(L^\infty)} + 1 \quad \text{(by (18.36a))},
\end{aligned}
$$

and

$$
\begin{aligned}
2\Delta t \|u_h^{k*}\|_{1,\infty} &\leqslant 2\Delta t(2\|u_h^k - \Pi_h u^k\|_{1,\infty} + \|u_h^{k-1} - \Pi_h u^{k-1}\|_{1,\infty} + \|\Pi_h u^{k*}\|_{1,\infty}) \\
&\leqslant 2\Delta t\{\alpha_{22} h^{-d/2}(2\|u_h^k - \Pi_h u^k\|_1 + \|u_h^{k-1} - \Pi_h u^{k-1}\|_1) + \|\Pi_h u^{k*}\|_{1,\infty}\} \\
&\leqslant 2\Delta t\big[\alpha_{22} h^{-d/2}\{2(\|u_h^k - \hat{u}_h^k\|_1 + \|\hat{u}_h^k - u^k\|_1 + \|u^k - \Pi_h u^k\|_1) \\
&\quad + (\|u_h^{k-1} - \hat{u}_h^{k-1}\|_1 + \|\hat{u}_h^{k-1} - u^{k-1}\|_1 + \|u^{k-1} - \Pi_h u^{k-1}\|_1)\} + \alpha_{23}\|u^{k*}\|_{1,\infty}\big] \\
&\leqslant 2\Delta t\Big[\alpha_{22} h^{-d/2}\Big\{2\Big(c_*(\Delta t^2 + h) + \alpha_{31} h \|(u,p)^k\|_{H^2\times H^1} + \alpha_{24} h \|u^k\|_2\Big) \\
&\quad + \Big(c_*(\Delta t^2 + h) + \alpha_{31} h \|(u,p)^{k-1}\|_{H^2\times H^1} + \alpha_{24} h \|u^{k-1}\|_2\Big)\Big\} + 3\alpha_{23}\|u\|_{C(W^{1,\infty})}\Big] \\
&\qquad\qquad \text{(by (18.38)}_k \text{ and (18.38)}_{k-1}) \\
&\leqslant 6\Delta t\big[\alpha_{22} h^{-d/2}\{c_*(\Delta t^2 + h) + (\alpha_{24} + \alpha_{31}) h \|(u,p)\|_{C(H^2\times H^1)}\} + \alpha_{23}\|u\|_{C(W^{1,\infty})}\big] \\
&\leqslant 6c_0\big[\alpha_{22}\{c_*(c_0^2 + h_0^{1-d/3}) + (\alpha_{24} + \alpha_{31}) h_0^{1-d/3} \|(u,p)\|_{C(H^2\times H^1)}\} \\
&\quad + \alpha_{23} h_0^{d/6}\|u\|_{C(W^{1,\infty})}\big] \quad \text{(by (18.12))} \\
&\leqslant \delta_1 \quad \text{(by (18.36b))},
\end{aligned}
$$

which lead to $(18.39)_k$.

We derive (F4). Since $A_i(\|u_h^{(k-1)*}\|_{0,\infty}) \leqslant a_i, i = 1, 2,$ hold, the inequality $(18.28)_k$ implies

$$
\overline{D}_{\Delta t} x_k + y_k \leqslant a_1(x_{k-1} + x_{k-2}) + \frac{1}{4} y_{k-1} + b_k,
$$

which is equivalent to $(18.40)_k$.

Step 2 (Induction): We prove P(n) for $n \in \{2, \dots, N_T\}$ by induction. Let us note that (F1) ensures (18.26) in the sequel.

Firstly, we prove P(2) and P(3) for given $(u_h^i, p_h^i), i = 0, 1$, satisfying Hypothesis 2. Since the last inequality of (18.11b) is nothing but $(18.25)_2$, there exists a unique solution (u_h^2, p_h^2) of (18.7) from Proposition 3-(i).

We prove P(2). (18.11c) is equivalent to $(18.27)_2$. Hence, $(18.28)_2$ holds from Proposition 3-(ii). The last inequality of (18.11a) and (F4) imply the estimate,

$$
\begin{aligned}
x_2 + \Delta t y_2 &\leqslant (1 + a_1 \Delta t) x_1 + a_1 \Delta t x_0 + \frac{1}{4} \Delta t y_1 + \Delta t b_2 \quad (\text{by } (18.40)_2) \\
&\leqslant (1 + a_1 \Delta t)(X_0 + \Delta t b_2) \leqslant \exp(a_1 \Delta t)(X_0 + \Delta t b_2),
\end{aligned}
$$

which is $(18.37)_2$, i.e., P(2).

Since $(18.38)_2$ and $(18.38)_1$ are obtained from (F2) and (18.11d) with $(\alpha_1/\sqrt{\nu})c_I \leqslant c_*$, respectively, we have $(18.39)_2$ and there exists a unique solution (u_h^3, p_h^3) of (18.7).

We prove P(3). $(18.27)_3$ is obtained from (18.7), and Proposition 3-(ii) implies $(18.28)_3$. From (F4) we have $(18.40)_3$ and the estimate,

$$
\begin{aligned}
x_3 + \frac{3}{4} \Delta t y_2 + \Delta t y_3 &\leqslant \left\{ (1 + a_1 \Delta t) x_2 + a_1 \Delta t x_1 + \frac{1}{4} \Delta t y_2 + \Delta t b_3 \right\} + \frac{3}{4} \Delta t y_2 \\
&\leqslant (1 + a_1 \Delta t)(x_2 + \Delta t y_2) + a_1 \Delta t x_1 + \Delta t b_3 \\
&\leqslant (1 + a_1 \Delta t) \exp(a_1 \Delta t)(X_0 + \Delta t b_2) + a_1 \Delta t X_0 + \Delta t b_3 \quad (\text{by P(2)}) \\
&\leqslant (1 + 2a_1 \Delta t) \exp(a_1 \Delta t) \Big(X_0 + \Delta t \sum_{i=2}^{3} b_i \Big) \leqslant \exp(3 a_1 \Delta t) \Big(X_0 + \Delta t \sum_{i=2}^{3} b_i \Big).
\end{aligned}
$$

Thus we get P(3).

Secondly, we prove the general step in the induction. Supposing that P($n-1$) and P($n-2$) hold true for an integer $n \in \{4, \dots, N_T\}$, we prove that P(n) also holds. P($n-1$) and P($n-2$) imply $(18.38)_{n-1}$ and $(18.38)_{n-2}$ from (F2), respectively. From (F3) we have $(18.39)_{n-1}$ and there exists a unique solution (u_h^n, p_h^n) of (18.7).

We prove P(n). $(18.39)_{n-1}$ implies $(18.40)_n$ from (F4). Noting P($n-1$) and P($n-2$), i.e.,

$$
x_{n-1} + \frac{3}{4} \Delta t \sum_{i=2}^{n-2} y_i + \Delta t y_{n-1} \leqslant \exp\big\{ a_1 (2n-5) \Delta t \big\} \Big(X_0 + \Delta t \sum_{i=2}^{n-1} b_i \Big), \tag{18.42a}
$$

$$
x_{n-2} + \frac{3}{4} \Delta t \sum_{i=2}^{n-3} y_i + \Delta t y_{n-2} \leqslant \exp\big\{ a_1 (2n-7) \Delta t \big\} \Big(X_0 + \Delta t \sum_{i=2}^{n-2} b_i \Big), \tag{18.42b}
$$

we have

$$
\begin{aligned}
& x_n + \frac{3}{4}\Delta t\sum_{i=2}^{n-1} y_i + \Delta t y_n = (x_n + \Delta t y_n) + \frac{3}{4}\Delta t \sum_{i=2}^{n-1} y_i \\
& \leqslant (1 + a_1\Delta t)x_{n-1} + a_1\Delta t x_{n-2} + \frac{1}{4}\Delta t y_{n-1} + \Delta t b_n + \frac{3}{4}\Delta t\sum_{i=2}^{n-1} y_i \quad \text{(by } (18.40)_n\text{)} \\
& \leqslant (1 + a_1\Delta t)\Big(x_{n-1} + \frac{3}{4}\Delta t\sum_{i=1}^{n-2} y_i + \Delta t y_{n-1}\Big) + a_1\Delta t x_{n-2} + \Delta t b_n \\
& \leqslant (1 + a_1\Delta t)\exp\big\{a_1(2n-5)\Delta t\big\}\Big(X_0 + \Delta t\sum_{i=2}^{n-1} b_i\Big) \\
& \quad + a_1\Delta t\exp\big\{a_1(2n-7)\Delta t\big\}\Big(X_0 + \Delta t\sum_{i=2}^{n-2} b_i\Big) + \Delta t b_n \quad \text{(by (18.42))} \\
& \leqslant (1 + 2a_1\Delta t)\exp\big\{a_1(2n-5)\Delta t\big\}\Big(X_0 + \Delta t\sum_{i=2}^{n-1} b_i\Big) + \Delta t b_n \\
& \leqslant \exp\big\{a_1(2n-3)\Delta t\big\}\Big(X_0 + \Delta t\sum_{i=2}^{n} b_i\Big),
\end{aligned}
$$

which is $(18.37)_n$, i.e., P(n). Thus, the induction is completed.

Step 3: Finally we derive the results (i) and (ii) of the theorem. The induction completed in the previous step implies that P(N_T) holds true. Hence, we have (i). The first inequality of (18.13) in (ii) is obtained from (F2) and the estimate,

$$
\|u_h - u\|_{l^\infty(H^1)} \leqslant \|e_h\|_{l^\infty(H^1)} + \|\eta\|_{l^\infty(H^1)} \leqslant \|e_h\|_{l^\infty(H^1)} + \alpha_{31} h\|(u,p)\|_{C(H^2\times H^1)}.
$$

We prove the second inequality of (18.13). We have

$$
\begin{aligned}
\|\varepsilon_h^n\|_0 & \leqslant \|(e_h^n, \varepsilon_h^n)\|_{V\times Q} \leqslant \frac{1}{\alpha_{30}} \sup_{(v_h,q_h)\in V_h\times Q_h} \frac{\mathscr{A}_h\big((e_h^n,\varepsilon_h^n),(v_h,q_h)\big)}{\|(v_h,q_h)\|_{V\times Q}} \\
& = \frac{1}{\alpha_{30}} \sup_{(v_h,q_h)\in V_h\times Q_h} \frac{\langle R_h^n, v_h\rangle - (\overline{D}_{\Delta t}^{(2)} e_h^n, v_h)}{\|(v_h,q_h)\|_{V\times Q}} \\
& \leqslant c_{(u,p)}\Big\{\Delta t^{3/2}\|u\|_{Z^3(t^{n-2},t^n)} + h\Big(\frac{1}{\sqrt{\Delta t}}\|(u,p)\|_{H^1(t^{n-2},t^n;H^2\times H^1)} + 1\Big) \\
& \qquad + \|e_h^{n-1}\|_1 + \|e_h^{n-2}\|_1 + \|\overline{D}_{\Delta t}^{(2)} e_h^n\|_0\Big\} \\
& \leqslant c_{(u,p)}\Big\{\Delta t^{3/2}\|u\|_{Z^3(t^{n-2},t^n)} + h\Big(\frac{1}{\sqrt{\Delta t}}\|(u,p)\|_{H^1(t^{n-2},t^n;H^2\times H^1)} + 1\Big) \\
& \qquad + \|e_h^{n-1}\|_1 + \|e_h^{n-2}\|_1 + \frac{3}{2}\|\overline{D}_{\Delta t} e_h^n\|_0 + \frac{1}{2}\|\overline{D}_{\Delta t} e_h^{n-1}\|_0\Big\} \qquad (18.43)
\end{aligned}
$$

for $n = 2, \ldots, N_T$, where we have used Lemmas 3 and 6 and the inequality $\|e_h^{n-1}\|_0 \leqslant \|e_h^{n-1}\|_1$. We obtain the result by combining (18.43), (18.41) and (18.11d) with the estimate

$$\|p_h - p\|_{l^2(L^2)} \leqslant \|\varepsilon_h\|_{l^2(L^2)} + \|\hat{p}_h - p\|_{l^2(L^2)} \leqslant \|\varepsilon_h\|_{l^2(L^2)} + \sqrt{T}\alpha_{31} h \|(u,p)\|_{C(H^2 \times H^1)}.$$

□

Remark 8 The former estimate of $(18.39)_k$ can be improved as

$$\left\|u_h^{k*}\right\|_{0,\infty} \leqslant \|u\|_{C(L^\infty)} + 1$$

under the additional condition $u \in C^1([0,T]; L^\infty(\Omega)^d)$ for $\Delta t \in (0, 1/\|u\|_{C^1(L^\infty)}]$.

18.4.4 Proof of Theorem 2

We use the next lemma, which is proved in Appendix.

Lemma 7 *Suppose Hypotheses 1 and 3 hold. Let $n \in \{2, \ldots, N_T\}$ be a fixed number and let u_h^{n-1} and $u_h^{n-2} \in V_h$ be known. Then, under the conditions $(18.25)_n$ and* (18.26) *we have*

$$\|R_{h2}^n\|_0 \leqslant c_{(u,p)} \left\{ \|e_h^{n-1}\|_0 + \|e_h^{n-2}\|_0 + h^2 \left(\frac{1}{\sqrt{\Delta t}} \|(u,p)\|_{H^1(t^{n-2},t^n; H^2 \times H^1)} + 1 \right) \right\}, \tag{18.44a}$$

$$\|R_{h3}^n\|_{V_h'} \leqslant c_{(u,p)} \left(\|e_h^{n-1}\|_0 + \|e_h^{n-2}\|_0 + h^2 \right), \tag{18.44b}$$

$$\|R_{h4}^n\|_{V_h'} \leqslant c_{(u,p)} \sum_{k=1}^{2} \left(1 + h^{-d/6} \|e_h^{n-k}\|_1\right) \left(\|e_h^{n-k}\|_0 + h^2 \right). \tag{18.44c}$$

Proof of Theorem 2. Since we have $\|e_h\|_{l^\infty(H^1)} \leqslant c_*(\Delta t^2 + h) \leqslant c_*(c_0^2 h_0^{d/6} + h_0^{1-d/6}) h^{d/6}$ from (F2) in the proof of Theorem 1, Hypothesis 2 and (18.12), (18.44c) implies

$$\|R_{h4}^n\|_{V_h'} \leqslant c_{(u,p)} \left(\|e_h^{n-1}\|_0 + \|e_h^{n-2}\|_0 + h^2 \right). \tag{18.45}$$

Substituting $(e_h^n, -\varepsilon_h^n)$ into (v_h, q_h) in (18.24) and using (18.17), (18.29a), (18.44a), (18.44b), (18.45) and the identity [19],

$$(\overline{D}_{\Delta t}^{(2)} e_h^n, e_h^n) = \frac{1}{\Delta t} \left\{ \frac{3}{4} \|e_h^n\|_0^2 - \|e_h^{n-1}\|_0^2 + \frac{1}{4} \|e_h^{n-2}\|_0^2 + \frac{1}{4} \|e_h^n - 2e_h^{n-1} + e_h^{n-2}\|_0^2 \right.$$
$$\left. + \frac{1}{2} \left(\|e_h^n - e_h^{n-1}\|_0^2 - \|e_h^{n-1} - e_h^{n-2}\|_0^2 \right) \right\},$$

we have

$$
\begin{aligned}
&\frac{1}{\Delta t}\Big\{\frac{3}{4}\|e_h^n\|_0^2 - \|e_h^{n-1}\|_0^2 + \frac{1}{4}\|e_h^{n-2}\|_0^2 + \frac{1}{2}\big(\|e_h^n - e_h^{n-1}\|_0^2 - \|e_h^{n-1} - e_h^{n-2}\|_0^2\big)\Big\} \\
&\quad + \frac{2\nu}{\alpha_1^2}\|e_h^n\|_1^2 + \delta_0|\varepsilon_h^n|_h^2 \leqslant \sum_{i=1}^{4}\langle R_{hi}^n, e_h^n\rangle \\
&\leqslant \Big(\sum_{i=1}^{4}\beta_i\Big)\|e_h^n\|_1^2 + c_{(u,p)}\Big(\sum_{i=2}^{4}\frac{1}{\beta_i}\Big)\big(\|e_h^{n-1}\|_0^2 + \|e_h^{n-2}\|_0^2\big) \\
&\quad + c'_{(u,p)}\Big[\frac{\Delta t^3}{\beta_1}\|u\|_{Z^3(t^{n-2},t^n)}^2 + h^4\Big\{\frac{1}{\beta_2\Delta t}\|(u,p)\|_{H^1(t^{n-2},t^n;H^2\times H^1)}^2 + \Big(\sum_{i=2}^{4}\frac{1}{\beta_i}\Big)\Big\}\Big]
\end{aligned}
$$

for any $\beta_i > 0\ (i = 1, \ldots, 4)$, where the inequality $\|e_h^n\|_0 \leqslant \|e_h^n\|_1$ has been employed. Hence, we have

$$
\begin{aligned}
&\frac{1}{\Delta t}\Big\{\frac{3}{4}\|e_h^n\|_0^2 - \|e_h^{n-1}\|_0^2 + \frac{1}{4}\|e_h^{n-2}\|_0^2 + \frac{1}{2}\big(\|e_h^n - e_h^{n-1}\|_0^2 - \|e_h^{n-1} - e_h^{n-2}\|_0^2\big)\Big\} \\
&\quad + \frac{\nu}{\alpha_1^2}\|e_h^n\|_1^2 + \delta_0|\varepsilon_h^n|_h^2 \\
&\leqslant c_{(u,p)}\big(\|e_h^{n-1}\|_0^2 + \|e_h^{n-2}\|_0^2\big) \\
&\quad + c'_{(u,p)}\Big\{\Delta t^3\|u\|_{Z^3(t^{n-2},t^n)}^2 + h^4\Big(\frac{1}{\Delta t}\|(u,p)\|_{H^1(t^{n-2},t^n;H^2\times H^1)}^2 + 1\Big)\Big\}
\end{aligned}
$$

by setting $\beta_i = \nu/(4\alpha_1^2)\ (i = 1, \ldots, 4)$. From Lemma 5 with

$$
\begin{aligned}
&x_n = \frac{1}{2}\|e_h^n\|_0^2, \quad y_n = \frac{1}{2}\|e_h^n - e_h^{n-1}\|_0^2, \quad z_n = \frac{\nu}{\alpha_1^2}\|e_h^n\|_1^2 + \delta_0|\varepsilon_h^n|_h^2, \\
&a_0 = 2c_{(u,p)}, \quad b_n = c'_{(u,p)}\Big\{\Delta t^3\|u\|_{Z^3(t^{n-2},t^n)}^2 + h^4\Big(\frac{1}{\Delta t}\|(u,p)\|_{H^1(t^{n-2},t^n;H^2\times H^1)}^2 + 1\Big)\Big\},
\end{aligned}
$$

and the inequality $y_1 \leqslant \|e_h^0\|_0^2 + \|e_h^1\|_0^2$ there exists a positive constant c_4 independent of h and Δt such that

$$
\|e_h\|_{l^\infty(L^2)} \leqslant c_4(\|e_h^1\|_0 + \|e_h^0\|_0 + \Delta t^2 + h^2) \tag{18.46}
$$

under the condition $\Delta t \leqslant 1/(2a_0)$. We note that there exists a positive constant $\tilde{c}_0 \leqslant c_0$ independent of h and Δt such that

$$
(\Delta t \leqslant)\tilde{c}_0 h_0^{d/6} \leqslant \frac{1}{2a_0}.
$$

Combining (18.46) and (18.14) with the inequality

$$\|u_h - u\|_{l^\infty(L^2)} \leqslant \|e_h\|_{l^\infty(L^2)} + \|\eta\|_{l^\infty(L^2)} \leqslant \|e_h\|_{l^\infty(L^2)} + \alpha_{32} h^2 \|(u,p)\|_{C(H^2\times H^1)},$$

we obtain (18.16). □

18.5 Numerical Results

In this section two- and three-dimensional test problems are computed by scheme (18.7) in order to recognize the theoretical convergence orders numerically.

We set u_h^0 as the first component of the Stokes projection of $(u^0, 0) \in V \times Q$ by (18.9). The approximation $(u_h^1, p_h^1) \in V_h \times Q_h$ is obtained by the stabilized LG scheme of first-order in time [14, 15, 17] with a time increment $\tau \equiv \Delta t^2$, i.e., (u_h^1, p_h^1) is defined by $(u_h^1, p_h^1) \equiv (\tilde{u}_h^{\tilde{N}_T}, \tilde{p}_h^{\tilde{N}_T})$ for the solution $\{(\tilde{u}_h^m, \tilde{p}_h^m)\}_{m=1}^{\tilde{N}_T} \subset V_h \times Q_h$ such that, for $m = 1, \ldots, \tilde{N}_T$,

$$\Big(\frac{\tilde{u}_h^m - \tilde{u}_h^{m-1}\circ X_1(\tilde{u}_h^{m-1}, \tau)}{\tau}, v_h\Big) + \mathscr{A}_h((\tilde{u}_h^m, \tilde{p}_h^m), (v_h, q_h)) = (\tilde{f}^m, v_h),$$
$$\forall (v_h, q_h) \in V_h \times Q_h, \qquad (18.47)$$

where $\tilde{u}_h^0 \equiv u_h^0$, $\tilde{N}_T \equiv \lfloor \Delta t/\tau \rfloor = \lfloor 1/\Delta t \rfloor$ and $\tilde{f}^m \equiv f(\cdot, m\tau)$. Then, Hypotheses 2 and 4 are satisfied. In the rest of this section we consider the above *complete* scheme, i.e., scheme (18.7) with u_h^0, the first component of the Stokes projection of $(u^0, 0) \in V \times Q$ by (18.9), and (u_h^1, p_h^1) obtained by (18.47). We simply call it *scheme* (18.7) *with* (18.47), and the error estimates (18.13) and (18.16) hold for the scheme.

Numerical quadrature formulae [22] of degree five for $d = 2$ (seven points) and 3 (fifteen points) are employed for the computation of the integrals of composite functions in (18.7) and (18.47), e.g.,

$$\int_K u_h^{n-1} \circ X_1(u_h^{(n-1)*}, \Delta t)(x) v_h(x)\, dx.$$

Theorems 1 and 2 hold for any fixed δ_0. Here we set $\delta_0 = 1$. The system of linear equations is solved by MINRES [2, 21].

Example 1 In problem (18.1) we set $\Omega = (0, 1)^d, T = 1$ and we consider four values of ν,

$$\nu = 10^{-k}, \quad k = 1, \ldots, 4.$$

The functions f and u^0 are given so that the exact solution is as follows:

for $d = 2$:

$$u(x,t) = \Big(\frac{\partial \psi}{\partial x_2}, -\frac{\partial \psi}{\partial x_1}\Big)(x,t), \quad p(x,t) = \sin\{\pi(x_1 + 2x_2 + t)\},$$
$$\psi(x,t) \equiv \frac{\sqrt{3}}{2\pi} \sin^2(\pi x_1)\sin^2(\pi x_2)\sin\{\pi(x_1 + x_2 + t)\},$$

for $d = 3$:

$$u(x,t) = \mathrm{rot}\,\Psi(x,t), \quad p(x,t) = \sin\{\pi(x_1 + 2x_2 + x_3 + t)\},$$
$$\Psi_1(x,t) \equiv \frac{8\sqrt{3}}{27\pi} \sin(\pi x_1)\sin^2(\pi x_2)\sin^2(\pi x_3)\sin\{\pi(x_2 + x_3 + t)\},$$
$$\Psi_2(x,t) \equiv \frac{8\sqrt{3}}{27\pi} \sin^2(\pi x_1)\sin(\pi x_2)\sin^2(\pi x_3)\sin\{\pi(x_3 + x_1 + t)\},$$
$$\Psi_3(x,t) \equiv \frac{8\sqrt{3}}{27\pi} \sin^2(\pi x_1)\sin^2(\pi x_2)\sin(\pi x_3)\sin\{\pi(x_1 + x_2 + t)\}.$$

These solutions are normalized so that $\|u\|_{C(L^\infty)} = \|p\|_{C(L^\infty)} = 1$.

Let N be the division number of each side of the domain. We set $N = 16$, 32, 64, 128, 256 and 512 for $d = 2$ and $N = 16$, 32 and 64 for $d = 3$, and (re)define $h \equiv 1/N$. The meshes of $N = 16$ are shown in Fig. 18.1 for $d = 2$ (left) and 3 (right). Example 1 is solved by scheme (18.7) with (18.47), and for the solution (u_h, p_h) we define the relative errors $Er1$ and $Er2$ by

$$Er1 \equiv \frac{\|u_h - \Pi_h u\|_{l^2(H^1)} + \|p_h - \Pi_h p\|_{l^2(L^2)}}{\|\Pi_h u\|_{l^2(H^1)} + \|\Pi_h p\|_{l^2(L^2)}}, \qquad Er2 \equiv \frac{\|u_h - \Pi_h u\|_{l^\infty(L^2)}}{\|\Pi_h u\|_{l^\infty(L^2)}},$$

where for the pressure we have used the same symbol Π_h as its scalar version, i.e., $\Pi_h : C(\bar{\Omega}) \to M_h$. We employ two relations between Δt and h, (i) $\Delta t = \lambda_1 h^{1/2}$ for $\lambda_1 = 0.25$ and (ii) $\Delta t = h$, in order to observe the convergence orders of (18.13) and (18.16), respectively. The orders are calculated as $O(\Delta t^2 + h) = O(\Delta t^2)$ and $O(\Delta t^2 + h^2) = O(\Delta t^2)$ for (i), and $O(\Delta t^2 + h) = O(\Delta t)$ and $O(\Delta t^2 + h^2) = O(\Delta t^2)$ for (ii). The results of the cases (i) and (ii) are shown in Figs. 18.2

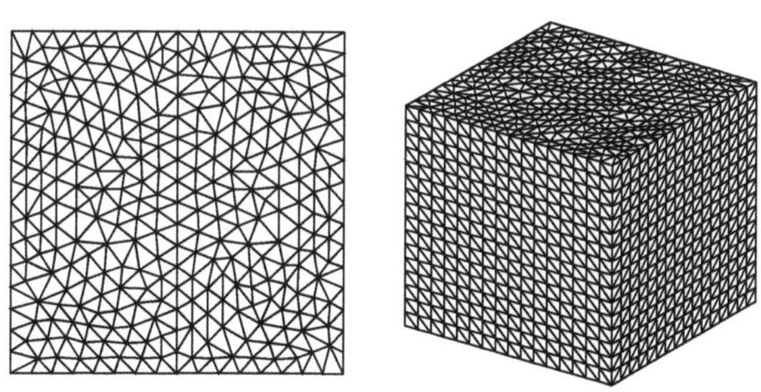

Fig. 18.1 Sample meshes for $d = 2$ (*left*) and 3 (*right*), $N = 16$

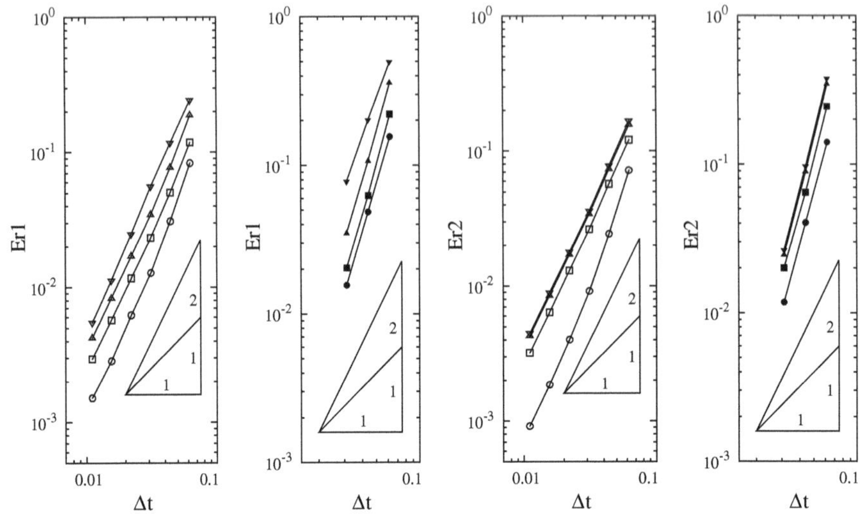

Fig. 18.2 $Er1$ versus Δt (the *left* two, *left*: $d = 2$, *right*: $d = 3$) and $Er2$ versus Δt (the *right* two, *left*: $d = 2$, *right*: $d = 3$) for $\Delta t = \lambda_1 h^{1/2}$ ($\lambda_1 = 0.25$)

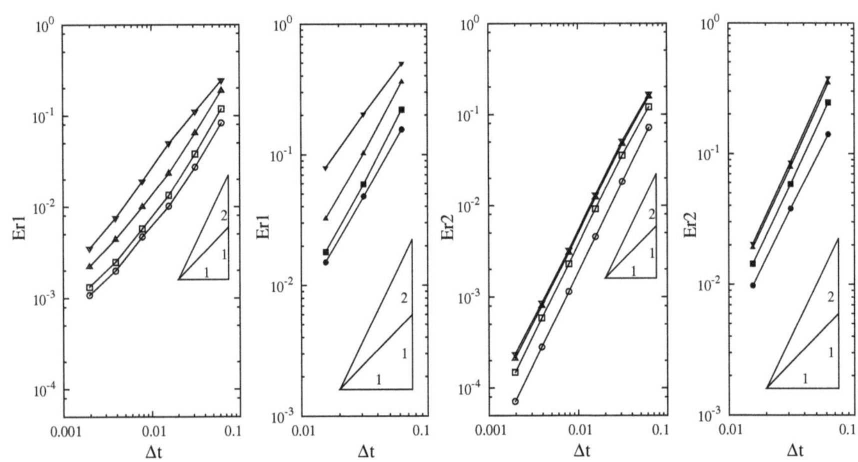

Fig. 18.3 $Er1$ versus Δt (the *left* two, *left*: $d = 2$, *right*: $d = 3$) and $Er2$ versus Δt (the *right* two, *left*: $d = 2$, *right*: $d = 3$) for $\Delta t = h$

and 18.3, respectively, and they exhibit the graphs of $Er1$ versus Δt (the left two) and $Er2$ versus Δt (the right two) in logarithmic scale for $d = 2$ and 3, where the symbols are summarized in Table 18.1. The values and the slopes of the graphs in Figs. 18.2 and 18.3 are given in Tables 18.2 and 18.3, respectively.

Table 18.1 Symbols used in Figs. 18.2 and 18.3

d	ν			
	10^{-1}	10^{-2}	10^{-3}	10^{-4}
2	○	□	△	▽
3	●	■	▲	▼

In Fig. 18.2 and Table 18.2, i.e., the results of (i), $Er1$ is almost of second order for $d = 2$ and is of better order than second one for $d = 3$, and the results of $Er2$ are similar to those of $Er1$. In Fig. 18.3 and Table 18.3, i.e., the results of (ii), $Er1$ is almost of first order for $d = 2$ and is of better order than first one for $d = 3$ and $Er2$ is almost of second order for both $d = 2$ and 3. All numerical results are consistent with Theorems 1 and 2.

18.6 Conclusions

We have proposed and analyzed a stabilized Lagrange–Galerkin scheme of second-order in time for the Navier–Stokes equations. The stabilization and the second-order accuracy in time have been realized by Brezzi–Pitkäranta's stabilization method and Adams-Bashforth's (two-step) method, respectively. Since it is a higher-order version in time of a stabilized LG scheme of first-order in time for the equations [14, 15, 17], the scheme has the same advantages of the first-order one, i.e., robustness for convection-dominated problems, symmetry of the resulting matrix and the small number of DOF. We note that the scheme is a fully discrete stabilized LG scheme in the sense that the exact solvability of ordinary differential equations describing the particle path is not required. Convergence with the optimal error estimates of order $O(\Delta t^2 + h)$ for the velocity in the H^1-norm and the pressure in the L^2-norm (Theorem 1) and of order $O(\Delta t^2 + h^2)$ for the velocity in the L^2-norm (Theorem 2) have been proved. The theoretical convergence orders have been recognized by two- and three-dimensional numerical results.

Acknowledgements This work was supported by JSPS (the Japan Society for the Promotion of Science) under the Japanese-German Graduate Externship (Mathematical Fluid Dynamics) and Grant-in-Aid for Scientific Research (S), No. 24224004. The authors are indebted to JSPS also for Grant-in-Aid for Young Scientists (B), No. 26800091 to the first author and for Grant-in-Aid for Scientific Research (C), No. 25400212 to the second author.

Table 18.2 Values of $Er1$, $Er2$ and their slopes of the graphs in Fig. 18.2

	$Er1$					$Er2$			
	N	$d=2$	Slope	$d=3$	Slope	$d=2$	Slope	$d=3$	Slope
$\nu = 10^{-1}$:	16	8.31×10^{-2}	–	1.55×10^{-1}	–	7.19×10^{-2}	–	1.40×10^{-1}	–
	32	3.09×10^{-2}	2.86	4.86×10^{-2}	3.36	2.43×10^{-2}	3.12	4.04×10^{-2}	3.59
	64	1.28×10^{-2}	2.54	1.56×10^{-2}	3.28	9.15×10^{-3}	2.82	1.18×10^{-2}	3.56
	128	6.23×10^{-3}	2.08	–	–	4.00×10^{-3}	2.38	–	–
	256	2.85×10^{-3}	2.25	–	–	1.86×10^{-3}	2.22	–	–
	512	1.51×10^{-3}	1.83	–	–	9.17×10^{-4}	2.04	–	–
$\nu = 10^{-2}$:	16	1.18×10^{-1}	–	2.20×10^{-1}	–	1.20×10^{-1}	–	2.45×10^{-1}	–
	32	5.03×10^{-2}	2.46	6.26×10^{-2}	3.63	5.69×10^{-2}	2.16	6.45×10^{-2}	3.85
	64	2.33×10^{-2}	2.22	2.04×10^{-2}	3.24	2.64×10^{-2}	2.22	2.00×10^{-2}	3.38
	128	1.17×10^{-2}	1.99	–	–	1.30×10^{-2}	2.04	–	–
	256	5.73×10^{-3}	2.05	–	–	6.38×10^{-3}	2.05	–	–
	512	2.96×10^{-3}	1.91	–	–	3.21×10^{-3}	1.98	–	–
$\nu = 10^{-3}$:	16	1.88×10^{-1}	–	3.59×10^{-1}	–	1.56×10^{-1}	–	3.47×10^{-1}	–
	32	7.71×10^{-2}	2.57	1.07×10^{-1}	3.50	7.34×10^{-2}	2.18	8.88×10^{-2}	3.93
	64	3.46×10^{-2}	2.31	3.48×10^{-2}	3.23	3.42×10^{-2}	2.20	2.44×10^{-2}	3.72
	128	1.70×10^{-2}	2.05	–	–	1.71×10^{-2}	2.00	–	–
	256	8.30×10^{-3}	2.07	–	–	8.49×10^{-3}	2.02	–	–
	512	4.24×10^{-3}	1.94	–	–	4.29×10^{-3}	1.97	–	–
$\nu = 10^{-4}$:	16	2.43×10^{-1}	–	4.95×10^{-1}	–	1.67×10^{-1}	–	3.77×10^{-1}	–
	32	1.17×10^{-1}	2.10	2.01×10^{-1}	2.61	7.76×10^{-2}	2.21	9.66×10^{-2}	3.93
	64	5.58×10^{-2}	2.14	7.76×10^{-2}	2.74	3.59×10^{-2}	2.23	2.63×10^{-2}	3.75
	128	2.48×10^{-2}	2.34	–	–	1.79×10^{-2}	2.01	–	–
	256	1.12×10^{-2}	2.29	–	–	8.89×10^{-3}	2.02	–	–
	512	5.49×10^{-3}	2.07	–	–	4.49×10^{-3}	1.97	–	–

Table 18.3 Values of $Er1$, $Er2$ and their slopes of the graphs in Fig. 18.3

	$Er1$					$Er2$			
	N	$d = 2$	Slope	$d = 3$	Slope	$d = 2$	Slope	$d = 3$	slope
$\nu = 10^{-1}$:	16	8.31×10^{-2}	–	1.55×10^{-1}	–	7.19×10^{-2}	–	1.40×10^{-1}	–
	32	2.73×10^{-2}	1.60	4.78×10^{-2}	1.70	1.84×10^{-2}	1.97	3.78×10^{-2}	1.89
	64	1.01×10^{-2}	1.43	1.50×10^{-2}	1.67	4.55×10^{-3}	2.01	9.80×10^{-3}	1.95
	128	4.69×10^{-3}	1.11	–	–	1.14×10^{-3}	1.99	–	–
	256	1.99×10^{-3}	1.24	–	–	2.80×10^{-4}	2.03	–	–
	512	1.08×10^{-3}	0.88	–	–	7.15×10^{-5}	1.97	–	–
$\nu = 10^{-2}$:	16	1.18×10^{-1}	–	2.20×10^{-1}	–	1.20×10^{-1}	–	2.45×10^{-1}	–
	32	3.81×10^{-2}	1.63	5.93×10^{-2}	1.89	3.57×10^{-2}	1.76	5.82×10^{-2}	2.07
	64	1.34×10^{-2}	1.51	1.80×10^{-2}	1.72	9.18×10^{-3}	1.96	1.44×10^{-2}	2.02
	128	5.76×10^{-3}	1.21	–	–	2.30×10^{-3}	2.00	–	–
	256	2.48×10^{-3}	1.22	–	–	5.89×10^{-4}	1.96	–	–
	512	1.32×10^{-3}	0.91	–	–	1.48×10^{-4}	1.99	–	–
$\nu = 10^{-3}$:	16	1.88×10^{-1}	–	3.59×10^{-1}	–	1.56×10^{-1}	–	3.47×10^{-1}	–
	32	6.46×10^{-2}	1.54	1.02×10^{-1}	1.81	4.75×10^{-2}	1.72	7.89×10^{-2}	2.14
	64	2.34×10^{-2}	1.46	3.24×10^{-2}	1.66	1.23×10^{-2}	1.96	1.91×10^{-2}	2.05
	128	9.99×10^{-3}	1.23	–	–	3.04×10^{-3}	2.01	–	–
	256	4.40×10^{-3}	1.18	–	–	7.99×10^{-4}	1.93	–	–
	512	2.23×10^{-3}	0.98	–	–	2.11×10^{-4}	1.92	–	–
$\nu = 10^{-4}$:	16	2.43×10^{-1}	–	4.95×10^{-1}	–	1.67×10^{-1}	–	3.76×10^{-1}	–
	32	1.11×10^{-1}	1.13	2.03×10^{-1}	1.28	5.12×10^{-2}	1.70	8.58×10^{-2}	2.13
	64	4.98×10^{-2}	1.15	7.94×10^{-2}	1.36	1.31×10^{-2}	1.96	2.05×10^{-2}	2.07
	128	1.91×10^{-2}	1.39	–	–	3.26×10^{-3}	2.01	–	–
	256	7.55×10^{-3}	1.33	–	–	8.60×10^{-4}	1.92	–	–
	512	3.50×10^{-3}	1.11	–	–	2.33×10^{-4}	1.88	–	–

Appendix

Proof of Lemma 5

From (18.22), there exists a non-negative sequence $\{\tilde{z}_n\}_{n\geqslant 2}$ such that

$$\frac{1}{\Delta t}\left(\frac{3}{2}x_n - 2x_{n-1} + \frac{1}{2}x_{n-2} + y_n - y_{n-1}\right) + \tilde{z}_n = a_0(x_{n-1} + x_{n-2}) + b_n, \quad \forall n \geqslant 2,$$

where $\tilde{z}_n$ satisfies

$$z_n \leqslant \tilde{z}_n, \quad \forall n \geqslant 2. \tag{18.48}$$

Let p and $q \in \mathbb{R}$ $(q < p)$ be the roots of the quadratic equation $(3/2)x^2 - 2x + 1/2 = a_0 \Delta t(x+1)$ and $\lambda \equiv 2/3$. We note that p and q satisfy

$$0 \leqslant q < 1 \leqslant p \tag{18.49}$$

from $\Delta t \in (0, 1/(2a_0)]$. Let any $n \geqslant 2$ be fixed. We have

$$\begin{aligned} x_n - px_{n-1} + \lambda(y_n - y_{n-1}) + \lambda\Delta t\tilde{z}_n &= q(x_{n-1} - px_{n-2}) + \lambda\Delta t b_n, \\ x_n - qx_{n-1} + \lambda(y_n - y_{n-1}) + \lambda\Delta t\tilde{z}_n &= p(x_{n-1} - qx_{n-2}) + \lambda\Delta t b_n, \end{aligned}$$

which lead to

$$\begin{aligned} x_n - px_{n-1} + \lambda\left\{\sum_{i=2}^{n} q^{n-i}y_i - \sum_{i=1}^{n-1} q^{n-1-i}y_i\right\} + \lambda\Delta t\sum_{i=2}^{n} q^{n-i}\tilde{z}_i \\ = q^{n-1}(x_1 - px_0) + \lambda\Delta t\sum_{i=2}^{n} q^{n-i}b_i, \end{aligned} \tag{18.50a}$$

$$\begin{aligned} x_n - qx_{n-1} + \lambda\left\{\sum_{i=2}^{n} p^{n-i}y_i - \sum_{i=1}^{n-1} p^{n-1-i}y_i\right\} + \lambda\Delta t\sum_{i=2}^{n} p^{n-i}\tilde{z}_i \\ = p^{n-1}(x_1 - qx_0) + \lambda\Delta t\sum_{i=2}^{n} p^{n-i}b_i. \end{aligned} \tag{18.50b}$$

Multiplying (18.50a) by q and (18.50b) by p and subtracting the first equation from the second, we get

$$
\begin{aligned}
&(p-q)x_n + \lambda\Big\{\sum_{i=2}^{n}(p^{n+1-i}-q^{n+1-i})y_i - \sum_{i=1}^{n-1}(p^{n-i}-q^{n-i})y_i\Big\}\\
&\qquad + \lambda\Delta t\sum_{i=2}^{n}(p^{n+1-i}-q^{n+1-i})\tilde{z}_i\\
&\qquad = (p^n-q^n)x_1 - pq(p^{n-1}-q^{n-1})x_0 + \lambda\Delta t\sum_{i=2}^{n}(p^{n+1-i}-q^{n+1-i})b_i.
\end{aligned}
\tag{18.51}
$$

The definition of p and q and (18.49) imply that

$$
(p-q)y_n - (p^n-q^n)y_1 \leqslant \sum_{i=2}^{n}(p^{n+1-i}-q^{n+1-i})y_i - \sum_{i=1}^{n-1}(p^{n-i}-q^{n-i})y_i,
\tag{18.52a}
$$

$$
\frac{2}{3} \leqslant p-q \leqslant p^{n+1-i}-q^{n+1-i} \leqslant p^{n-1}-q^{n-1} \leqslant p^n-q^n, \quad i\in\{2,\ldots,n\},
\tag{18.52b}
$$

$$
\begin{aligned}
p^n - q^n \leqslant p^n &= \Big\{\frac{1}{3}\Big(2 + a_0\Delta t + \sqrt{1+10a_0\Delta t + a_0^2\Delta t^2}\Big)\Big\}^n\\
&\leqslant \{1+2a_0\Delta t\}^n \leqslant \exp(2a_0 n\Delta t).
\end{aligned}
\tag{18.52c}
$$

Combining (18.51) with (18.52), we have

$$
\begin{aligned}
&(p-q)\Big(x_n + \lambda y_n + \lambda\Delta t\sum_{i=2}^{n}\tilde{z}_i\Big)\\
&\leqslant (p^n-q^n)x_1 - pq(p^{n-1}-q^{n-1})x_0 + \lambda(p^n-q^n)y_1 + \lambda\Delta t(p^n-q^n)\sum_{i=2}^{n}b_i\\
&\leqslant (p^n-q^n)\Big(x_1 + \lambda y_1 + \lambda\Delta t\sum_{i=2}^{n}b_i\Big)\\
&\leqslant \exp(2a_0 n\Delta t)\Big(x_1 + \lambda y_1 + \lambda\Delta t\sum_{i=2}^{n}b_i\Big),
\end{aligned}
\tag{18.53}
$$

and obtain the desired result as follows:

$$\text{LHS of (18.23)} \leqslant x_n + \lambda y_n + \lambda \Delta t \sum_{i=2}^{n} \tilde{z}_i \quad \text{(by (18.48))}$$
$$\leqslant \frac{3}{2} \exp(2a_0 n \Delta t)\Big(x_1 + \lambda y_1 + \lambda \Delta t \sum_{i=2}^{n} b_i\Big) \quad \text{(by (18.53), (18.52}b\text{))}.$$

□

Proof of Lemma 6

Let $t(s) \equiv t^{n-1} + s\Delta t$ $(s \in [0, 1])$. We prove (18.29a). Let $y(x, s) \equiv x - (1 - s) u^{(n-1)*}(x)\Delta t$. Using the identities

$$g'(1) - \Big\{\frac{3}{2}g(1) - 2g(0) + \frac{1}{2}g(-1)\Big\} = 2\int_0^1 sds \int_{2s-1}^{s} g'''(s_1)ds_1,$$
$$\tilde{g}(1) - 2\tilde{g}(0) + \tilde{g}(-1) = \int_0^1 ds \int_{s-1}^{s} \tilde{g}''(s_1)ds_1,$$

for $g(s) = u(y(\cdot, s), t(s))$ and $\tilde{g}(s) = u(\cdot, t(s))$, we have

$$\begin{aligned}
R_{h1}^n(x) &= \Big\{\Big(\frac{\partial}{\partial t} + u^n(x)\cdot\nabla\Big)u\Big\}(x, t^n) \\
&\quad - \frac{1}{2\Delta t}\big\{3u^n - 4u^{n-1} \circ X_1(u^{(n-1)*}, \Delta t) + u^{n-2} \circ X_1(u^{(n-1)*}, 2\Delta t)\big\}(x) \\
&= \Big\{\Big(\frac{\partial}{\partial t} + u^{(n-1)*}(x)\cdot\nabla\Big)u\Big\}(x, t^n) \\
&\quad - \frac{1}{2\Delta t}\big\{3u^n - 4u^{n-1} \circ X_1(u^{(n-1)*}, \Delta t) + u^{n-2} \circ X_1(u^{(n-1)*}, 2\Delta t)\big\}(x) \\
&\quad + \big\{\big((u^n - u^{(n-1)*})(x)\cdot\nabla\big)u^n\big\}(x) \\
&= 2\Delta t^2 \int_0^1 sds \int_{2s-1}^{s} \Big\{\Big(\frac{\partial}{\partial t} + u^{(n-1)*}(x)\cdot\nabla\Big)^3 u\Big\}\big(y(x, s_1), t(s_1)\big)ds_1 \\
&\quad + \Delta t^2 \int_0^1 ds \int_{s-1}^{s} \Big\{\Big(\frac{\partial^2 u}{\partial t^2}(x, t(s_1))\cdot\nabla\Big)u^n\Big\}(x)ds_1,
\end{aligned}$$

and

$$
\begin{aligned}
\|R_{h1}^n\|_0 &\leqslant 2\Delta t^2 \int_0^1 s ds \int_{2s-1}^s \Big\| \Big\{ \Big(\frac{\partial}{\partial t} + u^{(n-1)*}(\cdot)\cdot\nabla \Big)^3 u \Big\} \big(y(\cdot, s_1), t(s_1)\big) \Big\|_0 ds_1 \\
&\quad + \Delta t^2 \int_0^1 ds \int_{s-1}^s \Big\| \Big(\frac{\partial^2 u}{\partial t^2}(\cdot, t(s_1))\cdot\nabla \Big) u^n \Big\|_0 ds_1 \\
&\leqslant c_u \Delta t^2 \int_{-1}^1 \Big(\Big\| \Big\{ \Big(\frac{\partial}{\partial t} + \nabla \Big)^3 u \Big\} (\cdot, t(s_1)) \Big\|_0 + \Big\| \frac{\partial^2 u}{\partial t^2}(\cdot, t(s_1)) \Big\|_0 \Big) ds_1 \quad \text{(by (18.10))} \\
&\leqslant c_u' \Delta t^{3/2} \big(\|u\|_{Z^3(t^{n-2}, t^n)} + \|u\|_{H^2(t^{n-2}, t^n; L^2)} \big) \leqslant 2c_u' \Delta t^{3/2} \|u\|_{Z^3(t^{n-2}, t^n)},
\end{aligned}
$$

which implies (18.29a).

Inequality (18.29b) is obtained as follows:

$$
\begin{aligned}
\|R_{h2}^n\|_0 &\leqslant \alpha_{42} \|u^{(n-1)*} - u_h^{(n-1)*}\|_0 (2\|u^{n-1}\|_{1,\infty} + \|u^{n-2}\|_{1,\infty}) \\
&\leqslant 3\alpha_{42} \|u\|_{C(W^{1,\infty})} \|2(\eta^{n-1} - e_h^{n-1}) - (\eta^{n-2} - e_h^{n-2})\|_0 \\
&\leqslant c_u (\|e_h^{n-1}\|_0 + \|e_h^{n-2}\|_0 + \|\eta^{n-1}\|_0 + \|\eta^{n-2}\|_0) \qquad (18.54) \\
&\leqslant c_u \{ \|e_h^{n-1}\|_0 + \|e_h^{n-2}\|_0 + \alpha_{31} h (\|(u,p)^{n-1}\|_{H^2\times H^1} + \|(u,p)^{n-2}\|_{H^2\times H^1}) \} \\
&\leqslant c_{(u,p)} (\|e_h^{n-1}\|_0 + \|e_h^{n-2}\|_0 + h).
\end{aligned}
$$

We prove (18.29c). Let $y(x, s) \equiv x - (1-s) u_h^{(n-1)*}(x) \Delta t$. Since we have

$$
\begin{aligned}
R_{h3}^n &= \frac{1}{\Delta t} \Big\{ \frac{3}{2} \Big[\eta\big(y(\cdot, s), t(s)\big) \Big]_{s=0}^1 - \frac{1}{2} \Big[\eta\big(y(\cdot, s), t(s)\big) \Big]_{s=-1}^0 \Big\} \\
&= \frac{3}{2} \int_0^1 \Big\{ \Big(\frac{\partial}{\partial t} + u_h^{(n-1)*}(\cdot)\cdot\nabla \Big) \eta \Big\} \big(y(\cdot, s), t(s)\big) ds \\
&\quad - \frac{1}{2} \int_{-1}^0 \Big\{ \Big(\frac{\partial}{\partial t} + u_h^{(n-1)*}(\cdot)\cdot\nabla \Big) \eta \Big\} \big(y(\cdot, s), t(s)\big) ds,
\end{aligned}
$$

(18.29c) is obtained as follows:

$$
\begin{aligned}
\|R_{h3}^n\|_0 &\leqslant \frac{3}{2} \int_0^1 \Big\| \Big\{ \Big(\frac{\partial}{\partial t} + u_h^{(n-1)*}(\cdot)\cdot\nabla \Big) \eta \Big\} \big(y(\cdot, s), t(s)\big) \Big\|_0 ds \\
&\quad + \frac{1}{2} \int_{-1}^0 \Big\| \Big\{ \Big(\frac{\partial}{\partial t} + u_h^{(n-1)*}(\cdot)\cdot\nabla \Big) \eta \Big\} \big(y(\cdot, s), t(s)\big) \Big\|_0 ds \\
&\leqslant \frac{3}{2} \int_{-1}^1 \Big(\Big\| \frac{\partial \eta}{\partial t} \big(y(\cdot, s), t(s)\big) \Big\|_0 + \|u_h^{(n-1)*}\|_{0,\infty} \big\| \nabla \eta \big(y(\cdot, s), t(s)\big) \big\|_0 \Big) ds
\end{aligned}
$$

$$
\begin{aligned}
&\leqslant \frac{3}{\sqrt{2}}\int_{-1}^{1}\Big\{\Big\|\frac{\partial \eta}{\partial t}\big(\cdot, t(s)\big)\Big\|_0 + \|u_h^{n-1}\|_{0,\infty}\big\|\nabla\eta\big(\cdot, t(s)\big)\big\|_0\Big\}ds \quad \text{(by (18.10))}\\
&\leqslant \frac{3}{\sqrt{2\Delta t}}\Big(\Big\|\frac{\partial \eta}{\partial t}\Big\|_{L^2(t^{n-2},t^n;L^2)} + \|u_h^{(n-1)*}\|_{0,\infty}\big\|\nabla\eta\big\|_{L^2(t^{n-2},t^n;L^2)}\Big)\\
&\leqslant \frac{3\alpha_{31}h}{\sqrt{2\Delta t}}(\|u_h^{(n-1)*}\|_{0,\infty}+1)\|(u,p)\|_{H^1(t^{n-2},t^n;H^2\times H^1)}\\
&\leqslant \frac{ch}{\sqrt{\Delta t}}(\|u_h^{(n-1)*}\|_{0,\infty}+1)\|(u,p)\|_{H^1(t^{n-2},t^n;H^2\times H^1)}.
\end{aligned}
$$

We get (18.29d) from the estimate

$$
\begin{aligned}
\|R_{h4}^n\|_0 &= \frac{1}{2\Delta t}\big\|-4\{e_h^{n-1} - e_h^{n-1}\circ X_1(u_h^{(n-1)*}, \Delta t)\} + \{e_h^{n-2} - e_h^{n-2}\circ X_1(u_h^{(n-1)*}, 2\Delta t)\}\big\|_0\\
&\leqslant c\alpha_{40}\|u_h^{(n-1)*}\|_{0,\infty}(\|e_h^{n-1}\|_1 + \|e_h^{n-2}\|_1).
\end{aligned}
$$

□

Proof of Lemma 7

Inequality (18.44a) is obtained by combining (18.20b) with (18.54). For (18.44b) we divide R_{h3}^n into three terms,

$$
\begin{aligned}
R_{h3}^n = \overline{D}_{\Delta t}^{(2)}\eta^n &+ \frac{1}{2\Delta t}\big[4\{\eta^{n-1} - \eta^{n-1}\circ X_1(u^{(n-1)*}, \Delta t)\}\\
&\qquad - \{\eta^{n-2} - \eta^{n-2}\circ X_1(u^{(n-1)*}, 2\Delta t)\}\big]\\
&+ \frac{1}{2\Delta t}\big[4\{\eta^{n-1}\circ X_1(u^{(n-1)*}, \Delta t) - \eta^{n-1}\circ X_1(u_h^{(n-1)*}, \Delta t)\}\\
&\qquad - \{\eta^{n-2}\circ X_1(u^{(n-1)*}, 2\Delta t) - \eta^{n-2}\circ X_1(u_h^{(n-1)*}, 2\Delta t)\}\big]\\
&\equiv R_{h31}^n + R_{h32}^n + R_{h33}^n.
\end{aligned}
$$

We have, by virtue of (18.20b),

$$
\begin{aligned}
\|R_{h31}^n\|_{V_h'} &\leqslant \|\overline{D}_{\Delta t}^{(2)}\eta^n\|_0 \leqslant \frac{3}{2}\|\overline{D}_{\Delta t}\eta^n\|_0 + \frac{1}{2}\|\overline{D}_{\Delta t}\eta^{n-1}\|_0\\
&\leqslant \frac{3}{2\sqrt{\Delta t}}\Big\|\frac{\partial\eta}{\partial t}\Big\|_{L^2(t^{n-1},t^n;L^2)} + \frac{1}{2\sqrt{\Delta t}}\Big\|\frac{\partial\eta}{\partial t}\Big\|_{L^2(t^{n-2},t^{n-1};L^2)}\\
&\leqslant \frac{c\alpha_{32}h^2}{\sqrt{\Delta t}}\|(u,p)\|_{H^1(t^{n-2},t^n;H^2\times H^1)} \leqslant \frac{ch^2}{\sqrt{\Delta t}}\|(u,p)\|_{H^1(t^{n-2},t^n;H^2\times H^1)}, \qquad (18.55a)\\
\|R_{h32}^n\|_{V_h'} &\leqslant \alpha_{41}\|u^{(n-1)*}\|_{1,\infty}(2\|\eta^{n-1}\|_0 + \|\eta^{n-2}\|_0)\\
&\leqslant \alpha_{41}\|u^{(n-1)*}\|_{1,\infty}3\alpha_{32}h^2\|(u,p)\|_{C(H^2\times H^1)} \leqslant c_{(u,p)}h^2, \qquad (18.55b)
\end{aligned}
$$

$$
\begin{aligned}
\|R^n_{h33}\|_{V_h'} &= \sup_{v_h \in V_h} \frac{1}{\|v_h\|_1} \frac{1}{2\Delta t} \Big\{ 4\Big(\eta^{n-1} \circ X_1(u_h^{(n-1)*}, \Delta t) - \eta^{n-1} \circ X_1(u^{(n-1)*}, \Delta t), v_h\Big) \\
&\qquad - \Big(\eta^{n-2} \circ X_1(u_h^{(n-1)*}, 2\Delta t) - \eta^{n-2} \circ X_1(u^{(n-1)*}, 2\Delta t), v_h\Big)\Big\} \\
&\leqslant \sup_{v_h \in V_h} \frac{1}{\|v_h\|_1} \frac{1}{2\Delta t} \big(4\|\eta^{n-1} \circ X_1(u_h^{(n-1)*}, \Delta t) - \eta^{n-1} \circ X_1(u^{(n-1)*}, \Delta t)\|_{0,1} \\
&\qquad + \|\eta^{n-2} \circ X_1(u_h^{(n-1)*}, 2\Delta t) - \eta^{n-2} \circ X_1(u^{(n-1)*}, 2\Delta t)\|_{0,1}\big)\|v_h\|_{0,\infty} \\
&\leqslant 2\alpha_{43}\|u_h^{(n-1)*} - u^{(n-1)*}\|_0(\|\eta^{n-1}\|_1 + \|\eta^{n-2}\|_1)\alpha_{21}h^{-d/6} \qquad (18.55c) \\
&\leqslant ch^{-d/6}(\|\eta^{n-1}\|_1 + \|\eta^{n-2}\|_1)(\|e_h^{n-1}\|_0 + \|e_h^{n-2}\|_0 + \|\eta^{n-1}\|_0 + \|\eta^{n-2}\|_0) \\
&\leqslant c'\alpha_{32}h^{1-d/6}\|(u,p)\|_{C(H^2\times H^1)}\big(\|e_h^{n-1}\|_0 + \|e_h^{n-2}\|_0 + \alpha_{32}h^2\|(u,p)\|_{C(H^2\times H^1)}\big) \\
&\leqslant c_{(u,p)}\big(\|e_h^{n-1}\|_0 + \|e_h^{n-2}\|_0 + h^2\big). \qquad (18.55d)
\end{aligned}
$$

From (18.55a), (18.55b) and (18.55d) we obtain (18.44b).

For (18.44c) we use the bound on R^n_{h3}. R^n_{h4} is obtained by replacing η^{n-1} with $-e_h^{n-1}$ in $R^n_{h32} + R^n_{h33}$. Hence, from (18.55b) and (18.55c) we have

$$
\begin{aligned}
\|R^n_{h4}\|_{V_h'} &\leqslant \alpha_{41}\|u^{(n-1)*}\|_{1,\infty}(2\|e_h^{n-1}\|_0 + \|e_h^{n-2}\|_0) \\
&\quad + 2\alpha_{21}\alpha_{43}h^{-d/6}\|u_h^{(n-1)*} - u^{(n-1)*}\|_0(\|e_h^{n-1}\|_1 + \|e_h^{n-2}\|_1) \\
&\leqslant c\big\{\|u^{(n-1)*}\|_{1,\infty}(\|e_h^{n-1}\|_0 + \|e_h^{n-2}\|_0) + h^{-d/6}(\|e_h^{n-1}\|_1 + \|e_h^{n-2}\|_1) \\
&\qquad \times (\|e_h^{n-1}\|_0 + \|e_h^{n-2}\|_0 + \alpha_{32}h^2\|(u,p)\|_{C(H^2\times H^1)})\big\} \\
&\leqslant c_{(u,p)}\big\{1 + h^{-d/6}(\|e_h^{n-1}\|_1 + \|e_h^{n-2}\|_1)\big\}\big(\|e_h^{n-1}\|_0 + \|e_h^{n-2}\|_0 + h^2\big),
\end{aligned}
$$

which implies (18.44c). □

References

1. Achdou, Y., Guermond, J.-L.: Convergence analysis of a finite element projection/Lagrange-Galerkin method for the incompressible Navier-Stokes equations. SIAM J. Numer. Anal. **37**, 799–826 (2000)
2. Barrett, R., Berry, M., Chan, T.F., Demmel, J., Donato, J., Dongarra, J., Eijkhout, V., Pozo, R., Romine, C., van der Vorst, H.: Templates for the Solution of Linear Systems: Building Blocks for Iterative Methods. SIAM, Philadelphia (1994)
3. Boukir, K., Maday, Y., Métivet, B., Razafindrakoto, E.: A high-order characteristics/finite element method for the incompressible Navier-Stokes equations. Int. J. Numer. Methods Fluids **25**, 1421–1454 (1997)
4. Brezzi, F., Douglas Jr., J.: Stabilized mixed methods for the Stokes problem. Numer. Math. **53**, 225–235 (1988)
5. Brezzi, F., Pitkäranta, J.: On the stabilization of finite element approximations of the Stokes equations. In: Hackbusch, W. (ed.) Efficient Solutions of Elliptic Systems, pp. 11–19. Vieweg, Wiesbaden (1984)
6. Ciarlet, P.G.: The Finite Element Method for Elliptic Problems. North-Holland, Amsterdam (1978)

7. Douglas Jr., J., Russell, T.F.: Numerical methods for convection-dominated diffusion problems based on combining the method of characteristics with finite element or finite difference procedures. SIAM J. Numer. Anal. **19**, 871–885 (1982)
8. Duvaut, G., Lions, J.L.: Inequalities in Mechanics and Physics. Springer, Berlin (1976)
9. Ewing, R.E., Russell, T.F.: Multistep Galerkin methods along characteristics for convection-diffusion problems. In: Vichnevetsky, R., Stepleman, R.S. (eds.) Advances in Computer Methods for Partial Differential Equations, vol. IV, pp. 28–36. IMACS (1981)
10. Franca, L.P., Stenberg, R.: Error analysis of some Galerkin least squares methods for the elasticity equations. SIAM J. Numer. Anal. **28**, 1680–1697 (1991)
11. Girault, V., Raviart, P.-A.: Finite Element Methods for Navier-Stokes Equations Theory and Algorithms. Springer, Berlin (1986)
12. Heywood, J.G., Rannacher, R.: Finite element approximation of the nonstationary Navier-Stokes problem I. Regularity of solutions and second-order error estimates for spatial discretization. SIAM J. Numer. Anal. **19**, 275–311 (1982)
13. Heywood, J.G., Rannacher, R.: Finite element approximation of the nonstationary Navier-Stokes problem, Part IV: error analysis for second-order time discretization. SIAM J. Numer. Anal. **27**, 353–384 (1990)
14. Notsu, H.: Numerical computations of cavity flow problems by a pressure stabilized characteristic-curve finite element scheme. Trans Jpn. Soc. Comput. Eng. Sci. **20080032** (2008)
15. Notsu, H., Tabata, M.: A combined finite element scheme with a pressure stabilization and a characteristic-curve method for the Navier-Stokes equations. Trans. Jpn. Soc. Ind. Appl. Math. **18**, 427–445 (2008). (in Japanese)
16. Notsu, H., Tabata, M.: Error estimates of a pressure-stabilized characteristics finite element scheme for the Oseen equations. J. Sci. Comput. **65**, 940–955 (2015)
17. Notsu, H., Tabata, M.: Error estimates of a stabilized Lagrange–Galerkin scheme for the Navier–Stokes equations. ESAIM: M2AN, **50**, 361–380 (2016)
18. Pironneau, O.: On the transport-diffusion algorithm and its applications to the Navier-Stokes equations. Numer. Math. **38**, 309–332 (1982)
19. Ravindran, S.S.: Convergence of extrapolated BDF2 finite element schemes for unsteady penetrative convection model. Numer. Funct. Anal. Optim. **33**, 48–79 (2012)
20. Rui, H., Tabata, M.: A second order characteristic finite element scheme for convection-diffusion problems. Numer. Math. **92**, 161–177 (2002)
21. Saad, Y.: Iterative Methods for Sparse Linear Systems. SIAM, Philadelphia (2003)
22. Stroud, A.H.: Approximate Calculation of Multiple Integrals. Prentice-Hall, Englewood Cliffs, New Jersey (1971)
23. Süli, E.: Convergence and nonlinear stability of the Lagrange-Galerkin method for the Navier-Stokes equations. Numer. Math. **53**, 459–483 (1988)
24. Tabata, M., Uchiumi, S.: An exatly computable Lagrange–Galerkin scheme for the Navier–Stokes equations and its error estimates. Math. comput. (To appear)

Chapter 19
Chaotic Dynamics in an Integro-Differential Reaction-Diffusion System in the Presence of 0:1:2 Resonance

Toshiyuki Ogawa and Takashi Okuda Sakamoto

Abstract The dynamics and bifurcation structure of the normal form in the presence of 0:1:2 resonance are studied. It is proved that connecting orbits (heteroclinic cycles or homoclinic orbits) exist on the center manifold of the normal form. Moreover, to study the dynamics around the triple degeneracy of the normal form, we apply the results in Dumortier and Kokubu [4]. The sufficient conditions for the existence of heteroclinic cycles in a scaling family (blow-up vector field) of the 0:1:2 normal form are obtained. These results give a reasonable explanation for the behaviors of the solutions to an integro-reaction-diffusion system.

Keywords Normal form · 0:1:2 resonance · Connecting orbit · Heteroclinic loop

19.1 Introduction

The dynamics of patterns right after non-trivial instability can be characterized by the so-called normal form equations for critical modes. The dynamics observed in the standard normal form with O(2) symmetry can be classified by simple calculations (for instance, see [7, 8]). However, there are exceptional cases in the normal form with O(2) symmetry. One of the well-known example is the normal form with respect to 1 : 2 mode interaction. It has quadratic resonance terms, and moreover, if the sign of coefficients of the resonance terms are opposite, then the time-periodic solutions, traveling waves, and modulated traveling waves exist with a suitable choice of parameters [1, 14, 16]. The typical application of these results is the Kuramoto-Shivasinsky dynamics [2]. The O(2)-symmetric normal form (which has quadratic

T. Ogawa (✉)
Graduate School of Advanced Mathematical Science,
Meiji University, Nakano 164-8525, Japan
e-mail: togw@meiji.ac.jp

T.O. Sakamoto
School of Science and Technology, Meiji University,
Higashimita, Kawasaki 214-8571, Japan
e-mail: sakamoto@meiji.ac.jp

Y. Shibata and Y. Suzuki (eds.), *Mathematical Fluid Dynamics, Present and Future*, Springer Proceedings in Mathematics & Statistics 183,
DOI 10.1007/978-4-431-56457-7_19

terms) with respect to 0:1:2 mode interaction is also one of exceptional examples and studied in [15]:

$$\begin{cases} \dot{z}_0 = \mu_0 z_0 + 2(B_1|z_1|^2 + B_2|z_2|^2), \\ \dot{z}_1 = (\mu_1 - B_1 z_0)z_1 + \overline{z_1} z_2, \\ \dot{z}_2 = (\mu_2 - B_2 z_0)z_2 - z_1^2, \end{cases} \tag{19.1}$$

where $z_j(t) \in \mathbb{C}$ and $\overline{\cdot}$ denotes complex conjugate. This system has a "more rich structure": symmetry breaking pitchfork bifurcations, standing wave heteroclinic, chaotic heteroclinic cycles, torus bifurcations.

However, if the system has up-down symmetry, then the normal form does not have any quadratic term. In fact, O(2)-symmetric normal form with Z_2-symmetry in the presence of $0:1:2$ resonance is given by the following:

$$\begin{cases} \dot{z}_0 = (\mu_0 + a_1|z_0|^2 + a_2|z_1|^2 + a_3|z_2|^2)z_0 + a_4\overline{z_1}^2 z_2, \\ \dot{z}_1 = (\mu_1 + b_1|z_0|^2 + b_2|z_1|^2 + b_3|z_2|^2)z_1 + b_4 z_0 \overline{z_1} z_2, \\ \dot{z}_2 = (\mu_2 + c_1|z_0|^2 + c_2|z_1|^2 + c_3|z_2|^2)z_2 + c_4 z_0 z_1^2, \end{cases} \tag{19.2}$$

where $z_j(t) \in \mathbb{C}$. It has been revealed that the system exhibits another types of complex behaviors even when we consider the following system of Eqs. (19.3) on the invariant real sub-space of the system (19.2) (see [13]):

$$\begin{cases} \dot{z}_0 = (\mu_0 + a_1 z_0^2 + a_2 z_1^2 + a_3 z_2^2)z_0 + a_4 z_1^2 z_2, \\ \dot{z}_1 = (\mu_1 + b_1 z_0^2 + b_2 z_1^2 + b_3 z_2^2)z_1 + b_4 z_0 z_1 z_2, \\ \dot{z}_2 = (\mu_2 + c_1 z_0^2 + c_2 z_1^2 + c_3 z_2^2)z_2 + c_4 z_0 z_1^2, \end{cases} \tag{19.3}$$

where $z_j(t) \in \mathbb{R}$. Let us introduce, as an example, the following integro-differential reaction-diffusion system which is studied in [5, 12, 13]:

$$\begin{cases} u_t = D_1 u_{xx} + au + bv + F(u,v) + \dfrac{s}{L}\displaystyle\int_0^L u(t,x)\,dx, \ x \in (0,L), t > 0, \\ v_t = D_2 v_{xx} + cu + dv + G(u,v), \ x \in (0,L), t > 0, \\ u_x = v_x = 0 \text{ at } x = 0, L, \ t > 0. \end{cases} \tag{19.4}$$

It should be noted that the cases when the global feedback is negative ($s < 0$) and positive ($s > 0$) are studied in [5, 12] and [13], respectively. We consider the dynamical system (19.4) in a phase space

$$X := \left\{(u,v) \in [H^2(\Omega)]^2;\, u_x = v_x = 0 \text{ at } x = 0, L\right\},$$

where Ω denotes an interval $(0,L) \subset \mathbb{R}$. And we assume the following:

- (A1) The functions (higher order terms) F and G are sufficiently smooth;
- (A2) $F(u,v) \equiv -F(-u,-v)$ and $G(u,v) \equiv -G(-u,-v)$ hold;

- (A3) The coefficients of linear parts satisfy $a, c > 0,\ b, d < 0,\ a + d < 0$ and $\Delta := ad - bc > 0$;
- (A4) $\dfrac{bc}{d} + d < 0$ holds.

The integro-differential reaction-diffusion system (19.4) has a 0:1:2 - triple degenerate point around a trivial stationary solution $(u(t,x), v(t,x)) \equiv (0,0)$ by a suitable choice of the diffusion coefficients D_2, system size L and coefficients of global feedback s. And then the dynamics on the center manifold of (19.4) is given by the dynamics of (19.3). We briefly explain how to derive the system (19.3) from (19.4) in the Appendix A (see also [13]).

Definition Let $(U_\ell(x), V_\ell(x))$ be a stationary solution of (19.4). If there exist nonzero constants ξ_ℓ and η_ℓ such that the stationary solution $(U_\ell(x), V_\ell(x))$ has the form

$$U_\ell(x) = \xi_\ell \cos(k_0 \ell x) + o(3), \quad V_\ell(x) = \eta_\ell \cos(k_0 \ell x) + o(3),$$

then, we call $(U_\ell(x), V_\ell(x))$ is an ℓ-mode stationary solution (or ℓ -mode solution) of (19.4). Here, $o(3)$ denotes $o\left(\sqrt{\xi_\ell^2 + \eta_\ell^2}^{\,3}\right)$ and k_0 denotes π/L.

It is known that the system (19.3) has oscillating solutions around 1-mode stationary solutions [13]. Moreover, it has "chaotic" behavior around 1-mode as well (see Sect. 19.1.1 in more details). Therefore, the purpose of this paper is to study the detailed dynamics around 1-mode stationary solutions of the normal form (19.3). It should be noted that this behavior can not be observed in the quadratic case (19.1) since the system (19.1) dose not have 1-mode stationary solutions with any choice of parameters. While in (19.3), we will see that there are double zero degenerate point as a secondary bifurcation point on a branches of 1-mode stationary solutions by a suitable choice of parameters. Such a degenerate point include the Hopf instability point which has been studied in [13].

19.1.1 Numerical Results

In [13], not only the stationary bifurcation but also the Hopf bifurcation around 1-mode stationary solutions to the system (19.3) have been studied. In addition, the chaotic behavior was shown numerically. We can observe this fact numerically as shown in the Figs. 19.1, 19.2, 19.3 and 19.4 (Figs. 19.1 and 19.2 are numerical results for the normal form, while Fig. 19.3 and 19.4 are numerical results for the integro-differential reaction-diffusion system (19.4)). In those figures, the parameters in (19.4) are chosen as follows:

$$\begin{gathered} D_1 = 1/4, \quad a = 1, \quad b = -10, \quad c = 2, \quad d = -5, \\ F(u,v) = -u^3 \quad \text{and} \quad G(u,v) = -0.9u^3. \end{gathered} \tag{19.5}$$

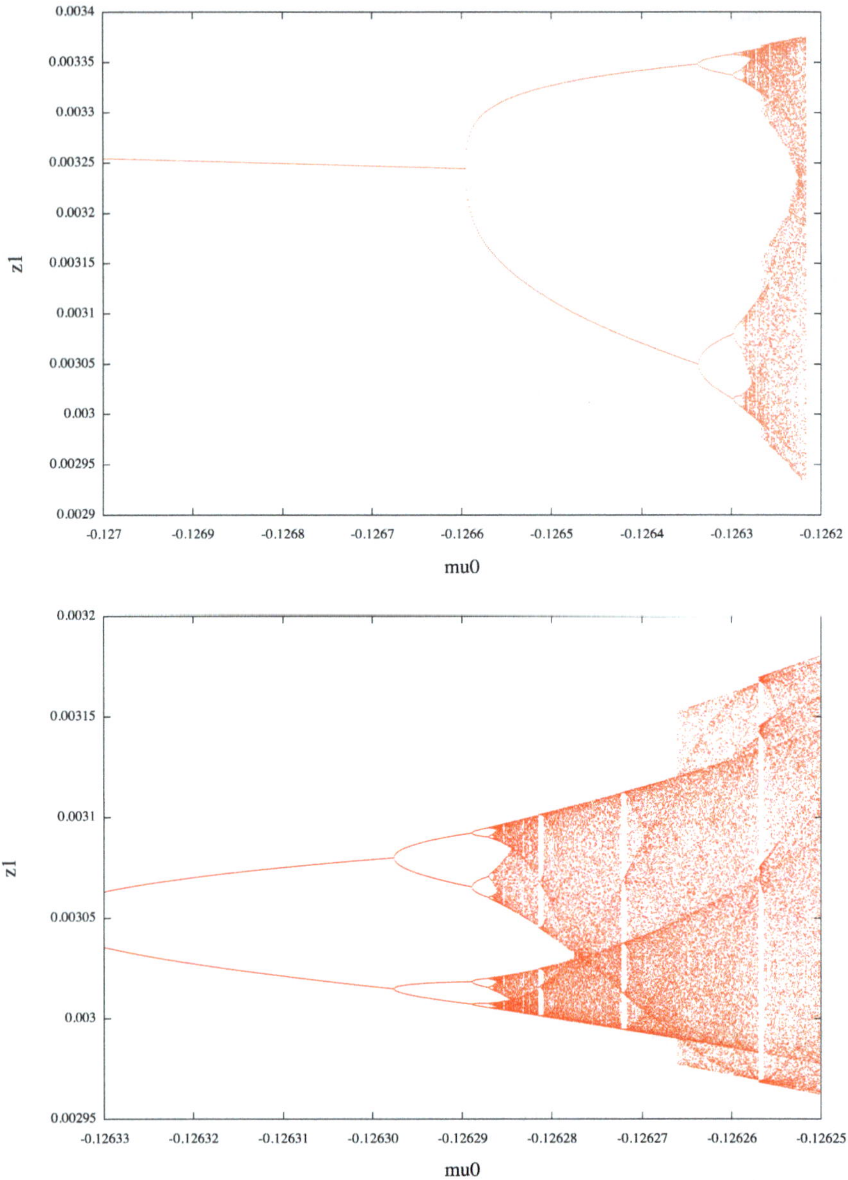

Fig. 19.1 Bifurcation diagram of the Poincaré map of (19.3) on the section $z_1 = 0$ with the coefficients in (19.6) and $\mu_1 = 1.524579805 \times 10^{-2}$, $\mu_2 = -10^{-3}$. The *vertical* and *horizontal* axes correspond to z_1 and μ_0, respectively. [*Above* $\mu_0 \in [-0.127, -0.12617]$] [*Below* Close-up view of the above figure in $\mu_0 \in [-0.12633, -0.12625]$]

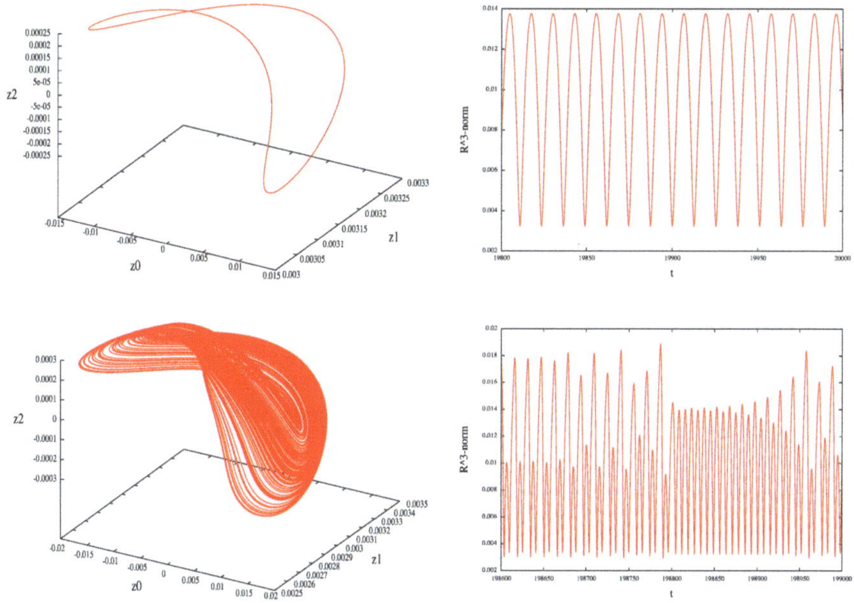

Fig. 19.2 The attractors in the phase space (*left*) and the $\mathbb{R}^3$-norm $\sqrt{z_0(t)^2 + z_1(t)^2 + z_2(t)^2}$ with respect to time (*right*) for (19.3) with the typical parameter values $\mu_0 = -0.126217$, $\mu_1 = 1.524579805 \times 10^{-2}$ and $\mu_2 = -0.001$, and the coefficients (19.6). The *above* and *below* figures correspond to the periodic orbit bifurcate from the 1-mode stationary solution and chaotic solution, respectively

Then we have $s^* = 3$ and $(k_0^{1,2}, D_2^{1,2}) \approx (0.87, 25.88)$, where s^* is the critical value for s so that the system (19.4) has 0:1:2 triple degeneracy at $(k_0, D_2) = (k_0^{1,2}, D_2^{1,2})$(see Appendix A for the details). Using the explicit forms of coefficients shown in the Appendix B, they can be obtained as follows.

$$\begin{aligned} &a_1 \approx 100.00,\ a_2 \approx 14644.17,\ a_3 \approx 1.69 \times 10^5,\ a_4 \approx 2.45 \times 10^5,\\ &b_1 \approx -49.29,\ b_2 \approx -1203.01,\ b_3 \approx -27695.10,\ b_4 \approx -1652.32,\\ &c_1 \approx -67.14,\ c_2 \approx -3277.22,\ c_3 \approx -18861.66,\ c_4 \approx -97.761. \end{aligned} \tag{19.6}$$

19.1.2 Main Results

Before stating the main results, let us classify the equilibria of (19.3): if $\mu_0 a_1 < 0$, $\mu_1 b_2 < 0, \mu_2 c_3 < 0$ then the system (19.3) has equilibria $\pm\mathbf{e}_0 := \pm(\sqrt{-\mu_0/a_1}, 0, 0)$, $\pm\mathbf{e}_1 := \pm(0, \sqrt{-\mu_1/b_2}, 0)$, $\pm\mathbf{e}_2 := \pm(0, 0, \sqrt{-\mu_2/c_3})$, respectively. And they correspond to the pure mode stationary solutions (the equilibria $\pm\mathbf{e}_\ell$ correspond to the ℓ-mode stationary solutions). If $(a_3\mu_2 - c_3\mu_0)(a_1c_3 - c_1a_3) > 0$ and $(c_1\mu_0 -$

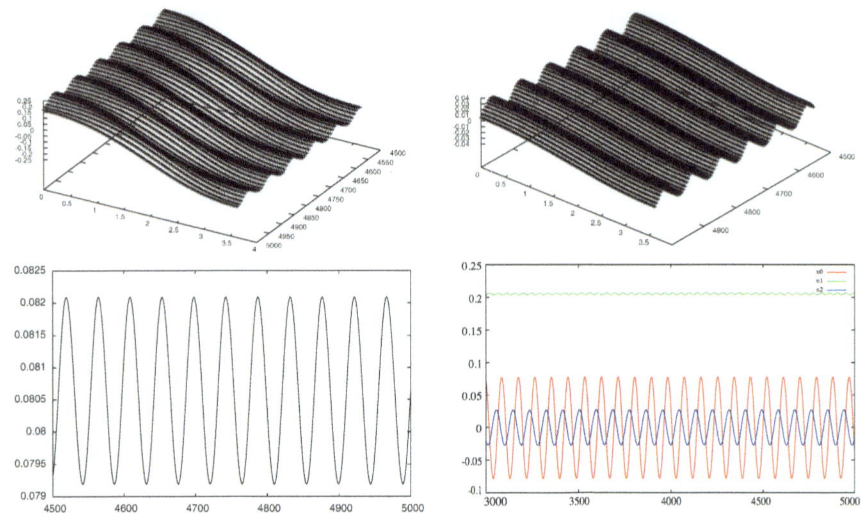

Fig. 19.3 Numerical solutions for the integro-differential reaction-diffusion system (19.4) with $D_2 = 27.13$ in the case of (19.5). The initial values are stable 1-mode stationary solutions at $D_2 = 27.0$. [*Above* The *left* and *right* figures correspond to the graph of $u(t, x)$ and $v(t, x)$, ($t \in [4500, 5000]$, $x \in [0, L]$), respectively.] [*Below left* The graph of $\|(u, v)\|_{L^2}(t)$, $t \in [4500, 5000]$. *Vertical axis* L^2 norm of u, *horizontal axis* t.] [*Below right* Graph of $u_0(t)$ (*black line*), $u_1(t)$ (*green line*) and $u_2(t)$ (*blue line*), where $u_j(t)$ denotes the j-th Fourier coefficients of $u(t, x)$.]

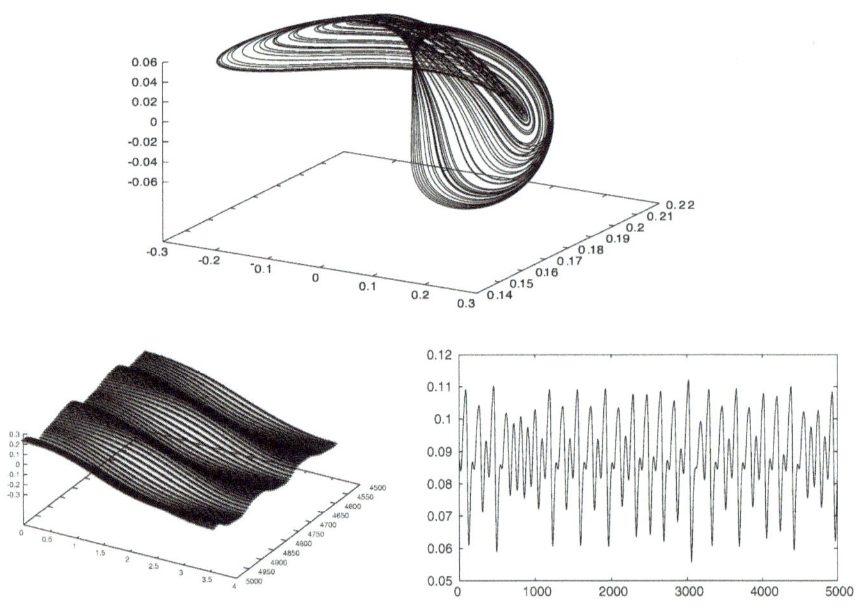

Fig. 19.4 Numerical solutions of (19.4) with $D_2 = 26.5$, $k_0 = 0.874919$, $s = 2.98212$, and the other parameters and nonlinear terms F and G are chosen as (19.5). (*Above*) The orbit of $(u_0(t), u_1(t), u_2(t))$ on Fourier space (Here, $u_j(t)$ denotes j-th Fourier mode of $u(t, x)$). (*Below left*) The profile of $u(t, x)$. (*Below right*) The graph of $\|(u, v)\|_{L^2}(t)$ with respect to time

$a_1\mu_2)(a_1c_3 - a_3c_1) > 0$, then the system (19.3) has equilibria $(z_0, z_1, z_2) = \pm(z_0^*, 0, z_2^*)$ and $(z_0, z_1, z_2) = \pm(z_0^*, 0, -z_2^*)$, where $z_0^* := [(a_3\mu_2 - c_3\mu_0)/(a_1c_3 - c_1a_3)]^{1/2}$ and $z_2^* := [(c_1\mu_0 - a_1\mu_2)/(a_1c_3 - a_3c_1)]^{1/2}$. They correspond to doubly mixed mode stationary solutions. In addition, if (z_{0*}, z_{1*}, z_{2*}), $z_{j_*} \neq 0$ is a root of

$$\begin{cases} \mu_0 + a_1z_0^2 + a_2z_1^2 + a_3z_2^2 + a_4\dfrac{z_1^2z_2}{z_0} = 0, \\ \mu_1 + b_1z_0^2 + b_2z_1^2 + b_3z_2^2 + b_4z_0z_2 = 0, \\ \mu_2 + c_1z_0^2 + c_2z_1^2 + c_3z_2^2 + c_4\dfrac{z_0z_1^2}{z_2} = 0, \end{cases} \tag{19.7}$$

then, it corresponds to a triple mixed mode stationary solution.

Now we can see that the onset of chaotic patters in (19.4) is the Hopf bifurcation around a stationary solution, and moreover, the leading Fourier mode of $u(t, x)$ in Fig. 19.4 seems to be 1-mode.

Based on this consideration, we study the detailed dynamics around the equilibrium $\mathbf{e}_1$ of the normal form (19.3). We can see that there are double zero degenerate point as a secondary bifurcation point on a branches of 1-mode stationary solutions by a suitable choice of parameters. Such a degenerate point include the Hopf instability point which has been studied in [13]. However, it is not sufficient to understand the whole dynamics. In fact, the analysis around the double degenerate point could be necessary to explain the "complex" patterns appearing in (19.4) near triple degenerate point. Our analysis in Sect. 19.2 yields the following results (see Fig. 19.5):

Proposition 19.1 *Let μ_1 be fixed so that $\mu_1b_2 < 0$ (then, 1-mode statically solutions exist). Generically, the followings hold with a suitable choice of parameters μ_0 and μ_2:*

(I) If $a_4c_4 \neq 0$, then there exist non-zero constants η_j, ξ_j, $j = 0, 1, 2$ (they depend on the parameters and coefficient of (19.4)) such that the integro-differential reaction-diffusion system (19.4) has the (small amplitude) triple mixed mode stationary solutions:

$$u(x) = \pm U_{mix}^{\pm} := \pm\left(\pm\,\xi_0 + \xi_1\cos\left(\frac{\pi}{L}x\right) \pm \xi_2\cos\left(2\frac{\pi}{L}x\right)\right) + o(3),$$
$$v(x) = \pm V_{mix}^{\pm} := \pm\left(\pm\,\eta_0 + \eta_1\cos\left(\frac{\pi}{L}x\right) \pm \eta_2\cos\left(2\frac{\pi}{L}x\right)\right) + o(3)$$

which bifurcate from the 1-mode stationary solutions $\pm(U_1(x), V_1(x))$ through the pitchfork bifurcations. Here, $o(3)$ denotes $o\left(\sqrt{\sum_{j=0}^{2}(\xi_j^2 + \eta_j^2)}^{\,3}\right)$.

(II) Moreover, if $a_4c_4 < 0$, then there exist time-periodic solutions $\Omega^{\pm}$ on the phase space X such that Ω^{+} and Ω^{-} bifurcate from the 1-mode stationary solutions $(U_1(x), V_1(x))$ and $(-U_1(x), -V_1(x))$ through the Hopf bifurcations, respectively. Moreover, one of the followings holds:

(i) There exist heteroclinic cycles Γ_1, Γ_2 on the phase space X such that Γ_1 connects (U^+_{mix}, V^+_{mix}) and (U^-_{mix}, V^-_{mix}), and Γ_2 connects $(-U^+_{mix}, -V^+_{mix})$ and $(-U^-_{mix}, -V^-_{mix})$;

(ii) There exist two pairs of homoclinic orbits $\Gamma_3^\pm$ and $\Gamma_4^\pm$ on the phase space X such that $\Gamma_3^\pm$ connects $(U_1(x), V_1(x))$ to itself, and $\Gamma_4^\pm$ connects $(-U_1(x), -V_1(x))$ to itself.

We also analyze the triple degenerate point of (19.3) in Sect. 19.3. The system (19.3) can be re-normalized as follows:

Proposition 19.2 *Let μ_1 be fixed so that $\mu_1 b_2 < 0$. Then, there are coordinate and parameter changes, and a scaling with respect to coordinate, parameters and time, such that the system (19.3) can be transformed into the following system:*

$$\begin{cases} \dot{x} = y, \\ \dot{y} = \lambda x + \mu y + axz + byz, \\ \dot{z} = \nu + x^2 + z^2. \end{cases} \tag{19.8}$$

This system is well studied by Dumortier and Kokubu [4]. They proved the existence of heteroclinic cycles and chaotic dynamics around the singularities of (19.8). We study the dynamics of (19.3) around the 1-mode equilibria by applying their results in Sect. 19.3.

This paper is organized as follows: in the next section, we study the dynamics around the 1-mode stationary solutions by using the center manifold theory and normalizing technic. The section is divided into three parts: in the first part, we state that the normal form (19.3) can exhibit double zero degenerate points around the 1-mode stationary solutions. In the second part (Sect. 19.2.1), we compute the bifurcation equations in the case when the system has simple 0-eigenvalue. The existence of periodic orbits and connecting orbits (heteroclinic cycles and homoclinic orbits) stated in Proposition 19.1 is the consequence of Theorem 19.3. Moreover, we will also discuss existence of the another types of time-periodic orbits and the stabilities of solutions. We also consider the blow-up vector fields associated with

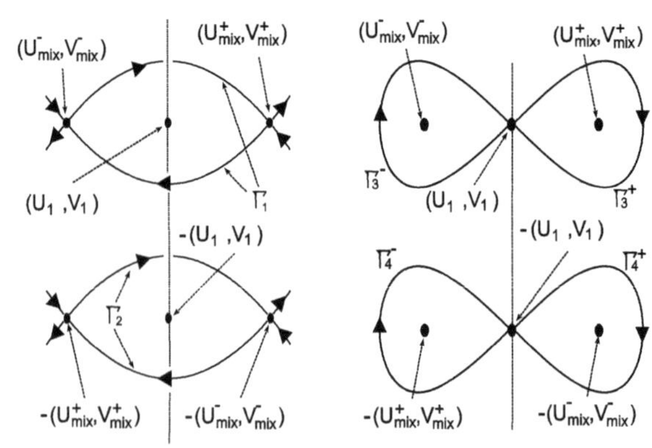

Fig. 19.5 Schematic pictures of heteroclinic cycles and homoclinic orbits stated in Proposition 19.1. The *left* and *right* figures correspond to the situation *(I)* and *(II)* of Proposition 19.1, respectively

the normal form (19.3) in Sect. 19.3. We can find a scaling family of (19.3) which have heteroclinic cycles around the 1-mode equilibria of (19.3).

19.2 Bifurcation Structure Around a 1-Mode Stationary Solutions

In this section, we study the bifurcation structure around a 1-mode equilibria

$$(z_0, z_1, z_2) = \pm\mathbf{e}_1 := (0, \pm\sqrt{-\mu_1/b_2}, 0).$$

Since the normal form (19.3) is invariant under the mapping

$$(z_0, z_1, z_2) \longrightarrow (z_0, -z_1, z_2),$$

it is sufficient to consider the bifurcation structure only around $\mathbf{e}_1$. Let $\tilde{M}_{\mathbf{e}_1}$ be a 2×2 matrix defined by

$$\tilde{M}_{\mathbf{e}_1} := \begin{pmatrix} \alpha & \beta \\ -c_4\mu_1/b_2 & \gamma \end{pmatrix},$$

where

$$\alpha = \mu_0 - a_2\mu_1/b_2, \quad \beta = -a_4\mu_1/b_2 \quad \text{and} \quad \gamma = \mu_2 - c_2\mu_1/b_2.$$

We also define the sets $\mathscr{S}_D$ and $\mathscr{S}_T$ as follows:

$$\mathscr{S}_D := \{(\mu_0, \mu_2)\,;\, \det\tilde{M}_{\mathbf{e}_1} = 0\}, \quad \mathscr{S}_T := \{(\mu_0, \mu_2)\,;\, \operatorname{tr}\tilde{M}_{\mathbf{e}_1} = 0\}.$$

Then we have the following lemma.

Lemma 19.1 *Let μ_1 be fixed so that $\mu_1 b_2 < 0$. Then the followings hold:*

- *If $a_4c_4 > 0$ then, then the linearized matrix around 1-mode equilibrium $\mathbf{e}_1$ has a zero eigenvalue if and only if $(\mu_0, \mu_2) \in \mathscr{S}_D$.*
- *If $a_4c_4 < 0$, then the linearized matrix around 1-mode equilibrium $\mathbf{e}_1$ has a zero eigenvalue if and only if $(\mu_0, \mu_2) \in \mathscr{S}_D \setminus \mathscr{S}_T$.*
- *If $a_4c_4 < 0$, then the linearized matrix of (19.3) around 1-mode equilibrium $\mathbf{e}_1$ has a double degenerate point*

$$(\mu_0, \mu_2) = P_+ := (\mu_0^+, \mu_2^+) \quad \text{and} \quad (\mu_0, \mu_2) = P_- := (\mu_0^-, \mu_2^-), \tag{19.9}$$

where

$$\mu_0^\pm := -\frac{\mu_1}{b_2}(-a_2 \pm \sqrt{-a_4c_4}) \quad \text{and} \quad \mu_2^\pm := \frac{\mu_1}{b_2}(c_2 \pm \sqrt{-a_4c_4}).$$

Proof Let us rewrite the system (19.3) as follows:

$$\begin{pmatrix} \dot{z}_1 \\ \dot{z}_0 \\ \dot{z}_2 \end{pmatrix} = \begin{pmatrix} \mu_1 & 0 & 0 \\ 0 & \mu_0 & 0 \\ 0 & 0 & \mu_2 \end{pmatrix} \begin{pmatrix} \tilde{z}_1 \\ z_0 \\ z_2 \end{pmatrix} + \begin{pmatrix} F_1(z_1, z_0, z_2) \\ F_0(z_1, z_0, z_2) \\ F_2(z_1, z_0, z_2) \end{pmatrix}, \tag{19.10}$$

where

$$\begin{aligned} F_0(z_1, z_0, z_2) &= (a_1 z_0^2 + a_2 z_1^2 + a_3 z_2^2) z_1 + a_4 z_1^2 z_2, \\ F_1(z_1, z_0, z_2) &= (b_1 z_0^2 + b_2 z_1^2 + b_3 z_2^2) z_1 + b_4 z_0 z_1 z_2, \\ F_2(z_1, z_0, z_2) &= (c_1 z_0^2 + c_2 z_1^2 + c_3 z_2^2) z_2 + c_4 z_0 z_1^2. \end{aligned}$$

Then, linearized matrix of $\mathbf{e}_1$ is given by

$$M_{\mathbf{e}_1} := \begin{pmatrix} -2\mu_1 & 0 & 0 \\ 0 & \alpha & \beta \\ 0 & -c_4\mu_1/b_2 & \gamma \end{pmatrix},$$

where

$$\alpha = \mu_0 - a_2\mu_1/b_2, \quad \beta = -a_4\mu_1/b_2 \quad \text{and} \quad \gamma = \mu_2 - c_2\mu_1/b_2.$$

We can see that if $a_4 c_4 > 0$, then the line

$$\mathscr{S}_T = \{(\mu_0, \mu_2)\,;\ \mathrm{tr}\, \tilde{M}_{\mathbf{e}_1}\} = \{(\mu_0, \mu_1)\,;\ \mu_0 + \mu_2 - (a_2 + c_2)\mu_1/b_2 = 0\}$$

and the set

$$\begin{aligned} \mathscr{S}_D &= \{(\mu_0, \mu_2)\,;\ \det \tilde{M}_{\mathbf{e}_1} = 0\} \\ &= \{(\mu_0, \mu_2)\,;\ (\mu_0 - a_2\mu_1/b_2)(\mu_2 - c_2\mu_1/b_2) - a_4 c_4 \mu_1^2/b_2^2 = 0\} \end{aligned}$$

don't intersect each other on μ_0-μ_2 plane. Therefore, the matrix $M_{\mathbf{e}_1}$ has a zero eigenvalue if and only if $\det \tilde{M}_{\mathbf{e}_1} = 0$. In addition, if $a_4 c_4 < 0$, then the matrix $M_{\mathbf{e}_1}$ has a zero eigenvalue if and only if $\det \tilde{M}_{\mathbf{e}_1} = 0$ and $\mathrm{tr}\, \tilde{M}_{\mathbf{e}_1} \neq 0$.

On the other hand, two sets $\mathscr{S}_D$ and $\mathscr{S}_T$ intersect each other at two points P_+ and P_- if $a_4 c_4 < 0$. Moreover, the Jordan block of the matrix M_{e_1} at the degenerate point $P_\pm$ is given by the following:

$$T^{-1} M_{\mathbf{e}_1} T = \begin{pmatrix} -2\mu_1 & 0 & 0 \\ 0 & 0 & 1 \\ 0 & 0 & 0 \end{pmatrix}, \quad \textit{where} \quad T = \begin{pmatrix} 1 & 0 & 0 \\ 0 & -2\beta & 0 \\ 0 & \alpha - \gamma & -2 \end{pmatrix}.$$

This completes the proof. □

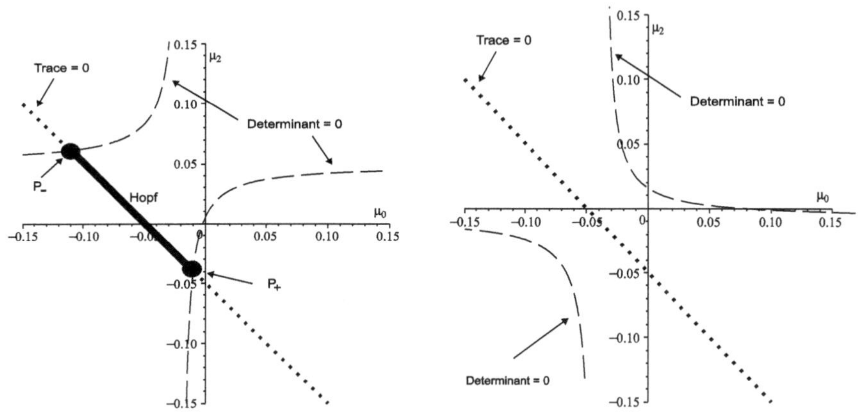

Fig. 19.6 Schematic picture of the set of critical points on the $\mu_0 - \mu_2$ plane. The *solid* and *doted lines* correspond to Tr $\tilde{M}_{\mathbf{e}_1} = 0$. The *dashed lines* correspond to det $\tilde{M}_{\mathbf{e}_1} = 0$. (*Left*) Two sets $\mathscr{S}_D$ and $\mathscr{S}_T$ intersect each other at P_+ and P_- when $a_2 > 0, b_2, c_2 < 0, a_2 + c_2 > 0$ and $a_4c_4 < 0$ hold. *Solid line* corresponds to the set of Hopf bifurcation points: $\{(\mu_0, \mu_2);\ \mathrm{tr}\,\tilde{M}_{\mathbf{e}_1} = 0 \text{ and } \det \tilde{M}_{\mathbf{e}_1} > 0\}$. (*Right*) Two sets $\mathscr{S}_D$ and $\mathscr{S}_T$ don't intersect each other when $a_2 > 0, b_2, c_2 < 0, a_2 + c_2 > 0$ and $a_4c_4 > 0$ hold

This lemma gives a set of critical points on the μ_0-μ_2 plane (see Fig. 19.6). We study the detailed dynamics and bifurcation structures on the center manifold of (19.3) in the case when $a_4c_4 > 0$ and $a_4c_4 < 0$ in Sects. 19.2.1 and 19.2.2, respectively.

19.2.1 The Case When $a_4c_4 > 0$

Let us study the bifurcations in the case when $a_4c_4 > 0$. We note again that the set of critical points in (μ_0, μ_2) space is given by

$$\mathscr{S}_D = \{(\mu_0, \mu_2);\ (\mu_0 - a_2\mu_1/b_2)(\mu_2 - c_2\mu_1/b_2) - a_4c_4\mu_1^2/b_2^2 = 0\}.$$

Let $\tilde{z}_1$ be a new variable defined by

$$\tilde{z}_1 = z_1 - z^*, \quad \text{where} \quad z^* = \sqrt{-\mu_1/b_2}.$$

Then, $(\tilde{z}_1(t), z_0(t), z_2(t))$ satisfies the following system:

$$\begin{pmatrix} \dot{\tilde{z}}_1 \\ \dot{z}_0 \\ \dot{z}_2 \end{pmatrix} = \begin{pmatrix} -2\mu_1 & 0 & 0 \\ 0 & \alpha & \beta \\ 0 & -c_4\mu_1/b_2 & \gamma \end{pmatrix} \begin{pmatrix} \tilde{z}_1 \\ z_0 \\ z_2 \end{pmatrix} + \begin{pmatrix} N_1(\tilde{z}_1, z_0, z_2) \\ N_0(\tilde{z}_1, z_0, z_2) \\ N_2(\tilde{z}_1, z_0, z_2) \end{pmatrix}, \qquad (19.11)$$

where

$$\begin{aligned} N_0(\tilde{z}_1, z_0, z_2) &= 2a_4 z^* \tilde{z}_1 z_2 + 2a_2 z^* \tilde{z}_1 z_0 + F_0(\tilde{z}_1, z_0, z_2), \\ N_1(\tilde{z}_1, z_0, z_2) &= b_1 z^* z_0^2 + 3b_2 z^* \tilde{z}_1^2 + b_3 z^* z_2^2 + b_4 z^* z_0 z_2 + F_1(\tilde{z}_1, z_0, z_2), \\ N_2(\tilde{z}_1, z_0, z_2) &= 2c_2 z^* \tilde{z}_1 z_2 + 2c_4 z^* \tilde{z}_1 z_0 + F_2(\tilde{z}_1, z_0, z_2). \end{aligned}$$

Let us introduce new variables as follows:

$$\begin{pmatrix} \tilde{z}_0 \\ \tilde{z}_2 \end{pmatrix} = \begin{pmatrix} \beta & \beta \\ -\alpha & \gamma \end{pmatrix}^{-1} \begin{pmatrix} z_0 \\ z_2 \end{pmatrix}.$$

If $(\mu_0^*, \mu_2^*) \in \mathscr{S}_D$, then $(\tilde{z}_1, \tilde{z}_0, \tilde{z}_2)$ satisfies the following system

$$\begin{pmatrix} \tilde{z}_1 \\ \tilde{z}_0 \\ \tilde{z}_2 \end{pmatrix} = \begin{pmatrix} -2\mu_1 & 0 & 0 \\ 0 & 0 & 0 \\ 0 & 0 & \alpha+\gamma \end{pmatrix} \begin{pmatrix} \tilde{z}_1 \\ \tilde{z}_0 \\ \tilde{z}_2 \end{pmatrix} + T_{(\mu_0^*, \mu_2^*)} \begin{pmatrix} N_1^{(\mu_0^*, \mu_2^*)}(\tilde{z}_1, \tilde{z}_0, \tilde{z}_2) \\ N_0^{(\mu_0^*, \mu_2^*)}(\tilde{z}_1, \tilde{z}_0, \tilde{z}_2) \\ N_2^{(\mu_0^*, \mu_2^*)}(\tilde{z}_1, \tilde{z}_0, \tilde{z}_2) \end{pmatrix}, \quad (19.12)$$

where $N^{(\mu_0^*, \mu_2^*)} = N_j(\tilde{z}_1, \beta(\tilde{z}_0 + \tilde{z}_2), -\alpha\tilde{z}_0 + \gamma\tilde{z}_2)$ and $T_{(\mu_0^*, \mu_2^*)}$ is a 3×3 matrix:

$$T_{(\mu_0^*, \mu_2^*)} = \begin{pmatrix} 1 & 0 & 0 \\ 0 & \beta & \beta \\ 0 & -\alpha & \gamma \end{pmatrix}.$$

Applying the center manifold theory for this system, we have the following theorem.

Theorem 19.1 *Let $(\mu_0^*, \mu_2^*) \in \mathscr{S}_D$ be a pair of constants. Then, there exists a positive constant ε such that if*

$$|(\mu_0, \mu_2) - (\mu_0^*, \mu_2^*)| + |(\tilde{z}_1, \tilde{z}_0, \tilde{z}_2)| < \varepsilon,$$

then the dynamics of (19.3) on the center manifold is topologically equivalent to the dynamics of the following system

$$\dot{\tilde{z}}_0 = (\nu + C_{(\mu_0^*, \mu_2^*)}\, \tilde{z}_0^2)\tilde{z}_0,$$

where

$$\nu = \begin{cases} \left(\alpha + \gamma - \sqrt{(\alpha+\gamma)^2 - 4(\alpha\gamma + \beta c_4 \mu_1 / b_2)}\right)/2, & \alpha + \gamma > 0, \\ \left(\alpha + \gamma + \sqrt{(\alpha+\gamma)^2 - 4(\alpha\gamma + \beta c_4 \mu_1 / b_2)}\right)/2, & \alpha + \gamma < 0, \end{cases}$$

$$\begin{aligned} C_{(\mu_0^*, \mu_2^*)} = \frac{1}{\beta(\alpha+\gamma)} \big[&\{\beta(c_4\beta - c_2\alpha) - \gamma(a_2\beta - a_4\alpha)\}(b_1\beta^2 + b_3\alpha^2 - b_4\alpha\beta)/b_2 \\ &+ \gamma(a_1\beta^3 + a_3\alpha^2\beta) + \beta(c_1\alpha\beta^2 + \alpha^3 c_3) \big]. \end{aligned}$$

Proof We can prove the theorem by applying the center manifold theory for (19.12) as follows: there exist functions $\tilde{z}_1 = h_1(\tilde{z}_0)$ and $\tilde{z}_2 = h_2(\tilde{z}_0)$ satisfying

$$\frac{dh_j}{d\tilde{z}_0}(0) = 0, \quad j = 1, 2$$

such that the center manifold of (19.12) is given by the graph on $\mathbb{R}^3$:

$$\{(\tilde{z}_1, \tilde{z}_0, \tilde{z}_2)\,;\ (\tilde{z}_1, \tilde{z}_0, \tilde{z}_0) = (h_1(\tilde{z}_0), \tilde{z}_0, h_2(\tilde{z}_0))\}.$$

Moreover, the dynamics on the center manifold is topologically equivalent to the dynamics of the ordinary differential equation

$$\tilde{z}_0 = \nu\tilde{z}_0 + N_0^{(\mu_0^*, \mu_2^*)}(h_1(\tilde{z}_0), \tilde{z}_0, h_2(\tilde{z}_0)).$$

To prove the theorem, it is necessary to compute the functions h_j approximately. Differentiating $\tilde{z}_j = h_j(\tilde{z}_0)$ with respect to t, we have

$$\lambda_j h_j + N_j^{(\mu_0^*, \mu_2^*)}(\tilde{z}_1, \tilde{z}_0, \tilde{z}_2) = \frac{dh_j}{d\tilde{z}_0}\frac{d\tilde{z}_0}{dt},$$

where

$$\lambda_j = \begin{cases} -2\mu_1 & (j = 1), \\ \alpha + \gamma & (j = 2). \end{cases}$$

Since $|\tilde{z}_0| < \varepsilon$ and $|(\mu_0, \mu_2) - (\mu_0^-, \mu_2^*)| < \varepsilon$, it holds that

$$\frac{dh_j}{d\tilde{z}_0}\frac{d\tilde{z}_0}{dt} = O(\varepsilon^3).$$

Substituting $h_j = H_j\tilde{z}_0^2 + \cdots$, and taking terms up to $O(\tilde{z}_0^2)$, we have

$$h_1(\tilde{z}_0) = \frac{z^*}{2\mu_1}(b_1\beta^2 - b_4\alpha\beta + b_3\alpha^2)z_0^2 + O(|z_0|^3),$$
$$h_2(\tilde{z}_0) = O(|z_0|^3).$$

Substituting these approximations into the second equation of (19.12), and taking terms up to the third order of $\tilde{z}_0$, we obtain the bifurcation equation stated in the theorem. □

Therefore, if $a_4c_4 > 0$, the bifurcation structure near the set of critical points on $\mathscr{S}_D$ is as follows: the stationary triple mixed mode solutions of (19.4) bifurcate from the 1-mode stationary solutions through the pitchfork bifurcation. These mixed mode solutions are stable if and only if $C_{(\mu_0^*, \mu_2^*)} < 0$ and $\alpha + \gamma < 0$ hold. Moreover, the leading term of the triple mixed mode stationary solutions are given by

$$u(x) \approx \pm 2\left[-d\beta\sqrt{\frac{-\nu}{C_{(\mu_0^*,\mu_2^*)}}} + B_1\sqrt{\frac{-\mu_1}{b_2}}\cos\left(\frac{\pi}{L}x\right) - B_2\alpha\sqrt{\frac{-\nu}{C_{(\mu_0^*,\mu_2^*)}}}\cos\left(2\frac{\pi}{L}x\right)\right],$$

$$v(x) \approx \pm 2c\left[\beta\sqrt{\frac{-\nu}{C_{(\mu_0^*,\mu_2^*)}}} + \sqrt{\frac{-\mu_1}{b_2}}\cos\left(\frac{\pi}{L}x\right) - \alpha\sqrt{\frac{-\nu}{C_{(\mu_0^*,\mu_2^*)}}}\cos\left(2\frac{\pi}{L}x\right)\right].$$

Remark 19.1 We can conclude that the bifurcation structure around the points $(\mu_0^*, \mu_2^*) \in \mathscr{S}_D$ even when $a_4c_4 < 0$ except the case when (μ_0^*, μ_2^*) is either P_+ or P_-.

19.2.2 The Case When $a_4c_4 < 0$

In this subsection, we study the bifurcation structures of (19.3) around the double degenerate points $P_\pm$ of (19.3) (see Fig. 19.6).

Let (z, x, y) be a coordinate defined by

$$\begin{pmatrix} z \\ x \\ y \end{pmatrix} = T^{-1}\begin{pmatrix} \tilde{z}_1 \\ z_0 \\ z_2 \end{pmatrix}.$$

Then, near the singular point, there exists constants p_1 and p_2 satisfying $|p_j| \ll 1$ such that the system (19.11) can be transformed into the following system:

$$\begin{pmatrix} \dot{z} \\ \dot{x} \\ \dot{y} \end{pmatrix} = \begin{pmatrix} -2\mu_1 & 0 & 0 \\ 0 & p_1 & 1 \\ 0 & 0 & p_2 \end{pmatrix}\begin{pmatrix} z \\ x \\ y \end{pmatrix} + \begin{pmatrix} \tilde{N}_1(z,x,y) \\ \tilde{N}_0(z,x,y) \\ \tilde{N}_2(z,x,y) \end{pmatrix}, \tag{19.13}$$

where

$$\begin{pmatrix} \tilde{N}_1(z,x,y) \\ \tilde{N}_2(z,x,y) \\ \tilde{N}_3(z,x,y) \end{pmatrix} = T^{-1}\begin{pmatrix} N_1(z,-2\beta x,(\alpha-\gamma)x-2y) \\ N_0(z,-2\beta x,(\alpha-\gamma)x-2y) \\ N_2(z,-2\beta x,(\alpha-\gamma)x-2y) \end{pmatrix}, \; T = \begin{pmatrix} 1 & 0 & 0 \\ 0 & -2\beta & 0 \\ 0 & \alpha-\gamma & -2 \end{pmatrix}$$

By applying the center manifold theory to the system (19.13), we can study detailed dynamics around the double degenerate point $P_\pm$.

Theorem 19.2 *If $|p_j| < 2\mu_1$ for $j = 1, 2$, then there exists a local invariant manifold $\mathscr{M}^c$ of (19.3). Moreover, the dynamics of (19.13) on $\mathscr{M}^c$ is given by the dynamics of the following system:*

$$\begin{pmatrix} \dot{x} \\ \dot{y} \end{pmatrix} = \begin{pmatrix} p_1 & 1 \\ 0 & p_2 \end{pmatrix}\begin{pmatrix} x \\ y \end{pmatrix} + \sum_{\substack{j,k\in\mathbb{N} \\ j+k=3}}\begin{pmatrix} f_{jk}\, x^j y^k \\ g_{jk}\, x^j y^k \end{pmatrix} + O(\|(x,y)\|^4), \tag{19.14}$$

where

$$
\begin{aligned}
f_{30} &= 2z^*(a_2 - a_4\alpha/\beta)H_{20} + 4(a_1\beta^2 + a_3\alpha^2),\\
f_{21} &= 2a_4z^*H_{20}/\beta + 2z^*(a_2 - a_4\alpha/\beta)H_{11} - 8a_3\alpha,\\
f_{12} &= 2a_4z^*H_{11}/\beta + 2z^*(a_2 - a_4\alpha/\beta)H_{02} + 4a_3,\\
f_{03} &= 2a_4z^*H_{02}/\beta,\\
g_{30} &= 2z^*[(a_2 - c_2)\alpha\beta - a_4\alpha^2 + c_4\beta^2]H_{20}/\beta + 4\alpha[(a_1 - c_1)\beta^2 + (a_3 - c_3)\alpha^2],\\
g_{21} &= 2z^*(a_4\alpha/\beta + c_2)H_{20} + 2z^*[\alpha(a_2 - c_2) - a_4\alpha^2/\beta + c_4\beta]H_{11}\\
&\quad + 4[\alpha^2(3c_3 - 2a_3) + \beta^2 c_1],\\
g_{12} &= 2z^*[(a_2 - c_2)\alpha - a_4\alpha^2/\beta + c_4\beta]H_{02} + 2z^*(a_4\alpha/\beta + c_2)H_{11} + 4\alpha(a_3 - 3c_3),\\
g_{03} &= 2z^*(a_4\alpha/\beta - c_2)H_{02} + 4c_3,\\
H_{20} &= 2z^*(\beta^2 b_1 + b_3\alpha^2 - \beta b_4\alpha)/\mu_1,\\
H_{11} &= [2z^*(\beta b_4 - 2\alpha b_3) - H_{20}]/\mu_1,\\
H_{02} &= z^*[2b_3(\mu_1 + \alpha) + b_4\beta]/\mu_1^2 - H_{11}/2\mu_1^2.
\end{aligned}
$$

Proof Proof is a simple application of the center manifold theory.

There exists a function $h(x, y)$ satisfying

$$h(0, 0) = h_x(0, 0) = h_y(0, 0) = 0$$

such that the $\mathcal{M}^c$ can be characterized as a graph on $\mathbb{R}^3$ near the origin:

$$\mathcal{M}^c = \{(z, x, y); z = h(x, y)\}.$$

Differentiating $z = h(x, y)$ with respect to t, we have

$$-2\mu_1 h(x, y) + \tilde{N}_1(h(x, y), x, y) = (\partial h/\partial x)\dot{x} + (\partial h/\partial y)\dot{y}.$$

Since $h(x, y) = O(|(x, y)|^2)$, we also have

$$
\begin{aligned}
&-2\mu_1 h(x, y) + \tilde{N}_1(h(x, y), x, y))\\
&\quad = (\partial h/\partial y)\dot{y} + O(|(x, y)|^3) + O(|p_1 + p_2| \cdot |(x, y)|^2)).
\end{aligned} \tag{19.15}
$$

Substituting $h(x, y) = h_{20}x^2 + h_{11}xy + h_{02}y^2$ into (19.15), we can conclude $h_{jk} = H_{jk}$ $(j + k = 2)$.

Then, the dynamics on the manifold $\mathcal{M}^c$ is given by the following system

$$\begin{pmatrix} \dot{x} \\ \dot{y} \end{pmatrix} = \begin{pmatrix} p_1 & 1 \\ 0 & p_2 \end{pmatrix} \begin{pmatrix} x \\ y \end{pmatrix} + \begin{pmatrix} \tilde{N}_0(h(x, y), x, y) \\ \tilde{N}_2(h(x, y), x, y) \end{pmatrix} \tag{19.16}$$

We obtain explicit form (19.14) by taking terms up to the third order in (19.16). □

Now, similarly to [11], we can normalize the system (19.16) by introducing the new variable (u, v):

$$u = x, \quad w = y + p_1 x + f_{30}\, x^3 + f_{21}\, x^2 y + f_{12}\, xy^2 + f_{03}\, y^3.$$

Then, (19.16) can be transformed into

$$\begin{aligned}\dot{u} &= w,\\ \dot{w} &= \tilde{p}_1 u + \tilde{p}_2 w + \sigma_1\, u^3 + \sigma_2 u^2 w + (g_{12} + 2f_{21}) uw^2\\ &\quad + (g_{03} + f_{12}) w^3 + O(|(x,y)|^5) + (|p_1 + p_2||(x,y)|^3),\end{aligned} \tag{19.17}$$

where

$$\tilde{p}_1 = -p_1 p_2, \quad \tilde{p}_2 = p_1 + p_2, \quad \sigma_1 = g_{30}, \quad \sigma_2 = (g_{21} + 3f_{30}).$$

We take the scaling for a small positive parameter ε as follows:

$$u = \varepsilon \frac{\sqrt{|\sigma_1|}}{|\sigma_2|} U, \ w = \varepsilon^2 \frac{|\sigma_1|^{3/2}}{\sigma_2^2} W, \ T = \varepsilon \left|\frac{\sigma_1}{\sigma_2}\right| t$$

$$\tilde{p}_1 = \varepsilon^2 \frac{\sigma_1^2}{\sigma_2^2} P_1, \ \tilde{p}_2 = \varepsilon^2 \left|\frac{\sigma_1}{\sigma_2}\right| P_2.$$

Then, we have

$$\begin{cases} U_T = W,\\ W_T = P_1 U + \varepsilon\, P_2 W + (\operatorname{sign}\sigma_1) U^3 + \varepsilon\, (\operatorname{sign}\sigma_2)\, U^2 W + O(\varepsilon^2).\end{cases}$$

If $\operatorname{sign}\sigma_2 = +1$, then by reversing the time and the variable W,

$$T \to -T, \quad W \to -W,$$

we have the following theorem.

Theorem 19.3 *Let μ_1 be fixed so that $b_2 \mu_1 < 0$. Then, there are coordinate and parameter changes and a time scaling so that the system (19.14) (which gives dynamics of (19.11) on the center manifold $\mathscr{M}^c$) is transformed into the following system*

$$\begin{cases} \dot{u} = w,\\ \dot{w} = p_1 u + p_2 w + \varsigma\, u^3 - \varepsilon\, u^2 w + O(\|(x,y)\|^4),\end{cases} \tag{19.18}$$

where $p_1 = P_1$ and $p_2 = \varepsilon P_2$ are bifurcation parameter, and ε is a positive small parameter and

$$\varsigma = \operatorname{sign}\sigma_1 = \operatorname{sign} g_{30}.$$

By direct computations, we have the following results:

Lemma 19.2 *At the critical point $P_\pm = (\mu_0^\pm, \mu_2^\pm)$, the coefficients σ_1 and σ_2 in (19.17) can be written by the Taylor series with respect to μ_1 as follows.*

$$\begin{aligned}
\sigma_1 &= 4a_4\mu_1^3[c_4a_4\{2(a_4b_1 - b_3c_4) + b_4(a_2 - c_2)\}\\
&\quad \pm \sqrt{-a_4c_4}\{a_4c_1b_2 - 2c_4a_4b_4 - a_1a_4b_2 + a_4b_1a_2 - a_4b_1c_2\\
&\quad + a_3b_2c_4 - c_3b_2c_4 - b_3a_2c_4 + b_3c_4c_2\}]/b_2^4 + O(\mu_1^4),\\
\sigma_2 &= 4\mu_1^2a_4[2c_4^2a_4b_3 - 3c_3c_4b_2^2 - a_3c_4b_2^2 - 2c_4a4^2b_1\\
&\quad - 3a_2a_4b_1b_2 - c_2a_4b_1b_2 + c_4a_4b_4 * c_2 + 3c_2c_4b_3b_2\\
&\quad - c_4a_4b_4a_2 + c_4a_2b_3b_2 + 3a_1a_4b_2^2 + c_1a_4b_2^2\\
&\quad \pm \sqrt{-a_4c_4}(-b_3c_4c_2 + b_3a_2c_4 + 2c_4a_4b_4 + 2a_2b_4b_2\\
&\quad + 2c_2b_4b_2 + 2c_4b_3b_2 - a_4b_1a_2 + a_4b_1c_2 + 2a_4b_1b_2)]/b_2^4 + O(\mu_1^3).
\end{aligned}$$

The dynamics of the system (19.18) is well studied: for instance, see Chap. 4 of [3], Chap. 7.3 of [6]. We show the phase diagram in Figs. 19.7 and 19.8.

These analysis yields that the non-generic conditions of proposition 1 are $C_{(\mu_0^*, \mu_2^*)} = 0$ and $\sigma_1 = 0$, for the case when $a_4c_4 > 0$ and $a_4c_4 < 0$, respectively. More precisely, we have the following results with respect to the existence and stabilities of the time-periodic solutions, heteroclinic cycles and Homoclinic orbits.

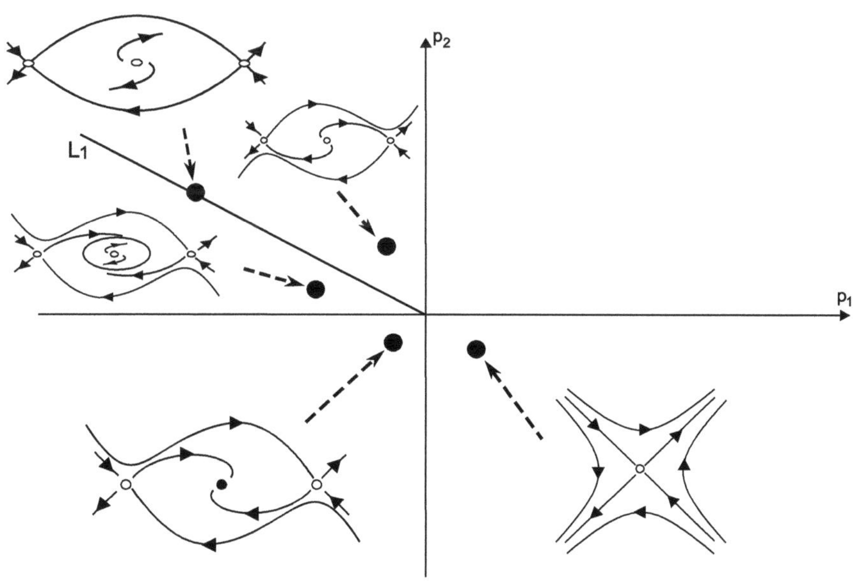

Fig. 19.7 Bifurcation diagram on phase plane when $\varsigma > 0$. The sets of critical points for the Hopf bifurcation, stationary bifurcation and heteroclinic bifurcation are $\{(p_1, p_2)\,;\, p_1 < 0, p_2 = 0\}$, $\{(p_1, p_2)\,;\, p_1 = 0\}$ and $L_1 := \{(p_1, p_2)\,;\, p_2 = -p_1/5 + O(p_1^2)\}$, respectively

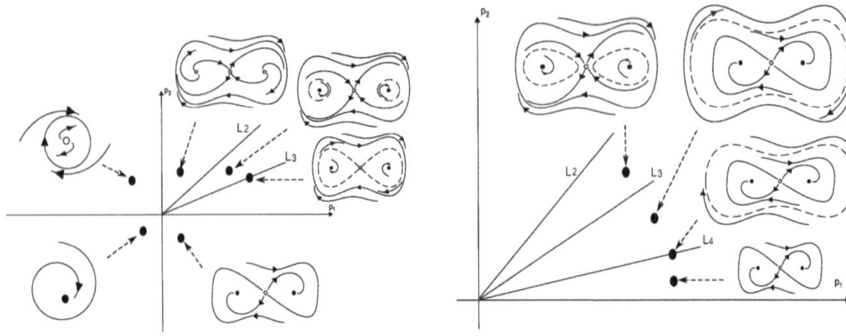

Fig. 19.8 Bifurcation diagram on phase plane when $\varsigma < 0$. The sets of critical points for Hopf and stationary bifurcations are the same as Fig. 19.7. The typical set of critical points is as follows: The set of critical points for the Hopf bifurcation around the nontrivial solutions of (19.18) is the curve $L_2 := \{(p_1, p_2)\,;\, p_2 = p_1\}$. The critical sets for the homoclinic bifurcation is the *curve* $L_3 := \{(p_1, p_2)\,;\, p_2 = 4p_1/5 + O(p_1^2)\}$. In the region between the p_1 axis and the *curve* L_3, there is the set of critical points $L_4 := \{(p_1, p_2)\,;\, p_2 \approx 0.752p_1\}$. Between L_3 and L_4, there are two different periodic orbits

Proposition 19.3 *If $b_2 < 0$, $a_4c_4 < 0$, $\sigma_2 < 0$ and $\sigma_1 \neq 0$, then, the time periodic solutions $\Omega^{\pm}$ of (19.4) are asymptotically locally stable, and bifurcate on the line $\{(p_1, p_2)\,;\, p_1 < 0, p_2 = 0\}$ through the Hopf bifurcation from the 1-mode stationary solutions $\pm(U_1(x), V_1(x))$. Moreover, if $\sigma_1 > 0$, then (II)–(i) of Proposition 19.1 holds by a suitable choice of parameters p_1 and p_2, and if $\sigma_1 < 0$, then (II)–(ii) of Proposition 19.1 holds by a suitable choice of parameters p_1 and p_2. In addition, the heteroclinic cycles Γ_1 and Γ_2 are locally stable, and the homoclinic orbits $\Gamma_3^{\pm}$ and $\Gamma_4^{\pm}$ are unstable.*

Remark 19.2 By phase plane analysis, if $b_2 < 0$, $a_4c_4 < 0$ and $\sigma_2 < 0$ hold, then we can obtain the stabilities of the stationary solutions as follows:

- If $\sigma_1 > 0$, then the unstable triple mixed mode stationary solutions bifurcate from unstable 1-mode stationary solutions through subcritical pitchfork bifurcations, and 1-mode stationary solutions become stable after the bifurcation.
- If $\sigma_1 < 0$, then the locally asymptotically stable triple mixed mode stationary solutions bifurcate form locally asymptotically stable 1-mode stationary solutions through super critical pitchfork bifurcation, and 1-mode stationary solutions lose its stability after the bifurcation.

Remark 19.3 If $\sigma_1 < 0$, then there exist not only the periodic orbits $\Omega^{\pm}$ but also another type of periodic orbits: for instance, there are periodic orbits which bifurcate from triple mixed mode stationary solutions through Hopf bifurcation (see Fig. 19.8).

Remark 19.4 If $\sigma_2 > 0$, then the dynamics of (19.18) is topologically equivalent to the dynamics of (19.10) by the time reverse transformation.

Remark 19.5 If $b_2 > 0$, then the center manifolds of (19.3) is not attractive, therefore, all solutions of (19.4) corresponding to the solutions of (19.18) are unstable on the phase space X of (19.4).

19.2.3 Case Study

Since it is complicated to determine the sign of coefficients σ_j, we compute them numerically. Let us take the constants in (19.4) as (19.5). Then, the coefficients of the normal form can be obtained as listed in (19.6). Take the parameter $\mu_1 = 0.01$, then we have

$$P_+ \approx (-0.081048,\ -0.013440) \quad \text{and} \quad P_- \approx (-0.162411,\ 0.067923).$$

Substituting then into the explicit form of f_{30}, g_{12} and g_{30}, we have

$$\sigma_1 \approx \begin{cases} 38.67 \text{ at } (\mu_0, \mu_2) = P_+, \\ 1287.74 \text{ at } (\mu_0, \mu_2) = P_-, \end{cases} \quad \sigma_2 \approx \begin{cases} 4221.37 \text{ at } (\mu_0, \mu_2) = P_+, \\ -1.84 \times 10^5 \text{ at } (\mu_0, \mu_2) = P_-. \end{cases}$$

Therefore, $\sigma_1 > 0$ for both double degenerate critical point $P_\pm$, and sign $\sigma_2 = +1$ at P_+, and sign $\sigma_2 = -1$ at P_- (Fig. 19.9).

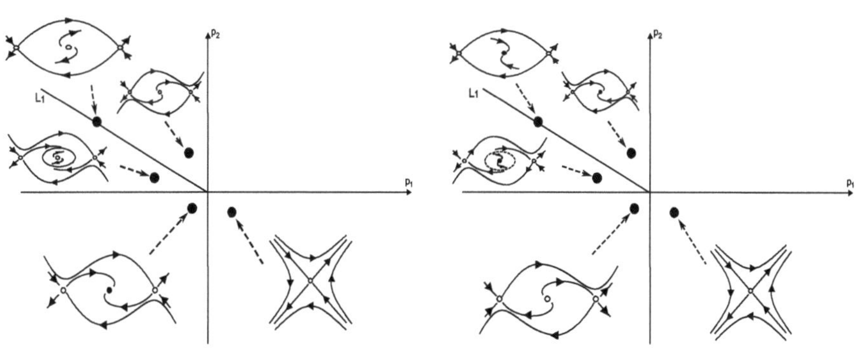

Fig. 19.9 Bifurcation structure of the system (19.10), where $p_1 = P_1$, $p_2 = \varepsilon P_2$. (*Left*) Bifurcation diagram on phase plane when $\sigma_1 > 0$, $\sigma_2 < 0$. (*Right*) Bifurcation diagram on phase plane when $\sigma_1 > 0$, $\sigma_2 > 0$

19.3 Heteroclinic Cycles on a Blow-Up Vector Field

In this section, we consider the blow-up vector field of the normal form (19.3) around the equilibrium $(z_0, z_1, z_2) = (0, z^*, 0)$, where $z^* = \sqrt{-\mu_1/b_2}$. To do that, we rescale the system (19.13) around an equilibrium $(0, 0, 0)$. We begin this section with the system (19.13):

$$\begin{pmatrix} \dot{z} \\ \dot{x} \\ \dot{y} \end{pmatrix} = \begin{pmatrix} -2\mu_1 & 0 & 0 \\ 0 & p_1 & 1 \\ 0 & 0 & p_2 \end{pmatrix} \begin{pmatrix} z \\ x \\ y \end{pmatrix} + \begin{pmatrix} \tilde{N}_1(z,x,y) \\ \tilde{N}_0(z,x,y) \\ \tilde{N}_2(z,x,y) \end{pmatrix},$$

where

$$\begin{aligned} \tilde{N}_1(z,x,y) &= B_1x^2 + B_2y^2 + B_3z^2 + B_4xy + B_5z^3 + B_6x^2z + B_7y^2z + B_8xyz, \\ \tilde{N}_0(z,x,y) &= A_1xz + A_2yz + A_3x^3 + A_4x^2y + A_5xy^2 + A_6xz^2 + A_7yz^2, \qquad (19.19) \\ \tilde{N}_2(z,x,y) &= C_1xz + C_2yz + C_3x^3 + C_4y^3 + C_5x^2y + C_6xy^2 + C_7xz^2 + C_8yz^2. \end{aligned}$$

The coefficients A_j, B_j and C_j are as follows:

$$A_1 = z^*\{2a_2\beta - a_4(\alpha - \gamma)\}/\beta, \quad A_2 = 2a_4z^*,$$

$$A_3 = 4a_1\beta^2 + a_3(\alpha - \gamma)^2, \quad A_4 = -4a_3, \quad A_5 = 4a_3,$$

$$B_1 = 4b_1z^*\beta^2 + b_3z^*(\alpha - \gamma)^2 - 2b_4z^*(\alpha - \gamma)\beta, \quad B_2 = 4b_3z^*, \quad B_3 = 3b_2z^*,$$

$$B_4 = 4z^*\{b_4\beta - b_3(\alpha - \gamma)\}, \quad B_5 = b_2, \quad B_6 = 4b_1\beta^2 + b_3(\alpha - \gamma)^2 - 2b_4\beta(\alpha - \gamma),$$

$$B_7 = 4b_3, \quad B_8 = -4b_3(\alpha - \gamma) + 4b_4\beta,$$

$$\begin{aligned} C_1 &= [z^*(\alpha - \gamma)\{2a_2\beta - a_4(\alpha - \gamma)\}]/(2\beta) + (2c_4\beta - c_2(\alpha - \gamma))z^*, \\ C_2 &= (\alpha - \gamma)a_4z^*/(\beta) + 2c_2z^*, \end{aligned}$$

$$C_3 = (\alpha - \gamma)\left\{2a_1\beta^2 + a_3(\alpha - \gamma)^2/2\right\} - \left\{2c_1\beta^2 + c_3(\alpha - \gamma)^2/2\right\}(\alpha - \gamma), \quad C_4 = 4c_3,$$

$$C_5 = (-2a_3 + 3c_3)(\alpha - \gamma)^2 + 4c_1\beta^2, \quad C_6 = 2(a_3 - 3c_3)(\alpha - \gamma),$$

$$C_7 = (\alpha - \gamma)\{2a_2\beta - a_4(\alpha - \gamma)\}/(4\beta) - \frac{1}{2}c_2(\alpha - \gamma) + c_4\beta, \quad C_8 = (\alpha - \gamma)a_4/(2\beta) + c_2.$$

19.3.1 Transformation to the Dumortier-Kokubu's Normal Form

In this subsection, we find the invertible coordinate, parameter changes, time reparameterization and scaling of variables and parameters such that the system (19.13) can be rescaled into the system (19.8):

$$\begin{cases} \dot{x} = y, \\ \dot{y} = \lambda x + \mu y + axz + byz, \\ \dot{z} = \nu + x^2 + z^2. \end{cases}$$

<Step 1: Near identity transformation>

We introduce new valuables by

$$u = x, \quad w = y + p_1 x + \tilde{N}_0(z, x, y).$$

Then, we have $\dot{x} = \dot{u} = w$ and

$$\begin{aligned}
\dot{w} &= \dot{y} + p_1\dot{x} + A_1(\dot{x}z + x\dot{z}) + A_2(\dot{y}z + y\dot{z}) + 3A_3x^2\dot{x} + A_4(2xy\dot{x} + x^2\dot{y}) \\
&\quad + A_5(\dot{x}y^2 + 2xy\dot{y}) + A_6(\dot{x}z^2 + 2xz\dot{x}) + A_7(\dot{y}z^2 + 2yz\dot{z}) \\
&= p_2y + C_1xz + C_2yz + C_3x^3 + C_4y^3 + C_5x^2y + C_6xy^2 + C_7xz^2 + C_8yz^2 \\
&\quad + p_1w + A_1[wz + (-2\mu_1zx + B_1x^3 + B_2xy^2 + B_3xz^2 + B_4x^2y) + O(4)] \\
&\quad + A_2[(p_2yz + C_1xz^2 + C_2yz^2) + (-2\mu_1yz + B_1x^2y + B_2y^3 + B_3z^2y + B_4xy^2) + O(4)] \\
&\quad + 3A_3wx^2 + A_4(2xyw + p_2x^2y + O(4)) + A_5(p_1xy^2 + y^3 + O(4)) \\
&\quad + A_6(p_1wz^2 - 4\mu_1xz^2 + O(4)) + A_7(p_2yz^2 - 4\mu_1z^2y + O(4))
\end{aligned}$$

Moreover, we obtain

$$\begin{aligned}
\dot{w} = &- p_1p_2u + (p_1 + p_2)w + (A_1 + C_2)wz + C_1uz + (C_3 + A_1B_1)u^3 \\
&+ h(p_1, p_2, \mu_1, u, w, z),
\end{aligned}$$

where

$$h(p_1, p_2, \mu_1, u, w, z) = O(\|(p_1, p_2, \mu_1)\| \|(u, w, z)\|^2) + O(\|(u, w, z)\|^3).$$

Here, we can assume that h does not contain the cubic term u^3. Therefore,

$$\begin{cases} \dot{u} = w, \\ \dot{w} = -p_1p_2u + (p_1 + p_2)w + (A_1 + C_2)wz + C_1uz + (C_3 + A_1B_1)u^3 + h(p_1, p_2, \mu_1, u, w, z), \\ \dot{z} = -2\mu_1z + B_1u^2 + B_2w^2 + B_3z^2 + B_4uw + O(\|(p_1, p_2, \mu_1)\| \|(u, w, z)\|^2) + O(\|(u, w, z)\|^3). \end{cases} \tag{19.20}$$

Let us use the following coordinate scaling:

$$u = \varrho \tilde{u}, \quad w = \varrho \tilde{w}, \quad z = \varrho \tilde{z}, \quad \varrho > 0 \tag{19.21}$$

and set

$$\tilde{A}_j = \varrho^2 A_j, \quad \tilde{B}_j = \varrho^2 B_j, \quad \tilde{C}_j = \varrho^2 C_j.$$

Then, the same system of (19.20) is obtained by dropping the tildes.

<Step 2: Changes of coordinates>

Substituting

$$z(t) = \zeta(t) - \delta$$

into 3rd equation of (19.20), we have

$$\dot{\zeta} = 2\mu_1\delta + B_3\delta^2 + (-2\mu_1 - 2B_3\delta)\zeta + B_3\zeta^2 + B_1u^2 + B_2w^2 + B_4uw.$$

By taking $\delta = -\mu_1/B_3$, then we have

$$\begin{cases} \dot{u} = w, \\ \dot{w} = \lambda u + \mu w + au\zeta + bwz + O(\varrho^3) + h, \\ \dot{\zeta} = \nu + B_1u^2 + B_2w^2 + B_3\zeta^2 + B_4uw + O(\|(p_1, p_2, \mu_1)\|\|(u, w, z)\|^2) + O(\|(u, w, \zeta)\|^3), \end{cases} \tag{19.22}$$

where

$$\lambda = \frac{a\mu_1}{B_3} - p_1p_2, \quad \mu = p_1 + p_2 + \frac{b\mu_1}{B_3},$$

$$a = C_1, \quad b = A_1 + C_2, \quad \nu = -\frac{\mu_1^2}{B_3}.$$

<Step 3: First scaling>

Let us scale the valuables and parameters as follows:

$$u = r^3x, \quad w = r^5y, \quad , \zeta = r^4z, \quad \lambda = r^4\tilde{\lambda}, \quad \mu = r^2\tilde{\mu}, \quad \nu = r^8\tilde{\nu}.$$

Then, we have

$$\begin{cases} x' = y, \\ y' = \tilde{\lambda}x + \tilde{\mu}y + axz + br^2yz + O(\varrho^3r^2) + O(r^4), \\ z' = r^2\tilde{\nu} + r^2B_3z^2 + B_1x^2 + r^2B_4xy + O(r^4). \end{cases} \tag{19.23}$$

Here, $\cdot\,'$ denotes the derivative $\dfrac{d}{d\tau}$ where $\tau = t/r^2$.

Remark 19.6 The coefficients of quadratic terms A_j, B_j and C_j in (19.19) are dependent on the parameters μ_j. Therefore, these coefficients in (19.23) depend on the scaling parameter r also. This means that if a quadratic term with the coefficient which is $O(r)$, we can consider it as a higher order term. However, we can take a scaling parameter $\varrho > 0$ in (19.21) so that the cubic term with respect to (x, y, z) goes to higher order terms, and the quadratic terms can be considered as leading terms.

<Step 4: Coordinate changes in the scaling family>

We can transform the 3rd equation of (19.23) as follows:

$$\frac{dz}{d\tau} = B_1\left(x + \frac{r^2B_4}{2B_1}y\right)^2 - \frac{r^4B_4^2}{4B_1}y^2 + r^2\tilde{\nu} + r^2B_3z^2 + O(r^4)$$

Let us put

$$X(\tau) = x(\tau) + r^2\frac{B_4}{2B_1}y(\tau)$$

then, we have

$$X' = x' + r^2\frac{B_4}{2B_1}y' = y + O(r^2),$$

and

$$y' = \tilde{\lambda}X + (\mu + O(r^2))y + aXz + r^2\left(b - \frac{B_4}{2B_1}\right)yz + r^2cx^3 + O(\varrho^3r^2) + O(r^4).$$

We rewrite $X = x$ and drop the tildes, we have

$$\begin{cases} x' = y + O(r^2), \\ y' = \lambda x + \mu y + axz + r^2\left(b - \dfrac{B_4}{2B_1}\right)yz + O(\varrho^3r^2) + O(r^4), \\ z' = r^2\nu + r^2B_3z^2 + B_1x^2 + O(r^4). \end{cases} \tag{19.24}$$

Remark 19.7 The system (19.24) can be normarized as follows: the $O(r^2)$ term in the first equation can be written as follows:

$$O(r^2)y + r^2\frac{B_4}{2B_1}\tilde{\lambda}X + r^2aXz + O(r^4).$$

Let us set

$$w = (1 + O(r^2))y + r^2\frac{B_4}{2B_1}\tilde{\lambda}X + r^2aXz + O(r^4).$$

Then we have

$$w' = (\mu + O(r^2))w + (1 + O(r^2)(\tilde{\lambda} + O(r^4))X + (1 + O(r^2))aXz$$
$$+ r^2\left(b - \frac{B_4}{2B_1}\right)zw + O(\varrho^3 r^2) + O(|\mu|r^2) + O(r^4) + O(\|(x, w, z)\|^3).$$

The system (19.24) now becomes

$$\begin{cases} x' = w, \\ w' = (\mu + O(r^2))w + (\tilde{\lambda} + O(r^2))x + (1 + O(r^2))axz \\ \qquad + r^2\left(b - \dfrac{B_4}{2B_1}\right)zw + O(\varrho^3 r^2) + O(|\mu|r^2) + O(r^4) + O(\|(x, w, z)\|^3), \\ z' = r^2\nu + r^2 B_3 z^2 + B_1 x^2 + O(r^4). \end{cases}$$

<Final step: Coordinate changes, re-parametarization of time, and final scaling>

Putting

$$r^2\left(b - \frac{B_4}{2B_1}\right) = \bar{b}, \quad r^2\nu = \bar{\nu}, \quad r^2 B_3 = \bar{B}_3,$$

and truncating the $O(\rho^3)$, $O(r^2)$ and $O(r^4)$ terms, the system (19.24) is

$$\begin{cases} x' = y, \\ y' = \lambda x + \mu y + axz + \bar{b}yz, \\ z' = \bar{\nu} + \bar{B}_3 z^2 + B_1 x^2. \end{cases} \tag{19.25}$$

Let us moreover introduce the following scaling:

$$x = |a||B_1\bar{B}_3^3|^{-\frac{1}{2}}\,\tilde{x}, \quad y = \delta\,|a||B_1\bar{B}_3^3|^{-\frac{1}{2}}\,\tilde{y}, \quad z = -\frac{a}{\bar{B}_3^2}\,\tilde{z}.$$

Then, we have

$$\begin{cases} \tilde{x}' = \delta\tilde{y}, \\ \delta\tilde{y}' = \lambda\tilde{x} + \mu\delta\tilde{y} - \dfrac{a^2}{\bar{B}_3^2}\tilde{x}\tilde{z} - \bar{b}\dfrac{\delta a}{\bar{B}_3^2}\tilde{y}\tilde{z}, \\ -\tilde{z}' = \bar{\nu} + \dfrac{a}{\bar{B}_3}\tilde{z}^2 + \dfrac{a}{\bar{B}_3}\tilde{x}^2. \end{cases}$$

Let $s = -\bar{B}_3\tau/a$ be a new time variable, and take $\delta = -\bar{B}_3\tau/a$, then we have

$$\begin{cases} \tilde{x}_s = \tilde{y}, \\ \tilde{y}_s = \tilde{\lambda}\,\tilde{x} + \tilde{\mu}\,\tilde{y} - \tilde{x}\tilde{z} + \tilde{b}\,\tilde{y}\tilde{z}, \\ \tilde{z}_s = \tilde{\nu} + \tilde{z}^2 + \tilde{x}^2, \end{cases} \tag{19.26}$$

where

$$\tilde{\lambda} = \lambda \frac{\bar{B}_3^2}{a^2}, \quad \tilde{\mu} = -\mu \frac{\bar{B}_3}{a}, \quad \tilde{b} = -\frac{\bar{b}}{\bar{B}_3}, \quad \tilde{\nu} = \bar{\nu} \frac{B_3}{a}.$$

This system is a system studied by Dumortier and Kokubu [4]. Let us rescale the system (19.26) by

$$\tilde{x} = \varepsilon^3 \bar{x}, \quad \tilde{y} = \varepsilon^5 \bar{y}, \quad \tilde{z} = \varepsilon^4 \bar{y},$$

$$\tilde{\lambda} = \varepsilon^4 \bar{\lambda}, \quad \tilde{\mu} = \varepsilon^2 \bar{\mu}, \quad \tilde{\nu} = \varepsilon^8 \bar{\nu}.$$

Then, we obtain

$$\begin{cases} \bar{x}' = \bar{y}, \\ \bar{y}' = \bar{\lambda} \bar{x} + \bar{\mu} \bar{y} - \bar{x}\bar{z} + O(\varepsilon^2) O(\bar{x}, \bar{y}), \\ \bar{z}' = \bar{x}^2 + \varepsilon^2 \bar{\nu} + \varepsilon^2 \bar{z}^2 + O(\varepsilon^4) O(\bar{x}, \bar{y}, \bar{z}). \end{cases} \tag{19.27}$$

Now we have the following theorem.

Theorem 19.4 ([4], Theorem 3.1) *Suppose* $\bar{\nu} < 0$. *There exists a smooth curve* $\bar{\lambda} = h(\bar{\mu})$ *for sufficiently small* $\bar{\mu} < 0$ *with* $h(\bar{\mu}) \in (-1, 1)$ *such that for any* $(\bar{\lambda}_0, \bar{\mu}_0) = (h(\bar{\mu}_0), \bar{\mu}_0)$ *and for any sufficiently small neighborhood* Λ *of* $(\bar{\lambda}_0, \bar{\mu}_0)$, *there exist* $\varepsilon_0 > 0$ *and a smooth function* $H(\bar{\mu}, \varepsilon)$ *defined for* $\varepsilon \in [0, \varepsilon_0]$ *and for* $\bar{\mu}_0 < 0$ *with* $H(\bar{\mu}_0, 0) = h(\bar{\mu})$ *for which the following statements are equivalent:*

(1) *there exists a connecting orbit from* $(0, 0, \bar{z}_-)$ *to* $(0, 0, \bar{z}_+)$ *in (19.27) with* $(\bar{\lambda}, \bar{\mu}, \varepsilon) \in \Lambda \times [0, \varepsilon_0]$, *where* $\bar{z}_\pm = \pm\sqrt{-\bar{\nu}} + O(\varepsilon^2)$.
(2) $\bar{\lambda} = H(\bar{\mu}, \varepsilon)$.

19.3.2 Heteroclinic Cycles in the Scaling Family

The parameters in (19.26) can be represented by the parameters in (19.13) as follows:

$$\bar{\mu} = -\varepsilon^{-2} B_3[\alpha + \gamma + b\mu_1/B_3]/C_1, \quad \bar{\nu} = -\varepsilon^{-8} r^{-4} \mu_1^2/C_1,$$

where

$$\alpha = \mu_0 - a_2\mu_1/b_2, \quad \beta = -a_4\mu_1/b_2 \quad \text{and} \quad \gamma = \mu_2 - c_2\mu_1/b_2.$$

For a given μ_1 satisfying $b_2\mu_1 < 0$, we can take $\bar{\mu} < 0$ with a suitable choice of parameters μ_0 and μ_2. We can compute sign of $\bar{\nu}$ as follows: we have sign $\bar{\nu} = -$sign C_1, where

$$C_1 = [z^*(\alpha - \gamma)\{2a_2\beta - a_4(\alpha - \gamma)\}]/(2\beta) + (2c_4\beta - c_2(\alpha - \gamma))z^*.$$

In addition, simple computation yields

$$C_1 = -\frac{1}{2z^*}[(\mu_0 - \mu_2)^2 - \{(a_2 - c_2)^2 + 4a_4c_4)\}(z^*)^4]. \tag{19.28}$$

It should be noted that $a_4c_4 < 0$ is necessary for the existence of double degenerate points $P_\pm$. At the degenerate points, we have

$$C_1 = \pm 2(z^*)^3(a_2 - c_2)\sqrt{-a_4c_4} \quad \text{at} \quad (\mu_0, \mu_2) = P_\pm,$$

where

$$P_\pm := (\mu_0^\pm, \mu_2^\mp) = \left(\frac{\mu_1}{b_2}(a_2 \mp \sqrt{-a_4c_4}), \frac{\mu_1}{b_2}(c_2 \pm \sqrt{-a_4c_4})\right).$$

Moreover, we have

$$\varepsilon^2\bar{\mu} = -2\mu_1(a_2 + c_2)/C_1 \quad \text{at} \quad (\mu_0, \mu_2) = P_\pm.$$

It is easy to see that if $b_2(a_2 + c_2) < 0$ and $b_2\mu_1 < 0$ hold, then $(a_2 + c_2)\mu_1 > 0$ and $b_2\mu_1 < 0$ hold, and vice versa. Therefore, we have the following proposition:

Proposition 19.4 *If $a_4c_4 < 0$ and $b_2(a_2 + c_2) < 0$, then for a given μ_1 satisfying $|\mu_1| \ll 1$ and $b_2\mu_1 < 0$, the following hold: if* sign $(a_2 - c_2) = +1$ *(resp. -1), then the scaling family (19.26) of (19.3) has the connecting orbit from $(0, 0, \bar{z}_-)$ to $(0, 0, \bar{z}_+)$ with a suitable choice of parameters around the degenerate point P_+ (resp. P_-).*

In [4], the following result was also given: there are infinitely many horseshoes for the Poincaré map around the hetroclinic cycles given in Theorem 19.4 (see Sect. 4 of [4] for the details). Therefore, we can conclude that Proposition 19.4 gives a reasonable explanation for the numerical results shown in introduction: chaotic dynamics in the integro-reaction-diffusion system (19.4). Indeed, the parameters given in (19.5) yield the following (see (19.6) and Sect. 19.2.3):

$$a_2 > 0, \quad c_2 < 0, \quad a_2 + c_2 > 0, \quad b_2 < 0, \quad a_4c_4 < 0, \quad \mu_0^+ < 0 \quad \text{and} \quad \mu_2^+ < 0.$$

Acknowledgements The authors would like to express their sincere gratitude to Professor Hiroshi Kokubu (Kyoto University) for valuable discussions and comments.

Appendix A

Here we show how to derive the normal form (19.3) from (19.4).

Let $\mathbf{u}(t,x) = (u(x,t), v(x,t))$ be a solution of (19.4). Define the function $\tilde{\mathbf{u}}(t,x)$, $t > 0, x \in (-L, L)$ by

$$\tilde{\mathbf{u}}(t,x) = \begin{cases} \mathbf{u}(t,x) & x \in (0,L), \\ \mathbf{u}(t,-x) & x \in (-L,0). \end{cases}$$

Then $\tilde{\mathbf{u}}$ is a solution of the system

$$\begin{cases} u_t = D_1 u_{xx} + au + bv + F(u,v) + \dfrac{s}{2L}\displaystyle\int_{-L}^{L} u(t,x)\,dx, \ x \in (-L,L), \ t > 0, \\ v_t = D_2 v_{xx} + cu + dv + G(u,v), \ x \in (-L,L), \ t > 0 \end{cases} \tag{19.29}$$

with Neumann boundary conditions at the boundary of interval $[-L, L] \subset \mathbb{R}$. Similarly, the solution $\tilde{\mathbf{u}}$ of (19.4) can be extended to the solution of (19.29) with periodic boundary conditions with period $2L$ on $\mathbb{R}$. Conversely, if the function $\mathbf{u}(t,x)$ satisfying $\mathbf{u}(t,x) = \mathbf{u}(t,-x)$ is a solution of (19.29) with period $2L$, then it satisfies (19.29) and Neumann boundary condition at $x = 0, L$. Therefore, the solution of (19.4) can be identified to the "even" solution of (19.29) with periodic boundary conditions with period $2L$. This implies the solution can be expressed by the Fourier series:

$$u(t,x) = \sum_{n\in\mathbb{Z}} u_m(t)e^{im\pi x/L}, \quad v(t,x) = \sum_{n\in\mathbb{Z}} v_m(t)e^{im\pi x/L}, \quad (u_m, v_m) = (u_{-m}, v_{-m}) \in \mathbb{R}^2.$$

Substituting into (19.29), we have

$$\frac{d}{dt}\begin{pmatrix} u_m \\ v_m \end{pmatrix} = M_m \begin{pmatrix} u_m \\ v_m \end{pmatrix} + \begin{pmatrix} f_m \\ g_m \end{pmatrix}, m \geq 0, \tag{19.30}$$

where

$$f_m = \sum_{\substack{m_1+m_2+m_3=m \\ m_1,m_2,m_3\in\mathbb{Z}}} (f_{30}u_{m_1}u_{m_2}u_{m_3} + f_{21}u_{m_1}u_{m_2}v_{m_3} + f_{12}u_{m_1}v_{m_2}v_{m_3} + f_{03}v_{m_1}v_{m_2}v_{m_3}),$$

$$g_m = \sum_{\substack{m_1+m_2+m_3=m \\ m_1,m_2,m_3\in\mathbb{Z}}} (g_{30}u_{m_1}u_{m_2}u_{m_3} + g_{21}u_{m_1}u_{m_2}v_{m_3} + g_{12}u_{m_1}v_{m_2}v_{m_3} + g_{03}v_{m_1}v_{m_2}v_{m_3}),$$

$$f_{j\ell} = \frac{1}{j!\ell!}\frac{\partial^{j\ell} F}{\partial u^j \partial v^\ell}(0,0), \qquad g_{j\ell} = \frac{1}{j!\ell!}\frac{\partial^{j\ell} G}{\partial u^j \partial v^\ell}(0,0), \quad j + \ell = 3, \, j, \ell \in \mathbb{N},$$

$$M_m = \begin{cases} \begin{pmatrix} a+s & b \\ c & d \end{pmatrix} & (m = 0), \\ \\ \begin{pmatrix} a - D_1 m^2 k_0^2 & b \\ c & d - D_2 m^2 k_0^2 \end{pmatrix} & (m \neq 0), \end{cases} \tag{19.31}$$

and $k_0 = \pi/L$. We consider the system (19.30) in a phase space

$$X_F := \Big\{\{(u_m, v_m)\}_{m\in\mathbb{Z}};\ (u_m, v_m) = (u_{-m}, v_{-m}) \in \mathbb{R}^2,$$
$$\|\{(u_m, v_m)\}_{m\in\mathbb{Z}}\|^2_{X_F} = \sum_{m\in\mathbb{Z}} (1+m^2)^2 |(u_m, v_m)|^2 < \infty\Big\}$$

Solving $\det M_0 = \det M_1 = \det M_2 = 0$ for s, k_0 and D_2, a triply degenerate point of 0:1:2-modes is given by the following:

$$k_0 = k_0^{1,2} := \left[\frac{1}{8dD_1}\left\{5\varDelta - \sqrt{25\varDelta^2 - 16ad\varDelta}\right\}\right]^{1/2},$$

$$D_2 = D_2^{1,2} := \frac{\{dD_1(k_0^*)^2 - \varDelta\}}{(k_0^*)^2\{D_1(k_0^*)^2 - a\}},$$

$$s = s^* := -\varDelta/d,$$

where $\varDelta = ad - bc$. Near this degenerate point, we can apply the center manifold theory (for instance, see [3, 9, 10]). To compute the dynamics on the center manifold of (19.4), we diagonalize the equations in (19.30) for $m = 0$, 1 and 2. Set $(k_0, D_2, s) = (k_0^{1,2}, D_2^{1,2}, s^*)$. Then changing variables ${}^t(u_m, v_m) = T_m\,{}^t(\tilde{u}_m, \tilde{v}_m)$, $(m = 0, 1, 2)$ by the matrix

$$T_0 = \begin{pmatrix} -d & bc/d \\ c & c \end{pmatrix},\ T_m = \begin{pmatrix} -d + D_2^{1,2} m^2 (k_0^{1,2})^2 & a - D_1 m^2 (k_0^{1,2})^2 \\ c & c \end{pmatrix}, m = 1, 2,$$

we have

$$\begin{pmatrix} \dot{\tilde{u}}_m \\ \dot{\tilde{v}}_m \end{pmatrix} = \begin{pmatrix} 0 & 0 \\ 0 & \mu_m^- \end{pmatrix} \begin{pmatrix} \tilde{u}_m \\ \tilde{v}_m \end{pmatrix} + T_m^{-1} \begin{pmatrix} \tilde{f}_m \\ \tilde{g}_m \end{pmatrix}, m = 0, 1, 2.$$

Here,

$$\mu_0^- := d + bc/d,\quad \mu_m^- := (a+d) - m^2(D_1 + D_2^{1,2})(k_0^{1,2})^2,$$
$$\tilde{f}_m := f_m|_{{}^t(u_{m_j}, v_{m_j}) = T_m\,{}^t(\tilde{u}_{m_j}, \tilde{v}_{m_j})},\quad \tilde{g}_m := g_m|_{{}^t(u_{m_j}, v_{m_j}) = T_m\,{}^t(\tilde{u}_{m_j}, \tilde{v}_{m_j})}.$$

Set

$$\rho := (k_0^{1,2}, D_2^{1,2}, s^*) - (k_0, D_2, s),\quad \mu_m^+ := \left\{\mathrm{tr}\, M_m + \sqrt{(\mathrm{tr}\, M_m)^2 - 4\det M_m}\right\}/2.$$

We define a neighborhood $\mathscr{U}_\varepsilon$ of $X_F \times \mathbb{R}^3$:

$$\mathscr{U}_\varepsilon := \left\{(\{(u_m, v_m)\}_{m\in\mathbb{Z}}, \rho) \in X_F \times \mathbb{R}^3;\ ||\{(u_m, v_m)\}_{m\in\mathbb{Z}}||_{X_F} + |\rho| < \varepsilon\right\}.$$

Then we have the following theorem.

Theorem A ([13]) *For given constants a, b, c, d, D_1, there exists a positive constant ε such that the local center manifold $\mathscr{W}^c_{loc}$ of* (19.30) *is contained in $\mathscr{U}_\varepsilon$. Moreover, the dynamics of* (19.30) *on the manifold $\mathscr{W}^c_{loc}$ is governed by the following system:*

$$\begin{cases} \dot{z}_0 = (\mu_0^+ + a_1 z_0^2 + a_2 z_1^2 + a_3 z_2^2) z_0 + a_4 z_1^2 z_2 + o(3), \\ \dot{z}_1 = (\mu_1^+ + b_1 z_0^2 + b_2 z_1^2 + b_3 z_2^2) z_1 + b_4 z_0 z_1 z_2 + o(3), \\ \dot{z}_2 = (\mu_2^+ + c_1 z_0^2 + c_2 z_1^2 + c_3 z_2^2) z_2 + c_4 z_0 z_1^2 + o(3). \end{cases} \tag{19.32}$$

Here, $z_j(t) \in \mathbb{R}$ denote $\tilde{u}_j(t)$ $(j = 0, 1, 2)$, and $o(3)$ denotes $o(|(z_0, z_1, z_2)|^3)$. In addition, the coefficients μ_j^+, a_j, b_j, c_j are dependent on the coefficients and parameters appearing in (19.30).

Proof The first statement of the theorem follows from standard center manifold theory. It also states that for $m \neq 0, 1, 2$, there exist functions

$$h^u_m(\tilde{u}_0, \tilde{u}_1, \tilde{u}_2; \rho), m \geq 3, \quad h^v_m(\tilde{u}_0, \tilde{u}_1, \tilde{u}_2; \rho), \ m \geq 0,$$

satisfying

$$\frac{\partial h^u_m}{\partial \tilde{u}_j}(0, 0, 0; 0) = \frac{\partial h^v_m}{\partial \tilde{u}_j}(0, 0, 0; 0) = 0, \ (j = 0, 1, 2)$$

and

$$\frac{\partial h^u_m}{\partial \rho}(0, 0, 0; 0) = \frac{\partial h^v_m}{\partial \rho}(0, 0, 0; 0) = 0$$

such that the local invariant manifold $\mathscr{W}^c_{loc}$ is expressed by

$$\begin{aligned} \mathscr{W}^c_{loc} = \big\{ &\{(\tilde{u}_\ell, \tilde{v}_\ell), (u_m, v_m)\}_{|\ell| \leq 2, \, |m| \geq 3} \in X_F; \, \tilde{v}_\ell = h^v_\ell(\tilde{u}_0, \tilde{u}_1, \tilde{u}_2; \rho), \\ &(u_m, v_m) = (h^u_m(\tilde{u}_0, \tilde{u}_1, \tilde{u}_2; \rho), h^v_m(\tilde{u}_0, \tilde{u}_1, \tilde{u}_2; \rho)), |\ell| \leq 2, \ |m| \geq 3 \big\}. \end{aligned}$$

We can check that if $|(\tilde{u}_0, \tilde{u}_1, \tilde{u}_2; \rho)| < \varepsilon$ then $h^u_m = h^v_m = o(\varepsilon^3)$. Then, the cubic truncated equations for $\tilde{u}_m$, $(m = 0, 1, 2)$ are given by the following:

$$\begin{aligned} \dot{\tilde{u}}_0 &= \mu_0^+ \tilde{u}_0 - \frac{1}{\mu_0^-} \left\{ \tilde{f}_0 - \frac{b}{d} \tilde{g}_0 \right\}, \\ \dot{\tilde{u}}_m &= \mu_m^+ \tilde{u}_m - \frac{1}{c \mu_m^-} \{ c \tilde{f}_m + (-a + D_1 m^2 (k_0^{1,2})^2) \tilde{g}_m \}, m = 1, 2, \end{aligned}$$

where

$$\begin{aligned} \tilde{f}_m := &\sum_{\substack{m_1 + m_2 + m_3 = m \\ m_j \in \{0, \pm 1, \pm 2\}}} (f_{30} B_{m_1} B_{m_2} B_{m_3} \tilde{u}_{m_1} \tilde{u}_{m_2} \tilde{u}_{m_3} \\ &+ c f_{21} B_{m_1} B_{m_2} \tilde{u}_{m_1} \tilde{u}_{m_2} \tilde{u}_{m_3} + c^2 f_{12} B_{m_1} \tilde{u}_{m_1} \tilde{u}_{m_2} \tilde{u}_{m_3} + c^3 f_{03} \tilde{u}_{m_1} \tilde{u}_{m_2} \tilde{u}_{m_3}) \end{aligned}$$

and

$$\tilde{g}_m := \sum_{\substack{m_1+m_2+m_3=m\\ m_j\in\{0,\pm1,\pm2\}}} (g_{30}B_{m_1}B_{m_2}B_{m_3}\tilde{u}_{m_1}\tilde{u}_{m_2}\tilde{u}_{m_3}$$
$$+cg_{21}B_{m_1}B_{m_2}\tilde{u}_{m_1}\tilde{u}_{m_2}\tilde{u}_{m_3} + c^2 g_{12}B_{m_1}\tilde{u}_{m_1}\tilde{u}_{m_2}\tilde{u}_{m_3} + c^3 g_{03}\tilde{u}_{m_1}\tilde{u}_{m_2}\tilde{u}_{m_3}).$$

Here, $B_j = -d + j^2 D_2^{1,2}(k_0^{1,2})^2$. This gives the cubic truncated system (19.32). We show the explicit form of coefficients in Appendix B.

Appendix B

We show the coefficients of system (19.3) explicitly. We put

$$\mu_0^- := d + bc/d, \qquad \mu_m^- := (a+d) - m^2(D_1 + D_2^{1,2})(k_0^{1,2})^2,$$
$$A_m := -a + D_1^{1,2}m^2(k_0^{1,2})^2, \quad B_m = -d + D_2^{1,2}m^2(k_0^{1,2})^2.$$

Then we have the following,

$$\mu_m := \left\{\operatorname{tr} M_m + \sqrt{(\operatorname{tr} M_m)^2 - 4\det M_m}\right\}/2,\ a_j = -\frac{1}{\mu_0^+}P_f^{a_j} + \frac{b}{d\mu_0^+}P_g^{a_j},\ j = 1\ldots4,$$

$$b_j = -\frac{1}{\mu_1^-}P_f^{b_j} - \frac{A_1}{c\mu_1^-}P_g^{b_j},\ j = 1\ldots4, \qquad c_j = -\frac{1}{\mu_2^-}P_f^{c_j} - \frac{A_2}{d\mu_2^-}P_g^{c_j},\ j = 1\ldots4.$$

Here,

$$\begin{pmatrix} P_f^{a_1} \\ P_g^{a_1}\end{pmatrix} := B_0^3\begin{pmatrix} F_{30} \\ G_{30}\end{pmatrix} + cB_0^2\begin{pmatrix} F_{21} \\ G_{21}\end{pmatrix} + c^2B_0\begin{pmatrix} F_{12} \\ G_{12}\end{pmatrix} + c^3\begin{pmatrix} F_{03} \\ G_{03}\end{pmatrix},$$

$$\begin{pmatrix} P_f^{a_2} \\ P_g^{a_2}\end{pmatrix} := 6B_1^2B_0\begin{pmatrix} F_{30} \\ G_{30}\end{pmatrix} + 2cB_1(B_1+2B_0)\begin{pmatrix} F_{21} \\ G_{21}\end{pmatrix} + 2c^2(B_0+2B_1)\begin{pmatrix} F_{12} \\ G_{12}\end{pmatrix} + 6c^3\begin{pmatrix} F_{03} \\ G_{03}\end{pmatrix}$$

$$\begin{pmatrix} P_f^{a_3} \\ P_g^{a_3}\end{pmatrix} := 6B_2^2B_0\begin{pmatrix} F_{30} \\ G_{30}\end{pmatrix} + 2cB_2(B_2+2B_0)\begin{pmatrix} F_{21} \\ G_{21}\end{pmatrix} + 2c^2(B_0+2B_2)\begin{pmatrix} F_{12} \\ G_{12}\end{pmatrix} + 6c^3\begin{pmatrix} F_{03} \\ G_{03}\end{pmatrix},$$

$$\begin{pmatrix} P_f^{a_4} \\ P_g^{a_4}\end{pmatrix} := 6B_1^2B_2\begin{pmatrix} F_{30} \\ G_{30}\end{pmatrix} + 2cB_1(B_1+2B_2)\begin{pmatrix} F_{21} \\ G_{21}\end{pmatrix} + 2c^2(B_2+2B_1)\begin{pmatrix} F_{12} \\ G_{12}\end{pmatrix} + 6c^3\begin{pmatrix} F_{03} \\ G_{03}\end{pmatrix},$$

$$\begin{pmatrix} P_f^{b_1} \\ P_g^{b_1} \end{pmatrix} := 3B_0^2B_1 \begin{pmatrix} F_{30} \\ G_{30} \end{pmatrix} + cB_0(B_0+2B_1) \begin{pmatrix} F_{21} \\ G_{21} \end{pmatrix} + c^2(B_1+2B_0) \begin{pmatrix} F_{12} \\ G_{12} \end{pmatrix} + 3c^3 \begin{pmatrix} F_{03} \\ G_{03} \end{pmatrix},$$

$$\begin{pmatrix} P_f^{b_2} \\ P_g^{b_2} \end{pmatrix} := 3B_1^3 \begin{pmatrix} F_{30} \\ G_{30} \end{pmatrix} + 3cB_1^2 \begin{pmatrix} F_{21} \\ G_{21} \end{pmatrix} + 3c^2B_1 \begin{pmatrix} F_{12} \\ G_{12} \end{pmatrix} + 3c^3 \begin{pmatrix} F_{03} \\ G_{03} \end{pmatrix},$$

$$\begin{pmatrix} P_f^{b_3} \\ P_g^{b_3} \end{pmatrix} := 6B_2^2B_1 \begin{pmatrix} F_{30} \\ G_{30} \end{pmatrix} + 2cB_2(B_2+2B_1) \begin{pmatrix} F_{21} \\ G_{21} \end{pmatrix} + 2c^2(B_1+2B_2) \begin{pmatrix} F_{12} \\ G_{12} \end{pmatrix} + 6c^3 \begin{pmatrix} F_{03} \\ G_{03} \end{pmatrix},$$

$$\begin{pmatrix} P_f^{b_4} \\ P_g^{b_4} \end{pmatrix} := 6B_0B_1B_2 \begin{pmatrix} F_{30} \\ G_{30} \end{pmatrix} + 2c(B_0B_1+B_1B_2+B_2B_0) \begin{pmatrix} F_{21} \\ G_{21} \end{pmatrix}$$

$$+2c^2(B_0+B_1+B_2) \begin{pmatrix} F_{12} \\ G_{12} \end{pmatrix} + 6c^3 \begin{pmatrix} F_{03} \\ G_{03} \end{pmatrix},$$

$$\begin{pmatrix} P_f^{c_1} \\ P_g^{c_1} \end{pmatrix} := 3B_0^2B_2 \begin{pmatrix} F_{30} \\ G_{30} \end{pmatrix} + cB_0(B_0+2B_2) \begin{pmatrix} F_{21} \\ G_{21} \end{pmatrix} + c^2(B_2+2B_0) \begin{pmatrix} F_{12} \\ G_{12} \end{pmatrix} + 3c^3 \begin{pmatrix} F_{03} \\ G_{03} \end{pmatrix},$$

$$\begin{pmatrix} P_f^{c_2} \\ P_g^{c_2} \end{pmatrix} := 6B_1^2B_2 \begin{pmatrix} F_{30} \\ G_{30} \end{pmatrix} + 2cB_1(B_1+2B_2) \begin{pmatrix} F_{21} \\ G_{21} \end{pmatrix} + 2c^2(B_2+2B_1) \begin{pmatrix} F_{12} \\ G_{12} \end{pmatrix} + 6c^3 \begin{pmatrix} F_{03} \\ G_{03} \end{pmatrix},$$

$$\begin{pmatrix} P_f^{c_3} \\ P_g^{c_3} \end{pmatrix} := 3B_2^3 \begin{pmatrix} F_{30} \\ G_{30} \end{pmatrix} + 3cB_2^2 \begin{pmatrix} F_{21} \\ G_{21} \end{pmatrix} + 3c^2B_2 \begin{pmatrix} F_{12} \\ G_{12} \end{pmatrix} + 3c^3 \begin{pmatrix} F_{03} \\ G_{03} \end{pmatrix},$$

$$\begin{pmatrix} P_f^{c_4} \\ P_g^{c_4} \end{pmatrix} := 3B_1^2B_0 \begin{pmatrix} F_{30} \\ G_{30} \end{pmatrix} + cB_1(B_1+2B_0) \begin{pmatrix} F_{21} \\ G_{21} \end{pmatrix} + c^2(B_0+2B_1) \begin{pmatrix} F_{12} \\ G_{12} \end{pmatrix} + 3c^3 \begin{pmatrix} F_{03} \\ G_{03} \end{pmatrix}.$$

References

1. Armbruster, D., Guckenheimer, J., Holmes, P.: Heteroclinic cycles and modulated travelling waves in system with O(2) symmetry. Physica **29D**, 257–282 (1988)
2. Armbruster, D., Guckenheimer, J., Holmes, P.: Kuramoto-Sivashinsky dynamics on the center-unstable manifold. SIAM J. Appl. Math. **49**(3), 676–691 (1988)
3. Carr, J.: Applications of center manifold theory. Springer (1981)
4. Dumortier, F., Kokubu, H.: Chaotic dynamics in $\mathbb{Z}_2$-equivalent unfoldings of codimension three singularities of vector fields in $\mathbb{R}^3$. Ergod. Theory Dyn. Syst. **20**, 85–107 (2000)
5. Fukushima, S., Nakanishi, S., Nakato, Y., Ogawa, T.: Selection principle for various modes of spatially non-uniform electrochemical oscillations. J. Chem. Phys. **128**, 014714 (2008)
6. Guckenheimer, J., Holmes, P.: Nonlinear Oscillations, Dynamical Systems, and Bifurcations of Vector Fields. Springer (1983)
7. Golubitsky, M., Stewart, I.: The Symmetry Perspective: From Equilibrium to Chaos in Phase Space and Physical Space. Brikhäuser (2003)
8. Golubitsky, M., Stewart, I., Schaeffer, G.D.: Singularities and Groups in Bifurcation Theory, vol. 1. Springer (1985). (vol. 2) (1988)

9. Haragus, M., Iooss, G.: Local Bifurcations, Center Manifolds, and Normal Forms in Infinite-dimensional Dynamical Systems. Springer (2010)
10. Henry, D.: Geometric Theory of Semilinear Parabolic Equations. Springer (1981)
11. Kuznetsov, Y.A.: Elements of Applied Bifurcation Theory, 3rd edn. Springer (1997)
12. Ogawa, T.: Degenerate Hopf Instability in Oscillatory Reaction-Diffusion Equations, DCDS Supplements, Special vol. 2007, pp. 784–793 (2007)
13. Ogawa, T., Okuda, T.: Oscillatory dynamics in a reaction-diffusion system in the presence of 0:1:2 resonance. Netw. Heterog. Media **7**(4), 893–926 (2012)
14. Proctor, M.R.E., Jones, C.A.: The interaction of two spatially resonant patterns in thermal convection, Part 1. Exact 1:2 resonance. J. Fluid Mech. **188**, 301–335 (1988)
15. Smith, T.R., Moehlis, J., Holmes, P.: Heteroclinic cycles and periodic orbits for the O(2)-equivariant 0:1:2 mode interaction. Physica **211D**, 347–376 (2005)
16. Porter, J., Knobloch, E.: New type of complex dynamics in the 1:2 spatial resonance. Physica **159D**, 125–154 (2001)

Chapter 20
L^∞-Stability of Discontinuous Traveling Waves in a Radiating Gas Model

Masashi Ohnawa

Abstract In the present article, we prove the L^∞-stability of discontinuous or supercritical shock waves which appear in a model system of radiating gases if the shock strength is greater than a certain critical value. The author has recently shown (SIAM J. Math. Anal. (2014), 2136–2159.) that all subcritical shock waves are stable to small perturbations while the critical shock wave blows up the first order derivative in a finite time if certain types of perturbations are added whatever small the perturbations may be. In the supercritical case, we show that the convection contributes to recover the stability by virtue of discontinuity in the asymptotic state compensating the insufficient smoothing effect of radiation.

Keywords Radiation gas system · Discontinuous shock wave · Asymptotic stability

20.1 Introduction

In the present paper, we study an initial value problem to a hyperbolic-elliptic coupled system called the Hamer model [3] arising in a gas dynamics including a radiation effect. The model system reads

$$u_t + uu_x + q_x = 0, \tag{20.1}$$
$$-q_{xx} + q + u_x = 0, \tag{20.2}$$

where $u(t,x)$ and $q(t,x)$ are scalar-valued functions satisfying

$$u(0,x) = u_0(x) \ \text{ with } \lim_{x\to\pm\infty} u_0(x) = u_\pm, \tag{20.3}$$
$$\lim_{x\to\pm\infty} q(t,x) = 0 \ \text{ for an arbitrary } t \geqslant 0. \tag{20.4}$$

M. Ohnawa (✉)
Faculty of Marine Science, Department of Ocean Sciences,
Tokyo University of Marine Science and Technology, Tokyo 108-8477, Japan
e-mail: ohnawa@m.kaiyodai.ac.jp

Y. Shibata and Y. Suzuki (eds.), *Mathematical Fluid Dynamics, Present and Future*, Springer Proceedings in Mathematics & Statistics 183,
DOI 10.1007/978-4-431-56457-7_20

We are particularly concerned with the stability of shock waves or traveling wave solutions to the Hamer model, which are expressed in the form of

$$(u, q)(t, x) = (U, Q)(\eta), \quad \eta = x - st$$

for a certain constant s. An interesting feature is that the traveling wave solutions are discontinuous if the shock strength:

$$\delta_S := u_- - u_+$$

is greater than a certain constant. Here we summarize conditions for the existence of shock waves and their properties clarified by Kawashima and Nishibata [5].

Proposition 20.1 [5] *If $\delta_S > 0$, there exists a traveling wave solution to (20.1)–(20.4) uniquely up to a shift with $s = (u_- + u_+)/2$. The function U is monotonically decreasing and is odd in a suitable coordinate. Moreover, it holds that*

$$|U(x) - u_S(x)| \leq \frac{1}{2}\delta_S e^{-c|x|}, \quad \text{where } u_S(x) := u_\pm \text{ for } \pm x > 0 \tag{20.5}$$

and c is a positive constant depending only on δ_S.

(i) (subcritical/critical cases) If $\delta_S \in (0, \sqrt{2}]$, the solution satisfies $(U, Q) \in C^1(\mathbb{R}) \times C^2(\mathbb{R})$, and

$$0 > U'(x) \geqslant U'(0) = \left(-1 + \sqrt{1 - \delta_S^2/2}\right)/2. \tag{20.6}$$

(ii) (supercritical case) If $\delta_S > \sqrt{2}$, the solution is discontinuous at only one point. Setting the point of discontinuity at $x = 0$, it holds that

$$0 > U'(x) > U'(\pm 0) = -1 \quad \text{for } x \in \mathbb{R}_0 := \mathbb{R} \setminus \{0\}. \tag{20.7}$$

The stability of traveling wave solutions to the Hamer model has been studied extensively in the last two decades. In L^1-topology, Serre [14, 15] showed that any shock wave is stable to arbitrary large L^1-initial perturbations. While in L^∞-topology, the earliest result was obtained in [5], which showed the L^∞-stability of shock waves with $\delta_S < \sqrt{6}/2 \approx 1.22$ using L^2 energy method. Extension of this work include multidimensional problem [2], general flux function problem [13], general systems [9, 10] to cite just a few. In a recent paper [12], the author showed that all subcritical shock waves (the case with $\delta_S < \sqrt{2}$) are L^∞-stable to small perturbations, while the critical shock wave (the case with $\delta_S = \sqrt{2}$) blows up the first order derivative in a finite time if certain types of perturbations are added whatever small the perturbations may be.

The difference in the behaviors between subcritical and critical shock waves indicates that the continuity of the profile of traveling waves might be related to the stability in C^1 framework. The objective of the present work is to show how the discontinuities in the supercritical shock waves help to recover the L^∞-stability. (Note that the result in [12] does not exclude the possibility of L^∞-stability of the critical shock wave, as inferred from [8, 14].) The importance of the appearance of discontinuous shock waves in the gas dynamics with radiation (in the present problem, the case with $\delta_S > \sqrt{2}$) has been addressed in [16] from the physical point of view.

Notation: For a constant $p \in [1, \infty]$, $|f|_p$ denotes the canonical L^p norm of a function f. For a nonnegative integer $m \geqslant 0$, H^m denotes the m-th order Sobolev space in the L^2 sense, equipped with the norm $\|\cdot\|_{H^m}$. We often simplify it as $\|\cdot\|_m$. For a function f which is continuous except for discontinuities of the first kind, $[\![f]\!](x)$ and $[\![f]\!](t,x)$ denote the jump amplitude in the spatial direction of f at x, i.e.,

$$[\![f]\!](x) := f(x-0) - f(x+0), \quad [\![f]\!](t,x) := f(t,x-0) - f(t,x+0)$$

and $\bar{f}(x)$ and $\bar{f}(t,x)$ denote the mean of the left and the right limit of f at x, i.e.,

$$\bar{f}(x) := (f(x-0) + f(x+0))/2, \quad \bar{f}(t,x) := (f(t,x-0) + f(t,x+0))/2.$$

Finally, c and C denote generic positive constants.

We consider smooth initial data except for a single discontinuity at $x = d_0$ and observe the pointwise behavior of solutions perturbed around discontinuous shock waves. In order to treat discontinuities, notion of an admissible solution is introduced in [5, 6] following the method of Kružkov [7].

If $u(t,x)$ is smooth except at $x = d(t)$, the entropy condition

$$[\![u]\!](t, d(t)) > 0 \tag{20.8}$$

and the Rankine-Hugoniot conditions

$$[\![q]\!](t, d(t)) = 0, \tag{20.9}$$

$$[\![u - q_x]\!](t, d(t)) = 0, \tag{20.10}$$

$$\dot{d}(t) = \tfrac{1}{2}\left(u(t, d(t)-0) + u(t, d(t)+0)\right) = \bar{u}(t, d(t)) \tag{20.11}$$

hold, where $\dot{d}(t)$ is a derivative of $d(t)$ in t.

To state our main results precisely, we define some quantities and functions. When

$$u_0 - u_S \in L^1, \tag{20.12}$$

is satisfied, we define the 'center of mass' of u_0 in the following way. Setting the shift of the traveling wave solution so that U is discontinuous at the origin, the center of mass x_0 of the initial data u_0 is given by

$$x_0 := \frac{1}{u_- - u_+}\int_{-\infty}^{\infty}(u_0(x) - U(x))\,dx,$$

which is equivalent to

$$\int_{-\infty}^{\infty}(u_0(x) - U(x - x_0))\,dx = 0.$$

Letting $s_0 = (u_- + u_+)/2$, we change variables as

$$\hat{u}(t,\hat{x}) = u(t,\hat{x} + x_0 + s_0 t) - s_0, \quad \hat{q}(t,\hat{x}) = q(t,\hat{x} + x_0 + s_0 t),$$

so that we may assume $s_0 = 0$ i.e. $u_- + u_+ = 0$ without loss of generality. In the new coordinate, the center of mass of the initial data is located at the origin. Hereafter we denote new variables $\hat{u}, \hat{q}, \hat{x}$ simply by u, q, x respectively and fix the shift of U so that $U(x)$ is discontinuous at $x = 0$.

The initial perturbation is

$$\phi_0(x) := u_0(d_0 + x) - U(x) \quad \text{for } x \in \mathbb{R}_0$$

and we define its potential or in other words anti-derivative by

$$\Phi_0(x) := \int_{\pm\infty}^{x}\phi_0(y)\,dy \quad \text{for} \quad \pm x > 0.$$

Φ_0 is well-defined by (20.5) and (20.12). In the case u_0 is odd, $u(t,\cdot)$ is also odd for an arbitrary $t \geqslant 0$ and $d(t)$ is identically zero. Our main theorem is the following.

Theorem 20.1 *Suppose u_0 is an odd function which is smooth except at $x = 0$, and satisfies the entropy condition*

$$[\![u_0]\!](0) > 0. \tag{20.13}$$

If

$$\phi_0 \in L^1(\mathbb{R}_0), \quad \text{and} \quad \Phi_0 \in H^3(\mathbb{R}_0) \tag{20.14}$$

are satisfied, where $\mathbb{R}_0 := \mathbb{R} \setminus \{0\}$, the initial value problem (20.1)–(20.4) with sufficiently small $\|\Phi_0\|_3$ has a unique global solution in the sense of Kružkov verifying

$$u(t,x) - U(x) \in \bigcap_{k=0}^{2} C^k([0,\infty); H^{2-k}(\mathbb{R}_0)), \tag{20.15}$$

and

$$q(t,x) - Q(x) \in \bigcap_{k=0}^{2} C^k([0,\infty); H^{3-k}(\mathbb{R}_0)). \tag{20.16}$$

Moreover, the solution converges uniformly to the shock wave:

$$\sup_{x\in\mathbb{R}_0} (u(t,x) - U(x),\ q(t,x) - Q(x)) \to 0 \ \text{as}\ t \to \infty. \tag{20.17}$$

Outline of the Paper

In Sect. 20.2, we prove the existence of a local solution and see that the solution exists arbitrarily long if the initial perturbation is sufficiently small. The first part of Sect. 20.3 is devoted to giving a bound to $|(\Phi, \phi)|_2$ and L^2-norms of ϕ_x and ϕ_{xx} in a domain away from the discontinuity in the usual manner. Finally, in the second part of Sect. 20.3, we bound $|(\phi_x, \phi_{xx})|_2$ noting that perturbations close to the discontinuity are 'swept away' thanks to the presence of the discontinuity and complete the proof of our main results.

20.2 Local Solvability

In order to construct Kružkov's admissible solutions to (20.1)–(20.4), we recall (see e.g. [1]) that a piecewise smooth data is admissible if it satisfies (20.1) and (20.2) classically where it is smooth while at the discontinuities it verifies the entropy condition (20.8) and the Rankine-Hugoniot conditions (20.9)–(20.11). The uniqueness of admissible solutions to (20.1)–(20.4) is shown in [4].

Besides stated in Proposition 20.1, we also recall facts about traveling wave solutions [5] that $(U, Q) \in C^\infty(\mathbb{R}_0) \times C^\infty(\mathbb{R}_0)$ and it satisfies

$$UU' + Q' = 0, \tag{20.18}$$
$$-Q'' + Q + U' = 0, \tag{20.19}$$

in the classical sense at $x \neq 0$ and

$$[\![Q]\!](0) = 0, \quad [\![Q' - U]\!](0) = 0, \quad Q(\pm\infty) = 0. \tag{20.20}$$

In the next lemma, we construct a local solution to (20.1)–(20.4) for initial value around (U, Q) in a suitable functional space. Due to the entropy condition at the discontinuities, this is proved by appealing to the Kato theory in a standard manner (see e.g. [11]).

Lemma 20.1 *Suppose $\phi_0 \in H^2(\mathbb{R}_0)$ is an odd function satisfying $[\![U + \phi_0]\!](0) > 0$. Then there exist a positive constant T depending only on $\|\phi_0\|_2$ and a unique $\phi \in \bigcap_{k=0}^{2} C^k([0, T]; H^{2-k}(\mathbb{R}_0))$ which solves*

$$\phi_t(t,x) + (U(x) + \phi(t,x))\phi_x(t,x) + U'(x)\phi(t,x) + \phi(t,x) - K * \phi(t,x) = 0, \tag{20.21}$$

$$\phi(0,x) = \phi_0(x). \tag{20.22}$$

The solution satisfies $\phi(t,-x) = -\phi(t,x)$ *for* $x \neq 0$, $\|\phi(t,\cdot)\|_2 \leqslant 2\|\phi_0\|_2$ *and*

$$[\![U + \phi]\!](t,0) > [\![U + \phi_0]\!](0)/2 \tag{20.23}$$

for an arbitrary $t \in [0,T]$. *Here,* $K(x) := e^{-|x|}/2$ *is a fundamental solution to the operator* $1 - \partial_{xx}$. *Defining* $\psi \in \bigcap_{k=0}^{2} C^k([0,T]; H^{3-k}(\mathbb{R}_0))$ *by* $\psi = -(K * \phi)_x$ *(the derivative is defined classically a.e.x since* $K * \phi$ *is Lipschitz continuous), (20.21) is equivalent to*

$$\phi_t(t,x) + (U(x) + \phi(t,x))\phi_x(t,x) + U'(x)\phi(t,x) + \psi_x(t,x) = 0, \tag{20.24}$$

$$-\psi_{xx}(t,x) + \psi(t,x) + \phi_x(t,x) = 0, \tag{20.25}$$

and

$$[\![\psi]\!](t,0) = 0, \quad [\![\psi_x - \phi]\!](t,0) = 0. \tag{20.26}$$

By the Sobolev embedding, we see that $(u,q)(t,x) := (\phi,\psi)(t,x) + (U,Q)(x)$ is piecewise smooth and satisfies (20.1)–(20.4) in the classical sense except at $x = 0$. The Rankine-Hugoniot conditions (20.9)–(20.11) and the entropy condition $[\![u]\!](t,0) > 0$ are also satisfied. Hence (u,q) is the unique admissible solution to (20.1)–(20.4).

In order to evaluate temporal evolution of integrals of certain time-dependent variables over domains which also change with time, we frequently use the so called Reynolds transport theorem stated as follows.

Lemma 20.2 *Suppose* $a(t)$ *and* $b(t)$ *are* C^1 *functions defined over* $t \in [t_0,t_1]$ *which satisfy* $a(t) < b(t)$. *If* f *is a* C^1 *function defined over* $\{(t,x) \mid t \in [t_0,t_1], x \in (a(t),b(t))\}$, *it holds*

$$\frac{d}{dt}\int_{a(t)}^{b(t)} f(t,x)dx = \int_{a(t)}^{b(t)} f_t(t,x)dx + f(b(t))\dot{b}(t) - f(a(t))\dot{a}(t).$$

The proof is elementary and we omit it.

Lemma 20.3 *Assuming (20.12), the solution* ϕ *obtained in Lemma 20.1 satisfies* $\phi(t,\cdot) \in L^1(\mathbb{R})$ *for* $t \in [0,T]$. *Defining the potential or the anti-derivative of* ϕ *by*

$$\Phi(t,x) := \int_{\pm\infty}^{x} \phi(t,y)dy \ \ for \ \pm x > 0, \tag{20.27}$$

it satisfies

$$\Phi_t(t,x) + U(x)\Phi_x(t,x) + \frac{1}{2}\phi^2(t,x) + \psi(t,x) = 0 \ \text{for}\ x \in \mathbb{R}_0. \tag{20.28}$$

Furthermore, if $\Phi_0 \in L^2(\mathbb{R}_0)$, *it holds that* $\Phi(t,\cdot) \in L^2(\mathbb{R}_0)$ *for* $t \in [0,T]$.

Proof The integrability of $\phi(t,\cdot)$ is proved by similar arguments to those in [5]. The governing equation for Φ and the final statement are obtained following arguments in [12] (the proof of Lemma 2.4) with the help of Lemma 20.2.

Lemma 20.4 *Define a positive constant* L_0 *by* $L_0 := |(U')^{-1}(-1/3)|$. *There exists a positive constant* δ_0 *such that* $\|\phi_0\|_2 \leqslant \delta_0$ *implies that the local solutions have a sufficiently long life span* T *so that characteristic curves emanating from* $x = \pm L_0$ *at* $t = 0$ *reach* $x = \pm 0$ *by* $t = T/2$.

Proof Let δ_1 be an arbitrary positive constant. By Lemma 20.1, we may assume that for an initial data satisfying $\|\phi_0\|_2 \leqslant \delta_1$, there exist a positive constant T_0 and a local solution over $[0, T_0]$ with $\sup_{t \leqslant T_0} \|\phi(t)\|_2 \leqslant 2\|\phi_0\|_2$. Therefore we can prolong the life span of the local solution arbitrary long by letting $\|\phi_0\|_2$ be suitably small.

By letting δ_1 further small if necessary, it holds that $\sup_{t \leqslant T_0} |\phi|_\infty(t) \leqslant |U(\pm 0)|/2$ thanks to the Sobolev embedding theorem. The conclusion follows easily from these considerations.

20.3 Asymptotic Stability

In this section, assuming the existence of a solution (ϕ, ψ) to (20.24) and (20.25) over $t \in [0, T]$ with the properties stated in Lemma 20.1, we give a-priori estimates to the solution. Let

$$N(T) := \sup_{t \in [0,T]} \|\Phi(t,\cdot)\|_3.$$

20.3.1 Energy Estimates Away from the Discontinuity

Lemma 20.5 *If* $N(T)$ *is sufficiently small, it holds that*

$$|\Phi, \phi|_2(t)^2 + \int_0^t |(\phi, \psi, \psi_x)|_2(s)^2 ds \leqslant C|\Phi_0, \phi_0|_2^2 \tag{20.29}$$

for an arbitrary $t \in [0, T]$, *where* C *is a positive constant independent of* T.

Proof By modifying the proof of Lemma 2.5 in [11] noting $\inf_x U'(x) = -1$ in the supercritical case, we arrive at the conclusion.

Lemma 20.6 *Let a domain Ω_0 be defined by $\Omega_0 := \{x \in \mathbb{R}_0 \mid U'(x) > -1/3\}$. If $N(T)$ is sufficiently small, it holds for an arbitrary $t \in [0, T]$ that*

$$|(\Phi, \phi)|_2(t)^2 + \int_{\Omega_0} \left(\phi_x^2 + \phi_{xx}^2\right)(t, x)dx \\ + \int_0^t \left(|(\phi, \psi, \psi_x)|_2(s)^2 + \int_{\Omega_0} (\phi_x^2 + \phi_{xx}^2)(s, x)dx \right) ds \leqslant C\|\Phi_0\|_3^2, \tag{20.30}$$

where C is a positive constant independent of T.

Proof Differentiate (20.24) in x and multiply the result by ϕ_x and use (20.25) to obtain

$$\partial_t \left(\frac{1}{2}\phi_x^2\right) + \partial_x \left(\frac{1}{2}(\phi + U)\phi_x^2\right) + \left(1 + \frac{3}{2}U' + \frac{1}{2}\phi_x\right) \phi_x^2 + \phi_x\psi + U''\phi\phi_x = 0. \tag{20.31}$$

Now we integrate (20.31) over Ω_0. Since $U'' \in L^\infty$, the integrals of the last two terms are estimated as

$$\left|\int_{\Omega_0} \left(\phi_x\psi + U''\phi\phi_x\right) dx\right| \leqslant \varepsilon \int_{\Omega_0} \phi_x^2 dx + C\varepsilon^{-1} \int_{\Omega_0} \left(\psi^2 + U''^2\phi^2\right) dx \\ \leqslant \varepsilon \int_{\Omega_0} \phi_x^2 dx + C\varepsilon^{-1}|(\phi, \psi)|_2^2, \tag{20.32}$$

where ε is an arbitrary positive constant. Noting $1 + 3U'(x)/2 > 1/2$ for $x \in \Omega_0$, if $N(T) \ll 1$ so that $|\phi|_\infty \ll |U(\pm 0)|$ $(\leqslant |U(\pm L_0)|)$ and $|\phi_x|_\infty \ll 1$ hold, letting ε in (20.32) suitably small and integrating in time yield

$$\int_{\Omega_0} \phi_x(t, x)^2 dx + \int_0^t \int_{\Omega_0} \phi_x(s, x)^2 dxds \leqslant C \int_{\Omega_0} \phi_0'(x)^2 dx + C \int_0^t |(\phi, \psi)|_2(s)^2 ds. \tag{20.33}$$

In the similar way, the second order derivative is estimated as

$$\int_{\Omega_0} \phi_{xx}(t, x)^2 dx + \int_0^t \int_{\Omega_0} \phi_{xx}(s, x)^2 dxds \\ \leqslant C \int_{\Omega_0} \phi_0''(x)^2 dx + C \int_0^t \left(|(\phi, \psi_x)|_2(s)^2 + \int_{\Omega_0} \phi_x(s, x)^2 dx \right) ds. \tag{20.34}$$

Combination of (20.29), (20.33), and (20.34) yields the desired estimate.

20.3.2 Energy Estimates over the Entire Domain

In this section we assume that $\|\Phi_0\|_3$ is sufficiently small so that the conclusion of Lemma 20.4 holds true.

For an arbitrary $s \geqslant 0$, we solve an ordinary differential equation

$$dX(t)/dt = U(X(t)) + \phi(t, X(t)) \ \text{ for } t > s \ \text{ with } X(s) = -L_0, \tag{20.35}$$

and define $T_1(y; s)$ for an arbitrary $y \in [-L_0, 0)$ so that $X(T_1(y; s)) = y$ holds. The solvability of (20.35) is assured by the boundedness of U' and ϕ_x. Due to Lemma 20.4, the smallness of $\|\Phi_0\|_3$ implies the existence of a limit $\lim_{y\to -0} T_1(y; 0)$, which we denote by T_0 (Fig. 20.1).

For an arbitrary $t > 0$, consider a characteristic curve subject to (20.35) arriving at $x = 0$ from left at time t, and denote its location at an arbitrary time $\tau \in [0, t)$ by $a_-(\tau; t)$. For an arbitrary $t(\geqslant T_0)$, we define $s_0(t)$ $(\geqslant 0)$ such that $a_-(s_0(t); t) = -L_0$ holds, and we set $s_0(t) = 0$ for $t \in [0, T_0)$. Letting T be the existence time of the solution, for an arbitrary $s \leqslant s_0(T)$ define $t_1(s)$ by $\lim_{y\to -0} T_1(y; s)$.

Lemma 20.7 *If $N(T)$ is sufficiently small, there exists a positive constant T_c independent of T such that*

$$t - s_0(t) < T_c \ \textit{ for an arbitrary } t \in (0, T],$$

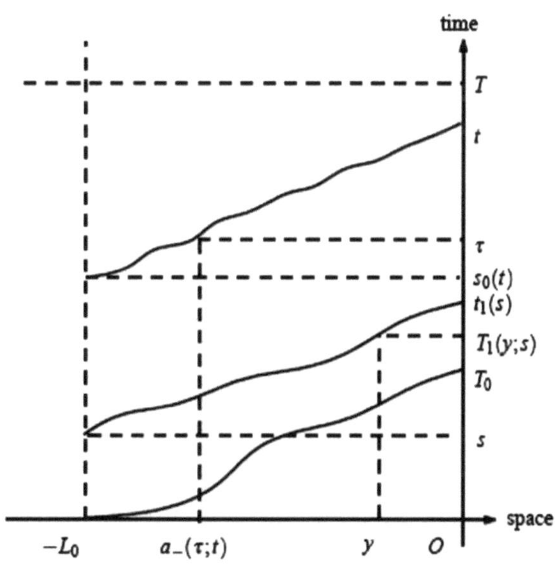

Fig. 20.1 Definitions of T_1, a_-, s_0 and t_1. *Solid lines* represent characteristic curves following (20.35)

and

$$t_1(s) - s < T_c \text{ for an arbitrary } s \in [0, s_0(T)]$$

hold. Moreover, the function $t_1(s)$ *is differentiable almost everywhere in* $s \in [0, s_0(T)/2]$ *and its derivative is bounded by a constant which is independent of* T.

Proof Similar arguments to those in the proof of Lemma 20.4 deduce the first two statements. By the definition of T_1, it holds for $y \in [-L_0, 0)$ that

$$T_1(y; s) = s + \int_{-L_0}^{y} \frac{dz}{U(z) + \phi(T_1(z; s), z)}.$$

Taking the difference quotient of this equality with respect to s, standard arguments lead us to the conclusions.

Lemma 20.8 *If* $N(T)$ *is sufficiently small, it holds that*

$$\|\Phi\|_3(t) \leqslant C\|\Phi_0\|_3 \text{ for an arbitrary } t \in [0, T],$$

where C *is a positive constant independent of* T.

Proof For arbitrary $t \in (0, T]$ and $\tau \in [0, t)$, define a time dependent domain $\Omega(\tau; t)$ by $\Omega(\tau; t) := \{x < a_-(\tau; t) | x \in \mathbb{R}\}$. Applying Lemma 20.2 to (20.31), we have

$$\begin{aligned} \frac{d}{d\tau} \int_{\Omega(\tau;t)} \frac{1}{2}\phi_x^2 dx = &- \int_{\Omega(\tau;t)} \partial_x \left(\frac{1}{2}(\phi + U)\phi_x^2\right) dx - \int_{\Omega(\tau;t)} \left(1 + \frac{3}{2}U' + \frac{1}{2}\phi_x\right) \phi_x^2 dx \\ &- \int_{\Omega(\tau;t)} (\phi_x \psi + U''\phi\phi_x)\, dx + \frac{1}{2}\phi_x^2(\tau, a_-(\tau; t))\partial_\tau a_-(\tau; t). \end{aligned} \tag{20.36}$$

Since $\partial_\tau a_-(\tau; t) = U(a_-(\tau; t)) + \phi(\tau, a_-(\tau; t))$, the first and the last terms in the right hand side cancel. Integrating (20.36) in τ over $[s_0(t), t]$ and using Lemma 20.7, we have

$$\begin{aligned} \int_{-\infty}^{0} \phi_x^2(t, x) dx &\leqslant e^{C(t - s_0(t))} \int_{\Omega(s_0(t),t)} \phi_x(s_0(t), x)^2 dx + C \int_{s_0(t)}^{t} e^{C(t-\tau)} |(\phi, \psi)|_2(\tau)^2 d\tau \\ &\leqslant C \int_{\Omega(s_0(t),t)} \phi_x(s_0(t), x)^2 dx + C \int_{s_0(t)}^{t} |(\phi, \psi)|_2(\tau)^2 d\tau \end{aligned} \tag{20.37}$$

for an arbitrary $t \in [0, T]$, where C is independent of T. In the same way, we have

$$\begin{aligned} &\int_{-\infty}^{0} (\phi_x^2 + \phi_{xx}^2)(t, x) dx \\ &\leqslant C \int_{\Omega(s_0(t),t)} \left(\phi_x^2 + \phi_{xx}^2\right)(s_0(t), x) dx + C \int_{s_0(t)}^{t} |(\phi, \psi, \psi_x)|_2(s)^2 ds. \end{aligned} \tag{20.38}$$

By Lemma 20.6, the last term in the right hand side of (20.38) is bounded above by $C\|\Phi_0\|_3^2$. In the case of $t \leqslant T_0$, then $s_0(t) = 0$ and $\Omega_0 \subset \Omega(s_0(t), t) \subset (-\infty, 0)$. Therefore the first term in the right hand side of (20.38) is not greater than $C\|\Phi_0\|_3^2$. In the case of $t > T_0$, this term appears in the left hand side of (20.30) because $a_-(s_0(t), t) = -L_0$ for $t > T_0$ and hence $\Omega(s_0(t), t) = \Omega_0$. In any case, $\int_{-\infty}^0 (\phi_x^2(t,x) + \phi_{xx}^2(t,x))dx \leqslant C\|\Phi_0\|_3^2$ holds for an arbitrary $t \in [0, T]$, where C is independent of T. Combined with the estimate for $|\Phi, \phi|_2(t)$ in Lemma 20.6, we have the desired estimate.

Lemma 20.9 *If $N(T)$ is sufficiently small, it holds for an arbitrary $t \in [0, T]$ that*

$$\|\Phi\|_3^2(t) + \int_0^t \|(\phi, \psi)\|_2(s)^2 ds \leqslant C\|\Phi_0\|_3^2,$$

where C is a positive constant independent of T.

Proof Choose an arbitrary interval $[0, t] \subset [0, T]$ and integrate (20.38) with t replaced by τ over $[0, t]$ to get

$$\begin{aligned}\int_0^t \int_{-\infty}^0 &(\phi_x^2(\tau, x) + \phi_{xx}^2(\tau, x))dxd\tau \\ &\leqslant C \int_0^t \int_{\Omega(s_0(\tau),\tau)} \left(\phi_x^2 + \phi_{xx}^2\right)(s_0(\tau), x)dxd\tau \\ &\qquad + C\int_0^t \int_{s_0(\tau)}^\tau |(\phi, \psi, \psi_x)|_2(s)^2 dsd\tau. \qquad (20.39)\end{aligned}$$

The first term in the right hand side of (20.39) is estimated as

$$\begin{aligned}&\int_0^t \int_{\Omega(s_0(\tau),\tau)} \left(\phi_x^2 + \phi_{xx}^2\right)(s_0(\tau), x)dxd\tau \\ &= \int_0^{T_0} \int_{\Omega(s_0(\tau),\tau)} \left(\phi_x^2 + \phi_{xx}^2\right)(s_0(\tau), x)dxd\tau + \int_{T_0}^t \int_{\Omega(s_0(\tau),\tau)} \left(\phi_x^2 + \phi_{xx}^2\right)(s_0(\tau), x)dxd\tau \\ &\leqslant T_0 \int_{-\infty}^0 \left((\phi_0')^2 + (\phi_0'')^2\right)(x)dx + \int_0^{s_0(t)} \int_{\Omega_0} \left(\phi_x^2 + \phi_{xx}^2\right)(s, x)dx \frac{dt_1(s)}{ds} ds \\ &\leqslant C\|\Phi_0\|_3^2, \qquad (20.40)\end{aligned}$$

where we used $s_0(\tau) = 0$ for $\tau \leqslant T_0$, $\Omega(s_0(\tau), \tau) = \Omega_0$ for $\tau > T_0$, $s_0^{-1}(s) = t_1(s)$ for $s > 0$, and Lemmata 20.6 and 20.7. The second term in the right hand side of (20.39) is estimated by using Lemmata 20.6 and 20.7 as

$$
\begin{aligned}
&\int_0^t \int_{s_0(\tau)}^{\tau} |(\phi,\psi,\psi_x)|_2(s)^2 ds d\tau \\
&= \int_0^{T_0}\int_0^{\tau} |(\phi,\psi,\psi_x)|_2(s)^2 ds d\tau + \int_{T_0}^t \int_{s_0(\tau)}^{\tau} |(\phi,\psi,\psi_x)|_2(s)^2 ds d\tau \\
&\leqslant T_0 \int_0^{T_0} |(\phi,\psi,\psi_x)|_2(s)^2 ds + \sup_{s\leqslant s_0(t)} (t_1(s)-s) \int_{T_0}^t |(\phi,\psi,\psi_x)|_2(s)^2 ds \\
&\leqslant (T_0+T_C)\int_0^t |(\phi,\psi,\psi_x)|_2(s)^2 ds \leqslant C\|\Phi_0\|_3^2. \qquad (20.41)
\end{aligned}
$$

Substituting (20.40) and (20.41) into (20.39), we obtain

$$
\int_0^t \int_{-\infty}^0 (\phi_x^2(\tau,x)+\phi_{xx}^2(\tau,x)) dx d\tau \leqslant C\|\Phi_0\|_3^2 \text{ for an arbitrary } t\in[0,T], \qquad (20.42)
$$

where C does not depend on T. Combining (20.25), Lemmata 20.6, 20.8, and (20.42), we complete the proof.

Proof of Theorem 20.1

Lemmata 20.1, 20.4 and 20.8 readily conclude the unique existence of the global solution. Passing $t\to\infty$, Lemma 20.9 gives $|\phi|_2(t)^2, |\phi_x|_2(t)^2 \in L^1(0,\infty)$. Also by Lemma 20.9, it is easy to see that

$$
\frac{d}{dt}|\phi|_2(t)^2,\ \frac{d}{dt}|\phi_x|_2(t)^2 \in L^1(0,\infty).
$$

Therefore we have

$$
|\phi|_2(t), |\phi_x|_2(t) \to 0 \text{ as } t\to\infty.
$$

This convergence and the boundedness of $\|\Phi\|_3$ given in Lemma 20.8 give

$$
|\Phi,\phi,\phi_x|_\infty(t) \to 0 \text{ as } t\to\infty
$$

with the help of the Sobolev embedding theorem.

The uniform convergence of q to Q is obtained using (20.25) with the help of the Young inequality. □

Acknowledgements The author wishes to express his sincere gratitude to the financial support from the Alliance for Breakthrough between Mathematics and Sciences (ABMS) supported by the Core Research for Evolutional Science and Technology (CREST) of the Japan Science and Technology Agency (JST), and to the Japan Society for the Promotion of Science (JSPS) for Grant-in-Aid for Young Scientists (B), No. 26800076. This work was done through a research project 'Study on Mathematical Fluid Mechanics' at Research Institute for Science and Engineering, Waseda University, and through the Japanese-German Graduate Externship (Mathematical Fluid Dynamics) supported by JSPS.

References

1. Bressan, A.: Hyperbolic Systems of Conservation Laws: The One Dimensional Cauchy Problem. Oxford University Press (2000)
2. Duan, R., Fellner, K., Zhu, C.: Energy method for multi-dimensional balance laws with non-local dissipation. J. Math. Pures Appl. **93**, 572–598 (2010)
3. Hamer, K.: Nonlinear effects on the propagation of sounds waves in a radiating gas. Q. J. Mech. Appl. Math. **24**, 155–168 (1971)
4. Ito, K.: BV-solutions of a hyperbolic-elliptic system for a radiating gas. Hokkaido Univ., Preprint, no. 368, p. 33 (1997)
5. Kawashima, S., Nishibata, S.: Shock waves for a model system of the radiating gas. SIAM J. Math. Anal. **30**, 95–117 (1998)
6. Kawashima, S., Nishibata, S.: Cauchy problem for a model system of the radiating gas: Weak solutions with a jump and classical solutions. Math. Models Meth. Sci. **9**, 69–91 (1999)
7. Kružkov, S.N.: First order quasilinear equations in several independent variables. Math. USSR Sbor. **10**, 217–243 (1970)
8. Lattanzio, C., Marcati, P.: Global well-posedness and relaxation limits of a model for radiating gas. J. Differ. Equ. **190**, 439–465 (2003)
9. Lattanzio, C., Mascia, C., Nguyen, T., Plaza, R.G., Zumbrun, K.: Stability of scalar radiative shock profiles. SIAM J. Math. Anal. **41**, 2165–2206 (2009)
10. Nguyen, T., Plaza, R.G., Zumbrun, K.: Stability of radiative shock profiles for hyperbolic-elliptic coupled systems. Phys. D **239**, 428–453 (2010)
11. Ohnawa, M.: Convergence rates towards the traveling waves for a model system of radiating gas with discontinuities. Kinet. Relat. Models **5**, 857–872 (2012)
12. Ohnawa, M.: L^∞-stability of continuous shock waves in a radiating gas model. SIAM J. Math. Anal. **46**(3), 2136–2159 (2014)
13. Schochet, S., Tadmor, E.: The regularized Chapman-Enskog expansion for scalar conservation laws. Arch. Ration. Mech. Anal. **119**, 95–107 (1992)
14. Serre, D.: L^1-stability of constants in a model for radiating gases. Commun. Math. Sci. **1**, 197–205 (2003)
15. Serre, D.: L^1-stability of nonlinear waves in scalar conservation laws. In: Dafermos, C.M., Feireisl, E. (eds.) Evolutionary Equations, vol.1 of Handbook of Differential Equations, pp. 473–553. Elsevier North Holland (2004)
16. Zel'dovich, Ya.B., Razier, Yu.P.: Physics of Shock Waves and High Temperature Hydrodynamic Phenomena: Volume II. Academic Press, New York (1967)

Chapter 21
Mathematical Analysis and Numerical Simulations for a Model of Atherosclerosis

Telma Silva, Jorge Tiago and Adélia Sequeira

Abstract Atherosclerosis is a chronic inflammatory disease that occurs mainly in large and medium-sized elastic and muscular arteries. This pathology is essentially caused by the high concentration of low-density-lipoprotein (LDL) in the blood. It can lead to coronary heart disease and stroke, which are the cause of around 17.3 million deaths per year in the world. Mathematical modeling and numerical simulations are important tools for a better understanding of atherosclerosis and subsequent development of more effective treatment and prevention strategies. The atherosclerosis inflammatory process can be described by a model consisting of a system of three reaction-diffusion equations (representing the concentrations of oxidized LDL, macrophages and cytokines inside the arterial wall) with non-linear Neumann boundary conditions. In this work we prove the existence, uniqueness and boundedness of global solutions, using the monotone iterative method. Numerical simulations are performed in a rectangle representing the intima, to illustrate the mathematical results and the atherosclerosis inflammatory process.

Keywords Atherosclerosis · Reaction-diffusion equations · Nonlinear boundary conditions · Upper and lower solutions · Monotone sequences · Existence-comparison theorem

21.1 Introduction

Atherosclerosis is a systemic disease affecting the entire arterial tree, but lesions involving the coronary, cerebral, and lower extremity circulations have the most clin-

T. Silva · J. Tiago · A. Sequeira (✉)
Department of Mathematics and CEMAT, IST, ULisboa, 1049-001 Lisbon, Portugal
e-mail: adelia.sequeira@math.tecnico.ulisboa.pt

T. Silva
e-mail: telma.da.silva@tecnico.ulisboa.pt

J. Tiago
e-mail: jftiago@math.tecnico.ulisboa.pt

Y. Shibata and Y. Suzuki (eds.), *Mathematical Fluid Dynamics, Present and Future*, Springer Proceedings in Mathematics & Statistics 183,
DOI 10.1007/978-4-431-56457-7_21

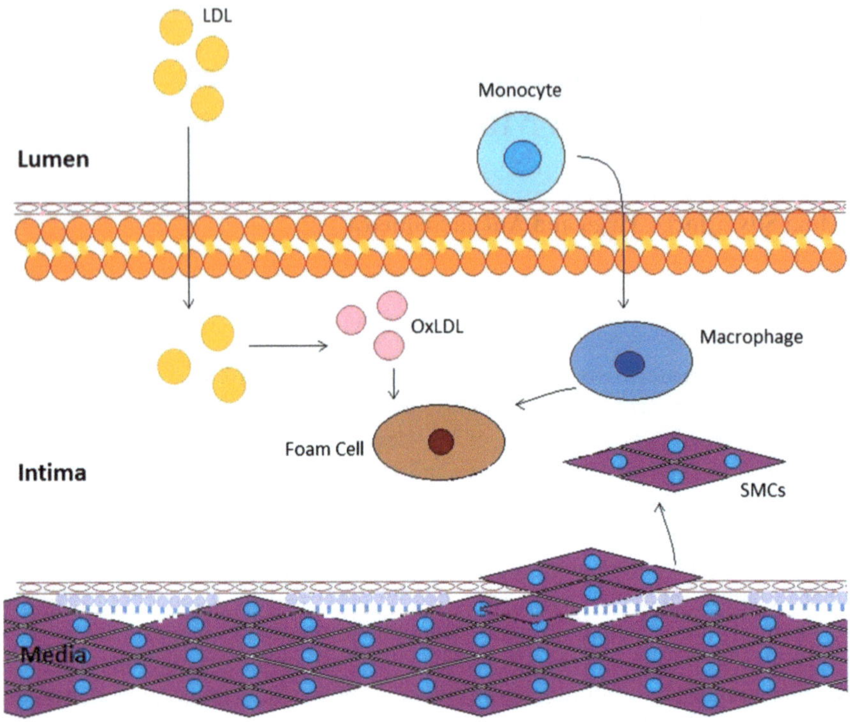

Fig. 21.1 Atherosclerosis schematics. The LDL penetrates the intima where it is oxidized. The oxLDL in the intima leads to monocytes recruitment. The monocytes penetrate the intima and differentiate into macrophages which phagocyte the ox-LDL leading to the formation of the foam cell and consequently to the chronic inflammatory reaction inside the intima. After a while, smooth muscle cells (SMCs) will migrate from media to intima creating a fibrous cap over the lipid deposit

ical significance for medical doctors. The pathogenesis of atherosclerosis involves a complex series of events, similar to a chronic inflammatory process, with the formation of atherosclerotic plaque as consequence [1, 2].

The genesis of atherosclerosis is still not known, but many researchers believe that the atherosclerosis starting point is the endothelial dysfunction, caused by high plasma concentration of cholesterol, in particular, the low-density-lipoprotein, (LDL), hyperglycemia, hypertension, infectious agents and/or smoking [1, 3]. High LDL concentration changes the permeability of the endothelial layer leading to subsequent deposition of lipids in the intima (the inner layer of the blood vessel) [4].

Intra intimal LDL undergoes oxidation (oxLDL) by oxidant mechanisms. Oxidized LDL is considered as a dangerous agent, hence an inflammatory reaction is launched: monocytes adhere to the endothelium, then they penetrate into the intima, where they differentiate into active macrophages (Fig. 21.1).

Active macrophages recognize and absorb oxLDL in the intima by the phagocytosis process. The ingestion of large amounts of oxLDL transforms the fatty macrophages into foam cells (lipid-laden cells) which in turn have to be removed by the immune system. Hence, they set up a chronic inflammatory reaction (auto-amplification phenomenon) by secreting pro-inflammatory cytokines that promote the recruitment of new monocytes and the production of new pro-inflammatory cytokines.

The inflammation process involves the proliferation (growth or production of cells by multiplication of their parts) and the migration of smooth muscle cells (SMCs) to create a fibrous cap over the lipid deposit. This fibrous cap changes the geometry of the vessel and consequently modifies the blood flow.

Although atherosclerosis is asymptomatic in the beginning, with time can lead to cardiovascular diseases, such as coronary artery diseases, cerebrovascular diseases or peripheral arterial diseases, which are responsible for around 17.3 million deaths per year in the world [5]. Therefore, a deep understanding of this pathogenesis and subsequent development of more effective treatment and prevention strategies are essential. Mathematical modeling and numerical simulations are two powerful tools which have a key role in this framework.

Mathematical modeling of the atherosclerosis processes leads to complex systems of flow, transport, chemical reactions, interactions of fluid and elastic structures, movement of cells, coagulation and growth processes and additional complex dynamics of the vessel walls.

Partial differential equations have been used to model this complex process. As an example found in the literature, we can cite [6], where the authors present a model consisting of reaction-diffusion equations, describing how the concentration of macrophages and cytokines in the intima (a vessel layer) leads to an inflammatory disease. A model leading to the atherosclerotic plaque formation and the early atherosclerotic lesions was suggested in [7] and [8], respectively. Systems of convection-reaction-diffusion equations were used to describe the transport and the concentration of oxidized low densities lipoproteins (LDL), macrophages, foam cells and the pro-inflammatory signal emission in the intima. Recently, a more complex and realistic model was presented in [9]. The authors used reaction-diffusion equations to describe the distribution of substance in the intima, such as LDL, high densities lipoproteins (HDL), oxidized LDL, and free radicals, among others, and convection-reaction-diffusion equations for each species of cells, such as macrophages, T cells or foam cells.

Many works have been devoted to the understanding of the atherosclerosis process through numerical simulations [7–12]. Nevertheless, concerning the mathematical analysis there are still many open problems. In 2009, Khatib et. al. presented results of existence of traveling waves for a system with two reaction-diffusion equations in a strip with nonlinear boundary conditions [13] and in 2012, for the same model, they proved the existence and uniqueness of global solutions in Hölder spaces, [6]. Results of existence, uniqueness and boundedness of global solutions, based on the monotone sequences method, as well as the analysis of stability and the long time behavior of the solutions for a system of three reaction-diffusion equations in 1D with homogeneous Neumann boundary conditions was presented in [14].

The main contribution of the present paper consists in extending the results given in [14] for the two-dimensional case. Based on the monotone sequences method, we prove the existence, uniqueness and boundedness of global solutions for a system of three reaction-diffusion equations in 2D with non-linear Neumann boundary conditions. In fact, this result can directly be applied to the 3D case, without any additional restriction. The monotone iterative method consists in using an upper or a lower solution as the initial iteration in a suitable iterative process, to obtain a monotone sequence that converges to a solution of the problem [15].

To illustrate the mathematical results, we perform numerical simulations for the concentration of oxLDL, macrophages and cytokines in a 2D geometry representing the intima.

This paper is organized as follows. In Sect. 21.2, we present the description of a 2D atherosclerosis model imposing some mathematical assumptions. In Sect. 21.3 we describe the core results of this work. We start by introducing some notations and rewriting the mathematical model as a parabolic problem with nonlinear boundary conditions. These will be used to describe the monotone iterative method, that appears thereafter. We then prove the existence, uniqueness and boundedness of global solutions and we make some comments about the simplified model with linear boundary conditions. Finally, in Sect. 21.4, numerical results are presented to illustrate the mathematical model.

21.2 Atherosclerosis Mathematical Modeling

Let the inner layer of the blood vessel (the intima) be defined as a two-dimensional domain, $\Omega = (0, L) \times (0, h)$, where L and h are respectively, the length and the height of the intima. Let the boundary of Ω be denoted by, $\partial\Omega = \Gamma_{in} \cup \Gamma_{end} \cup \Gamma_{med} \cup \Gamma_{out}$, where Γ_{end} represents the interface between the intima and the lumen (the endothelium layer), Γ_{med} is the interface between the intima and the media, Γ_{in} and Γ_{out} are respectively the proximal and distal sections (see Fig. 21.2).

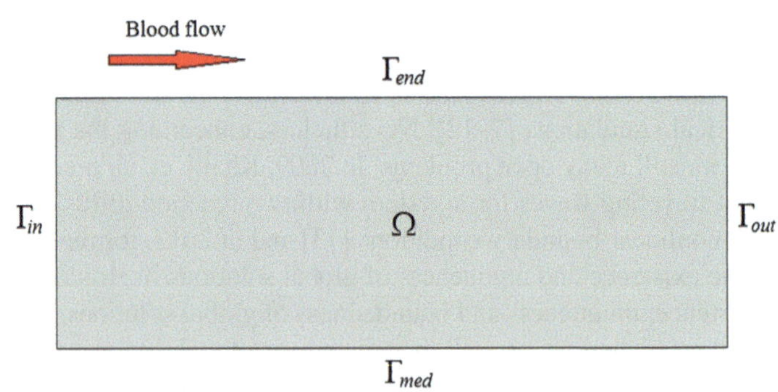

Fig. 21.2 A rectangular representation of the intima layer

The atherosclerosis inflammatory process can be described by the following system of three reaction-diffusion equations,

$$\partial_t Ox - d_{ox} \Delta Ox = -\beta Ox \cdot M \quad (21.1a)$$
$$\partial_t M - d_M \Delta M = -\beta Ox \cdot M \quad (21.1b)$$
$$\partial_t S - d_S \Delta S = \beta Ox \cdot M - \lambda S + \gamma \left(Ox - Ox^{th} \right) \quad (21.1c)$$

in Ω, for all $t \in \mathbb{R}^+$, with the boundary conditions

$$\nabla Ox \cdot \mathbf{n} = \tau(x) C_{LDL} \quad \text{on } \Gamma_{end} \quad \text{and} \quad \nabla Ox \cdot \mathbf{n} = 0 \quad \text{on } \partial\Omega \backslash \Gamma_{end} \quad (21.1d)$$
$$\nabla M \cdot \mathbf{n} = g(S) \quad \text{on } \Gamma_{end} \quad \text{and} \quad \nabla M \cdot \mathbf{n} = 0 \quad \text{on } \partial\Omega \backslash \Gamma_{end} \quad (21.1e)$$
$$\nabla S \cdot \mathbf{n} = 0 \quad \text{on } \partial\Omega \quad (21.1f)$$

for all $t \in \mathbb{R}^+$, where $\mathbf{n}$ is the outward unit normal vector to $\partial\Omega$, and with the initial conditions

$$Ox(x, 0) = Ox_0(x), \; M(x, 0) = M_0(x), \; S(x, 0) = S_0(x) \; \text{ in } \Omega. \quad (21.1g)$$

The functions Ox, M and S are, respectively, the concentrations of oxidized LDL, macrophages and inflammatory signal (a generic chemoattractant which gathers the cytokines), which are continuously differentiable in t and twice continuously differentiable in x.

The second term on the left-hand side of each reaction-diffusion equation of system (21.1), represents the diffusion term, and d_{ox}, d_M and d_S are the diffusion coefficients, which are positive constants. The first term on the right-hand side represents the ingestion of oxLDL by the macrophages. The parameter β is a positive constant of proportionality.

The second term in the Eq. (21.1c) denotes the natural death of the cells and λ is the degradation rate. The starting point of the signal emission is assumed to be a too high oxidized LDL concentration in the intima. This is described by the third term $\gamma \left(Ox - Ox^{th} \right)$, where Ox^{th} corresponds to a given oxLDL quantity and γ is the activation rate. In order to have an inflammatory response Ox should be greater than Ox^{th}.

In the boundary conditions on Γ_{end}, we assume that LDL and monocytes enter the tunica intima through the endothelial layer. The function τ in the boundary condition (21.1d) is the permeability of the vessel wall, which depends on the wall shear stress (WSS), the mechanical force imposed on the endothelium by the flowing blood. As we know, low WSS favors the penetration of both LDL and monocytes [2]. The permeability function τ is defined to be a smooth and nonnegative function of WSS.[1] The parameter C_{LDL} is a given LDL-cholesterol concentration, which is positive. We

[1] The wall shear stress function, WSS, is computed using the solution of the blood flow model (for instance a generalized Navier-Stokes model).

assume that the incoming monocytes immediately differentiate into macrophages and the recruitment of new monocytes depends on a general pro-inflammatory signal which gathers both chemokines and cytokines. The boundary condition (21.1e) considers that this signal acts through the function g, which is defined to impose a limit in the macrophages recruitment. There are many ways to define the function g, see for instance [6, 8, 15]. Here, for simplicity, we have, $g(S) = S/(1+S)$ and

$$g(S) > 0 \text{ for } S > 0, \quad g(0) = 0, \text{ and } g(S) \to 1 \text{ as } S \to \infty.$$

The functions $Ox_0(x)$, $M_0(x)$ and $S_0(x)$ defined in (21.1g) are smooth and nonnegative functions satisfying the boundary conditions (21.1d)–(21.1f) at $t = 0$.

21.3 Existence, Uniqueness and Boundedness of Solutions

System (21.1) is coupled through the boundary conditions, as well as through the differential equations themselves and, in this sense, the analysis becomes more complex than the one performed in [14] for homogeneous Neumann boundary conditions. Nevertheless, the monotone iterative method used to establish the existence-comparison theorem in [14] can be extended to system (21.1), which is a two-dimensional model with nonlinear boundary conditions. But in this case, we should require the quasimonotone property of the boundary function $\boldsymbol{G} = (G_1, G_2, G_3)$ together with the quasimonotone property of the reaction function $\boldsymbol{\Phi} = (\Phi_1, \Phi_2, \Phi_3)$. For the 3D case, we can use the same argument as in 2D, without any additional requirement.

The results in this section are based in the general theory of parabolic problems presented in [15].

21.3.1 Notations

Let Ω be an open set in $\mathbb{R}^d$. We denote by $\partial\Omega$ the boundary of Ω and by $\overline{\Omega}$ its closure. For each $T > 0$, let $\Omega_T = \Omega \times (0, T]$ be a domain in $\mathbb{R}^{d+1}$, $\partial\Omega_T = \partial\Omega \times (0, T]$ and $\overline{\Omega}_T$ the closure of Ω_T.

We denote by $C^{2,1}(\Omega_T)$ the space of all functions that are twice continuously differentiable in x and once continuously differentiable in t, for all (x, t) in Ω_T.

The product function space $\left(C^k(Q)\right)^3$, where Q can be Ω_T or $\overline{\Omega}_T$ is denoted by $\mathscr{C}^k(Q)$. For any vector function $\mathbf{U} = (U_1, U_2, U_3)$ in $\mathscr{C}^k(Q)$ the components U_1, U_2, U_3 are each one in $C^k(Q)$, which is the set of all continuous functions whose partial derivatives up to the kth order are continuous in Q. When $Q = \Omega_T$ we denote by $\mathscr{C}^{2,1}(\Omega_T)$ the product space whose components are in $C^{2,1}(\Omega_T)$.

21.3.2 Parabolic Problem with Nonlinear Boundary Conditions

Let $\Omega_T = \Omega \times (0, T]$ and $\partial\Omega_T = \partial\Omega \times (0, T]$, for an arbitrary finite $T > 0$.

We define the concentrations

$$(C_1, C_2, C_3) = (Ox, M, S), \tag{21.2}$$

the diffusion coefficients

$$(d_1, d_2, d_3) = (d_{ox}, d_M, d_S), \tag{21.3}$$

the reaction functions

$$\Phi_1 = \Phi_2 = -\beta C_1 \cdot C_2 \text{ and } \Phi_3 = \beta Ox \cdot M - \lambda S + \gamma \left(Ox - Ox^{th}\right) \tag{21.4}$$

and the boundary functions

$$G_1 = \tau(x) C_{LDL} \psi(x), \quad G_2 = g(C_3)\psi(x) \text{ and } \quad G_3 = 0 \text{ on } \partial\Omega, \tag{21.5}$$

where $\psi(x)$ is a bump function.[2]

With these notations, system (21.1) with the boundary conditions (21.1d)–(21.1f) and the initial conditions (21.1g) can be rewritten, as follows

$$\begin{aligned} \partial_t C_i - d_i \Delta C_i &= \Phi_i && \text{in } \Omega_T \\ \partial_{\mathbf{n}} C_i &= G_i && \text{on } \partial\Omega_T \\ C_i(x, 0) &= C_{i,0}(x) && \text{in } \Omega \end{aligned} \tag{21.6}$$

for $i = 1, 2, 3$, where the functions $\left(C_{1,0}, C_{2,0}, C_{3,0}\right) = (Ox_0, M_0, S_0)$ and $\partial_{\mathbf{n}} C_i$ denotes the outward normal derivative of C_i on $\partial\Omega_T$.

Let

$$\mathbf{C} = \left(C_i, [\mathbf{C}]_{a_i}, [\mathbf{C}]_{b_i}\right) \tag{21.7}$$

be the split notation of the vector function $\mathbf{C}$, where, $[\mathbf{C}]_{a_i}$ and $[\mathbf{C}]_{b_i}$ denote, respectively, the a_i and b_i-components of the vector $\mathbf{C}$. We rewrite the function Φ_i as

$$\Phi_i(\mathbf{C}) = \Phi_i\left(C_i, [\mathbf{C}]_{a_i}, [\mathbf{C}]_{b_i}\right), \quad \text{for } i = 1, 2, 3, \tag{21.8}$$

[2]Let K be an arbitrary compact set and U an open subset of Γ_{end}, taken as a very small neighbourhood of K, containing K. There exists a bump function $\psi(x)$ which is equal to 1 on K and falls off rapidly to 0 outside of K, while still being smooth.

and the function G_i as

$$G_i\left(\mathbf{C}\right) = G_i\left(\cdot, C_i, [\mathbf{C}]_{\alpha_i}, [\mathbf{C}]_{\rho_i}\right), \quad \text{for } i = 1, 2, 3, \tag{21.9}$$

where a_i, b_i, α_i and ρ_i are nonnegative integers with $a_i + b_i = \alpha_i + \rho_i = 2$.

Considering the split notations (21.7), (21.8) and (21.9), the reaction-diffusion system (21.6) can be written as

$$\begin{aligned} \partial_t C_i - d_i \Delta C_i &= \Phi_i\left(C_i, [\mathbf{C}]_{a_i}, [\mathbf{C}]_{b_i}\right) && \text{in } \Omega_T \\ \partial_{\mathbf{n}} C_i &= G_i\left(\cdot, C_i, [\mathbf{C}]_{\alpha_i}, [\mathbf{C}]_{\rho_i}\right) && \text{on } \partial\Omega_T \\ C_i\left(x, 0\right) &= C_{i,0}\left(x\right) && \text{in } \Omega \end{aligned} \tag{21.10}$$

for $i = 1, 2, 3$.

Since the initial conditions $C_{i,0}$ and the boundary functions G_i, for $i = 1, 2, 3$, are nonnegative, and for all $C_1, C_2, C_3 \geqslant 0$ we have

$$\begin{aligned} \Phi_1\left(0, C_2\right) &= 0, \\ \Phi_2\left(C_1, 0, C_3\right) &= 0, \\ \Phi_3\left(C_1, C_2, 0\right) &= \beta C_1 \cdot C_2 + \gamma\left(C_1 - C_{ox}^{th}\right) \geqslant 0, \end{aligned}$$

the nonnegativity of the solutions of (21.6) is preserved in time [15, 16].

21.3.3 Monotone Iterative Method

The monotone iterative method consists in using an upper or a lower solution as the initial iteration in a suitable iterative process, in order to obtain a monotone sequence which converges to a solution of the problem.

The definition of upper and lower solutions and the construction of monotone sequences depend on the quasimonotone property of the reaction function $\boldsymbol{\Phi}$ and the boundary function $\boldsymbol{G}$.

A function $\mathbf{F} = (f_1, f_2, ..., f_n)$ is said to possess the quasimonotone property if for each i there exist nonnegative integers a_i, b_i with $a_i + b_i = n - 1$ such that $f_i\left(C_i, [\mathbf{C}]_{a_i}, [\mathbf{C}]_{b_i}\right)$ is monotone nondecreasing in $[\mathbf{C}]_{a_i}$ and monotone nonincreasing in $[\mathbf{C}]_{b_i}$, (see [15]).

We need to see if $\boldsymbol{\Phi} = (\Phi_1, \Phi_2, \Phi_3)$ and $\boldsymbol{G} = (G_1, G_2, G_3)$ possess the quasimonotone property. Considering the reaction functions Φ_i, for $i = 1, 2, 3$, and for all $\mathbf{C} \geqslant 0$, we have

$$\begin{aligned} [\mathbf{C}]_{a_1} &= 0 && \text{and } [\mathbf{C}]_{b_1} = C_2 \\ [\mathbf{C}]_{a_2} &= 0 && \text{and } [\mathbf{C}]_{b_2} = C_1 \\ [\mathbf{C}]_{a_3} &= (C_1, C_2) && \text{and } [\mathbf{C}]_{b_3} = 0. \end{aligned}$$

Looking at the boundary functions G_i (with $i = 1, 2, 3$), since G_1 is linear and $G_3 = 0$, we just need to take into account G_2. Therefore, for all $\mathbf{C} \geqslant 0$,

$$[\mathbf{C}]_{\alpha_2} = C_3 \text{ and } [\mathbf{C}]_{\rho_2} = 0.$$

Hence, we conclude that $\boldsymbol{\Phi}$ and $\boldsymbol{G}$ defined in (21.6), are quasimonotone in $\mathbf{C}$, for all $\mathbf{C} \geqslant 0$.

Based on the quasimonotone property of $\boldsymbol{\Phi} = (\Phi_1, \Phi_2, \Phi_3)$ and $\boldsymbol{G} = (G_1, G_2, G_3)$ we have the following definition of upper and lower solutions.

Definition 21.1 Two smooth functions $\widetilde{\mathbf{C}} = \left(\widetilde{C}_1, \widetilde{C}_2, \widetilde{C}_3\right)$, $\widehat{\mathbf{C}} = \left(\widehat{C}_1, \widehat{C}_2, \widehat{C}_3\right)$ in $\mathscr{C}\left(\overline{\Omega}_T\right) \cap \mathscr{C}^{2,1}\left(\Omega_T\right)$ are called a pair of **coupled upper and lower solutions** of (21.6) if $\widetilde{\mathbf{C}} \geqslant \widehat{\mathbf{C}}$ and if they satisfy the inequalities

$$\partial_t \widetilde{C}_i - d_i \Delta \widetilde{C}_i \geqslant \Phi_i\left(\widetilde{C}_i, \left[\widetilde{\mathbf{C}}\right]_{a_i}, \left[\widehat{\mathbf{C}}\right]_{b_i}\right) \quad \text{in } \Omega_T \tag{21.11a}$$

$$\partial_t \widehat{C}_i - d_i \Delta \widehat{C}_i \leqslant \Phi_i\left(\widehat{C}_i, \left[\widehat{\mathbf{C}}\right]_{a_i}, \left[\widetilde{\mathbf{C}}\right]_{b_i}\right) \quad \text{in } \Omega_T \tag{21.11b}$$

$$\partial_{\mathbf{n}} \widetilde{C}_i \geqslant G_i\left(\cdot, \widetilde{C}_i, \left[\widetilde{\mathbf{C}}\right]_{\alpha_i}, \left[\widehat{\mathbf{C}}\right]_{\rho_i}\right) \quad \text{on } \partial\Omega_T \tag{21.11c}$$

$$\partial_{\mathbf{n}} \widehat{C}_i \leqslant G_i\left(\cdot, \widehat{C}_i, \left[\widehat{\mathbf{C}}\right]_{\alpha_i}, \left[\widetilde{\mathbf{C}}\right]_{\rho_i}\right) \quad \text{on } \partial\Omega_T \tag{21.11d}$$

$$\widetilde{C}_i\left(x, 0\right) \geqslant C_{i,0}\left(x\right) \geqslant \widehat{C}_i\left(x, 0\right) \quad \text{in } \Omega \tag{21.11e}$$

for $i = 1, 2, 3$.

The differential inequalities (21.11a) and (21.11b) can be written explicitly as

$$\partial_t \widetilde{C}_1 - d_1 \Delta \widetilde{C}_1 \geqslant -\beta \widetilde{C}_1 \cdot \widehat{C}_2 \tag{21.12a}$$
$$\partial_t \widetilde{C}_2 - d_2 \Delta \widetilde{C}_2 \geqslant -\beta \widehat{C}_1 \cdot \widetilde{C}_2 \tag{21.12b}$$
$$\partial_t \widetilde{C}_3 - d_3 \Delta \widetilde{C}_3 \geqslant \beta \widetilde{C}_1 \cdot \widetilde{C}_2 - \lambda \widetilde{C}_3 + \gamma\left(\widetilde{C}_1 - Ox^{th}\right) \tag{21.12c}$$

$$\partial_t \widehat{C}_1 - d_1 \Delta \widehat{C}_1 \leqslant -\beta \widehat{C}_1 \cdot \widetilde{C}_2 \tag{21.12d}$$
$$\partial_t \widehat{C}_2 - d_2 \Delta \widehat{C}_2 \leqslant -\beta \widetilde{C}_1 \cdot \widehat{C}_2 \tag{21.12e}$$
$$\partial_t \widehat{C}_3 - d_3 \Delta \widehat{C}_3 \leqslant \beta \widehat{C}_1 \cdot \widehat{C}_2 - \lambda \widehat{C}_3 + \gamma\left(\widehat{C}_1 - Ox^{th}\right) \tag{21.12f}$$

in Ω_T and the boundary inequalities (21.11c) and (21.11d) as

$$\partial_{\mathbf{n}} \widetilde{C}_1 \geqslant \tau\left(x\right) C_{LDL} \psi(x) \tag{21.13a}$$
$$\partial_{\mathbf{n}} \widetilde{C}_2 \geqslant g\left(\widetilde{C}_3\right) \psi(x) \tag{21.13b}$$
$$\partial_{\mathbf{n}} \widetilde{C}_3 \geqslant 0 \tag{21.13c}$$

$$\partial_{\mathbf{n}} \widehat{C}_1 \leqslant \tau\,(x)\, C_{LDL} \psi(x) \tag{21.13d}$$

$$\partial_{\mathbf{n}} \widehat{C}_2 \leqslant g\left(\widehat{C}_3\right) \psi\,(x) \tag{21.13e}$$

$$\partial_{\mathbf{n}} \widehat{C}_3 \leqslant 0 \tag{21.13f}$$

on $\partial \Omega_T$.

For a given pair of coupled upper and lower solutions $\widetilde{\mathbf{C}}, \widehat{\mathbf{C}}$, the sector $\langle \widehat{\mathbf{C}}, \widetilde{\mathbf{C}} \rangle$ is defined by the functional interval

$$\langle \widehat{\mathbf{C}}, \widetilde{\mathbf{C}} \rangle \equiv \left\{ \mathbf{C} \in \mathscr{C}\left(\overline{\Omega_T}\right) : \widehat{\mathbf{C}} \leqslant \mathbf{C} \leqslant \widetilde{\mathbf{C}} \right\} \tag{21.14}$$

where the inequalities between vectors should be interpreted in the componentwise sense.

The reaction function $\boldsymbol{\Phi}$ and the boundary function $\boldsymbol{G}$ are continuous in $\Omega_T \times \langle \widehat{\mathbf{C}}, \widetilde{\mathbf{C}} \rangle$ and in $\partial \Omega_T \times \langle \widehat{\mathbf{C}}, \widetilde{\mathbf{C}} \rangle$, respectively, and continuously differentiable in $\left(\mathbb{R}^+\right)^3$ with respect to $\mathbf{C}$. Therefore, they satisfy the Lipschitz condition,

$$\begin{aligned} \left|\Phi_i\,(\mathbf{C}) - \Phi_i\left(\mathbf{C}'\right)\right| &\leqslant R_i \left|\mathbf{C} - \mathbf{C}'\right|, \quad \text{for } \mathbf{C}, \mathbf{C}' \in \langle \widehat{\mathbf{C}}, \widetilde{\mathbf{C}} \rangle \\ \left|G_i\,(\cdot, \mathbf{C}) - G_i\left(\cdot, \mathbf{C}'\right)\right| &\leqslant R_i \left|\mathbf{C} - \mathbf{C}'\right|, \quad \text{for } \mathbf{C}, \mathbf{C}' \in \langle \widehat{\mathbf{C}}, \widetilde{\mathbf{C}} \rangle \end{aligned} \tag{21.15}$$

where $\widehat{C}_i \leqslant C_i' \leqslant C_i \leqslant \widetilde{C}_i$ and the Lipschitz constant, R_i, are given by

$$R_i = \sup \left\{ \left| \frac{\partial \Phi_i}{\partial C_i} \right| : \widehat{C}_i \leqslant C_i \leqslant \widetilde{C}_i, \right\}, \quad \text{for } i = 1, 2, 3. \tag{21.16}$$

Consequently the one-sided Lipschitz conditions

$$\begin{aligned} \Phi_i\left(C_i, [\mathbf{C}]_{a_i}, [\mathbf{C}]_{b_i}\right) - \Phi_i\left(C_i', [\mathbf{C}]_{a_i}, [\mathbf{C}]_{b_i}\right) &\geqslant -R_i\left(C_i - C_i'\right) \\ G_i\left(\cdot, C_i, [\mathbf{C}]_{\alpha_i}, [\mathbf{C}]_{\rho_i}\right) - G_i\left(\cdot, C_i', [\mathbf{C}]_{\alpha_i}, [\mathbf{C}]_{\rho_i}\right) &\geqslant -R_i\left(C_i - C_i'\right) \end{aligned} \tag{21.17}$$

are also satisfied.

Below, we describe the construction of monotone sequences by an iterative process, choosing as initial iterations

$$\overline{\mathbf{C}}^{(0)} = \left(\widetilde{C}_1, \widetilde{C}_2, \widetilde{C}_3\right) \quad \text{and} \quad \underline{\mathbf{C}}^{(0)} = \left(\widehat{C}_1, \widehat{C}_2, \widehat{C}_3\right). \tag{21.18}$$

The upper and lower sequences

$$\left\{\overline{\mathbf{C}}^{(k)}\right\} = \left\{\overline{C}_1^{(k)}, \overline{C}_2^{(k)}, \overline{C}_3^{(k)}\right\} \quad \text{and} \quad \left\{\underline{\mathbf{C}}^{(k)}\right\} = \left\{\underline{C}_1^{(k)}, \underline{C}_2^{(k)}, \underline{C}_3^{(k)}\right\} \tag{21.19}$$

are obtained through the following iterative process

$$\partial_t \overline{C}_i^{(k)} - d_i \Delta \overline{C}_i^{(k)} + R_i \overline{C}_i^{(k)} = R_i \overline{C}_i^{(k-1)} + \Phi_i \left(\overline{C}_i^{(k-1)}, \left[\overline{\mathbf{C}}^{(k-1)} \right]_{a_i}, \left[\underline{\mathbf{C}}^{(k-1)} \right]_{b_i} \right) \text{ in } \Omega_T \tag{21.20a}$$

$$\partial_t \underline{C}_i^{(k)} - d_i \Delta \underline{C}_i^{(k)} + R_i \underline{C}_i^{(k)} = R_i \underline{C}_i^{(k-1)} + \Phi_i \left(\underline{C}_i^{(k-1)}, \left[\underline{\mathbf{C}}^{(k-1)} \right]_{a_i}, \left[\overline{\mathbf{C}}^{(k-1)} \right]_{b_i} \right) \text{ in } \Omega_T \tag{21.20b}$$

and

$$\partial_{\mathbf{n}} \overline{C}_i^{(k)} + R_i \overline{C}_i^{(k)} = R_i \overline{C}_i^{(k-1)} + G_i \left(\cdot, \overline{C}_i^{(k-1)}, \left[\overline{\mathbf{C}}^{(k-1)} \right]_{\alpha_i}, \left[\underline{\mathbf{C}}^{(k-1)} \right]_{\rho_i} \right) \text{ on } \partial\Omega_T \tag{21.21a}$$

$$\partial_{\mathbf{n}} \underline{C}_i^{(k)} + R_i \underline{C}_i^{(k)} = R_i \underline{C}_i^{(k-1)} + G_i \left(\cdot, \underline{C}_i^{(k-1)}, \left[\underline{\mathbf{C}}^{(k-1)} \right]_{\alpha_i}, \left[\overline{\mathbf{C}}^{(k-1)} \right]_{\rho_i} \right) \text{ on } \partial\Omega_T \tag{21.21b}$$

with initial conditions

$$\overline{C}_i^{(k)} (x, 0) = C_{i,0} (x) = \underline{C}_i^{(k)} (x, 0) \qquad \text{in } \Omega \tag{21.22}$$

for $k = 1, 2, \ldots$ and $i = 1, 2, 3$. Here, R_i (with $i = 1, 2, 3$) are the Lipschitz constants defined in (21.16) and in this case they are given by

$$\begin{aligned} R_1 &= \sup_{\widehat{C}_2 \leqslant C_2 \leqslant \widetilde{C}_2} \{\beta C_2\}, \\ R_2 &= \sup_{\widehat{C}_1 \leqslant C_1 \leqslant \widetilde{C}_1} \{\beta C_1\}, \\ R_3 &= \lambda. \end{aligned}$$

Since for each k, (21.20)–(21.22) are uncoupled scalar linear problems (which have a unique solution in Ω_T) and by the properties of Φ_i and G_i, the sequences $\left\{ \overline{\mathbf{C}}^{(k)} \right\}$ and $\left\{ \underline{\mathbf{C}}^{(k)} \right\}$ are well defined (for the proof see [15] pp. 58, 493). The following lemma gives the monotone property of these sequences.

Lemma 21.1 *The upper and the lower sequences* $\left\{ \overline{\mathbf{C}}^{(k)} \right\}$, $\left\{ \underline{\mathbf{C}}^{(k)} \right\}$ *given by the iterative process (21.20)–(21.22), with* $\overline{\mathbf{C}}^{(0)} = \widetilde{\mathbf{C}}$ *and* $\underline{\mathbf{C}}^{(0)} = \widehat{\mathbf{C}}$, *possess the monotone property*

$$\widehat{\mathbf{C}} \leqslant \underline{\mathbf{C}}^{(k)} \leqslant \underline{\mathbf{C}}^{(k+1)} \leqslant \overline{\mathbf{C}}^{(k+1)} \leqslant \overline{\mathbf{C}}^{(k)} \leqslant \widetilde{\mathbf{C}} \quad \text{in } \overline{\Omega}_T \tag{21.23}$$

for every k.

Proof Let $\mathbf{U}^{(0)} = \overline{\mathbf{C}}^{(0)} - \overline{\mathbf{C}}^{(1)} = \widetilde{\mathbf{C}} - \overline{\mathbf{C}}^{(1)}$. By the property of upper solutions (21.11a) and (21.11c), and using the sequences (21.20), we have

$$\partial_t U_i^{(0)} - d_i \Delta U_i^{(0)} + R_i U_i^{(0)} = \partial_t \widetilde{C}_i - d_i \Delta \widetilde{C}_i - \Phi_i \left(\widetilde{C}_i, [\widetilde{\mathbf{C}}]_{a_i}, [\widehat{\mathbf{C}}]_{b_i} \right) \geqslant 0 \text{ in } \Omega_T$$

$$\partial_{\mathbf{n}} \widetilde{U}_i^{(0)} + R_i U_i^{(0)} = \partial_n \widetilde{C}_i - G_i \left(\cdot, \widetilde{C}_i, [\widetilde{\mathbf{C}}]_{\alpha_i}, [\widehat{\mathbf{C}}]_{\rho_i} \right) \geqslant 0 \text{ on } \partial\Omega_T$$

for $i = 1, 2, 3$.

From the initial conditions (21.11e), we obtain

$$U_i^{(0)}(x, 0) = \widetilde{C}_i(x, 0) - C_{i,0}(x) \geqslant 0.$$

By the maximum principle [15, 16], $U_i^{(0)} \geqslant 0$ or equivalently

$$\overline{\mathbf{C}}^{(1)} \leqslant \overline{\mathbf{C}}^{(0)}.$$

Similarly, using the property of lower solutions (21.11b) and (21.11d), and by the sequences (21.21), we have

$$\underline{\mathbf{C}}^{(1)} \geqslant \underline{\mathbf{C}}^{(0)}.$$

Now, let $\mathbf{U}^{(1)} = \overline{\mathbf{C}}^{(1)} - \underline{\mathbf{C}}^{(1)}$, then, by the sequences (21.20)–(21.21), the one-sided Lipschitz condition (21.17) and the monotone property of Φ_i and G_i, we have

$$\begin{aligned} &\partial_t U_i^{(1)} - d_i \Delta U_i^{(1)} + R_i U_i^{(1)} = R_i \left(\overline{C}_i^{(0)} - \underline{C}_i^{(0)} \right) + \\ &+ \left[\Phi_i \left(\overline{C}_i^{(0)}, \left[\overline{\mathbf{C}}^{(0)} \right]_{a_i}, [\underline{\mathbf{C}}^{(0)}]_{b_i} \right) - \Phi_i \left(\underline{C}_i^{(0)}, [\underline{\mathbf{C}}^{(0)}]_{a_i}, \left[\overline{\mathbf{C}}^{(0)} \right]_{b_i} \right) \right] \geqslant 0 \text{ in } \Omega_T \end{aligned}$$

and

$$\begin{aligned} &\partial_{\mathbf{n}} U_i^{(1)} + R_i U_i^{(1)} = R_i \left(\overline{C}_i^{(0)} - \underline{C}_i^{(0)} \right) + \\ &+ \left[G_i \left(\overline{C}_i^{(0)}, \left[\overline{\mathbf{C}}^{(0)} \right]_{\alpha_i}, [\underline{\mathbf{C}}^{(0)}]_{\rho_i} \right) - G_i \left(\underline{C}_i^{(0)}, [\underline{\mathbf{C}}^{(0)}]_{\alpha_i}, \left[\overline{\mathbf{C}}^{(0)} \right]_{\rho_i} \right) \right] \geqslant 0 \text{ on } \partial\Omega_T \end{aligned}$$

for $i = 1, 2, 3$.

It follows from the initial conditions $U_i^{(1)}(x, 0) = 0$ that $U_i^{(1)} > 0$ for each $i = 1, 2, 3$. This shows that

$$\underline{\mathbf{C}}^{(0)} \leqslant \underline{\mathbf{C}}^{(1)} \leqslant \overline{\mathbf{C}}^{(1)} \leqslant \overline{\mathbf{C}}^{(0)} \text{ in } \overline{\Omega}_T. \tag{21.24}$$

For a fixed $k \in \mathbb{N}$, the function $U_i^{(k+1)} = \overline{C}_i^{(k)} - \overline{C}_i^{(k+1)}$ satisfies the relations

$$\partial_t U_i^{(k)} - d_i \Delta U_i^{(k)} + R_i U_i^{(k)} = R_i \left(\overline{C}_i^{(k-1)} - \overline{C}_i^{(k)}\right) +$$
$$+ \left[\Phi_i \left(\overline{C}_i^{(k-1)}, \left[\overline{\mathbf{C}}^{(k-1)}\right]_{a_i}, \left[\underline{\mathbf{C}}^{(k-1)}\right]_{b_i}\right) - \Phi_i \left(\overline{C}_i^{(k)}, \left[\overline{\mathbf{C}}^{(k)}\right]_{a_i}, \left[\underline{\mathbf{C}}^{(k)}\right]_{b_i}\right)\right] \geqslant 0 \text{ in } \Omega_T$$

$$\partial_{\mathbf{n}} U_i^{(k)} + R_i U_i^{(k)} = R_i \left(\overline{C}_i^{(k-1)} - \overline{C}_i^{(k)}\right) +$$
$$+ \left[G_i \left(\overline{C}_i^{(k-1)}, \left[\overline{\mathbf{C}}^{(k-1)}\right]_{\alpha_i}, \left[\underline{\mathbf{C}}^{(k-1)}\right]_{\rho_i}\right) - G_i \left(\overline{C}_i^{(k)}, \left[\overline{\mathbf{C}}^{(k)}\right]_{\alpha_i}, \left[\underline{\mathbf{C}}^{(k)}\right]_{\rho_i}\right)\right] \geqslant 0 \text{ on } \partial\Omega_T$$

for $i = 1, 2, 3$. These relations and $U_i^{(k)}(x, 0) = 0$ ensure that $\overline{\mathbf{C}}^{(k+1)} \leqslant \overline{\mathbf{C}}^{(k)}$.

A similar argument yields $\underline{\mathbf{C}}^{(k+1)} \geqslant \underline{\mathbf{C}}^{(k)}$ and $\overline{\mathbf{C}}^{(k+1)} \geqslant \underline{\mathbf{C}}^{(k)}$.

By induction, we prove that the sequence $\left\{\overline{\mathbf{C}}^{(k)}\right\}$ is monotone nonincreasing and $\{\underline{\mathbf{C}}^{(k)}\}$ is monotone nondecreasing. And so, the monotone property

$$\widehat{\mathbf{C}} \leqslant \underline{\mathbf{C}}^{(k)} \leqslant \underline{\mathbf{C}}^{(k+1)} \leqslant \overline{\mathbf{C}}^{(k+1)} \leqslant \overline{\mathbf{C}}^{(k)} \leqslant \widetilde{\mathbf{C}} \text{ in } \overline{\Omega}_T$$

follows, for every $k = 0, 1, 2, \ldots$.

21.3.4 Existence of Upper and Lower Solutions

The main condition for the existence of a unique solution to problem (21.6) is the existence of a pair of coupled upper and lower solutions when the reaction function Φ_i and the boundary function G_i are quasimonotone.

Since

$$\begin{aligned} \Phi_1 \left(0, [\mathbf{0}]_{a_i}, [\mathbf{C}]_{b_i}\right) &= 0 \\ \Phi_2 \left(0, [\mathbf{0}]_{a_i}, [\mathbf{C}]_{b_i}\right) &= 0 \\ \Phi_3 \left(0, [\mathbf{0}]_{a_i}, [\mathbf{C}]_{b_i}\right) &= 0 \end{aligned} \quad (21.25)$$

the function Φ_i satisfies the additional condition

$$\Phi_i \left(0, [\mathbf{0}]_{a_i}, [\mathbf{C}]_{b_i}\right) \geqslant 0 \text{ when } [\mathbf{C}]_{b_i} \geqslant 0, \quad (21.26)$$

for $i = 1, 2, 3$, where $[\mathbf{C}] \geqslant 0$ stands for $\mathbf{C} \geqslant 0$.

Remark 21.1 The last equality in (21.25) becomes true due to the relation $C_1 \geqslant Ox^{th}$, considered as an assumption in order to have an inflammatory process. So, if $C_1 = 0$ then Ox^{th} also vanishes.

Since G_1 is linear and nonnegative, $G_3 = 0$ and $G_2\left(0, [\mathbf{0}]_{a_i}, [\mathbf{C}]_{b_i}\right) = 0$, each boundary function G_i satisfies the condition

$$G_i\left(\cdot, 0, [\mathbf{0}]_{\alpha_i}, [\mathbf{C}]_{\rho_i}\right) \geqslant 0 \ \text{ when } \ [\mathbf{C}]_{\rho_i} \geqslant 0 \tag{21.27}$$

for $i = 1, 2, 3$.

From (21.26) and (21.27), we conclude that the trivial function $\mathbf{C} = 0$ is a lower solution.

Any positive function $\widetilde{\mathbf{C}} = \left(\widetilde{C}_1, \widetilde{C}_2, \widetilde{C}_3\right)$ satisfying

$$\begin{aligned} \partial_t \widetilde{C}_i - d_i \Delta \widetilde{C}_i &\geqslant \Phi_i\left(\widetilde{C}_i, \left[\widetilde{\mathbf{C}}\right]_{a_i}, [\mathbf{0}]_{b_i}\right) && \text{in } \Omega_T \\ \partial_{\mathbf{n}} \widetilde{C}_i &\geqslant G_i\left(\cdot, \widetilde{C}_i, \left[\widetilde{\mathbf{C}}\right]_{\alpha_i}, [\mathbf{0}]_{\rho_i}\right) && \text{on } \partial\Omega_T \\ \widetilde{C}_i(x, 0) &\geqslant C_{i,0}(x) && \text{in } \Omega \end{aligned} \tag{21.28}$$

for $i = 1, 2, 3$, also satisfies the upper and lower inequalities (21.11). Therefore, the requirement of an upper solution is reduced to

$$\begin{aligned} \partial_t \widetilde{C}_1 - d_1 \Delta \widetilde{C}_1 &\geqslant 0 && \text{in } \Omega_T \\ \partial_t \widetilde{C}_2 - d_2 \Delta \widetilde{C}_2 &\geqslant 0 && \text{in } \Omega_T \\ \partial_t \widetilde{C}_3 - d_3 \Delta \widetilde{C}_3 &\geqslant \beta \widetilde{C}_1 \cdot \widetilde{C}_2 - \lambda \widetilde{C}_3 + \gamma\left(\widetilde{C}_1 - Ox^{th}\right) && \text{in } \Omega_T \\ \partial_{\mathbf{n}} \widetilde{C}_1 &\geqslant \tau(x)\, C_{LDL} \psi(x) && \text{on } \partial\Omega_T \\ \partial_{\mathbf{n}} \widetilde{C}_2 &\geqslant g\left(\widetilde{C}_3\right) \psi(x) && \text{on } \partial\Omega_T \\ \partial_{\mathbf{n}} \widetilde{C}_3 &\geqslant 0 && \text{on } \partial\Omega_T \\ \widetilde{C}_i(x, 0) &\geqslant C_{i,0} \end{aligned} \tag{21.29}$$

in Ω for $i = 1, 2, 3$.

21.3.5 Existence-comparison Theorem

By the monotone property (21.23), the pointwise and componentwise limits

$$\lim_{k\to\infty} \overline{\mathbf{C}}^{(k)}(x, t) = \overline{\mathbf{C}}(x, t), \quad \lim_{k\to\infty} \underline{\mathbf{C}}^{(k)}(x, t) = \underline{\mathbf{C}}(x, t)$$

exist and satisfy the relation

$$\widehat{\mathbf{C}} \leqslant \underline{\mathbf{C}} \leqslant \overline{\mathbf{C}} \leqslant \widetilde{\mathbf{C}} \text{ in } \overline{\Omega}_T. \tag{21.30}$$

Due to the quasimonotone property of Φ_i and G_i, the one-sided Lipschitz condition (21.17), the Lipschitz condition (21.15) and the monotone property relation

(21.23) we can conclude, by applying the existence-comparison theorem for parabolic problems proved in [15] (pp. 494), that the limits of $\left\{\overline{\mathbf{C}}^{(k)}\right\}$ and $\{\underline{\mathbf{C}}^{(k)}\}$ coincide and yield to a unique solution of problem (21.6).

These conclusions can be summarized in the following result.

Theorem 21.1 *Let* $\widetilde{\mathbf{C}} = (\widetilde{C}_1, \widetilde{C}_2, \widetilde{C}_3)$ *and* $\widehat{\mathbf{C}} = (\widehat{C}_1, \widehat{C}_2, \widehat{C}_3)$ *be a pair of nonnegative coupled upper and lower solutions of* (21.6) *and let* $\boldsymbol{\Phi} = (\Phi_1, \Phi_2, \Phi_3)$ *and* $\mathbf{G} = (G_1, G_2, G_3)$ *be quasimonotone functions satisfying the global Lipschitz condition* (21.15)*. Then, the upper and lower sequences* $\left\{\overline{\mathbf{C}}^{(k)}\right\}$, $\{\underline{\mathbf{C}}^{(k)}\}$ *given by* (21.20)–(21.22)*, converge monotonically to a unique solution* $\mathbf{C} = (C_1, C_2, C_3)$ *with*

$$\left(\widehat{C}_1, \widehat{C}_2, \widehat{C}_3\right) \leqslant (C_1, C_2, C_3) \leqslant \left(\widetilde{C}_1, \widetilde{C}_2, \widetilde{C}_3\right) \text{ in } \Omega_T. \tag{21.31}$$

The existence-comparison theorem can directly be applied to the 3D case, imposing similar conditions.

21.3.6 The Case of Linear Boundary Conditions

In the literature it has also been suggested to describe atherosclerosis using mathematical models with linear boundary conditions [7, 11]. In the case of homogeneous Neumann boundary conditions, such models can be seen as simplified versions of the parabolic problem with nonlinear boundary conditions (21.6), presented in Sect. 21.2.

The mathematical model of the atherosclerosis inflammatory process, under homogeneous Neumann boundary conditions, can be described as follows

$$\begin{aligned} \partial_t C_i - d_i \Delta C_i &= F_i && \text{in } \Omega_T \\ \partial_{\mathbf{n}} C_i &= 0 && \text{on } \partial\Omega_T \\ C_i(x,0) &= C_{i,0}(x) && \text{in } \Omega \end{aligned} \tag{21.32}$$

for $i = 1, 2, 3$, where the reaction functions (F_1, F_2, F_3) are defined as

$$F_1(x, C_1, C_2) = -\beta C_1 \cdot C_2 + \tau(x) C_{LDL}, \tag{21.33a}$$

$$F_2(C_1, C_2, C_3) = -\beta C_1 \cdot C_2 + g(C_3), \tag{21.33b}$$

$$F_3(C_1, C_2, C_3) = \beta C_1 \cdot C_2 - \lambda C_3 + \gamma\left(C_1 - Ox^{th}\right),. \tag{21.33c}$$

If the term $\tau(x)C_{LDL}$ in the function F_1 is defined only on the boundary $\Gamma_T^{end} = \Gamma_{end} \times (0, T]$, we have

$$\begin{aligned} \partial_t C_1 - d_1 \Delta C_1 &= -\beta C_1 \cdot C_2 && \text{in } \Omega_T \\ \partial_{\mathbf{n}} C_1 &= \tau(x) C_{LDL} && \text{on } \Gamma_T^{end} \\ \partial_{\mathbf{n}} C_1 &= 0 && \text{on } \partial\Omega_T \setminus \Gamma_T^{end} \\ C_1(x,0) &= C_{1,0}(x) && \text{in } \Omega \end{aligned} \tag{21.34}$$

Problem (21.34) with C_2 and C_3 defined as in (21.32) leads to a 2D model, with linear Neumann boundary conditions.

The existence, uniqueness and boundedness of global solutions of these problems can be proved by the monotone iterative method following the same reasoning as in the one dimensional case presented in [14], since the boundary conditions are independent of C_i and the function $\mathbf{F} = (F_1, F_2, F_3)$ is quasimonotone.

21.4 Numerical Simulations

Numerical simulations are an important tool to better understand the atherosclerosis mechanism and to improve the mathematical models. In this section we present numerical results concerning the concentrations of oxidized low density lipoproteins, macrophages and signal path between cells inside the intima. To represent the intima, we consider a rectangle $(0, L) \times (0, h)$, with length $L = 0.5\,\text{cm}$ and height $h = 0.167\,\text{cm}$ and the following physical and biological parameters, taken from [1]:

$$
\begin{aligned}
&d_{ox} = d_S = 10^2 \times d_M = 10^{-3}\ cm/s\\
&\lambda = 10\ s^{-1}\\
&\beta = 1\ cm/(g \cdot s)\\
&\gamma = 1\ s^{-1}
\end{aligned}
$$

The model (21.1a)–(21.1g) presented in Sect. 21.2 assumes that atherosclerosis starts with the penetration of LDL into the intima, through the endothelial cells, where they are oxidized (oxLDL). Choosing a region of LDL penetration, Γ^p_{end}, we can define the permeability function τ as a smooth step function that is equal to one in Γ^p_{end} and zero otherwise. We consider that $C_{LDL} = 0.1 g/cm^3$ is a given quantity of LDL that enters in the region where $\tau(x) = 1$. Figure 21.3 shows the evolution in time of oxLDL concentration in the intima.

As we expected, in the region Γ^p_{end}, the concentration of Ox is higher and will spread out in time across the domain.

The model also considers that if the concentration of oxLDL (Ox) exceeds a threshold Ox^{th}, an inflammatory reaction will set up promoting the recruitment of monocytes, which will transform in active macrophages (M) to phagocyte the dangerous product, oxLDL. In Fig. 21.4 we can observe how the macrophages concentration depends on the concentration of oxLDL. Moreover, we can also observe the effect of diffusion over time.

The recruitment of monocytes depends on a general pro-inflammatory signal (S) which acts through the function g. The time evolution of signal concentration in the intima is presented in Fig. 21.5.

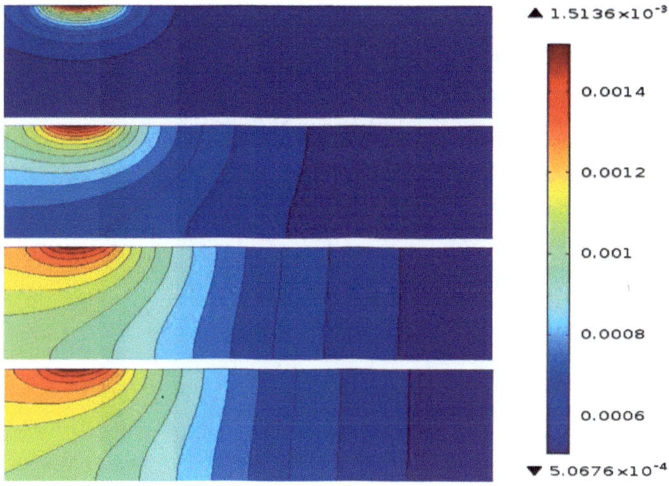

Fig. 21.3 The concentration of oxLDL in the intima for time values T = 1, 10, 50 and 100 s (respectively, fist, second, third and last image starting from the *top*). Defining a region of LDL penetration, Γ^{p}_{end}, the concentration of Ox will spread out in time through all the domain

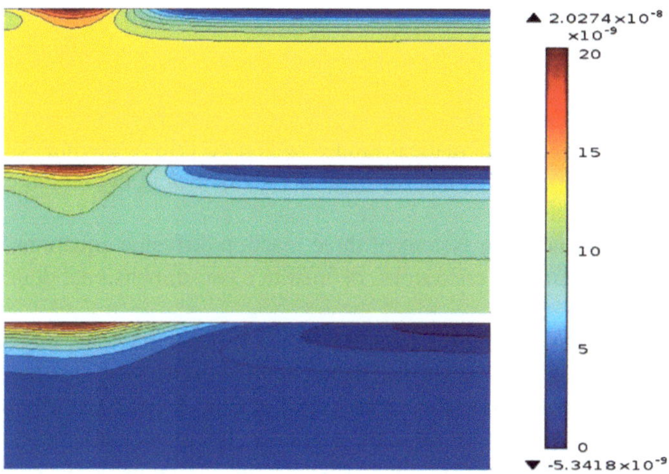

Fig. 21.4 The concentration of macrophages in the intima for time values T = 30, 50 and 100 s (respectively, fist, second, and last image starting from the *top*). A high concentration of Ox in the Γ^{p}_{end} leads to an high value of M in the same region

Comparing the results of Figs. 21.3 and 21.5 we can notice a strong relation between the oxLDL and the signal near the endothelium, as well as the diffusion effect over time in all the domain.

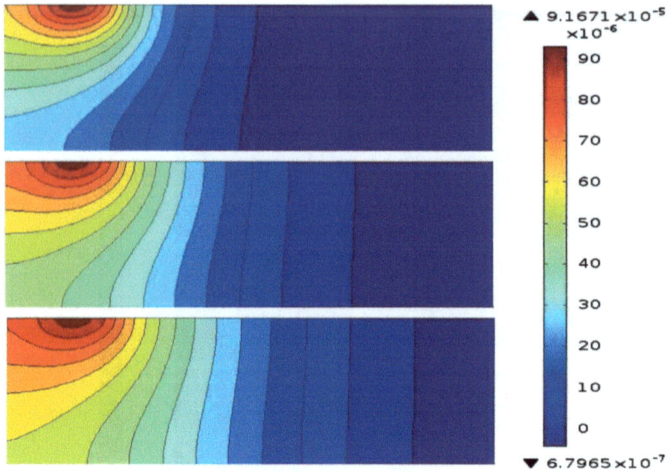

Fig. 21.5 The concentration of cytokines in the intima for time value T = 30, 50 and 100 s (respectively, fist, second and last image starting from the *top*). An elevated value of Ox in the Γ_{end}^{p} contributes to high concentration of S in the same region

21.5 Conclusions

In this work, we presented the existence, uniqueness and boundedness of solutions for an atherosclerosis mathematical model, which describes how the variations in the concentration of oxLDL, macrophages and cytokines in the intima can lead to an inflammatory disease.

The model consists of a system of three reaction-diffusion equations with nonlinear Neumann boundary conditions, defined in a two-dimensional domain, representing the intima.

Since the reaction and the boundary functions are quasimonotone, we could define a pair of upper and lower solutions and use an iterative process to construct monotone sequences. The smoothness of the reaction and boundary functions and monotonicity arguments have been used to prove the existence, uniqueness and boundedness of global solutions in 2D. However, the existence-comparison theorem can directly be applied to the 3D case, without any additional condition.

Numerical simulations have been performed to better understand the atherosclerosis mechanism described by the model.

Acknowledgements FCT (Fundação para a Ciência e a Tecnologia, Portugal) through the grant SFRH/BPD/66638/2009, the project EXCL/MAT-NAN/0114/2012 and the Research Center CEMAT-IST are gratefully acknowledged.

References

1. Mitchell, M.E., Sidawy, A.N.: The pathophysiology of atherosclerosis. Semin. Vasc. Surg. **3**(11), 134–141 (1998)
2. Ross, R.: Atherosclerosis—an inflammatory disease. Mass. Med. Soc. **340**(2), 115–126 (1999)
3. Mitrovska, S., Jovanova, S., Matthiesen, I., Libermans, C.: Atherosclerosis: Understanding Pathogenesis and Challenge for Treatment. Nova Science Publishers Inc., New York (2009)
4. Guretzki, H.J., Gerbitza, K.D., Olgemöller, B., Schleicher, E.: Atherogenic levels of low density lipoprotein alter the permeability and composition of the endothelial barrier. Elsevier **107**(1), 15–24 (1994)
5. Mendis, S., Puska, P., Norrving, B.: Global Atlas on Cardiovascular Disease Prevention and Control. World Health Organization, Geneva (2011)
6. El Khatib, N., Génieys, S., Kazmierczak, B., Volpert, V.: Reaction-diffusion model of atherosclerosis development. J. Math. Biol. **65**, 349–374 (2012)
7. Calvez, V., Ebde, A., Meunier, N., Raoult, A.: Mathematical and numerical modeling of the atherosclerotic plaque formation. ESAIM Proc. **28**, 1–12 (2009)
8. Calvez, V., Houot, J., Meunier, N., Raoult, A., Rusnakova, G.: Mathematical and numerical modeling of early atherosclerotic lesions. ESAIM Proc. **30**, 1–14 (2010)
9. Hao, W., Friedman, A.: The LDL-HDL profile determines the risk of atherosclerosis- a mathematical model. PLoS ONE **9**(3), 1–15 (2014)
10. Liu, B., Tang, D.: Computer simulations of atherosclerosis plaque growth in coronary arteries. Mol. Cell. Biomech. **7**(4), 193–202 (2010)
11. Filipovic, N., Nikolic, D., Saveljic, I., Milosevic, Z., Exarchos, T., Pelosi, G., Parodi, O.: Computer simulation of three-dimensional plaque formation and progression in the coronary artery. Comput. Fluids **88**, 826–833 (2013)
12. Silva, T., Sequeira, A., Santos, R., Tiago, J.: Mathematical modeling of atherosclerotic plaque formation coupled with a non-Newtonian model of blood flow. In: Hindawi Publishing Corporation Conference Papers in Mathematics (2013)
13. El Khatib, N., Génieys, S., Kazmierczak, B., Volpert, V.: Mathematical modeling of atherosclerosis as an inflammatory disease. Philos. Trans. R. Soc. A **367**, 4877–4886 (2009)
14. Silva, T., Sequeira, A., Santos, R., Tiago, J.: Existence, uniqueness, stability and asymptotic behavior of solutions for a model of atherosclerosis, DCDS-S, AIMS, **9**(1), 343–362 (2016)
15. Pao, C.V.: Nonlinear Parabolic and Elliptic Equations. Plenum Press, New York (1992)
16. Fife, P.C.: Mathematical Aspects of Reacting and Diffusing Systems. Springer-Verlag, Berlin Heidelberg (1979)

Chapter 22
Regularity for the Solution of a Stochastic Partial Differential Equation with the Fractional Laplacian

Satoshi Yokoyama

Abstract We study the regularity properties for the mild solution of a stochastic partial differential equation in $\mathbf{R}^d$ with the fractional Laplacian $-(-\Delta)^{\frac{\gamma}{2}}$ under the condition where its solution exists uniquely as a function valued process. To show its regularity, we estimate the fundamental solution and use the Kolmogorov-Centsov theorem. Due to the unboundedness of the domain, we need to check the behavior of the fundamental solution for sufficiently large $|x|$, $x \in \mathbf{R}^d$.

Keywords Stochastic partial differential equations · Mild solutions · Regularity · Fractional Laplacian

22.1 Introduction

Stochastic partial differential equations (SPDEs) appear in many fields such as physics, biology and economy. For example, in the field of fluid dynamics, Navier-Stokes and Euler equations with random forces such as an additive or a multiplicative colored noise have been studied. In this paper, we consider the solution $u = u(t, x)$ of the initial value problem of the following stochastic partial differential equation (22.1):

$$\frac{\partial u}{\partial t} = -(-\Delta)^{\frac{\gamma}{2}} u + R(t, u(t,x)) + F(t, u(t,x))\dot{w}(t,x), \quad (t,x) \in \mathbf{R}_+ \times \mathbf{R}^d, \tag{22.1}$$

where $-(-\Delta)^{\frac{\gamma}{2}}$ is the fractional Laplacian which will be defined later in Sect. 22.2 and $\dot{w}(t,x)$ is a two-parameter white noise defined on a probability space $(\Omega, \mathscr{F}, P)$. We assume that both R and F are bounded Lipschitz continuous functions on $\mathbf{R}_+ \times \mathbf{R}$. In the case $\gamma = 2m$, $m \in \mathbf{N}$, $2m > d$, it is known that a unique solution exists. Unless $2m > d$, its solution does not live in a function space, see also Funaki [6] or

S. Yokoyama (✉)
Graduate School of Mathematical Sciences, The University of Tokyo,
Komaba, Tokyo 153-8914, Japan
e-mail: satoshi2@ms.u-tokyo.ac.jp

Y. Shibata and Y. Suzuki (eds.), *Mathematical Fluid Dynamics, Present and Future*, Springer Proceedings in Mathematics & Statistics 183,
DOI 10.1007/978-4-431-56457-7_22

Kotelenez [9] for the details. It is known that u is a random field which has a continuous modification. Concerning the spatial variable x, it is known that its α-th differential $D_x^\alpha u(t,\cdot) = \frac{\partial^{|\alpha|}}{\partial_{x_1}^{\alpha_1}\dots\partial_{x_d}^{\alpha_d}} u(t,\cdot)$, where $\alpha = (\alpha_1, \dots, \alpha_d) \in (\mathbf{N}_+)^d$, $\mathbf{N}_+ = \mathbf{N} \cup \{0\}$, $|\alpha| = \alpha_1 + \cdots + \alpha_d$, belongs to $D_x^\alpha u(t,\cdot) \in \bigcap_{\varepsilon>0} C^{\frac{2m-d-2|\alpha|}{2}-\varepsilon}(\mathbf{R}^d)$, a.s. The regularity of the solution of SPDEs has been discussed by many authors. In the case $\alpha = 2$ with Neumann boundary condition, the regularity of the unique solution is discussed by Walsh [13]. Funaki [6] considers the case of whole region in $\mathbf{R}^d$ and shows that the solution exists uniquely and its spacial regularity. Peszat and Szymon [12] studies SPDEs in a bounded domain in $\mathbf{R}^d$ and [9] discusses SPDEs in an unbounded domain with a pseudo-differential operator with symbol of order γ, $\gamma > d$.

In addition, Debbi and Dozzi [3] and Niu and Xie [11] study SPDEs in one dimension with a certain fractional operator which is different from our paper. Burgers' equation with a fractional Laplacian is studied by Biler, Funaki and Woyczynski [1]. Furthermore, Chang and Lee [2] study the existence and uniqueness of the solution of SPDEs in the sense of generalized solutions in $L^2(\mathbf{R}^d)$ with the fractional Laplacian $(-\Delta)^{\frac{\gamma}{2}}$ with $\gamma \in (0, 2)$ and investigate the regularity of the solution by using the Sobolev space theory. As discussed in [6], for the second order SPDEs with the space-time white noise to be well-posed, the space dimension d must be equal to 1. In the case of $d \geqslant 2$, the solution is no longer a function and exists in only distribution sense. These facts are related to the condition $2m > d$ ($m = 1$ in the second order). On the other hand, if we consider higher order differential operators, the associated solution has better regularity. In this paper, we define the solution of (22.1) as the mild solution which lives in a class of $L^2_\rho(\mathbf{R}^d)$, which will be stated in Sect. 22.2, then discuss the regularity of the solution with respect to the spatial variable and the relationship between the regularity of the solution of the case $2m$-th order and that of the fractional order. To find the function spaces where the solution lives, we investigate the differentiability for the associate fundamental solution and apply Kolmogorov-Centsov theorem for our solution (see also Appendix). Recently, Hairer [7] established an innovative method for the solutions of SPDEs such as KPZ equation by using the framework of so called the regularity structure. His method brings us to another concept for solutions to the equation of the second order differential operator with space-time white noise in higher space dimension. The paper is organized as follows: In Sect. 22.2, we formulate our problem and state our main result. In Sect. 22.3, we give the proof of the main result.

22.2 Formulation of Our Problem

22.2.1 *Notations*

Let us denote by $\mathscr{S} = \mathscr{S}(\mathbf{R}^d)$ a family of Schwartz's rapidly decreasing functions on $\mathbf{R}^d$. Set

$$\hat{u}(\xi) = \int_{\mathbf{R}^d} e^{-\sqrt{-1}x\cdot\xi} u(x)dx, \qquad u \in \mathscr{S}. \tag{22.2}$$

We define the fractional Laplacian $-(-\Delta)^{\frac{\gamma}{2}}$ as

$$-(-\Delta)^{\frac{\gamma}{2}} u(x) = \frac{1}{(2\pi)^d} \int_{\mathbf{R}^d} e^{\sqrt{-1}x\cdot\xi} (-|\xi|^\gamma) \hat{u}(\xi) d\xi. \tag{22.3}$$

Note that if $\gamma = 2$, the above definition coincides with the usual Laplacian. Furthermore, set

$$G(t, r, q) = \frac{1}{(2\pi)^d} \int_{\mathbf{R}^d} e^{\sqrt{-1}(r-q)\cdot\xi} g(t, \xi) d\xi, \qquad t > 0,\ r, q \in \mathbf{R}^d, \tag{22.4}$$

where

$$g(t, \xi) = e^{-t|\xi|^\gamma}, \qquad \gamma > d, \tag{22.5}$$

and $\sqrt{-1}$ is i with $i^2 = -1$ and $a \cdot b$, (resp. $|a|$), $a, b \in \mathbf{R}^d$ denotes the inner product on $\mathbf{R}^d$ (resp. the Euclidean norm on $\mathbf{R}^d$). Note that if $\gamma = 2$, $G(t, r, q)$ is the Gaussian kernel. To use later, we set $\lambda(x) = (1 + |x|^2)^{\frac{1}{2}}$, $x \in \mathbf{R}^d$ and

$$L^2_\rho(\mathbf{R}^d) = \left\{ u : \mathbf{R}^d \to \mathbf{R} \,\middle|\, |u|^2_\rho := \int_{\mathbf{R}^d} |u(x)|^2 \lambda^{-\rho}(x) dx < \infty \right\}, \qquad \rho \geqslant 0. \tag{22.6}$$

22.2.2 *Definition of Solutions*

Here, we formulate the solution of (22.1). We denote by $\mathbf{X}(\mathbf{R}^d)$ the family of $\mathscr{F}_t$-adapted stochastic processes $X_t \equiv \{X(t, x), x \in \mathbf{R}^d\}$ defined on some probability space $(\Omega, \mathscr{F}, P)$ on which the $\mathscr{F}_t$-cylindrical Brownian motion on $L^2(\mathbf{R}^d)$ $w = \{w_t(\psi),\ t \geqslant 0,\ \psi \in L^2(\mathbf{R}^d)\}$ with its covariance operator $Q = I_d$(=identity operator) is defined and the map $(t, \omega) \mapsto X(t, \cdot, \omega) \in L^2_\rho(\mathbf{R}^d)$ is measurable. We say $X \in \mathbf{X}_T(L^2_\rho(\mathbf{R}^d))$ if X belongs to $\mathbf{X}(\mathbf{R}^d)$ and satisfies

$$\sup_{0 \leqslant t \leqslant T} \mathbf{E}\Big[|X(t)|^2_\rho \Big] < \infty. \tag{22.7}$$

Definition 22.2.1 We say u is a solution of (22.1) with an initial value u_0 with $|u_0|_\rho < \infty$ if
(s.1) $u \in \mathbf{X}_T(L^2_\rho(\mathbf{R}^d))$ for some $T > 0$ and $\rho > d$.
(s.2) For almost every $t \in (0, T]$ and $x \in \mathbf{R}^d$,

$$\begin{aligned} u(t,x) &= \int_{\mathbf{R}^d} G(t,x,q)u_0(q)dq + \int_0^t \int_{\mathbf{R}^d} G(t-s,x,q)R(s,u(s,q))dsdq, \quad (22.8) \\ &+ \int_0^t \int_{\mathbf{R}^d} G(t-s,x,q)F(s,u(s,q))w(dsdq), \\ &\equiv S_1(t,x) + S_2(t,x) + S_3(t,x), \qquad a.s., \end{aligned}$$

holds.

22.2.3 Existence and Uniqueness of Solutions of (22.1)

The solutions defined in Definition 22.2.1 are usually called mild solutions. It is known that the generalized solution, namely, as the definition by the duality with test functions, is equivalent to the mild solution. We assume

$$\gamma > d. \tag{22.9}$$

Concerning $G(t, r, q)$, the following proposition holds:

Proposition 22.2.1 *There exist a Borel measurable function $P(r)$ on $\mathbf{R}^d$ and $\rho > d$ such that the following three inequalities*

$$|G(t,r,q)| \leqslant t^{-\frac{d}{\gamma}} P((q-r)t^{-\frac{1}{\gamma}}), \quad t > 0, \tag{22.10}$$

$$\int_{\mathbf{R}^d} P(r)\lambda^\rho(r)dr < \infty, \tag{22.11}$$

$$\sup_{0 \leqslant s < t \leqslant T} \sup_{r \in \mathbf{R}^d} P(r(t-s)^{-\frac{1}{\gamma}})\lambda^{2\rho}(r) < \infty, \quad T > 0, \tag{22.12}$$

hold. In particular, ρ can be chosen as $d < \rho < \frac{\gamma+3d}{4} (< \frac{\gamma+d}{2})$.

Proof Concerning (22.10), from change of variables, we have

$$|G(t,r,q)| = (\frac{1}{2\pi})^d t^{-\frac{d}{\gamma}} \left| \int_{\mathbf{R}^d} e^{\sqrt{-1}(r-q)t^{-\frac{1}{\gamma}} \cdot \xi} e^{-|\xi|^\gamma} d\xi \right|. \tag{22.13}$$

Thus if we set

$$P(r) = (\frac{1}{2\pi})^d \left| \int_{\mathbf{R}^d} e^{\sqrt{-1}r \cdot \xi} e^{-|\xi|^\gamma} d\xi \right|, \tag{22.14}$$

(22.10) holds indeed as an equality.

Next we will show (22.11). Clearly, $P(r)$ and $\lambda^\rho(r)$ are continuous in r. Since $P(r)$ and $\lambda^\rho(r)$ are bounded in $|r| \leqslant 1$, it suffices to check the integral over $|r| \geqslant 1$ satisfies the estimate like (22.11). As will be shown later in the proof the Theorem 22.3.1, there exists a constant $c > 0$ such that $|P(r)| \leqslant c|r|^{-k}$, $|r| \geqslant 1$ holds for $k < \gamma + d$ and k can be chosen as $k = \gamma + d - \varepsilon_0$ for arbitrarily small $\varepsilon_0 > 0$ (see (22.53) with $|\alpha| = 0$). Using $\lambda^\rho(r) \leqslant (\sqrt{2}|r|)^\rho$, we have

$$|\int_{|r|\geqslant 1} P(r)\lambda^\rho(r)dr| \leqslant 2^{\frac{\rho}{2}} c \int_{|r|\geqslant 1} |r|^{\rho-k} dr, \tag{22.15}$$

which is finite if $\rho < k - d$, that is, $\rho < \gamma - \varepsilon_0$. Let us set $\varepsilon_0 = \frac{\gamma-d}{2}$, which is positive by (22.9), then it follows that ρ should satisfy $\rho < \frac{\gamma+d}{2}$. From this and using (22.9) again, we can find ρ satisfying

$$d < \rho < \frac{\gamma + d}{2}. \tag{22.16}$$

Thus, (22.11) holds for such ρ.

Concerning (22.12), we only need to see $|r| \geqslant 1$ since $P(r(t-s)^{-\frac{1}{\gamma}})$ and $\lambda^\rho(r)$ are bounded in $|r| \leqslant 1$ for each $0 \leqslant s < t \leqslant T$. For $|r| \geqslant 1$, by similar argument, we obtain $P(r(t-s)^{-\frac{1}{\gamma}}) \leqslant c|r(t-s)^{-\frac{1}{\gamma}}|^{-k}$ for $k < \gamma + d$, where c is some constant. Recall $k = \gamma + d - \varepsilon_0$ and $\varepsilon_0 = \frac{\gamma-d}{2}$, namely, $k = \frac{\gamma+3d}{2}$. Using $\lambda^\rho(r) \leqslant (\sqrt{2}|r|)^\rho$ again, we have

$$P(r(t-s)^{-\frac{1}{\gamma}})\lambda^{2\rho}(r) \leqslant cT^{\frac{\gamma+3d}{2\gamma}}|r|^{2\rho-(\frac{\gamma+3d}{2})}, \quad |r| \geqslant 1. \tag{22.17}$$

Choose ρ as

$$d < \rho \leqslant \frac{\gamma + 3d}{4}. \tag{22.18}$$

Indeed, this set is not empty due to (22.9) and $2\rho - (\frac{\gamma+3d}{2}) \leqslant 0$ holds for such ρ. Thus, (22.12) holds and (22.11) also holds for ρ satisfying (22.18). The proof is complete.

Combined Proposition 22.2.1 with Theorem 3.1 in [9], we have the unique existence of solutions of (22.1), that is,

Theorem 22.2.2 *Let*

$$\gamma > d. \tag{22.19}$$

Suppose that there exists some ρ with $d < \rho < \frac{\gamma+d}{2}$ such that (22.10), (22.11) *and* (22.12) *are satisfied and that for each $T > 0$, there exist K_0 and $K > 0$ such that*

$$|R(t,0)| + |F(t,0)| \leqslant K_0, \quad \forall t \in [0,T], \tag{22.20}$$

$$|F(t,x) - F(t,y)| + |R(t,x) - R(t,y)| \leqslant K|x-y|, \quad \forall t \in [0,T], \quad x, y \in \mathbf{R}. \tag{22.21}$$

Then, there exists a unique solution $u \in X_T(L^2_\rho(\mathbf{R}^d))$ *with the initial value* u_0 *satisfying* $|u_0|_\rho < \infty$.

Remark 22.2.1 In the above theorem, even when F is constant, it is essential to introduce the weighted L^2-space, that is, $L^2_\rho(\mathbf{R}^d)$ as the state space. On the other hand, it is not necessary to introduce such function space in the setting of [8].

22.3 Regularity of the Solution

In this section, we will study the regularity with respect to the spatial variable $x \in \mathbf{R}^d$ of the solution $u(t,x)$ of (22.1). We assume that F and R are bounded in addition to (22.19)–(22.21). By Theorem 22.2.2, a unique solution $u \in \mathbf{X}_T(L^2_\rho(\mathbf{R}^d))$ exists. Under these settings, we will show the following result:

Theorem 22.3.1 *If* $\gamma \notin 2\mathbf{N}$,

$$D^\alpha_x u(t) \in \bigcap_{\varepsilon>0} C^{\left(\frac{\gamma-2|\alpha|-d}{2}\wedge(1-\frac{d}{\gamma+|\alpha|+d})\right)-\varepsilon}(\mathbf{R}^d), \quad |\alpha| = [\frac{\gamma-d}{2}], \quad a.s., \tag{22.22}$$

where $\alpha = (\alpha_1, \dots, \alpha_d) \in (\mathbf{N}_+)^d$, $|\alpha| = \alpha_1 + \cdots + \alpha_d$, $D^\alpha_x = \frac{\partial^{|\alpha|}}{\partial^{\alpha_1}_{x_1}\dots\partial^{\alpha_d}_{x_d}}$ *and the notation* $[a]$ *represents the maximal integer which is not over* a.

Proof First, we will study the regularity of the stochastic integral part $S_3(t,x)$ by computing its $2m$-th moment. In what follows, for simplicity, C denotes a positive constant whose value might change from one formula to another. Concerning $S_3(t,x)$, we have

$$\begin{aligned}
&\mathbf{E}\Big[\Big|D^\alpha_x(S_3(t,x+h) - S_3(t,x))\Big|^{2m}\Big] \qquad (22.23)\\
&= \mathbf{E}\Big[\Big|\int_0^t \int_{\mathbf{R}^d} \big(D^\alpha_x G(t-s,x+h,q) - D^\alpha_x G(t-s,x,q)\big)F(s,u(s,q))w(dsdq)\Big|^{2m}\Big]\\
&\leqslant C\Big(\mathbf{E}\Big[\int_0^t \int_{\mathbf{R}^d} \big(D^\alpha_x G(t-s,x+h,q) - D^\alpha_x G(t-s,x,q)\big)^2 F(s,u(s,q))^2 dsdq\Big]\Big)^m\\
&\leqslant C\Big(\int_0^t \int_{\mathbf{R}^d} \big(D^\alpha_x G(t-s,x+h,q) - D^\alpha_x G(t-s,x,q)\big)^2 dsdq\Big)^m,
\end{aligned}$$

where we have used the Burkholder-Davis-Gundy's inequality at the first inequality and the boundedness of F at the last line. Since

$$D^\alpha_x G(t-s,x,q) = \frac{1}{(2\pi)^d}\int_{\mathbf{R}^d} e^{\sqrt{-1}(x-q)\cdot\xi}\sqrt{-1}^{|\alpha|}\xi^\alpha g(t-s,\xi)d\xi, \tag{22.24}$$

for $t > s > 0,\ x, q \in \mathbf{R}^d$, where $\xi = (\xi_1, \dots, \xi_d)$ and $\xi^\alpha = \xi_1^{\alpha_1} \dots \xi_d^{\alpha_d}$,

$$\begin{aligned}
&\left|D_x^\alpha G(t-s, x+h, q) - D_x^\alpha G(t-s, x, q)\right| \qquad (22.25)\\
&= C\left|\int_{\mathbf{R}^d} (e^{\sqrt{-1}(x+h-q)\cdot\xi} - e^{\sqrt{-1}(x-q)\cdot\xi})\xi^\alpha g(t-s,\xi)d\xi\right|,\\
&\leqslant C\int_{\mathbf{R}^d} \left|e^{\sqrt{-1}h\cdot\xi} - 1\right||\xi|^{|\alpha|}|g(t-s,\xi)|d\xi,\\
&= C(t-s)^{-\frac{d+|\alpha|}{\gamma}}\int_{\mathbf{R}^d} \left|e^{\sqrt{-1}h(t-s)^{-\frac{1}{\gamma}}\cdot\xi} - 1\right||\xi|^{|\alpha|}|g(t-s,(t-s)^{-\frac{1}{\gamma}}\xi)|d\xi.
\end{aligned}$$

Using $|e^{\sqrt{-1}x\cdot\xi} - 1| \leqslant C|x||\xi|$ and $\int_{\mathbf{R}^d} |\xi|^{1+|\alpha|}|g(t-s,(t-s)^{-\frac{1}{\gamma}}\xi)|d\xi$ is finite since

$$g(t-s,(t-s)^{-\frac{1}{\gamma}}\xi) = e^{-|\xi|^\gamma}, \qquad (22.26)$$

we have

$$\left|D_x^\alpha G(t-s, x+h, q) - D_x^\alpha G(t-s, x, q)\right| \leqslant C(t-s)^{-\frac{d+|\alpha|+1}{\gamma}}|h|. \qquad (22.27)$$

On the other hand, for every $p \in [0, 1]$, we have

$$\begin{aligned}
&\left(\left|D_x^\alpha G(t-s, x+h, q)\right| + \left|D_x^\alpha G(t-s, x, q)\right|\right)^{2(1-p)} \qquad (22.28)\\
&= \frac{1}{(2\pi)^{2d(1-p)}}(t-s)^{-\frac{2(d+|\alpha|)}{\gamma}(1-p)}\left(\left|\int_{\mathbf{R}^d} e^{\sqrt{-1}(x+h-q)(t-s)^{-\frac{1}{\gamma}}\cdot\xi}\xi^\alpha e^{-|\xi|^\gamma}d\xi\right|\right.\\
&\quad \left. + \left|\int_{\mathbf{R}^d} e^{\sqrt{-1}(x-q)(t-s)^{-\frac{1}{\gamma}}\cdot\xi}\xi^\alpha e^{-|\xi|^\gamma}d\xi\right|\right)^{2(1-p)}\\
&\leqslant C(t-s)^{-\frac{2(d+|\alpha|)}{\gamma}(1-p)}\left(\left|\int_{\mathbf{R}^d} e^{\sqrt{-1}(x+h-q)(t-s)^{-\frac{1}{\gamma}}\cdot\xi}\xi^\alpha e^{-|\xi|^\gamma}d\xi\right|^{2(1-p)}\right.\\
&\quad \left. + \left|\int_{\mathbf{R}^d} e^{\sqrt{-1}(x-q)(t-s)^{-\frac{1}{\gamma}}\cdot\xi}\xi^\alpha e^{-|\xi|^\gamma}d\xi\right|^{2(1-p)}\right),
\end{aligned}$$

where we have used $(a+b)^k \leqslant C(a^k + b^k),\ a, b \geqslant 0,\ k \geqslant 0$ and (22.26). From (22.27) and (22.28), we have

$$\int_0^t \int_{\mathbf{R}^d} \Big(D_x^\alpha\big(G(t-s,x+h,q)-G(t-s,x,q)\big)\Big)^2 dsdq \tag{22.29}$$
$$= \int_0^t \int_{\mathbf{R}^d} \Big(D_x^\alpha\big(G(t-s,x+h,q)-G(t-s,x,q)\big)\Big)^{2p}$$
$$\Big(D_x^\alpha\big(G(t-s,x+h,q)-G(t-s,x,q)\big)\Big)^{2(1-p)} dsdq$$
$$\leqslant C|h|^{2p} \int_0^t (t-s)^{-\frac{d+|\alpha|+1}{\gamma}2p-\frac{2(d+|\alpha|)}{\gamma}(1-p)} ds$$
$$\times \int_{\mathbf{R}^d} \Big|\int_{\mathbf{R}^d} e^{\sqrt{-1}(x+h-q)(t-s)^{-\frac{1}{\gamma}}\cdot\xi} \xi^\alpha e^{-|\xi|^\gamma} d\xi\Big|^{2(1-p)}$$
$$+ \Big|\int_{\mathbf{R}^d} e^{\sqrt{-1}(x-q)(t-s)^{-\frac{1}{\gamma}}\cdot\xi} \xi^\alpha e^{-|\xi|^\gamma} d\xi\Big|^{2(1-p)} dq.$$

Furthermore, by change of variables $(x+h-q)(t-s)^{-\frac{1}{\gamma}} = z$ and $(x-q)(t-s)^{-\frac{1}{\gamma}} = z$, we obtain from (22.29),

$$\int_0^t \int_{\mathbf{R}^d} \Big(D_x^\alpha\big(G(t-s,x+h,q)-G(t-s,x,q)\big)\Big)^2 dsdq \tag{22.30}$$
$$\leqslant C_p|h|^{2p} \int_0^t (t-s)^{-\frac{d+|\alpha|+1}{\gamma}2p-\frac{2(d+|\alpha|)}{\gamma}(1-p)+\frac{d}{\gamma}} ds$$
$$\times \int_{\mathbf{R}^d} \Big|\int_{\mathbf{R}^d} e^{\sqrt{-1}z\cdot\xi} \xi^\alpha e^{-|\xi|^\gamma} d\xi\Big|^{2(1-p)} dz.$$

Set

$$\phi(z) = \Big|\int_{\mathbf{R}^d} e^{\sqrt{-1}z\cdot\xi} \xi^\alpha e^{-|\xi|^\gamma} d\xi\Big|. \tag{22.31}$$

In particular, for each $n \in \mathbf{N}$, $j = 1, \ldots, d$ and $|z| \geqslant 1$, we have

$$\Big|\int_{\mathbf{R}^d} e^{\sqrt{-1}z\cdot\xi} \xi^\alpha e^{-|\xi|^\gamma} d\xi\Big| = \frac{1}{|z_j|^n}\Big|\int_{\mathbf{R}^d} (D_{\xi_j}^n e^{\sqrt{-1}z\cdot\xi}) \xi^\alpha e^{-|\xi|^\gamma} d\xi\Big|, \tag{22.32}$$

where $D_{\xi_j}^n = \frac{\partial^n}{\partial \xi_j^n}$. Let us decompose ϕ into two parts as follows:

$$\phi(z) = \Big|\int_{\mathbf{R}^d} e^{\sqrt{-1}z\cdot\xi} \xi^\alpha e^{-|\xi|^\gamma} d\xi\Big| 1_{\{|z|<1\}} \tag{22.33}$$
$$+ \frac{1}{|z_j|^n}\Big|\int_{\mathbf{R}^d} (D_{\xi_j}^n e^{\sqrt{-1}z\cdot\xi}) \xi^\alpha e^{-|\xi|^\gamma} d\xi\Big| 1_{\{|z|\geqslant 1\}} \equiv \phi_1(z) + \phi_2(z),$$

for $j = 1, \dots, d$. We will check the integrability of the integral:

$$\int_{\mathbf{R}^d} |\phi(z)|^{2(1-p)} dz. \tag{22.34}$$

To this end, since $\phi(z)$ is continuous and $\int_{|z|<1} |\phi_1(z)|^{2(1-p)} dz < \infty$ holds for every $p \in [0, 1]$, let us check the behaviour of $\phi_2(z)$ for large $|z|$. Note that in the case of $\gamma \notin 2\mathbf{N}$ and $k > \gamma$, $D^k_{\xi_j} e^{-|\xi|^\gamma}$ diverges as $|\xi|$ tends to 0. Note also that for any $k \in \mathbf{N} \cup \{0\}$, there is no diverging terms in $D^k_{\xi_j} e^{-|\xi|^\gamma}$ in a neighborhood of $\xi = 0$. Indeed, no terms of ξ with negative exponent appear. When $\gamma \notin 2\mathbf{N}$,

$$\int_{\mathbf{R}^d} |D^k_{\xi_j} (\xi^\alpha e^{-|\xi|^\gamma})| d\xi < \infty, \qquad k \in \mathbf{N} \cup \{0\}, \tag{22.35}$$

holds if k satisfies $k < \gamma + |\alpha| + d$. To see this, it suffices to check the behavior of the integrand of (22.35) near $\xi = 0$. Indeed, using the polar coordinate $(r, \theta_1, \dots, \theta_{d-1})$ of $(\xi_1, \dots, \xi_d)$ with $\xi_1 = r \cos\theta_1$, $\xi_i = r\big(\prod_{j=1}^{i-1} \sin\theta_j\big) \cos\theta_i$, $i = 2, \dots, d-1$, $\xi_d = r \prod_{j=1}^{d-1} \sin\theta_j$ and noting that its Jacobian is given by $r^{d-1} g(\theta_1, \dots, \theta_d)$, where $g(\theta_1, \dots, \theta_d)$ is some bounded function, the condition of k for which (22.35) holds is obtained by direct computation.

Let us assume

$$\int_{\mathbf{R}^d} \left| D^k_{\xi_j} (\xi^\alpha e^{-|\xi|^\gamma}) \right| d\xi < \infty, \tag{22.36}$$

for some $k \in \mathbf{N}_+$. Then,

$$\begin{aligned}
&\int_{\mathbf{R}^d} D_{\xi_j} e^{\sqrt{-1}z\cdot\xi} D^k_{\xi_j} (\xi^\alpha e^{-|\xi|^\gamma}) d\xi \qquad (22.37)\\
&= \lim_{\substack{e\to 0\\ N\to\infty}} \int_{\mathbf{R}^d} D_{\xi_j} e^{\sqrt{-1}z\cdot\xi} D^k_{\xi_j} (\xi^\alpha e^{-|\xi|^\gamma}) 1_{[-N,-\varepsilon]\cup[\varepsilon,N]}(\xi_j) 1_{B^j_{\varepsilon,N}}(\tilde{\xi}) d\xi\\
&= \lim_{\substack{\varepsilon\to 0\\ N\to\infty}} \int_{\mathbf{R}^{d-1}} e^{\sqrt{-1}\tilde{z}\cdot\tilde{\xi}} 1_{B^j_{\varepsilon,N}}(\tilde{\xi}) d\tilde{\xi}\\
&\quad \times \int_{\mathbf{R}} (D_{\xi_j} e^{\sqrt{-1}z_j\xi_j}) D^k_{\xi_j} (\xi^\alpha e^{-|\xi|^\gamma}) 1_{[-N,-\varepsilon]\cup[\varepsilon,N]}(\xi_j) d\xi_j\\
&= \lim_{\substack{\varepsilon\to 0\\ N\to\infty}} \int_{\mathbf{R}^{d-1}} e^{\sqrt{-1}\tilde{z}\cdot\tilde{\xi}} \Big[e^{\sqrt{-1}z_j\xi_j} D^k_{\xi_j} (\xi^\alpha e^{-|\xi|^\gamma}) \Big]_{\xi_j=-N}^{\xi_j=-\varepsilon} 1_{B^j_{\varepsilon,N}}(\tilde{\xi}) d\tilde{\xi}\\
&\quad + \int_{\mathbf{R}^{d-1}} e^{\sqrt{-1}\tilde{z}\cdot\tilde{\xi}} \Big[e^{\sqrt{-1}z_j\xi_j} D^k_{\xi_j} (\xi^\alpha e^{-|\xi|^\gamma}) \Big]_{\xi_j=\varepsilon}^{\xi_j=N} 1_{B^j_{\varepsilon,N}}(\tilde{\xi}) d\tilde{\xi}\\
&\quad - \int_{\mathbf{R}^{d-1}} e^{\sqrt{-1}\tilde{z}\cdot\tilde{\xi}} \Big(\int_{\mathbf{R}} e^{\sqrt{-1}z_j\xi_j} D^{k+1}_{\xi_j} (\xi^\alpha e^{-|\xi|^\gamma}) 1_{[-N,-\varepsilon]\cup[\varepsilon,N]}(\xi_j) d\xi_j \Big) 1_{B^j_{\varepsilon,N}}(\tilde{\xi}) d\tilde{\xi},
\end{aligned}$$

where $\tilde{z} = (z_1, \ldots, z_{j-1}, z_{j+1}, \ldots, z_d)$, $\tilde{\xi} = (\xi_1, \ldots, \xi_{j-1}, \xi_{j+1}, \ldots, \xi_d) \in \mathbf{R}^{d-1}$, $1_{[a,b]}$ is the indicator function on $[a, b]$, $\big[a(\xi)\big]_{\xi_j=\varepsilon}^{\xi_j=N} = a(\xi)\Big|_{\xi_j=N} - a(\xi)\Big|_{\xi_j=\varepsilon}$ and $B^j_{\varepsilon,N} = \{\tilde{\xi} \mid \varepsilon \leqslant |\tilde{\xi}| \leqslant N\}$, $|\tilde{\xi}| = \big(\sum_{1\leqslant i\leqslant d,\, i\neq j} \xi_i^2\big)^{\frac{1}{2}}$. On the other hand, since

$$\begin{aligned}
&\Big[e^{\sqrt{-1}z_j\xi_j} D^k_{\xi_j}\big(\xi^\alpha e^{-|\xi|^\gamma}\big)\Big]_{\xi_j=\varepsilon}^{\xi_j=N} \\
&= \frac{1}{N-\varepsilon}\int_{\mathbf{R}} \Big[e^{\sqrt{-1}z_j\xi_j} D^k_{\xi_j}\big(\xi^\alpha e^{-|\xi|^\gamma}\big)\Big]_{\xi_j=\varepsilon}^{\xi_j=N} 1_{[\varepsilon,N]}(\xi_j)d\xi_j,
\end{aligned} \tag{22.38}$$

(22.37) is rewritten as

$$\begin{aligned}
&\lim_{\substack{\varepsilon\to 0\\ N\to\infty}} \Big[\int_{\mathbf{R}^{d-1}} e^{\sqrt{-1}\tilde{z}\cdot\tilde{\xi}} \\
&\times\Big(\frac{1}{N-\varepsilon}\int_{\mathbf{R}} \Big[e^{\sqrt{-1}z_j\xi_j} D^k_{\xi_j}\big(\xi^\alpha e^{-|\xi|^\gamma}\big)\Big]_{\xi_j=-N}^{\xi_j=-\varepsilon} 1_{[-N,-\varepsilon]}(\xi_j)d\xi_j\Big) 1_{B^j_{\varepsilon,N}}(\tilde{\xi})d\tilde{\xi} \\
&+\int_{\mathbf{R}^{d-1}} e^{\sqrt{-1}\tilde{z}\cdot\tilde{\xi}}\Big(\frac{1}{N-\varepsilon}\int_{\mathbf{R}} \Big[e^{\sqrt{-1}z_j\xi_j} D^k_{\xi_j}\big(\xi^\alpha e^{-|\xi|^\gamma}\big)\Big]_{\xi_j=\varepsilon}^{\xi_j=N} 1_{[\varepsilon,N]}(\xi_j)d\xi_j\Big) 1_{B^j_{\varepsilon,N}}(\tilde{\xi})d\tilde{\xi} \\
&-\int_{\mathbf{R}^{d-1}} e^{\sqrt{-1}\tilde{z}\cdot\tilde{\xi}}\Big(\int_{\mathbf{R}} e^{\sqrt{-1}z_j\xi_j} D^{k+1}_{\xi_j}\big(\xi^\alpha e^{-|\xi|^\gamma}\big) 1_{[-N,-\varepsilon]\cup[\varepsilon,N]}(\xi_j)d\xi_j\Big) 1_{B^j_{\varepsilon,N}}(\tilde{\xi})d\tilde{\xi}\Big] \\
&\equiv \lim_{\substack{\varepsilon\to 0\\ N\to\infty}} \Big(\frac{1}{N-\varepsilon}(I) + \frac{1}{N-\varepsilon}(II) + (III)\Big).
\end{aligned} \tag{22.39}$$

Suppose that $k+1 < \gamma + |\alpha| + d$ holds. Then, (22.36) holds for $k+1$. Concerning (II), using the polar coordinate $(r, \theta_1, \ldots, \theta_{d-2})$ for $\tilde{\xi}$, it is easy to check $\lim_{N\to\infty}(N-\varepsilon)^{-1}(II) = 0$. For sufficiently small r, we obtain that $r^{\gamma+|\alpha|-k+(d-2)}$ is the fastest divergent term in the terms contained in (II). From $k+1 < \gamma + |\alpha| + d$, we get $\gamma + |\alpha| - k + (d-2) > -1$, which implies that $r^{\gamma+|\alpha|-k+(d-2)}$ is integrable near 0. So $\lim_{\varepsilon\to 0, N\to\infty}(N-\varepsilon)^{-1}(II) = 0$ for $\tilde{z}$. Similarly, $\lim_{\varepsilon\to 0, N\to\infty}(N-\varepsilon)^{-1}(I) = 0$ for $\tilde{z}$.

Concerning (III), when $k+1 < \gamma + |\alpha| + d$, we obtain

$$\lim_{\substack{\varepsilon\to 0\\ N\to\infty}} \frac{1}{N-\varepsilon}\Big|(III)\Big| = \Big|\int_{\mathbf{R}^d} e^{\sqrt{-1}z\cdot\xi} D^{k+1}_{\xi_j}(\xi^\alpha e^{-|\xi|^\gamma})d\xi\Big|, \tag{22.40}$$

for z. As a result, when $k+1 < \gamma + |\alpha| + d$, we have

$$\Big|\int_{\mathbf{R}^d} D_{\xi_j} e^{\sqrt{-1}z\cdot\xi} D^k_{\xi_j}(\xi^\alpha e^{-|\xi|^\gamma})d\xi\Big| = \Big|\int_{\mathbf{R}^d} e^{\sqrt{-1}z\cdot\xi} D^{k+1}_{\xi_j}(\xi^\alpha e^{-|\xi|^\gamma})d\xi\Big|. \tag{22.41}$$

Suppose that an integer l satisfies $l < \gamma + |\alpha| + d$ and $l + 1 \geqslant \gamma + |\alpha| + d$. By proceeding similarly to the previous argument, for all $m \geqslant l$,

$$\left| \int_{\mathbf{R}^d} D_{\xi_j}^m e^{\sqrt{-1}z\cdot\xi} (\xi^\alpha e^{-|\xi|^\gamma}) d\xi \right| = \left| \int_{\mathbf{R}^d} D_{\xi_j}^{m-l} e^{\sqrt{-1}z\cdot\xi} D_{\xi_j}^l (\xi^\alpha e^{-|\xi|^\gamma}) d\xi \right|. \tag{22.42}$$

However, (22.42) does not hold for $l + 1$. Set

$$n = \begin{cases} \gamma + |\alpha| + d - 1, & \gamma \in 2\mathbf{N} - 1, \\ [\gamma + |\alpha| + d], & \gamma \notin \mathbf{N}. \end{cases} \tag{22.43}$$

Applying (22.42) with $m = l = n$ to $\phi_2(z)$, we have

$$\phi_2(z) = \frac{1}{|z_j|^n} \left| \int_{\mathbf{R}^d} e^{\sqrt{-1}z\cdot\xi} D_{\xi_j}^n (\xi^\alpha e^{-|\xi|^\gamma}) d\xi \right|. \tag{22.44}$$

Since the integral of the right hand side of (22.44) is finite, we have $\int_{|z|\geqslant 1} \phi_2(z)^{2(1-p)} dz < \infty$ if $2(1-p)(-n) < -d$.

Thus, $\int_{\mathbf{R}^d} |\phi(z)|^{2(1-p)} dz < \infty$ holds for p satisfying $2(1-p)n > d$, that is, $p \in \left(0, 1 - \frac{d}{2n}\right)$.

Although (22.44) does not hold for $n + 1$, we can obtain faster decay of $\phi_2(z)$ as $|z_j| \to \infty$. The definition of ϕ_2 of (22.33) is written as

$$\frac{1}{|z_j|^{n+1}} \left| \int_{\mathbf{R}^d} (D_{\xi_j}^{n+1} e^{\sqrt{-1}z\cdot\xi}) \xi^\alpha e^{-|\xi|^\gamma} d\xi \right| 1_{\{|z|\geqslant 1\}}. \tag{22.45}$$

For the time being, we omit the notation $1_{\{|z|\geqslant 1\}}$ for simplicity. Then, (22.45) is rewritten to

$$\begin{aligned} \psi(z) &\equiv \frac{1}{|z_j|^{n+1}} \left| \int_{\mathbf{R}^d} (D_{\xi_j} e^{\sqrt{-1}z\cdot\xi}) D_{\xi_j}^n (\xi^\alpha e^{-|\xi|^\gamma}) d\xi \right| \qquad (22.46) \\ &= \frac{1}{|z_j|^{n+1}} \left| \int_{\mathbf{R}^{d-1}} e^{\sqrt{-1}\tilde{z}\cdot\tilde{\xi}} \left(\int_{\mathbf{R}} D_{\xi_j} (e^{\sqrt{-1}z_j\xi_j} - 1) D_{\xi_j}^n (\xi^\alpha e^{-|\xi|^\gamma}) d\xi_j \right) d\tilde{\xi} \right| \\ &= \frac{1}{|z_j|^{n+1}} \lim_{\substack{\varepsilon\to 0 \\ N\to\infty}} \left| \int_{\mathbf{R}^{d-1}} e^{\sqrt{-1}\tilde{z}\cdot\tilde{\xi}} \left(\int_{\mathbf{R}} D_{\xi_j} (e^{\sqrt{-1}z_j\xi_j} - 1) D_{\xi_j}^n (\xi^\alpha e^{-|\xi|^\gamma}) \right.\right. \\ &\qquad \left.\left. 1_{[-N,-\varepsilon]\cup[\varepsilon,N]}(\xi_j) d\xi_j \right) 1_{B_{\varepsilon,N}^j}(\tilde{\xi}) d\tilde{\xi} \right|. \end{aligned}$$

By integration by parts and recalling (22.38), we have

$$\begin{aligned}\psi(z) = {} & \frac{1}{|z_j|^{n+1}} \lim_{\substack{\varepsilon\to 0\\ N\to\infty}} \Big[\int_{\mathbf{R}^{d-1}} e^{\sqrt{-1}\tilde{z}\cdot\tilde{\xi}} \\ & \times \Big(\frac{1}{N-\varepsilon} \int_{\mathbf{R}} \Big[(e^{\sqrt{-1}z_j\xi_j} - 1) D^n_{\xi_j} (\xi^\alpha e^{-|\xi|^\gamma}) \Big]_{\xi_j=-N}^{\xi_j=-\varepsilon} 1_{[-N,-\varepsilon]}(\xi_j) d\xi_j \Big) 1_{B^j_{\varepsilon,N}}(\tilde{\xi}) d\tilde{\xi} \\ & + \int_{\mathbf{R}^{d-1}} e^{\sqrt{-1}\tilde{z}\cdot\tilde{\xi}} \\ & \times \Big(\frac{1}{N-\varepsilon} \int_{\mathbf{R}} \Big[(e^{\sqrt{-1}z_j\xi_j} - 1) D^n_{\xi_j} (\xi^\alpha e^{-|\xi|^\gamma}) \Big]_{\xi_j=\varepsilon}^{\xi_j=N} 1_{[\varepsilon,N]}(\xi_j) d\xi_j \Big) 1_{B^j_{\varepsilon,N}}(\tilde{\xi}) d\tilde{\xi} \\ & - \int_{\mathbf{R}^{d-1}} e^{\sqrt{-1}\tilde{z}\cdot\tilde{\xi}} \\ & \times \Big(\int_{\mathbf{R}} (e^{\sqrt{-1}z_j\xi_j} - 1) D^{n+1}_{\xi_j} (\xi^\alpha e^{-|\xi|^\gamma}) 1_{[-N,-\varepsilon]\cup[\varepsilon,N]}(\xi_j) d\xi_j \Big) 1_{B^j_{\varepsilon,N}}(\tilde{\xi}) d\tilde{\xi} \Big] \\ \equiv {} & \frac{1}{|z_j|^{n+1}} \lim_{\substack{\varepsilon\to 0\\ N\to\infty}} \Big(\frac{1}{N-\varepsilon}(IV) + \frac{1}{N-\varepsilon}(V) + (VI) \Big).\end{aligned}$$

Note that n satisfies (22.43). Concerning (V), by the polar coordinate $(r, \theta_1, \dots, \theta_{d-2})$ for $\tilde{\xi}$, using $|e^{\sqrt{-1}z_j\xi_j} - 1| \leqslant C|z_j||\xi_j|$ and similar argument done in (22.39), we have $\lim_{\varepsilon\to 0, N\to\infty}(N-\varepsilon)^{-1}(IV) = 0$ and $\lim_{\varepsilon\to 0, N\to\infty}(N-\varepsilon)^{-1}(V) = 0$ for $\tilde{z}$. Concerning (VI), if the function $(e^{\sqrt{-1}z_j\xi_j} - 1) D^{n+1}_{\xi_j}(\xi^\alpha e^{-|\xi|^\gamma})$ is integrable on $\mathbf{R}^d$, we have

$$\lim_{\substack{\varepsilon\to 0\\ N\to\infty}} |(VI)| = \Big| \int_{\mathbf{R}^{d-1}} e^{\sqrt{-1}\tilde{z}\cdot\tilde{\xi}} \Big(\int_{\mathbf{R}} (e^{\sqrt{-1}z_j\xi_j} - 1) D^{n+1}_{\xi_j}(\xi^\alpha e^{-|\xi|^\gamma}) d\xi_j \Big) d\tilde{\xi} \Big|. \tag{22.47}$$

Using the polar coordinate $(r, \theta_1, \dots, \theta_{d-1})$ and

$$|e^{z_j\xi_j} - 1| \leqslant C|z_j|^\delta |\xi_j|^\delta, \ \delta \in [0, 1], \tag{22.48}$$

for some constant $C > 0$, we obtain that the right hand side of (22.47) is bounded from above by

$$C|z_j|^\delta \int_0^\infty r^{\gamma+|\alpha|-(n+1)+(d-1)+\delta} e^{-r^\gamma} dr\, h(\theta_1, \dots, \theta_{d-1}), \tag{22.49}$$

where $h(\theta_1, \dots, \theta_{d-1})$ is some bounded function. If $\gamma + |\alpha| - n + (d-1) + \delta > 0$, the integral of (22.49) is finite. Thus, we need to choose $\delta \in [0, 1]$ satisfying

$$1 - (\gamma + |\alpha| + d) + n < \delta \leqslant 1. \tag{22.50}$$

As a result, for such δ, we have

$$\psi(z) = \frac{1}{|z_j|^{n+1}}\Big|\int_{\mathbf{R}^{d-1}} e^{\sqrt{-1}\tilde{z}\cdot\tilde{\xi}}\Big(\int_{\mathbf{R}}(e^{\sqrt{-1}z_j\xi_j}-1)D_{\xi_j}^{n+1}(\xi^\alpha e^{-|\xi|^\gamma})d\xi_j\Big)d\tilde{\xi}\Big| \leqslant \frac{C}{|z_j|^{n+1-\delta}}. \tag{22.51}$$

Therefore, we set

$$\delta = 1-(\gamma+|\alpha|+d)+n+\varepsilon_0, \tag{22.52}$$

for arbitrary small $\varepsilon_0 > 0$ and n the integer defined as (22.43). As a result, we have

$$\psi(z) \leqslant \frac{C}{|z_j|^{\gamma+|\alpha|+d-\varepsilon_0}}, \quad |z| \geqslant 1. \tag{22.53}$$

Consequently, we obtain better estimate: $\int_{\mathbf{R}^d}|\phi(z)|^{2(1-p)}dz < \infty$ holds for p satisfying $-2(1-p)(\gamma+|\alpha|+d-\varepsilon_0) < -d$, that is,

$$p \in \Big(0, 1-\frac{d}{2(\gamma+|\alpha|+d-\varepsilon_0)}\Big). \tag{22.54}$$

Thus, since ε_0 is arbitrary, from (22.30) and (22.31), we have

$$\int_0^t\int_{\mathbf{R}^d}\big(D_x^\alpha G(t-s,x+h,q)-D_x^\alpha G(t-s,x,q)\big)^2 dsdq \leqslant C|h|^{2p}\int_0^t (t-s)^{-\frac{d+|\alpha|+1}{\gamma}2p-\frac{2(d+|\alpha|)}{\gamma}(1-p)+\frac{d}{\gamma}}ds, \tag{22.55}$$

for $p \in \big(0, 1-\frac{d}{2(\gamma+|\alpha|+d)}\big]$. Furthermore, the integral of the right hand side of (22.55) is finite if $p \in (0, \frac{\gamma-2|\alpha|-d}{2})$. From (22.23) and (22.30), as a result, for all $p \in \Big(0, \frac{\gamma-2|\alpha|-d}{2} \wedge (1-\frac{d}{2(\gamma+|\alpha|+d)})\Big)$,

$$\mathbf{E}\Big[\Big|D_x^\alpha\Big(S_3(t,x+h)-S_3(t,x)\Big)\Big|^{2m}\Big] \leqslant C|h|^{2pm}, \tag{22.56}$$

holds for any $m \in \mathbf{N}$, thus we have obtained the evaluate for $S_3(t,x)$.

Next, we will compute the $2m$-th moment of $S_2(t,x)$. Indeed, we have

$$\begin{aligned}
&\mathbf{E}\Big[\Big|D_x^\alpha(S_2(t,x+h)-S_2(t,x))\Big|^{2m}\Big] \qquad (22.57)\\
&\leqslant C\mathbf{E}\Big[\Big(\int_0^t\int_{\mathbf{R}^d}\Big|D_x^\alpha G(t-s,x+h,q)-D_x^\alpha G(t-s,x,q)\Big|dsdq\Big)^{2m}\Big]\\
&\leqslant C\mathbf{E}\Big[\Big(\int_0^t ds\int_{\mathbf{R}^d}\Big|D_x^\alpha\Big(G(t-s,x+h,q)-G(t-s,x,q)\Big)\Big|^r\\
&\quad\times\Big(\Big|D_x^\alpha G(t-s,x+h,q)\Big|+\Big|D_x^\alpha G(t-s,x,q)\Big|\Big)^{1-r}dq\Big)^{2m}\Big],
\end{aligned}$$

for $r\in[0,1]$. By proceeding similarly to the estimate of S_3, we obtain that the right hand side of (22.57) is bounded from above by

$$C|h|^{2rm}\Big(\int_0^t (t-s)^{-\frac{d+|\alpha|+1}{\gamma}r-\frac{(d+|\alpha|)}{\gamma}(1-r)+\frac{d}{\gamma}}ds \times\int_{\mathbf{R}^d}\Big|\int_{\mathbf{R}^d}e^{\sqrt{-1}z\cdot\xi}\xi^\alpha e^{-|\xi|^\gamma}d\xi\Big|^{(1-r)}dz\Big)^{2m}. \qquad (22.58)$$

Recall that $\phi(z)$ is defined as (22.33). Similarly to the argument for S_3, $\int_{\mathbf{R}^d}|\phi(z)|^{1-r}dz < \infty$ holds if $-(1-r)(\gamma+|\alpha|+d-\varepsilon_0)<-d$. Noting that ε_0 is arbitrarily small, r should be

$$r\in\Big(0,1-\frac{d}{\gamma+|\alpha|+d}\Big]. \qquad (22.59)$$

On the other hand, if $r\in(0,\gamma-|\alpha|)$,

$$\int_0^t (t-s)^{-\frac{d+|\alpha|+1}{\gamma}r-\frac{(d+|\alpha|)}{\gamma}(1-r)+\frac{d}{\gamma}}ds<\infty. \qquad (22.60)$$

Clearly, $(\gamma-|\alpha|)>1-\frac{d}{(\gamma+|\alpha|+d)}$. Therefore, we have

$$\mathbf{E}\Big[\Big|D_x^\alpha\Big(S_2(t,x+h)-S_2(t,x)\Big)\Big|^{2m}\Big]\leqslant C_{p,m}|h|^{2rm}, \qquad (22.61)$$

for any $m\in\mathbf{N}$ and $r\in\Big(0,1-\frac{d}{\gamma+|\alpha|+d}\Big]$. Equations (22.56) and (22.61) hold for each $x, x+h\in\mathbf{R}^d$. Finally, $S_1(t,x)$ is clearly of C^∞-class. As a result, the assertion follows from Kolmogorov-Centsov theorem.

Remark 22.3.1 Our result implies better estimate than that shown in [9]. Indeed, [9] shows

$$D_x^\alpha u(t,\cdot)\in\bigcap_{\varepsilon>0}C^{\left(\frac{\gamma-2|\alpha|-d}{2}\wedge\frac{1}{4}\right)-\varepsilon}(\mathbf{R}^d),\qquad |\alpha|=[\frac{\gamma-d}{2}],\qquad a.s., \qquad (22.62)$$

Remark 22.3.2 In the case $\gamma = 2m$, $m \in \mathbf{N}$, $2m > d$ with (22.1), we have

$$D_x^\alpha u(t,\cdot) \in \bigcap_{\varepsilon>0} C^{\frac{2m-2|\alpha|-d}{2}-\varepsilon}(\mathbf{R}^d), \qquad |\alpha| = [\frac{2m-d}{2}], \qquad a.s.. \tag{22.63}$$

This is obtained by changing slightly the proof of Theorem 22.3.1. Indeed, (22.35) holds for any $k \in \mathbf{N} \cup \{0\}$. Thus, (22.42) holds for all m, l with $m \geqslant l$. By proceeding similarly to the case of $\gamma \notin 2\mathbf{N}$, we can choose arbitrary integer n in (22.43) if $\gamma \in 2\mathbf{N}$, hence, we have $\int_{\mathbf{R}^d} |\phi(z)|^{2(1-p)} dz < \infty$ for all $p \in (0, 1 - \frac{1}{2n})$. It suffices to choose n arbitrary large in such a way that $\frac{\gamma-2|\alpha|-d}{2} < 1 - \frac{1}{2n}$ holds, which leads to (22.63).

Remark 22.3.3 Funaki [6] shows

$$D_x^\alpha u(t) \in \bigcap_{\varepsilon>0} C^{\frac{2m-2|\alpha|-d}{2}-\varepsilon}(\mathbf{R}^d), \qquad |\alpha| = [\frac{2m-d}{2}], \qquad a.s., \tag{22.64}$$

by using the estimate of the fundamental solution $q(t,x,y)$ of $\partial_t + \mathscr{A}$, where $\mathscr{A}$ is a $2m$-th order elliptic differential operator, namely,

$$|\frac{\partial^{j+\alpha+\beta}}{\partial t^j \partial x^\alpha \partial y^\beta} q(t,x,y)| \leqslant t^{-\frac{|\alpha|+|\beta|}{2m}-j} \bar{q}(t,x,y),$$

for $t \in (0,T]$, $x, y \in \mathbf{R}^d$, $j \in \mathbf{Z}_+$ and $\alpha, \beta \in \mathbf{Z}_+^d$, where

$$\bar{q}(t,x,y) = K_1 t^{-\frac{d}{2m}} e^{-K_2 (\frac{|x-y|^{2m}}{t})^{\frac{1}{2m-1}}},$$

where K_1 and K_2 are positive constants depending on T, j, α and β and they are taken uniformly in (j, α, β) such that $0 \leqslant j, |\alpha|, |\beta| \leqslant c$, for any $c \in \mathbf{Z}_+$, see also [4] or Chap. 9 in [5]. In our case ($\gamma \notin 2\mathbf{N}$), the estimate for the fundamental solution of the $2m$-th order parabolic type as above is not obvious, however, our result in the case $\gamma = 2m$ coincides with that of [6].

Finally, we consider more special case:

$$\frac{\partial u}{\partial t} = -(-\Delta)^{\frac{\gamma}{2}} u + f(x) R(t, u(t,x)) + g(x) \dot{w}(t,x) \qquad (t,x) \in \mathbf{R}_+ \times \mathbf{R}^d, \tag{22.65}$$

where $f, g \in C_0^\infty(\mathbf{R}^d)$, where $C_0^\infty(\mathbf{R}^d)$ is a family of functions whose support are compact in $\mathbf{R}^d$. In this case, we have

Corollary 22.3.2

$$D_x^\alpha u(t) \in \bigcap_{\varepsilon>0} C^{\frac{\gamma-2|\alpha|-d}{2}-\varepsilon}(\mathbf{R}^d), \qquad |\alpha| = [\frac{\gamma-d}{2}], \qquad a.s. \tag{22.66}$$

Proof The solution u of (22.66) is given by

$$u(t,x) = \int_{\mathbf{R}^d} G(t,x,q)u_0(q)dq + \int_0^t \int_{|q|\leqslant K} G(t-s,x,q)R(s,u(s,q))f(q)dsdq + \int_0^t \int_{|q|\leqslant K} G(t-s,x,q)g(q)w(dsdq), \quad a.s. \tag{22.67}$$

In this case, it suffices to take the consideration of

$$\int_{|x|\leqslant K} |\psi(x)|^l dx, \tag{22.68}$$

into account, for $l = 2(1-p)$ or $1-r$, where

$$\psi(x) = \int_{\mathbf{R}^d} e^{\sqrt{-1}z\cdot\xi}\xi^\alpha e^{-|\xi|^\gamma} d\xi. \tag{22.69}$$

Seeing (22.68), it follows that (22.55) and (22.58) hold for $p, r \in [0,1]$ by the argument in the proof of Theorem 22.3.1. Thus the assertion is obtained.

Acknowledgements This work is supported by Japan Society for the Promotion of Science (JSPS) "Construction of mathematical theory to investigate the macro structure and the meso structure of the fluid motion" (KAKENHI, No. 24224004) and the Core Research for Evolutional Science and Technology (CREST) of the Japan Science and Technology Agency (JST). The author would like to thank Professor Yoshihiro Shibata in Waseda university for giving him an opportunity to prepare for this manuscript and the financial support.

Appendix

For the convenience of the readers, we briefly recall the Kolmogorov-Centsov theorem for random fields without its proof (see e.g. Theorem 1.4.4 in [10]).

Theorem 22.3.3 *Let $X(x,y)$, $x \in D_1$, $y \in D_2$ be a random field with values in a Banach space B with its norm $\|\cdot\|$, where D_1, D_2 are domains in $\mathbf{R}^{d_1}$ and $\mathbf{R}^{d_2}$, respectively. Suppose that there exist constants $C, \gamma > 0$ and $\alpha_1 > d_1$, $\alpha_2 > d_2$ such that*

$$\mathbf{E}\Big[||(X(x', y') - X(x, y')) - (X(x', y) - X(x, y))||^{\gamma}\Big] \leqslant C|x - x'|^{\alpha_1}|y - y'|^{\alpha_2}, \tag{22.70}$$

$$\mathbf{E}\Big[||X(x', y) - X(x, y)||^{\gamma}\Big] \leqslant C|x - x'|^{\alpha_1}, \tag{22.71}$$

$$\mathbf{E}\Big[||X(x, y') - X(x, y)||^{\gamma}\Big] \leqslant C|y - y'|^{\alpha_2} \tag{22.72}$$

hold for any $x, x' \in D_1$ *and* $y, y' \in D_2$*. Then,* $X(x, y)$ *has a continuous modification* $Y(x, y)$*, which is* (β_1, β_2)*-Hölder continuous for any* $\beta_1 \in (0, \frac{\alpha_1 - d_1}{\gamma})$ *and* $\beta_2 \in (0, \frac{\alpha_2 - d_2}{\gamma})$.

References

1. Biler, P., Funaki, T., Woyczynski, W.A.: Fractal Burgers equations. J. Differ. Equ. **148**, 9–46 (1998)
2. Chang, T., Lee, K.: On a stochastic partial differential equation with a fractional Laplacian operator. Stoch. Process. Appl. **122**, 3288–3311 (2012)
3. Debbi, L., Debbi, M.: On the solutions of nonlinear stochastic fractional partial differential equations in one spatial dimension. Stoch. Process. Appl. **115**, 1764–1781 (2005)
4. Eidel'man, S.D.: Parabolic Systems. London North-Holland Publishing Co., Groningen (1969)
5. Friedman, A.: Partial Differential Equations of Parabolic Type. Prentice-Hall Inc, Englewood Cliffs (1964)
6. Funaki, T.: Regularity properties for stochastic partial differential equations of parabolic type. Osaka J. Math. **28**, 495–516 (1991)
7. Hairer, M.: A theory of regularity structures. Invent. Math. **198**, 269–504 (2014)
8. Kim, K.-H., Kim, P.: An L_p-theory of a class of stochastic equations with the random fractional Laplacian driven by Levy processes. Stoch. Process. Appl. **122**, 3921–3952 (2012)
9. Kotelenez, P.: Existence, uniqueness and smoothness for a class of function valued stochastic partial differential equations. Stoch. Stoch. Rep. **41**, 177–199 (1992)
10. Kunita, H.: Stochastic Flows and Stochastic Differential Equations. Cambridge University Press, Cambridge (1997)
11. Niu, M., Xie, B.: Regularity of a fractional partial differential equation driven by space-time white noise. Proc. Am. Math. Soc. **138**, 1479–1489 (2010)
12. Peszat, S., Szymon, J.: Existence, uniqueness and smoothness for a class of function valued stochastic partial differential equations. Math. Bohem. **123**, 7–32 (1998)
13. Walsh, J.B.: An introduction to stochastic partial differential equations. Lecture Notes in Mathematics, vol. 1180, pp. 265–439. Springer, Berlin (1986)